Stem Cell Therapeutics

Scrivener Publishing
100 Cummings Center, Suite 541J
Beverly, MA 01915-6106

Publishers at Scrivener
Martin Scrivener (martin@scrivenerpublishing.com)
Phillip Carmical (pcarmical@scrivenerpublishing.com)

Stem Cell Therapeutics

Edited by

Ayush Madan

Sumel Ashique

Deepika Arora

and

Mantosh Kumar Satapathy

Scrivener Publishing

WILEY

This edition first published 2025 by John Wiley & Sons, Inc., 111 River Street, Hoboken, NJ 07030, USA and Scrivener Publishing LLC, 100 Cummings Center, Suite 541J, Beverly, MA 01915, USA

For more information about Scrivener publications please visit www.scrivenerpublishing.com.

Wiley Global Headquarters
111 River Street, Hoboken, NJ 07030, USA

For details of our global editorial offices, customer services, and more information about Wiley products visit us at www.wiley.com.

The manufacturer's authorized representative according to the EU General Product Safety Regulation is Wiley-VCH GmbH, Boschstr. 12, 69469 Weinheim, Germany, e-mail: Product_Safety@wiley.com.

Library of Congress Cataloging-in-Publication Data

ISBN 9781394313754

Cover image: Generated with AI using Adobe Firefly
Cover design by Russell Richardson

Set in size of 11pt and Minion Pro by Manila Typesetting Company, Makati, Philippines

Printed in the USA

10 9 8 7 6 5 4 3 2 1

Contents

Preface xxix

1 Unraveling the Mechanisms: Insights Into Stem Cell Behavior in Disease Microenvironments 1
Pranshul Sethi, Aniruddha Sen, Sonima, Ayush Madan, Syed Mohsin Waheed and Bibhas Kumar Bhunia
1.1 Introduction 2
1.2 Stem Cell Biology and the Stem Cell Niche 4
1.2.1 Characteristics of Stem Cells 4
1.2.1.1 Self-Renewal 4
1.2.1.2 Differentiation Potential 5
1.2.2 The Stem Cell Niche 6
1.2.2.1 Normal Physiological Conditions 6
1.2.2.2 Key Components and Their Roles 8
1.3 Disease Microenvironments and Their Impact on Stem Cells 9
1.3.1 Alterations in the Extracellular Matrix 9
1.3.2 Role of Inflammation 10
1.3.3 Effects of Hypoxia 11
1.3.4 Disease-Specific Factors 11
1.4 Molecular and Cellular Mechanisms 12
1.4.1 Signaling Pathways Involved 12
1.4.1.1 Wnt Signaling Pathway 12
1.4.1.2 Notch Signaling Pathway 13
1.4.1.3 Hedgehog Signaling Pathway 13
1.4.2 Molecular Interactions 14
1.4.3 Mechanical and Biochemical Cues 15
1.5 Techniques for Studying Stem Cell Behavior in Diseased States 16
1.5.1 Advanced Imaging Techniques 16
1.5.2 Single-Cell RNA Sequencing 17

1.5.3 *In Vitro* Models 18
1.5.4 *In Vivo* Models 18
1.6 Stem Cell Dynamics in Various Diseases 19
1.6.1 Cancer 19
1.6.1.1 Tumor Microenvironment 19
1.6.1.2 Stem Cell Behavior in Cancer 20
1.6.2 Neuro-Degenerative Diseases 22
1.6.2.1 Microenvironmental Changes 22
1.6.2.2 Stem Cell Responses 22
1.6.3 Cardiovascular Diseases 23
1.6.3.1 Ischemic Conditions 23
1.6.3.2 Stem Cell Behavior and Repair Mechanisms 24
1.7 Therapeutic Implications 24
1.7.1 Potential Therapeutic Targets 24
1.7.2 Strategies to Enhance Stem Cell Function 25
1.7.3 Tailoring Treatments to Microenvironmental Contexts 26
1.8 Case Studies 27
1.8.1 Successful Regenerative Therapies 27
1.8.2 Challenges and Lessons Learned 28
1.9 Future Directions 29
1.9.1 Emerging Research Areas 29
1.9.2 Innovations in Therapeutic Approaches 29
1.9.3 Long-Term Goals and Visions for Stem Cell Therapy 30
References 31

2 CRISPR-Cas9 and Beyond: Gene Editing Frontiers in Stem Cell Therapeutics 39
Pranshul Sethi, Sonali Rastogi, Kajal Sherawat, Neetu Panwar, Ayush Madan and Mojahidul Islam
2.1 Introduction 40
2.1.1 Background of Gene Editing 40
2.2 Significance of CRISPR-Cas9 Technology 41
2.3 Overview of Stem Cell Therapeutics 42
2.4 Fundamentals of CRISPR-Cas9 Technology 42
2.4.1 Mechanism of CRISPR-Cas9 42
2.4.2 Advances in CRISPR-Cas9 Precision and Efficiency 44
2.5 Ethical and Regulatory Considerations 50
2.6 Applications in Stem Cell Research 51
2.7 Stem Cells: A Therapeutic Frontier 52
2.7.1 Types of Stem Cells: Embryonic, Adult, and Induced Pluripotent Stem Cells (iPSCs) 53

2.7.2 Current and Emerging Stem Cell Therapies 54
2.7.2.1 Current Stem Cell Therapies 54
2.7.2.2 Emerging Applications in Tissue Regeneration and Organ Repair 60
2.7.3 Challenges in Stem Cell Therapeutics 60
2.8 CRISPR-Cas9 in Stem Cell Therapeutics 62
2.8.1 Genetic Modification of Stem Cells Using CRISPR-Cas9 63
2.8.2 Case Studies: Successful Applications in Disease Models 65
2.8.3 Limitations and Challenges in Clinical Applications 71
2.8.4 Safety and Long-Term Implications 77
2.9 Beyond CRISPR-Cas9: Emerging Gene Editing Technologies 79
2.9.1 CRISPR-Cas12 and CRISPR-Cas13 Systems 79
2.9.1.1 CRISPR-Cas12 79
2.9.1.2 CRISPR-Cas13 79
2.9.2 Base Editing and Prime Editing 80
2.9.2.1 Base Editing 80
2.9.2.2 Prime Editing 80
2.9.3 Other Gene Editing Tools (TALENs, ZFNs) 81
2.9.3.1 TALENs 81
2.9.3.2 ZFNs 82
2.9.3.3 Comparing TALENs and ZFNs with CRISPR-Cas9 82
2.9.4 Comparative Analysis with CRISPR-Cas9 in Stem Cell Therapy 82
2.9.4.1 Precision and Specificity 82
2.9.4.2 Versatility 83
2.9.4.3 Safety and Off-Target Effects 83
2.9.4.4 Applications in Stem Cell Therapy 83
2.9.5 Future Roles 84
2.10 Ethical, Legal, and Social Implications (ELSI) 84
2.10.1 Ethical Concerns in Gene Editing and Stem Cell Research 84
2.10.2 Public Perception and Misconceptions 86
2.10.2.1 Public Engagement and Education 87
2.10.3 Legal and Regulatory Frameworks Globally 88
2.10.3.1 Regulatory Approaches in Different Countries 88
2.10.3.2 Impact on Research and Clinical Applications 92
2.10.3.3 Challenges in Creating International Guidelines 92

2.10.4 Future Directions in Ethical Governance 93
2.10.4.1 The Need for Balanced Policies 93
2.10.4.2 Public Engagement and Stakeholder Involvement 93
2.10.4.3 International Cooperation and Harmonization 94
2.10.4.4 Adaptive Governance 94
2.11 Future Prospects and Challenges 95
2.11.1 Advances in Precision Medicine through Gene Editing 95
2.11.1.1 CRISPR-Cas9 in Precision Medicine 95
2.11.1.2 Ongoing Research and Anticipated Breakthroughs 95
2.11.2 Overcoming Technical Barriers in Stem Cell Editing 96
2.11.3 Potential for Personalized Stem Cell Therapies 97
2.11.3.1 Personalized Therapies Using iPSCs 98
2.11.3.2 Current State of Research and Clinical Applications 98
2.11.3.3 Timeline for Clinical Application 98
2.11.4 The Road Ahead: Speculative Futures and Technological Innovations 99
2.12 Conclusion 100
2.12.1 Implications for Future Research and Clinical Practice 101
2.12.2 Final Thoughts on the Integration of Gene Editing and Stem Cell Therapeutics 101
References 102

3 Future Vistas: Revolutionary Technologies and Paradigms in Stem Cell Therapeutics 107

Dattatraya M. Shinkar, Sharayu P. Rathod, Priyanka D. Dabir, Sandip R. Purkar, Deepali D. Bhandari and Sunil V. Amrutkar

3.1 Introduction 108
3.2 Characteristics of Several Stem Cell Types Employed in Therapeutic Applications 110
3.2.1 Embryonic Stem Cells 111
3.2.2 Totipotent Stem Cells 113
3.2.3 Pluripotent Stem Cells 113
3.2.4 Multipotent Stem Cells 114
3.2.5 Oligopotent Stem Cells 114
3.2.6 Unipotent Stem Cells 114
3.2.7 Adult Stem Cells 114

3.2.7.1 Hematopoietic Stem Cells 116
3.2.7.2 Mesenchymal Stem Cells 116
3.2.8 Neutral Stem Cells 118
3.2.9 Epithelial Stem Cells 118
3.2.10 Skin Stem Cells 119
3.2.11 Embryonic Stem Cells (ESCs) 119
3.2.12 Induced Pluripotent Stem Cells (iPSCs) 119
3.3 Recent Advancements and Applications of Stem Cell Therapy 120
3.3.1 HSC Transplantation 120
3.3.2 HSC Therapy (HSCT) 120
3.3.3 Placental Stem Cell Therapy 121
3.3.4 Autologous Transplantation of Limbal Embryonic Stem Cells 121
3.3.5 Advancements in the Field of Synthetic Organ Engineering 121
3.3.6 Hollow Organ Engineering 121
3.3.7 Anti-Aging Effects 122
3.3.8 Minimizing Mitochondrial Injury 122
3.3.9 Treatment of Diabetes 122
3.3.10 Modeling Diseases and Researching Differentiation 122
3.3.11 Cell-Free Therapy 123
3.3.12 Wound Healing 123
3.3.13 Treatment of Burn Wounds 123
3.3.14 Significance in Research 123
3.4 Current Limitations 123
3.5 Innovative Isolation and Culture Techniques and Their Use in Research on Stem Cells 125
3.5.1 Conventional Methods of Separation 126
3.5.2 New Approaches 129
3.5.3 Prospects and Difficulties 133
3.6 Future Prospects 134
3.6.1 Stem Cell in Gene Editing 134
3.6.2 Stem Cell for Autoimmune Illnesses 135
3.6.3 Regenerative Medicine 137
3.6.3.1 ESCs in Regenerative Medicine 138
3.6.3.2 TSPSCs in Regenerative Medicine 138
3.6.3.3 Stem Cells and MSCs in Regenerative Medicine 139
3.6.3.4 UCSCs in Regenerative Medicine 139
3.6.3.5 BMSCs in Bioregenerative Therapy 140

3.6.4 Personalized Medicine 141
3.6.5 Tissue Engineering 141
3.6.6 Artificial Organs and Body Parts 146
3.6.6.1 Artificial Oxygen Carriers 146
3.6.6.2 Cardiac Support and Blood Pumps 146
3.6.6.3 Membrane Oxygenation and Cardiopulmonary Support 147
3.6.6.4 Functional Electrical Stimulation and Neuromuscular Support 148
3.6.6.5 Pancreatic Support 148
3.6.6.6 Pulmonary Support 148
3.6.6.7 Renal Support and Dialysis 148
3.6.6.8 Vascular and Valve Support 149
3.6.6.9 Visual Support 149
3.6.7 Use in Research 149
References 150

4 Nanobiomaterials at the Nexus: Integrating Stem Cells and Nanotechnology for Precision Therapeutics 159
Aditya Banerjee and Deepika Arora
4.1 Introduction 160
4.1.1 The Rise of Precision Therapy 160
4.1.2 Challenges in Conventional Therapy 161
4.1.3 Integration of Nanotechnology and Stem Cell Therapy 162
4.2 Nanobiomaterials: Tailoring the Cellular Microenvironment 163
4.2.1 Revolutionizing Medical Applications with Nanobiomaterials 163
4.2.2 Tailoring the Nanoworld: Properties Engineering Toward Precision Medicine 164
4.2.2.1 Biocompatibility 165
4.2.2.2 Drug Delivery Efficiency 165
4.2.2.3 Targeting Capabilities 166
4.2.2.4 Novel Approaches and Intelligent Materials 166
4.2.3 Design Strategies for Stem Cell–Nanobiomaterial Integration 167
4.2.3.1 Biocompatibility and Surface Modification 168
4.2.3.2 Scaffold Design and Tissue Engineering 168
4.2.3.3 Control of Biologically Active Particle Release 169
4.2.3.4 Advanced Functionalities 169
4.2.3.5 Immunomodulation 170
4.2.3.6 Integration with Smart Technologies 170

4.3 Stem Cells for Regenerative Medicine 171
4.3.1 Adult vs. Pluripotent Stem Cells 171
4.3.1.1 Sources 171
4.3.1.2 Characteristics 171
4.3.1.3 Therapeutic Potentials 172
4.3.1.4 Comparative Advantages and Challenges 173
4.3.1.5 Can Umbilical Cord-Derived Mesenchymal Stem Cells Pave a New Wave in Regenerative Medicine? 173
4.3.2 Differentiation and Therapeutic Potential of Stem Cells 174
4.3.2.1 Mechanisms of Differentiation 174
4.3.2.2 Therapeutic Potential in Regenerative Medicine 175
4.3.2.3 Future Direction 176
4.3.3 Challenges in Stem Cell Transplantation 176
4.3.3.1 Immune Rejection 176
4.3.3.2 Ensuring Cell Viability 177
4.3.3.3 Homing and Engraftment 178
4.3.3.4 Safety Concerns 178
4.3.3.5 Ethical and Regulatory Issues 178
4.3.3.6 Cost and Accessibility 179
4.4 Exosomes: Natural Nanocarriers for Intercellular Communication 179
4.4.1 Biogenesis and Properties of Exosomes 179
4.4.1.1 Biogenesis 179
4.4.1.2 Structure of Exosomes 180
4.4.1.3 Natural Functions of Exosomes in Cellular Communication 181
4.4.2 Targeted Delivery Using Engineered Exosomes 182
4.4.2.1 Surface Alteration for Targeting 183
4.4.2.2 Increasing Cargo Capacity 183
4.4.2.3 Enhancing Stability and Circulation 184
4.4.2.4 Controlled Release Mechanisms 185
4.4.3 Exosome-Based Therapeutics 186
4.4.3.1 Drug Delivery 186
4.4.3.2 Immune Modulation 187
4.4.3.3 Regenerative Medicine 188
4.4.3.4 Applications for Diagnosis 188
4.5 Reprogramming Immunity for Cancer Through CAR-T Cell Therapy 189
4.5.1 CAR-T Cell Design and Engineering 189

4.5.1.1 CAR-T Cell Design 189
4.5.1.2 Genetic Engineering of CAR-T Cells 190
4.5.2 Clinical Applications and Efficacy of CAR-T Cell Therapy 191
4.5.2.1 Hematologic Malignancies 191
4.5.2.2 Solid Tumors 192
4.5.2.3 Efficacy and Long-Term Outcomes 192
4.5.3 Challenges and Future Directions of CAR-T Cell Therapy 192
4.5.3.1 Toxicity 192
4.5.3.2 Resistance and Relapse 193
4.5.3.3 Manufacturing and Accessibility 193
4.5.3.4 Future Directions 194
4.6 Integrating Nanobiomaterials with Stem Cells and CAR-T Cells 194
4.6.1 Nanobiomaterials for Enhanced Stem Cell Therapy 194
4.6.1.1 Proliferation Enhancement 195
4.6.1.2 Differentiation Enhancement 195
4.6.1.3 Improvement of Homing and Engraftment 196
4.6.2 Nanocarriers for Delivery and Activation of CAR-T Cells 197
4.6.2.1 Enhanced Delivery to Tumor Sites 197
4.6.2.2 Improved Activation and Expansion 198
4.6.2.3 Enhancement of Persistence and Survival 199
4.6.3 Combination Treatments for Improved Therapeutic Effectiveness 200
4.6.3.1 Nanobiomaterials and Stem Cells 200
4.6.3.2 CAR-T Cells and Nanobiomaterials 201
4.6.3.3 CAR-T Cells and Stem Cells 202
4.7 Challenges and Future Perspectives 204
4.7.1 Regulatory Considerations for Nanobiomaterial-Based Therapeutics 204
4.7.1.1 Approval Processes 204
4.7.1.2 Safety Standards 205
4.7.1.3 Challenges in Regulation 205
4.7.1.4 Future Perspectives 206
4.7.2 Conquering Biocompatibility and Safety Issues 206
4.7.2.1 Material Selection and Surface Engineering 206
4.7.2.2 Incorporating Biodegradability 206
4.7.2.3 Mitigating Immune Responses 207
4.7.2.4 Evaluating Long-Term Safety 207

4.7.2.5 Regulatory Guidelines and Standardization 207
4.7.2.6 Interdisciplinary Collaboration 207
4.7.3 Personalized Medicine and Patient-Specific Therapeutics 208
4.7.3.1 Customized Drug Delivery 208
4.7.3.2 Stem Cell Treatments for Particular Patients 208
4.7.3.3 Genomic and Proteomic Profiling 209
4.7.3.4 Customized Cancer Immunotherapy 209
4.7.3.5 Precision Diagnostics 209
4.7.3.6 Real-Time Monitoring and Feedback 209
4.7.3.7 Ethical and Social Considerations 210
4.8 Conclusion 210
4.8.1 Stem Cells, Nanotechnology, Exosomes, and CAR-T Cells: What the Future Holds 210
4.8.1.1 Stem Cell Treatment Improvement 211
4.8.1.2 Targeted Release Mechanisms 211
4.8.1.3 Therapeutics Based on Exosome 211
4.8.1.4 CAR-T Cell Engineering 211
4.8.1.5 Combination Therapy 212
4.8.1.6 Personalized Medicine 212
4.8.1.7 Regulatory and Ethical Considerations 212
4.8.2 The Potential of Nanobiomaterials for Precision Therapeutics 213
4.8.2.1 Improved Drug Delivery 213
4.8.2.2 More Accurate Diagnosis 213
4.8.2.3 Regenerative Medicine 213
4.8.2.4 Cancer Treatment 214
4.8.2.5 Gene Therapy 214
4.8.2.6 Individualized Approach 214
4.8.2.7 Future Outlook 215
4.9 Summary 215
References 216

5 Exosomes from Stem Cells as Therapeutic Agents for Cardiovascular Regeneration 227
Maryam Taghavi Narmi, Nikoo Baghal Darbandi, Abdulwahab Teflischi Gharavi, Ali Babaeizad and Mohsen Sheykhhasan
5.1 Introduction 228
5.1.1 Background on Cardiovascular Diseases 228
5.1.2 The Role of Stem Cells in Regenerative Medicine 229

5.1.3 Overview of Exosomes and their Biological Functions 237
5.2 Biogenesis and Characteristics of Stem-Cell-Derived Exosomes 242
5.2.1 Exosome Formation and Release 242
5.2.2 Molecular Composition and Functional Properties 244
5.3 Mechanism of Cardiovascular Regeneration Mediated by Exosomes 249
5.3.1 Modulation of Inflammation 251
5.3.2 Promotion of Angiogenesis 254
5.3.3 Enhancement of Cell Survival and Proliferation 261
5.3.4 Reduction of Fibrosis 263
5.4 Therapeutic Applications of Exosomes in Cardiovascular Diseases 269
5.4.1 Myocardial Infarction 269
5.4.2 Heart Failure 271
5.4.3 Peripheral Artery Disease 271
5.5 Preclinical and Clinical Studies 272
5.5.1 Animal Models of Cardiovascular Disease 272
5.5.2 Human Clinical Trials 277
5.6 Challenges and Future Directions 277
5.7 Conclusion 279
References 279

6 Cardiovascular Renaissance: Stem Cell Strategies for Heart Diseases 309

Apurva, Pracheta Janmeda, Divya Kumari, Narotam Sharma and Priya Chaudhary

6.1 Introduction 310
6.2 Stem Cell Types Considered for the Treatment of Heart Disease 311
6.2.1 Skeletal Myoblast 311
6.2.2 Bone-Marrow-Derived Mononuclear Cells 312
6.2.3 BM-Derived Endothelial Progenitor Cells (EPCs) 312
6.2.4 BM-Derived Mesenchymal Stem Cells (MSCs) 313
6.2.5 Mobilized Stem and Progenitor Cells 313
6.2.6 Adipose-Derived Stem and Progenitor Cells (ASPCs) 314
6.2.7 Cardiac Stem and Progenitor Cells 315
6.2.8 Embryonic Stem Cells 316
6.2.9 Induced Pluripotent Stem Cells 316
6.2.10 Adult Stem Cell 317
6.3 Stem Cell Therapy Against Different Heart Diseases 317

6.3.1 Acute and Chronic Coronary Syndromes 318
6.3.2 Myocardial Infarction 322
6.3.3 Ischemia–Reperfusion Injury 322
6.3.4 Ischemic Cardiomyopathy 323
6.3.5 Ischemic Heart Failure 324
6.3.6 Modeling Heart Diseases on a Chip 325
6.4 Mechanism of Stem-Cell Based Therapy for Heart Disease 325
6.4.1 Differentiation of Transplanted Stem Cells into Cardiac Cells 325
6.4.2 Formation of New Blood Vessels from Transplanted Stem Cells 326
6.4.3 Paracrine Effect 326
6.4.4 Cell Fusion 327
6.4.5 Treatment by Extracellular Vesicle-Derived Stem Cells 327
6.5 Gene-Modified Stem Cells 328
6.5.1 Increase the Capacity of Antiapoptotic 328
6.5.2 Promote Migration 329
6.5.3 Anti-Inflammatory 330
6.5.4 Multigene-Modified MSC Exploration 330
6.6 Limitations of Stem Cell Strategies for Cardiovascular Diseases 331
6.6.1 Heterogeneity of Stem Cells 331
6.6.2 Survival of Transplanted Stem Cell *In Vivo* 331
6.6.3 Migration and Homing of Stem Cell *In Vivo* 332
6.6.4 Immunogenicity of Stem Cells After Transplantation 333
6.6.5 Uncontrolled Differentiation of Stem Cells *In Vivo* Resulting in Undesired Phenotype 334
6.6.6 Acquisition of Adequate Stem Cells 334
6.6.7 *Ex Vivo* Expansion Efficiency 335
6.7 Strategies for Improving Stem Cell Therapy 335
6.7.1 Stem Cell Delivery 335
6.7.2 Homing 335
6.7.3 Survival/Engraftment 336
6.8 Bioengineering Strategies: New Era 336
6.8.1 Cell-Free Therapy 336
6.8.2 Improving Target Ability 337
6.8.3 Overcoming Low Cell Retention 337
6.8.4 Biomimetic Strategies in Stem-Cell-Based Therapy 338
6.9 Conclusion 338
References 339

7 Neurological Disorders: Advancements in Stem-Cell-Based Treatments **351**
Dheeraj Sharma and Shriyansh Srivastava
7.1 Introduction 352
7.2 Types of Stem Cells 353
7.2.1 Embryonic Stem Cells 354
7.2.2 Adult Stem Cells 355
7.2.3 Induced Pluripotent Stem Cells (iPSCs) 355
7.2.4 Neural Stem Cells 356
7.3 Mechanisms of Action 357
7.3.1 Differentiation and Integration 357
7.3.2 Neuroprotection 357
7.3.3 Immunomodulation 358
7.3.4 Paracrine Effects 358
7.4 Technological Advancements in Stem Cell Research 358
7.4.1 CRISPR and Gene Editing 359
7.4.2 Organoids and 3D Bioprinting 359
7.4.3 Biomaterials and Scaffolding 360
7.4.4 Single-Cell Sequencing 360
7.5 Stem-Cell-Based Treatments for Specific Neurological Disorders 362
7.5.1 Parkinson's Disease 362
7.5.2 Alzheimer's Disease 362
7.5.3 Multiple Sclerosis 363
7.5.4 Stroke 363
7.5.5 Spinal Cord Injuries 364
7.5.6 Amyotrophic Lateral Sclerosis (ALS) 364
7.6 Clinical Trials and Case Studies 365
7.6.1 Overview of Current Clinical Trials 365
7.6.2 Success Stories and Outcomes 365
7.6.3 Challenges and Setbacks 366
7.7 Ethical Considerations and Regulatory Challenges 367
7.7.1 Ethical Issues in Stem Cell Research 367
7.7.2 Regulatory Landscape 368
7.7.3 Patient Consent and Safety 369
7.8 Future Directions and Potential Innovations 370
7.8.1 Potential for Preventive Therapies 371
7.8.2 The Future of Stem-Cell-Based Treatments in Neurology 371
7.9 Conclusion 372
References 373

8 Stem Cell Therapy for Metabolic Disorders: Harnessing the Power of Stem Cells to Treat Diabetes, Obesity, and Other Metabolic Disorders 381
Sindhu D. Bali, Anand Kumar Shukla, Rosaleen Sahoo, Varsha A. Mahadik and Narendra Kadoo
8.1 Introduction 382
8.1.1 Overview of Metabolic Disorders and Their Current Treatment Options 382
8.1.2 Significance of Stem Cell Therapy in Metabolic Disorders 386
8.1.3 Fundamental Principles of Stem Cell Therapy 386
8.1.3.1 Stem Cell Types 386
8.1.3.2 Unique Characteristics of Stem Cells 388
8.2 Clinical Applications of Stem Cell Therapy for Metabolic Disorders 389
8.2.1 Mesenchymal Stem Cells (MSCs) 389
8.2.2 Induced Pluripotent Stem Cells (iPSCs) 389
8.2.3 Hematopoietic Stem Cells (HSCs) 390
8.2.4 Embryonic Stem Cells (ESCs) 391
8.2.5 Neural Stem Cells (NSCs) 391
8.3 Molecular Mechanisms Underlying the Action of Stem Cells in Metabolic Disorders 391
8.3.1 Homing to Target Tissues 392
8.3.2 Tissue Repair and Regeneration 392
8.3.3 Modulation of Immune Response 393
8.3.4 Signaling Pathways Involved in Metabolic Disorders 394
8.3.4.1 The mTOR Signaling Pathway 394
8.3.4.2 The JNK Signaling Pathway 395
8.3.4.3 The AMPK Pathway 395
8.3.4.4 Inflammatory Signaling Pathways 395
8.3.4.5 Signaling by Leptin and Ghrelin 395
8.4 Challenges and Limitations of Stem Cell Therapy for Metabolic Disorders 396
8.4.1 Off-Target Effects 396
8.4.2 Immune Responses 397
8.4.3 Tissue Heterogeneity 397
8.4.4 Stem Cell Exhaustion 398
8.5 Future Directions for Stem Cell Therapy in Metabolic Disorders 399
8.5.1 Advancements in Sources of Stem Cells 399

8.5.2 Enhancing Treatment Safety and Efficacy 399
8.5.3 Personalized Medicine Approaches 400
8.5.4 Combination Therapies 400
8.5.5 Longitudinal Studies and Clinical Trials 400
8.5.6 Regulatory and Ethical Considerations 400
8.6 Conclusions 401
Acknowledgments 402
References 402

9 Immunomodulation in Autoimmune Disorders: Harnessing the Power of Stem Cell 409
Neha Kamboj, Rahul Kumar and Debasis Mitra
9.1 Introduction 410
9.1.1 Importance of Immunomodulation in Managing Autoimmune Diseases 411
9.1.2 Role of SCs in Medical Research and Treatment 411
9.2 SCs: Types and Characteristics 412
9.3 Mechanisms of SC-Based Immunomodulation 414
9.3.1 SC Interactions with the Immune System 415
9.3.2 Anti-Inflammatory Effects of SCs 419
9.3.3 Paracrine Signaling and Secretion of Immunomodulatory Factors 421
9.3.4 Regulation of Immune Cell Populations (e.g., T Cells, B Cells, Dendritic Cells) 422
9.4 Clinical Applications of SCs in Autoimmune Disorders 427
9.4.1 Hematopoietic SC Transplantation (HSCT) 428
9.4.2 Mechanism of Action and Therapeutic Effects 430
9.4.3 Clinical Outcomes and Case Studies 432
9.4.4 Mesenchymal SCs (MSCs) 434
9.4.5 Mechanism of Action and Therapeutic Effects 436
9.4.6 Clinical Trials and Outcomes in Autoimmune Diseases (e.g., Rheumatoid Arthritis, Multiple Sclerosis) 438
9.4.7 Induced Pluripotent SCs (iPSCs) 439
9.4.8 Challenges and Future Directions 441
9.5 Challenges and Considerations in SC Therapy 442
9.5.1 Safety Concerns and Risk of Tumorigenesis 444
9.5.2 Immunogenicity and Rejection Issues 445
9.5.3 Standardization and Quality Control in SC Production 446
9.5.4 Regulatory and Clinical Trial Design Challenges 447

9.5.5 Advances in SC Engineering and Gene Editing 449
9.5.6 Combination Therapies with SCs and Other Immunomodulatory Agents 451
9.6 Conclusion 453
References 454

10 The Microbiome Connection: Exploring Interactions with Stem Cell Therapies 463
Rizwan Ahamad, Asad Ali, Priya Manna, Neha Quadri, Zuber Khan, Radheshyam Pal, Mohammad Muztaba, Shakira Ghazanfar, Anas Islam, Asif Iqbal and Sumel Ashique
10.1 Introduction 464
10.2 The Function of Microbiota in Gut Health 464
10.3 The Intersection of Microbiome and Stem Cell Research 465
10.4 Microbiota Metabolites and Stem Cell Differentiation 466
10.5 Modulation of Mucosal Immunity and Intestinal Inflammation 468
10.5.1 Fungal Colonization in the Gut 469
10.5.2 Fungi's Role in Gut Immune Balance 469
10.5.3 Fungal–Immune Interactions Extend Beyond the Gut: Gut Fungi Influence Lung Health 469
10.5.4 Gut Fungi Influence Overall Immune Defense 470
10.5.5 Impact of Different Fungi on Immune Response 470
10.6 Influences on Stem Cell Engraftment 471
10.6.1 Factors Influencing Stem Cell Engraftment 471
10.6.2 Graft-Versus-Host Disease (GVHD) 472
10.6.3 Additional Factors 472
10.6.4 A Novel Approach to Improving Outcomes 472
10.7 Role in Neurological Diseases 474
10.7.1 Gut Microbiota–Brain Axis in Neuro-Inflammation 474
10.7.2 Gut–Brain Axis 474
10.7.3 Gut Dysbiosis, an Imbalance of Gut Bacteria, is Linked to Brain Problems 474
10.7.4 The Gut Microbiota Influences Brain Function 475
10.7.5 The Gut–Brain Axis: An Intricate Connection 475
10.8 Emerging Significance of Gut Microbiota in Neurogenesis 475
10.8.1 Gut Bacteria Significantly Impact Brain Health 475
10.8.2 Neurological Pathway in the Gut–Brain Axis (GBA) 476

10.8.3 Neurological Disorders and the Gut–Brain Axis (GBA) 476
10.8.4 The Gut–Brain Axis and Neurological Disorders 477
10.9 Clinical Trials 477
10.10 Role in Gastrointestinal Diseases 477
10.10.1 Inflammatory Bowel Disease (IBD) 477
10.10.2 COVID-19 Vaccines in IBD Patients 478
10.10.3 Bivalent Vaccines are Suggested for IBD Patients Due to Higher Breakthrough Infections 478
10.10.4 SARS-CoV-2 Vaccines are Safe and Effective in IBD Patients, with No Increased Risk of Disease Flares 479
10.10.5 COVID-19 Complications in IBD Patients Include Mental Health Issues and Potential Long-Term Effects (Long COVID) 479
10.10.6 COVID-19 and IBD: Key Findings 479
10.10.7 Patients' Perspectives on COVID-19 and IBD 479
10.10.8 Significance of the Microbiota to Liver Pathophysiology 480
10.10.9 Gut–Liver Axis: A Bidirectional Relationship 480
10.11 Microbiome Influence on Tumor Microenvironment 482
10.11.1 Microbiome-Supported Stem Cell Treatment 483
10.11.2 Microbiome Diversity 484
10.12 Challenges of Stem Cell Therapy 485
10.12.1 Inadequate Source of Stem Cells 485
10.12.2 Presence of Optimal Time Window 486
10.12.3 Inherited Limitation of Stem Cells 486
10.12.4 Adverse Effect 486
10.13 Microbiome Engineering and Integration of Multi-Omics for Personalized Medicine 487
References 487

11 Cellular Synchrony: Coordinated Strategies for Effective Tissue Regeneration Using Stem Cells 505
Amit Anand, Santhepete Nanjundiah Manjula and Mruthunjaya Kenganora
Abbreviations 506
11.1 Introduction to Cellular Synchrony and Tissue Regeneration 506
11.1.1 Overview of Tissue Regeneration 506
11.1.2 Importance of Cellular Synchrony 507

11.1.3 Function of Stem Cells in Tissue Repair and Regeneration 508
11.2 Fundamentals of Stem Cells 509
11.2.1 Types of Stem Cells 510
11.2.1.1 Embryonic Stem Cells 510
11.2.1.2 Adult Stem Cells (ASCs) 511
11.2.2 Potency of Stem Cells 512
11.2.2.1 Totipotent 512
11.2.2.2 Pluripotent 513
11.2.2.3 Unipotent 514
11.2.2.4 Multipotent 514
11.2.2.5 Oligopotent 515
11.2.3 Properties of Stem Cells 515
11.2.3.1 Differentiation of Stem Cells 515
11.2.3.2 Self-Renewal of Stem Cells 516
11.2.3.3 Clonogenicity of Stem Cells 517
11.3 Mechanisms of Cellular Synchrony 518
11.3.1 Signaling Pathways Involved in Synchrony 518
11.3.1.1 MAPK Signaling Pathways 518
11.3.1.2 P13K/AKT Signaling Pathway 518
11.3.1.3 JAK-STAT Signaling Pathways 519
11.3.1.4 Wnt Signaling Pathway 519
11.4 Methods of Investigating Cell Synchrony 520
11.4.1 Gene Regulation Model 520
11.4.2 Simulation Studies 520
11.4.3 Biological Methods 521
11.4.3.1 Serum Starvation 521
11.4.3.2 Double Thymidine Block 521
11.5 Strategies for Enhancing Cellular Synchrony in Tissue Regeneration 522
11.5.1 Overexpression of Cell Cycle Regulators 522
11.5.1.1 Cyclins and Cyclin-Dependent Kinase 522
11.5.2 Controlled Release of Growth Factors 524
11.5.3 Electrical and Mechanical Stimulation 525
11.6 Applications of Stem Cell-Based Tissue Regeneration 527
11.6.1 Bone and Cartilage Tissue Regeneration 527
11.6.2 Cardiac Tissue Regeneration 527
11.6.3 Regeneration of Neural Tissue 528
11.6.4 Regeneration of Sensory Organ 529
11.7 Challenges and Future Directions 530
Conclusion 531

Acknowledgments 532
References 532

12 Tissue Engineering Paradigms: Creating Functional Replacements with Stem Cells 541
Karan Goel, Isha Chawla, Garima, Sarita Sharma and Sumeet Gupta
12.1 Introduction to Tissue Engineering 542
12.2 Fundamentals of Stem Cells 542
12.2.1 Types of Stem Cells 542
12.2.2 Characteristics and Properties 543
12.2.2.1 Pluripotency and Multipotency 543
12.2.2.2 Self-Renewal and Differentiation Potential 544
12.2.2.3 Regenerative Medicine Applications 544
12.3 Biomaterials and Components in Tissue Engineering 544
12.3.1 Types of Biomaterials 544
12.3.2 Components in Tissue Engineering 544
12.3.3 Properties and Selection Criteria 546
12.3.4 Biomaterial Scaffolds: Designing the Extracellular Matrix 546
12.3.4.1 Scaffold Architecture 546
12.4 Cell-Seeding Techniques 547
12.4.1 Scaffold-Based Methods 547
12.4.1.1 Scaffold Fabrication Techniques 547
12.4.1.2 Seeding Efficiency 547
12.4.1.3 Dynamic Seeding Methods 548
12.4.2 Bioprinting Technologies 548
12.4.2.1 Bioprinting Principles 548
12.4.2.2 Types of Bioprinting 548
12.5 Bioreactor Systems for Tissue Culture 549
12.5.1 Types of Bioreactors 549
12.5.1.1 Static Bioreactors 549
12.5.1.2 Dynamic Bioreactors 549
12.5.1.3 Microfluidic Bioreactors 549
12.5.2 Applications in Tissue Engineering 550
12.6 Clinical Applications of Tissue Engineering 552
12.7 Regulatory and Ethical Considerations 556
12.8 Future Directions in Tissue Engineering 564
12.9 Conclusion and Future Perspectives 568
References 569

13 Personalized Medicine Frontiers: Interrelating Stem Cell Microenvironments with Stem Cell Therapies for Individual Patients **575**
Sagar Mondal, Swati Priya, Jutishna Bora, Smita Lata, Ajay Kumar Mahalka and Sumira Malik
13.1 Introduction 576
13.2 Stem Cells and their Microenvironments as Personalized Medicine 578
13.3 The Disease Microenvironment: A Complex Ecosystem 580
13.3.1 Intercellular Interaction 580
13.3.2 Extracellular Matrix Interaction 581
13.3.3 Role of Soluble Bioactive Agents 584
13.4 Stem Cell Plasticity and Adaptation 585
13.4.1 Influence of Signaling Pathways 587
13.5 Stem Cell Dysfunction in Disease 588
13.5.1 Stem Cell-Derived Cancer Stem Cells 590
13.5.2 Therapeutic Implications 591
13.6 Stem Cell Therapy for Human Diseases 594
13.7 Conclusion 595
References 595

14 Stem Cell Therapy for Cancer Treatment: Current Advances and Future Directions **599**
Anand Kumar Shukla, Sindhu D. Bali, Varsha A. Mahadik, Rosaleen Sahoo and Narendra Kadoo
14.1 Introduction 599
14.1.1 Importance of Stem Cell Therapy in Cancer Treatment 604
14.1.1.1 Enhanced Targeting of Tumors 604
14.1.1.2 Overcoming Therapy Resistance 604
14.1.1.3 Regenerative Potential 605
14.1.1.4 Immune Modulation 605
14.1.2 Fundamental Principles of Stem Cell Therapy 605
14.1.3 Types of Stem Cells 606
14.1.3.1 Mesenchymal Stem Cells (MSCs) 606
14.1.3.2 Neural Stem Cells (NSCs) 607
14.1.3.3 Hematopoietic Stem Cells (HSCs) 607
14.1.4 Unique Properties of Stem Cells 608
14.2 Current Clinical Applications of Stem Cell Therapy for Cancer 608
14.2.1 Mesenchymal Stem Cells (MSCs) 608

14.2.2 Neural Stem Cells (NSCs) 609
14.2.3 Hematopoietic Stem Cells (HSCs) 610
14.2.4 Induced Pluripotent Stem Cells (iPSCs) 610
14.2.5 Embryonic Stem Cells (ESCs) 611
14.3 Molecular Mechanisms Underlying the Action of Stem Cells 611
14.3.1 Homing to Bone Marrow 614
14.3.2 Tumor-Tropic Effect 614
14.3.3 Secretion of Paracrine Factors and Differentiation Capacity 614
14.3.4 Signaling in Cancer Stem Cells 615
14.4 Challenges and Limitations of Stem Cell Therapy for Cancer 615
14.5 Future Directions for Stem Cell Therapy in Cancer Treatment 617
14.5.1 Integration with Other Therapeutic Approaches 618
14.5.2 Development of New Stem Cell Sources 619
14.5.3 Gene Editing and Gene Therapy 619
14.5.4 Clinical Trials and Regulatory Mechanisms 620
14.6 Summary and Conclusions 621
Acknowledgments 621
References 622

15 Challenges and Breakthroughs: Lessons from Stem-Cell-Based Cancer Therapies 629

Priya Chaudhary, Ayush Madan, Divya Kumari, Apurva and Pracheta Janmeda

15.1 Introduction 630
15.2 Current Hurdles in Stem-Cell-Based Treatment for Cancer 631
15.2.1 Ethical Considerations 631
15.2.2 Technical Constraints 632
15.2.3 Regulatory Restrictions 632
15.2.4 Unanticipated Clinical Results 633
15.3 Stem Cell Classification and their Properties 634
15.4 Mechanism Underlying the Action of Stem Cells in Cancer 636
15.4.1 Homing to Bone Marrow 636
15.4.2 Tumor-Tropic Effect 636
15.4.3 Paracrine Factor Secretion and Differentiation Capacity 636
15.4.4 Signaling in Cancer Stem Cells 637

15.5 Potential Application of Stem Cell Therapy in Cancer Treatment 638
15.5.1 HSC Transplantation 638
15.5.2 MSC Transplantation After Cancer Treatment 638
15.5.3 Enzyme Therapy 640
15.5.4 Secreted Agents 641
15.5.5 Viral Therapy 642
15.5.6 Nanoparticle Carrier 642
15.5.7 Exosomes as Therapeutic Carrier 642
15.6 Other Applications of Stem-Cell-Based Cancer Therapy 643
15.6.1 Inhibition of Angiogenesis 643
15.6.2 Regenerative Medicine 644
15.6.3 Immunotherapy 644
15.6.4 Anticancer Drug Screening 645
15.6.5 Targeting Cancer Stem Cells (CSCs) 645
15.7 Factors Influencing Stem Cell Therapy 645
15.7.1 Stem Cell Type 645
15.7.2 Transplantation Route 646
15.7.3 Cell Number and Transplanting Time 646
15.8 Diapause and Hibernation Mechanism in Cancer Stem Cells 647
15.9 Cancer Stem Cell Models 648
15.9.1 Cell-Line-Derived Models 648
15.9.2 Patient-Derived Xenografts 648
15.9.3 3D Culture Systems 649
15.9.4 Genetically Engineered Mouse Models (GEMMs) 649
15.9.5 Spheroids 650
15.10 Biomarkers Associated with Cancer 650
15.11 Side Effects Related with Stem Cell Therapy 652
15.11.1 Tumorigenesis 653
15.11.2 Drug Toxicity and Resistance 654
15.11.3 Adverse Events in Allogenic HSC Transplantation 654
15.11.4 Increased Immune Responses and Autoimmunity 654
15.11.5 Viral Infection 655
15.12 Challenges to Stem Cell Therapy 655
15.12.1 Treatment Durability 655
15.12.2 Potential Tumorigenesis Concerns 656
15.12.3 Epithelial Mesenchymal Transition (EMT) 656
15.12.4 High Levels of Multidrug Resistance or Detoxification Proteins 656
15.12.5 Tumor Microenvironment 657

15.13 Strategies to Improve the Anticancer Efficacy of Stem-Cell-Based Therapy 657
15.13.1 Improvement of Stem Cell Immunotolerance 657
15.13.2 Enhancement of the Stem Cell Tumor Tropism 658
15.13.3 Optimal Choice of the Stem Cell Administration Route 658
15.13.4 Combinational Approaches 659
15.14 Conclusion 659
References 660

16 Clinical Trials and Triumphs: Navigating the Path from Bench to Bedside **673**
Abdelmonem Siddiq, Al-Hassan Soliman Wadan, Abdalla Ali, Shubham Shrestha and Ejaz Ahmad Khan
16.1 Introduction 674
16.2 Current and Ongoing Clinical Trials 676
16.3 Approved Stem Cell Therapies 687
16.4 Challenges in Clinical Trials 693
16.5 Conclusion 695
References 696

17 Beyond Borders: Global Perspectives on Stem Cell Therapeutics **705**
N.L. Swathi, Praveen Mallari, Mihir Dabhi and Shweta Shrivastava
17.1 Introduction 706
17.1.1 Overview of Stem Cell Therapies 706
17.2 Global Advances in Stem Cell Research 707
17.2.1 Breakthroughs in North America 707
17.2.2 Innovations in Europe 712
17.2.3 Progress in Asia-Pacific 714
17.2.4 Developments in Latin America 714
17.2.5 Research in Africa and the Middle East 716
17.3 Regional Regulatory Frameworks 717
17.3.1 Regulatory Environment in the USA 717
17.4 Ethical Considerations and Public Perception 719
17.4.1 Ethical Dilemmas in Stem Cell Research 719
17.4.1.1 Embryonic Stem Cells (ESCs) 719
17.4.1.2 Safety and Efficacy of Therapy 720
17.4.1.3 Animal Models and Chimeras 721
17.4.1.4 Responsible and Ethical Research 721

17.4.2 Cultural Influences on Public Perception 722
17.4.2.1 Religious Viewpoints 722
17.4.3 Case Studies of Public Engagement 722
17.4.3.1 UK Stem Cell Network (UKSCN) 722
17.4.3.2 Stem Cell Network (SCN) Canada 723
17.4.3.3 California Institute for Regenerative Medicine (CIRM) 723
17.5 Challenges and Barriers 724
17.5.1 Scientific and Technical Challenges 724
17.5.1.1 Cell Source and Differentiation 724
17.5.1.2 Cell Homing, Survival, and Integration 725
17.5.1.3 Immune Tolerance and Tumorigenicity 725
17.5.2 Logistical and Financial Barriers 726
17.5.2.1 Cost of Therapy 726
17.5.2.2 Copyright and Competition 726
17.5.3 Overcoming Obstacles: Best Practices 727
17.6 Future Directions 729
17.6.1 Emerging Technologies in Stem Cell Research 729
17.6.2 Predictions for the Next Decade 729
17.6.3 Global Impact and Potential Developments 730
17.7 Conclusion 733
References 733

18 Ethical Considerations in Stem Cell Therapeutics: Striking a Balance 739

Rahul Kumar, Ayush Madan, Debasis Mitra and Jyotsana Singh Chandravanshi

18.1 Introduction 740
18.2 Properties of Stem Cells 741
18.3 Sources of Stem Cell 742
18.4 Research on Stem Cells 744
18.5 Source and Ethical Concerns of Deriving Stem Cells 744
18.5.1 Utilizing Human Embryonic Stem Cells 744
18.5.2 Utilizing Extra/Spare Embryos Created During IVF Applications 745
18.5.3 Utilization of Fetal Tissue Received From Termination of Pregnancy (Intentionally Induced or Miscarriages) 745
18.5.4 Producing Embryos for Use in Stem Cell Studies 746
18.5.5 Somatic Cell Uses 746
18.6 Research Ethics Concerns and Process 747

18.6.1 Concerns of Ethics in Relation to Women's Reproductive Rights 747
18.6.2 Informed Consent During the Research Process to Volunteers 747
18.6.3 Privacy Confidentiality and Information Masking During the Research Process 748

18.7 Ethical Concerns Throughout the Clinical Phase and After 748
18.7.1 Problems Caused by Business Interests Taking the Initiative and Persistently Pursuing Research 748
18.7.2 Promotion, Deception of the Public, Amazing Therapies, and Sharing of Research Findings with Unproven Theories 749
18.7.3 During the Research Process, Several Treatment Methods That Have Not Been Proven to be Effective Were Reflected as Miraculous Treatments 749
18.7.4 Copyright and Patent Issues 749
18.7.5 Issues with Treatment Accessibility and Costs to be Developed as a Result of Stem Cell Research 750
18.7.6 Disregarding Other Regional and Worldwide Health Issues Unrelated to Stem Cell Research 750

18.8 Fundamental Methods of Stem Cell Research in Different Nations 750
18.8.1 International Law and National Strategies for Research on Stem Cells 751
18.8.2 Oviedo Convention on Human Rights and Biomedicine, Article 18 752

References 753

Index **757**

Preface

Stem cell research stands at the forefront of modern biomedical science, offering profound possibilities for revolutionizing the treatment of a wide range of diseases and disorders. Over the past few decades, remarkable strides have been made in understanding stem cell biology, engineering novel therapeutic strategies, and translating research innovations from the laboratory to clinical applications. This progress has illuminated new avenues in regenerative medicine, personalized therapeutics, and the management of complex chronic conditions.

"*Stem Cell Therapeutics*" emerges from the collective efforts of numerous researchers and clinicians passionate about the transformative potential of stem cells. This volume is designed to provide a comprehensive, multidisciplinary perspective on the current advancements, underlying mechanisms, and therapeutic applications of stem cells across various fields of medicine. It delves into critical topics including disease microenvironments, gene editing technologies, nanotechnology integration, cardiovascular and neurological regeneration, autoimmune modulation, metabolic disorders, cancer therapy, and the pivotal role of the microbiome.

Each chapter brings together expertise from diverse disciplines, ensuring a rich and up-to-date presentation of scientific concepts, clinical applications, challenges, and future directions. Special attention has been given to ethical, regulatory, and global perspectives, acknowledging that the journey from discovery to clinical success must be guided by responsibility, inclusivity, and innovation.

We hope this book will serve as a valuable reference for students, researchers, healthcare professionals, and academicians who are engaged or interested in the expanding universe of stem cell research and therapeutics. We believe that the integration of novel technologies, deeper understanding of stem cell behavior, and global collaborations will continue to accelerate the transition of stem cell therapies from experimental treatments to mainstream healthcare solutions.

We are deeply grateful to all the contributors for sharing their knowledge, and to the readers who continually inspire the pursuit of scientific excellence and meaningful innovation. It is our aspiration that *Stem Cell Therapeutics* will ignite curiosity, inspire new research endeavors, and ultimately contribute to improving human health and quality of life.

Editors
Ayush Madan
Sumel Ashique
Deepika Arora
Mantosh Kumar Satapathy

1

Unraveling the Mechanisms: Insights Into Stem Cell Behavior in Disease Microenvironments

Pranshul Sethi[1,2], Aniruddha Sen[3], Sonima[4], Ayush Madan[5,6], Syed Mohsin Waheed[7] and Bibhas Kumar Bhunia[8*]

[1]*Chitkara College of Pharmacy, Chitkara University, Punjab, India*
[2]*College of Pharmacy, Shri Venkateshwara University, Gajraula, Uttar-Pradesh, India*
[3]*Department of Biochemistry, All India Institute of Medical Sciences, Gorakhpur, Uttar Pradesh, India*
[4]*Department of Pharmacology, University Institute of Pharma Science and Research, Chandigarh University Gharuan, Punjab, India*
[5]*Department of Biotechnology, School of Research and Technology, People's University, Bhanpur, Bhopal, Madhya Pradesh, India*
[6]*Institute of Tropical Aquaculture and Fisheries, Universiti Malaysia Terengganu, Kuala Nerus, Terengganu, Malaysia*
[7]*Divacc Research Laboratories Pvt. Ltd. AIC-JNUFI, Jawaharlal Nehru University, New Delhi, India*
[8]*Department of Life Sciences, Parul Institute of Applied Sciences & Research and Development Cell, Parul University, Vadodara, Gujarat, India*

Abstract

Stem cell research holds great promise for the development of regenerative medicine, with potential cures and treatments for a wide range of diseases. This is still a subject area of stem cells interacting with the pathological microenvironments and their behavior that remains a core point of investigation.

This chapter offers general insight into the mechanisms by which stem cell behavior is governed within these complex settings, many of which tend to be hostile. This chapter will begin focusing on the basic features of stem cell biology and how a stem cell niche is typically organized under physiological conditions.

**Corresponding author*: bibhasmicro@gmail.com

Ayush Madan, Sumel Ashique, Deepika Arora and Mantosh Kumar Satapathy (eds.) Stem Cell Therapeutics, (1–38) © 2025 Scrivener Publishing LLC

This is accompanied by changes in the microenvironment within these niches, driven by diseases, including modifications to the extracellular matrix, oxygen levels, and the presence of inflammatory signals. These are relevant since they will largely influence the fates of stem cells vis-à-vis the processes of differentiation, migration, and self-renewal. This goes deep into the critical signaling pathways and molecular interactions pivotal in determining stem cell behavior in disease contexts, such as Wnt, Notch, Hedgehog, and so on. This chapter also discusses how stem cells respond to the mechanical and biochemical cues of their altered microenvironments and the implication of such responses for disease progression and therapy.

Moreover, the chapter discusses some of the most recent methodological developments in studying stem cell–disease interactions. Advanced imaging techniques, single-cell RNA sequencing, and innovative *in vitro* and *in vivo* models are discussed in this context, the ways those are being applied to shed new light upon disease-induced dynamics of stem cells. The chapter finishes with an outlook on potential therapeutic strategies built based on understanding derived from stem cell behavior within a diseased microenvironment. Pursuing targeted molecular pathways and microenvironmental factors allows the potential for optimal stem-cell-based therapies, leading to better clinical outcomes. This chapter will not only improve our understanding of the biology of stem cells in disease but it is also set to pave the way toward targeted regenerative therapeutics.

Keywords: Stem cells, disease microenvironments, regenerative medicine, stem cell niche, therapeutic strategies, molecular signaling pathways

1.1 Introduction

Human stem cell research has substantially developed over the past few decades and is beginning to emerge as a critical pillar for regenerative medicine. Stem cells are relevant because of their self-renewal and differentiation ability into different cells; therefore, they can help study the development and mechanisms of diseases with potential therapeutic applications. This continued in full swing after the 1981 discovery of embryonic stem cells from mice. Martin Evans and Matthew Kaufman went on to show that those cells could differentiate into any cell type [1]. The process was carried even further by James Thomson, who derived human embryonic stem cells in 1998, thus making a breakthrough in biomedical research [2].

Early research primarily focused on the fundamental properties of stem cells, such as pluripotency and multipotency. Pluripotent stem cells comprise embryonic and induced pluripotent stem (iPSC) cells. Since they can differentiate to all body cell types, this is more generalized for fundamental research as well as for clinical use [3]. Notably, this field was revolutionized by the discovery of iPS cells in 2006 (which can be derived from adult somatic cells and reprogrammed to an embryonic-like state) that avoid the ethical issues associated with embryonic stem (ES) cell lines [4]. Our knowledge of adult stem cells, i.e., hematopoietic stem cells (HSCs) and mesenchymal stem cells (MSCs), has also blossomed. Symporemonic cells are so named because they are quasistepwise-insoluble virtually in whole tissues, and they abode an moment character to network retrial and helium/popper. This led to our understanding of HSCs, the cells responsible for lifelong renewal of all blood cell types and successful bone marrow transplantation (now utilized worldwide for leukemia and other blood diseases) [5]. Conversely, because of their immunomodulating and regenerative properties [6], MSCs have also become promising tools for the treatment of various diseases including inflammatory diseases or tissue injuries.

However, how stem cells function for a long term in disease microenvironments will be a key factor in developing successful stem cell therapies. When in a pathogenic setting, the way stem cells act is very different and usually more negatively impacted than under normal conditions. For example, in cancer, the unique tumor microenvironment with hypoxic conditions, inflammation, and differences in the extracellular matrix components generally may precondition stem cell fate, leading to tumorigenesis and metastasis in a more interactive way [7]. In neurodegenerative diseases such as Alzheimer's and Parkinson's, the inhibitory effects are due to the inflammatory milieu and neurotoxic factors acting on neural stem cells' regenerative capacity, which precludes them from developing their regenerative potential for repair purposes [8].

Several reasons argue for studying these interactions. It may be helpful for understanding how certain disease factors modify the function of stem cells in terms of differentiation, migration, and self-renewal. Such a knowledge would point exactly toward the optimization of the stem cell therapy procedure in a way that enhances its effectiveness and safety. For example, a comprehension of the precise signaling pathways and molecular cues (e.g., Wnt, Notch, and Hedgehog) might permit to develop measures that would alter the signals, leading to less bountiful therapeutic manifestations [9]. Understanding these mechanisms in a disease context, e.g., within endothelial stem cells, is the key to designing new targeted therapies aimed at addressing the root cause of a diseases rather than merely masking symptoms. For instance, in heart diseases, stem cell therapy is adopted to repair and regenerate the damaged heart muscle cells. How hypoxia influences the properties of stem cells has been largely elucidated, and this information can further us in developing improved approaches to enhance their therapeutic potential [10].

The aim of this chapter is to provide an insight into the possible ways that alterations in the stem cell niche within disease microenvironments can regulate the behavior of a stem cell. This chapter discusses ways in which stem cells are influenced by altered biochemical and biophysical components of the niche (e.g., ECM, oxygen, inflammatory signals) with a focus on how changes to these factors affect fate selections of stem cells. This chapter provides a synopsis of the key signaling pathways and molecular crosstalk that condition stem cell function, within the context of different diseases including Wnt, Notch, and Hedgehog signaling. In addition, the chapter reviews recent methodological advances—advanced imaging, single-cell RNA sequencing, and new *in vitro* and *in vivo* models—that are transforming research into stem cell–disease interactions. While this review is not aimed at documenting the development and utility of each, these tools have been extremely valuable in advancing our knowledge regarding stem cell biology on disease states and only deserve mention here. Lastly, we will discuss potential therapeutic interventions arising from our increasing understanding of how stem cells may behave in diseased microenvironments. We seek to provide the strategy's details concerning molecular targets and the microenvironment on how to realize enhanced effectiveness of stem cell-based therapies toward ensuring better clinical outcomes overall.

1.2 Stem Cell Biology and the Stem Cell Niche

1.2.1 Characteristics of Stem Cells

Stem cells are unique cells with two defining characteristics: self-renewal and differentiation potential. These properties are essential for their role in development, tissue repair, and regenerative medicine.

1.2.1.1 Self-Renewal

A stem cell can also replicate itself without differentiation for many divisions. This phenomenon is known as self-renewal. Self-renewal is a mechanism to ensure stem cell numbers maintain a consistent quantity over the life of an organism. However, it is to be balanced against the opposing activities of differentiation. Intrinsic and extrinsic regulators finely control it.

Among the inherent, the most important are the transcription factor network and signaling pathways. Among these, Wnt, Notch, and Hedgehog pathways are included. Major signaling pathways that help maintain the undifferentiated state of stem cells include Wnt signaling, which promotes the expression of genes associated with stemness and blocks differentiation signals [9]. Another major pathway is Notch signaling, which controls cell fate by determining the balance between differentiation and self-renewal, often context-dependent [11]. The Hedgehog pathway is thus vital in the modulation of stem cell proliferation and differentiation, mainly for tissue repair and regeneration [12].

Extrinsically regulated components of the stem cell niche comprise other cellular and molecular elements providing signals for self-renewal; these include growth factors, cytokines, and constituents of the extracellular matrix interacting with stem cells' cell membrane receptors that modulate their behavior [13]. The niche not only supply the necessary factors for self-renewal but also provides a protective environment that excludes stem cells from differentiation cues and extrinsic damage.

1.2.1.2 Differentiation Potential

The ability of stem cells to differentiate is often referred to as potency, which is a measure of the potential for generating different cell types. The capacity changes with the kind of stem cell and is typically divided into totipotent cells, pluripotent cells, and multipotent cells.

Cells at these stages of development have totipotency, which is the ability to give rise to all types of cells in the body and extraembryonic tissues [14]. Pluripotent stem cells, including embryonic stem cells (ESCs) and induced pluripotent stem cells (iPSCs), have the potential to develop into any one of the cell types that are derived from three germ layers: ectoderm, mesoderm, and endoderm. ESCs are isolated from the inner cell mass of a blastocyst. At the same time, iPSCs are generated through the reprogramming of an adult somatic cell with forced expression of a particular group of transcription factors [2, 4].

Multipotent stem cells, such as hematopoietic (HSCs) and mesenchymal stem cells (MSCs), can only differentiate into cell types of one lineage. HSCs are multipotent stem cells residing in the bone marrow, capable of differentiation into all kinds of blood cells and representing important sources for treating diseases of hematological origin, such as bone marrow transplantation [15]. Most tissue-resident MSCs can differentiate into osteoblasts, chondrocytes, and adipocytes; therefore, these cells are widely researched in regenerative therapy because of their immunomodulatory properties (Figure 1.1).

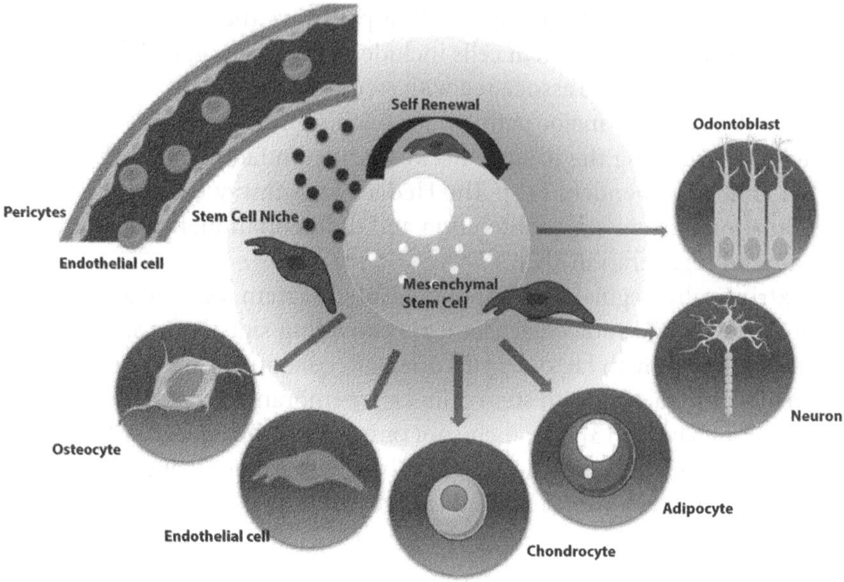

Figure 1.1 MSCs have a multipotential ability to differentiate into cells that depend on certain signaling pathways such as osteoblasts (bone-forming cells), endothelial cells (lining blood vessels), chondrocytes (cartilage forming cells), adipocytes (fat-tissue-forming cells), odontoblasts (dentin produced in teeth), and neural lineages involved in the nervous system.

It is vital to determine the mechanisms by which differentiation is regulated to utilize stem cells for therapeutic purposes. Cells undergo differentiation through the activation and repression of specific gene networks regulated by signaling pathways such as transforming growth factor beta (TGF-β), bone morphogenetic proteins (BMP), and fibroblast growth factor (FGF), among others [16]. In contrast, these pathways act to external signals from the niche and guide stem cells into specific lineages.

1.2.2 The Stem Cell Niche

1.2.2.1 Normal Physiological Conditions

A stem cell niche is such a specialized microenvironment that supports, controls, and helps convey signals necessary for stem cell maintenance with stemness properties and guides its function. In fact, under normal physiological conditions, the niche regulates the balance of the self-renewal capacity of the stem cells against differentiation to guarantee tissue homeostasis, thereby permitting the repair and regeneration of tissues effectively [17]. This microniche comprises several cellular and acellular components: stromal cells, extracellular matrix (ECM) proteins, blood vessels, and a milieu of signaling molecules (Figure 1.2). Each plays a critical role in creating an enabling environment for the stem cells [18].

On the other hand, HSCs are located in two different niches of the bone marrow: the osteoblastic niche and the vascular niche. The osteoblastic niche, located at the endosteal surface of the bone, mainly supports HSC quiescence and long-term self-renewal by supporting osteoblast differentiation. Osteoblasts secrete factors within the HSC, such as osteopontin and angiopoietin-1, which activate adhesion molecules through the interaction of their receptors to keep them in a quiescent state [19].

The vascular niche, on the other hand, is associated with blood vessels and supports HSC proliferation and differentiation by being rich in growth factors and oxygen supply [20]. In essence, therefore, a dual-niche system for the dynamic balancing of maintenance and mobilization against physiological needs is set. Other cell types residing within the neural stem cell niche are located mainly in the subventricular zone (SVZ) and the hippocampus and include astrocytes, ependymal cells, and endothelial cells, which also participate in the regulatory environment [21]. Astrocytes secrete essential factors for the growth and differentiation of neural stem

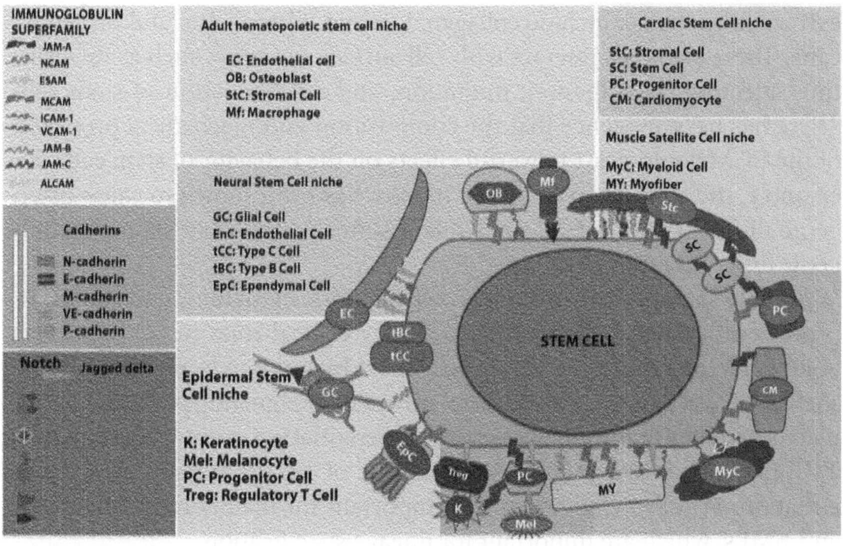

Figure 1.2 Heterocellular molecular contacts in the mammalian stem cell niche.

cells (NSCs), such as brain-derived neurotrophic factor and fibroblast growth factor (FGF) [22]. Endothelial cells forming the blood–brain barrier secrete signals that influence neurogenesis and facilitate the integration of new neurons into mature networks [23].

The depiction of the stem cell niche showcases stromal cells, adhesion molecules, and the extracellular matrix, highlighting how these components regulate stem cell behavior and maintain tissue homeostasis.

1.2.2.2 Key Components and Their Roles

The key components comprise cellular constituents, extracellular matrix components, and soluble factors within a stem cell niche—all of which interact to control the behavior of stem cells.

Cellular constituents: The niche consists of different cell types directly interacting with stem cells to provide support and regulatory signals. As such, stromal cells produce ECM proteins and growth factors required for the fate decisions of the interacting stem cells [24]. For instance, in bone marrow, osteoblasts and endothelial cells are needed for the maintenance of HSCs by producing certain niche-specific factors such as CXCL12, which retains HSCs in a quiescent state [25].

It has been reported to contain complex proteomic and polysaccharide networks, making it responsible for the mechanical support of the stem cells and holding other biochemical signals. The main factors in the extracellular matrix include collagen, laminin, fibronectin, and proteoglycans. These elements interact with cell surface receptors such as integrins, thus affecting the adhesion, migration, and differentiation of stem cells. There has been evidence that the composition and mechanical properties of the ECM could exert powerful effects on the behavior of stem cells. For example, the stiffness of the ECM *in vitro* has been shown to cause osteogenic differentiation of MSCs, while a softer matrix may support adipogenic differentiation [26].

Soluble factors—this means growth factors, cytokines, and hormones produced by niche cells themselves—play critical roles in SC regulation. Soluble factors in general exert their effect by binding to receptors on the surface of SCs and thereby activate intracellular signaling pathways to regulate self-renewal, proliferation, and differentiation events. These include TGF-β, which is involved in the regulation of cell proliferation and differentiation; Wnt proteins, which are important for maintaining "stemness"; and BMPs, which are important for lineage specification.

The complex interactions of these elements were essential in unraveling whether the niche can regulate stem cell responses in homeostasis. Building on this knowledge might be essential one day for designing therapeutic strategies to intervene in the niche in order to increase the effectiveness of stem-cell-based therapies across a range of diseases.

1.3 Disease Microenvironments and Their Impact on Stem Cells

1.3.1 Alterations in the Extracellular Matrix

The stem cell niche, which includes inanimate and living components surrounding the stem cells, provides a structural support as well as biochemical signals that contribute to sustaining the functions of embedded stem cells; one typical component is called extracellular matrix (ECM). Under normal physiological conditions, the ECM serves to maintain tissue integrity and cell-to-cell communication. However, in diseased states, particularly where the ECM is compositionally and mechanically altered, there may be an aberrant influence on the behavior of stem cells.

During diseases such as cancer, fibrosis, and chronic inflammation, the ECM undergoes remodeling that alters many of its features including stiffness, composition, and presence of bioactive molecules. ECM composition and stiffness have been reported to be altered in the tumor microenvironment, leading to increased proliferation and migration of cancer cells [27]. Such changes will also alter the characteristics of resident stem cells. One of these are cancer-associated fibroblasts (CAFs), which secrete extracellular matrix components and remodeling enzymes that generate a pro-tumorigenic niche, leading to cancer stem cells (CSC) stemness and survival [7].

Fibrotic diseases, characterized by excessive ECM deposition, similarly affect stem cell function. In liver fibrosis, for example, the dense and stiff ECM impairs the regenerative capacity of hepatic progenitor cells, leading to compromised tissue repair [28]. Similarly, in pulmonary fibrosis, the altered ECM microenvironment hinders the differentiation potential of lung progenitor cells, contributing to disease progression [29]. Table 1.1 outlines how different disease conditions alter the extracellular matrix in stem cell niches and the subsequent impact on stem cells, alongside potential therapeutic strategies.

Table 1.1 Changes in stem cell niches under disease conditions.

Disease condition	Normal ECM components	Altered ECM components	Impact on stem cells	Therapeutic strategies
Cancer	Collagen, laminin	Increased stiffness, altered composition	Enhanced stemness, survival of CSCs	ECM modification, CAF targeting
Fibrosis	Balanced ECM proteins	Excessive ECM deposition	Impaired regeneration, increased stiffness	Anti-fibrotic therapies
Chronic inflammation	Normal cytokine levels	Elevated TNF-alpha, IL-6	Induced apoptosis, senescence	Anti-inflammatory drugs

1.3.2 Role of Inflammation

Inflammation is considered to be a hallmark of many diseases and exerts a profound impact on stem cell behavior. Although inflammatory signals, such as cytokines, chemokines, and growth factors, can either positively or negatively regulate stem cell (SC) functions depending on the context and duration of the inflammatory response, the inflammation itself might have dual effects on SC-mediated tissue regeneration.

Chronic inflammation, a hallmark of autoimmune disorders, infections, and cancer, represents a somewhat unfavorable microenvironment for normal SC activities. Proinflammatory cytokines—tumor necrosis factor-alpha (TNF-α), interleukin-6 (IL-6), and IL-1β—may induce SC apoptosis, senescence, or aberrant differentiation [30]. For example, in the context of inflammatory bowel disease, chronic inflammation impairs the regenerative capacity of intestinal stem cells, leading to persistent epithelial damage and disrupted gut homeostasis [31].

Acute inflammation after injury can, conversely, directly activate stem cells and promote tissue repair. In the initial phases of wound healing, for example, inflammatory cells have been demonstrated to secrete trophic signals, thereby linking the recruitment and activation of stem cells at the site of injury that enables regeneration [32]. Knowing exactly how inflammation exerts these opposing roles in the regulation of stem cells will be important in the design of therapies intended for translation into clinical application by tipping the balance toward beneficial inflammatory signals while dampening harmful effects.

1.3.3 Effects of Hypoxia

Hypoxia, or low oxygen levels, is another significant factor influencing stem cell activity in disease microenvironments. Hypoxic situations are often described in ischemic tissues, malignant tissues, and inflammatory areas. Hypoxia-inducible factors (HIFs) are crucial regulators of the physiological response to low oxygen levels, as well as the maintenance and differentiation of stem cells. Tumor microenvironmental hypoxia activates HIF-1α and HIF-2α, which increase the lifetime of cancer stem cells by activating genes related to survival, stemness, and metabolic flexibility [33]. Because of their more stem-like character, hypoxic cancer cells are more resistant to treatment and are more able to spread [34].

As demonstrated by myocardial infarction and stroke, hypoxia in ischemic conditions affects local stem cells and their ability to support tissue restoration. For instance, in myocardial infarction, hypoxia diminishes the potential of cardiac progenitor cells to repair damaged cardiac tissue [10]. One of the most important ways to improve outcomes in ischemic disorders is to enhance stem cell survival and function in low oxygen environments and to modulate hypoxia responses.

1.3.4 Disease-Specific Factors

Different illnesses effect stem cell function in different ways by changing the microenvironment. Because of the accumulation of toxic proteins and neuroinflammatory signals in their environment, neural stem cells in neurodegenerative diseases like Parkinson's and Alzheimer's are unable to proliferate and differentiate into functional neurons [8]. This results in a slow loss of neuronal populations and a reduced capacity to effectively repair the damaged nervous system. In metabolic diseases like diabetes, high glucose levels and oxidative stress alter the stem cell niche, which affects the capacity of several stem cell types, such as mesenchymal stem cells and pancreatic progenitor cells, to regenerate [35]. These modifications make it more difficult for stem cells to return normal tissue function, which accelerates the course of the illness.

The immune system targets healthy tissues in autoimmune disorders, causing an inflammatory environment that disturbs the stem cell niche. For example, in multiple sclerosis, the immune system's destruction of the myelin sheaths surrounding neurons produces an environment that prevents oligodendrocyte progenitor cells from differentiating, which hinders remyelination and neural repair [36].

1.4 Molecular and Cellular Mechanisms

1.4.1 Signaling Pathways Involved

Stem cell behavior is tightly regulated by a network of signaling pathways that maintain their self-renewal and differentiation capabilities. Among these, the Wnt, Notch, and Hedgehog pathways play pivotal roles in various physiological and pathological contexts as summarized in Table 1.2.

1.4.1.1 Wnt Signaling Pathway

The control of stem cell self-renewal, proliferation, and differentiation depends on the Wnt signaling pathway. It is highly conserved and has a role in tissue homeostasis, regeneration, and embryonic development. Different cellular outputs are mediated by both canonical (β-catenin dependent) and non-canonical (β-catenin independent) routes of wnt signaling [37].

Within the canonical Wnt pathway, β-catenin is stabilized and undergoes nuclear translocation as a result of Wnt ligands binding to Frizzled receptors and co-receptors LRP5/6. After entering the nucleus, β-catenin binds to TCF/LEF transcription factors to trigger target genes that hinder differentiation and encourage cell division [38]. The preservation of stem cells' undifferentiated status depends on this route. For instance, Wnt signaling in the intestinal crypts maintains the population of stem cells that are in charge of the ongoing regeneration of the intestinal epithelium [39].

Table 1.2 Key signaling pathways and their roles in stem cell regulation.

Signaling pathway	Key components	Role in stem cell regulation	Disease contexts	Therapeutic implications
Wnt	Beta-catenin, Frizzled receptors	Promotes stemness, inhibits differentiation	Cancer, Neurodegenerative diseases	Potential target for cancer therapy
Notch	Notch receptors, delta/jagged ligands	Controls cell fate decisions, maintains stem cell populations	Hematopoietic disorders, T-cell leukemia	Target for Notch-dependent cancers
Hedgehog	Ptch, Smo, Gli transcription factors	Regulates stem cell proliferation and differentiation	Basal cell carcinoma, medulloblastoma	Target for Hedgehog pathway inhibitors

Cancer is among the many disorders linked to the dysregulation of Wnt signaling. The proliferation and survival of cancer stem cells in colorectal cancer have been linked to hyperactivation of the Wnt/β-catenin pathway [40]. Gaining knowledge on the molecular processes that underlie Wnt signaling in stem cells might help identify new therapeutic targets for cancer and regenerative medicine.

1.4.1.2 Notch Signaling Pathway

Another extremely conserved system that governs stem cell maintenance, differentiation, and cell fate decisions is the Notch signaling pathway. When a Notch receptor on one cell surface attaches to a ligand (Delta or Jagged) on a neighboring cell, Notch signaling is set off. This contact induces proteolytic cleavages that release the Notch intracellular domain (NICD), which translocates to the nucleus and interacts with the CSL (CBF1/Su(H)/Lag-1) transcription factor to control gene expression [41].

A key function of notch signaling is seen in many stem cell systems. Notch signaling is essential for the upkeep of hematopoietic stem cells (HSCs) throughout hematopoiesis and affects how these cells differentiate into various blood cell lineages [42]. Notch signaling controls the ratio of neural stem cell proliferation to differentiation throughout neural development, promoting healthy brain growth and function [43]. Gain-of-function mutations in Notch1 cause unchecked cell proliferation in T-cell acute lymphoblastic leukemia (T-ALL), one malignancy where aberrant Notch signaling is linked [44]. The possibility of therapeutic approaches that target Notch signaling, including γ-secretase inhibitors, to treat malignancies that are dependent on Notch is being investigated [45].

1.4.1.3 Hedgehog Signaling Pathway

For both adult tissue homeostasis and embryonic development, the Hedgehog (Hh) signaling system is essential. In order to relieve the inhibition of the Smoothened (Smo) receptor, it entails the binding of Hh ligands (Sonic, Desert, and Indian Hedgehog) to the Patched (Ptch) receptor. The activation of Gli transcription factors, which control the expression of target genes involved in cell proliferation, differentiation, and survival, is the result of a signaling cascade started by activated Smo [46].

Hedgehog signaling is critical for maintaining stem cell populations and regulating stem cell habitats in the setting of stem cells. For example, Hedgehog signaling promotes the growth and maintenance of hematopoietic stem cells in the bone marrow and neural stem cells in the brain [47]. It is also important for the regeneration of tissues including bones, intestines, and skin.

Cancer and developmental problems are linked to Hedgehog signaling dysregulation. For instance, medulloblastoma and basal cell cancer can result from mutations in pathway components [48]. Evidence that inhibition of Hedgehog signaling was therapeutic, at least in the treatment of these malignancies with drugs similar to vismodegib, reinforces the viability of targeting this system [49].

1.4.2 Molecular Interactions

Stem cell behavior and function are highly regulated by molecular interactions within the stem cell niche. These interactions are mediated by an intricate network of signaling pathways, cell–cell contacts, and with the extracellular matrix (ECM). This knowledge is essential to elucidating how stem cells perceive their biochemical surroundings and how these responses are altered in disease.

The extracellular matrix (ECM) is a complex molecule interaction which provides the structural support and biochemical cues needed to maintain adhesions between cells, but stem cell behavior is among the most important. The extracellular matrix (ECM) surrounds cells and is composed of proteins such as collagen, fibronectin, laminin, and proteoglycans that interact with integrins and other cell surface receptors. These interactions also trigger intracellular signaling cascades that determine the fate choices of stem cells [13]. Integrin engagement to the extracellular matrix (ECM) activates focal adhesion kinase (FAK) and subsequent downstream MAPK/ERK pathway, leading to stem cell survival and proliferation; for example, this is essential for stemness maintenance of CSCs from human glioblastoma cells [50].

Moreover, stem cell regulation by the niche is substantially mediated through cell–cell interactions. Two of the more understudied molecular signaling pathways that govern the destiny of stem cells are cell–cell interactions, and Notch signaling is one of the quintessential examples for this process. In response to one of its cell ligands (Jagged or Delta) directly expressed on a neighboring stem cell, Notch expressed on a stem cell connects with the membrane-bound receptor which in turn triggers proteolytic cleavages that release the Notch intra-cellular domain (NICD). In order to manage stem cell differentiation and preserve stem cell populations, the NICD translocates to the nucleus and interacts with CSL transcription factors to regulate gene expression [41].

Moreover, stem cell control depends on paracrine communication, in which cells release chemicals that influence neighboring cells. Growth factors that are released by niche cells, including transforming growth factor-beta (TGF-β), epidermal growth factor (EGF), and fibroblast growth factor (FGF), bind to receptors on stem cells and initiate signaling pathways that control cell proliferation, differentiation, and death [16].

1.4.3 Mechanical and Biochemical Cues

In their surroundings, stem cells respond very well to mechanical and pharmacological stimuli. These signals are essential for controlling the migration, differentiation, and proliferation of stem cells.

Mechano-transduction processes allow stem cells to detect mechanical signals such cyclic strain, shear stress, and substrate stiffness. For example, substrate stiffness can control stem cell differentiation. It has been shown recently that osteoblasts are the main type of cells differentiated from mesenchymal stem cells (MSCs) when they are cultured on a stiff substrate, whereas adipocytes develop more frequently from MSCs cultured within a soft matrix [26]. Specifically, mechanosensitive ion channels and cytoskeletal links from the ECM to the nucleus that allow it to directly influence gene expression are involved [51]. An important component for this to happen is the activation of YAP/TAZ pathway when mechanical stress occurs, which regulates cell division and proliferation gene expression [52].

Endothelial cells in blood vessels experience shear stress too, which also affect stem cell activity. It has been discovered that shear stress, for example, stimulates the differentiation of endothelial progenitor cells to mature ECs, thereby promoting vascular regeneration and repair [53]. The application of mechanical forces such as stretch (and strain) to specific tissues, like muscle and tendon, induces signaling pathways which coordinate the activity of resident stem cells within that tissue, thus contributing to tissue repair and maintenance [54].

Influence of chemical signals to the activity with stem cells is also a prime importance. Soluble molecules, such as growth factors, hormones, and cytokines, bind to cell surface receptors, leading to intracellular signaling cascades that have an impact on the fate of stem cells. TGF-β signaling, for example, is critical to maintain stemness in a variety of different stem cell types and to promote differentiation [55]. As mentioned above, wnt signaling is the key to maintain stem cells in an undifferentiated way and also stimulate growth [37].

In addition, how mechanical and biochemical stimuli combine to control stem cell responses is still not well understood. The extracellular matrix (ECM) not only stiffens the structure when the stem cell retains physical support. For example, integrins transduce environmental signals and mechanical stimuli from extracellular matrix to direct cell decisions [56]. All the factors discussed above, including molecular signaling pathways, molecular interactions, and biomechanical cues, interact to modulate stem cell behavior, ultimately influencing therapeutic treatment outcomes.

1.5 Techniques for Studying Stem Cell Behavior in Diseased States

1.5.1 Advanced Imaging Techniques

Advanced imaging techniques provide thorough spatial and temporal information on cellular processes, so they are required to study the behavior of stem cells undergoing pathological conditions. Live-cell imaging, two-photon microscopy, and confocal microscopy can be used by researchers to observe stem cells as well as their interactions with the microenvironment on real time.

Confocal microscopy uses spatial pinhole and point illumination in order to exclude out-of-focus light and delivers high resolution, three dimensional images. This method is also particularly powerful for investigating stem cell motility and localization within tissues. For example, confocal microscopy has been employed to track migration and differentiation of neural stem cells in brain damage models, revealing insights into regenerative medicine [57].

Two-photon microscopy excites fluorescent molecules with near-infrared light to achieve deeper tissue penetration without photodamage. This technique enables *in vivo* imaging of stem cells, which allows for visualization of cellular behavior at their physiological location. Two-photon microscopy has also recently shown that mesenchymal stem cells (MSCs) colonize injury sites and how they affect surrounding tissue during tissue repair [58]. Live-cell imaging techniques including fluorescent lifetime imaging microscopy (FLIM) and fluorescence resonance energy transfer (FRET) provide dynamic insights into molecular interactions and signaling networks in stem cells. These approaches have been applied for investigations of pharmacological stimuli [59] and monitoring real-time activation of signaling cascades, such as Wnt and Notch in stem cells [40]. This section lists examples of advanced approaches for investigating stem

Table 1.3 Advanced techniques for studying stem cell–disease interactions.

Technique	Principle	Applications	Advantages	Limitations
Confocal microscopy	Spatial pinhole, point illumination	Studying stem cell motility, localization	High-resolution 3D imaging	Limited tissue penetration
Two-photon microscopy	Near-infrared light, deep tissue penetration	Imaging stem cells in living animals	Minimizes photodamage	Expensive equipment
Single-cell RNA sequencing	Sequencing RNA from individual cells	Identifying stem cell subpopulations	High-resolution gene expression data	High cost, complex data analysis

cells and their interactions in disease mechanisms, shown in Table 1.3, including the basic principles used, their applications as well as advantages and limitations.

1.5.2 Single-Cell RNA Sequencing

Single-cell RNA sequencing (sc-RNA-seq) offers by far the greatest potential for detailed exploration of single cell gene expression. The technique allows an analysis of DNA variations in single cells. Scientists automatically controlled cellular response to chemical signals.

ScRNA-seq accomplishes this by isolating individual cells, reverse transcribing (creating cDNA) the RNA from these cells, and sequencing each of these cDNAs to measure the expression levels of specific genes. This technique has been a real game-changer in our understanding of stem cell biology as it showed just how complex the environment of stem cell niches can be and just how different the properties of a single population of stem cells in healthy tissue is from those found among many states directly related to disease. For instance, single-cell transcriptomic analyses have revealed heterogeneity in cancer stem cell populations with varying degrees of stemness and drug resistance between malignant subtypes [60].

In addition, scRNA-seq has also offered valuable new knowledge on the molecular mechanisms specific for stem cell plasticity and differentiation. Alternatively, a lower level of stem cell expression in sick versus healthy situations could lead to the identification of critical regulatory pathways and potential therapeutic targets for intervention. When applied to an example, scRNA-seq analysis of hematopoietic stem cells in leukemia, dysregulated signaling pathways promoting disease progression and drug resistance were identified [61].

1.5.3 *In Vitro* Models

Therefore, in order to test the function of stem cells in a regulated environment, we need suitable methods—such as those used with an *in vitro* model. These models include less complex two-dimensional (2D) cultures, organoid and tissue engineering models in three-dimensions.

2D cultures are also the simplest, most inexpensive way to study many basic biological questions such as differentiation, proliferation and drug response. One way to do this is by doping pellets in one of several types of hydrogel and follow 2-hydroxyethyl methacrylate (HEMA) and 2-hydroxyethyl acrylate (HEA) polymerization for 3D growth, or grow as a monolayer on flat surfaces, which constitutes more standard 2D culture. Nevertheless, the 2D culture does not provide the intricate natural milieu of tissue that is associated with *in vivo* conditions [62]. Organoids, which have been derived from stem cells and grown using 3D culture, reproduce the cellular heterogeneity and tissue architecture, making them a more physiologically relevant model. The brain, liver, and intestines are among several organs for which organoids have been created as the artificial versions provide significant insights into how drugs work, progress of a certain illness, and also discovery. For instance, intestinal organoids models have been employed to interrogate the effects—inflammatory cues—on stem cell function and epithelial regeneration in the context of inflammatory bowel disease [63].

Tissue engineering approaches that exploit advances in biophysical and mechanical properties of matrices with bioengineered scaffolds or hydrogels offer an opportunity to reproduce the extracellular environment *in vitro* by coupling chemical cues and physical forces similar to the stem cell niche. These models inherently provide a platform for research into diseased states where understanding the interactions between stem cells and their environment in context with disease will assist in the discovery of new approaches to treatment [64].

1.5.4 *In Vivo* Models

In vivo models are mandatory to investigate stem cell activity at the level of the whole organism. These models have been useful in delineating how stem cells respond to physiological and pathologic stimuli, and interact within their endogenous milieu.

There are genetic ways to manipulate mouse physiology, and because of its homology with human physiology, mouse models often find applications in stem cell research. Transgenic and knockout animal models can also be used to investigate how genes, or signaling pathways, function in the regulation of stem cells as well as pathological development. Wnt and Notch likewise have a specific role in stem cell maintenance and development among other lineages which has been possible to clarify by using conditional ablation of components to these pathways within normal mice suggesting that the requirement for these pathways is more general [65].

The transparency and quick growth of zebrafish embryos have made them appealing models to study stem cell dynamics as they occur in real time. For example, studies have analyzed the migration and proliferation of hematopoietic stem cells in zebrafish [66] or whether spinal cord injury also induces regeneration of neural stem cells [67]. Another approach to examine cancer stem cells and determine potential therapeutic agents is to use patient-derived xenografts (PDX). To use PDXs, immunodeficient mice have to be transplanted with human tumor tissues. The PDX models have led to the discovery of potential treatment targets and biomarkers, significantly contributing to personalized cancer therapy [67].

1.6 Stem Cell Dynamics in Various Diseases

1.6.1 Cancer

Cancer arises due to unrestrained cell proliferation and the potential to infiltrate surrounding tissues and metastasize to remote organs. Both the tumor microenvironment (TME) and behavior of cancer stem cells (CSCs) play pivotal roles in the advancement of malignancy and therapeutic resistance.

1.6.1.1 Tumor Microenvironment

The tumor microenvironment is characterized by a complex and dynamic milieu of immune cells, stromal cells, cancer cells, extracellular matrix (ECM) and blood vessels. It is important for tumor growth and metastasis as well as response to therapy. Such elements together create a niche that benefits clonal growth and survival of cancer stem cells [68].

CAFs (cancer-associated fibroblasts) are one of the primary constituents of the TME. The release of extracellular matrix proteins, cytokines and growth factors that may promote tumor development. For example, activated CAFs are known to secrete transforming growth factor-beta (TGF-β), which promotes an epithelial-mesenchymal transition (EMT) in cancer cells causing increased invasion of the tumor cells [69]. CAF-induced modification of the extracellular matrix (ECM) has also been described and is thought to be one way in which a desmoplastic stroma exerts its effect, by promoting the growth and invasion of tumor cells into a stiffer ECM [27]. Immune cells are present in the tumor microenvironment (TME) and also one of the main regulatory factors for tumor behavior, including macrophages, T-cells and myeloid-derived suppressor cells (MDSCs). Tumor-associated macrophages (TAMs) can switch into a tumoricidal phenotype [70, 71] or pro-tumorigenic phenotype by secreting of cytokines such vascular endothelial growth factor (VEGF) and interleukin-10 (IL-10) that both promote angiogenesis and attenuate anti-tumor immune responses. Regulatory T cells (Tregs) help cancer cells escape the immune system by inhibiting effector T cell function [71].

Low oxygen levels, or hypoxia—a hallmark of the TME—profoundly impact the activity of cancer cells. Upon the establishment of hypoxic conditions, hypoxia-inducible factors (HIFs) induce the transcription of angiogenesis metabolic genes and cell survival. Furthermore, hypoxia plays the important role in maintaining CSCs by improving CSCs' stemness characteristics and resistance to therapy [33].

1.6.1.2 Stem Cell Behavior in Cancer

A fraction of tumor cells that is able to self-renew and differentiate into distinct cells types within the tumor are called cancer stem cell (CSCs). They are believed to contribute to tumor initiation, growth and spread, as well as relapse. GSCs share several common properties with normal stem cells, including upregulation of some stem cell markers and activation of key signaling pathways [72]. One of the most critical features attributed to CSCs is their ability to withstand traditional cancer therapeutics. This resistance can be achieved through various mechanisms such as the activation of anti-apoptotic pathway, upregulation of DNA repair capacities or expression of drug efflux pumps [3]. CSCs often express high levels of multidrug resistance-associated proteins, p-glycoprotein and other ATP-binding cassette (ABC) transporters which efflux many chemotherapeutic agents from the cell leading to reduced intracellular concentrations and decreased efficacy [73].

The TME supports CSC survival and self-renewal or tumor dedifferentiation through diverse molecular pathways. The Wnt/β-catenin pathway is frequently active in CSCs and crucial for protecting their stemness, including proliferation [74]. In addition to control of CSCs cell fate choices, Notch signaling system is also a critical transcriptional regulator that maintains the CSC inside the tumor [41]. Hedgehog signaling is also attributed to CSCs control. It exerts a function of promoting tumor formation and maintaining the stem cell properties of CSCs. In fact, the inhibition of Hedgehog signaling has been demonstrated to reduce both the number of CSCs, as well as the progression of tumor formation in many studies/animals with automated citation [75]. One advantage of CSCs is their plasticity, allowing them to transition between different states, as required by fluctuating environmental and treatment pressures. The TME controls this plasticity by delivering signals that might trigger EMT, a mechanism that amplifies CSCs' capacity for invasion and metastasis [76]. Incorporating virus-like particles (VLPs) as a novel cancer therapy strategy holds significant promise, as they enhance drug delivery and imaging capabilities in cancer treatment, providing a dual-functional approach that can improve therapeutic outcomes [77]. Improving the results of cancer treatment may be achieved by focusing on CSCs and how they interact with the TME. There is current research being done on therapies that interfere with the beneficial interactions that CSCs have with the TME, block important signaling pathways, or target CSC-specific markers only [78]. Table 1.4 presents how stem cells respond to different disease microenvironments, such as in Alzheimer's Disease, Myocardial Infarction, and Diabetes, highlighting key microenvironmental changes, stem cell responses, outcomes, and potential interventions.

Table 1.4 Stem cell responses in various disease microenvironments.

Disease	**Key microenvironmental changes**	**Stem cell response**	**Outcome**	**Potential interventions**
Alzheimer's disease	Amyloid-beta plaques, neuroinflammation	Impaired proliferation and differentiation	Neuronal loss, cognitive decline	Anti-inflammatory agents, stem cell transplantation
Myocardial infarction	Hypoxia, fibrosis	Reduced regenerative capacity	Heart failure	Preconditioning of stem cells, growth factor delivery
Diabetes	High glucose, oxidative stress	Impaired stem cell function	Compromised tissue repair	Antioxidants, improved glycemic control

1.6.2 Neuro-Degenerative Diseases

Neurodegenerative diseases, such as Alzheimer's, Parkinson's, and Huntington's disease, are characterized by progressive loss of neuronal function and structure. Understanding the microenvironmental changes and stem cell responses in these diseases is crucial for developing effective therapies.

1.6.2.1 Microenvironmental Changes

Neurodegenerative illnesses cause profound changes to the microenvironment that affect how resident neural stem cells (NSCs) behave and how the disease progresses. Within the steady state brain, NSCs reside in restricted microenvironments, termed niches: these correspond to sites including the hippocampus dentate gyrus and the subventricular zone (SVZ), where NSCs are nurtured by an intricate mixture of growth factors, ECM elements as well as cellular interactions [79]. Neurodegeneration is often associated with alterations of the ECM composition. Pathologic changes of Alzheimer's disease include the development of amyloid-beta plaques and neurofibrillary tangles, which result in neuronal damage as well as a deviation to structural integrity and signaling environment of the NSC niches [80]. This disturbed ECM may reduce the brain's capacity for recompensation and reduced NSCs proliferation/differentiation.

In addition, another hallmark of all neurodegenerative disorders is chronic inflammation. Upon activation, microglia and astrocytes start secreting pro-inflammatory cytokines, including TNF-α, IL-1β and IL-6 that result in an unfavorable microenvironment for NSCs. Inflammatory signals may additionally impair the potential for brain regeneration and result in apoptosis or senescence of NSCs [8]. Oxidative stress, in turn, exacerbates neuronal damage and impairs NSC function [81], with the additional generation of reactive oxygen species (ROS) also implicated.

1.6.2.2 Stem Cell Responses

Neuronal stem cells (NSCs) maintain their restorative capabilities and adaptation to disease conditions in the harsh diseased brain environment. However, but in the presence of progressing disease too often they do not regenerate and repair injured brain tissue adequately. In reply for neurodegradation due to insults, NSCs attempts to expand and differentiate into new glial and neuronal cells. However, it comes with a huge sacrifice of efficiency in the mentioned methods. In Parkinson's disease that translates to the death of dopaminergic neurons in the substantia nigra and subsequent decreased motor function. Despite the capacity of NSCs to differentiate into dopaminergic neurons, the harsh environment and lack of endogenous instructive signals in the SVZ make therapeutic exposure difficulty [82].

Stem cell therapies due to their unprecedented tissue regenerative potential may serve as a significant modulatory adjunct for promoting the natural process of repair by NSCs. Approaches such as implantation of iPSCs or exogenous NSCs have been considered to achieve this. These cells are able to integrate in the host tissue, differentiate into neurons and produce trophic factors supporting endogenous healing processes after transplantation [83]. Finally, the introduction of CRISPR/Cas9 and other gene editing approaches allows for the correction of genetic abnormalities associated with neurodegenerative diseases to improve the function of either local or transplanted stem cells [84].

1.6.3 Cardiovascular Diseases

Cardiovascular diseases (CVD) are one of the largest causes of morbidity and mortality globally, with myocardial infarction (MI), including the subsequent chronic ischemic heart disease, among the most prevalent and debilitating [1]. The study of stem cells in the setting of acute ischemia and their capacity to participate in repair has become an important focus for regenerative therapies.

1.6.3.1 Ischemic Conditions

Due to the reduced blood supply, inadequate oxygen and nutrients are supplied to the cardiac tissue leading to ischemic conditions in the heart. Succinctly, this hypoxic environment triggers a cascade of events including inflammation, fibrosis and apoptosis leading to heart failure [85].

In response to ischemia the heart attempts to activate its own healing responses. The new study further adds that hypoxia-inducible factors (HIFs) are the ones activating the cardiac stem cells (CSCs) and progenitor cells in their niche inside the heart to seek out a low-oxygen environment. These stimuli induce transcriptional control of gene expression that enhances angiogenesis, metabolic flexibility and cell survival [86]. However, the severity of the ischemic insult and the extent of fibrotic remodeling in the heart tissue that follows often limits how well these endogenous repair mechanisms can work.

1.6.3.2 Stem Cell Behavior and Repair Mechanisms

Heart research has heavily focused on the regenerative capacity of stem cells to heal and replace scarred heart tissue. Work in both preclinical and clinical settings supports the promise and potency of endogenous CSCs or exogenously administered stem cells like MSCs and iPSCs [87]. Endogenous CSCs repair tissues and it can give rise to smooth muscle, endothelial, and cardiomyocyte through healing. This return in the quantity of these cells is insufficient to fully restore cardiac function after a major ischemia episode, however. Genetic modification of CSCs to overexpress prosurvival and proangiogenic genes or exogenous administration of growth factors such as VEGF, FGF have been shown to enhance the survival advantage for providing a higher chance of rendered myocardial repair [88].

Transplanting stem cells to a heart with local disease is referred to as exogenous stem cell therapy. MSCs derived from bone marrow or adipose tissue are thus attractive candidates due to their ability to secrete various paracrine factors that assists in the repair of tissues and their immunomodulatory properties. They are able to restore heart function by decreasing inflammation, inhibiting formation of fibrotic tissue, and increasing angiogenesis [89]. In contrast, iPSCs could undergo differentiation into cardiomyocytes and other cardiac cell types *in vitro* prior to transplantation to provide a source of functional cells that are able to replace damaged tissue [90]. Although promising, stem cell treatment faces hurdles with regards to the survival and integration of transplanted cells, immunological rejection, and tumor generation. Current research focuses on new cell survival, engraftment optimization and better cell integration as well as improved distribution of cells and long-term impact on heart function with the use of stem cell treatment for myocardial tissue [91].

1.7 Therapeutic Implications

1.7.1 Potential Therapeutic Targets

The identification of robust new therapeutic targets within stem cell biology is essential to the disentanglement of novel and effective treatments for a variety of diseases. It is anticipated that interfering with cancer stem cells (CSCs) involving in the genesis, invasion, dissemination and relapse of tumors maybe a promising therapeutic target for cancer [5]. As many standard therapies demonstrate resistance in CSCs, unique approaches are necessary to target them. The Wnt/β-catenin signaling, an important pathway in maintenance of the CSC phenotype is one such target. Inhibition of this mechanism by, small compounds and monoclonal antibodies have proved to be efficient in decreasing the CSC populations, as it could help affected conventional chemotherapies more effective [9]. The Notch signaling system, which is an evolutionarily conserved cell–cell communication mechanism and maintains CSC populations through controlling lineage specification have been targeted by inhibitors of γ-secretase or neutralizing antibodies [41].

The oxidative stress and inflammatory cellular processes are potential targets for treatment in neurodegenerative diseases. Neural stem cells (NSCs) under conditions in which the pro-inflammatory cytokines, secreted by microglia and astrocytes have been as much possible modulated8 can better exert their reparative and regenerative activity. Moreover, oxidative stress is attenuated by upregulating antioxidants' pathways that protect NSCs and neurons [81]. The listed therapeutic targets for heart repair or angiogenesis related signaling pathways against cardiovascular disorders. VEGF and FGF processes are needed for restoring heart function and promoting new blood vessel growth. Thus, targeting these pathways through the use of therapeutic strategies can offer increased blood supply to ischemic tissues and preserve the cell survival, and functionality of CSCs [86].

1.7.2 Strategies to Enhance Stem Cell Function

Maximizing stem cell activity is critical to unlocking the full therapeutic capabilities of these cells. A number of strategies have been developed to improve the survival, proliferation, differentiation, and engraftment of stem cells into host tissues. One such is preconditioning of stem cells during harvesting before transplantation. This can be done by exposing stem cells to hypoxia or treating them with pharmacological conditions mimicking those found in the tissue environment. This kind of preconditioning enhances the antioxidant and metabolic support of stem cells, that may help them to survive and function better after transplantation [92].

One other approach to augmenting the therapeutic power of the stem cells is gene modification. Although to augment that stem cells can be furthermore genetically engineered to resist the hostile environments of diseased tissues as well by over-expressing genes aiding angiogenesis, differentiation or cell survival. Specifically, upregulation of the anti-apoptotic gene Bcl-2 in MSCs enhances resistance to apoptosis and improves therapeutic efficacy in repairing the infarcted heart [93]. One of the reasons why biomaterials and scaffolds improve stem cell function may be related to their ability to mimic the extracellular matrix (ECM), in terms of both structure and physical support. These substances can be engineered to release cytokines and growth factors which help the cells to differentiate and proliferate. In addition to these properties, such scaffolds can be further configured for incorporation of mechanical and pharmacological signals creating a favorable microenvironment that enhances stem cell activity [64].

1.7.3 Tailoring Treatments to Microenvironmental Contexts

To best translate benefits of stem cell therapies to the clinic, therapies that use stem cells must be tailored to the specific microenvironmental contexts of individual diseases. Comprehending these connections will lead to the development of stronger therapies. A major modulator of stem cell function is the microenvironment. Hypoxia, inflammation, and changes in the extracellular matrix are some of those characteristics within the tumor microenvironment (TME) that are present in cancers. Inflammation can be reduced with anti-inflammatory drugs, blood vessel normalization increases oxygen delivery, and ECM remodeling blocks cancer cell invasion and metastasis [94]. Here are some ways to change the TME to reduce support for CSCs.

In neurodegenerative disorders, the balance between a reduction in oxidative stress and chronic inflammation can favor NSC niches. This can be done by using neuroprotective drugs, antioxidants, and anti-inflammatory drugs so as to improve the environment to make it more favorable for NSC differentiation and proliferation [95]. Improving the cardiovascular disorder microenvironment means reducing fibrosis and vascularization. Direct administration or gene therapy can promote angiogenesis by assisting the blood flow to ischemic regions via growth factors like VEGF and FGF. Similarly, pathological fibrogenesis is reduced by antifibrotic drugs, which in turn promote engraftment of implanted stem cells into the host tissue [87]. So, developing specific methods considering differently microenvironmental obstacles of diseases can also help to optimize stem cell activity and enhanced results after treatment. By understanding the environment of each disease, stem cell treatments can be better tailored for clinical application.

1.8 Case Studies

1.8.1 Successful Regenerative Therapies

This area has seen significant innovation especially over the last few years where very effective treatments have emerged in both restoring function and causing tissue regeneration! One of the most successful examples is the application of stem cell therapy for hematological disorders. This transplantation of hematopoietic stem cells (HSCT) has been adopted as a curative therapy for patients with leukemia, lymphoma, and other blood diseases. High-dose chemotherapy or radiation therapy followed by hematopoietic stem cell transplantation (HSCT) with peripheral blood, bone marrow, or umbilical cord blood is used to reconstitute the hematopoietic system in patients when it has been devastated [96], and while this medication is now archaic in the world of oncology, it has brought several patients way back and given them their fighting chance. Regenerative medicine used to treat osteoarthritis with mesenchymal stem cells (MSCs). MSCs are excellent candidates for regenerating damaged cartilage due to their well-demonstrated immunomodulatory and reparative properties. It has been demonstrated in clinical trials that intra-articular MSCs injections can be effective to help people with osteoarthritis feel and function better on their weighted-bearing joints. It may be a viable alternative to established therapies like corticosteroids and hyaluronic acid injections [97].

In the field of cardiology, stem cell therapy has been proven to be effective in treating myocardial infarction (MI). Autologous bone-marrow-derived stem cells that are being deposited in myocardium previously infarcted have been shown to promote both cardiac function and decrease the extent of scar, leading to better mortality rates for those with heart failure [98]. In addition, a potential treatment for heart disease involves the generation of cardiomyocytes derived from induced pluripotent cells (iPSCs) to ultimately transplant these into the host. One advantage of using iPSCs is that these types can be individualized to the patient and hence may not elicit immune responses, leading to rejection [1].

The production of bioengineered tissues and organs is a milestone in the field of regenerative medicine. Based on a combination of scaffolds and stem cells, researchers have been building functional tissue structures like bladder patches and skin transplants. One good example is designed skin grafts in which autologous keratinocytes and fibroblasts are applied for the definitive treatment of patients suffering from extensive burns that have offered a skin replacement of high durability and usefulness in these cases [99]. On the other hand, long-term survival with function has been achieved following transplantation of bioengineered bladder tissues in patients with bladder dysfunction [100].

1.8.2 Challenges and Lessons Learned

As such, there are several problems of regenerative medicine that need to be addressed to reach its utilization at the maximum level. SCoPE-2 could provide new insights, but there are thousands of experiments to perform and the supply and scalability of cells is one of the biggest problems. Generating sufficient numbers of high-quality stem cells for therapeutic needs is difficult, particularly in a way that allows production on an industrial scale while maintaining both the potency and activity of these cells. For example, *in vitro* expansion of MSCs can lead to cellular senescence and decrease their regeneration capacity and therefore limit the versatility as a therapeutic tool [101].

Furthermore, the complexity of tissue microenvironment and need for accurate replication *in vitro* adds on the additional challenges. The microenvironment plays an important role in regulating stem cell activity and is very difficult to reproduce in a laboratory. While 3D culture systems and bioreactors that mimic the real tissue environment have been described, further advancements are required to optimize these models for therapeutic applications [64]. The major issues with allogeneic stem cell therapy remain rejection and immunogenicity. Although autologous cells lower the possibility of immunological rejection, their use is not always possible because of the patient's condition or the need for prompt therapy. These difficulties may be addressed by creating universal donor cells using gene editing methods or by devising ways to stimulate immunological tolerance [9].

Regenerative medicine has a complicated and dynamic regulatory environment. Strict adherence to regulatory guidelines and extensive clinical testing are necessary to guarantee the safety and effectiveness of stem cell treatments. Unfortunately, the creation of suitable regulatory frameworks frequently lags behind the quick speed of invention in this area, which causes delays in the transfer of promising treatments from the lab to the clinic [102].

Furthermore, regeneration therapy costs are a major deterrent to their broad use. Stem-cell-based therapy can be costly to produce, store, and administer; therefore, it can be difficult to balance affordability with strict adherence to quality and safety regulations. Creating affordable manufacturing procedures and payment plans is essential to enabling a larger patient base to get these treatments [103].

1.9 Future Directions

1.9.1 Emerging Research Areas

Stem cell research is still developing quickly, and a number of new directions seem likely to have a big influence on regenerative medicine. Among the most short-term prospecting was in fact stems can be enhanced for therapy using CRISPR/Cas9 and other gene editing techniques. Because of their ability to be genome-edited *in vitro*, patient-derived induced pluripotent stem cells (iPSCs) have quickly emerged as appealing for personalized therapy. For example, CRISPR/Cas9 has been used to correct the genetic defect in iPSCs from sickle cell anemia patients [104], which might be a promising cure for this devastating disease.

Another relatively new area is the application of artificial intelligence (AI) to stem cell research. Using AI and machine learning algorithms, large datasets can be analyzed to identify trends and predict the behavior of stem cells. This facilitates the optimization of techniques on stem cell differentiation and culture conditions. This approach can potentially speed stem cell therapies to the clinic and make stem cell-based treatments applicable on a broader scale [62]. A second interesting area is organoids—3D cell culture systems that mimic the structure and function of real organs. These organoids can then be developed out of stem cells for multiple purposes, such as testing drugs, modeling diseases and studying the development of organs *in vitro*. Liver, gut, brain and other organoids have been generated in high success rate, providing crucial information about the pathophysiology of diseases as well as enabling unprecedented therapeutic interventions [105].

1.9.2 Innovations in Therapeutic Approaches

Further strategies to optimize stem cell therapies are emerging as new therapeutic options. One approach is to use biomaterials that mimic the native extracellular matrix (ECM) supporting scaffolds, which characterize most tissues. The bioactive molecules could also promote survival of cells and tissue regeneration, recruit stem cells into the injured area and for structural support. For instance, hydrogel-based scaffolds containing growth-promoting molecules have been used to enhance both engraftment and functional contribution of transplanted stem cells in cardiac repair [106].

Extracellular vesicles (EVs) are small membrane-enclosed particles which contain proteins, lipids and nucleic acids that stem from cells. Stem-cell-derived EVs become a promising alternative for cell based treatments due to their ability to modulate the immune system, reduce inflammation, and promote tissue repair. To indirectly verify the safety and efficacy of EVs for the treatment of various diseases, including neurodegenerative syndromes or cardiovascular abnormalities [107], clinical trials are currently ongoing. Even more are under investigation, including combination treatments, which use stem cells and another type of therapy together. For example, use of stem cell therapy combined with immunotherapy could potentially enhance the anti-tumor immune response and overall outcomes in cancer treatment. Conversely, gene therapy and the use of stem cells may complement each other to further enhance gene-based therapeutic approaches for genetic diseases as well as enhance the regenerative potential of stem cells [108].

1.9.3 Long-Term Goals and Visions for Stem Cell Therapy

The long-term objectives of stem cell therapy are lofty and include transforming the way that many different illnesses are treated. One of the main objectives is to restore damaged tissues and organs in a safe and efficient manner, offering a remedy for diseases for which there are presently few therapeutic choices. This covers uncommon genetic conditions and injuries in addition to prevalent illnesses like diabetes and heart disease.

A significant additional goal for stem cell treatment in the future is personalized medicine. Therapy may be customized to a patient's genetic composition by employing patient-specific stem cells, which lowers the chance of immunological rejection and enhances therapy results. The creation of tailored cell-based therapeutics that can be utilized to treat a range of illnesses is being made possible by developments in iPSC technology and gene editing [109].

The potential effects of climate change on human health and disease progression can also influence the future directions of stem cell research, as shifting environmental conditions exacerbate diseases and create new health challenges [110]. Understanding these impacts is crucial for developing responsive and effective stem cell therapies.

The creation of easily accessible stem cell products that may be used right away in emergency scenarios like trauma or acute organ failure is another long-term objective. To preserve their therapeutic potential, these allogeneic stem cell products would be extracted from donors who have been carefully chosen and vetted, processed, and kept under ideal circumstances. One important step in reaching this objective is the creation of universal donor cells, which are designed to be immunologically compatible with any recipient [8].

Another crucial goal is making sure stem cell treatments are accessible and affordable. Significant hurdles include the high cost of stem cell therapies and the intricate manufacturing procedures. Aims are being made to create scalable manufacturing procedures that can generate huge volumes of high-quality stem cells while also streamlining production and cutting costs. To overcome these obstacles and make stem cell treatments available to patients everywhere, cooperation between academic institutions, business, and regulatory bodies is crucial [103].

References

1. Evans, M.J. and Kaufman, M.H., Establishment in culture of pluripotential cells from mouse embryos. *Nature.*, 292, 5819, 154–6, 1981 Jul 9.
2. Thomson, J.A., Itskovitz-Eldor, J., Shapiro, S.S., Waknitz, M.A., Swiergiel, J.J., Marshall, V.S., *et al.*, Embryonic stem cell lines derived from human blastocysts. *Science.*, 282, 5391, 1145–7, 1998 Nov 6.
3. Ying, Q.L., Nichols, J., Chambers, I., Smith, A., BMP induction of Id proteins suppresses differentiation and sustains embryonic stem cell self-renewal in collaboration with STAT3. *Cell.*, 115, 3, 281–92, 2003 Oct 31.
4. Takahashi, K. and Yamanaka, S., Induction of pluripotent stem cells from mouse embryonic and adult fibroblast cultures by defined factors. *Cell.*, 126, 4, 663–76, 2006 Aug 25.
5. Thomas, E.D., Lochte, H.L., Lu, W.C., Ferrebee, J.W., Intravenous infusion of bone marrow in patients receiving radiation and chemotherapy. *N. Engl. J. Med.*, 257, 11, 491–6, 1957 Sep 12.
6. Pittenger, M.F., Mackay, A.M., Beck, S.C., Jaiswal, R.K., Douglas, R., Mosca, J.D., *et al.*, Multilineage potential of adult human mesenchymal stem cells. *Science.*, 284, 5411, 143–7, 1999 Apr 2.
7. Kalluri, R. and Zeisberg, M., Fibroblasts in cancer. *Nat. Rev. Cancer*, 6, 5, 392–401, 2006 May.
8. Heneka, M.T., Golenbock, D.T., Latz, E., Innate immunity in Alzheimer's disease. *Nat. Immunol.*, 16, 3, 229–36, 2015 Mar.
9. Clevers, H., Wnt/beta-catenin signaling in development and disease. *Cell*, 127, 3, 469–80, 2006 Nov 3.
10. Murry, C.E., Reinecke, H., Pabon, L.M., Regeneration gaps: observations on stem cells and cardiac repair. *J. Am. Coll. Cardiol.*, 47, 9, 1777–85, 2006 May 2.
11. Artavanis-Tsakonas, S., Rand, M.D., Lake, R.J., Notch signaling: cell fate control and signal integration in development. *Science.*, 284, 5415, 770–6, 1999 Apr 30.
12. Ingham, P.W. and McMahon, A.P., Hedgehog signaling in animal development: paradigms and principles. *Genes. Dev.*, 15, 23, 3059–87, 2001 Dec 1.

13. Hynes, R.O., The extracellular matrix: not just pretty fibrils. *Science.*, 326, 5957, 1216–9, 2009 Nov 27.
14. Su, Y.H., Tang, L.P., Zhao, X.Y., Zhang, X.S., Plant cell totipotency: Insights into cellular reprogramming. *J. Integr. Plant Biol.*, 63, 1, 228–43, 2021 Jan.
15. Weissman, I.L., Stem cells: units of development, units of regeneration, and units in evolution. *Cell.*, 100, 1, 157–68, 2000 Jan 7.
16. Massagué, J., TGFβ signalling in context. *Nat. Rev. Mol. Cell Biol.*, 13, 10, 616–30, 2012 Oct.
17. Scadden, D.T., The stem-cell niche as an entity of action. *Nature.*, 441, 7097, 1075–9, 2006 Jun 29.
18. Yin, T. and Li, L., The stem cell niches in bone. *J. Clin. Invest.*, 116, 5, 1195–201, 2006 May.
19. Stier, S., Ko, Y., Forkert, R., Lutz, C., Neuhaus, T., Grünewald, E., *et al.*, Osteopontin is a hematopoietic stem cell niche component that negatively regulates stem cell pool size. *J. Exp. Med.*, 201, 11, 1781–91, 2005 Jun 6.
20. Haug, J.S., He, X.C., Grindley, J.C., Wunderlich, J.P., Gaudenz, K., Ross, J.T., *et al.*, N-cadherin expression level distinguishes reserved versus primed states of hematopoietic stem cells. *Cell Stem Cell*, 2, 4, 367–79, 2008 Apr 10.
21. Doetsch, F., Caillé, I., Lim, D.A., García-Verdugo, J.M., Alvarez-Buylla, A., Subventricular zone astrocytes are neural stem cells in the adult mammalian brain. *Cell.*, 97, 6, 703–16, 1999 Jun 11.
22. Song, H., Stevens, C.F., Gage, F.H., Neural stem cells from adult hippocampus develop essential properties of functional CNS neurons. *Nat. Neurosci.*, 5, 5, 438–45, 2002 May.
23. Palmer, T.D., Takahashi, J., Gage, F.H., The Adult Rat Hippocampus Contains Primordial Neural Stem Cells. *Mol. Cell. Neurosci.*, 8, 6, 389–404, 1997 Jan 1.
24. Watt, F.M. and Hogan, B.L., Out of Eden: stem cells and their niches. *Science*, 287, 5457, 1427–30, 2000 Feb 25.
25. Sugiyama, T., Kohara, H., Noda, M., Nagasawa, T., Maintenance of the hematopoietic stem cell pool by CXCL12-CXCR4 chemokine signaling in bone marrow stromal cell niches. *Immunity.*, 25, 6, 977–88, 2006 Dec.
26. Engler, A.J., Sen, S., Sweeney, H.L., Discher, D.E., Matrix Elasticity Directs Stem Cell Lineage Specification. *Cell.*, 126, 4, 677–89, 2006 Aug 25.
27. Levental, K.R., Yu, H., Kass, L., Lakins, J.N., Egeblad, M., Erler, J.T., *et al.*, Matrix crosslinking forces tumor progression by enhancing integrin signaling. *Cell.*, 139, 5, 891–906, 2009 Nov 25.
28. Wells, R.G., The role of matrix stiffness in regulating cell behavior. *Hepatol. Baltim. Md.*, 47, 4, 1394–400, 2008 Apr 1.
29. Zhou, Y., Hagood, J.S., Lu, B., Merryman, W.D., Murphy-Ullrich, J.E., Thy-1-integrin alphav beta5 interactions inhibit lung fibroblast contraction-induced latent transforming growth factor-beta1 activation and myofibroblast differentiation. *J. Biol. Chem.*, 285, 29, 22382–93, 2010 Jul 16.
30. Balkwill, F., Tumour necrosis factor and cancer. *Nat. Rev. Cancer*, 9, 5, 361–71, 2009 May.

31. Sturm, A. and Dignass, A.U., Epithelial restitution and wound healing in inflammatory bowel disease. *World J. Gastroenterol.*, 14, 3, 348–53, 2008 Jan 21.
32. Eming, S.A., Krieg, T., Davidson, J.M., Inflammation in wound repair: molecular and cellular mechanisms. *J. Invest. Dermatol.*, 127, 3, 514–25, 2007 Mar.
33. Keith, B. and Simon, M.C., Hypoxia-inducible factors, stem cells, and cancer. *Cell.*, 129, 3, 465–72, 2007 May 4.
34. Mohyeldin, A., Garzón-Muvdi, T., Quiñones-Hinojosa, A., Oxygen in stem cell biology: a critical component of the stem cell niche. *Cell Stem Cell*, 7, 2, 150–61, 2010 Aug 6.
35. King, G.L., The role of inflammatory cytokines in diabetes and its complications. *J. Periodontol.*, 79, 8 Suppl, 1527–34, 2008 Aug.
36. Kotter, M.R., Zhao, C., van Rooijen, N., Franklin, R.J.M., Macrophage-depletion induced impairment of experimental CNS remyelination is associated with a reduced oligodendrocyte progenitor cell response and altered growth factor expression. *Neurobiol. Dis.*, 18, 1, 166–75, 2005 Feb.
37. Clevers, H. and Nusse, R., Wnt/β-catenin signaling and disease. *Cell*, 149, 6, 2012 Jun 8.
38. MacDonald, B.T., Tamai, K., He, X., Wnt/beta-catenin signaling: components, mechanisms, and diseases. *Dev. Cell*, 17, 1, 9–26, 2009 Jul.
39. Sato, T. and Clevers, H., Growing self-organizing mini-guts from a single intestinal stem cell: mechanism and applications. *Science.*, 340, 6137, 1190–4, 2013 Jun 7.
40. Polakis, P., Wnt signaling and cancer. *Genes. Dev.*, 14, 15, 1837–51, 2000 Aug 1.
41. Kopan, R. and Ilagan, M.X.G., The canonical Notch signaling pathway: unfolding the activation mechanism. *Cell.*, 137, 2, 216–33, 2009 Apr 17.
42. Lampreia, F.P., Carmelo, J.G., Anjos-Afonso, F., Notch Signaling in the Regulation of Hematopoietic Stem Cell. *Curr. Stem Cell Rep.*, 3, 3, 202–9, 2017.
43. Ables, J.L., Breunig, J.J., Eisch, A.J., Rakic, P., Not(ch) just development: Notch signalling in the adult brain. *Nat. Rev. Neurosci.*, 12, 5, 269–83, 2011 May.
44. Weng, A.P., Ferrando, A.A., Lee, W., Morris, J.P., Silverman, L.B., Sanchez-Irizarry, C., *et al.*, Activating mutations of NOTCH1 in human T cell acute lymphoblastic leukemia. *Science.*, 306, 5694, 269–71, 2004 Oct 8.
45. Wu, Y., Cain-Hom, C., Choy, L., Hagenbeek, T.J., de Leon, G.P., Chen, Y., *et al.*, Therapeutic antibody targeting of individual Notch receptors. *Nature.*, 464, 7291, 1052–7, 2010 Apr 15.
46. Jiang, J. and Hui, C.C., Hedgehog signaling in development and cancer. *Dev. Cell*, 15, 6, 801–12, 2008 Dec.

47. Palma, V. and Altaba, AR., Hedgehog-GLI signaling regulates the behavior of cells with stem cell properties in the developing neocortex. *Development*, 131, 2, 337–45, 2004 Jan 15.
48. Scales, S.J. and de Sauvage, F.J., Mechanisms of Hedgehog pathway activation in cancer and implications for therapy. *Trends Pharmacol. Sci.*, 30, 6, 303–12, 2009 Jun.
49. Sekulic, A., Migden, M.R., Oro, A.E., Dirix, L., Lewis, K.D., Hainsworth, J.D., *et al.*, Efficacy and safety of vismodegib in advanced basal-cell carcinoma. *N. Engl. J. Med.*, 366, 23, 2171–9, 2012.
50. Schlaepfer, D.D. and Mitra, S.K., Multiple connections link FAK to cell motility and invasion. *Curr. Opin. Genet. Dev.*, 14, 1, 92–101, 2004 Feb 1.
51. Discher, D.E., Mooney, D.J., Zandstra, P.W., Growth factors, matrices, and forces combine and control stem cells. *Science.*, 324, 5935, 1673–7, 2009 Jun 26.
52. Dupont, S., Morsut, L., Aragona, M., Enzo, E., Giulitti, S., Cordenonsi, M., *et al.*, Role of YAP/TAZ in mechanotransduction. *Nature.*, 474, 7350, 179–83, 2011 Jun 8.
53. Lehoux, S. and Tedgui, A., Cellular mechanics and gene expression in blood vessels. *J. Biomech.*, 36, 5, 631–43, 2003 May 1.
54. Gilbert, P.M., Havenstrite, K.L., Magnusson, K.E.G., Sacco, A., Leonardi, N.A., Kraft, P., *et al.*, Substrate elasticity regulates skeletal muscle stem cell self-renewal in culture. *Science.*, 329, 5995, 1078–81, 2010 Aug 27.
55. Moustakas, A. and Heldin, C.H., Mechanisms of TGFβ-Induced Epithelial-Mesenchymal Transition. *J. Clin. Med.*, 5, 7, 63, 2016 Jun 29.
56. Geiger, B., Spatz, J.P., Bershadsky, A.D., Environmental sensing through focal adhesions. *Nat. Rev. Mol. Cell Biol.*, 10, 1, 21–33, 2009 Jan.
57. Elliott, A.D., Confocal Microscopy: Principles and Modern Practices. *Curr. Protoc. Cytom.*, 92, 1, e68, 2020 Mar.
58. Driskell, R.R. and Watt, F.M., Understanding fibroblast heterogeneity in the skin. *Trends Cell Biol.*, 25, 2, 92–9, 2015 Feb.
59. Akerboom, J., Rivera, J.D.V., Guilbe, M.M.R., Malavé, E.C.A., Hernandez, H.H., Tian, L., *et al.*, Crystal Structures of the GCaMP Calcium Sensor Reveal the Mechanism of Fluorescence Signal Change and Aid Rational Design*. *J. Biol. Chem.*, 284, 10, 6455–64, 2009 Mar 6.
60. Tirosh, I., Venteicher, A.S., Hebert, C., Escalante, L.E., Patel, A.P., Yizhak, K., *et al.*, Single-cell RNA-seq supports a developmental hierarchy in human oligodendroglioma. *Nature.*, 539, 7628, 309–13, 2016 Nov 10.
61. van Galen, P., Hovestadt, V., Wadsworth Ii, M.H., Hughes, T.K., Griffin, G.K., Battaglia, S., *et al.*, Single-Cell RNA-Seq Reveals AML Hierarchies Relevant to Disease Progression and Immunity. *Cell*, 176, 6, 1265–1281.e24, 2019 Mar 7.
62. Kim, J., Koo, B.K., Knoblich, J.A., Human organoids: model systems for human biology and medicine. *Nat. Rev. Mol. Cell Biol.*, 21, 10, 571–84, 2020 Oct.

63. Sato, T., Vries, R.G., Snippert, H.J., van de Wetering, M., Barker, N., Stange, D.E., *et al.*, Single Lgr5 stem cells build crypt-villus structures *in vitro* without a mesenchymal niche. *Nature.*, 459, 7244, 262–5, 2009 May 14.
64. Murphy, S.V. and Atala, A., 3D bioprinting of tissues and organs. *Nat. Biotechnol.*, 32, 8, 773–85, 2014 Aug.
65. Nusse, R. and Clevers, H., Wnt/β-Catenin Signaling, Disease, and Emerging Therapeutic Modalities. *Cell*, 169, 6, 985–99, 2017 Jun 1.
66. Kizil, C., Kyritsis, N., Brand, M., Effects of inflammation on stem cells: together they strive? *EMBO Rep.*, 16, 4, 416–26, 2015 Apr.
67. Hidalgo, M., Amant, F., Biankin, A.V., Budinská, E., Byrne, A.T., Caldas, C., *et al.*, Patient-derived xenograft models: an emerging platform for translational cancer research. *Cancer Discov.*, 4, 9, 998–1013, 2014 Sep.
68. Hanahan, D. and Weinberg, R.A., Hallmarks of cancer: the next generation. *Cell.*, 144, 5, 646–74, 2011 Mar 4.
69. Massagué, J., TGFbeta in Cancer. *Cell.*, 134, 2, 215–30, 2008 Jul 25.
70. DeNardo, D.G., Johansson, M., Coussens, L.M., Immune cells as mediators of solid tumor metastasis. *Cancer Metastasis Rev.*, 27, 1, 11–8, 2008 Mar.
71. Zou, W., Regulatory T cells, tumour immunity and immunotherapy. *Nat. Rev. Immunol.*, 6, 4, 295–307, 2006 Apr.
72. Visvader, J.E. and Lindeman, G.J., Cancer stem cells: current status and evolving complexities. *Cell Stem Cell*, 10, 6, 717–28, 2012 Jun 14.
73. Phi, L.T.H., Sari, I.N., Yang, Y.G., Lee, S.H., Jun, N., Kim, K.S., *et al.*, Cancer Stem Cells (CSCs) in Drug Resistance and their Therapeutic Implications in Cancer Treatment. *Stem Cells Int.*, 2018, 5416923, 2018 Feb 28.
74. Reya, T., Duncan, A.W., Ailles, L., Domen, J., Scherer, D.C., Willert, K., *et al.*, A role for Wnt signalling in self-renewal of haematopoietic stem cells. *Nature.*, 423, 6938, 409–14, 2003 May 22.
75. Takebe, N., Miele, L., Harris, P.J., Jeong, W., Bando, H., Kahn, M., *et al.*, Targeting Notch, Hedgehog, and Wnt pathways in cancer stem cells: clinical update. *Nat. Rev. Clin. Oncol.*, 12, 8, 445–64, 2015 Aug.
76. Brabletz, T., Jung, A., Spaderna, S., Hlubek, F., Kirchner, T., Opinion: migrating cancer stem cells - an integrated concept of malignant tumour progression. *Nat. Rev. Cancer*, 5, 9, 744–9, 2005 Sep.
77. S, V., Kumar, P., Sen, A., Anand, A., Kaushik, M., Bhowmick, M., *et al.*, Unlocking the Role of Virus-like Particles (VLPs) for Cancer Treatment. *Curr. Cancer Ther. Rev.*, 20, 1–27, 2024 May.
78. Nassar, D. and Blanpain, C., Cancer Stem Cells: Basic Concepts and Therapeutic Implications. *Annu. Rev. Pathol.*, 11, 47–76, 2016 May 23.
79. Ming, G.L. and Song, H., Adult neurogenesis in the mammalian brain: significant answers and significant questions. *Neuron.*, 70, 4, 687–702, 2011 May 26.
80. Selkoe, D.J. and Hardy, J., The amyloid hypothesis of Alzheimer's disease at 25 years. *EMBO Mol. Med.*, 8, 6, 595–608, 2016 Jun.

81. McCord, J.M., Oxygen-derived free radicals in postischemic tissue injury. *N. Engl. J. Med.*, 312, 3, 159–63, 1985 Jan 17.
82. Dauer, W. and Przedborski, S., Parkinson's disease: mechanisms and models. *Neuron.*, 39, 6, 889–909, 2003 Sep 11.
83. Goldman, S., Stem and progenitor cell-based therapy of the human central nervous system. *Nat. Biotechnol.*, 23, 7, 862–71, 2005 Jul.
84. Hsu, P.D., Lander, E.S., Zhang, F., Development and applications of CRISPR-Cas9 for genome engineering. *Cell.*, 157, 6, 1262–78, 2014 Jun 5.
85. Hu, X., Dai, S., Wu, W.J., Tan, W., Zhu, X., Mu, J., *et al.*, Stromal cell derived factor-1 alpha confers protection against myocardial ischemia/reperfusion injury: role of the cardiac stromal cell derived factor-1 alpha CXCR4 axis. *Circulation.*, 116, 6, 654–63, 2007 Aug 7.
86. Carmeliet, P. and Jain, R.K., Molecular mechanisms and clinical applications of angiogenesis. *Nature.*, 473, 7347, 298–307, 2011 May 19.
87. Menasché, P., Cell therapy trials for heart regeneration - lessons learned and future directions. *Nat. Rev. Cardiol.*, 15, 11, 659–71, 2018 Nov.
88. Sanganalmath, S.K. and Bolli, R., Cell therapy for heart failure: a comprehensive overview of experimental and clinical studies, current challenges, and future directions. *Circ. Res.*, 113, 6, 810–34, 2013 Aug 30.
89. Gnecchi, M., Zhang, Z., Ni, A., Dzau, V.J., Paracrine mechanisms in adult stem cell signaling and therapy. *Circ. Res.*, 103, 11, 1204–19, 2008 Nov 21.
90. Zhang, J., Klos, M., Wilson, G.F., Herman, A.M., Lian, X., Raval, K.K., *et al.*, Extracellular matrix promotes highly efficient cardiac differentiation of human pluripotent stem cells: the matrix sandwich method. *Circ. Res.*, 111, 9, 1125–36, 2012 Oct 12.
91. Sluijter, J.P.G., Davidson, S.M., Boulanger, C.M., Buzás, E.I., de Kleijn, D.P.V., Engel, F.B., *et al.*, Extracellular vesicles in diagnostics and therapy of the ischaemic heart: Position Paper from the Working Group on Cellular Biology of the Heart of the European Society of Cardiology. *Cardiovasc. Res.*, 114, 1, 19–34, 2018 Jan 1.
92. Cencioni, C., Capogrossi, M.C., Napolitano, M., The SDF-1/CXCR4 axis in stem cell preconditioning. *Cardiovasc. Res.*, 94, 3, 400–7, 2012 Jun 1.
93. Phillips, M.I. and Tang, Y.L., Genetic modification of stem cells for transplantation. *Adv. Drug Deliv. Rev.*, 60, 2, 160–72, 2008 Jan 14.
94. Jain, R.K., Antiangiogenesis strategies revisited: from starving tumors to alleviating hypoxia. *Cancer Cell.*, 26, 5, 605–22, 2014 Nov 10.
95. Yamasaki, T.R., Holmes, B.B., Furman, J.L., Dhavale, D.D., Su, B.W., Song, E.S., *et al.*, Parkinson's disease and multiple system atrophy have distinct α-synuclein seed characteristics. *J. Biol. Chem.*, 294, 3, 1045–58, 2019 Jan 18.
96. Appelbaum, F.R., Hematopoietic-cell transplantation at 50. *N. Engl. J. Med.*, 357, 15, 1472–5, 2007 Oct 11.
97. Freitag, J., Bates, D., Boyd, R., Shah, K., Barnard, A., Huguenin, L., *et al.*, Mesenchymal stem cell therapy in the treatment of osteoarthritis: reparative

pathways, safety and efficacy - a review. *BMC Musculoskelet. Disord.*, 17, 230, 2016 May 26.

98. Laflamme, M.A. and Murry, C.E., Heart regeneration. *Nature.*, 473, 7347, 326–35, 2011 May 19.
99. Atiyeh, B.S., Hayek, S.N., Gunn, S.W., New technologies for burn wound closure and healing–review of the literature. *Burns J. Int. Soc. Burn. Inj.*, 31, 8, 944–56, 2005 Dec.
100. Atala, A., Bauer, S.B., Soker, S., Yoo, J.J., Retik, A.B., Tissue-engineered autologous bladders for patients needing cystoplasty. *Lancet Lond. Engl.*, 367, 9518, 1241–6, 2006 Apr 15.
101. Wagner, W. and Ho, A.D., Mesenchymal stem cell preparations–comparing apples and oranges. *Stem Cell Rev.*, 3, 4, 239–48, 2007 Dec.
102. Lysaght, T. and Campbell, A.V., Regulating autologous adult stem cells: the FDA steps up. *Cell Stem Cell*, 9, 5, 393–6, 2011 Nov 4.
103. Mason, C. and Dunnill, P., A brief definition of regenerative medicine. *Regen Med.*, 3, 1, 1–5, 2008 Jan.
104. DeWitt, M.A., Corn, J.E., Carroll, D., Genome editing via delivery of Cas9 ribonucleoprotein. *Methods San Diego Calif.*, 121–122, 9–15, 2017 May 15.
105. Clevers, H., Modeling Development and Disease with Organoids. *Cell*, 165, 7, 1586–97, 2016 Jun 16.
106. Kloxin, A.M., Kloxin, C.J., Bowman, C.N., Anseth, K.S., Mechanical properties of cellularly responsive hydrogels and their experimental determination. *Adv. Mater. Deerfield Beach Fla.*, 22, 31, 3484–94, 2010 Aug 17.
107. Andaloussi S, E.L., Mäger, I., Breakefield, X.O., Wood, M.J.A., Extracellular vesicles: biology and emerging therapeutic opportunities. *Nat. Rev. Drug Discov.*, 12, 5, 347–57, 2013 May.
108. June, C.H. and Sadelain, M., Chimeric Antigen Receptor Therapy. *N. Engl. J. Med.*, 379, 1, 64–73, 2018 Jul 5.
109. Takahashi, K. and Yamanaka, S., Induced pluripotent stem cells in medicine and biology. *Dev. Camb. Engl.*, 140, 12, 2457–61, 2013 Jun.
110. Sen, A. and Chatterjee, A., Confronting the Health Impacts of Climate Change: A Comprehensive Exploration, in: *Preserving Health, Preserving Earth: The Path to Sustainable Healthcare [Internet]*, P.K. Prabhakar and W. Leal Filho (Eds.), Springer Nature Switzerland, Cham, 2024, [cited 2024 Jun 19]. p. 1–9. Available from: https://doi.org/10.1007/978-3-031-60545-1_1.

2

CRISPR-Cas9 and Beyond: Gene Editing Frontiers in Stem Cell Therapeutics

Pranshul Sethi[1,2], Sonali Rastogi[2,3], Kajal Sherawat[4], Neetu Panwar[5], Ayush Madan[6,7] and Mojahidul Islam[8]*

[1]*College of Pharmacy, Shri Venkateshwara University, Gajraula, UP, India*
[2]*Chitkara College of Pharmacy, Chitkara University, Rajpura, Punjab, India*
[3]*Department of Pharmacy, Venkateshwara Group of Institutions, Jatoli, Meerut, UP, India*
[4]*Department of Pharmacy, Noida Institute of Engineering and Technology, Greater Noida, UP, India*
[5]*School of Applied Sciences, Shri Venkateshwara University, Gajraula, UP, India*
[6]*Department of Biotechnology, School of Research and Technology, People's University, Bhanpur, Bhopal, Madhya Pradesh, India*
[7]*Institute of Tropical Aquaculture and Fisheries, Universiti Malaysia Terengganu, Kuala Nerus, Terengganu, Malaysia*
[8]*Department of Pharmacy, Orlean College of Pharmacy, Greater Noida, UP, India*

Abstract

The advent of CRISPR-Cas9 technology has revolutionized the field of gene editing, opening new avenues for therapeutic interventions, particularly in stem cell research. This chapter delves into the transformative potential of CRISPR-Cas9 and its subsequent advancements in the realm of stem cell therapeutics. It explores the underlying mechanisms of CRISPR-Cas9, its application in correcting genetic disorders, enhancing regenerative medicine, and overcoming the challenges associated with gene editing technologies. Furthermore, the paper investigates the ethical considerations and future prospects of integrating CRISPR-Cas9 with other emerging gene editing tools to create more precise and efficient therapeutic strategies. The goal is to provide a comprehensive overview of the current state and future directions of gene editing in stem cell therapeutics, highlighting the profound implications for personalized medicine and disease treatment.

**Corresponding author*: mojahiduii@gmail.com

Ayush Madan, Sumel Ashique, Deepika Arora and Mantosh Kumar Satapathy (eds.) Stem Cell Therapeutics, (39–106) © 2025 Scrivener Publishing LLC

Keywords: CRISPR-Cas9, gene editing, stem cell therapeutics, regenerative medicine, genetic disorders

2.1 Introduction

2.1.1 Background of Gene Editing

Gene editing is a powerful tool that allows scientists to make precise alterations to the DNA of living organisms, providing a method for understanding gene function and correcting genetic disorders [1]. The concept of gene editing emerged from early discoveries in molecular biology, particularly the understanding of how DNA functions as the blueprint of life. Over the past several decades, scientists have developed technologies capable of modifying specific genes, providing unprecedented control over biological systems [2].

The historical development of gene editing technologies dates back to the late 20th century, with early techniques like zinc finger nucleases (ZFNs) and transcription activator-like effector nucleases (TALENs) leading the way [3]. ZFNs, discovered in the 1990s, used zinc finger proteins to recognize and bind to specific DNA sequences. These proteins were fused to a nuclease, an enzyme that could cut DNA at specific locations, enabling researchers to modify the gene at the target site [4]. However, ZFNs had significant limitations, including the complexity of designing and producing zinc finger proteins for each target, and the potential for off-target effects that could damage other parts of the genome [5].

TALENs, developed in the 2000s, represented an improvement over ZFNs. They were more flexible and allowed researchers to target a broader range of DNA sequences. Despite these advances, TALENs also presented challenges, such as their labor-intensive construction and the relatively high cost of their application [6].

While these early methods paved the way for gene editing, their limitations—such as the difficulty of design, off-target effects, and restricted accessibility—motivated scientists to seek new, more efficient technologies. The development of these early techniques laid the groundwork for the discovery of CRISPR-Cas9, a revolutionary gene-editing tool that has transformed the field of genetic research and medicine [7].

2.2 Significance of CRISPR-Cas9 Technology

The advent of CRISPR-Cas9 (Clustered Regularly Interspaced Short Palindromic Repeats-CRISPR associated protein 9) marked a pivotal moment in the history of gene editing [8]. Discovered by Jennifer Doudna, Emmanuelle Charpentier, and their colleagues in 2012, the CRISPR-Cas9 system originates from a bacterial immune defense mechanism, wherein bacteria use CRISPR sequences and Cas proteins to recognize and destroy the DNA of invading viruses [9]. By harnessing this natural system, researchers developed a method to edit genes with unparalleled precision and efficiency.

The discovery and development of CRISPR-Cas9 revolutionized gene editing due to its simplicity, efficiency, and versatility [10]. Unlike ZFNs and TALENs, which required the design of complex proteins for each new target, CRISPR-Cas9 relies on a short RNA molecule (called guide RNA or gRNA) to guide the Cas9 enzyme to a specific location in the genome [11]. This allows researchers to easily target virtually any gene by simply changing the sequence of the guide RNA, significantly reducing the time and cost associated with gene editing [12].

CRISPR-Cas9 also offers several key advantages over previous gene-editing methods:

- Simplicity: Designing a new CRISPR-Cas9 system requires only altering the sequence of a guide RNA, rather than engineering new proteins, making the process significantly simpler and more cost-effective [13].
- Efficiency: CRISPR-Cas9 can target genes with higher precision and fewer off-target effects compared to earlier methods. The high degree of specificity minimizes unwanted changes to the genome [7].
- Versatility: CRISPR-Cas9 can be used to make a wide variety of genetic modifications, from knocking out genes to adding or replacing genes [8].
- Multiplexing: One of the key innovations of CRISPR-Cas9 is its ability to target multiple genes simultaneously, something that was previously challenging with ZFNs and TALENs [9].

2.3 Overview of Stem Cell Therapeutics

Stem cells are undifferentiated cells capable of developing into various specialized cell types. They hold immense potential in regenerative medicine and disease treatment due to their ability to self-renew and differentiate into multiple cell lineages [10]. There are several types of stem cells, each with unique characteristics and therapeutic applications [10].

Stem cells are classified into:

- Embryonic stem cells (ESCs): Derived from the inner cell mass of blastocysts, these pluripotent stem cells can give rise to nearly any cell type in the body [11].
- Adult (somatic) stem cells: These multipotent cells are found in various tissues, such as bone marrow, skin, and fat, and are responsible for maintaining and repairing the tissues in which they are found [12].
- Induced pluripotent stem cells (iPSCs): Discovered by Shinya Yamanaka in 2006, iPSCs are adult cells reprogrammed to a pluripotent state through the introduction of specific genes [13].

2.4 Fundamentals of CRISPR-Cas9 Technology

2.4.1 Mechanism of CRISPR-Cas9

Derived from the adaptive immune system found in bacteria, the CRISPR-Cas9 system functions as a robust genome editing instrument through a basic but remarkably efficient mechanism. Fundamentally, the CRISPR-Cas9 method is dependent on two crucial elements: the guiding RNA molecule and the Cas9 enzyme. Working together in a synchronized biological dance, the gRNA initiates base-pairing with the target DNA sequence while Cas9 provides the enzymatic might necessary to enact the desired genomic alteration. In the CRISPR-Cas9 system, the guide RNA (gRNA) directs the Cas9 enzyme to a target DNA sequence, where it induces a double-strand break (DSB), subsequently repaired *via* pathways such as non-homologous end joining (NHEJ) or homology-directed repair (HDR) (Figure 2.1). Though straightforward in design, the CRISPR-Cas9 platform demonstrates immense transformative capacity, promising revolutionary potential for directed manipulations of the genetic blueprint underpinning all life.

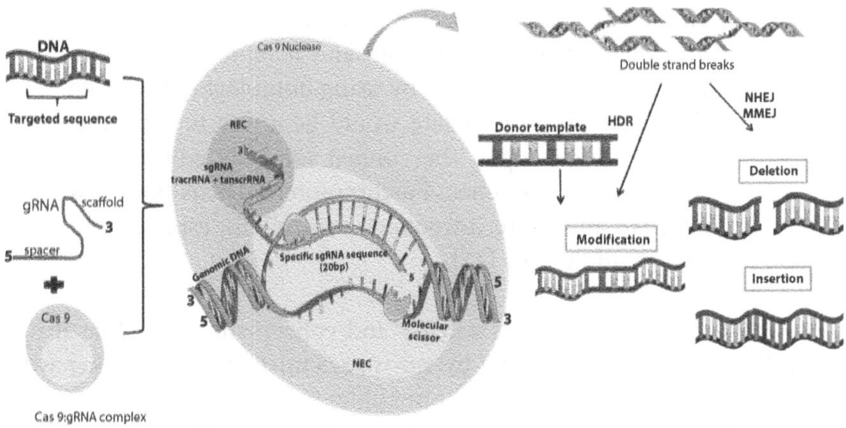

Figure 2.1 Mechanisms of CRISPR-Cas9.

a. Guide RNA (gRNA): The gRNA is a small synthetic RNA that has two components to it – the scaffold sequence which allows for Cas9 binding, and then 20 nucleotides that complement their respective target DNA sequences. This 20- bp sequence on the crRNA guides the Cas9 protein to cut at that specific position in genome [1, 14].

b. Cas9 protein: Cas9 is an endonuclease enzyme that functions like molecular scissors and can cleave double-stranded DNA. If the gRNA is complementary with a target DNA sequence and thus able to bind, then it will association with Cas9 protein. The complex validates the DNA for a protospacer adjacent motif (PAM) sequence and then activates Cas9. Following PAM recognition, the Cas9 protein cleaves DNA in which sgRNAs bind, thereby generating a double-strand break (DSB) [15].

c. DNA repair pathways: Following the cut of the DNA, repair machinery within a cell will try to mend this break using primarily one or two pathways — non-homologous end joining (NHEJ) and homology-directed repair (HDR).

- Non-homologous end joining (NHEJ): It is an error-prone repair process that proteins directly ligates both broken DNA ends. This typically results in small insertions or deletions (indels) at the nuclease cleavage site that will disrupt gene function and is widely used to generate knockouts [16].

- Homology-directed break repair (HDR): HDR is the second major repair mechanism and it represents a more accurate way to fix break by using homologous template DNA. The DNA template would be a way to hack this same pathway, open the context for gene editors putting in or fixing genes. Furthermore, HDR is both less effective and more infrequent than NHEJ, especially in non-dividing cells [17].

CRISPR-Cas9 technology is so easy and convenient to use that it has taken the world of gene editing by storm; scientists can now tinker with genes in a way never before thought possible.

The gRNA binds with cas9 to form the cas9 complex. Once the complex is formed, cas9 specifically targets the DNA. The Cas9 nuclease, in conjunction with sgRNA complexes, possesses the capacity to produce precise double-strand breaks (DSBs) in the DNA of the host organism. The specific DNA double-strand breaks (DSBs) have the capability to trigger the inherent repair systems within the cell. DSBs can be repaired using three distinct mechanisms: homology-directed repair (HDR), non-homologous end joining (NHEJ), and microhomology-mediated end joining (MMEJ). High-fidelity homology-directed repair (HDR) entails utilizing the sister chromatid as a template to precisely incorporate genomic alterations into the DNA of the host organism. Conversely, non-homologous end joining (NHEJ) refers to the process of connecting DNA ends without requiring any similarity between them. Finally, MMEJ entails the alignment of DNA ends by utilizing a limited number of nucleotides with homologous sequences, referred to as microhomology. The presence of insertions and deletions is most likely attributed to non-homologous end joining (NHEJ) and microhomology-mediated end joining (MMEJ). [Abbreviations—DSBs: double-strand breaks, HDR: homology-directed repair, NHEJ: non-homologous end joining, MMEJ: microhomology-mediated end joining]

2.4.2 Advances in CRISPR-Cas9 Precision and Efficiency

While CRISPR-Cas9 is a powerful tool, its initial iterations had limitations, particularly concerning off-target effects—where the Cas9 protein cuts DNA at unintended sites. These off-target effects pose significant risks in therapeutic applications, as they can lead to unintended mutations and genomic instability. Over the years, researchers have made significant advancements to enhance the precision and efficiency of CRISPR-Cas9. A detailed comparison of CRISPR-Cas9 with other gene-editing technologies, including TALENs and ZFNs, highlighting their respective strengths, limitations, and applications is presented in Table 2.1.

Table 2.1 Comprehensive comparison of gene editing technologies.

Aspect	CRISPR-Cas9	TALENs	ZFNs	References
Discovery	2012, derived from a bacterial immune system mechanism	2010, based on TALE proteins found in *Xanthomonas* bacteria	2005, based on zinc finger proteins naturally occurring in transcription factors	[1–3]
Mechanism of DNA recognition	RNA-guided: A guide RNA (gRNA) directs the Cas9 protein to a specific DNA sequence by complementary base pairing	Protein–DNA interaction: TALE proteins recognize and bind to specific DNA sequences through modular repeats	Protein–DNA interaction: zinc finger motifs each recognize a 3-nucleotide DNA sequence	[4–6]
Nuclease component	Cas9 protein, which introduces a double-strand break (DSB) at the target site	FokI nuclease domain, which dimerizes to create a DSB when two TALENs bind to adjacent DNA sequences	FokI nuclease domain, similar to TALENs, requires dimerization to create a DSB	[7–9]
Targeting specificity	High, but potential for off-target effects, especially with less optimized gRNAs. High-fidelity Cas9 variants can reduce off-target activity	Very high, with minimal off-target effects due to precise protein–DNA interactions	Moderate to high, but higher risk of off-target effects compared to TALENs, especially if zinc fingers are not well-designed	[10, 11]

(*Continued*)

Table 2.1 Comprehensive comparison of gene editing technologies. (*Continued*)

Aspect	CRISPR-Cas9	TALENs	ZFNs	References
Design complexity	Simple and easily customizable: designing a new gRNA is straightforward and cost-effective	Moderately complex: requires the construction of new TALE repeats for each target, but still more flexible than ZFNs	Highly complex: requires the design of a unique set of zinc fingers for each target sequence, which is labor-intensive and costly	[12–14]
Editing efficiency	High: Generally high efficiency in both *in vitro* and *in vivo* systems, especially with optimized delivery and repair templates	Moderate to high: efficiency depends on the target and the design of TALENs, but generally effective	Moderate: efficiency is often lower than CRISPR-Cas9 and TALENs, with variability depending on target and cell type	[1, 15, 16]
Delivery methods	Versatile: Compatible with various methods including viral vectors (AAV, lentivirus), lipid nanoparticles, electroporation, and direct injection	Limited by size: Larger size makes TALENs more challenging to deliver *via* viral vectors, but non-viral methods are feasible	Challenging: Large and complex, making delivery *via* common vectors like AAV difficult; often requires electroporation or other non-viral methods	[8, 17, 18]
Applications	Broad: used in research, therapeutics, agriculture, and synthetic biology. Suitable for gene knockout, knock-in, base editing, and epigenetic modifications	Research and some therapeutic applications, particularly in creating isogenic cell lines and targeted gene correction	Primarily research, with limited therapeutic use due to complexity; historically used in early gene therapy trials and functional genomics	[19–21]

(*Continued*)

Table 2.1 Comprehensive comparison of gene editing technologies. (*Continued*)

Aspect	CRISPR-Cas9	TALENs	ZFNs	References
Off-target effects	Manageable: Off-target effects can occur but are mitigated with high-fidelity Cas9 variants, truncated gRNAs, or paired Cas9 nickases	Minimal: Highly specific, with low risk of off-target effects due to precise protein-DNA interactions	Significant: Higher risk compared to CRISPR and TALENs, particularly if zinc finger arrays are not perfectly designed	[22–24]
Ease of use	High: user-friendly, with simple design and implementation. Supported by many open-source tools and commercial kits	Moderate: more complex than CRISPR due to the need for protein engineering, but still accessible	Low: requires extensive protein engineering and optimization for each target, making it less accessible and user-friendly	[25–27]
Cost	Low to moderate: inexpensive to design and produce gRNAs, but costs increase with delivery methods and large-scale applications	Moderate: higher costs due to the need for custom protein engineering, but less expensive than ZFNs	High: significant costs associated with protein engineering and optimization for each specific target	[23, 28, 29]
Regulatory and IP landscape	Mixed: widely used, but subject to ongoing patent disputes and complex IP landscape. Many tools are available for research use under open licenses	Less restrictive: fewer patents and IP disputes compared to CRISPR, though some aspects are patented	Heavily patented: complex and restrictive IP landscape, with significant patent coverage limiting accessibility	[30–32]

(*Continued*)

Table 2.1 Comprehensive comparison of gene editing technologies. (*Continued*)

Aspect	CRISPR-Cas9	TALENs	ZFNs	References
Ethical considerations	Similar to other gene editing technologies, but with greater public awareness due to its widespread use. Ethical debates often focus on germline editing potential	Ethical considerations are similar to CRISPR, but TALENs have seen fewer high-profile debates	Similar to CRISPR and TALENs, but less public focus due to reduced use in clinical settings	[24, 33]
Advantages	High precision, versatility, ease of design and use, broad applicability across organisms and cell types	High specificity, minimal off-target effects, suitable for complex genome modifications	Historically significant, with strong foundational research; effective in contexts requiring high specificity	[34–36]
Limitations	Off-target effects, potential immune responses, IP disputes, and ethical concerns related to germline editing	Delivery challenges due to size, more complex design and production compared to CRISPR	Complexity in design, cost, off-target effects, and limited flexibility compared to newer technologies	[28, 30, 37]
Notable applications	Treatment of genetic disorders (e.g., sickle cell disease, β-thalassemia), cancer immunotherapy (CAR-T cells), agricultural biotech, gene drives	Disease modeling, gene correction in stem cells, functional genomics, limited therapeutic applications	Early gene therapy trials, functional genomics, research in model organisms	[38–40]

(*Continued*)

Table 2.1 Comprehensive comparison of gene editing technologies. (*Continued*)

Aspect	CRISPR-Cas9	TALENs	ZFNs	References
Overall flexibility	Very high: adaptable to a wide range of applications, from basic research to clinical trials	High: particularly useful in contexts requiring high specificity and minimal off-target effects	Moderate: useful in specific research contexts but less adaptable due to complexity and cost	[41–43]

A. High-fidelity Cas9 variants: One of the major advancements in improving CRISPR-Cas9 specificity has been the development of high-fidelity Cas9 variants. These engineered versions of Cas9, such as SpCas9-HF1 and eSpCas9, have been modified to reduce non-specific interactions with DNA, thereby minimizing off-target effects. These variants retain their cutting efficiency while drastically reducing the likelihood of unintended DNA cleavage [17].
B. Guide RNA engineering: Another critical area of improvement has been the optimization of gRNA design. Researchers have developed algorithms and tools to predict and design gRNAs with high on-target activity and minimal off-target potential. Additionally, truncated gRNAs (tru-gRNAs) have been introduced, which have shorter target sequences and are less likely to bind to off-target sites, further enhancing the precision of CRISPR-Cas9 [18].
C. Base editing and prime editing: Base editing and prime editing represent significant advancements beyond traditional CRISPR-Cas9. Base editors, such as cytosine base editors (CBEs) and adenine base editors (ABEs), enable the conversion of one DNA base pair into another without inducing a double-strand break. Prime editing, a more recent development, combines the specificity of CRISPR with a reverse transcriptase enzyme, allowing precise insertions, deletions, and all 12 possible base-to-base conversions without requiring a donor DNA template or double-strand breaks. These innovations offer higher precision and efficiency, expanding the range of possible genetic modifications [19].

These enhancements have not only improved the safety and efficacy of CRISPR-Cas9 but have also broadened its applicability, particularly in clinical settings where precision is paramount. As we explore the advancements and mechanisms of CRISPR-Cas9, it is essential to compare it with other existing gene editing technologies. The Table 2.1 provides a comprehensive comparison of CRISPR-Cas9 with alternative gene editing methods such as TALENs and ZFNs, highlighting their respective strengths, limitations, and applications.

2.5 Ethical and Regulatory Considerations

The rapid development of CRISPR-Cas9 technology has outpaced the establishment of ethical guidelines and regulatory frameworks, particularly concerning its use in human germline editing. Germline editing involves making changes to the DNA of embryos, sperm, or eggs that can be passed on to future generations, raising profound ethical and moral questions.

A. Ethical implications of germline editing: The prospect of editing the human germline has sparked significant debate. Proponents argue that germline editing could potentially eliminate genetic diseases, improving the quality of life for future generations. However, critics caution against unintended consequences, such as off-target effects that could introduce new genetic problems or exacerbate social inequalities by enabling "designer babies" [20]. The potential for eugenics and the loss of genetic diversity are additional concerns that have been raised in ethical discussions [21].
B. Regulatory landscape: Globally, the regulation of CRISPR-Cas9 and gene editing varies widely. In many countries, germline editing is either banned or heavily restricted. For example, the United States prohibits the use of federal funds for research involving germline editing, although private research is not explicitly banned. In contrast, countries like China have more permissive regulations, which has led to controversial developments, such as the birth of the first gene-edited babies in 2018, an event that drew widespread condemnation from the global scientific community [22].

C. Debates and future directions: The debate over CRISPR-Cas9 and germline editing is ongoing, with calls for a global consensus on the ethical use of gene editing technologies. International bodies such as the World Health Organization (WHO) have recommended establishing a registry for human gene-editing research and developing global standards. The challenge lies in balancing the potential benefits of gene editing with the ethical and societal implications, ensuring that the technology is used responsibly and equitably [23].

These ethical and regulatory considerations are crucial as the technology advances and becomes more integrated into medical practice. The need for robust and universally accepted guidelines is paramount to prevent misuse and ensure that gene editing is used to benefit humanity as a whole.

2.6 Applications in Stem Cell Research

CRISPR-Cas9 has revolutionized stem cell research by providing a precise and efficient means to edit the genome, enabling scientists to study gene function, create disease models, and develop therapeutic applications.

a. Studying gene function in stem cells: CRISPR-Cas9 has been instrumental in advancing our understanding of gene function in stem cells. By knocking out specific genes, researchers can observe the resulting phenotypic changes, thereby elucidating the role of those genes in stem cell biology. This approach has been used to identify key regulators of stem cell pluripotency and differentiation, providing insights into the molecular mechanisms that govern stem cell fate [24].
b. Creating disease models: One of the most significant applications of CRISPR-Cas9 in stem cell research is the creation of disease models. By introducing disease-associated mutations into human pluripotent stem cells (PSCs), researchers can generate cellular models of various genetic disorders. These models are invaluable for studying disease pathology, identifying potential drug targets, and testing therapeutic interventions. For example, CRISPR-Cas9 has been used to create models for conditions such as cystic fibrosis, sickle cell anemia, and Huntington's disease [25].

c. Correcting genetic defects in stem cells: CRISPR-Cas9 also holds promise for therapeutic applications, particularly in correcting genetic defects in stem cells before they are used in regenerative medicine. This approach has been explored for treating genetic disorders such as β-thalassemia and muscular dystrophy. In these cases, CRISPR-Cas9 is used to correct the disease-causing mutations in patient-derived iPSCs, which are then differentiated into the desired cell type and transplanted back into the patient. This strategy offers the potential for personalized medicine, where patients receive treatments tailored to their genetic makeup [26].
d. Challenges and future prospects: Despite its potential, the application of CRISPR-Cas9 in stem cell research faces several challenges. These include ensuring the safety and efficacy of edited cells, particularly with regard to off-target effects and the long-term stability of the edits. Additionally, ethical concerns surrounding the use of human embryos in research and the potential for germline modifications must be carefully navigated. Future research is likely to focus on improving the precision and efficiency of CRISPR-Cas9, as well as developing strategies to overcome these challenges, paving the way for broader clinical applications [27].

CRISPR-Cas9 has undeniably transformed stem cell research, providing powerful tools to investigate gene function, model diseases, and develop novel therapies. As the technology continues to evolve, it will play an increasingly central role in advancing our understanding of biology and improving human health.

2.7 Stem Cells: A Therapeutic Frontier

Stem cells hold immense promise in regenerative medicine due to their unique ability to differentiate into various cell types. This section explores the different types of stem cells, their therapeutic potential, current and emerging therapies, and the challenges that must be overcome for their widespread clinical application.

2.7.1 Types of Stem Cells: Embryonic, Adult, and Induced Pluripotent Stem Cells (iPSCs)

Stem cells are categorized based on their origin and potential to differentiate into different cell types. The three primary types of stem cells—embryonic stem cells (ESCs), adult stem cells, and induced pluripotent stem cells (iPSCs)—each have distinct characteristics and therapeutic potential.

a. Embryonic stem cells (ESCs): Embryonic stem cells are derived from the inner cell mass of a blastocyst, an early-stage embryo. ESCs are pluripotent, meaning they can differentiate into any cell type in the body, making them incredibly versatile for research and therapeutic applications [28]. Their ability to give rise to all cell types offers the potential to generate cells and tissues for regenerative medicine. For example, ESCs have been explored for generating pancreatic beta cells to treat diabetes, cardiomyocytes for heart disease, and neurons for neurodegenerative disorders [29].

   However, the use of ESCs is fraught with ethical concerns, primarily because their derivation involves the destruction of human embryos. This ethical dilemma has led to significant controversy and regulatory restrictions in many countries, limiting the availability and use of ESCs in research and therapy [24]. Additionally, there is a risk of immune rejection when ESC-derived cells are transplanted into patients, as they may be recognized as foreign by the recipient's immune system [30].

b. Adult stem cells: Adult stem cells, also known as somatic or tissue-specific stem cells, are found in various tissues throughout the body, including the bone marrow, brain, and skeletal muscle. Unlike ESCs, adult stem cells are typically multipotent, meaning they can only differentiate into a limited range of cell types related to their tissue of origin. For example, hematopoietic stem cells (HSCs) found in the bone marrow can give rise to various blood cells, while mesenchymal stem cells (MSCs) can differentiate into bone, cartilage, and fat cells [31].

   Adult stem cells are less controversial than ESCs because they can be obtained from the patient's own body, eliminating the ethical concerns associated with embryo destruction. Additionally, using a patient's own stem cells (autologous transplantation) reduces the risk of immune rejection, as the cells are genetically identical to those of the recipient. However, adult stem cells have limitations, including their restricted differentiation potential and the fact that they are often present in small quantities, making them challenging to isolate and expand in sufficient numbers for therapeutic use [32].

c. Induced pluripotent stem cells (iPSCs): Induced pluripotent stem cells represent a significant breakthrough in stem cell research. iPSCs are generated by reprogramming adult somatic cells (such as skin or blood cells) back into a pluripotent state, similar to that of ESCs. This reprogramming is achieved by introducing a specific set of genes (often through viral vectors) that reset the cell's developmental clock, allowing it to differentiate into any cell type.

   The most significant advantage of iPSCs is that they offer the therapeutic versatility of ESCs without the associated ethical concerns. Since iPSCs can be generated from a patient's own cells, they are also less likely to be rejected by the immune system, making them ideal for personalized medicine. For example, iPSCs can be used to create patient-specific models of diseases, enabling researchers to study disease mechanisms and test potential treatments in a laboratory setting [24]. Additionally, iPSCs hold promise for generating tissues and organs for transplantation, potentially overcoming the shortage of donor organs.

Despite their potential, iPSCs are not without challenges. The reprogramming process can introduce genetic and epigenetic abnormalities, raising concerns about the safety of iPSC-based therapies. Moreover, the use of viral vectors to deliver reprogramming factors can increase the risk of insertional mutagenesis and tumorigenicity [33]. Ongoing research aims to address these challenges, such as developing safer and more efficient methods for reprogramming cells.

2.7.2 Current and Emerging Stem Cell Therapies

2.7.2.1 Current Stem Cell Therapies

Stem cell therapies have advanced significantly over the past few decades, with several treatments now in clinical use and many more under investigation. Several ongoing clinical trials have explored the potential of CRISPR-Cas9 in correcting genetic disorders, with promising outcomes detailed in Table 2.2.

Table 2.2 Summary of clinical trials involving CRISPR-Cas9.

Trial identifier	Condition/disease	Target gene(s)	Phase	Location	Status	Results/outcomes	References
NCT03745287	Sickle cell disease	BCL11A	Phase 1/2	United States	Recruiting	Early results show increased fetal hemoglobin levels; ongoing assessment	[37]
NCT03872479	β-thalassemia	BCL11A	Phase 1/2	Europe	Active	Patients have become transfusion-independent; promising efficacy observed	[48]
NCT03399448	Multiple myeloma	BCMA	Phase 1	China	Completed	Demonstrated safety and preliminary efficacy; continuing follow-up for long-term effects	[39]

(Continued)

Table 2.2 Summary of clinical trials involving CRISPR-Cas9. (*Continued*)

Trial identifier	Condition/ disease	Target gene(s)	Phase	Location	Status	Results/outcomes	References
NCT04037566	Leber congenital amaurosis	CEP290	Phase 1/2	United States	Active	Initial safety confirmed; visual improvements observed in treated patients	[40]
NCT04441606	COVID-19 pneumonia	ACE2	Phase 1	United States	Withdrawn	Trial withdrawn due to funding issues; no results available	[41]
NCT04244656	HIV infection	CCR5	Phase 1/2	China	Active	Ongoing; preliminary data suggest potential reduction in viral load	[42]
NCT04134325	Transfusion-dependent β-thalassemia	HBB	Phase 1/2	Australia	Active	Early results indicate promising hemoglobin production; trials continuing	[43]

(*Continued*)

Table 2.2 Summary of clinical trials involving CRISPR-Cas9. *(Continued)*

Trial identifier	Condition/disease	Target gene(s)	Phase	Location	Status	Results/outcomes	References
NCT04146052	Sickle Cell disease	HBB	Phase 2	United States	Recruiting	Focused on correcting the sickle cell mutation; early phase results pending	[44]
NCT03655678	T-cell acute lymphoblastic leukemia (T-ALL)	TRAC, CD52	Phase 1	United States	Recruiting	Investigating the safety and efficacy of CRISPR-edited T-cells in treating T-ALL	[45]
NCT04002026	Relapsed/refractory multiple myeloma	BCMA, CD19	Phase 1/2	China	Recruiting	Early results show promising anti-tumor activity; ongoing recruitment and data collection	[46]

(Continued)

Table 2.2 Summary of clinical trials involving CRISPR-Cas9. (*Continued*)

Trial identifier	Condition/ disease	Target gene(s)	Phase	Location	Status	Results/outcomes	References
NCT04557490	Duchenne muscular dystrophy (DMD)	Dystrophin	Phase 1	United States	Recruiting	Aimed at restoring dystrophin expression in DMD patients; early results pending	[47]
NCT03690011	Sickle Cell disease	HBB	Phase 2	Canada	Recruiting	Exploring the safety and efficacy of CRISPR-Cas9 gene editing in SCD patients	[48]

a. Hematopoietic stem cell transplantation (HSCT): Hematopoietic stem cell transplantation is one of the most established stem cell therapies and has been used for decades to treat blood-related disorders such as leukemia, lymphoma, and aplastic anemia. HSCT involves the transplantation of HSCs, usually derived from the bone marrow, peripheral blood, or umbilical cord blood, to restore the blood and immune systems after they have been damaged by chemotherapy or radiation therapy [34]. Autologous HSCT, where the patient's own HSCs are used, is preferred, when possible, as it reduces the risk of graft-versus-host disease (GVHD). Allogeneic HSCT, using donor HSCs, is used when the patient's own cells are not viable, but it carries a higher risk of complications [35].
b. Stem cell therapies for cardiovascular diseases: Cardiovascular diseases, including heart failure and myocardial infarction, are leading causes of death worldwide. Stem cell therapy holds promise for regenerating damaged heart tissue and improving heart function. Clinical trials have investigated the use of various stem cell types, including MSCs, cardiac stem cells, and iPSC-derived cardiomyocytes, for treating heart disease. Although early results have been promising, challenges such as low cell engraftment rates, limited cell survival, and potential arrhythmogenic risks need to be addressed [36].
c. Stem cell therapies for neurodegenerative disorders: Neurodegenerative diseases such as Parkinson's disease, Alzheimer's disease, and amyotrophic lateral sclerosis (ALS) are characterized by the progressive loss of neurons, leading to debilitating symptoms. Stem cell-based therapies aim to replace the lost neurons and restore function. For example, clinical trials are exploring the transplantation of dopaminergic neurons derived from ESCs or iPSCs into the brains of patients with Parkinson's disease [30]. While preclinical studies have shown potential, translating these therapies into safe and effective treatments remains a significant challenge due to the complexity of the central nervous system and the risks of tumorigenicity and immune rejection [28].

As CRISPR-Cas9 technology continues to advance, its application in clinical settings has become increasingly prominent. Numerous clinical trials are underway to evaluate the safety and efficacy of CRISPR-Cas9-based therapies, particularly in the context of genetic disorders and stem cell research.

2.7.2.2 Emerging Applications in Tissue Regeneration and Organ Repair

Beyond the treatment of specific diseases, stem cells are being explored for broader applications in tissue regeneration and organ repair. For instance, MSCs have been investigated for their ability to promote tissue repair and modulate the immune response in conditions such as osteoarthritis, spinal cord injury, and chronic wounds. Additionally, advances in tissue engineering, combined with iPSCs, have led to the development of organoids—miniature, simplified versions of organs grown *in vitro*. Organoids provide a platform for studying organ development, disease modeling, and drug testing, and they hold potential for future organ replacement therapies [29].

2.7.3 Challenges in Stem Cell Therapeutics

The development and clinical application of stem cell therapies face numerous technical, ethical, and logistical challenges. Addressing these challenges is crucial for the successful translation of stem cell research into effective treatments.

a. Immune rejection: One of the primary challenges in stem cell therapy is the risk of immune rejection, particularly in allogeneic transplants where the stem cells are derived from a donor rather than the patient. The recipient's immune system may recognize the transplanted cells as foreign and mount an immune response, leading to the destruction of the cells and the failure of the therapy. To mitigate this risk, immunosuppressive drugs are often used, but they can have significant side effects and increase the risk of infections and cancer. The development of iPSCs, which can be generated from the patient's own cells, offers a potential solution to this problem by reducing the risk of immune rejection [33, 34].

b. Tumorigenicity: Another major concern in stem cell therapy is the risk of tumorigenicity, particularly with pluripotent stem cells such as ESCs and iPSCs. These cells have the potential to form teratomas—tumors that contain a variety of tissue types—if their differentiation is not carefully controlled. The use of iPSCs is particularly challenging in this regard, as the reprogramming process can introduce genetic and epigenetic abnormalities that may increase the risk of tumor formation. To address this issue, researchers are developing methods to improve the safety of stem cell therapies, such as using safer reprogramming techniques, thoroughly screening cells for abnormalities before transplantation, and developing strategies to ensure the complete differentiation of pluripotent cells before they are used in therapy [35].
c. Directing stem cell differentiation: A critical challenge in stem cell therapy is ensuring that the transplanted stem cells differentiate into the desired cell type and integrate properly into the target tissue. For example, in therapies aimed at regenerating heart tissue, it is essential that the stem cells differentiate into cardiomyocytes and no other cell types. This requires a deep understanding of the signals that guide stem cell differentiation and the development of protocols to direct this process *in vitro* before transplantation. Moreover, once transplanted, the cells must respond appropriately to the local environment and integrate with the surrounding tissue to restore function [31].
d. Scalability and manufacturing: For stem cell therapies to become widely available, it is essential to develop scalable and cost-effective methods for producing large quantities of high-quality stem cells. This includes developing robust protocols for cell culture, expansion, and differentiation, as well as ensuring the consistency and reproducibility of the final cell products. Manufacturing challenges also include the need for strict quality control measures to ensure that the cells meet safety and efficacy standards. Additionally, the logistics of storing, transporting, and delivering stem cells to patients must be carefully managed to maintain cell viability and function [29].

e. Ethical considerations: The ethical challenges associated with stem cell therapy, particularly the use of ESCs, continue to be a topic of debate. The destruction of embryos to obtain ESCs raises significant ethical concerns, leading to restrictions on ESC research in many countries. The development of iPSCs has alleviated some of these concerns, as these cells do not require the destruction of embryos. However, the ethical implications of creating and using iPSCs, particularly in the context of potential germline modifications, remain an area of active discussion. It is essential to continue engaging with ethicists, policymakers, and the public to navigate these ethical challenges and develop guidelines that balance scientific progress with respect for human dignity [23].
f. Regulatory and clinical translation: The translation of stem cell research into clinical therapies requires navigating complex regulatory environments. Regulatory agencies such as the U.S. Food and Drug Administration (FDA) and the European Medicines Agency (EMA) have established frameworks for evaluating the safety and efficacy of stem cell therapies. However, the regulatory landscape is continually evolving, and researchers must stay informed about the latest requirements and guidelines. Additionally, the clinical translation of stem cell therapies involves conducting rigorous preclinical and clinical trials to demonstrate their safety and efficacy. This process can be time-consuming and expensive, and many promising therapies face significant challenges in moving from the laboratory to the clinic [33].

Addressing these challenges is crucial for the successful development and widespread adoption of stem cell therapies. Ongoing research and innovation, coupled with careful ethical and regulatory oversight, will be essential in overcoming these barriers and realizing the full potential of stem cell-based treatments.

2.8 CRISPR-Cas9 in Stem Cell Therapeutics

CRISPR-Cas9 technology has revolutionized the field of gene editing, offering precise and efficient tools to modify the genome of stem cells.

2.8.1 Genetic Modification of Stem Cells Using CRISPR-Cas9

CRISPR-Cas9 has become the tool of choice for genetic modification of stem cells due to its simplicity, precision, and versatility. The process begins with the design of a guide RNA (gRNA) that matches the specific DNA sequence targeted for modification. The gRNA, in complex with the Cas9 protein, guides the Cas9 enzyme to the target location within the genome, where it induces a double-strand break (DSB) [1]. Once the break is introduced, the cell's natural repair mechanisms, non-homologous end joining (NHEJ) or homology-directed repair (HDR), come into play.

A. Non-homologous end joining (NHEJ): NHEJ is an error-prone repair process that often results in small insertions or deletions (indels) at the site of the DSB. This can disrupt the gene's function, making NHEJ useful for creating gene knockouts. In stem cell research, NHEJ has been used to study the effects of gene loss on cell differentiation and function, providing insights into gene function and disease mechanisms [15].
B. Homology-directed repair (HDR): HDR, on the other hand, is a more precise repair mechanism that uses a homologous DNA template to guide the repair process. Researchers can exploit HDR to introduce specific genetic changes, such as correcting a mutation or inserting a therapeutic gene. In stem cells, HDR is often used to correct disease-causing mutations, making it a powerful tool for developing therapies for genetic disorders [16].

Examples of Successful Modifications in iPSCs

Induced pluripotent stem cells (iPSCs) have been a primary focus for CRISPR-Cas9-mediated gene editing due to their pluripotency and potential for personalized medicine. For example, researchers have successfully used CRISPR-Cas9 to correct mutations in iPSCs derived from patients with genetic disorders such as sickle cell anemia and β-thalassemia. In these cases, the corrected iPSCs were differentiated into hematopoietic stem cells, which were then transplanted into disease models, demonstrating the potential to cure these blood disorders [25].

The ability to introduce or correct genetic mutations in iPSCs has significant implications for treating a wide range of genetic disorders. By generating patient-specific iPSCs, correcting the genetic defect, and then differentiating these cells into the required cell type, researchers can develop personalized therapies that offer a potential cure for otherwise untreatable conditions. The role of CRISPR-Cas9 in modifying induced pluripotent stem cells (iPSCs) for studying disease models and developing cell-based therapies is illustrated in Figure 2.2.

Genome editing has the potential to treat both monogenic illnesses and complicated conditions that are linked to a specific gene target. The utilization of human induced pluripotent stem cells (hiPSCs) has been employed to test several candidate treatments for diverse diseases, hence obtaining a disease model. hiPSCs are often utilized in conjunction with gene editing techniques to specifically target disease-causing genes. This allows for a more comprehensive investigation of the underlying mechanisms of the disease, facilitates drug screening, and enhances the possibility for cell-based therapies. The most potent gene editing method is represented by clustered, regularly interspaced, short palindromic repeats (CRISPR) and CRISPR-associated protein (CRISPR/Cas9). This method can be regarded as one of the most potent and adaptable tools for both gene editing and transcriptional control, as well as for epigenetic regulation.

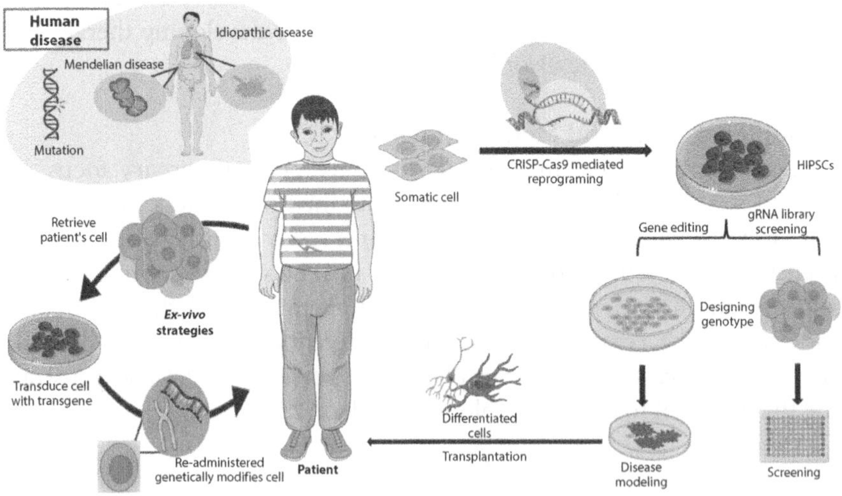

Figure 2.2 CRISPR-Cas9 in stem cell therapeutics.

2.8.2 Case Studies: Successful Applications in Disease Models

CRISPR-Cas9 has been applied in numerous studies to correct genetic mutations in stem cells, offering promising results in disease models. These case studies highlight the potential of CRISPR-modified stem cells to move from bench to bedside. A comprehensive overview of CRISPR-Cas9 applications in treating genetic conditions is summarized in Table 2.3.

a. Sickle cell anemia: Sickle cell anemia is caused by a single point mutation in the β-globin gene, leading to the production of abnormal hemoglobin that causes red blood cells to assume a sickle shape. This shape impairs the cells' ability to transport oxygen, leading to severe health complications. Researchers have used CRISPR-Cas9 to correct this mutation in iPSCs derived from patients with sickle cell anemia. The corrected iPSCs were then differentiated into hematopoietic stem cells, which produced normal red blood cells when transplanted into animal models [49]. This approach demonstrated the feasibility of using CRISPR-Cas9 to correct genetic defects and restore normal function in diseased tissues.
b. Muscular dystrophy: Duchenne muscular dystrophy (DMD) is a severe genetic disorder characterized by progressive muscle degeneration due to mutations in the dystrophin gene. CRISPR-Cas9 has been used to target and correct these mutations in iPSCs derived from DMD patients. In one study, researchers used CRISPR-Cas9 to excise the mutated exons from the dystrophin gene, allowing the production of a functional, albeit shorter, version of dystrophin. The corrected iPSCs were then differentiated into muscle cells that expressed the corrected dystrophin protein, offering a potential therapeutic approach for DMD [26].
c. Cystic fibrosis: Cystic fibrosis (CF) is caused by mutations in the CFTR gene, leading to the production of defective chloride channels in epithelial cells. This results in thick mucus accumulation in the lungs and other organs, causing chronic infections and respiratory failure. Researchers have used CRISPR-Cas9 to correct CFTR mutations in iPSCs derived from CF patients. The corrected iPSCs were differentiated into lung epithelial cells, which exhibited restored chloride channel function *in vitro*, providing proof-of-concept for a potential gene therapy approach to CF [50].

Table 2.3 Applications of CRISPR-Cas9 in disease treatment.

Disease	Target gene	CRISPR-Cas9 approach	Model/system used	Current status	References
Sickle Cell anemia	HBB	Correction of the point mutation in the β-globin gene in iPSCs derived from patients	iPSCs differentiated into hematopoietic stem cells, transplanted into animal models	Preclinical studies demonstrate restored normal red blood cell production	[49]
Duchenne muscular dystrophy (DMD)	Dystrophin	Excision of mutated exons in the dystrophin gene using CRISPR-Cas9	iPSCs derived from DMD patients, differentiated into muscle cells	Functional dystrophin protein expression observed in muscle cells, showing potential for therapeutic application	[26]
Cystic fibrosis	CFTR	Correction of CFTR gene mutations using CRISPR-Cas9 in iPSCs derived from cystic fibrosis patients	iPSCs differentiated into lung epithelial cells	Chloride channel function restored in corrected cells, offering proof-of-concept for gene therapy	[50]

(Continued)

Table 2.3 Applications of CRISPR-Cas9 in disease treatment. (*Continued*)

Disease	Target gene	CRISPR-Cas9 approach	Model/system used	Current status	References
β-thalassemia	HBB	Correction of mutations in the HBB gene using CRISPR-Cas9	iPSCs differentiated into hematopoietic stem cells, transplanted into animal models	Demonstrates potential to cure β-thalassemia by producing normal hemoglobin in animal models	[25]
Huntington's disease	HTT	Introduction of disease-associated mutations into PSCs to create disease models	Human pluripotent stem cells (PSCs)	Valuable for understanding disease pathology and aiding drug discovery	[25]
Retinal diseases	Various	Correction of mutations in retinal cells using CRISPR-Cas9, followed by differentiation into retinal pigment epithelium cells	Gene-edited iPSCs differentiated into retinal cells	Preclinical models show potential for treating retinal diseases by restoring retinal function	[22]
Cancer (immunotherapy)	PD-1, CAR-T targets	Engineering of T cells using CRISPR-Cas9 to target tumor antigens, creating individualized cancer immunotherapies	T cells modified to recognize and attack specific tumors	Clinical trials underway; shows promise in enhancing the efficacy of cancer immunotherapy	[77]

(*Continued*)

Table 2.3 Applications of CRISPR-Cas9 in disease treatment. (*Continued*)

Disease	Target gene	CRISPR-Cas9 approach	Model/system used	Current status	References
Blood disorders (e.g., β-thalassemia, sickle cell disease)	HBB, BCL11A	Editing of hematopoietic stem cells to correct genetic mutations, focusing on hemoglobin production	iPSCs differentiated into hematopoietic stem cells	Ongoing clinical trials show promise in curing genetic blood disorders, with several trials in advanced stages	[78]
HIV/AIDS	CCR5	Knockout of CCR5 co-receptor gene in T cells using CRISPR-Cas9 to confer resistance to HIV infection	Patient-derived T cells, mouse models	Early clinical trials show safety and partial efficacy; ongoing trials aim to improve T cell persistence and HIV resistance	[15]
Liver disorders (e.g., hemophilia B)	F9	CRISPR-Cas9 used to correct mutations in the F9 gene in hepatocytes to restore clotting factor production	iPSCs differentiated into hepatocytes, mouse models	Preclinical studies demonstrate restored clotting function; gene therapy approaches moving towards clinical trials	[16]
Amyotrophic lateral sclerosis (ALS)	SOD1	CRISPR-Cas9 used to knock out mutant SOD1 gene associated with familial ALS	iPSCs differentiated into motor neurons, mouse models	Proof-of-concept studies show potential to slow disease progression; early stage research ongoing	[51]

(*Continued*)

Table 2.3 Applications of CRISPR-Cas9 in disease treatment. (*Continued*)

Disease	Target gene	CRISPR-Cas9 approach	Model/system used	Current status	References
Cardiovascular diseases	PCSK9, MYBPC3	Editing of genes involved in cholesterol metabolism and cardiomyopathies to prevent cardiovascular diseases	Human pluripotent stem cells, animal models	Preclinical studies show promise in preventing disease onset; moving towards human trials for heart disease prevention	[18]
Leber congenital amaurosis	CEP290	CRISPR-Cas9 used to correct the CEP290 mutation in retinal cells to treat blindness caused by LCA	iPSCs differentiated into retinal cells, mouse models	Preclinical models demonstrate restoration of retinal function; early-stage clinical trials initiated for vision restoration	[27]
Alzheimer's disease	APOE4	CRISPR-Cas9 used to knock out or modify the APOE4 allele associated with increased risk of Alzheimer's disease	Human iPSCs, brain organoids, mouse models	Research ongoing to understand the therapeutic potential and safety; still in early stages of development	[20]
Type 1 diabetes	Insulin	Editing pancreatic beta cells to enhance insulin production or protect against autoimmune destruction in type 1 diabetes	iPSCs differentiated into pancreatic beta cells, mouse models	Preclinical studies show potential for restoring insulin production; early research into gene therapy for diabetes management	[17]

(*Continued*)

Table 2.3 Applications of CRISPR-Cas9 in disease treatment. (*Continued*)

Disease	Target gene	CRISPR-Cas9 approach	Model/system used	Current status	References
Hemophilia A	F8	CRISPR-Cas9-mediated gene correction in the F8 gene to restore factor VIII production in patients with Hemophilia A	iPSCs differentiated into hepatocytes, mouse models	Proof-of-concept studies show promising results; clinical trials in early stages for hemophilia gene therapy	[52]
Parkinson's disease	SNCA	CRISPR-Cas9 used to model Parkinson's disease by introducing mutations in the SNCA gene in pluripotent stem cells	iPSCs, differentiated into dopaminergic neurons, mouse models	Ongoing research to understand disease mechanisms and explore potential gene therapy approaches for Parkinson's disease	[27]
Tay–Sachs disease	HEXA	Gene editing to correct mutations in the HEXA gene responsible for Tay–Sachs disease	iPSCs, mouse models	Preclinical studies indicate potential for therapeutic intervention; still in early research phases	[19]

These case studies illustrate the potential of CRISPR-Cas9 to correct genetic mutations in stem cells and provide a foundation for developing new treatments for genetic disorders. The ability to generate patient-specific stem cells, correct the underlying genetic defect, and then differentiate these cells into the required tissue offers a promising path towards personalized medicine.

As stem cell research continues to evolve, CRISPR-Cas9 has emerged as a powerful tool for addressing genetic disorders by enabling precise gene modifications. The Table 2.3 showcases various applications of CRISPR-Cas9 in treating diseases, detailing the target genes, approaches used, model systems, and the current status of these research efforts. This table provides concrete examples of how CRISPR-Cas9 is being utilized to advance stem cell therapeutics.

2.8.3 Limitations and Challenges in Clinical Applications

While the application of CRISPR-Cas9 in stem cell therapeutics is promising, several challenges must be addressed before these technologies can be translated into clinical practice. A comparative evaluation of various CRISPR-Cas9 delivery methods, emphasizing their efficiency, accuracy, and potential drawbacks, is presented in Table 2.4.

Table 2.4 Efficiency and accuracy of different delivery methods for CRISPR-Cas9.

Delivery method	Efficiency	Accuracy	Advantages	Disadvantages	Current status	References
Viral vectors (lentivirus, AAV)	High	High	Efficient at delivering CRISPR components into a wide range of cell types; stable integration into the genome	Risk of insertional mutagenesis, potential immunogenicity, limited cargo size	Widely used in preclinical and clinical studies; ongoing research to improve safety and efficiency	[51, 68]
Lipid nanoparticles	Moderate	Moderate	Non-viral, reduces risk of immunogenicity and insertional mutagenesis; can carry larger cargos than viral vectors	Lower delivery efficiency compared to viral vectors, potential cytotoxicity	Increasingly used in preclinical studies; research focused on enhancing targeting specificity and delivery efficiency	[51, 68]

(Continued)

Table 2.4 Efficiency and accuracy of different delivery methods for CRISPR-Cas9. (*Continued*)

Delivery method	Efficiency	Accuracy	Advantages	Disadvantages	Current status	References
Electroporation	High	High	Direct delivery of CRISPR components into cells with high efficiency, especially *in vitro*	Can cause significant cell damage and death; less efficient *in vivo*	Commonly used for *in vitro* experiments and *ex vivo* gene editing; ongoing research to adapt for *in vivo* applications	[51]
Gold nanoparticles	Moderate	High	Low toxicity, non-viral, can be functionalized for targeted delivery	Limited to small cargo sizes, lower delivery efficiency compared to other methods	Experimental stage; research ongoing to optimize delivery and expand therapeutic applications	[68]

(*Continued*)

Table 2.4 Efficiency and accuracy of different delivery methods for CRISPR-Cas9. (*Continued*)

Delivery method	Efficiency	Accuracy	Advantages	Disadvantages	Current status	References
Physical methods (microinjection)	High	High	Precise delivery of CRISPR components directly into the nucleus of individual cells	Technically challenging, not scalable for large numbers of cells or *in vivo* applications	Primarily used in research settings for precise gene editing; limited clinical application due to scalability issues	[51, 64]
Peptide-based delivery	Moderate	Moderate	Flexible and can be modified for targeted delivery; low immunogenicity	Lower efficiency compared to viral and lipid-based methods; stability and delivery challenges	Emerging method with potential for targeted *in vivo* applications; research ongoing to improve stability and efficiency	[51, 68]

(*Continued*)

Table 2.4 Efficiency and accuracy of different delivery methods for CRISPR-Cas9. (*Continued*)

Delivery method	Efficiency	Accuracy	Advantages	Disadvantages	Current status	References
Hydrodynamic injection	Moderate	Moderate	Simple technique for delivering large volumes, useful for liver targeting	Potential tissue damage, limited to specific organs (e.g., liver), low efficiency	Primarily experimental, used in animal models; research focused on refining technique for broader application	[64]
Polymeric nanoparticles	Moderate	High	Biocompatible, can carry large payloads, potential for targeted delivery	Complex synthesis, potential variability in delivery efficiency, slower translation to clinical use	Increasingly explored for therapeutic delivery, particularly in cancer and genetic diseases; research ongoing to enhance targeting and efficiency	[21, 68]

a. Delivery methods: One of the significant hurdles in applying CRISPR-Cas9 therapeutically is the efficient and safe delivery of the CRISPR components to the target cells. Current delivery methods include viral vectors (such as lentivirus and adeno-associated virus), lipid nanoparticles, and electroporation. Each method has its advantages and limitations. Viral vectors, for example, can efficiently deliver CRISPR components but may integrate into the host genome, raising safety concerns. Non-viral methods like lipid nanoparticles avoid genomic integration but may have lower efficiency and specificity [51]. Developing delivery methods that can target specific tissues, ensure efficient gene editing, and minimize off-target effects is critical for the success of CRISPR-based therapies.
b. Off-target effects: Off-target effects, where CRISPR-Cas9 cuts DNA at unintended sites, pose a significant risk in therapeutic applications. Such unintended modifications can lead to the disruption of essential genes or the activation of oncogenes, potentially causing harmful side effects or contributing to cancer development. Despite advances in improving CRISPR-Cas9 specificity, off-target effects remain a concern, particularly in clinical settings where safety is paramount. Techniques such as high-fidelity Cas9 variants and improved gRNA design have been developed to reduce off-target effects, but further refinement is needed [18].
c. Long-term stability of genetic modifications: The long-term stability and safety of CRISPR-induced genetic modifications are another area of concern. In stem cell therapies, it is essential that the genetic modifications are stably maintained as the cells differentiate and proliferate in the body. There is also a need to ensure that the modified cells do not acquire additional mutations over time that could lead to malignant transformation or other adverse effects. Long-term studies in animal models and eventually in humans are required to assess the durability and safety of these genetic modifications [27].
d. Ethical and regulatory challenges: The ethical and regulatory challenges associated with using CRISPR-Cas9 in stem cell therapeutics cannot be overlooked. Germline editing, where genetic changes are heritable and passed on to future generations, raises significant ethical concerns. The potential

for creating "designer babies" or introducing unintended genetic changes with long-term consequences has led to calls for strict regulation and oversight of germline editing. Even in somatic cell editing, ethical considerations related to informed consent, access to therapy, and the potential for misuse must be carefully addressed. Regulatory agencies must establish clear guidelines to ensure that CRISPR-based therapies are developed and applied responsibly [20].

Overcoming these limitations and challenges is essential for the successful translation of CRISPR-Cas9-modified stem cells into clinical therapies. Continued research and innovation in delivery methods, specificity, safety, and ethical considerations will be critical to realizing the full potential of these technologies in medicine.

2.8.4 Safety and Long-Term Implications

The use of CRISPR-Cas9 in stem cell therapeutics presents several safety and long-term implications that must be carefully considered to ensure the responsible application of this technology.

a. Unintended genetic changes: One of the primary safety concerns with CRISPR-Cas9 is the potential for unintended genetic changes, including off-target effects and mosaicism (where not all cells carry the same genetic modification). These unintended changes can have unpredictable consequences, such as disrupting essential genes or activating oncogenes, leading to cancer or other health issues [17]. The risk of unintended genetic changes underscores the importance of developing more precise gene-editing tools and implementing robust screening methods to detect and minimize these effects.
b. Immune responses: The introduction of CRISPR-Cas9 components into the body can trigger immune responses, particularly against the bacterial Cas9 protein. Immune responses can reduce the efficiency of gene editing and, in severe cases, lead to inflammatory reactions or anaphylaxis. Managing immune responses is a critical consideration in developing CRISPR-based therapies, particularly for repeated or systemic administration. Strategies such as using Cas9 orthologs from non-pathogenic bacteria or engineering humanized Cas9 variants may help mitigate immune reactions [52].

c. Long-term consequences of genetic modifications: The long-term consequences of CRISPR-induced genetic modifications remain largely unknown, particularly in the context of stem cell therapeutics, where edited cells are expected to proliferate and differentiate over time. There is a need to ensure that the genetic modifications do not have deleterious effects on the cells' function, longevity, or potential to form tumors. Long-term studies in animal models and careful monitoring of patients in clinical trials are necessary to assess the safety and stability of these modifications over time [27].
d. Strategies to mitigate risks: To mitigate the risks associated with CRISPR-Cas9, several strategies are being developed. These include the use of high-fidelity Cas9 variants that reduce off-target effects, base editors that make precise nucleotide changes without inducing DSBs, and prime editors that offer even greater precision and versatility in gene editing. Additionally, researchers are exploring the use of transient CRISPR systems, where the CRISPR components are quickly degraded after achieving the desired edit, reducing the potential for off-target effects [19].
e. Regulatory and ethical considerations: The regulatory and ethical landscape surrounding CRISPR-Cas9 is evolving as the technology advances. Regulatory agencies must establish clear guidelines for the clinical use of CRISPR-modified stem cells, ensuring that therapies are safe, effective, and ethically sound. Ethical considerations include issues of informed consent, equitable access to therapies, and the potential for germline modifications. Public engagement and transparent dialogue between scientists, ethicists, and policymakers are essential to navigate these challenges responsibly [23].

Addressing these safety and long-term implications is crucial for the successful and responsible development of CRISPR-Cas9 in stem cell therapeutics. As research progresses, a focus on precision, safety, and ethical considerations will be key to ensuring that these powerful technologies are used to benefit patients while minimizing risks.

2.9 Beyond CRISPR-Cas9: Emerging Gene Editing Technologies

As powerful as CRISPR-Cas9 has proven to be, the field of gene editing continues to evolve with the development of new technologies that offer enhanced precision, versatility, and efficiency.

2.9.1 CRISPR-Cas12 and CRISPR-Cas13 Systems

CRISPR-Cas12 and CRISPR-Cas13 are part of the expanding CRISPR family, offering unique advantages over the Cas9 system, particularly in terms of targeting different types of nucleic acids.

2.9.1.1 CRISPR-Cas12

Cas12 is a class 2 CRISPR-associated protein that, like Cas9, targets DNA. However, it has distinct features that make it a valuable tool for gene editing. Cas12 is known for its ability to create staggered cuts in DNA, leaving overhangs that can be more easily repaired through homology-directed repair (HDR), potentially improving the precision of gene editing [53]. Additionally, Cas12 has a unique trans-cleavage activity, where it can cleave single-stranded DNA nonspecifically once activated. This property has been harnessed for diagnostic purposes, such as in the development of the DETECTR system for detecting viral infections like COVID-19 [54].

In the context of stem cell research, CRISPR-Cas12 offers potential advantages for creating more precise genetic modifications. Its ability to generate overhangs might improve the efficiency of HDR in stem cells, which is crucial for therapeutic applications that require precise gene correction. Moreover, its versatility in creating various types of DNA modifications could expand the range of possible genetic edits in stem cells [55].

2.9.1.2 CRISPR-Cas13

Cas13 is unique among CRISPR-associated proteins because it targets RNA rather than DNA. This RNA-targeting capability opens new avenues for gene editing, particularly in applications where temporary or reversible modifications are desired. Cas13 can be used to knock down gene expression at the RNA level, providing a tool for studying gene function without altering the underlying DNA sequence [56].

Cas13's RNA-editing capabilities have significant implications for stem cell research. It allows for the modulation of gene expression without permanent changes to the genome, which can be particularly useful in studies where transient gene expression changes are needed to guide stem cell differentiation or to study the effects of gene knockdown in stem cell-derived tissues. Additionally, Cas13 could be used to correct RNA-level defects in diseases caused by splicing errors or other RNA-related mutations, offering a new therapeutic avenue for conditions that are not easily addressed by DNA-level editing [57].

The CRISPR-Cas12 and Cas13 systems expand the toolkit available for gene editing, offering distinct mechanisms and potential advantages over Cas9. Their applicability in stem cell research and therapy could lead to more refined and versatile approaches to genetic modification.

2.9.2 Base Editing and Prime Editing

While traditional CRISPR-Cas9 introduces double-strand breaks (DSBs) in DNA, newer technologies like base editing and prime editing allow for precise nucleotide changes without the need for DSBs, reducing the risk of off-target effects and unwanted mutations.

2.9.2.1 Base Editing

Base editing is a powerful technique that enables the direct conversion of one DNA base pair into another without causing a DSB. Developed by David Liu and colleagues, base editors consist of a deactivated Cas9 (dCas9) or a nickase (Cas9n) fused to a deaminase enzyme. This enzyme catalyzes the conversion of specific bases; for example, cytosine base editors (CBEs) convert C•G base pairs into T•A base pairs, while adenine base editors (ABEs) convert A•T base pairs into G•C base pairs [58].

In stem cell research, base editing offers a precise method for correcting point mutations, which are a common cause of genetic diseases. For instance, base editors have been used to correct the mutation responsible for sickle cell anemia in iPSCs. The edited iPSCs were then differentiated into hematopoietic stem cells that produced normal hemoglobin, demonstrating the potential of base editing for developing curative therapies [59].

2.9.2.2 Prime Editing

Prime editing is a more recent advancement that combines the precision of base editing with the flexibility of traditional CRISPR-Cas9. Prime editing uses a fusion protein consisting of Cas9 nickase and a reverse transcriptase enzyme. This system is guided by a prime editing guide RNA (pegRNA), which not only directs the Cas9 to the target site but also encodes the desired edit. The reverse transcriptase then synthesizes the new DNA strand directly at the target site, enabling the insertion, deletion, or substitution of nucleotides without creating a DSB [19].

Prime editing offers several advantages over both traditional CRISPR-Cas9 and base editing, including the ability to perform a wider range of genetic modifications with higher precision and lower off-target effects. In stem cell research, prime editing holds promise for correcting a broader spectrum of genetic mutations, including those that cannot be addressed by base editing alone. This makes it an attractive tool for developing therapies for complex genetic disorders that involve multiple types of mutations [60].

Base editing and prime editing represent significant advancements in gene editing technology, offering more precise and versatile tools for modifying stem cell genomes. These technologies have the potential to address a wider range of genetic disorders with greater accuracy and safety than traditional CRISPR-Cas9.

2.9.3 Other Gene Editing Tools (TALENs, ZFNs)

Before the rise of CRISPR-Cas9, other gene editing technologies such as Transcription Activator-Like Effector Nucleases (TALENs) and Zinc Finger Nucleases (ZFNs) were the main tools used for targeted genetic modifications. Although CRISPR-Cas9 has largely overshadowed these methods due to its simplicity and efficiency, TALENs and ZFNs still have important roles in specific contexts, particularly in stem cell research.

2.9.3.1 TALENs

TALENs are engineered nucleases that consist of a DNA-binding domain derived from transcription activator-like effectors (TALEs) and a FokI nuclease domain. Each TALE is engineered to recognize a specific DNA sequence, allowing TALENs to target and cut DNA at precise locations. TALENs are particularly useful for creating gene knockouts or introducing specific mutations [61].

One of the advantages of TALENs is their high specificity, which can be superior to CRISPR-Cas9 in certain contexts. TALENs have been used successfully in stem cell research to create isogenic cell lines, where only a single gene is altered while the rest of the genome remains unchanged. This is particularly important in studying gene function and disease mechanisms in stem cells, where high specificity is critical [62].

2.9.3.2 ZFNs

Zinc finger nucleases (ZFNs) are another class of engineered nucleases that consist of a DNA-binding domain composed of zinc finger motifs and a FokI nuclease domain. Each zinc finger recognizes a specific triplet of DNA bases, and by combining multiple zinc fingers, ZFNs can be designed to target specific DNA sequences for cleavage [14].

ZFNs were among the first tools used for targeted gene editing and have been employed in a variety of applications, including the modification of stem cells. For example, ZFNs have been used to disrupt the CCR5 gene in hematopoietic stem cells to make them resistant to HIV infection, demonstrating the therapeutic potential of ZFNs in treating viral diseases [63].

2.9.3.3 Comparing TALENs and ZFNs with CRISPR-Cas9

While TALENs and ZFNs offer high specificity, they are more challenging to design and use compared to CRISPR-Cas9. CRISPR-Cas9's ease of use, lower cost, and versatility have made it the preferred tool for most gene editing applications. However, in cases where the highest possible specificity is required, or where CRISPR-Cas9 is less effective, TALENs and ZFNs remain valuable alternatives. These tools are also less prone to off-target effects in certain genomic contexts, making them suitable for applications where precision is paramount [5].

TALENs and ZFNs continue to be important tools in the gene editing toolbox, particularly in scenarios where CRISPR-Cas9 may not be the best option. Their continued development and refinement could lead to new applications in stem cell research and therapy.

2.9.4 Comparative Analysis with CRISPR-Cas9 in Stem Cell Therapy

To understand the potential future roles of these emerging gene editing technologies in stem cell therapy, it is essential to compare them with CRISPR-Cas9 in terms of their strengths, weaknesses, and applicability.

2.9.4.1 Precision and Specificity

CRISPR-Cas9 is known for its ease of use and versatility but can sometimes lead to off-target effects, which can be problematic in therapeutic applications. High-fidelity versions of Cas9 and improved gRNA design have mitigated these issues, but the risk remains. In contrast, TALENs and ZFNs offer higher specificity in certain contexts, albeit at the cost of greater complexity in design and use. Base editing and prime editing, by avoiding DSBs, offer even higher precision, making them particularly suited for correcting single nucleotide mutations with minimal risk of off-target effects [17].

2.9.4.2 Versatility

CRISPR-Cas9 is unparalleled in its versatility, capable of targeting virtually any sequence in the genome by simply altering the gRNA. This flexibility is unmatched by TALENs and ZFNs, which require extensive protein engineering to target new sequences. Base editing and prime editing extend the versatility of CRISPR-based systems by allowing for a wider range of genetic modifications, including precise nucleotide substitutions and small insertions or deletions, without the need for DSBs [58].

2.9.4.3 Safety and Off-Target Effects

Safety is a critical concern in gene editing, particularly in clinical applications. TALENs and ZFNs have a track record of high specificity, which can be advantageous in therapeutic contexts. However, their complexity and the potential for unintended interactions with the genome make them less user-friendly. CRISPR-Cas9, while more accessible, carries a higher risk of off-target effects, although advancements like high-fidelity Cas9 have significantly improved its safety profile. Base and prime editing, by minimizing the creation of DSBs, further reduce the risk of off-target effects and genomic instability, making them attractive options for therapeutic applications [19].

2.9.4.4 Applications in Stem Cell Therapy

CRISPR-Cas9 has already demonstrated its potential in stem cell therapy through various successful applications, such as correcting genetic mutations in iPSCs. Base editing and prime editing could offer even more precise tools for stem cell therapy, particularly in cases where correcting a single nucleotide could cure a disease. TALENs and ZFNs, while less commonly used today, remain valuable for applications requiring the highest specificity, such as generating isogenic cell lines or modifying difficult-to-target regions of the genome [5].

2.9.5 Future Roles

As gene editing technologies continue to evolve, it is likely that CRISPR-Cas9 will remain a cornerstone of the field due to its simplicity and adaptability. However, base editing and prime editing are poised to play increasingly significant roles in both research and clinical settings, offering unparalleled precision and safety. TALENs and ZFNs will continue to be important in niche applications where their specific advantages are required. The future of stem cell therapy will likely involve a combination of these tools, each chosen based on the specific needs of the therapeutic application [60].

The comparative analysis highlights the complementary nature of these gene editing technologies, each offering unique advantages that can be leveraged in different contexts. As stem cell therapy continues to advance, these tools will play a crucial role in developing more effective and precise treatments for a wide range of diseases.

2.10 Ethical, Legal, and Social Implications (ELSI)

The advent of gene editing technologies, particularly CRISPR-Cas9, has revolutionized biomedical research and therapeutics, offering unprecedented opportunities for treating genetic disorders. However, these advances come with significant ethical, legal, and social implications (ELSI) that must be carefully considered to ensure responsible use.

2.10.1 Ethical Concerns in Gene Editing and Stem Cell Research

The ethical implications of gene editing, especially in the context of stem cell research, are profound and multifaceted. Key concerns include the potential for eugenics, issues of informed consent, and the consequences of germline editing.

a. Eugenics and designer babies: One of the most controversial ethical concerns is the potential for gene editing to be used for eugenics—the idea of improving the human species by selecting desirable traits. The ability to edit the human genome raises the possibility of creating "designer babies," where parents could select physical, intellectual, or behavioral traits for their offspring. This prospect has sparked fears of a new era of genetic inequality, where access to gene editing technologies could exacerbate social divides and lead to discrimination against individuals who do not possess certain engineered traits [21].

The ethical dilemma here revolves around the distinction between therapeutic and enhancement applications of gene editing. While therapeutic uses, such as correcting genetic defects to prevent serious diseases, are widely accepted, the use of gene editing for enhancement purposes—such as increasing intelligence or physical abilities—remains highly controversial. The line between these two applications is often blurred, raising questions about where society should draw the line [20].

b. Informed consent: Informed consent is a cornerstone of ethical medical practice, ensuring that individuals are fully aware of the risks and benefits before undergoing any medical procedure. However, gene editing presents unique challenges in this regard, particularly in the context of germline editing. Germline edits are heritable, meaning they will affect not only the individual receiving the therapy but also their descendants. Obtaining informed consent in this context is problematic, as future generations cannot consent to the genetic changes being made on their behalf [23].

Furthermore, the complexity of gene editing technologies and the potential long-term consequences make it difficult for individuals to fully understand the implications of the procedures. This raises concerns about whether truly informed consent can be obtained, particularly when the risks and benefits of these technologies are not yet fully understood.

c. Germline editing: Germline editing, which involves making genetic changes that are passed on to future generations, is perhaps the most ethically contentious issue in gene editing. While somatic cell editing, which affects only the individual treated, is generally considered acceptable, germline editing raises profound ethical questions. The potential for unintended consequences, such as off-target effects or unforeseen impacts on future generations, makes germline editing a highly risky endeavor. Moreover, germline editing could have irreversible effects on the human gene pool, potentially leading to unintended consequences that could affect entire populations [64].

Many ethicists argue that the risks of germline editing outweigh the potential benefits, at least until more is known about the long-term effects of gene editing. Others believe that with stringent regulation and oversight, germline editing could be used responsibly to prevent serious genetic diseases. The ethical debate surrounding germline editing continues to evolve as the technology advances, and it remains one of the most challenging aspects of gene editing governance.

The ethical concerns surrounding gene editing and stem cell research underscore the need for careful consideration and responsible governance. As these technologies continue to develop, it will be crucial to establish clear ethical guidelines that protect human rights and ensure that gene editing is used for the benefit of all.

2.10.2 Public Perception and Misconceptions

Public perception plays a crucial role in shaping the development and acceptance of new technologies, including gene editing and stem cell research. Understanding how the public views these technologies, as well as addressing common misconceptions, is essential for fostering informed public discourse and guiding policy decisions.

a. Public attitudes toward gene editing: Public attitudes toward gene editing are diverse and often influenced by a variety of factors, including cultural beliefs, religious views, and the perceived risks and benefits of the technology. Surveys have shown that while there is general support for the use of gene editing to treat serious diseases, there is significant apprehension about its use for non-therapeutic purposes, such as enhancement or altering traits like intelligence or appearance [65].

   The public's concerns often center around the ethical implications of gene editing, particularly the potential for unintended consequences and the fear of "playing God" by altering the human genome. Additionally, there is widespread concern about the accessibility and equity of gene editing technologies, with many fearing that these advancements could lead to greater social inequality if only the wealthy have access to them [22].

b. Common misconceptions: There are several common misconceptions about gene editing and stem cell research that contribute to public apprehension. One prevalent misconception is the belief that all gene editing involves germline modifications, which would permanently alter the human gene pool. In reality, most current gene editing research focuses on somatic cells, which do not affect future generations [66].

Another misconception is the idea that gene editing technologies are already fully developed and ready for widespread use. In fact, while CRISPR-Cas9 and other gene editing tools have shown great promise, they are still in the early stages of development, and many technical, ethical, and regulatory challenges remain. Public understanding of the limitations and current state of the technology is essential to temper expectations and avoid unrealistic hopes or fears [67].

c. The role of media: The media plays a significant role in shaping public perceptions of gene editing and stem cell research. Media coverage often focuses on the most sensational aspects of these technologies, such as the potential for "designer babies" or the ethical controversies surrounding germline editing. While such coverage can raise important ethical questions, it can also contribute to fear and misunderstanding if not balanced with accurate information about the science and its potential benefits [65].

To address these issues, there is a need for more accurate and balanced reporting on gene editing and stem cell research. Scientists and ethicists should engage with the media to provide clear, factual information that helps the public understand the complexities of these technologies. Public education campaigns and outreach efforts can also play a crucial role in dispelling myths and fostering informed debate.

2.10.2.1 Public Engagement and Education

Given the ethical and social implications of gene editing, it is essential to involve the public in discussions about the development and regulation of these technologies. Public engagement initiatives can help ensure that diverse perspectives are considered in policy decisions and that the technologies are developed in a way that reflects societal values [64].

Educational efforts are also critical for empowering the public to make informed decisions about gene editing. This includes providing accessible information about how gene editing works, its potential applications, and the ethical considerations involved. Schools, universities, and community organizations can play a key role in these efforts, helping to build a more scientifically literate and ethically aware society.

Understanding and addressing public perceptions and misconceptions about gene editing and stem cell research is essential for fostering informed public discourse and guiding the responsible development of these technologies. Public engagement and education will be critical components of this effort.

2.10.3 Legal and Regulatory Frameworks Globally

The rapid development of gene editing technologies has outpaced the creation of comprehensive legal and regulatory frameworks, leading to a patchwork of regulations that vary widely from country to country. Understanding these frameworks is essential for navigating the global landscape of gene editing and ensuring that research and clinical applications are conducted responsibly.

2.10.3.1 Regulatory Approaches in Different Countries

Regulatory frameworks for gene editing and stem cell research differ significantly across the world, reflecting varying cultural, ethical, and political perspectives on these technologies. In the United States, the Food and Drug Administration (FDA) regulates gene editing therapies under its framework for biologics, which includes stringent requirements for safety and efficacy before a product can enter the market. However, the use of federal funds for research involving human embryos is restricted, limiting the scope of research that can be conducted in certain areas, such as germline editing [67]. A detailed overview of the regulatory frameworks governing gene editing technologies across different countries is provided in Table 2.5.

In contrast, the European Union (EU) has adopted a more precautionary approach to gene editing, particularly in the context of germline modifications. The EU's regulatory framework emphasizes the need for thorough ethical review and public consultation before gene editing technologies can be approved for clinical use. The European Medicines Agency (EMA) plays a key role in regulating gene editing therapies, with a focus on ensuring patient safety and ethical compliance [22].

Countries like China have adopted more permissive regulatory frameworks, allowing for a broader range of research and clinical applications. This has led to significant advancements in gene editing research, but it has also raised concerns about the potential for ethical lapses and the need for stronger oversight. The case of the world's first gene-edited babies, which was widely condemned by the international community, highlights the risks of inadequate regulation and the importance of establishing clear guidelines [68].

Table 2.5 Regulatory frameworks across different countries.

Country/ region	Regulatory body	Permitted applications	Prohibited applications	Regulatory approach	References
United States	Food and Drug Administration (FDA)	Somatic gene editing for therapeutic purposes, research on CRISPR-Cas9, preclinical and clinical trials	Germline editing (with federal funds), non-therapeutic genetic modifications, editing of embryos for reproductive use	Rigorous approval process for gene therapies; focus on safety and efficacy; federal funding restrictions on embryo research	[69]
European Union	European Medicines Agency (EMA)	Somatic gene editing for disease treatment, clinical trials under strict ethical oversight, research on gene editing	Germline editing, embryo editing for reproductive purposes, non-therapeutic genetic enhancements	Precautionary approach; emphasizes ethical review and public consultation; strict regulation of clinical applications	[70]
China	National Health Commission (NHC)	Research and clinical applications of gene editing, including somatic and some germline editing under strict oversight	Unregulated germline editing leading to heritable changes, unethical experiments (e.g., non-consensual human trials)	More permissive regulatory environment with rapid advancements in gene editing; growing calls for stricter oversight post-2018 incidents	[71]

(Continued)

Table 2.5 Regulatory frameworks across different countries. (*Continued*)

Country/ region	Regulatory body	Permitted applications	Prohibited applications	Regulatory approach	References
United Kingdom	Human Fertilisation and Embryology Authority (HFEA)	Research on gene editing, including CRISPR-Cas9, particularly for disease treatment and understanding genetic disorders	Germline editing for reproductive purposes, non-therapeutic genetic enhancements	Strong ethical framework governing research, particularly on embryos; stringent licensing for gene editing research	[72]
Japan	Ministry of Health, Labour and Welfare (MHLW)	Somatic gene editing, preclinical and clinical research, particularly in disease treatment	Germline editing, embryo editing for reproductive purposes	Focus on safety and ethical concerns; strict guidelines for clinical trials and research applications	[73]
Australia	National Health and Medical Research Council (NHMRC)	Gene editing for therapeutic purposes, research on CRISPR-Cas9	Germline editing, embryo editing for non-therapeutic purposes	Prohibits germline editing while allowing therapeutic research; strict ethical and safety guidelines	[74]
Canada	Health Canada	Somatic gene editing for therapeutic and research purposes, including clinical trials	Germline editing, use of CRISPR-Cas9 for non-therapeutic modifications, embryo editing for reproductive purposes	Conservative approach with strict regulations; focus on ensuring safety and ethical compliance in all gene editing applications	[75]

(*Continued*)

Table 2.5 Regulatory frameworks across different countries. (*Continued*)

Country/ region	Regulatory body	Permitted applications	Prohibited applications	Regulatory approach	References
South Korea	Ministry of Food and Drug Safety (MFDS)	Somatic gene editing for therapeutic uses, clinical research, and trials	Germline editing, non-therapeutic genetic enhancements, embryo editing	Highly regulated environment with a strong emphasis on safety and ethical oversight; germline editing is strictly prohibited	[76]
India	Indian Council of Medical Research (ICMR)	Research on gene editing for therapeutic purposes, including somatic editing and preclinical studies	Germline editing, reproductive cloning, embryo editing for non-therapeutic purposes	Focus on ethical research practices; regulation is still evolving, with increasing attention to the implications of gene editing technologies	[77]
Brazil	National Health Surveillance Agency (ANVISA)	Somatic gene editing for therapeutic purposes and research, clinical trials under strict ethical review	Germline editing, non-therapeutic genetic modifications, embryo editing for reproductive purposes	Similar to EU regulations, emphasizing ethical review and patient safety in clinical trials; germline editing remains prohibited	[78]

Given the diverse ethical and legal perspectives on gene editing worldwide, regulatory frameworks vary significantly across countries. Understanding these differences is crucial for navigating the global landscape of gene editing research and applications. The following table provides a comparison of the regulatory frameworks across different countries, highlighting key differences and commonalities in their approaches to governing gene editing technologies.

2.10.3.2 Impact on Research and Clinical Applications

The variation in regulatory frameworks has a significant impact on the progress of gene editing research and the development of clinical applications. In countries with more restrictive regulations, researchers may face challenges in securing funding, obtaining ethical approval, and conducting clinical trials. This can slow the pace of innovation and limit the availability of new therapies to patients.

Conversely, more permissive regulatory environments may facilitate rapid advancements in gene editing, but they also carry the risk of ethical breaches and insufficient oversight. The lack of harmonization in global regulations can lead to "regulatory tourism," where researchers or companies seek to conduct trials in countries with less stringent regulations, potentially undermining ethical standards and patient safety [22].

2.10.3.3 Challenges in Creating International Guidelines

The development of international guidelines for gene editing and stem cell research is a complex and challenging task, given the diverse ethical, cultural, and legal perspectives that exist worldwide. While there have been efforts to establish global norms, such as the World Health Organization's (WHO) advisory committee on human genome editing, achieving consensus on key issues remains difficult [68].

One of the main challenges is balancing the need for scientific progress with the protection of human rights and ethical standards. Different countries have different views on issues such as germline editing, informed consent, and the commercialization of gene editing technologies, making it challenging to develop guidelines that are acceptable to all. Additionally, there is a need to ensure that any international guidelines are flexible enough to accommodate future developments in the rapidly evolving field of gene editing [64].

The lack of international consensus also raises concerns about enforcement and accountability. Even if global guidelines are established, ensuring compliance across different jurisdictions will require strong international cooperation and the development of mechanisms for monitoring and enforcement.

Navigating the global legal and regulatory landscape for gene editing and stem cell research requires a nuanced understanding of the diverse approaches taken by different countries. Efforts to create international guidelines will be essential for ensuring the responsible development and use of these technologies while respecting the diverse ethical perspectives that exist worldwide.

2.10.4 Future Directions in Ethical Governance

As gene editing technologies continue to advance, there is an urgent need to develop robust ethical governance frameworks that can guide their responsible development and use. These frameworks must balance the protection of human rights with the promotion of scientific progress, ensuring that the benefits of gene editing are realized while minimizing the risks.

2.10.4.1 The Need for Balanced Policies

One of the key challenges in ethical governance is striking the right balance between enabling scientific innovation and protecting individual rights and societal values. Policies must be designed to allow for the development of new therapies that can address serious medical needs while also safeguarding against potential abuses, such as the use of gene editing for non-therapeutic enhancement or eugenics [21].

Balanced policies should also consider issues of accessibility and equity. As gene editing technologies become more widely available, there is a risk that they could exacerbate existing social inequalities if only the wealthy have access to them. Ethical governance frameworks should include measures to ensure that the benefits of gene editing are distributed fairly and that vulnerable populations are not exploited [65].

2.10.4.2 Public Engagement and Stakeholder Involvement

Future ethical governance must involve meaningful public engagement and the inclusion of diverse stakeholder perspectives. This includes not only scientists and ethicists but also patients, advocacy groups, religious leaders, and policymakers. Engaging the public in discussions about gene editing and its implications can help ensure that policies reflect the values and priorities of society as a whole [64].

Stakeholder involvement is also crucial for addressing the ethical challenges associated with germline editing and other contentious issues. By bringing together diverse perspectives, it is possible to develop more nuanced and ethically robust policies that take into account the potential long-term consequences of gene editing technologies.

2.10.4.3 International Cooperation and Harmonization

Given the global nature of gene editing research, there is a need for greater international cooperation and harmonization of ethical standards. This includes developing shared guidelines for the conduct of research, the approval of clinical applications, and the management of ethical concerns such as informed consent and the protection of vulnerable populations [22].

International cooperation will also be essential for addressing the risks associated with germline editing and other technologies that have the potential to affect the global gene pool. By working together, countries can develop a coordinated approach to managing these risks and ensuring that the benefits of gene editing are realized without compromising ethical standards.

2.10.4.4 Adaptive Governance

The rapid pace of technological development in gene editing requires an adaptive approach to governance that can respond to new challenges as they arise. This means that ethical governance frameworks must be flexible and capable of evolving in response to new scientific discoveries, technological advancements, and societal changes [64].

Adaptive governance also involves ongoing monitoring and evaluation of the impacts of gene editing technologies, both in research and in clinical practice. By continuously assessing the ethical implications of these technologies and making adjustments to policies as needed, it is possible to ensure that gene editing is used in a way that aligns with societal values and ethical principles.

The future of ethical governance in gene editing and stem cell research will depend on the development of balanced, inclusive, and adaptive frameworks that can guide the responsible use of these powerful technologies. By engaging with diverse stakeholders, promoting international cooperation, and remaining flexible in the face of new challenges, it is possible to ensure that the benefits of gene editing are realized while protecting human rights and ethical standards.

2.11 Future Prospects and Challenges

The rapid advancements in gene editing technologies, particularly CRISPR-Cas9, have opened up new frontiers in biomedical research and medicine. This section explores the future prospects and challenges of gene editing, focusing on its role in precision medicine, the technical barriers in stem cell editing, the potential for personalized therapies, and the speculative future of these technologies.

2.11.1 Advances in Precision Medicine through Gene Editing

Precision medicine aims to tailor medical treatment to the individual characteristics of each patient, particularly their genetic profile. Gene editing technologies like CRISPR-Cas9 hold the potential to revolutionize precision medicine by enabling the precise modification of disease-causing genes, allowing for highly personalized therapeutic approaches.

2.11.1.1 CRISPR-Cas9 in Precision Medicine

CRISPR-Cas9 has already shown promise in correcting genetic mutations that lead to diseases such as cystic fibrosis, sickle cell anemia, and muscular dystrophy. By directly targeting and repairing the defective genes, CRISPR-Cas9 can offer curative treatments that are tailored to the specific genetic makeup of the patient [67]. For instance, ongoing research is exploring the use of CRISPR-Cas9 to correct mutations in hematopoietic stem cells, which can then be reintroduced into the patient to treat blood disorders like β-thalassemia and sickle cell disease.

The potential of CRISPR-Cas9 extends beyond single-gene disorders. Researchers are investigating its use in more complex diseases, such as cancer, where it could be used to modify the genetic profile of tumors, making them more susceptible to existing treatments or less aggressive. Additionally, CRISPR-Cas9 could be used to create individualized cancer immunotherapies by engineering T cells that are specifically targeted to the patient's tumor antigens [67].

2.11.1.2 Ongoing Research and Anticipated Breakthroughs

Several clinical trials are underway to test the safety and efficacy of CRISPR-Cas9-based therapies in humans. One of the most notable trials involves the use of CRISPR-Cas9 to treat sickle cell disease by editing the patient's hematopoietic stem cells to produce fetal hemoglobin, which can compensate for the defective adult hemoglobin [68]. Early results from these trials are promising, indicating that gene editing could become a cornerstone of precision medicine.

Looking forward, anticipated breakthroughs include the development of CRISPR-based therapies that can target multiple genes simultaneously, which could be crucial for treating complex diseases with polygenic causes. Additionally, advancements in delivery methods, such as nanoparticle-based delivery systems, could enhance the precision and efficiency of gene editing *in vivo*, making it a more viable option for a wider range of diseases [19].

The future of precision medicine is closely tied to the continued development and refinement of gene editing technologies. As these tools become more precise, efficient, and accessible, they have the potential to transform how we diagnose, treat, and prevent disease on an individual basis.

2.11.2 Overcoming Technical Barriers in Stem Cell Editing

Despite the significant progress made in gene editing, several technical challenges remain, particularly in the context of editing stem cells for therapeutic purposes. These challenges include improving delivery methods, ensuring precise targeting, and achieving efficient correction rates.

a. Improving delivery methods: One of the major hurdles in stem cell editing is the efficient and safe delivery of CRISPR components into the target cells. Viral vectors, such as lentiviruses and adeno-associated viruses (AAVs), are commonly used to deliver CRISPR-Cas9 into stem cells. However, these methods have limitations, including potential immunogenicity, insertional mutagenesis, and limitations in cargo size [51]. Non-viral delivery methods, such as electroporation, lipid nanoparticles, and gold nanoparticles, are being explored as alternatives that may offer greater safety and efficiency, but they also present challenges, such as lower delivery efficiency and potential cytotoxicity.
b. Ensuring precise targeting: Achieving precise targeting of the genome is crucial to avoid off-target effects, which can lead to unintended genetic modifications that could compromise the safety and efficacy of the therapy. While advances in gRNA design and the development of high-fidelity Cas9

variants have reduced the incidence of off-target effects, further improvements are needed to achieve the level of precision required for clinical applications. New strategies, such as the use of CRISPR-Cas12a (Cpf1) and base editors, which offer greater specificity and lower off-target activity, are being investigated as potential solutions to this challenge [58].

c. Achieving efficient correction rates: Another challenge in stem cell editing is achieving high correction rates without compromising cell viability or function. HDR, the preferred pathway for precise gene correction, is inherently less efficient than NHEJ, which can limit the effectiveness of CRISPR-Cas9 in therapeutic settings. Researchers are exploring various strategies to enhance HDR efficiency, including the use of small molecules that inhibit NHEJ or promote HDR, the optimization of donor DNA templates, and the timing of CRISPR delivery relative to the cell cycle [59].
d. Emerging solutions and technologies: Emerging technologies and approaches are being developed to address these technical barriers. For example, prime editing, which does not rely on DSBs, offers a promising alternative to traditional CRISPR-Cas9 that could achieve higher precision and efficiency in stem cell editing. Additionally, the use of synthetic biology to engineer more precise and controllable gene editing systems, as well as the development of programmable delivery vehicles that can target specific cell types, are areas of active research that could significantly improve the feasibility of stem cell therapies [60].

Overcoming these technical barriers is essential for the successful translation of gene editing technologies into clinical practice. Continued innovation and refinement of these tools will be crucial for realizing the full potential of stem cell editing in regenerative medicine.

2.11.3 Potential for Personalized Stem Cell Therapies

The combination of gene editing technologies with induced pluripotent stem cells (iPSCs) offers the potential for highly personalized therapies tailored to the individual patient's genetic makeup. This approach could revolutionize the treatment of a wide range of diseases, from genetic disorders to cancer.

2.11.3.1 Personalized Therapies Using iPSCs

iPSCs are derived from the patient's own somatic cells, which are reprogrammed to a pluripotent state, allowing them to differentiate into any cell type. By using CRISPR-Cas9 or other gene editing technologies to correct disease-causing mutations in iPSCs, researchers can generate patient-specific stem cells that can be used to treat the underlying cause of the disease. This approach not only addresses the genetic defect but also minimizes the risk of immune rejection, as the cells are autologous [30].

For example, researchers are exploring the use of gene-edited iPSCs to treat retinal diseases by correcting mutations in the patient's retinal cells and then differentiating these cells into retinal pigment epithelium (RPE) cells for transplantation. Similar approaches are being investigated for a range of other conditions, including neurodegenerative diseases, cardiovascular diseases, and diabetes, where the corrected iPSCs can be used to generate functional cells that replace the damaged or diseased tissues [33].

2.11.3.2 Current State of Research and Clinical Applications

While significant progress has been made in the development of personalized stem cell therapies, most of this research is still in the preclinical or early clinical trial stages. Challenges such as ensuring the safety and efficacy of the edited cells, optimizing differentiation protocols, and scaling up production for clinical use remain to be addressed. However, early results are promising, and several clinical trials are underway to test the feasibility of using gene-edited iPSCs in humans [68].

2.11.3.3 Timeline for Clinical Application

The timeline for the widespread clinical application of personalized stem cell therapies will depend on the resolution of the technical and regulatory challenges mentioned earlier. It is likely that we will see the first approvals for gene-edited iPSC therapies within the next decade, particularly for conditions where there are no existing effective treatments. However, broader adoption will require continued advances in gene editing precision, delivery methods, and manufacturing processes [67].

The potential for personalized stem cell therapies is immense, offering the possibility of curative treatments for a wide range of diseases. As research continues to progress, these therapies could become a mainstay of precision medicine, providing tailored solutions that address the root cause of genetic disorders.

2.11.4 The Road Ahead: Speculative Futures and Technological Innovations

As we look to the future, the possibilities for gene editing and stem cell therapies are vast and exciting. However, with these possibilities come new challenges, both technological and ethical, that will need to be navigated carefully.

a. Potential technological innovations: The field of gene editing is likely to continue evolving rapidly, with new technologies emerging that could surpass CRISPR-Cas9 in precision, efficiency, and versatility. For example, developments in synthetic biology could lead to the creation of entirely new gene editing tools that are programmable at a much finer level, allowing for complex genetic modifications that are currently beyond our reach [60]. One area of particular interest is the development of gene writing technologies, which could allow for the insertion of large sequences of DNA with high precision. This could enable the correction of more complex genetic disorders that involve large deletions or duplications, which are currently difficult to address with existing gene editing tools [19]. Another potential innovation is the use of artificial intelligence (AI) and machine learning to optimize gene editing strategies. AI could be used to design more effective gRNAs, predict off-target effects with greater accuracy, and tailor gene editing protocols to individual patients, further enhancing the precision and safety of these technologies [58].
b. Societal impacts and ethical dilemmas: As gene editing technologies become more powerful, they will also raise new ethical dilemmas and societal impacts that will need to be carefully considered. For example, the ability to edit the human germline could lead to profound changes in how we think about human reproduction, inheritance, and identity. While germline editing has the potential to prevent serious genetic disorders, it also raises concerns about unintended consequences, social inequality, and the potential for misuse [21]. The societal impacts of gene editing will also extend to issues of access and equity. As these technologies become more advanced, there is a risk that they could exacerbate existing health disparities if only certain populations have

access to the benefits. Ensuring that gene editing therapies are accessible to all, regardless of socioeconomic status, will be a key challenge for policymakers and healthcare providers [65].

c. The ethical governance of emerging technologies: The future of gene editing will require robust ethical governance frameworks that can adapt to new challenges as they arise. This will involve ongoing dialogue between scientists, ethicists, policymakers, and the public to ensure that the development and application of these technologies are guided by ethical principles and respect for human rights [64].

As new technologies emerge, there will be a need for continuous evaluation of their ethical implications and the development of policies that can address the potential risks while maximizing the benefits. This will require a flexible and adaptive approach to governance that can keep pace with the rapid advancements in the field.

2.12 Conclusion

Throughout this chapter, we have explored the revolutionary impact of CRISPR-Cas9 and its applications in stem cell therapeutics. CRISPR-Cas9 has emerged as a powerful tool for precise genetic modifications, enabling researchers to correct genetic mutations, study gene function, and develop novel therapies for a wide range of diseases. Its application in stem cell research has opened new avenues for personalized medicine, offering the potential to generate patient-specific treatments that address the root cause of genetic disorders.

Key advancements in CRISPR-Cas9 technology, such as the development of high-fidelity Cas9 variants, base editing, and prime editing, have significantly improved the precision and safety of gene editing. These innovations have enhanced our ability to edit stem cells with greater accuracy, reducing the risk of off-target effects and enabling more efficient correction of genetic mutations. Additionally, the emergence of other CRISPR systems, such as Cas12 and Cas13, has expanded the gene editing toolkit, offering new possibilities for targeting both DNA and RNA.

Despite these advancements, several challenges remain in translating CRISPR-Cas9-modified stem cells from bench to bedside. Technical barriers such as efficient delivery methods, precise targeting, and achieving high correction rates continue to pose obstacles. Moreover, the ethical and regulatory landscape surrounding gene editing is complex and evolving, with concerns about germline editing, informed consent, and equitable access to these technologies.

Overall, while significant progress has been made, the field of CRISPR-Cas9 and stem cell therapeutics is still in its early stages, with much work to be done to fully realize its potential.

2.12.1 Implications for Future Research and Clinical Practice

The findings discussed in this chapter have profound implications for future research and clinical practice. As gene editing technologies continue to advance, they will likely play an increasingly central role in the development of precision medicine. Researchers must continue to innovate and refine these technologies to overcome existing challenges, such as improving the specificity and efficiency of gene editing and developing safer and more effective delivery methods.

The clinical translation of CRISPR-Cas9 and other gene editing technologies will require rigorous preclinical testing and carefully designed clinical trials to ensure their safety and efficacy. Regulatory frameworks must be adaptive and responsive to the rapid pace of technological development, providing clear guidelines for the use of gene editing in clinical settings while ensuring that ethical considerations are not overlooked.

Ethical considerations must remain at the forefront of future developments in gene editing and stem cell therapeutics. As these technologies become more powerful and accessible, it is crucial to engage in ongoing dialogue with stakeholders, including ethicists, policymakers, patients, and the public, to ensure that gene editing is used responsibly and equitably. Issues such as germline editing, informed consent, and access to therapies must be carefully navigated to prevent potential misuse and to protect human rights.

2.12.2 Final Thoughts on the Integration of Gene Editing and Stem Cell Therapeutics

The integration of gene editing with stem cell therapy holds immense promise for addressing currently untreatable diseases. By combining the precision of CRISPR-Cas9 with the regenerative potential of stem cells, it is possible to develop therapies that not only treat symptoms but also correct the underlying genetic causes of disease. This approach could lead to curative treatments for a wide range of conditions, from genetic disorders to neurodegenerative diseases and beyond.

Achieving the full potential of gene editing and stem cell therapeutics will require interdisciplinary collaboration across fields such as molecular biology, genetics, bioinformatics, and clinical medicine. Researchers, clinicians, and ethicists must work together to develop safe and effective therapies that are grounded in a deep understanding of both the science and the ethical implications of these technologies.

The future of gene editing and stem cell therapy is bright, with the potential to transform medicine and improve the lives of millions of people. However, realizing this potential will require careful navigation of the scientific, ethical, and societal challenges that lie ahead. By fostering collaboration, innovation, and ethical governance, we can ensure that these powerful technologies are used to benefit all of humanity.

References

1. Jinek, M., Chylinski, K., Fonfara, I., Hauer, M., Doudna, J.A., Charpentier, E., A programmable dual-RNA–guided DNA endonuclease in adaptive bacterial immunity. *Science, 337*, 6096, 816–821, 2012.
2. Urnov, F.D., Miller, J.C., Lee, Y.L., Beausejour, C.M., Rock, J.M., Augustus, S., Rebar, E.J., Highly efficient endogenous human gene correction using designed zinc-finger nucleases. *Nature, 435*, 7042, 646–651, 2005.
3. Carroll, D., Genome engineering with zinc-finger nucleases. *Genetics, 188*, 4, 773–782, 2011.
4. Kim, H. and Kim, J.S., A guide to genome engineering with programmable nucleases. *Nat. Rev. Genet., 15*, 5, 321–334, 2014.
5. Joung, J.K. and Sander, J.D., TALENs: a widely applicable technology for targeted genome editing. *Nat. Rev. Mol. Cell Biol., 14*, 1, 49–55, 2013.
6. Gaj, T., Gersbach, C.A., Barbas, C.F., ZFN, TALEN, and CRISPR/Cas-based methods for genome engineering. *Trends Biotechnol., 31*, 7, 397–405, 2013.
7. Cong, L., Ran, F.A., Cox, D., Lin, S., Barretto, R., Habib, N., Zhang, F., Multiplex genome engineering using CRISPR/Cas systems. *Science, 339*, 6121, 819–823, 2013.
8. Ran, F.A., Hsu, P.D., Wright, J., Agarwala, V., Scott, D.A., Zhang, F., Genome engineering using the CRISPR-Cas9 system. *Nat. Protoc., 8*, 11, 2281–2308, 2013.
9. Mali, P., Yang, L., Esvelt, K.M., Aach, J., Guell, M., DiCarlo, J.E., Church, G.M., RNA-guided human genome engineering *via* Cas9. *Science, 339*, 6121, 823–826, 2013.
10. Takahashi, K. and Yamanaka, S., Induction of pluripotent stem cells from mouse embryonic fibroblasts by four transcription factors. *Cell, 126*, 4, 663–676, 2006.

11. Zakrzewski, W., Dobrzyński, M., Szymonowicz, M., Rybak, Z., Stem cells: past, present, and future. *Stem Cell Res. Ther.*, *10*, 1, 1–22, 2019.
12. Bianco, P. and Robey, P.G., Stem cells in tissue engineering. *Nat. Biotechnol.*, *24*, 7, 800–801, 2015.
13. Stadtfeld, M. and Hochedlinger, K., Induced pluripotency: history, mechanisms, and applications. *Genes Dev.*, *24*, 20, 2239–2263, 2010.
14. Carroll, D., Progress and Prospects: Zinc-Finger Nucleases as Gene Therapy Agents. *Gene Ther.*, 15, 22, 1463–1468, 2008.
15. Barrangou, R., *et al.*, CRISPR Provides Acquired Resistance Against Viruses in Prokaryotes. *Science*, 315, 5819, 1709–1712, 2007.
16. Hsu, P.D., *et al.*, Development and Applications of CRISPR-Cas9 for Genome Engineering. *Cell*, 157, 6, 1262–1278, 2014.
17. Kleinstiver, B.P., *et al.*, High-Fidelity CRISPR-Cas9 Nucleases with No Detectable Genome-Wide Off-Target Effects. *Nature*, 529, 7587, 490–495, 2016.
18. Slaymaker, I.M., *et al.*, Rationally Engineered Cas9 Nucleases with Improved Specificity. *Science*, 351, 6268, 84–88, 2016.
19. Anzalone, A.V., *et al.*, Search-and-Replace Genome Editing Without Double-Strand Breaks or Donor DNA. *Nature*, 576, 7785, 149–157, 2019.
20. Bosley, K.S., *et al.*, CRISPR Germline Engineering—the Community Speaks. *Nat. Biotechnol.*, 33, 5, 478–486, 2015.
21. Baylis, F., Human Germline Genome Editing: A Game of CRISPR-Cas or Russian Roulette? *Nature*, 571, 7764, 31–33, 2019.
22. Isasi, R.M. and Knoppers, B.M., Regulatory Approaches for Genome Editing: State of the Art in Human Applications. *Eur. J. Hum. Genet.*, 28, 10, 1391–1400, 2020.
23. Hyun, I., Ethics of Stem Cell Research and Therapy. *J. Clin. Invest.*, 120, 1, 71–75, 2010.
24. Hockemeyer, D. and Hamilton, G.A., CRISPR-Cas Technology and Human Induced Pluripotent Stem Cells. *Nat. Rev. Mol. Cell Biol.*, 16, 9, 507–517, 2015.
25. Xie, F., *et al.*, Seamless Gene Correction of β-Thalassemia Mutations in Patient-Specific iPSCs Using CRISPR-Cas9 and PiggyBac. *Genome Res.*, 24, 9, 1526–1533, 2014.
26. Long, C., *et al.*, Prevention of Muscular Dystrophy in Mice by CRISPR/Cas9–Mediated Editing of Germline DNA. *Science*, 345, 6197, 1184–1188, 2014.
27. Wu, Y., *et al.*, Highly Efficient Therapeutic Gene Editing of Human Hematopoietic Stem Cells. *Nat. Med.*, 25, 5, 776–783, 2019.
28. Trounson, A. and McDonald, J.C., Stem Cell Therapies in Clinical Trials: Progress and Challenges. *Cell Stem Cell*, 17, 1, 11–22, 2015.
29. Mao, A.S. and Mooney, D.J., Regenerative Medicine: Current Therapies and Future Directions. *Proc. Natl. Acad. Sci.*, 112, 47, 14452–14459, 2015.
30. Takahashi, K. and Yamanaka, S., Induction of Pluripotent Stem Cells from Adult Human Fibroblasts by Defined Factors. *Cell*, 131, 5, 861–872, 2007.

31. Bianco, P. and Robey, P.G., Stem Cells in Tissue Engineering. *Nature*, 414, 6859, 118–121, 2001.
32. Yu, J., *et al.*, Induced Pluripotent Stem Cell Lines Derived from Human Somatic Cells. *Science*, 318, 5858, 1917–1920, 2007.
33. Yamanaka, S., Pluripotent Stem Cell-Based Regenerative Medicine: Progress in iPS Cell Research. *Philos. Trans. R. Soc. B Biol. Sci.*, 370, 1680, 20140368, 2015.
34. Copelan, E.A., Hematopoietic Stem-Cell Transplantation. *New Engl. J. Med.*, 354, 17, 1813–1826, 2006.
35. Appelbaum, F.R., Hematopoietic-Cell Transplantation at 50. *New Engl. J. Med.*, 357, 15, 1472–1475, 2007.
36. Wollert, K.C. and Drexler, H., Clinical Applications of Stem Cells for the Heart. *Circ. Res.*, 96, 2, 151–163, 2005.
37. ClinicalTrials.gov, *Study of CRISPR-Cas9 Gene Editing in Patients with Sickle Cell Disease (NCT03745287)*, Sickle Cell Disease Gene Therapy Trial. Bethesda, MD: U.S. National Library of Medicine, 2019, Available at: https://clinicaltrials.gov/ct2/show/NCT03745287.
38. ClinicalTrials.gov, *CRISPR-Cas9 Gene Editing in β-Thalassemia Patients (NCT03872479)*, β-Thalassemia Gene Therapy Trial. Sponsor: CRISPR Therapeutics, Cambridge, MA, USA. U.S. National Library of Medicine, 2019, Available at: https://clinicaltrials.gov/ct2/show/NCT03872479.
39. ClinicalTrials.gov, *A Study of BCMA-Targeted CAR T Cells in Multiple Myeloma (NCT03399448)*, Multiple Myeloma Gene Therapy Trial. Sponsor: Shenzhen University, Shenzhen, China, U.S. National Library of Medicine, 2018, Available at: https://clinicaltrials.gov/ct2/show/NCT03399448.
40. ClinicalTrials.gov, *Gene Therapy for Leber Congenital Amaurosis Using CRISPR-Cas9 (NCT04037566)*, Leber Congenital Amaurosis Trial. Bethesda, MD: U.S. National Library of Medicine, 2019, Available at: https://clinicaltrials.gov/ct2/show/NCT04037566.
41. ClinicalTrials.gov, *CRISPR/Cas9-Based Therapeutic Approach for COVID-19 Pneumonia (NCT04441606)*, COVID-19 Pneumonia Trial. Bethesda, MD: U.S. National Library of Medicine, 2020, Available at: https://clinicaltrials.gov/ct2/show/NCT04441606.
42. ClinicalTrials.gov, *CRISPR-Cas9 Gene Editing in HIV-Infected Patients (NCT04244656)*, HIV Infection Gene Therapy Trial. Sponsor: Beijing YouAn Hospital, Beijing, China. U.S. National Library of Medicine, 2020, Available at: https://clinicaltrials.gov/ct2/show/NCT04244656.
43. ClinicalTrials.gov, *Gene Editing of HBB Gene in Transfusion-Dependent β-Thalassemia (NCT04134325)*, β-Thalassemia Transfusion-Dependent Trial. Sponsor: Murdoch Children's Research Institute, Melbourne, Australia. U.S. National Library of Medicine, 2019, Available at: https://clinicaltrials.gov/ct2/show/NCT04134325.
44. ClinicalTrials.gov, *CRISPR-Cas9 for Sickle Cell Disease (NCT04146052)*, Sickle Cell Disease Mutation Correction Trial. Bethesda, MD: U.S. National

Library of Medicine, 2019, Available at: https://clinicaltrials.gov/ct2/show/NCT04146052.
45. ClinicalTrials.gov, *A Study of CRISPR-Cas9 Edited T Cells in Patients with T-Cell Acute Lymphoblastic Leukemia (T-ALL) (NCT03655678)*, T-Cell Acute Lymphoblastic Leukemia Trial. Bethesda, MD: U.S. National Library of Medicine, 2019, Available at: https://clinicaltrials.gov/ct2/show/NCT03655678.
46. ClinicalTrials.gov, *CRISPR-Cas9 Edited T Cells Targeting BCMA and CD19 in Relapsed/Refractory Multiple Myeloma (NCT04002026)*, Relapsed/Refractory Multiple Myeloma Trial. Sponsor: Peking University, Beijing, China. U.S. National Library of Medicine, 2019, Available at: https://clinicaltrials.gov/ct2/show/NCT04002026.
47. ClinicalTrials.gov, *Gene Editing to Restore Dystrophin Expression in Duchenne Muscular Dystrophy Patients (NCT04557490)*, Duchenne Muscular Dystrophy Gene Therapy Trial. Bethesda, MD: U.S. National Library of Medicine, 2020, Available at: https://clinicaltrials.gov/ct2/show/NCT04557490.
48. ClinicalTrials.gov, *Safety and Efficacy of CRISPR-Cas9 in Sickle Cell Disease Patients (NCT03690011)*, Sickle Cell Disease CRISPR-Cas9 Gene Editing Trial. Sponsor: The Hospital for Sick Children, Toronto, Canada. U.S. National Library of Medicine, 2019, Available at: https://clinicaltrials.gov/ct2/show/NCT03690011.
49. DeWitt, M.A., *et al.*, Selection-Free Genome Editing of the Sickle Mutation in Human Adult Hematopoietic Stem Cells. *Sci. Transl. Med.*, 8, 360, 360ra134, 2016.
50. Schwank, G., *et al.*, Functional Repair of CFTR by CRISPR/Cas9 in Intestinal Stem Cell Organoids of Cystic Fibrosis Patients. *Cell Stem Cell*, 13, 6, 653–658, 2013.
51. Yin, H., *et al.*, Non-Viral Vectors for Gene-Based Therapy. *Nat. Rev. Genet.*, 15, 8, 541–555, 2014.
52. Wagner, D.L., *et al.*, High Prevalence of Streptococcus pyogenes Cas9-Reactive T Cells within the Adult Human Population. *Nat. Med.*, 25, 2, 242–248, 2019.
53. Chen, J.S., *et al.*, CRISPR-Cas12a Target Binding Unleashes Indiscriminate Single-Stranded DNase Activity. *Science*, 360, 6387, 436–439, 2018.
54. Gootenberg, J.S., *et al.*, Nucleic Acid Detection with CRISPR-Cas13a/C2c2. *Science*, 356, 6336, 438–442, 2017.
55. Li, L., *et al.*, CRISPR-Cas12a-Mediated Precise Genome Editing *via* Targeted DNA DSB and HDR in Human Cells. *Nat. Commun.*, 9, 1, 2162, 2018.
56. Abudayyeh, O.O., *et al.*, C2c2 is a Single-Component Programmable RNA-Guided RNA-Targeting CRISPR Effector. *Science*, 353, 6299, aaf5573, 2016.
57. Cox, D.B.T., Gootenberg, J.S., Zhang, F., RNA Editing with CRISPR-Cas13. *Science*, 358, 6366, 1019–1027, 2017.
58. Komor, A.C., *et al.*, Programmable Editing of a Target Base in Genomic DNA without Double-Stranded DNA Cleavage. *Nature*, 533, 7603, 420–424, 2016.

59. Gaudelli, N.M., *et al.*, Programmable Base Editing of A•T to G•C in Genomic DNA without DNA Cleavage. *Nature*, 551, 7681, 464–471, 2017.
60. Liu, L., *et al.*, Prime Editing in Plants and Mammals: Efficiency, Specificity, and Applications. *Nat. Biotechnol.*, 38, 8, 889–897, 2020.
61. Boch, J., *et al.*, Breaking the Code of DNA Binding Specificity of TAL Effectors. *Science*, 326, 5959, 1509–1512, 2009.
62. Schmid-Burgk, J.L., *et al.*, A Scalable Approach for High-Throughput Quantification of Genetic Interactions. *Nat. Methods*, 13, 10, 1001–1005, 2016.
63. Perez, E.E., *et al.*, Establishment of HIV-1 Resistance in CD4+ T Cells by Genome Editing Using Zinc-Finger Nucleases. *Nat. Biotechnol.*, 26, 7, 808–816, 2008.
64. Jasanoff, S., *et al.*, Human Genome Editing: Science, Ethics, and Governance. *Science*, 356, 6341, 148–150, 2017.
65. Nerlich, B., *et al.*, The 'CRISPR' Bug in the Popular Imagination: Accounts of the Cultural Impact of Gene Editing. *Public Understanding Sci.*, 26, 6, 714–730, 2017.
66. Doudna, J.A. and Sternberg, S.H., *A Crack in Creation: Gene Editing and the Unthinkable Power to Control Evolution*, Houghton Mifflin Harcourt, Boston, 2017.
67. Ledford, H., CRISPR-Cas Debates Ignite. *Nature*, 528, 7580, 319–320, 2015.
68. Cyranoski, D., What Next for Gene-Edited Babies? *Nature*, 566, 7745, 440–442, 2019.
69. U.S. Food and Drug Administration, *Human Gene Therapy for Rare Diseases: Guidance for Industry*, FDA, Silver Spring, MD, 2020.
70. European Medicines Agency, *Guidelines on Human Gene Therapy Clinical Trials. EMA/CHMP/GTWP/125459/2006*, EMA, London, 2018.
71. National Health Commission of the People's Republic of China, *Ethical Guidelines for Human Embryonic Stem Cell Research*, NHC, Beijing, 2017.
72. Human Fertilisation and Embryology Authority, *Code of Practice*, 9th Edition, HFEA, London, 2019.
73. Ministry of Health, Labour and Welfare, *Guidelines on Clinical Research Using Human Stem Cells*, MHLW, Tokyo, 2015.
74. National Health and Medical Research Council, *National Statement on Ethical Conduct in Human Research*, NHMRC, Canberra, 2018.
75. Government of Canada, *Assisted Human Reproduction Act. S.C. 2004, c. 2*, Department of Justice, Ottawa, 2004.
76. Ministry of Health and Welfare, *Bioethics and Safety Act*, MOHW, Seoul, 2016.
77. Indian Council of Medical Research, *National Guidelines for Stem Cell Research*, ICMR, New Delhi, 2017.
78. National Health Surveillance Agency, *Resolution RDC No. 260/2018: Regulations on Clinical Research with Advanced Therapy Medicinal Products*, ANVISA, Brasília, 2018.

3

Future Vistas: Revolutionary Technologies and Paradigms in Stem Cell Therapeutics

Dattatraya M. Shinkar[1]*, Sharayu P. Rathod[1], Priyanka D. Dabir[1], Sandip R. Purkar[2], Deepali D. Bhandari[2] and Sunil V. Amrutkar[2]

[1]Department of Pharmaceutics, GES's Sir Dr. M. S. Gosavi College of Pharmaceutical Education and Research, Nashik, Maharashtra, India
[2]Department of Pharmaceutical Chemistry, GES's Sir Dr. M. S. Gosavi College of Pharmaceutical Education and Research, Nashik, Maharashtra, India

Abstract

New approaches to healing injured or diseased organs, tissues, or cells could completely transform healthcare through advances in regenerative medicine. Owing to its ability to provide treatments for illnesses that were once thought to be incurable, this area of medicine has grown in popularity. Crucial to the area of healing with regeneration requires discovering and employing stem cells. Because of their innate flexibility and aptitude for self-renewal, stem cells have the ability to differentiate into a wide variety of cell types. They are beneficial regarding regenerative medicine due to their exceptional and one-of-a-kind characteristics. Breakthroughs in tissue engineering and treatments with stem cells possess the capacity to change healthcare. Stem cell therapy is employed to treat a number of sickness, such as cancer and Parkinson's disease. Regenerative medicine confronts ethical concerns, but it is gaining favor. It is almost clear that the areas of medical regeneration and stem cell therapy will expand rapidly. Due to continuous research and technological advancements, the way we approach and treat a wide range of diseases may change significantly over the next few decades. Regenerative medicine is projected to grow even more potent as biotechnology, nanotechnology, and tissue engineering combine, paving the way for previously unthinkable treatment alternatives. This therapy and regeneration in medicine embodies a paradigm change in contemporary healthcare. Their ability to restore and revitalize signals the dawn of a new era in medical science and offers hope for healing a multitude of illnesses. As research progresses, it is imperative to maintain a delicate equilibrium between scientific advancement and

**Corresponding author*: dattashinkar@gmail.com

Ayush Madan, Sumel Ashique, Deepika Arora and Mantosh Kumar Satapathy (eds.) Stem Cell Therapeutics, (107–158) © 2025 Scrivener Publishing LLC

moral issues to ensure that the area of regenerative medicine is distinguished by inventiveness, responsibility, and revolutionary healing.

Keywords: Stem cells, therapeutics, regenerative medicine, personalized medicine, tissue engineering, applications, healthcare

3.1 Introduction

Medical research relies heavily on stem cells due to their unique characteristics and ability to treat various illnesses. These are distinct cells with the propensity to grow into any sort of every bodily cell, including brain and muscle cells. For regenerative medicine and therapy, this makes them an indispensable tool. The two main mechanisms of working of stem cells are that they can replicate exactly by constantly renewing themselves, and they are the only cells capable of differentiating into specialized kinds of cells that are suitable for substitution or repair particular cell types within the body [1].

With their special capacity to regenerate themselves and give birth to specific types of cells, stem cells are a rare and flexible form of cell that can divide indefinitely. While the majority of body cells, such those found in the skin or heart, are committed to carrying out certain tasks, stem cells are not, and they stay that way until they receive a signal to change into a different type of cell. The remarkable capacity of stem cells to reproduce and differentiate makes them extraordinary [1].

Diverse varieties of stem cells exist, each possessing distinct capabilities and uses. The primary kinds consist of the following: Different tissues can provide Mesenchymal Stem Cell (MSCs), which are resultant from the early legs of mesoderm progress. MSCs derived from human umbilical cord (HUC), bone marrow (BM), and adipose (AD) are the most often used types of MSCs in clinical practice. The blood–brain barrier (BBB) can be crossed; BM-MSCs are readily obtained, possess the ability to relocate to the wounded site, and are well-known for their therapeutic implications. Research shows that administering BM-MSCs to rats who have experienced intracranial hemorrhage (ICH) can improve overall neurological function recovery, mitigate neurodegenerative impairments, and restore blood–brain barrier function [2].

HUC-MSCs have been used to treat several neurological conditions, which includes ICH, since they may be obtained in high quantities with little invasion. The combination of HUC-MSC transplantation and minimally invasive hematoma aspiration has demonstrated potential in minimizing brain injury and enhancing neurological functions. Because of their accessibility, abundance, and low apoptotic rates, AD-MSCs—which are derived from adipose tissue—are well-suited for large-scale production. Studies concentrating the administration of ADMSCs reveal their capacity to develop into cells resembling neurons and astrocytes in the vicinity of the lesion, resulting in enhanced functionality and a reduction in the progression of chronic brain deterioration [3].

A very undifferentiated kind of cell that is separated from the original or early embryos' gonads is recognized as embryonic stem cells or ESCs. Since ESCs can multiply endlessly and they show great promise for the therapy of diseases since they can differentiate in various directions *in vitro* affecting the central nervous system. Although their ability to differentiate into a range of cell types—including glial and neuronal cells—makes them desirable for medicinal uses, there are a number of issues to be carefully considered, including the unidentified chemical processes underlying cell migration and cytokine regulation, as well as ethical questions surrounding their use [4, 5].

Hematopoietic stem cells (HSCs) come from umbilical cord and bone marrow, among other hematopoietic organs. Research indicates that improved functional outcomes are linked to increased levels of circulating HSCs following ICH [6, 7]. The promotion of functional recovery in both ischemic stroke and ICH has been demonstrated by the mobilization of HSCs *via* G-CSF. HSCs are a possible contender for widespread clinical application due to their easy acquisition in huge quantities [8, 9]. Within the neurological system are neural stem cells (NSCs), which can proliferate into any glial cells/neurons and self-renew. In mice models with ICH and brain ischemia, both endogenous and exogenous NSC transplantation have proven to be able to support functional recovery [10].

Neurorestoration is aided by the migration of endogenous NSCs to damage sites and their activation in pathological settings. Because of their strong histocompatibility with brain tissue, minimal immunogenicity, and adaptability, NSCs offer a capable alternative for the stroke therapy. A revolutionary development, induced stem cells with pluripotency are used in stem cell studies (iPSCs) are produced through somatic cell reprogramming with particular transcription factors [11]. Research has documented positive outcomes in rat models of cerebral ischemia and ICH. iPSCs have the ability to release neuro-trophic substances that develop into stem cells that resemble and function like neuroepithelium. Although difficulties like low reprogramming efficiency require more work, the iPSC method shows potential for therapeutic applications by overcoming ethical and immunological rejection concerns [12].

Additionally, stem cells' production of cytokines and growth factors promotes angiogenesis, neurogenesis, and neurogenesis—all of which improve the circumstances for brain repair. Furthermore, through managing the immune system's reaction to the shock, they lower inflammation—a vital step in preventing more stroke damage [13]. Stem cells can also enhance synaptic plasticity and remodeling of the brain network, which are necessary for reversing the functional deficits brought on by the stroke. To maximize stem cell interventions and increase their effectiveness in therapeutic settings for stroke rehabilitation, it is critical to understand these pathways [14].

For many years, researchers have been looking for ways to repair diseased or damaged cells and tissues with stem cells. Recently, stem cells have garnered a lot of attention. Since this discovery a few years ago, evidence has surfaced suggesting that these stem cells possess the ability to differentiate into nearly every specialized cell in the body. They might thus be able to create new cells for various tissues and organs, such as the heart, pancreas, and nervous system. Therefore, many of our most severe diseases and disorders may cause harm to or destruction of cells or tissues that this fraction of human stem cells has the capacity to replace and restore [2].

3.2 Characteristics of Several Stem Cell Types Employed in Therapeutic Applications

Three major categories can be employed to organize various stem cells after a thorough examination. Embryonic stem cells and ESCs are the primary and most basic kind of stem cells. Due to their totipotency, these cells have the capacity to develop into new individuals. However, their therapeutic application and study are constrained as a result of ethical disputes. ASCs are the second class, and they are the most diverse. In the field of medical regeneration, these stem cells are the cells that are used the most. iPSC is the third category. Because of the ESC-like traits they possess, treatment options for these cells have been investigated of different types of neurodegenerative diseases.

Human embryonic stem cells, or ESCs, are immature cells which can develop into any type of body cell. Because these cells are capable of differentiating into any type of bodily cell, they can also be called as human pluripotent stem cells, or iPSCs. These are collections of cells found 5 to 7 days following zygote development in the blastocyst [15].

At birth, the umbilical cord yields cord blood stem cells, which can develop into different types of body cells. They can grow forever in culture, unlike adult stem cells (ASCs); however, they have limited differentiation capacities. They are a never-ending supply of illness and tissue models.

Intentionally reprogramming adult stem cells (ASCs) to resemble embryos results in induced pluripotent stem cells. This phenomenon is known as the "reverse direction of time." While they differ, induced stem cells share many characteristics with ESCs. This innovative method will not lessen the necessity for ESCs; rather, it will be helpful in the creation of disease models.

The best human embryonic stem cells (ESCs) for stem cell research are those obtained using somatic cell nuclear transfer, or SCNT. Using this method, the nucleus of an adult cell—such as a skin or blood cell—replaces the nucleus of an egg cell, which is then chemically or electrically encouraged to proliferate.

Potential uses of stem cells for modeling and treating diseases have been impacted by varying viewpoints among researchers regarding their potential. These perspectives are based on the stem cells' location and our comprehension of the biology of the production of stem cells [16].

3.2.1 Embryonic Stem Cells

Embryonic cells have totipotent potential, which means they can develop into every kind of tissue or stem cell that is present in the human body. These cells are produced when the sperm and egg fuse together. These cells can now be derived from, differentiated into, and multiplied under favorable conditions thanks to technological advancements. They are also referred to as iPSCs since they originate from the ICM from the embryonic blastocyst and possess the capacity to differentiate into each of the three germ layers both *in vivo* and *in vitro*. Their phenotype is undifferentiated and may be sustained for a very long duration (about 1 to 2 years, with cell division occurring every 36–48 h). They can be separated by immunological dissection mediated by complement or physical microdissection [17].

Techniques such as vitrification or fast freezing (shock) are employed to stop unintended early natural differentiation. For therapeutic and research purposes, the ESCs should be cultured for an extended length of time, as they need special attention during this time. Conversely, somatic stem cells and MSCs are incapable of dividing and have restricted capacity for regeneration. Cell type, donor type, donor age, and telomerase activity all affect how long cells live. The structures known as telomeres shield the length of chromosomes. The enzyme telomerase promotes cell division by appending unique, brief sequences to the ends of chromosomes. In ESCs, telomerase activity is relatively high.

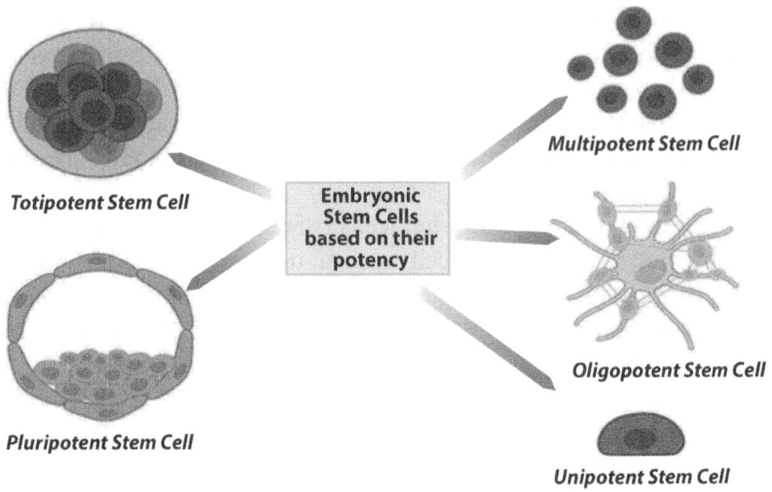

Figure 3.1 Types of embryonic stem cells based on their potency.

Table 3.1 Potency of different stem cells.

Potency of stem cell		Examples
Totipotent	Has the ability to create extraembryonic and embryonic cell lines. Noted during the morula stage's growth, starts from a single cell zygote	The initial blastomeres and a fertilized oocyte
Pluripotent	Possesses the ability to differentiate into any one of the three germ layer cells. Confines to the inner cell mass of the blastocyst stage until the formation of the trilaminar germ layer	Stem cells from embryos
Multipotent	Can differentiate into several types of cell lines, but not from three germ layers. These persist until adulthood and beyond the trilaminar developmental stage	Neural stem cells (NSCs), mesenchymal stem cells (MSCs)
Oligopotent	Can proliferate into a few distinct but related cell types	Leukemic or lymphoid stem cells
Unipotent	Has the ability to proliferate into a single cell	Sperm-producing germ line stem cells and skin-producing epidermal stem cells

Numerous countries have studies utilizing embryonic stem cells is prohibited or severely limited. South Korea and Singapore, on the other hand, rank among the least restrictive nations [18].

The fertilized egg proceeds to proliferate instantaneously. In the emerging embryo, all the cells are totipotent. In a few days, these cells develop a hollow structure. Clustering of cells in an area forms the inner cell mass. Pluripotent cells, which comprise the growing foetus, are present in this.

Additional types for the embryonic stem cells based on their potency depicted in Figure 3.1 and Table 3.1 shows the potency of different stem cells.

3.2.2 Totipotent Stem Cells

A single cell's totipotency refers to its capacity to mature into any kind of adult somatic cell as well as tissues found in the placenta and fetal membranes. The zygote up to the 16-cell stage, which is formed after fertilization of the ovum, is the greatest example of totipotent stem cells. After which, it divides into placental cytotrophoblast and syncytiotrophoblast and any of the three germ layers: endoderm, ectoderm, and mesoderm. At the 16-cell stage, 4 days postfertilization, cells start to shed their totipotency and transform into pluripotent embryonic stem cells [19].

One kind of embryonic stem cell is totipotent stem cell. Cell which possess the capacity to proliferate into all components of a conceived organism, including the embryo and extraembryonic tissues including fetal membranes and placental tissues. In the case of humans, a zygote is created when the sperm and egg gametes combine. The freshly created zygote's cells begin to divide. During the morula stage of development of the embryo they maintain their totipotency. In humans, the morula stage of embryonic development happens approximately at E4. Day 5 sees the formation of a blastocyst from the morula, which has a fluid-filled blastocoel, a trophoblast layer, and a clearly defined core cell mass. During the blastocyst phase, totipotent cells become extinct [20].

3.2.3 Pluripotent Stem Cells

The ability of pluripotent stem cells to differentiate into any embryonic cell line from any of the three cell layers is what makes them unique among embryonic stem cells. Extraembryonic tissues, however, cannot be produced by them. The embryonic blastocyst stage is when the pluripotent inner cell mass cells begin, which is traditionally when they so-called immature embryonic cells are detected. They have the ability to develop a complete fetus because they can produce any of the three cell lines that make up the germ layers developing pluripotent adult somatic cell lines, such as fibroblasts [19–21].

3.2.4 Multipotent Stem Cells

Although these adolescent stem cells are located in the same germ layer, they have the ability to differentiate into several distinct cell lines that are not always related. They can be found in many different "adult" tissues and organs. Organs such as bone marrow, adipose tissue, and Wharton's jelly include mesenchymal stromal cells (MSC) and hematopoietic stem cells, which are precursors of lymphocytes, neutrophils, and monocytes. MSC can differentiate into osteoblasts, chondrocytes, and adipocytes in the correct circumstances [19, 20].

3.2.5 Oligopotent Stem Cells

These can differentiate into a few closely similar cell lines and are present in a variety of "adult" tissues. For instance, several B and T cell lines can be produced using lymphocyte progenitors. They cannot, however, create neutrophils or platelets.

3.2.6 Unipotent Stem Cells

Only one cell line can be differentiated from unipotent stem cells, as the name indicates. Precursors is the commonly used term for them. Hepatoblasts, which are found in liver tissues, can help with hepatocyte regeneration [20, 21].

3.2.7 Adult Stem Cells

During the process of cell division, undifferentiated stem cells, or adult stem cells (ASCs) can regenerate tissue that has been harmed or has expired. They are present in particular bodily tissues. Adult stem cells are also referred to as "somatic stem cells." Sperm and eggs are examples of non-reproductive cells; these cells are called "somatic." Because ASCs are uncommon in native tissues, it has been difficult to obtain and analyze them for scientific research [19, 20].

Within the mainstream of bodily tissues, ASCs create new cells to replenish those that have been lost due to sickness, injury, or natural repair. ASCs can be found in many different tissues throughout a person's life, including the skin, stomach, brain, muscle, placenta, adipose tissue, and neck. Blood was produced in 1948 using the first ASCs that were separated. Using adult bone marrow cells for therapeutic purposes of blood diseases was a major step forward for this strategy in 1968.

Adult stem cells are prevalent. Their principal duties are to keep a cell's homeostasis intact, or stable condition of functioning, and, to the greatest extent feasible, to replace cells that die from illness or injury. Furthermore, adult stem cells behave differently depending on their surroundings and are spread throughout the tissues of a fully formed animal. For example, bone marrow is a continual source of HSCs, which grow into dissimilar types of blood cells. It is true that HSCs' primary role is to restore blood cells. In contrast, stem cells in digestive tract are stationary and physically separated from the adult cell types they generate. Gut epithelial stem cells, also known as progenitors, arise at the base of vaults, which are profound breaches among the mature, differentiated epithelial cells lining the lumen of the intestine. These epithelial crypt cells divide frequently, yet remain part of the immobile group of cells they produce [21].

Adult stem cells are incompetent of being identified in same manner as embryonic stem cells are identifiable *via* their source- the inward cell cluster of blastocyst. The origins of adult stem cells in any maturing tissue are unknown. Adult tissues containing stem cells are increasingly being discovered, plus bone marrow, peripheral blood, the brain, blood vessels, spinal cord, dental pulp, skeletal muscle, skin and digestive system epithelia, retina, liver, cornea, and pancreas [22]. To be termed an adult, a stem cell essentially has the capacity to replenish itself throughout the organism's lifetime. It is difficult to demonstrate this prerequisite *in vivo*, despite the fact that it is critical to stem cell biology.

Conflicting research has been done to validate the particularity of creating ASCs; some indicate that the cells that ASCs develop from their resident tissue are the only cell types they can produce, while other research suggests that ASCs can form different tissue sorts.

Types of Adult Stem Cells

3.2.7.1 Hematopoietic Stem Cells

Hematopoietic stem cells, or HSCs, remain stem cells that develop into numerous types of blood cells. This progression is known as hemopoiesis. The first conclusive HSCs in animals originate from the adaxial endothelium the aortic wall in embryonic form inside the (midgestational) aorta–gonad–mesonephros area through a method called endothelial-to-hematopoietic transition. Hemostasis occurs in the hematopoietic system. of adults, which is located in the middle of the bones. There is development of red bone marrow in the embryo layer known as the mesoderm [23].

Hematopoiesis is the mechanism that creates all mature blood cells. An individual's daily creation of over 500 billion blood cells necessitates tremendous manufacturing needs, but it also needs to balance the need to regulate the amount of each type of blood cell in the body. In most vertebrates, the bone marrow is the site of hematopoiesis. It is generated by a small number of highly self-renewing, multipotent hematopoietic stem cells [24].

Hematopoietic stem cells develop into lymphoid and myeloid blood cell lines. Dendritic cells develop from lymphoid and myeloid lineages. Myeloid cell types comprise monocytes, eosinophils, neutrophils, macrophages, erythrocytes, basophils, megakaryocytes, and platelets. B and T cells are examples of lymphoid cells, along with innate lymphoid cells also ordinary killer cells.

The phrase "hematopoietic stem cells" has evolved since their discovery in 1961. The hematopoietic tissue contains committed oligopotent, unipotent, and multipotent progenitors, in addition to cells capable of both immediate and long-term restoration. In myeloid tissue, hematopoietic stem cells account for one in every 10,000 cells [25].

Because HSC transplants have regenerative properties, they are used to treat immune system disorders including cancer.

3.2.7.2 Mesenchymal Stem Cells

Mesoderm germ layer cells are especially referred to as "mesenchymal" due to their embryonic origin. At the beginning of the embryo's development, the mesoderm germ layer forms, which is one amongst the three basic germ layers. The urogenital system, blood vessels, and the cells that make them up are all derived from the mesoderm, along with a variety of connective tissues including muscle, bone, cartilage, and fat.

Stem cells, or MSCs, can differentiate into multiple lineages while also self-renewing. MSCs is separate from a range of tissues, including bone marrow, adipose tissue, menstrual blood, endometrial polyps, and the umbilical cord. This is due to the sources' great utility for experimental and possible medicinal applications, as well as their ease of harvesting and yield. Menstrual blood and endometrium are two unique sources of MSCs that have recently been identified shown in Figure 3.2.

Among the mesenchymal stem cells' (MSCs') many varied functions is their capacity to discriminate into numerous cell categories, including chondrocytes (cartilage cells), adipocytes (fat cells), and osteoblasts (bone cells). Furthermore, mesenchymal stem cells (MSCs) partake the ability to suppress the immune scheme and reassure tissue regeneration.

Due to their capacity to suppress immunological responses, lower inflammation, and modify immune system activity, mesenchymal stem cells (MSCs) are renowned for their ability to modulate immunity. Their ability to do this gives them a viable option for cell-based treatments for a number of illnesses, including graft-versus-host disease, transplant rejection and autoimmune diseases.

By releasing growth aspects and cytokines, mesenchymal stem cells (MSCs) can aid in tissue regeneration by attracting other cells to the wounded area. These cytokines and growth factors can encourage the creation of novel blood vessels, which are necessary for tissue healing.

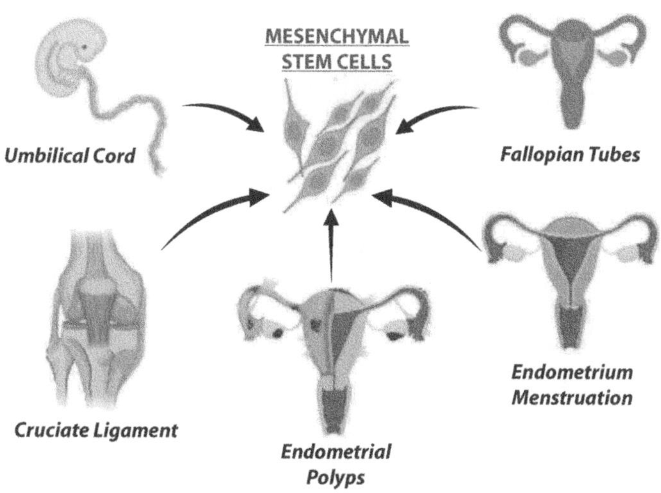

Figure 3.2 Various new sources of MSCs.

Preclinical and clinical trials are also being conducted to assess the potential of mesenchymal stem cells (MSCs) to treat a variety of illnesses, including multiple sclerosis, Parkinson's disease, osteogenesis imperfecta, osteoarthritis, and Crohn's disease [26, 27].

3.2.8 Neutral Stem Cells

Additionally, stem cells can be discovered in the brain as shown in Figure 3.3. Although hardly any new neurons are created in mammals after birth, a few olfactory bulb neurons and hippocampal regions are always growing [26]. These neurons are derived neural stem cells, which could be cultured *in vitro* as neurospheres—small stem cell-based cell clusters, as well as some of their progeny. For use in cells, this kind of stem cell is being researched for use in cell therapy to treat Parkinson disease, other neurodegenerative ailments, and harm incurred by the nervous system's centers [27].

3.2.9 Epithelial Stem Cells

Layers of skin cells known as keratinocytes are found within the skin. Separating cells can only be discovered in the basal layer, which is next to the dermis. These cells are mostly transit amplifying cells; however, some are stem cells. As they develop, the keratinocytes gradually peel off at the skin's surface after dying and moving slowly outward through the epidermis. Small holes known as crypts are scattered throughout the projections

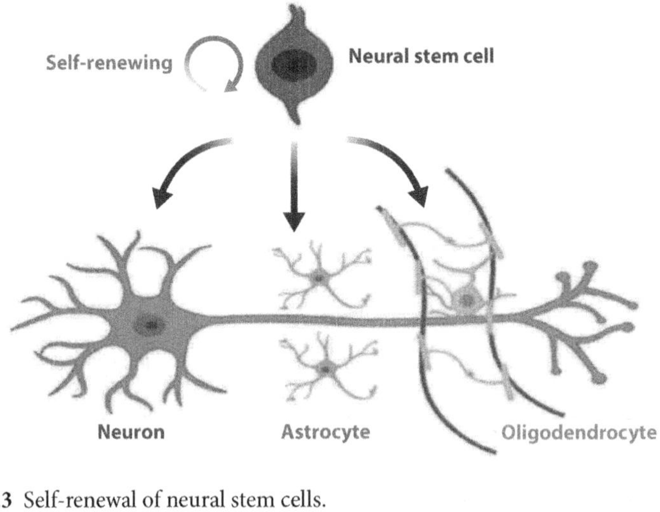

Figure 3.3 Self-renewal of neural stem cells.

known as villi that the small intestine's epithelium makes [27]. There are crypts containing the dividing cells, and around the foundation of each crypt are the stem cells. From the crypts, cells are constantly being created. They then move onto the villi and, at some point, are expelled into the intestinal lumen.

3.2.10 Skin Stem Cells

Skin stem cells can be detected in the dermis, hair follicle bulge area, and epidermis. Stem cells from skin and hair follicles are the most prevalent and convenient source of pluripotent adult stem cells that are actively growing. Because bulge stem cells can produce both skin and hair epithelia, whereas epidermal stem cells can only develop into the epidermis in a physiological context, bulge stem cells are alleged to be more stem-like than epidermal stem cells [26, 27]. Many adult stem cell types, including mesenchymal, hematopoietic, and the skin, have vast stores of neural stem cells. Skin stem cells have the capacity to be used in the following research and treatment areas: skin cancer, burns, ulcers, and injuries, genetic disorders and congenital defects, autoimmune diseases, hair loss and alopecia, hair transplantation and methods for anti-aging, cosmetic surgery, and cosmeceuticals.

3.2.11 Embryonic Stem Cells (ESCs)

The embryo, known as a blastocyst at this point, has a central volume of cells that may differentiate into either of the specialized tissue types that make up the human body in the days to 3 to 5 days following conception and before to implantation. Embryonic stem cells (ESCs) are created from the inner cell mass of an embryo after it has been *in vitro* fertilized and voluntarily donated for study. ESCs are not produced from fertilized eggs inside a woman's body. Apparently pluripotent stem cells can differentiate into absolutely any type of cell and are only found in the very beginning phases of development.

3.2.12 Induced Pluripotent Stem Cells (iPSCs)

Adolescent somatic cells could be genetically reconfigured to resemble embryonic stem (ES) cells by forcing the regulation of genes and other elements required to maintain the unique properties of ES cells. With this technique, an engineered pluripotent stem is created. Regenerative medicine appears to have great potential with pluripotent stem cells. The liver is one site where cells from damage or disease might be found to replenish lost tissue, but liver cells can also divide indefinitely to are precursors to all other body cell types, such as liver, heart, pancreas, and neuronal cells. iPSCs can be created directly from adult tissues, eliminating the requirement for embryos. Furthermore, they can be generated according to patient matching, which denotes that every individual may have their own line of pluripotent stem cells. These endless sources of autologous cells allow transplants might be performed without the risk of immunological rejection.

3.3 Recent Advancements and Applications of Stem Cell Therapy

The applications of stem cells are numerous in addition include mitochondrial therapy as well as the creation of artificial organs for study and transplantation. Among the changes that have occurred recently are discussed below.

3.3.1 HSC Transplantation

Patients with leukemia, lymphoma, and other blood illnesses, as well as those with tumors, can get healthy HSC transplants to replace their damaged bone marrow cells. Allogeneic (cell taken from an autonomous person), syngeneic (cells from identical twins), and autologous (cells from the patient) can all be used in the transfer. With a long history, bone marrow transplants are now considered standard medical practice [28, 29].

3.3.2 HSC Therapy (HSCT)

In clinical trials, HSCT was used to treat multiple sclerosis. An autoimmune condition that has an influence on central nervous system is multiple sclerosis. Therapy aimed at modifying disease is the conventional method of treating multiple sclerosis (DMT). DMT works on the immune scheme by modifying it, changing the way immune cells are trafficked, or lowering the number of immune cells. It must be taken consistently, though, and there may be harmful side effects. Compared to DMT, HSCT has yielded better results in clinical trials [30].

3.3.3 Placental Stem Cell Therapy

Promising outcomes and potential for treating a variety of disorders, including Alzheimer's, pancreatic, hepatic, and myocardial infarction, muscular lung fibrosis, dystrophy, with substantial bone lytic lesions, have been demonstrated by placental stem cells. They are useful in tissue engineering as well [31].

3.3.4 Autologous Transplantation of Limbal Embryonic Stem Cells

Patients who have lost their corneal epithelium may benefit from culturing of autologous limbal cells, which encompasses stem cells called holoclones. Burns to the eye can cause blindness if the limbus is destroyed or there is insufficient limbal stem cell activity. In Europe, Holoclar has received formal approval for people with medium to severe deficiencies in their limbal stem cells [32–34].

3.3.5 Advancements in the Field of Synthetic Organ Engineering

Stem cells proliferate and differentiate into structures similar to their source when they are grown in a habitat with three dimensions and favorable growth circumstances without the need for outside input. These structures are referred to as "organoids" because they resemble organs and serve as a home for stem cells. Large lytic lesions in organoids vary in size, shape, and cell mass composition, and other characteristics from culture to culture, but they also display an organization level that is unreplicable with existing technology. Numerous investigations make use of these organoids [35].

3.3.6 Hollow Organ Engineering

When it comes to the creation of dull organs like the vagina and trachea, stem cells provide encouraging outcomes. A case study demonstrates how autologous stem cells can be successfully accustomed to produce trachea. For those who were afflicted with mustard gas, new trachea can be created using stem cells. Many disorders that go untreated or have ineffective treatments can be treated by constructing an artificial vagina using MSCs and the right scaffold [36].

3.3.7 Anti-Aging Effects

Molecular factors such as telomere shortening, DNA damage, mitochondrial malfunction, loss of proteostasis, and stem cell depletion are what define the aging process. Adipose-derived stem cells gradually alter cell metabolism to look like youthful cells by promoting mitophagy, increasing mitochondrial synthesis, and lowering reactive oxygen species. There is also enrichment in the pathways connected to mitochondria and nucleotide metabolism [37].

3.3.8 Minimizing Mitochondrial Injury

Numerous diseases that appear unconnected, including neurological disorders, heart diseases, sepsis, cancer, diabetes, and fluoroquinolone-associated impairment, are largely caused by mitochondrial malfunction. MSCs help cells with damaged or dysfunctional mitochondria recover faster, encourage mitophagy, and trigger the transfer of healthy mitochondria [38].

3.3.9 Treatment of Diabetes

In clinical trials, a multitude of stem cells have been used to treat diabetes. Diabetes mellitus may result from autoimmune disorders (type 1 diabetes), lifestyle decisions and genetic predispositions (type 2 diabetes), and even shifts in hormone levels brought on by pregnancy (gestational diabetes). Insulin injections are a common part of diabetes treatment; however, they come with drawbacks because of the expense of the injections and their short-term effects. This issue can be resolved by stem cells directly repairing pancreatic cells. Diabetes-related conditions such not-healing wounds are also preserved with stem cells [39–42].

3.3.10 Modeling Diseases and Researching Differentiation

Since its creation, researchers have employed iPSCs in many different contexts, including the pathophysiology of diseases, the development of new iPSC-forming techniques, the investigation of genetic disease inheritance, the study of neurodegenerative disorders, etc. [43].

3.3.11 Cell-Free Therapy

Extracellular vesicles (EVs) that originate from stem cells, which include exosomes, have been shown to have effects on neuroregeneration, neural protection, neurological growth, including enhanced neural function. Because EVs are non-invasive, blood–brain barrier-crossing, and non-tumorous, they reduce the risks and limitations of cell-based therapy [44].

3.3.12 Wound Healing

Because they secrete antimicrobial substances, stem cells have antibacterial capabilities, aid in the control over the immune system, and encourage cell division and proliferation at the site of the wound. Immune rejection is eliminated with the usage of autologous stem cells [45].

3.3.13 Treatment of Burn Wounds

As it relates to treating burn wounds, stem cells exhibit greater promise and outcomes than the techniques used today. The direct injection of stem cells, exosome treatment, tissue-engineered grafts or exhibit encouraging outcomes in the healing of blister wounds [46].

3.3.14 Significance in Research

Research into disorders like congenital heart disease, neurodevelopmental issues, as well as how the environment affects the development of cells and tissues is being conducted with stem cells.

3.4 Current Limitations

Stem cell research is now moving gradually due to an assortment of factors. Some of these factors include the inability to successfully culture the majority of stem cells, the high cost and inefficiency of conventional 2D culturing techniques, the difficulty in accurately replicating the stem cell role, the loss of differentiation ability to culture, the absence of defined methods for 3D culturing, the deficiency of appropriate scale-up practices, and so on [46]. Table 3.2 depicts various advantages and disadvantages of different types of stem cells.

When PSCs are injected into the body, they can become teratomas, which are benign tumors that contain tissues from all germ layers. Studies using placental stem cells cultivate a wide variety of cell types; yet, acquiring extremely pure, specialized cells in a sustainable manner is difficult.

Table 3.2 Pros and cons of different stem cells.

	Adult stem cells	**Embryonic stem cells**	**Induced pluripotent stem cells**
Pros	These cells can be reprogrammed and transdifferentiated, albeit this has not been thoroughly investigated. Believed to have a lower chance of rejection when used in transplants. Numerous clinical uses have previously shown accomplishment.	Able to endure and grow in popularity for an entire year or more. Recognized regulations for conserving culture. Pluripotent embryonic stem cells are able to form majority of cell types (ESCs). It is possible to learn more about the expansion course by seeking ESCs.	Giver somatic cells in large measures can be employed. With donor/beneficiary transplants, histocompatibility glitches can be prevented. Incredibly helpful for progressive research and medication advance. The acquaintance gained after the "reprogramming" technique could be pragmatic to *in vivo* treatments aimed at reprogramming sick or damaged cells or tissues.
Cons	Although the limitations of ASC differentiation are still unknown, multi- or unipotentiation is now thought to exist. Cannot be cultivated for long periods of time in culture. They are difficult to find and separate since they are frequently extremely uncommon in each tissue. As of yet, no technique exists for mass-producing stem cells in culturing.	The procedure of generating an ESC line is inefficient. Uncertain if their transplantation would be accepted. Much more research and testing are needed because ESC pathway therapeutics are still in their infancy. Using ESC homogeneous culture prep directly for tissue transplants can lead to teratomas or tumor growth.	Practices to pledge maintenance and reproducibility because discriminated tissues are inexact. Currently, viruses are laboring to supplement genes into embryos; in experiments on mice, viruses have been found to induce cancer.
Ethical concerns	Significant ethical issues have not been brought up.	To access the inner cell mass, the embryo must be destroyed. Peril of giving permission to female donors.	When given the correct circumstances, induced pluripotent stem cells can develop into embryos.

3.5 Innovative Isolation and Culture Techniques and Their Use in Research on Stem Cells

Research on stem cells presents previously unattainable prospects for the advancement of medical therapeutics for severe diseases, as well as a new channel for digging into fundamental biological problems.

Stem cells' distinct properties, in particular, their ability to differentiate and refresh oneself into several cell types offers promising predictions for the development of innovative therapeutics. The capability to significantly alter the way crippling diseases are treated is among the most crucial uses of stem cell research [47]. There is hope for illnesses for which there are currently few or no viable treatments by means of stem cells to restore or repair injured tissues and organs. Repairs could be possible using stem cell therapy. Heart attack-related cardiac tissue destruction, regrow neurons in neurodegenerative ailments like Alzheimer's and Parkinson's, and even restore function to spinal cord injuries. Additionally, the study of stem cells advances our comprehension of basic biological processes. Examining the systems that govern stem cell function can teach scientists more about the initial phases of growth and the intricate mechanisms underlying tissue regeneration. This information not only aids in the improvement of stem cell-based treatments but also broadens our understanding of the complexities of human biology [48].

Regenerative medicine has shifted its focus toward developing effective stem cell isolation technologies, especially when it comes to diabetic therapy. The potential to generate functional islet clusters from stem cells holds great promise for addressing the problems related to conventional diabetes medications. Meanwhile, adipose tissue is emerging as a crucial a source of stem cells for uses in regeneration, with advantages such as the capacity to distinguish between several lineages and self-renewal. However, the multiplicity of separation techniques—from cutting-edge non-enzymatic technologies to traditional enzymatic methods—introduces problems that necessitate careful consideration [47, 48].

Stem cell rehabilitation requires large-scale stream of highly viable, pure stem cells. and translational stem cell research, which has prompted the creation of stem cell separation algorithms. It is commonly recognized that stem cells with the aptitude to regenerate are present in tissues, blood, and bone marrow. With varying degrees of effectiveness, numerous stem cell for the previous ten years, treatments have been employed to aid in the regeneration of these tissues and organs. For example, autologous stem cell implantation has been employed in the clinical therapy of myocardial infarction and macular degeneration. The progressions in this field have directed to a growing demand for stem cell seclusion methods that are extra effective, straightforward, readily applicable, and capable of producing highly functioning stem cells from wide range of tissue types. Despite the widespread use of numerous commercially available stem cell isolation techniques, their efficiency and specificity remain inadequate [49].

3.5.1 Conventional Methods of Separation

A crucial component of both basic and clinical research is the extraction of pure populations of stem cells from a diverse solution. As a result, numerous commercially available stem cell separation procedures have been created. Table 3.3 elaborates conventional methods of separation with benefits and drawbacks. Methods for separating cells are broadly alienated into two groups: strategies based on empathy (electrical, chemical or magnetic couplings) and methods based on physical factors (size/density). The first category's techniques capitalize on the reality that stem cells often fluctuate from other cells in the same tissue in terms of size and density. Density gradient centrifugation is the simplest example; a test cell sample is layered on top of a tube that has been filled with a density gradient made longitudinally [50].

Cells are propelled down a density gradient by centrifugal force, gathering at the intersection of the medium's and the cells' densities. The desired cell type density must be known in advance in order to use this procedure. Solutions with the potential to create a density gradient for stem cell isolation—such as RosetteSep™, Percoll™, and Ficoll-paque™—are employed as a medium [50, 51].

One sorting technique that does not involve stem cell labeling is field flow fractionation (FFF). Depending on their size and form, cells in FFF are moved to collectors at varying times by inertial or non-inertial forces. Thus, FFF is able to distinguish between various cell types based just on physical characteristics. Because of the nonexistence of tagging, the magnitude and density of the stem cell population are notably different from those of the other cell populations, FFF is an incredibly effective method for separating cells. To distinguish between human mesenchymal stem cells and those produced from both amniotic and fetal membranes, the researchers employed a non-equilibrium gravitational force field (FFF) [49–51].

Table 3.3 Benefits and drawbacks of conventional methods of separation.

Method	Benefits	Drawbacks	Examples
Centrifugation using density gradient	Extensive compiling of volume, simple scaling up	Reduced chastity, increased output differentiality	Percoll; Ficoll- Paque
Pre-plating	Lots of processing, simple scaling up, and ease of use	Reduced purity, increased output heterogeneity	Both human adipose-derived and human embryonic stem cells
Centrifugation-negative selection using density gradient	Extensive volume of processing, excellent purity, simple scaling up, and ease of use	Use solely whole blood	Stem cell technology: Rosettesep
Dielectrophoresis (DEP)	Sort cells straightly, label free on stem cells, without diluting them in fluid, and arrange cells according to viability	Low purity is the consequence of cross-reactivity.	Adipose-derived stem cells in living
Conditioned expansion media	Lots of processing, simple scaling up, and ease of use	Reduced purity, increased output heterogeneity	Human bone marrow-derived mesenchymal stem cells

(Continued)

Table 3.3 Benefits and drawbacks of conventional methods of separation. (*Continued*)

Method	Benefits	Drawbacks	Examples
Flow cytometry (FACS)	Strong, extremely accurate, sensitive, and high-resolution	Needs expensive, specialized equipment, and well-trained personnel to operate; not suitable for extensive processing	Hematopoietic stem and primordial cells, as well as neural cell populations derived from human embryonic stem cells
Field flow fractionation (FFF)	No label for stem cells	Only able to differentiate stem cells according to variations in size and density	Human fetal membrane-derived mesenchymal stem cells
Magnetic activated cell sorting (MACS)	Less complicated than FACS	Magnetic particles are present in target stem cells	Easysep (stem cell technology) without columns; Macs (Miltenyi)

Dielectrophoresis (DEP) does not require antibody labeling because it separates cells according to their inherent electro-physical characteristics. When a cell is in an electric field, it develops polarized and forms a dipole. Conditional on the surface charge, size, nucleic acid content, and other factors of the cells, a net force is applied to them in an irregular electric field, which causes motion in a single direction. DEP has extracted CD34+ bone marrow or blood-derived hematopoietic stem cells [52]. DEP has two advantages over most other methods: Cells can be directly concentrated using it during isolation, and it doesn't require labels. Splitting non-stem cells from stem cells is generally difficult due to differences in dielectrophoretic potential and cell size, and DEP-purified stem cells are frequently tainted with undesirable cells [52, 53].

When comparing stem cells to non-stem cells, there is some variation in their size and density, the main disadvantage of these density- and size-based methods is that they are not very precise. Furthermore, with these methods, the majority of other stem cells that are similar to the surrounding non-stem cells in size and density cannot be isolated. To get around these restrictions, techniques for affinity-based differentiation were created, which use antibodies to identify surface indicators unique to stem cells. These antibodies are associated with a tag that is recognizable to automated devices. Binding fluorescently dye-labeled antibodies to mixtures of cell suspensions is the process of fluorescence-activated cell categorization, or FACS. After that, each cell is sorted separately using light scattering and fluorescence. Using FACS, cell populations that are 95% pure or more can be obtained. However, the equipment needed is costly and the throughput is restricted to about 107 cells per hour. Separation using magnetic beads layered with antibodies is an alternative to FACS. Target cells can be handled in parallel utilizing magnet-activated cell sorting (MACS) to produce a quicker separation rate of about 1,011 cells per hour. About 75% of pure stem cells can be recovered using MACS, and one of the main sources of interference is the presence of magnetic beads in the pure stem cell population [54–56].

Pre-plating, often referred to as static adhesion separation of stem cells, differs from labeling techniques in that it relies on the observation that stem cells have a greater propensity than other cell types to cling to dishes and culture plates. It has been demonstrated that this technique can separate human adipose-derived stem cells from lipoaspirate—tissue taken out during liposuction—and concentrate mesenchymal stem cells from bone marrow. However, stem cells extracted by pre-plating frequently contain undesirable cell types because some non-stem cells also attach to the growth plate, though not as successfully. A stem cell stimulating medium can be utilized to increase the stem cell population in a targeted manner. Consequently, a target stem cell population can be efficiently enriched by cultivating the beginning population in a selective medium [57–59].

3.5.2 New Approaches

There are now new separation methods available in the field with the objectives of producing samples in large quantities, simplifying the separation procedure, and attaining superior purity when separating stem cells. The new approaches are either new theories or enhanced versions of the current procedure. Benefits and drawbacks of new approaches of separation depicted in Table 3.4.

Temperature-sensitive polymer poly(N-isopropylacrylamide) (PNIPAAm) is water soluble at ambient temperature (20°C), but at 32–35°C, it precipitates out of aqueous solutions, is one of the revolutionary stem cell separation techniques. This justifies the use of PNIPAm-conjugated antibodies, which identify markers specific to stem cells, increasing the temperature

Table 3.4 Benefits and drawbacks of new methods of separation.

Method	Benefits	Drawbacks	Examples
Temperature-responsive polymer in an aqueous two-phase system	Easy to use, adaptable, and reusable antibodies	Low rate of recuperation and purity	CD34+ KG1 cells
SELEX	Reduce the cost of *in vitro* testing, select stem cells without choose stem cells in the absence of any molecular information and target non-protein targets	Aptamer identification screening is ineffective, with low specificity and low processing volume	Human and non-human mesenchymal stem cells
Microfluidics	Miniaturization, automation, reduced operator handling, cells without labels and several operations on the same chip, and mass processing	Long operation duration, not extremely pure, with sporadic high recovery rates	Microfluidic apparatus utilizing droplet, cell affinity chromatography, and negative DEP chromatography

from 20 to 32°C–35°C will enable the extraction of stem cells from an aqueous solution. Centrifugation in a two-phase polyethylene glycol (superior phase) and dextran (inferior phase) density gradient can be used to separate the precipitated cells from the remaining cells. As the tainted cells settle to the bottom, target cells at the point of intersection between the two periods create bands of sediment. The temperature-sensitive polymer technique of this large-scale separation can be accomplished with ease and quickly using an aqueous two-phase system. Nevertheless, KG-1 cell sediment band had a considerable number of contaminating Jurkat cells when assessed on isolating CD34+ KG-1 cells in Jurkat cells with a preliminary population of 10%. The presence of non-target Jurkat cells in the target KG-1 cell band may have been caused primarily by the concentration of NaCl supplied to the lower phase, which affects cell partition. Purity can be increased by raising the NaCl content in the phase below, encouraging the aqueous two-phase separation procedure twice on a pure cell combination, or attaching a sufficient quantity of antibody conjugates to cells [60].

Cell Target cells can be selectively captured using ssDNA, RNA, or modified nucleic acids as aptamers in methodical ligand evolution by exponential enrichment (SELEX). Aptamer libraries—oligonucleotides that are not found naturally are cultured with a population of stem cells to initiate SELEX-based cell selection. Any aptamers that remain unbound are eliminated. Following their release from the stem cell surfaces, the bound aptamers undergo further amplification by RT-PCR in preparation for development in successive SELEX cycles, also known as progression of aptamers. After several cycles of grasp and unleash, there will be a generation of aptamers that have a high kinship for the target stem cells. The SELEX approach has been used to isolate stem cells extracted from bone marrow that are mesenchymal, both from humans and other species [61].

The primary benefit of SELEX lies in its empirically generated aptamers for target stem cell enrichment. This benefit makes up for the fact that many target stem cells' surface marker characteristics are unknown beforehand. Furthermore, SELEX can detect non-protein markers on stem cells and does not require expensive antibodies. SELEX is a screening method that examines a lot of potential aptamers. Nevertheless, SELEX has a limited throughput and necessitates drawn-out aptamer selection and amplification procedures. Eventually, the specificity may not always be sufficient to extract high-purity stem cells, even after multiple iterations of aptamer development [62].

The topic of microfluidics is a recent development that focuses on the downsizing of devices and processes. The separation of stem cells is one of its uses. The majority of approaches covered have been modified for use with microfluidic platforms. Microfluidics uses less reagent and sample. The whole process of injecting, sorting, and collecting samples is carried out continuously on a single chip. There are techniques that can be automated to reduce the amount of operator handling. However, throughput could be a problem for applications with large volumes. Some researchers came up with inventive techniques to parallelize the microfluidic devices, boost flow rates, or enhance the surface area of a single device. Here are some instances of cell separation methods that have been scaled down for use on microfluidic platforms.

For instance, appropriate aptamers were chosen using SELEX microfluidics to enable a quicker and more effective screening procedure. Microfluidics based on negative DEP (nDEP) has been utilized to isolate Stro-1+ bone marrow-derived human endothelial precursor cells (EPCs) and mesenchymal stem cells derived from white blood cell models. To locate and accumulate uncommon precursor cells derived from autologous periosteal tissue, single cells are encapsulated in dissimilar field beads and screened on a microfluidic stage using droplet microfluidics. Microfluidic cell sorting systems that are label-free and affinity-based concentrate on immobilized antibodies instead of pre-labeling cells using magnetic beads or fluorescent dyes. CD34+ cell lines (KG1a cells) are coated on the luminal surface of a analogous range of hollow filaments in order to immobilize antibodies and capture them [63].

Target cell detachment was carried out in a fluid flow under a predetermined shear stress. Research has demonstrated that Mice's adipose-derived stem cells and human whole blood EPCs can be separated from a heterogeneous suspension using antibody-coated microfluidic devices. Antibodies are fixed on alginate hydrogel covering for the EPC selection investigation. Following the capture of target cells in antibody-gel complexes, the alginate hydrogel can be dissolved and very pure EPCs can be collected by rinsing with EDTA. Promising applications as mentioned above demonstrated the microfluidics' advantages for multiplying devices although the separation process can be streamlined and conducted in parallel, more work still has to be done to achieve higher stem cell recovery, improved purity, shorter operating times, and a decrease in the requirement for overpriced antibodies in adhesion-based parting [63].

While every clinical example and tissue engineering are different, there exist a few industry standards that can be universally applied to efficaciously use microfluidics in technical and medical settings. The target cell inhabitants must be consistently secluded and free of labels or tags that can potentially obstruct post-separation analysis or negatively alter cell activity.

A microfluidic platform needs to outperform state-of-the-art conventional platforms in terms of efficiency of 95% or higher, 1,011 cells or more throughput per hour at minimum, and purity of greater than 95% in order to advance as a platform for stem cell isolation. Regarding purity, throughput, efficiency, and total cell yield, no exact quantitative benchmark has been established. Such large-scale separation criteria have not yet been reached by any microfluidics platform. Numerous stem cell separation microfluidics platforms that meet or exceed one or two requirements have been documented in the literature. As far as the authors are aware, though, no platform has been able to separate stem cells at throughputs that are on par with MACS without compromising on efficiency or purity. Furthermore, the optimal strategy for the precise separation of these delicate cells remains unknown due to the paucity of research on the impression of the parting procedure on the function and phenotypic of stem cells after separation. It should be mentioned that even if the majority of research still aims for high purity cell isolation from the original heterogeneous cell suspension, most studies still want ultra-high purity when it comes to stem and progenitor cells. Because supporting cells resemble the micro-cellular atmosphere for stem cell culture, the absence of supporting non-stem/predecessor cells may eventually affect the survivability and/or purposeful properties of the stem/predecessor cells in culture.

3.5.3 Prospects and Difficulties

In the preceding decade, significant advancements have been made in enrichment and seclusion of rare stem cells. Because specificity of antibodies that identify antigens on the outward of stem cells is great, affinity-based approaches have proven to be the most efficient and dependable among the various stem cell separation techniques. FACS has a remarkable purity of >95%, although MACS is portable. In the future, affinity-based methods for stem cell separation will continue to be highly effective due to the increasing understanding of more effective stem cell indicators and the production of extra targeted aptamers *via* SELEX. Higher purity should be pursued in tandem with other critical goals, such as fewer disruption of stem cells, speed, and cost-effectiveness. Microfluidics presents intriguing answers to these problems by providing an automated, compact, multiplex, and portable way to separate stem cells. It will be highly useful to combine conventional techniques on a microfluidic platform, even though the quality of stem cell isolation by microfluidic devices still has to be significantly enhanced. Last but not least, a significant obstacle to tissue engineering and its therapeutic use is large-scale stem cell isolation. The throughput of most separation techniques can be increased. Among them FFF, density gradient centrifugation, and DEP have the most promise for large-scale manufacturing. Future breakthroughs in stem cell isolation will have a major impact on the advancement of translational biology research as well as clinical stem cell therapy.

Since mesenchymal stem cells in various tissues share traits that make it challenging to identify an exact stem cell group, differentiating stem cell types from human tissues and isolating them, including adipose tissue, has proven to be challenging. Political and principled concerns have limited the usage of embryonic stem cells in research since many current techniques necessitate the death of human embryos in order to excerpt embryonic stem cells. The the capacity to isolate and cultivate stem cells of adults for use in medicine is still limited. Furthermore, the nonexistence of particular indicators for isolation makes it difficult to identify hepatic stem cells in the future. While there has been optimism over the therapy of end-stage liver disease due to the extraction of liver stem/progenitor cells, methods for maintaining and isolating these cells are still being developed. Furthermore, Adult somatic stem cells are less able to differentiate than embryonic stem cells, which poses challenges for their therapeutic use. Although there have been encouraging advances in the methods for isolating and cultivating human pancreatic stem/progenitor cells, more developments are need to identify and isolate these cells. Furthermore, despite extensive research, stem cell applications for standardized therapy in veterinary medicine are merely beginning their growth.

3.6 Future Prospects

3.6.1 Stem Cell in Gene Editing

The integration of stem cell technology with CRISPR-Cas9 gene editing techniques unbolts new avenues for the conduct of genetic defects and inherited illnesses. It may be possible to heal hereditary illnesses with gene-edited stem cells and reduce the likelihood that harmful mutations would be inherited by subsequent generations [64].

In regenerative medicine, CRISPR-based genome editing of stem cells represents a new frontier. Stem cell therapy may benefit from this technology's capacity to precisely alter DNA and RNA. This method offers several applications such as base editing, knocking-in, knocking-out, and RNA expression modification. By gaining a deeper understanding of the role played by crucial genes and biological cascades, researchers were able to investigate and enhance stem cell activity. Stem cell research is anticipated to be greatly advanced by this technology, which will also make stem cell therapy more accessible [64].

Mutations at the genome level give birth to a diverse array of hereditary diseases. The usefulness of stem cell-based healings in the management of particular illnesses has been shown *via* clinical trials. One intriguing method is the autologous implantation of heritably engineered stem cells. These heritably engineered stem cells have multiple benefits, including the ability to self-renew, an extended lifespan for the patient, and the skill to divide into daughter cells that can develop into resident cells in countless tissues [29]. An increasing amount of studies has established that the CRISPR/Cas9 genome excision technology is a convenient tool for stem cell-based medicine and research. HPSCs can develop into a diversity of cell types and have a high skill for proliferation. Using CRISPR/Cas9 gene editing techniques, these cells (mesenchymal, hematopoietic, and brain stem cells) can be altered and used therapeutically [65, 66].

3.6.2 Stem Cell for Autoimmune Illnesses

Immunological deficits caused by AIDS may be healed by stem cells. They are capable of treating autoimmune ailments because they normalize the invulnerable system and encourage tissue regeneration. Subsequent research endeavors will center toward devising safe and effective stem cell therapies for ailments like rheumatoid arthritis, multiple sclerosis, and type 1 diabetes [67].

Diabetes type 1 in children: The autoimmune condition known as type 1 diabetes is typified by the demise of the insulin-producing pancreatic cells. There are currently very few pancreas donated annually, and the immunosuppressive drug dealings obligatory to avert graft denial are extremely toxic. This severely limits the ability to treat these affected with human islet replacement in an attempt to reinstate insulin secretory purpose (gained from human pancreas). Pluripotent stem cells might be able to solve the scarcity of therapeutically effective material available for transplantation if given the instruction to discern into a precise type of pancreatic cell known as a beta cell. They also provide the chance to design these cells so that they can successfully fend off immunological assault and graft rejection [68].

Illnesses of the nervous system: Loss of nerve cells is the cause of many disorders of the nervous system. Adult nerve cells are unable to proliferate in order to replace missing ones. Thus, there are no treatment options available in the absence of a "new" source of functional nerve tissue. With Parkinson's illness, the nerve cells that produce dopamine are destroyed.

The cells responsible for producing specific neurotransmitters die in Alzheimer's disease. The motor nerve cells in people with amyotrophic lateral sclerosis responsible for firing muscles die. Multiple cell types are lost or killed in brain trauma, strokes, and spinal cord injury. Trauma to the brain, stroke, and spinal cord injuries cause the loss or death of several cell types. Multiple sclerosis causes the loss of glia, which are cells that protect nerve fibers. The skill to use pluripotent stem cells to generate new nerve tissue that can restore function may be the only chance for treating these people [68].

Surprisingly, human clinical trials have shown that this therapy technique may be beneficial. Fetal cells have been surgically implanted into Parkinson's sufferers' brains, with moderate success. Though probably not entirely successful due to insufficient numbers of cells secreting dopamine, comparable studies employing adequately differentiated stem cells ought to get beyond those challenges.

Using mouse models of Parkinson's disease, more complex research has already been conducted successfully. Replicable techniques might be developed to replenish the impaired or deceased neurons in the hippocampal and cortical brain regions affected by Alzheimer's disease.

Diseases of primary immunodeficiency: It is possible to treat almost all primary immunodeficiency illnesses with pluripotent stem cells. Currently, immune system defects that are hereditary or congenital have been identified in over 70 different kinds. These illnesses have some of the worst prognoses and are among the most difficult to treat. This group of illnesses includes Wiskott-Aldrich Syndrome, lupus, and extreme combination immune deficiency illness (often known as the "bubble boy" condition). Relevant in this context are the immunological deficits brought on by acquired immune deficiency syndrome (AIDS), which develops after viral infection. These conditions are distinguished by an atypical vulnerability to infection and are frequently linked to anemia, rheumatism, diarrhea, and some types of cancer. Nonetheless, the implantation of stem cells that have been selectively restored may restore immunological function, as well as a functional standardization of life expectancy and life quality, for these patients [67, 68].

Illnesses of the bone and cartilage: When correctly generated, a wide range of illnesses can be treated by stem cells and degenerative conditions marked by a deficiency of healthy cartilage or bone cells. This could be beneficial for treating genetic diseases such as chondrodysplasias and osteogenesis imperfecta. In a similar vein, cells may be grown and injected into big bone gaps left by surgery or fractures, or into osteoarthritis-related damage to some areas of the joint cartilage.

Cancer: Patients undergoing high-dose chemotherapy are currently saved by using bone marrow stem cells, which are a more committed kind of stem cells. Sadly, the proficiency of these restored cells in order to completely recover immunological function is restricted in this particular context. It is envisaged that patients receiving bone marrow transplants would regain full immune response with injections of properly-differentiated stem cells. For instance, full and functional healing will be necessary for the efficacy of immune/vaccine anticancer therapy. More crucially, a successful outcome would allow use of extremely toxic and efficient chemotherapy routines that are presently not possible due to their inability to rebuild immunological and bone marrow function [67, 68].

3.6.3 Regenerative Medicine

"Regenerative medicine" is the newest and fastest-growing area of medicine it focuses on helping patients with certain diseases restore their ability to use particular tissues or organs of the patients suffering from serious wounds or long-term medical issues in which the natural healing powers of the body are inadequate [69].

Finding alternatives has become more urgent as a result of the current situation, which prevents donated tissues and organs from meeting the transplant needs of aging and ill populations. Thanks to their limitless the capacity to divide cells and change their differentiation into distinct cell types, stem cells have arisen as a major tool in regenerative medication, aiding in the restoration of damaged organs and tissues caused by disease, aging, and birth defects [69, 70].

Stem cells provide the building blocks for every organ and tissue system in body and play a variety of roles in the formation, progression, and healing of diseases in the host. A common way to categorize stem cells is as (1) multipotent, (2) unipotent, (3) totipotent, or (4) pluripotent based on their capacity for transdifferentiation. The human body contains only one totipotent stem cell, the zygote, which can differentiate into a whole creature through a process called transdifferentiation. The embryo's internal cell mass, or ICM, on the other hand, contains pluripotent cells that can discern into extraembryonic tissue cells but cannot distinguish into cells that characterize the three germ layers [71].

Stem cells can be characterized as bone marrow stem cells (BMSCs), mesenchymal stem cells (MSCs), umbilical cord stem cells (UCSCs), embryonic stem cells (ESCs), and progenitor stem cells specific to tissues (TSPSCs) based on their potential for regenerative uses. Pathogen or cancerous cell immunolysis and tissue regeneration can be achieved *via* autologous, allogenic, and syngeneic stem cell transplantation. Immunosuppressive medications and tissue typing utilizing human leucocyte antigens (HLA) are recommended to minimize the deleterious penalties of host-versus-graft rejections. Tissue tolerance at the graft site is made possible by stem cells' reduced MHC receptor expression and their secretion of chemokines, which attract immune and endothelial cells [72, 73].

The current methods of stem cell regenerative medicine rely on technology for tissue engineering, which combine the concepts of material science, organoid development, and cell transplantation to create organoids that is suitable for repairing damaged tissue and organs physiologically. Transplanted stem cells need to thrive, proliferate, undergo site-specific differentiation, and get integrated into the host's circulatory system in order for regenerative outcomes to be effective [72–74].

3.6.3.1 ESCs in Regenerative Medicine

In 1998, Thomson successfully extracted human embryonic stem cells (hESCs). ESCs have great promise in the amelioration of all diseases due to their pluripotency, which allows them to develop into over 200 different cell types. The transcription factors' functional dynamics referred to as pluripotency factors, such as SOX2, OCT4, NANOG, and others, determine the pluripotency fate of ESCs. One important regulatory switch for ESC lineage commitment has been identified as the homologous pairing phase throughout the embryogenesis and transdifferentiation processes. In ESCs, the pluripotent state maintains the two OCT4 alleles apart. ESCs are a prime model for regenerative medicine for treating diseases and tissue abnormalities because of their varied lineage commitment capability [75, 76].

3.6.3.2 TSPSCs in Regenerative Medicine

Through continued cell division, TSPSCs preserve tissue homeostasis; in contrast to ESCs, TSPSCs preserve the tissues-specific elasticity and diversification of stem cells, resulting in the creation of a limited variety of cell types. Given that the population's TSPSC count is very low in relation to all cells, it will be difficult to collect them and manipulate them *in vitro* in order to investigate their potential application in rehabilitation.

The foundation of the human body is made up of many kinds of TSPSCs; one example is the potential therapeutic use of pancreatic progenitor cells (PPCs). PPCs result in β-cells that produce insulin throughout embryogenesis. Insulin has a negative regulatory effect on PPC differentiation into β-cells. In order to support the functional importance of the niche impact in processes connected to the transdifferentiation and self-renewal, PPCs necessitate active FGF + Notch signaling. PPCs grow more quickly in communities than in single cell populations. In a three-dimensional scaffold culture system, PPCs derived from mouse embryos develop into spheres of porous organoids, which then discern into clusters of β cells that generate insulin [77, 78].

3.6.3.3 Stem Cells and MSCs in Regenerative Medicine

Multilineage stem cells, or MSCs, can only differentiate into mesodermal tissue, which contains muscles, ligaments, tendons, neurons, and cartilage. MSCs are defined as cells expressing a alliance of the following indicators: CD73+, HLA-DR, CD90+, CD11b–, CD105+, CD34–, CD14–, CD45–, CD19–, and CD79a–. The physiological and anatomical traits of the recipient and donor both influence the course of therapy. Bone marrow-derived MSCs from baboons (BMDMSCs) are comparable to bladder stem cells in both morphology and phenotype, and they can be employed to regenerate bladder tissue. Within ten weeks following transplantation, the GFP-expressing BMDMSCs (CD73+, CD105+, CD34–, and CD45–) stimulated by (SIS) small intestinal submucosa scaffolds promote the repair of damaged bladder tissue. Combinatorial CD characterization of donor MSCs appears to produce better regeneration results, as the MACs identified by CD are functionally active at the transplant site [79, 80].

Retroviral administration, which is used in *ex vivo* transdifferentiation as of MSCs, has the potential to cause cancer in cells. MSCs are used in regenerative medicine in two different ways: either directly through transplantation or by transdifferentiating the cells first and then re-transplanting [80].

3.6.3.4 UCSCs in Regenerative Medicine

The primary source of stem cells is the umbilical cord, which is typically thrown away after childbirth and may be obtained effectively with less ethical restrictions than embryonic stem cells. Myeloid stem cells are abundant in the umbilical cord (MSCs), which have an amazing ability to regenerate, and hematopoietic stem cells (HSCs). The HSCs in cord blood are accountable for the continuous restitution of all blood types and immune-supporting cells. HSC proliferation is controlled by the decrease signaling of the aryl hydrocarbon receptor (AHR) in stem cells, which is controlled by the Musashi-2 protein. Cryopreserved UCSCs are available from stem cell banks [81].

Five HSCs have received FDA approval to treat blood and other immune-related issues [97]. When amniotic fluid is extracted during pregnancy for routine diagnostic testing, it is typically thrown away without taking its potential for vasculogenesis into account. The greatest substitute are UCSCs for patients without fully coordinated HLA donors for bone marrow, PBMCs and peripheral blood [98]. Then, *ex vivo* UCSC multiplication can meet the need for the intended results. The number of cells in a transplant is a significant issue with UCSCs; fewer cells in a transplant increase the danger of infection and death and require longer for engraftment to mature. Different procedures are available to expand UCSCs *ex vivo* [81, 82].

3.6.3.5 BMSCs in Bioregenerative Therapy

All peripheral blood is created in the spongy, soft bone marrow, consisting of stem cells with hematopoiesis, which create stromal cells and blood cells, which create fat, bones, and cartilage. Aesthetically, bone marrow is divided into two types: yellow marrow, which produces fat cells and some WBC, and red marrow, which is myeloid tissue and produces RBC, platelets, and most WBC [83, 84].

An imbalance in the makeup of the bone marrow may lead to the illness. Bone marrow transplantation has gained widespread acceptance since 1980 as a cancer therapy. HLA type of donors is necessary to prevent rejection of graft, but the availability of fully matched donors is restricted to members of the family, which complicates the use of allogenic transplantation. Determining the key antigens for patients using a haploidentical allogenic donor who are unable to identify a fully matched donor may alleviate donor restrictions in this situation, as matching of all HLA antigens is not strictly necessary. Patients with aplastic anemia and hematological malignancies can benefit from the reconstruction of their host immune systems through the two-step administration of myeloid and lymphoid (BMSCs) bone marrow stromal cells derived from haploidentical donors. The results are nearly identical to those of fully matched transplants, suggesting that HLA profiling is adequate to ensure successful BMSC transplantation.

The main procedure for minorities and others without access to matched donors is the haploidentical HLA matching approach. Moreover, BMSC-based therapies involve other considerations besides antigen profiling [85].

3.6.4 Personalized Medicine

Since the development of iPSCs, it has been possible to separate embryonic stem cells off a patient's cells, providing an endless supply of unique cells that can be used for therapeutic purposes. Given that IPSCs and patients might be ethnically matched, the chance of immunological denial can be greatly decreased, and customized treatments for a range of illnesses can be made possible.

A key element in the quest for customized therapy is stem cells. These are an important instrument for regenerative healthcare due to their bulk to develop into numerous types of cells. Physicians can customize therapy for their patients by utilizing stem cells that belong to the patient. This strategy may completely change the way we treat various illnesses.

Stem cells have countless applications in customized therapy. Treatments using stem cells are presently being explored by researchers for a number of illnesses, such as diabetes, arthritis, and cancer. In regenerative medicine, where they can be utilized to replace damaged tissue and reestablish organ function, stem cells have also demonstrated tremendous potential.

Treating cancer is one of the utmost fascinating uses of stem cells in personalized medicine. Doctors can develop a targeted therapy that targets cancer cells while sparing healthy cells by utilizing the individual's stem cells. This strategy could transform way that cancer is treated and enhance patient outcomes.

Multiple sclerosis and lupus are two autoimmune illnesses that are being treated with stem cells. Instead of only treating the symptoms of a sickness, doctors can develop a therapy that addresses the underlying cause of the condition by employing the patient's own stem cells.

3.6.5 Tissue Engineering

One increasingly popular strategy to address these issues is tissue engineering; it is a crucial element of regenerative medication. Tissue engineering, as the name implies, is a multifaceted discipline that uses the ideas and techniques of bioengineering, material science, and life sciences to produce biological replacements with the goal of maintaining, restoring, and improving tissue functioning after trauma or disease-related damage. The fundamental concepts of tissue engineering involve the fusion of live cells with a biodegradable natural or synthetic scaffold to form a 3-D living structure that aims to mimic or surpass the function, structure, and mechanical properties of the tissue that is being replaced.

Four essential elements must be carefully chosen in order to establish such a construct, namely:

1. scaffold
2. extracellular matrix (ECM)
3. growth factors
4. cells

Three-dimensional tissue structures known as scaffold materials direct the cell arrangement, differentiation and development. Scaffolds have to be biocompatible and made to fulfill the dietary and physiological requirements of the precise cell type. Evolution features are peptides that bind to cellular receptors and cause cells to respond by permitting or inhibiting tissue proliferation and/or differentiation. Through controlling variables including pH, temperature, oxygen saturation, and mechanical forces, the extracellular matrix (ECM) must establish an environment within the structure that promotes cell adhesion, proliferation, and differentiation. These conditions depend on the specific cell types and the characteristics of the scaffold. Lastly, constructing a viable framework requires a sufficient supply of cells that are nonimmunogenic, highly capable of proliferation, easy to harvest, and can differentiate into various specialized cell types.

Harvesting and cultivating cells can be done in two primary ways. When direct cell harvesting is not practical—as in the case of individuals with severe terminal organ disease or cells with restricted ability for proliferation in culture—stem cells are considered a backup source of cells. Tissue engineering's ultimate objective is to reconstruct a patient's tissues and restore organ function by implanting cells and matrix. Currently, there are three types of tissue engineering approaches: substitutive, histioconductive, and histioinductive. Histioconductive techniques (*ex vivo*) replace damaged or absent organ tissue segments with *ex vivo* constructs; substitutive procedures (*ex vivo*) replace entire organs. The histioinductive methods, on the other hand, promote self-healing and may include gene therapy through the delivery of DNA using plasmid vectors or growth factors.

For wounded tissues to be effectively and permanently repaired, a number of requirements must be satisfied, namely:

1. For the deficiency to be filled, enough cells must be produced.
2. Cells should be capable of transforming into the desired cell types.
3. In order to generate ECM, cells must adapt to an appropriate three-dimensional structural support or scaffold.

4. The generated cells must have a structure and mechanical properties that align with natural cells.
5. The ability of cells to successfully merge with inherent cells and go past the danger of immunological denial is essential.
6. Any associated biological risks should be minimal.

Additional classifications for cell sources include mature (non-stem) cells, embryonic stem cells (ESCs), adult or somatic stem cells, and zygotes or totipotent stem cells. Mature cells have limited proliferative and differentiation capability, which limits their utility. Adult stem cells are resident stem cells that are essential to the integrity of tissues like skin, blood, and bone. They can be located in particular niches or tissue compartments. These cells are pluripotent, meaning they can be genetically engineered to specialize into specific tissue types. In mammals, somatic stem cells have been classified into at least 20 primary groups. They are present in the GI tract, bone marrow, blood, cornea, retina, tooth pulp, liver, and pancreas. The pre-implantation blastocyst's inner mass of cell is the resource of ESCs. These are young cells that are not differentiated but exhibit a tendency to distinguish and self-renew indefinitely [86].

With the use of biodegradable biomaterial scaffolds, tissue manufacturing has the tendency to encourage the rejuvenation of tissues, together with cartilage, bone, and neural tissues. For example, the therapeutic application of tubular collagen nerve guides has demonstrated efficacy in dealing of peripheral nerve damages, including the management of critical gap lengths more than 10 mm in primates. *Via* embedding fibers or cellular hydrogel, this efficacy can be further increased. Furthermore, bone-specific tissue-engineered constructions have shown osteoconductive qualities, which encourage bone cell adhesion, proliferation, and migration.

Additionally, stem cells are employed in the arena of tissue engineering, which includes the use of scaffolds, biomaterials, and stem cells in the tissue engineering of the heart, bone, liver, brain and other organs. A few instances of these applications are covered here.

1. Cardiac tissue engineering

Congestive heart failure is among the leading causes of death in the United States, which results from myocardial infarction (MI) and ischemia, or the loss of purposeful cardiac cells (CMs). The intricate processes involved in the loss of ischemic myocardial cells and the post-MI remodeling that follows, which results in heart failure, are not given adequate consideration by the available treatments. Tissue restoration and stem cell applications could be highly beneficial for heart repair. Using natural acellular hydrogel scaffolds to support the infarcted heart is a promising strategy because of their ability to minimize wall stress, compensate for retrenchment function, and limit ventricular restoration due to their mechanical structure [87].

Basic fibroblast growth factor (bFGF) was demonstrated to be vital to angiogenesis by being encapsulated in heparin-alginate microspheres in a chronic MI pig model, which resulted in a considerable improvement in myocardial function *in vivo*. Hydrogel frameworks have been explored *in vitro* to promote cardiogenic differentiation and cell proliferation. For instance, the enhanced interaction between clusters of CMs generated from stem cells generated from muscular tissue (MDSCs) resulted in the upregulation of the countenance of cardiac genes, such as connexin 43 and cardiac troponin-T, in collagen scaffolds. This suggests that the incorporation of MDSCs into collagen scaffolds provides a valuable three-dimensional culture medium for directly evaluating the inflexible characteristics of *in vitro* developed CMs [87, 88].

Furthermore, a continuous flow bioreactor evaluation of a hybrid support structure made of collagen and PGA—two natural and synthetic biomaterials—showed enhanced cardiac stem cell (CSC) adhesion. Additionally, physical influences such as mechanical pressure resulted in a twofold increase in CMs, along with alignment of matrix fibers, myofibrillogenesis, and sarcomeric banding. Meanwhile, repeated mechanical pressure leads to a 2.2-fold increase in CM hypertrophy and a 21% rise in proliferation rates, compared to controls without mechanical pressure [89].

2. Neural tissue engineering

The mind and the cervical spine make up the body's central nervous system. The extracellular matrix (ECM) structure of this unique tissue network gives it unique soft physical qualities; for example, natural brain tissue has an elastic modulus of approximately 500 Pa, while bone has an elastic modulus of 109–1,010 Pa. Many illnesses and injuries can affect the CNS, such as multiple sclerosis, traumatic brain injury, stroke, Parkinson's disease and spinal cord injury. Since the composition, structure, and mechanical characteristics of the extracellular matrix (ECM) function as cell activity effectors, soft hydrogel scaffolds are employed in CNS claims to mimic the mechanical and biochemical characteristics of the CNS. For example, arginine glycine-aspartic acid (RGD)-containing PEG-and acrylamide-based hydrogel scaffolds can control the behaviors of cells, including such as neural stem cell (NSC) differentiation, adhesion, and cell renewal. It has been shown that PDGF-AA immobilized agarose scaffolds can promote the differentiation of neural progenitor cells and neural stem cells into oligodendrocytes [90–93].

3. Bone tissue engineering

($Ca_{10}[PO_4]6[OH]_2$) and other ions, together with hydroxyapatite, are in charge of regulating the homeostasis, metabolism, and high content and endurance of bone, which are extensively mineralized in the collagenous extracellular matrix (ECM) that makes up bone. Bone tissue engineering uses scaffolds made of amalgam materials, which blend synthetic and natural components, because collagen lacks structural stability. For instance, it has been demonstrated that fibrous scaffolds composed of PLGA, collagen, and hydroxyapatite enhance human MSC (hMSC) binding while also offering mechanical integrity and structural stability. Collagen scaffolds' structural stability may be improved by inorganic substances like (ACP) amorphous calcium phosphate or calcium phosphate. Furthermore, bone is a tissue with a high metabolism. that depends on a plentiful vascular supply for growth, remodeling, and homeostasis throughout its structure. Consequently, studies of simultaneous angiogenesis/vasculogenesis and osteogenesis have been conducted using 3D co-culture systems based on biomaterials [94–97].

4. Liver tissue engineering

With regard to organ size, the liver is the largest in humans, is essential for detoxification, metabolism, and protein synthesis. The majority of the liver's metabolism, synthesis, and storage processes are carried out by hepatocytes, which is a primary cell type in the liver. The way non-parenchymal cells interact with hepatocytes also impacts how well the liver functions. Hepatocytes have a tendency to become dysfunctional in *in vitro* culture conditions, indicating that stem cells could serve as a substitute cell basis for liver tissue engineering when combined with additional variables such differentiation-inducing factors. It was observed that rat ESCs cultivated in a 3D culture scheme may develop into hepatic-like cells possessing certain traits of healthy, matured hepatocytes, in existence of other components such dexamethasone, insulin, transferrin, selenium, oncostatin, transgenic mice hepatic development factor and fibroblast proliferation factor. Furthermore, the rESCs continued to be alive after being placed in chronically immunodeficient mice, where they underwent *in vivo* hepatic-like cell maturation and subsequent differentiation [98, 99].

Together with the previously mentioned uses for cardiac muscle, bone, nerve, and liver, stem cells and tissue engineering may also be able to renew other tissue sorts like eyes, skin, cartilage, and bladder. Clinical trials are now in progress for methods to restore bladder, kidney, and urethra tissue utilizing a fusion of regenerative cells and tissue regeneration [100–102].

3.6.6 Artificial Organs and Body Parts

We might see the use of stem cells to create completely functional elements of the body that are artificial in the not too distant future. This may help with the organ scarcity for transplants and offer people in need of options that are specifically tailored to their needs.

An essential component in developing biocompatible, durable, and functional medical devices are artificial organs and organoids. These developments have great promise for personalized therapy, drug testing, and as workable substitutes for organ donation. Now withstanding the noteworthy progress, the field still faces formidable challenges. One of the primary challenges is to replicate the intricate microarchitecture and wide range of functions of local tissues. The challenges of maintaining long-term biocompatibility and lowering the likelihood of immune system rejection continue to be very important [103].

3.6.6.1 Artificial Oxygen Carriers

Preclinical investigations conducted by Ijad Kao *et al.* investigated the ability of the oxygen carrier based on hemoglobin HbMP-700 to cause vasoconstriction, nitric oxide (NO) scavenging and oxidative tissue damage. Because of the high oxygen affinity this product offered, tiny blood vessel vasoconstriction *in vitro* was avoided along with premature oxygen overstock. These units are 100–1,000 nm in size to minimize phagocytosis of big particles, NO-scavenging, and penetration through the endothelial gaps in the blood channel [104].

3.6.6.2 Cardiac Support and Blood Pumps

Following their departure from inpatient cardiac rehabilitation (CR), Thomas Schmidt *et al.* looked back at the functional performance and exercise ability of 68 heart failure patients receiving assistance from a left ventricular assist device. The findings demonstrate that cardiopulmonary exercise ability is still significantly limited even after left ventricular assist device (LVAD) implantation. On the other hand, functional performance, as determined by the walk distance of six minutes, is within acceptable bounds. Patients appear to be realistically capable of completing light daily duties, which opens the door to their exit from inpatient rehabilitation.

A comparison study of six of the utmost advanced physiological managers for blood pumps that rotate was conducted *in vitro* by Anastasios Petrou *et al.* A natural circulation and a LVAD running at constant speed were used to compare the controller's performances. All controllers produced a shift in pump flow during preload fluctuations that was similar to the physiological circulation's cardiac output response. The controllers' responses varied in response to afterload fluctuations; some showed a high susceptibility to drift in the sensor or contractility, which resulted in suction and overload of the left ventricle. Preload-based physiological controllers outperformed the CS operation, demonstrating their clinical significance and offering numerous advantages for LVAD therapy [105, 106].

3.6.6.3 Membrane Oxygenation and Cardiopulmonary Support

Retrospectively, Lin Lyu *et al.* evaluated the occurrence, contributing variables, and consequences of hyperbilirubinemia in 89 adult patients with heart problems undergoing veno-arterial extracorporeal membrane oxygenation. A total of 73% of people had hyperbilirubinemia (>3 mg/dL). In affected role with extremely high bilirubin, there was a low in-hospital survival rate and low platelets during ECMO. Patients on VA ECMO continue to frequently have hyperbilirubinemia, which is linked to low platelets and increased in-hospital impermanence. Risk factors for hyperbilirubinemia include basic high bilirubin levels, hemolysis, and liver failure during ECMO.

Elizabeth Profeta *et al.* evaluated a different a circuit for continuous renal replacement treatment (CRRT) and newborn extracorporeal life support (ECLS) in a pediatric patient simulation. Any circuit with a combination of ECLS and CRRT performed hemodynamically no differently from the circuit without CRRT when a centrifugal pump was employed. For some CRRT settings, The performance parameters of a circuit with a roller pump are changed when CRRT is added. Recommendations are provided for neonatal combination CRRT circuits with both blood pumps and extracorporeal membrane oxygenation, taking hemodynamic performance into account.

In order to replicate human artery waveforms in a typical oxygenation circuit, Gengo Sunagawa *et al.* studied the VentriFlo True Pulse Pump in piglets. With an arterial pulsation pressure of 24.6 ± 5.7mm Hg, the VentriFlo showed normal arterial waveforms. During a prolonged cardiac bypass, this pilot study showed the VentriFlo device that maintained appropriate perfusion of all organs and created pulsatile flow [107–109].

3.6.6.4 *Functional Electrical Stimulation and Neuromuscular Support*

A total of 11 Sprague Dawley rats' somatosensory cortex electrocorticograms and intracortical signals were used by Cristian Sevcencu *et al.* to study a model of prolonged seizures. Pentylenetetrazol (PTZ) was infused intravenously under strict supervision as part of a technique designed to cause spike-and-wave (SW) seizures and keep them steady during the whole experiment. All rats had SW discharges that were initiated and were steady for 2 h. The required quantity of animals to optimize treatments using electrical stimulation designed to prevent seizures may be significantly reduced by taking advantage of this 2-h time period, which allows for repeated experiments with varying stimulation parameters in each animal [110].

3.6.6.5 *Pancreatic Support*

Iwata Hiroo *et al.* provided an update on their ongoing investigation of a bioartificial pancreas (BAP). To offer insights into logical layout of three different BAP kinds, they derived a number of equations taking into account the oxygen supply in BAPs (cylindrical capsules, spherical microcapsules, and planar capsules). The formulas provide (i) the capsule membrane thickness that, depending on the initial magnitude of the encapsulated cell amassed, precludes central necrosis; (ii) the effect of encasing a group of cells on the release of insulin; (iii) oxygen concentration patterns in BAPs; (iv) number of enclosed cells required to get a patient's blood sugar levels back to normal; and (v) the combined sizes and volumes of BAPs [111].

3.6.6.6 *Pulmonary Support*

Current study on a oxygenator design with parallel plates centered around proof-of-concept studies with silicon micropores shows adequate oxygen transmission which establishes the tools and knowledge required to mature silicon micropore membrane oxygenators over new generations [112].

3.6.6.7 *Renal Support and Dialysis*

A relationship between colloid osmotic pressure (COP) and hydrostatic transmembrane pressure (TMPh) was investigated in high-flux (HF) and low-flux (LF) dialyzers. Dialyzers are used to measure blood flow without ultrafiltration can be connected to TMPh through a relationship that can be characterized by the slope associated with internal filtration and the intercept associated with observed COP. This relationship might be helpful in *in vivo* conditions to calculate COP and internal filtration [113].

3.6.6.8 Vascular and Valve Support

Florian Uhle *et al.* examined inflammation and its effects following transcatheter (TAVR) versus surgical (SAVR) aortic valve repair in 38 patients in a prospective, single-center, nonrandomized trial. Following either SAVR or transfemoral (TA)-TAVR, patients have a strong pro-inflammatory response that is only counterbalanced in SAVR patients on the cytokine level. Their findings suggest that TA-TAVR has a stronger effect on neurocognitive performance and that some patients may have a harmful platelet activation following SAVR [114].

3.6.6.9 Visual Support

In order to help determine the optimal instantaneous processing of images technique for restricted sum of electrodes and to provide the utmost valuable visual evidence for prosthetic implants is reported [115].

3.6.7 Use in Research

Every facet of human biology still has a great deal to be learned and discovered. There has been a persistent lack of resources needed to either make the initial discoveries or utilize the knowledge obtained from discoveries to understand complex systems. In basic and clinical biology, stem cells may be the key to solving some of the most challenging problems [115, 116].

Fresh perspective on the biology of human development: Practical and ethical constraints notably affect the study of human developmental biology. It could help scientists understand more concerning the early stages of human cells perpetrate to the main ancestries that comprise the body, how such lineages establish the fundamental structures of the tissues and organs in the body, and how the cells within these fundamentals differentiate to generate multiple functional cell types that support normal function in adults by using human ES cells. The information acquired will affect numerous fields. For example, the study of cancer ecology will gain a great deal from this, since it is already established that defects in normal embryonic processes often serve as the source of cancers. The availability of human embryonic stem cells would also enable the understanding of congenital defects and, in turn, open the door to possible prevention [115, 116].

Models of human illness limited by the available animal and cell-based models: *In vitro* model availability is a major barrier to research into many human diseases. Many harmful viruses, such as the hepatitis C virus and the HIV-related human immunodeficiency, can only proliferate in chimpanzee or human cells. Research on these and other viral ailments may be substantially accelerated by the cell and tissue types that ES cells may supply. The process of neurodegenerative disorders like Alzheimer's disease is only partially reflected in the animal models that are now available [116].

Transplantation: Without the need for harmful immunosuppression or consideration for tissue compatibility, pluripotent stem cells might be useful to generate an infinite number of tissues, organs, or even cells that might be utilized to return functionality. These cells would effectively be appropriate for "universal" donation when they were utilized in transplantation treatments. The risky, costly, and challenging process of bone marrow transplantation may one day be safe, affordable, and accessible for the management of numerous clinical ailments, such as aplastic anemia and some hereditary blood disorders. This would be crucial for those whose exposure to harmful substances, such as radiation or toxic compounds, resulted in the loss of bone marrow function. It is predictable that more tissues, such as the heart, skin, nervous system components, and other key organs, will grow and be transplanted in case of illness or accident [114–116].

Gene therapy: When using gene therapy, genetic material is placed inside an organ to have a medicinal impact, either by replacing a absent or essential protein or by causing a biochemical activity that is clinically important. For gene-based therapies to develop slowly and have a therapeutic impact, the target gene must be transferred into organ stem cells. (more especially, those involving DNA sequences). There have been significant advancements in the delivery of therapeutic DNA since the initial gene therapy procedure about a decade ago, but actual breakthroughs have not yet been achieved. In addition to issues with transportation, one significant barrier to the effective use of gene therapy is absence of expression or inadequate expression, which can be addressed by putting genetic material into stem cells [116, 117].

References

1. Zakrzewski, W., Dobrzyński, M., Szymonowicz, M., Rybak, Z., Stem cells: past, present, and future. *Stem Cell Res. Ther.*, 10, 1, 1–22, 2019 Dec.

2. Kumar, S. and Singh, N.P., Stem Cells: A new paradigm. *Indian J. Hum. Genet.*, 12, 1, 4–10, 2006.
3. Pittenger, M.F., Discher, D.E., Péault, B.M., Phinney, D.G., Hare, J.M., Caplan, A.I., Mesenchymal stem cell perspective: cell biology to clinical progress. *npj Regener. Med.*, 4, 1, 22, 2019 Dec 2.
4. Wang, C., Fei, Y., Xu, C., Zhao, Y., Pan, Y., Bone marrow mesenchymal stem cells ameliorate neurological deficits and blood-brain barrier dysfunction after intracerebral hemorrhage in spontaneously hypertensive rats. *Int. J. Clin. Exp. Path.*, 8, 5, 4715, 2015.
5. Chen, J., Tang, Y.X., Liu, Y.M., Chen, J., Hu, X.Q., Liu, N., Wang, S.X., Zhang, Y., Zeng, W.G., Ni, H.J., Zhao, B., Transplantation of adipose-derived stem cells is associated with neural differentiation and functional improvement in a rat model of intracerebral hemorrhage. *CNS Neurosci. Ther.*, 18, 10, 847–54, 2012 Oct.
6. Martello, G. and Smith, A., The nature of embryonic stem cells. *Annu. Rev. Cell Dev. Biol.*, 30, 1, 647–75, 2014 Oct 6.
7. Sobrino, T., Arias, S., Pérez-Mato, M., Agulla, J., Brea, D., Rodríguez-Yáñez, M., Castillo, J., Cd34+ progenitor cells likely are involved in the good functional recovery after intracerebral hemorrhage in humans. *J. Neurosci. Res.*, 89, 7, 979–85, 2011 Jul.
8. England, T.J., Abaei, M., Auer, D.P., Lowe, J., Jones, D.R., Sare, G., Walker, M., Bath, P.M., Granulocyte-colony stimulating factor for mobilizing bone marrow stem cells in subacute stroke: the stem cell trial of recovery enhancement after stroke 2 randomized controlled trial. *Stroke.*, 43, 2, 405–11, 2012 Feb.
9. Tang, T., Li, X.Q., Wu, H., Luo, J.K., Zhang, H.X., Luo, T.L., Activation of endogenous neural stem cells in experimental intracerebral hemorrhagic rat brains. *Chin. Med. J.*, 117, 09, 1342–7, 2004 Sep 1.
10. Liu, H., Jiang, C., Peng, J., Hu, X., Xia, Y., Transplantation of Neural Stem Cells-Overexpressed Ku70 Improves Neurological Deficits in a Mice Model of Cerebral Ischemia Stroke. *Neurochem. Res.*, 49, 3, 718–31, 2024 Mar.
11. Qin, J., Ma, X., Qi, H., Song, B., Wang, Y., Wen, X., Wang, Q.M., Sun, S., Li, Y., Zhang, R., Liu, X., Transplantation of induced pluripotent stem cells alleviates cerebral inflammation and neural damage in hemorrhagic stroke. *PLoS One*, 10, 6, e0129881, 2015 Jun 18.
12. Cherkashova, E., Namestnikova, D., Leonov, G., Gubskiy, I., Sukhinich, K., Melnikov, P., Chekhonin, V., Yarygin, K., Goldshtein, D., Salikhova, D., Comparative study of the efficacy of intra-arterial and intravenous transplantation of human induced pluripotent stem cells-derived neural progenitor cells in experimental stroke. *PeerJ.*, 11, e16358, 2023 Nov 9.
13. Behdarvandy, M., Karimian, M., Atlasi, M.A., Azami Tameh, A., Heat shock protein 27 as a neuroprotective biomarker and a suitable target for stem cell therapy and pharmacotherapy in ischemic stroke. *Cell Biol. Int.*, 44, 2, 356–67, 2020 Feb.

14. Chrostek, M.R., Fellows, E.G., Crane, A.T., Grande, A.W., Low, W.C., Efficacy of stem cell-based therapies for stroke. *Brain Res.*, 1722, 146362, 2019 Nov 1.
15. Daley, G.Q., Stem cells and the evolving notion of cellular identity. *Philos. Trans. R. Soc. B Biol. Sciences.*, 370, 1680, 20140376, 2015 Oct 19.
16. Lodi, D., Iannitti, T., Palmieri, B., Stem cells in clinical practice: applications and warnings. *J. Exp. Clin. Cancer Res.*, 30, 1–20, 2011 Dec.
17. Abu-Dawud, R., Graffmann, N., Ferber, S., Wruck, W., Adjaye, J., Pluripotent stem cells: induction and self-renewal. *Philos. Trans. R. Soc. B Biol. Sciences.*, 373, 1750, 20170213, 2018 Jul 5.
18. Priddle, H., Jones, D.R.E., Burridge, P.W., Patient, R., Hematopoiesis from human embryonic stem cells: overcoming the immune barrier in stem cell therapies. *Stem Cells*, 24, 4, 815–824, 2006.
19. Monga, I., Kaur, K., Dhanda, S.K., Revisiting hematopoiesis: applications of the bulk and single-cell transcriptomics dissecting transcriptional heterogeneity in hematopoietic stem cells. *Briefings Funct. Genomics*, 21, 3, 159–76, 2022 May.
20. Birbrair, A. and Frenette, P.S., Niche heterogeneity in the bone marrow. *Ann. N. Y. Acad. Sci.*, 1370, 1, 82–96, 2016 Apr.
21. Dzierzak, E. and Bigas, A., Blood development: hematopoietic stem cell dependence and independence. *Cell Stem Cell*, 22, 5, 639–51, 2018 May 3.
22. Ottersbach, K., Endothelial-to-haematopoietic transition: an update on the process of making blood. *Biochem. Soc. Trans.*, 47, 2, 591–601, 2019 Apr 30.
23. Till, J.E. and McCulloch, E.A., A direct measurement of the radiation sensitivity of normal mouse bone marrow cells. *Radiat. Res.*, 14, 2, 213–22, 1961 Feb 1.
24. Eaves, C.J., Hematopoietic stem cells: concepts, definitions, and the new reality. *Blood J. Am. Soc. Hematology.*, 125, 17, 2605–13, 2015 Apr 23.
25. Müller, A.M., Huppertz, S., Henschler, R., Hematopoietic stem cells in regenerative medicine: astray or on the path? *Transfus. Med. Hemotherapy.*, 43, 4, 247–54, 2016 Jul 29.
26. Ullah, I., Subbarao, R.B., Rho, G.J., Human mesenchymal stem cells-current trends and future prospective. *Biosci. Rep.*, 35, 2, e00191, 2015 Apr 28.
27. Ding, D.C., Shyu, W.C., Lin, S.Z., Mesenchymal stem cells. *Cell Transplant.*, 20, 1, 5–14, 2011 Feb.
28. Henig, I. and Zuckerman, T., Hematopoietic stem cell transplantation - 50 years of evolution and future perspectives. *Rambam Maimonides Med. J.*, 2014 Oct 29, 5, 4, e0028.
29. Merrick, T.L., *Program Evaluation of Post-Transplant Care Program Options: Inpatient Versus Short Stay Unit Outpatient and Advanced Care at Home Approaches*, PhD diss., Jacksonville University, 2023.
30. Atkins, H., Stem cell transplantation to treat multiple sclerosis. *Jama.*, 321, 2, 153–5, 2019 Jan 15.
31. Oliveira, M.S. and Barreto-Filho, J.B., Placental-derived stem cells: Culture, differentiation and challenges. *World J. Stem Cells*, 7, 4, 769, 2015 May 5.

32. Trounson, A. and McDonald, C., Stem cell therapies in clinical trials: progress and challenges. *Cell Stem Cell*, 17, 1, 11–22, 2015 Jul 2.
33. Rama, P., Matuska, S., Paganoni, G., Spinelli, A., De Luca, M., Pellegrini, G., Limbal stem-cell therapy and long-term corneal regeneration. *N. Engl. J. Med.*, 363, 2, 147–55, 2010 Jul 8.
34. Gupta, D.R. and Singh, S., Stem cells: Current applications and future prospects. *Indian J. Med. Sciences.*, 1, 1–5, 2024 Jan.
35. Brassard, J.A. and Lutolf, M.P., Engineering stem cell self-organization to build better organoids. *Cell Stem Cell*, 24, 6, 860–76, 2019 Jun 6.
36. Henckes, N.A., Faleiro, D., Chuang, L.C., Cirne-Lima, E.O., Scaffold strategies combined with mesenchymal stem cells in vaginal construction: a review. *Cell Regener.*, 10, 1–1, 2021 Dec.
37. Lv, M., Zhang, S., Jiang, B., Cao, S., Dong, Y., Cao, L., Guo, S., Adipose-derived stem cells regulate metabolic homeostasis and delay aging by promoting mitophagy. *FASEB J.*, 35, 7, e21709, 2021 Jul.
38. Zhao, L., Hu, C., Zhang, P., Jiang, H., Chen, J., Mesenchymal stem cell therapy targeting mitochondrial dysfunction in acute kidney injury. *J. Transl. Med.*, 17, 1–9, 2019 Dec.
39. Päth, G., Perakakis, N., Mantzoros, C.S., Seufert, J., Stem cells in the treatment of diabetes mellitus—Focus on mesenchymal stem cells. *Metabolism*, 90, 1–5, 2019 Jan 1.
40. Mihardja, L., Delima, D., Massie, R.G., Karyana, M., Nugroho, P., Yunir, E., Prevalence of kidney dysfunction in diabetes mellitus and associated risk factors among productive age Indonesian. *J. Diabetes Metab. Disord.*, 17, 53–61, 2018 Jun.
41. Hunt, L.M., Arndt, E.A., Bell, H.S., Howard, H.A., Are corporations redefining illness and health? The diabetes epidemic, goal numbers, and blockbuster drugs. *J. Bioethical Inq.*, 18, 477–97, 2021 Sep.
42. Lopes, L., Setia, O., Aurshina, A., Liu, S., Hu, H., Isaji, T., Liu, H., Wang, T., Ono, S., Guo, X., Yatsula, B., Stem cell therapy for diabetic foot ulcers: a review of preclinical and clinical research. *Stem Cell Res. Ther.*, 9, 1–6, 2018 Dec.
43. Singh, V.K., Kalsan, M., Kumar, N., Saini, A., Chandra, R., Induced pluripotent stem cells: applications in regenerative medicine, disease modeling, and drug discovery. *Front. Cell Dev. Biol.*, 3, 2, 2015 Feb 2.
44. Nasiry, D. and Khalatbary, A.R., Stem cell-derived extracellular vesicle-based therapy for nerve injury: A review of the molecular mechanisms. *World Neurosurgery X.*, 19, 100201, 2023 Jul 1.
45. Ayavoo, T., Murugesan, K., Gnanasekaran, A., Roles and mechanisms of stem cell in wound healing. *Stem Cell Invest.*, 8, 1–9, 2021.
46. Tian, C.M., Zhang, Y., Yang, M.F., Xu, H.M., Zhu, M.Z., Yao, J., Wang, L.S., Liang, Y.J., Li, D.F., Stem cell therapy in inflammatory bowel disease: a review of achievements and challenges. *J. Inflammation Res.*, 31, 2089–119, 2023 Dec.

47. Radisic, M., Iyer, R.K., Murthy, S.K., Micro-and nanotechnology in cell separation. *Int. J. Nanomed.*, 1, 1, 3–14, 2006 Dec 1.
48. Brakke, M.K., Density gradient centrifugation: a new separation technique1. *J. Am. Chem. Soc.*, 73, 4, 1847–8, 1951 Apr.
49. Gothard, D., Tare, R.S., Mitchell, P.D., Dawson, J.I., Oreffo, R.O., In search of the skeletal stem cell: isolation and separation strategies at the macro/micro scale for skeletal regeneration. *Lab Chip*, 11, 7, 1206–20, 2011.
50. Pappas, D. and Wang, K., Cellular separations: a review of new challenges in analytical chemistry. *Anal. Chim. Acta*, 601, 1, 26–35, 2007 Oct 3.
51. Roda, B., Lanzoni, G., Alviano, F., Zattoni, A., Costa, R., Di Carlo, A., Marchionni, C., Franchina, M., Ricci, F., Tazzari, P.L., Pagliaro, P., A novel stem cell tag-less sorting method. *Stem Cell Rev. Rep.*, 5, 420–7, 2009 Dec.
52. Burt, R.K., Loh, Y., Pearce, W., Beohar, N., Barr, W.G., Craig, R., Wen, Y., Rapp, J.A., Kessler. J., Clinical applications of blood-derived and marrow-derived stem cells for nonmalignant diseases. *JAMA*, 299, 8, 925–36, 2008 Feb 27.
53. Pethig, R., Dielectrophoresis: Status of the theory, technology, and applications. *Biomicrofluidics*, 4, 2, 1–5, 2010 Jun 1.
54. Stephens, M., Talary, M.S., Pethig, R., Burnett, A.K., Mills, K.I., The dielectrophoresis enrichment of CD34+ cells from peripheral blood stem cell harvests. *Bone Marrow Transplant.*, 18, 4, 777–82, 1996 Oct 1.
55. Thiel, A., Scheffold, A., Radbruch, A., Immunomagnetic cell sorting—pushing the limits. *Immunotechnology.*, 4, 2, 89–96, 1998 Oct 1.
56. Putnam, D.D., Namasivayam, V., Burns, M.A., Cell affinity separations using magnetically stabilized fluidized beds: Erythrocyte subpopulation fractionation utilizing a lectin-magnetite support. *Biotechnol. Bioeng.*, 81, 6, 650–65, 2003 Mar 20.
57. Bauer, J., Advances in cell separation: recent developments in counterflow centrifugal elutriation and continuous flow cell separation. *J. Chromatogr. B Biomed. Sci. Appl.*, 722, 1-2, 55–69, 1999 Feb 5.
58. Locke, M., Windsor, J., Dunbar, P.R., Human adipose-derived stem cells: isolation, characterization and applications in surgery. *ANZ J. Surgery.*, 79, 4, 235–44, 2009 Apr.
59. Zhang, Y., Fan, Y., Wang, Z., Wan, Y., Zhou, Z., Zhong, B., Wang, L., Wang, F., Isolation, characterization, and gene modification of dairy goat mesenchymal stem cells from bone marrow. *In Vitro Cell. Dev. Biology-Animal*, 48, 418–25, 2012 Aug.
60. Conget, P.A. and Minguell, J.J., Phenotypical and functional properties of human bone marrow mesenchymal progenitor cells. *J. Cell. Physiol.*, 181, 1, 67–73, 1999 Oct.
61. Kamihira, M. and Kumar, A., Development of separation technique for stem cells. *Cell Separation Fundamentals Anal. Prep. Methods*, 106, 173–93, 2007.
62. Guo, K.T., SchÄfer, R., Paul, A., Gerber, A., Ziemer, G., Wendel, H.P., A new technique for the isolation and surface immobilization of mesenchymal stem

cells from whole bone marrow using high-specific DNA aptamers. *Stem Cells*, 24, 10, 2220–31, 2006 Oct 1.

63. Didar, T.F. and Tabrizian, M., Adhesion based detection, sorting and enrichment of cells in microfluidic Lab-on-Chip devices. *Lab Chip*, 10, 22, 3043–53, 2010.
64. Srisa-Art, M., Bonzani, I.C., Williams, A., Stevens, M.M., Demello, A.J., Edel, J.B., Identification of rare progenitor cells from human periosteal tissue using droplet microfluidics. *Analyst.*, 134, 11, 2239–45, 2009.
65. Valenti, M.T., Serena, M., Dalle Carbonare, L., Zipeto, D., CRISPR/Cas system: An emerging technology in stem cell research. *World J. Stem Cells*, 11, 11, 937, 2019 Nov 11.
66. Matano, M., Date, S., Shimokawa, M., Takano, A., Fujii, M., Ohta, Y., *et al.*, Modeling colorectal cancer using CRISPR-Cas9–mediated engineering of human intestinal organoids. *Nat. Med.*, *21*, 3, 256–262, 2015.
67. Fessler, E., Drost, J., van Hooff, S.R., Linnekamp, J.F., Wang, X., Jansen, M., De Sousa E Melo, F., Prasetyanti, P.R., IJspeert, J.E., Franitza, M., Nürnberg, P., TGFβ signaling directs serrated adenomas to the mesenchymal colorectal cancer subtype. *EMBO Mol. Med.*, 8, 7, 745–60, 2016 Jul.
68. Chapman, A., Frankel, M., Garfinkel, M., Stem cell research and applications: monitoring the frontiers of biomedical research, in: *American Association for the Advancement of Science*, 1999.
69. Schierle, G.S., Hansson, O., Leist, M., Nicotera, P., Widner, H., Brundin, P., Caspase inhibition reduces apoptosis and increases survival of nigral transplants. *Nat. Med.*, 5, 1, 97–100, 1999 Jan.
70. Mahla, R.S., Stem cells applications in regenerative medicine and disease therapeutics. *Int. J. Cell Biol.*, 2016, 1, 6940283, 2016.
71. Mason, C. and Dunnill, P., A brief definition of regenerative medicine. *Regen. Med.*, 3, 1, 1–5, 2008 Jan 1
72. Fortier, L.A., Stem cells: classifications, controversies, and clinical applications. *Vet. Surgery.*, 34, 5, 415–23, 2005 Sep.
73. Petersdorf, E.W., Malkki, M., Gooley, T.A., Martin, P.J., Guo, Z., MHC haplotype matching for unrelated hematopoietic cell transplantation. *PLoS Med.*, 4, 1, e8, 2007 Jan.
74. Suman, S., Domingues, A., Ratajczak, J., Ratajczak, M.Z., Potential clinical applications of stem cells in regenerative medicine, in: *Stem Cells. Advances in Experimental Medicine and Biology*, Ratajczak, M. (eds), 1201, 1–22, 2019.
75. Jin, Y., Li, S., Yu, Q., Chen, T., Liu, D., Application of stem cells in regeneration medicine. *MedComm.*, 4, 4, e291, 2023 Aug.
76. JA, T., Embryonic stem cell lines, derived from human blastocysts. *Science*, 282, 1145–7, 1998.
77. Hogan, M.S., Parfitt, D.E., Zepeda-Mendoza, C.J., Shen, M.M., Spector, D.L., Transient pairing of homologous Oct4 alleles accompanies the onset of embryonic stem cell differentiation. *Cell Stem Cell*, 16, 3, 275–88, 2015 Mar 5.

78. Greggio, C., De Franceschi, F., Figueiredo-Larsen, M., Gobaa, S., Ranga, A., Semb, H., Lutolf, M., Grapin-Botton, A., Artificial three-dimensional niches deconstruct pancreas development *in vitro*. *Development.*, 140, 21, 4452–62, 2013 Nov 1.
79. Ye, L., Robertson, M.A., Mastracci, T.L., Anderson, R.M., An insulin signaling feedback loop regulates pancreas progenitor cell differentiation during islet development and regeneration. *Dev. Biol.*, 409, 2, 354–69, 2016 Jan 15.
80. Dominici, M.L., Le Blanc, K., Mueller, I., Slaper-Cortenbach, I., Marini, F.C., Krause, D.S., Deans, R.J., Keating, A., Prockop, D.J., Horwitz, E.M., Minimal criteria for defining multipotent mesenchymal stromal cells. The International Society for Cellular Therapy position statement. *Cytotherapy.*, 8, 4, 315–7, 2006 Jan 1.
81. Sharma, A.K., Bury, M.I., Marks, A.J., Fuller, N.J., Meisner, J.W., Tapaskar, N., Halliday, L.C., Matoka, D.J., Cheng, E.Y., A nonhuman primate model for urinary bladder regeneration using autologous sources of bone marrow-derived mesenchymal stem cells. *Stem Cells*, 29, 2, 241–50, 2011 Feb 1.
82. Rentas, S., Holzapfel, N.T., Belew, M.S., Pratt, G.A., Voisin, V., Wilhelm, B.T., Bader, G.D., Yeo, G.W., Hope, K.J., Musashi-2 attenuates AHR signalling to expand human haematopoietic stem cells. *Nature.*, 532, 7600, 508–11, 2016 Apr 28.
83. Rosemann, A., Why regenerative stem cell medicine progresses slower than expected. *J. Cell. Biochem.*, 115, 12, 2073–6, 2014 Oct 15.
84. Travlos, G.S., Normal structure, function, and histology of the bone marrow. *Toxicol. Pathol.*, 34, 5, 548–65, 2006 Aug.
85. Gschweng, E., De Oliveira, S., Kohn, D.B., Hematopoietic stem cells for cancer immunotherapy. *Immunol. Rev.*, 257, 1, 237–49, 2014 Jan.
86. Gaballa, S., Palmisiano, N., Alpdogan, O., Carabasi, M., Filicko-O'Hara, J., Kasner, M., Kraft, W.K., Leiby, B., Martinez-Outschoorn, U., O'Hara, W., Pro, B., A two-step haploidentical versus a two-step matched related allogeneic myeloablative peripheral blood stem cell transplantation. *Biol. Blood Marrow Transplant.*, 22, 1, 141–8, 2016 Jan 1.
87. Kim, K.M. and Evans, G.R., Tissue engineering: the future of stem cells. *Top Tissue Eng.*, 2, 1–21, 2005.
88. Hong, J.K. and Kwon, S.M., Application of tissue engineering in stem cell therapy. *J. Biomed. Sci. Eng.*, 7, 2, 67–74, 2014 Jan 29.
89. Hosseinkhani, H., Hosseinkhani, M., Hattori, S., Matsuoka, R., Kawaguchi, N., Micro and nano-scale *in vitro* 3D culture system for cardiac stem cells. *J. Biomed. Mater. Res. Part A An Off. J. Soc. Biomaterials, Jpn. Soc. Biomaterials, Aust. Soc. Biomater. Korean Soc. Biomaterials.*, 94, 1, 1–8, 2010 Jul.
90. Tulloch, N.L., Muskheli, V., Razumova, M.V., Korte, F.S., Regnier, M., Hauch, K.D., Pabon, L., Reinecke, H., Murry, C.E., Growth of engineered human myocardium with mechanical loading and vascular coculture. *Circ. Res.*, 109, 1, 47–59, 2011 Jun 24.

91. Choi, J.S., Choi, Y.C., Kim, J.D., Kim, E.J., Lee, H.Y., Kwon, I.C., Cho, Y.W., Adipose tissue: A valuable resource of biomaterials for soft tissue engineering. *Macromol. Res.*, 22, 932–47, 2014 Sep.
92. Leipzig, N.D. and Shoichet, M.S., The effect of substrate stiffness on adult neural stem cell behavior. *Biomaterials.*, 30, 36, 6867–78, 2009 Dec 1.
93. Li, X., Katsanevakis, E., Liu, X., Zhang, N., Wen, X., Engineering neural stem cell fates with hydrogel design for central nervous system regeneration. *Prog. Polym. Sci.*, 37, 8, 1105–29, 2012 Aug 1.
94. Saha, K., Irwin, E.F., Kozhukh, J., Schaffer, D.V., Healy, K.E., Biomimetic interfacial interpenetrating polymer networks control neural stem cell behavior. *J. Biomed. Mater. Res. Part A*, 81, 1, 240–9, 2007 Apr.
95. Kwon, S.G., Kwon, Y.W., Lee, T.W., Park, G.T., Kim, J.H., Recent advances in stem cell therapeutics and tissue engineering strategies. *Biomater. Res.*, 22, 1–8, 2018 Dec.
96. Rao, R.R. and Stegemann, J.P., Cell-based approaches to the engineering of vascularized bone tissue. *Cytotherapy.*, 15, 11, 1309–22, 2013 Nov 1.
97. Jose, M.V., Thomas, V., Xu, Y., Bellis, S., Nyairo, E., Dean, D., Aligned bioactive multi-component nanofibrous nanocomposite scaffolds for bone tissue engineering. *Macromol. Biosci.*, 10, 4, 433–44, 2010 Apr 8.
98. Maas, M., Guo, P., Keeney, M., Yang, F., Hsu, T.M., Fuller, G.G., Martin, C.R., Zare, R.N., Preparation of mineralized nanofibers: collagen fibrils containing calcium phosphate. *Nanoletters.*, 11, 3, 1383–8, 2011 Mar 9.
99. Ordovás, L., Park, Y., Verfaillie, C.M., Stem cells and liver engineering. *Biotechnol. Adv.*, 31, 7, 1094–107, 2013 Nov 15.
100. Wang, Y., Zhang, Y., Zhang, S., Peng, G., Liu, T., Li, Y., Xiang, D., Wassler, M.J., Shelat, H.S., Geng, Y., Rotating microgravity-bioreactor cultivation enhances the hepatic differentiation of mouse embryonic stem cells on biodegradable polymer scaffolds. *Tissue Eng. Part A*, 18, 21–22, 2376–85, 2012 Nov 1.
101. Chen, H., Huang, J., Yu, J., Liu, S., Gu, P., Electrospun chitosan-graft-poly (ε-caprolactone)/poly (ε-caprolactone) cationic nanofibrous mats as potential scaffolds for skin tissue engineering. *Int. J. Biol. Macromol.*, 48, 1, 13–9, 2011 Jan 1.
102. Rizk, A. and Rabie, A.B., Human dental pulp stem cells expressing transforming growth factor β3 transgene for cartilage-like tissue engineering. *Cytotherapy.*, 15, 6, 712–25, 2013 Jun 1.
103. Jin, G., Prabhakaran, M.P., Kai, D., Ramakrishna, S., Controlled release of multiple epidermal induction factors through core–shell nanofibers for skin regeneration. *Eur. J. Pharm. Biopharm.*, 85, 3, 689–98, 2013 Nov 1.
104. Jung, C., Westenfeld, R., Kapur, N.K., Transaxillary Impella support: Bridging the gap of powerful left ventricular support. *Artif. Organs*, 43, 11, 1053–1055, 2019 Nov 1.
105. Kao, I., Xiong, Y., Steffen, A., Smuda, K., Zhao, L., Georgieva, R., Pruss, A., Bäumler, H., Preclinical *in vitro* safety investigations of submicron sized

hemoglobin based oxygen carrier HbMP-700. *Artif. Organs*, 42, 5, 549–59, 2018 May.
106. Schmidt, T., Bjarnason-Wehrens, B., Bartsch, P., Deniz, E., Schmitto, J., Schulte-Eistrup, S., Willemsen, D., Reiss, N., Exercise capacity and functional performance in heart failure patients supported by a left ventricular assist device at discharge from inpatient rehabilitation. *Artif. Organs*, 42, 1, 22–30, 2018 Jan.
107. Petrou, A., Lee, J., Dual, S., Ochsner, G., Meboldt, M., Schmid Daners, M., Standardized comparison of selected physiological controllers for rotary blood pumps: *in vitro* study. *Artif. Organs*, 42, 3, E29–42, 2018 Mar.
108. Lyu, L., Yao, J., Gao, G., Long, C., Hei, F., Ji, B., Liu, J., Yu, K., Hu, Q., Hu, J., Incidence, risk factors, and outcomes of hyperbilirubinemia in adult cardiac patients supported by veno-arterial ECMO. *Artif. Organs*, 42, 2, 148–54, 2018 Feb.
109. Profeta, E., Shank, K., Wang, S., O'Connor, C., Kunselman, A.R., Woitas, K., Myers, J.L., Ündar, A., Evaluation of hemodynamic performance of a combined ECLS and CRRT circuit in seven positions with a simulated neonatal patient. *Artif. Organs*, 42, 2, 155–65, 2018 Feb.
110. Sunagawa, G., Karimov, J.H., Dessoffy, R., Byram, N., Grady, P., Sinkewich, M., Naber, J.P., Vincent, D., Okano, S., Stewart, R.D., Sale, S.M., New technology mimics physiologic pulsatile flow during cardiopulmonary bypass. *Artif. Organs*, 42, 2, 231–5, 2018 Feb.
111. Sevcencu, C., Jiao, J., Harreby, K.R., Jensen, W., A new rat model of seizures suitable for screening antiepileptic electrical stimulation therapies. *Artif. Organs*, 42, 1, 94–9, 2018 Jan.
112. Iwata, H., Arima, Y., Tsutsui, Y., Design of bioartificial pancreases from the standpoint of oxygen supply. *Artif. Organs*, 42, 8, E168–85, 2018 Aug.
113. Dharia, A., Abada, E., Feinberg, B., Yeager, T., Moses, W., Park, J., Blaha, C., Wright, N., Padilla, B., Roy, S., Silicon micropore-based parallel plate membrane oxygenator. *Artif. Organs*, 42, 2, 166–73, 2018 Feb.
114. Schneditz, D., Sarikakis, G., Kontodima, M., Sauseng, N., The influence of colloid osmotic pressure on hydrostatic pressures in high-and low-flux hemodialyzers. *Artif. Organs*, 42, 5, 525–32, 2018 May.
115. Uhle, F., Castrup, C., Necaev, A.M., Grieshaber, P., Lichtenstern, C., Weigand, M.A., Böning, A., Inflammation and its consequences after surgical versus transcatheter aortic valve replacement. *Artif. Organs*, 42, 2, E1–2, 2018 Feb.
116. Zhao, Y., Geng, X., Li, Q., Jiang, G., Gu, Y., Lv, X., Recognition of a virtual maze scene using simulated prosthetic vision. *Artif. Organs*, 42, 5, 560–8, 2018 May.
117. Singh, M.P., Bhaduaria, R.S., Sharma, C.S., Stem cells: An "Aladdin's lamp": An Overview. *Pharmacologyonline*, 3, 261–273, 2010.

4

Nanobiomaterials at the Nexus: Integrating Stem Cells and Nanotechnology for Precision Therapeutics

Aditya Banerjee and Deepika Arora*

Regenerative Cell Research Institute, Genome Valley, Hyderabad (T.S.), India

Abstract

The healthcare industry has been transformed by nanotechnology and cellular therapy, making precision medicine a reality. In this chapter, the impact of biomaterials such as nanopolymers and nanoparticles on enhancement of treatment efficiency is discussed. Biocompatibility, targeted delivery, and engineered functionalities are among the specific properties that improve drug efficacy. The merger of stem cells with new-edge bio-materials controls the specificity and targeted cellular microenvironments via maintenance of stemness, enhancement of sustainability, tacking, and cells' differentiating potential for their use in regenerative medicine and inflammatory conditions. The extracellular released exosomes can be modified to function as nanocarriers for cell-to-cell communication that have capability to deliver drugs to specific sites within the body. Through nanobiomaterial's modification, CAR-T cell therapy stands to benefit from improved distribution and activation. This revolution in personalized medicine is poised to result from the convergence of nanotechnology with stem cells, exosomes as well as CAR-T cell therapy. With better accuracy, efficiency, and safety due to individualizing treatments based on genetic and biological makeup of patients, a new era of therapeutic interventions may emerge.

Keywords: Nanobiomaterials, stem cell therapy, exosomes, CAR-T cell therapy, precision therapeutics

**Corresponding author*: deepikaarora.23@rediffmail.com

Ayush Madan, Sumel Ashique, Deepika Arora and Mantosh Kumar Satapathy (eds.) Stem Cell Therapeutics, (159–226) © 2025 Scrivener Publishing LLC

4.1 Introduction

4.1.1 The Rise of Precision Therapy

Precision therapy or precision medicine is an innovative healthcare method that individualizes the treatment of patients based on their personal characteristics [1]. In contrast to traditional "one-size-fits-all" interventions, precision therapy considers genetic, environmental and lifestyle factors that influence the course of disease and response to medications [2]. Precision therapy has evolved following advances in genomics, bioinformatics, and biotechnology offering a more focused approach to patient management [3].

The concept of precision therapy has roots in early twentieth century but gained significant ground with completion of Human Genome Project in 2003 [4]. The project provided a global map of genes in humans hence enabling understanding of genetic basis for most diseases by scientists. Consequently, genomics became more developed which facilitated the identification of genetic mutations as well as biomarkers associated with different ailments [5]. These findings carved the way for targeted therapies especially in oncology whereby treatment could be personalized to attack only cancer cells with specific genetic changes [6].

Precision therapy, as a medical model, entails individualized medicine where decisions, treatments, practices, and products that a patient gets are tailored to that person. This means requires categorizing individuals into subclasses having different susceptibilities to certain diseases, biology or prognosis of these diseases, or responses to specific therapies [7]. Precision therapy attempts to integrate genetic, environmental, and lifestyle information in order to accurately predict which treatment and prevention strategies for a given disease will be effective in certain groups of people [2]. With the present direction taken by precision therapy, it is focused on integrating multi-omics approaches such as genomics, proteomics, metabolomics, and transcriptomics in an attempt to obtain a molecular-level understanding of disease mechanisms [8]. This holistic perspective allows the identification of new biomarkers and therapeutic targets, thereby accelerating personalized treatment strategies. Moreover the use of artificial intelligence (AI) together with machine learning advances the capability to analyze complex datasets to predict disease outcomes and suggest individualized treatment options [9–11].

In oncology, targeted therapies and immunotherapies involve monoclonal antibodies and CAR-T-cell therapy among others, which have had astonishing success in the treatment of some cancers [12]. Consequently, monoclonal antibodies like trastuzumab (Herceptin) are used to treat HER2-positive breast cancer, while pembrolizumab (Keytruda) is effective in selective genetically profiled malignancies [13, 14]. Additionally, companion diagnostics are increasingly being used as tests that can identify those patients who will benefit most from a particular therapy.

Apart from oncology, there are other medical fields where precision therapy is making headway such as cardiology, neurology and infectious diseases [15]. For instance, personalized medicine prescription based on genetics is currently being practiced using pharmacogenomics which reduces the effects of adverse drug reactions associated with drug administration [16].

The introduction of genomics, bioinformatics, and AI into medicine provides new possibilities for improving patient care via ensuring that various forms of treatment can be matched to individuals, thereby enhancing their outcomes while reducing adverse effects [3, 4]. Further inclusion of multi-omics data integration and novel therapeutic approaches during these transitions will enhance precision and efficacy in medical interventions going forward [16].

4.1.2 Challenges in Conventional Therapy

Standardized treatments, which are often synonymous with the conventional therapies, have been highly significant in controlling various diseases. Nevertheless, these methods have their various limitations in that they may have an effect on efficacy of the overall treatment and patient outcomes. Lack of specificity is among the primary impediments, possible adverse effects, as well as lack of efficacy in certain patients' groups [17].

One-size-fits-all is usually a principle underpinning most conventional therapies. This does not take into account individual genetic variations, biology differences, or how diseases manifest themselves differently within people [17]. For example, when treating cancer by chemotherapy method, the use of cytotoxic drug compounds that hinder rapid cell division is adopted. These drugs effectively kill cancer cells but cause harm to other healthy cells as well, which divide rapidly like those in the bone morrow, gastrointestinal tract, and hair follicles. Hence, this aspect of non-specificity can lead to a major collateral damage, thus reducing the efficiency of therapy and compromising the life quality of patients [18].

Due to their non-specific nature, a wide range of side effects comes with conventional therapies. These kinds of reactions could be mild such as nausea and fatigue or go up to severe ones that range from organ damage to secondary cancers and immune suppression. Limiting their dosage and duration, the toxicity of these treatments can reduce the patients' tolerability to them, leading to inadequate treatment outcomes. Also, this often requires additional medication and intervention in order to manage them, which complicates patient care as well as increases healthcare costs [19].

In certain patients, traditional therapies may offer no relief since success rates among individuals may vary considerably. Drug metabolism, response to interventions and side effect tolerance are all subject to genetic, epigenetic and environmental influences. For instance, a number of cancer patients have specific genetic mutations or resistance mechanisms that make them resistant to standard chemotherapy regimens. Consequently, certain chronic illnesses like diabetes or high blood pressure may still require other forms of therapy beyond or supplementary to regular treatments [20].

4.1.3 Integration of Nanotechnology and Stem Cell Therapy

Nanotechnology's integration with stem cell therapy offers new avenues for surmounting limitations intrinsic to conventional approaches. It brings together nanotechnology's precision delivery advantages and the capacity of stem cells to regenerate, thus offering alternative means of improving therapeutic efficacy, specificity and patient outcomes [21]. This impactful amalgamation gives the great benefits for stem cell therapy via allowing the targeted delivery of therapeutic agents. Nanoparticles may be designed in such a way as to carry drugs, growth factors, or genetic material and directed specifically to disease sites [21]. This approach helps minimize off-target effects and maximize therapeutic impact on the target site. Nanoparticles may be conjugated with ligands that bind specifically to tumor cell markers which allows that encapsulated drug only released in their desired areas. This can reduce systemic toxicity associated with conventional chemotherapy and improve the therapeutic index [22, 23].

Nanotechnology can also improve the viability and function of transplanted stem cells thus enhancing their sustainability. The use of nanomaterials to encapsulate stem cells offers protection from adverse *in vivo* conditions, thereby raising cell survival rates after transplantation. Furthermore, nanoparticles have the potential to transport bioactive substances that encourage stem cell growth, diversification, and assimilation into host tissue. In particular, the controlled release of these molecules can be used for the introduction of growth factors or signaling molecules, which promotes the use of stem cells by tissue engineering grafts for repairing tissues [24]. The incorporation of nanotechnology also enables advanced monitoring and imaging of stem cell therapies. For instance, nano-particles can be engineered as contrast agents for MRI, PET, or fluorescence imaging, hence allowing real-time visualization of stem cell distribution, migration, and differentiation *in vivo* [25]. This capability allows for precise tracking of therapeutic outcomes resulting in timely adjustments to treatment protocols, thus achieving the best efficacy possible and safety at all times.

Overcoming the immunogenicity or rejection process is a significant challenge for stem cell therapy, especially with allogeneic transplants. Nanotechnology can be used to solve these problems by changing the surface properties of stem cells or nanoparticles to avoid the immune system. The immunogenic profile of transplanted cells can be decreased by adding biocompatible polymers on the surfaces or through incorporation of immune modulatory molecules thereby promoting acceptance and reducing the risk of rejection [21, 24]. The versatility of nanotechnology allows for development of multifunctional therapeutic platforms. These can be designed to simultaneously deliver multiple therapeutic agents, combining regenerative potential of stem cells with anti-inflammatory, anti-cancer, or gene editing therapies. This multi-pronged strategy addresses the complexity and multiple factors associated with most diseases, hence offering a comprehensive treatment plan [11, 23].

When implemented together, nanotechnology and stem cell therapy offer the prospect of completely changing the face of treatments since the conventional therapies have proven to have limitations [24]. In terms of precise targeting, improved cell survival, real-time monitoring of the bio-effect, immunomodulation aspect, and construction of the multifunctional theragnostic system, this integrated concept opens a new opportunity for the next generation of medicine with higher efficiency, specificity, and safety. Thus, further development of research in this field is expected to help develop new approaches to the treatment of various diseases, resulting in enhanced patient outcomes as well as the improvement of their quality of life [21].

4.2 Nanobiomaterials: Tailoring the Cellular Microenvironment

4.2.1 Revolutionizing Medical Applications with Nanobiomaterials

There are many different types of nanobiomaterials that have been created at the nanoscale which can be used for a wide variety of medical applications;

these include polymers, nanoparticles, hydrogels, and lipid-based nanomaterials. What sets them apart is their ratio of surface area to volume as well as how they interact with molecules in living systems—this allows doctors to find unique solutions within medicine itself [26].

Polymer-based examples like polylactic acid (PLA), poly(lactic-co-glycolic acid) (PLGA), or even polyethylene glycol (PEG) are often preferred because they are biocompatible, biodegradable, and versatile, too [27]. Particles made from these materials could then be used for drug delivery systems which would slowly release medication into specific parts of the body over time; this is commonly employed when treating cancerous tumors through chemotherapy sessions where healthy tissues need protecting from harmful toxins being released all at once throughout someone's bloodstream. PEGylation refers primarily toward attaching chains consisting only out of PEG units onto therapeutic agents so as to increase their stability, solubility, and half-life in circulation, thereby making them more effective over longer periods; it provides an outer protective layer around drugs or other biological substances [1].

Gold nanoparticles can be used for killing cancerous cells when heated up by light; magnetic resonance imaging (MRI) uses magnets placed around someone's body along with injected contrast agents containing iron oxide particles which then become attracted toward areas where there are higher concentrations of cells being actively dividing due to tumor growths, etc. [28]. Another thing worth mentioning is that these materials are often made in the form of hydrogels, which are basically water-based gels that have been cross-linked so they don't dissolve or fall apart easily. They are commonly used as scaffolds for growing tissues such as skin, bone, and cartilage; this allows doctors to repair damaged organs without having any negative effects on the patient themselves [29].

4.2.2 Tailoring the Nanoworld: Properties Engineering Toward Precision Medicine

Creating nanobiomaterials suitable for particular medical applications requires specific properties to be engineered such as biocompatibility, drug delivery efficiency, or targeting ability. These factors are crucial since they ensure that materials work well within a human body offering therapy where needed while minimizing any negative side effect. In this article, we will look into how various properties are engineered for different medical applications with recent advances being highlighted [30].

4.2.2.1 Biocompatibility

For nanobiomaterials' design, consideration should be made to ensure that they do not cause any harmful immune responses or toxicities when introduced into the body. In other words, these materials have to be non-toxic, non-immunogenic, and able to blend with biological tissues without being rejected by the latter [30]. The biocompatible nature would depend on two criteria, namely:

(a) Material selection: Natural polymers like collagen, gelatin, and alginate can be used due to their inherently high levels of biocompatibility and biodegradability. Additionally, synthetic polymers such as poly(ethylene glycol) (PEG) or poly(lactic-co-glycolic) acid (PLGA) may also find wide application but their chemical reactivity needs careful controls during the synthesis steps [27].

(b) Surface modification: Coating nanoparticles with biocompatible polymers or biomolecules increases their compatibility with host cells, thus enhancing their therapeutic potentialities, too, where necessary instead of only acting like foreign objects inside living organisms. A good example here is using zwitterionic coatings which mimic cell membrane surfaces, thereby reducing protein adsorption—hence immune recognition—thus improving biocompatibility [31].

4.2.2.2 Drug Delivery Efficiency

To enhance efficiency in drug delivery systems, it involves designing nanobiomaterials so they can transport therapeutic agents directly into the target organs or cells, release them under a controlled manner while keeping them stable and effective throughout their journey within the patient's body [26]. These drug delivery systems broadly include the following:

(a) Nanocarriers: Liposomes, polymeric nanoparticles, and dendrimers are designed to encapsulate drugs, thereby protecting against degradation as well as increasing solubility/bioavailability [32].

(b) Controlled release systems: PLGA nanoparticles can be made to degrade at specific rates which allows for a sustained release of drug over long periods of time [33].

(c) Stimuli-responsive systems: Some nanocarriers respond to different stimuli like pH value, temperature changes, or presence certain enzymes so that they only discharge drug when appropriate conditions have been met such as a site inflammation being detected. This approach gives better control over where drugs should act so as to reduce their toxicity levels elsewhere in the body while maximizing local efficacy [34].

4.2.2.3 Targeting Capabilities

The ability to hit diseased cells with drugs without affecting the healthy ones is very important in medicine if high cure rates are be achieved against various ailments. However, this is only possible when the nanobiomaterials used have adequate targeting capabilities.

(a) Active targeting: Nanobiomaterials may be modified by attaching ligands, e.g., antibodies, peptides, or small molecules which will recognize and bind specifically onto certain biomarkers found on abnormal (cancerous) cells but not normal ones. For instance, gold nanoparticles functionalized anti-HER2 antibodies selectively bind HER2 positive breast cancer cell surface receptors [35].

(b) Passive targeting: Here particles take advantage of leaky blood vessels within solid tumors coupled with poor lymphatic drainage system around them, whereby most nanoscale-range objects can enter easily due to their size alone without requiring active uptake mechanisms, engineering, shapes, sizes, etc., would thus optimize passive targeting toward particular types/locations of malignancies [36].

(c) Multifunctional nanocarriers: These have been developed recently; they possess dual functions like active targeting ability in addition to being capable of passively localizing at sites where higher levels or more complex diseases such as cancer, involving multiple cell types, are present [37].

4.2.2.4 Novel Approaches and Intelligent Materials

The design of nanobiomaterials for specific functions in medicine is being improved constantly by emerging technologies and novel approaches.

(a) Biomimetic nanocarriers: These are designed to resemble natural biological systems so as to improve targeting and delivery. As an example, nanoparticles coated with cell membranes take advantage of the inherent ability of immune cells, platelets, or cancer cells to target diseases, thereby escaping recognition by the immune system while homing onto these sick areas [37].

(b) CRISPR-Cas9 delivery: This involves the use of nanoparticles as carriers for CRISPR-Cas9 gene editing systems which allows accurate modifications of genomes used in treating genetic disorders [38].

(c) Smart nanobiomaterials: Such materials can react to external stimuli like light, magnetic fields, and ultrasound among others, thus enabling on-demand drug release, improving imaging capabilities as well as providing therapies with high precision. For instance, magnetic nanoparticles can be steered toward targets by outside magnetic forces before releasing drugs or generating heat for hyperthermia treatment of tumors [34].

Biocompatibility, drug delivery efficiency, and targeting abilities are some aspects considered when engineering nanobiomaterials for specific medical applications. These components should therefore be precise, efficient, and safe in therapy provision while still being designed through new techniques plus advanced equipment. With increased investigations into this area, it becomes clear that nanobiomaterials have the potential to transform healthcare, hence leading to better patient outcomes.

4.2.3 Design Strategies for Stem Cell–Nanobiomaterial Integration

Effective therapeutic outcomes require strategic designs when integrating stem cells with nanobiomaterials. When combined stem cells–nanobiomatrices have been shown to enhance cell viability, targeting specificity at diseased sites, differentiation potency, and overall regenerative potentiality among others. In this section, we will discuss different design principles as well as methods employed during fabrication process according to their relevance such as biocompatible factors, scaffold architecture, controlled release mechanisms, and multifunctional materials.

4.2.3.1 *Biocompatibility and Surface Modification*

Integration of stem cells into nanobiomaterials demands that biocompatibility is taken care of. Modification surface characteristics nanomaterials often leads to increased adhesion, proliferation, and differentiation of these types to a particular kind of tissue.

(a) Surface coatings: Coating surfaces with extracellular matrix proteins like fibronectin collagen laminin or using bioactive peptides mimicking extracellular matrix (ECM) components improves cellular interaction with them.
(b) Functionalization: When growth factors are attached onto substrate molecules which can signalize nearby cells, then it becomes possible for the surrounding environment to support survival functions necessary for differentiated status expression among others.
(c) Advanced coatings: Zwitterionic or hydrophilic coatings reduce immune responses to biomaterials, thus making them more compatible with stem cells while enhancing their interaction properties [39].

4.2.3.2 *Scaffold Design and Tissue Engineering*

For successful integration, regeneration needs scaffolds that support guide such process. Natural ECM ought to be imitated through various forms of nanobiomatrices that provide a conducive milieu for the growth of living organisms, especially those derived from humans.

(a) Hydrogels: These materials should have injectable capability so as to allow easy delivery into hard-to-reach areas within body systems without causing any damage; besides that, they need to exhibit mechanical strength similar to native tissues being regenerated around, while at same time allowing appropriate chemical cues to promote desired cell functions like proliferation, differentiation, etc.
(b) Nanofibrous scaffolds: In order to enhance viability, functionality, electrospinning, high porosity, and large surface area, nano-sized fibers can be used to create appropriate conditions for nutrient transport, waste removal, support, adhesion, differentiation, growth, and migration of stem cells during the tissue formation stage [40].

(c) Nanoscale cues: The physical features of native ECM-like topographies may also play an important role in determining the fate and behavior of multipotent cells by providing the right signals through the incorporation of nanosized patterns on surfaces that mimic these structures.

4.2.3.3 Control of Biologically Active Particle Release

Synthesizing stem cells with nanobiomaterials enables biologically active molecules to be released at will, which is necessary for improving therapeutic outcomes.

(a) Nanocarriers: For example, polymeric nanoparticles, liposomes, and silica nanoparticles can be used to wrap growth factors, cytokines, or any other therapeutic agents. They can be integrated into scaffolds or delivered together with stem cells, hence ensuring extended release time as well as localization of biologically active particles [40].
(b) Stimuli-responsive systems: Nanocarriers that react with certain environmental changes such as pH levels, temperature, or enzymes by releasing their contents give better control over when and where drugs are released, thus increasing cure rates.

4.2.3.4 Advanced Functionalities

Designing stem cell–nanobiomaterial systems with additional functions could greatly improve treatment outcomes.

(a) Magnetic nanoparticles: These can guide stem cells to specific sites using external magnetic fields, thereby improving targeting as well as retention at the injury site. Also, they facilitate real-time imaging, thus enabling the monitoring of stem cell migration and engraftment.
(b) Gene delivery: Genetic materials that promote differentiation among stem cells or enhance regenerative potentials could be delivered through nanocarriers. For instance, nanoparticles carrying CRISPR-Cas9 components have the ability to genetically modify stem cells so that they acquire more powerful therapeutic properties [34].

4.2.3.5 Immunomodulation

The success rate of allogeneic transplantation in stem cell therapy largely depends on how well we modulate immune responses with nanobiomaterials.

(a) Surface modifications: This involves decorating surfaces either with immunomodulatory molecules or coatings designed specifically to escape from immune system detection, thereby minimizing chances of rejection while increasing integration between host tissues and implanted stems cells.
(b) Anti-inflammatory agents: Nanocarrier-mediated delivery of immunosuppressive drugs/anti-inflammatory agents creates a favorable environment for survival and proper functioning of infused stem cells [41].

4.2.3.6 Integration with Smart Technologies

Smart technologies can be used to make stem cell–nanobiomaterial systems responsive, thus gaining more control over therapeutic procedures [42].

(a) Biosensors: Biosensors embedded in scaffolds are capable of monitoring local biochemical conditions which would then help in adjusting real-time release rates for bioactive molecules.
(b) Responsive materials: Physiological conditions that enhance differentiation as well tissue regeneration among stem cells may result from dynamic environments created through material properties changing under influence external stimuli like light, electrical fields, or mechanical stress.

This is a multidimensional approach that combines stem cells with nanobiomaterials based on biocompatibility, scaffold design, controlled release, and advanced functionalities all aimed at improving treatment outcomes. With these design strategies, regenerative medicine will advance rapidly while providing better targeted therapies against numerous diseases and injuries. The integrated systems could revolutionize medical treatments as they promise significant improvements in patient care, thus expanding their potential with ongoing research having already shown great results in various areas of application.

4.3 Stem Cells for Regenerative Medicine

4.3.1 Adult vs. Pluripotent Stem Cells

Regenerative medicine utilizes stem cells which are capable of self-renewal and differentiation into various types of cells. Stem cells have two categories: adult stem cells and pluripotent stem cells, each with unique sources, properties, and therapeutic possibilities. In this comparative analysis, these variations will be considered vis-à-vis medical applications [43].

4.3.1.1 Sources

These somatic or tissue-specific stem cells also known as adult stem cells can be found in multiple parts of the body such as the brain, bone marrow, and adipose tissues among others. Generally, these are either autologous (obtained from the same individual) or allogeneic (donated by someone else) and they are multipotent, meaning they can differentiate only into a few cell lines within their related tissue types, e.g., all blood cell lineages may arise from hematopoietic stem cells residing in the bone marrow while mesenchymal ones from adipose tissue can differentiate to form cartilage, fat, or bone [44].

The other type is pluripotent stem cells that include embryonic stem (ES) cells and induced pluripotent stem (IPS) cells. ES comes from the inner cell mass of blastocysts at early stages of embryogenesis capable of differentiation to any kind of somatic or germ-line derived lineage thus expressing true pluripotency. On the other hand, IPS are reprogrammed by specific transcription factors into a pluripotent state starting from adult somatic counterparts like skin fibroblasts; this was first developed by Shinya Yamanaka and co-workers, allowing the generation of patient-tailored pluripotent stem cells but without the issues accompanying ESCs involving donor eggs or embryos. Furthermore, ethical concerns regarding ESC use coupled with allogenic IPSC therapy have potential immunogenicity [44].

4.3.1.2 Characteristics

Adult stem cells on their part have limited capacity for differentiation and thus mostly function in maintaining their resident tissues. Unlike other cells, they are less likely to grow into tumors which is a big plus in terms of medical treatment. On the other hand, they are usually scarce and not easily isolated or maintained in lab culture conditions. Furthermore, their differentiation ability is limited, thus restricting their usability in tissue repair.

Pluripotent stem cells such as ESCs and iPSCs can differentiate into any type of cell; hence, they provide excellent options with regards to regenerative medicine. They can be cultivated indefinitely, providing an inexhaustible supply for numerous applications. Nevertheless, pluripotent stem cells have a higher risk of developing teratomas (tumors composed of several cell types) that pose great safety concern [44].

4.3.1.3 Therapeutic Potentials

Adult stem cells have shown promise as a tool for regenerative medicine especially in autologous therapies where immune rejection risks are minimalized. Hematopoietic stem cell transplantation is already a well-established treatment method for various blood disorders, e.g., leukemia and lymphoma. Other uses include mesenchymal stem cells which may be used to cure osteoarthritis, myocardial infarction, and neurodegenerative diseases among others, while their immune modulation also makes them good candidates for managing autoimmunity like rheumatoid arthritis as well as inflammation. Good examples here are hematopoietic stem cell transplantation that has been widely used in treating blood-related diseases like leukemia and mesenchymal stem cells which exhibit therapeutic potential against osteoarthrosis (OA), myocardial infarction (MI), neurodegenerative diseases, etc. Their immunomodulatory properties make these MSCs attractive agents for the treatment of autoimmune diseases such as RA or inflammatory conditions that occur in patients with chronic infections or septicemia. Subsequently, it can be concluded that mesenchymal stem cells are suitable for patients suffering from rheumatoid arthritis who may not be responding to conventional therapies [43].

Stem cells that can turn into any type of cell hold great promise for the treatment of many diseases and conditions. For example, there have been preclinical and clinical trials with ESCs targeting macular degeneration, spinal cord injury, and diabetes. One advantage of iPSCs is that they allow for personalized therapy, thereby reducing chances of immune rejection. There are now studies going on to see how iPSCs could be applied in terms of cardiac repair, neurodegenerative disorders, and individualized medication. They also represent valuable tools for disease modeling, drug discovery screening, or understanding the underlying mechanisms behind a multitude of disease processes at the cellular level [43].

4.3.1.4 *Comparative Advantages and Challenges*

On one hand, safety and moral acceptability of adult stem cells are at a better advantage over other stem cells; however, they have limited capacity for differentiation and lack availability. iPSCs, a type of pluripotent stem cell, has the potential to be shaped into any cell type (pluripotency) and also can be used as an ideal patient-specific treatment option; nevertheless, these methods present limitations in terms of safety concerns, ethical considerations, and technical challenges. It will require continued research that seeks to improve efficiency as well as secure protocols for differentiation process among others [45].

4.3.1.5 *Can Umbilical Cord-Derived Mesenchymal Stem Cells Pave a New Wave in Regenerative Medicine?*

Umbilical cord-derived mesenchymal stem cells (UC-MSCs) hold much therapeutic potential because they are uniquely placed in their characteristics and advantages relative to their counterpart MSCs. The non-invasive harvesting allows UC-MSCs derived from Wharton's jelly of the umbilical cord to bypass ethical issues associated with embryonic MSCs or invasive procedures like bone marrow extraction. They have strong immunosuppressive effects which are useful for treating autoimmune diseases such as multiple sclerosis or rheumatoid arthritis among various conditions characterized by inflammation.

As compared to other sources, these types have higher mitotic rate offering a greater supply for treatment purposes alone; but they can differentiate into osteoblasts and chondrocytes, called adipocytes, hence ideal for regenerative medicine. To further elaborate this point, low allogenicity minimizes danger of immune rejection, thus allowing more allogenic therapies, namely, transplants, that may be performed on them. This makes them valuable for treating cardiovascular diseases, neurodegenerative disorders, musculoskeletal injuries, and inflammatory conditions, thereby increasing the chances for successful regenerative therapies in a range of diseases [46].

Each of these types of stem cells presents unique benefits as well as challenges that make them different from the other. The choice between them depends on the specific medical application, with adult stem cells, including umbilical cord derived mesenchymal stem cells, currently leading in certain therapies due to their safety profile, while pluripotent stem cells offer broader treatment options. This field remains promising for overcoming those problems through technological developments [43, 45].

4.3.2 Differentiation and Therapeutic Potential of Stem Cells

Stem cells have an amazing ability to differentiate into definite cell types which is crucial for their function in regenerative medicine. This ability to develop a range of cell types makes them capable of repairing or replacing damaged tissues, thus offering hope for the treatment of many diseases and injuries [43]. This chapter focuses on aspects related to the differentiation of stem cells and its applications in regenerative medicine.

4.3.2.1 Mechanisms of Differentiation

Stem cell differentiation is highly controlled by both intrinsic and extrinsic factors. Intrinsic factors include such things as the genetic makeup and epigenetic characteristics of the cell that determine which type it can transform itself into. Extrinsic factors are the niche or microenvironment where stem cells are found; they consist of growth factors, cytokines, extracellular matrix components, and cell–cell interactions [47].

4.3.2.1.1 Intrinsic Factors

(a) Gene expression: A specific gene is activated or deactivated during differentiation. Transcription factors play a huge role in this process by maintaining gene expression patterns that identify a cell's identity and function properly [48].

(b) Epigenetic modifications: Methylation, acetylation among others, result from chemical changes on DNA or histones which may affect chromatin structures thereby affecting gene expression positively or negatively. These alterations are critical for maintaining pluripotency in stem cells and guiding differentiation [48].

4.3.2.1.2 Extrinsic Factors

(a) Growth factors and cytokines: Soluble molecules like bone morphogenetic protein (BMPs), fibroblast growth factor (FGF), transforming growth factor-beta (TGF-β), etc., enhance or inhibit lineage-specific differentiations, e.g., BMPs contribute toward osteogenic while TGF-β determines chondrogenic differentiations, etc. [48, 49].

(b) Extracellular matrix (ECM): Components present within the ECM interact with membrane receptors on cells, activating signaling pathways governing cell development through providing structural support and biochemical signals. They include collagen, laminin, and fibronectin.

(c) Cell-to-cell interactions: Differentiation is regulated by direct cell contact through gap junctions, cadherins, and integrins between stem cells and their adjacent cells [48].

4.3.2.2 *Therapeutic Potential in Regenerative Medicine [50, 51]*

(a) Cardiovascular diseases: In the treatment of myocardial infarction, for example, cardiac tissue regeneration has been achieved with the use of mesenchymal stem cells (MSCs) and induced pluripotent stem cells (iPSCs). These cells can become cardiomyocytes as well as endothelial and smooth muscle cells, thus enhancing repair processes within the heart and resulting in its improved function [52].

(b) Neurodegenerative disorders: Neural stem cells (NSCs) and iPSCs have potential for treating neurodegenerative diseases like Parkinson's, Alzheimer's etc., as well as spinal cord injuries. Such neural precursors are capable of being transformed into mature neurons, astrocytes, or oligodendrocytes replacing lost or damaged neural constituents, leading to restoration of normal neuronal functions [53].

(c) Musculoskeletal disorders: MSCs are extensively studied for their ability to regenerate bone, cartilage, and muscle tissues due to different musculoskeletal disorders like osteoarthritis or osteoporosis. The differentiation of MSCs into chondrocytes or osteoblast in these conditions thereby helps in skeletal tissue repair/regeneration [54].

(d) Diabetes: Stem cell therapy for diabetes is aimed at inducing the differentiation of stem cells into insulin-producing beta cells. Consequently, these functional beta-cells were successfully derived from both embryonic stem cell (ESC) and iPSC sources offering a promising means for restoring insulin production to diabetic individuals through cellular replacement therapy [55].

(e) Liver diseases: For example, iPSCs are being looked into as potential candidates for generating hepatocytes which represent the functional cells of the liver. By making functional hepatocytes via this approach, it would be possible to treat illnesses such as liver cirrhosis and liver failure since these could replace the damaged hepatic tissue [56].

(f) Diseases associated with the eye: In research, mesenchymal cells are being used to treat age-related macular degeneration (AMD). By replacing damaged RPE cells and encouraging retinal health, these cells may help regain vision [57].

4.3.2.3 Future Direction

Though stem cells have significant therapeutic potential, there are many challenges. This requires ensuring that stem cell therapy is safe and effective in humans. Some of the risks include tumorigenicity, improper differentiation and immune rejection which need to be addressed through extensive preclinical and clinical research. Also, it would be necessary to develop scalable and reproducible approaches for differentiating and delivering stem cells in order to translate these therapies into clinical use [43, 50, 51].

There are promising developments in understanding the molecular mechanisms governing stem cell differentiation as well as innovations in biomaterials, tissue engineering and gene editing technologies which can help overcome these challenges [58–60]. Regenerative medicine could revolutionize the treatment of a range of diseases and injuries by exploiting all aspects of stem cells resulting new hope for patients worldwide [61].

4.3.3 Challenges in Stem Cell Transplantation

Regenerative medicine has a vast potential to treat many diseases and injuries with the help of stem cells. The successful delivery and transplantation of these cells are, however, subject to challenges such as immune rejection, cell viability assurance, and others.

4.3.3.1 Immune Rejection [62, 63]

One of the greatest obstacles in stem cell therapy is the problem of immune rejection. This is when the patient's immune system identifies transplanted stem cells as foreign bodies and mounts an immune response toward them. This can be particularly problematic in case of allogeneic stem cell transplantation, where the donor cells come from a different person. To tackle this challenge, several approaches are being considered by researchers, namely:

(a) Immunosuppressive therapy: Administering immunosuppressive drugs to the patient can reduce the immune response and improve graft survival. However, prolonged use of these drugs can increase the risk of infections and other complications.
(b) Immune-matching: Like organ transplantation, matching donor and recipient for human leukocyte antigen (HLA) diminishes rejection chances; unfortunately it is hard to find perfect matches within short periods.
(c) Engineering hypoimmunogenic stem cells: Genetic engineering techniques can modify stem cells so that they become less likely to stimulate immune reactions. For example, CRISPR/Cas9 technology can be used to delete specific genes involved in immune recognition.

4.3.3.2 Ensuring Cell Viability [64, 65]

The preservation of viable functional stem cells during their transplantation is extremely important for therapeutic success. Several factors determine whether a cell will remain alive or not, namely:

(a) Storage and transport: In order to keep them alive, stem cells must be stored under proper conditions while being moved around from one place to another. Cryopreservation is frequently used; however, if not done properly, it may damage living tissues, resulting in cellular problems. Hence, there is a need for better cryopreservation techniques and protocols.
(b) Delivery methods: The method of delivering stem cells to the target site affects their viability and therapeutic efficacy. Direct injection, intravenous infusion, and scaffold-based delivery are some methods used. Each has its advantages and limitations. For instance, direct injection may cause local tissue damage, while intravenous infusion can lead to cell trapping in non-target tissues.
(c) Microenvironment: The survival and integration of stem cells heavily depends on the microenvironment at the

transplant site. Oxygen levels, nutrient availability, and presence of inflammatory signals among others determine cell viability. Biological matrices like scaffolds made from biomaterials or hydrogels serve as useful supportive environments that promote cell survival and functionality.

4.3.3.3 Homing and Engraftment [66]

For effective stem cell therapy, transplanted cells must migrate to target sites (homing) and integrate with host tissue (engraftment). This is influenced by several factors, namely:

(a) Chemotactic signals: Natural molecules in the body attract stem cells to injured sites. Enhancing such signals or making them more responsive to stem cells will improve homing efficiency.
(b) Cell surface modifications: Modifying the surface of a stem cell so that it expresses certain adhesive molecules can increase its ability to adhere to and become part of the target tissue.

4.3.3.4 Safety Concerns [67]

Safety should be prioritized when carrying out stem cell therapies since there is potential risks associated with them:

(a) Tumorigenicity: In pluripotent stem cells (PSCs), there is a danger of abnormal growth leading up to tumor formation. Therefore it is important to develop approaches ensuring complete differentiation of those cells as well as elimination of undifferentiated ones.
(b) Genetic changelessness: It is very important that stem cells are not damaged by genetic mutation during culture and expansion. There must be regular screening and quality control measures to ensure that genetic constancy is maintained.

4.3.3.5 Ethical and Regulatory Issues [68]

The ethical issues and regulatory frameworks for stem cell therapies are intricate and differ from one country to another. Making sure that the guidelines provided by regulatory bodies are adhered to, as well as solving all ethical matters connected with the use of embryonic stem cells, is key in moving ahead with stem cell treatments.

4.3.3.6 Cost and Accessibility

The high costs of stem cell therapies together with their technical complexity can be a major hindrance to accessibility. Efforts aimed at cutting down on costs, making production processes smoother and developing standard protocols need to be made so that these therapies become more available [69].

In summary, there remain several barriers in delivery and transplantation that need to be resolved before the full potential of stem cell therapies can be realized. Addressing immune rejection; ensuring viable cells; homing and engraftment enhancement, safety assurance, ethical regulation, cost reduction; these represent crucial steps toward successful application of such treatments on a large scale. With ongoing researches and technological improvements, this area is progressing toward effective regenerative approaches that are accessible to everyone.

4.4 Exosomes: Natural Nanocarriers for Intercellular Communication

Exosomes, which are small extracellular vesicles ranging in diameter from 30 to 150 nm, are vital for intercellular communication. They perform a variety of important functions during different physiological and pathological processes, with their natural ability to act as nanocarrier for delivering biomolecules between cells [70].

4.4.1 Biogenesis and Properties of Exosomes

4.4.1.1 Biogenesis

The exosomes develop through intracellular events that involve inward budding of the endosomal membrane to form Extravesicular bodies (EVBs). These EVBs contain Intraluminal vesicles (ILVs), which include proteins, lipids and nucleic acids. Upon fusion of EVBs with the plasma membranes, ILVs are released into the extracellular environment as exosomes [71].

Important molecules and pathways involved in exosome biogenesis include the following [72]:

(a) Endosomal sorting complex required for transport (ESCRT): The ESCRT machinery is made up of four complexes known as ESCRT-0, -I,-II,-III that sorts cargo into ILVs, thus facilitating membrane bending.
(b) Non-ESCRT pathways: Not all exosome biogenesis processes depend on ESCRT machinery. These utilize tetraspanins (e.g., CD9, CD63, CD81), which are integral membrane proteins that aid in sorting cargo or regulating membrane dynamics.
(c) Lipid rafts: Lipid microdomains present on endosomal membrane contributes to creation of ILV as hence effecting lipid composition on exosome.

4.4.1.2 Structure of Exosomes [73]

Exosomes possess specific lipids, glycoproteins, and proteins enriched lipid bilayered membrane that give them unique characteristics. These aspects consist of the following:

(a) Surface proteins: Various kinds of surface proteins are found on exosomes including tetraspanins (CD9, CD63, CD81), integrins as well as major histocompatibility complex (MHC) molecules. This class of proteins facilitates cell targeting adhesion immune modulation.
(b) Molecular cargo: Exosomes carry a variety of molecular cargo including proteins, lipids, mRNA, microRNA (miRNA), long non-coding RNA (lncRNA), and DNA. This is selectively packaged in these organelles during their formation with the exact content reflecting the cell of origin, making it a "fingerprint" for cell type and state.
(c) Lipid composition: The lipid bilayer of exosome consists of sphingomyelin, cholesterol, phosphatidylserine, and ceramide, which give it stability and functionality.

4.4.1.3 *Natural Functions of Exosomes in Cellular Communication [74–76]*

Exosomes are involved in several physiological and pathological processes which include the following:

(a) Mediating cell–cell communication: Exosomes mediate multi-cell communication by delivering proteins, RNAs and lipids among others to other cells. The behavior and function of recipient cells can be altered through this intercellular exchange that affects processes such as cell multiplication as well as differentiation apoptosis.
(b) Immune modulation: Exosome takes part in immune regulation via antigen presentation to immune cells leading to immunomodulation. This happens when dendritic cell-derived exosome antigens are presented to T cells which stimulate or hinder immune response. Tumor-derived exosomes also bear immunosuppressive molecules that help cancer cells evade immune surveillance.
(c) Tumor progression and metastasis: In terms of cancer, exosomes contribute significantly to tumor progression and metastases through transfer of oncogenic factors initiation of angiogenesis plus modification of the tumor microenvironment. Cancer cell-derived exosome prepared distant tissues for tumor colonization through formation premalignant niches.
(d) Tissue repair and regeneration: There have been reports that exosomes derived from stem cells specifically mesenchymal stem cells (MSCs) promote tissue repair and regeneration. Wound healing can be improved by bioactive molecules such as growth factors, cytokines miRNAs carried by these exosomes, thus reducing inflammation and stimulating tissue regeneration.
(e) Biomarkers in disease diagnosis and monitoring: Exosomes can be used as biomarkers of diseases such as cancer, neurodegenerative disorders, cardiac conditions, or infections. The analysis of exosomal RNA, DNA, and proteins from patient samples (e.g., blood) may give hints about what stage a disease is at or how it is progressing or even how it responds to a given treatment.

(f) Therapeutic delivery vehicles: The ability of exosomes to encapsulate cargo protect it from degradation upon its release within target cells also eases them across biological barriers. This is why scientists are currently engaged in experiments aimed at developing special exosomes for the purpose of drug delivery to some specific places using gene therapy and immune modulation among other things.

(g) Others: Neurotransmission and synaptic plasticity are interconnected processes within the nervous system, where exosomes play a crucial role. These vesicles facilitate the uptake of neurotransmitter receptors, synaptic proteins, and RNAs, thereby modulating synaptic function and plasticity, which in turn influences learning and memory. Pathogens, such as viruses and bacteria, exploit the exosome pathway to disseminate infections and evade the immune system. For instance, HIV-1 incorporates viral proteins and RNA into exosomes, enhancing infection propagation and modulating host immune responses, thereby aiding in immune evasion and disease progression.

Exosomes mediate various physiological and pathological processes that occur between cells fostering intercommunication; these vesicles are indispensable for life. Therefore, thorough knowledge on their origin as well as structure together with their natural activities forms a basis essential in determining whether they can potentially contribute toward diagnosis or therapeutics, thereby leading to innovative precision medicine approaches along with regenerative therapies.

4.4.2 Targeted Delivery Using Engineered Exosomes

Owing to characteristics like biocompatibility, stability, and biomolecule carrying capability, exosomes are a good method of achieving targeted drug delivery. There are many techniques used in engineering exosomes for specific delivery, all with the aim of improving their targeting specificity, loading capacity and therapeutic efficacy [77]. This part describes various methods and approaches which are employed to modify exosomes for individualized medicine.

4.4.2.1 *Surface Alteration for Targeting*

One way to engineer an exosome is by changing its surface so that it becomes more specific in targeting certain tissues or cells [78]. This can be done through the processes discussed below.

4.4.2.1.1 Use of Ligands

(a) Antibodies and peptides: Antibodies or peptides can be attached onto the surface of an exosome thereby enabling selective recognition of antigens expressed on cells' surfaces. For example, HER2-positive breast cancer cells could be targeted by anti-HER2 adorned vesicles [78].

(b) Aptamers: These small single-stranded DNA/RNA molecules have high affinity binding sites against particular proteins found only at the cell membrane level hence useful when attached to exosomes as homing devices [78].

4.4.2.1.2 Genetic Manipulation of Donor Cells

Protein expression: In order to have personalized vesicles, genetic engineers need them express specific proteins on their membranes through modification of parent cell's genome; this allows for production tailored-made particles which will bind selectively with target tissues or organs. For instance, Lamp2b fused together with homing peptides could be expressed in such a way that resultant exosomes home either neuron or cancer cell depending on where should treatment take place [79].

4.4.2.2 *Increasing Cargo Capacity*

The more drugs an exosome can contain inside it the better because those medications will cause stronger effects when they reach their destinations within human body systems. Therefore, efficient packaging mechanisms should be developed into these tiny transporters. One good thing about them is that there exists several techniques meant to achieve high levels of efficiency [78].

4.4.2.2.1 Passive Loading

This involves incubating drugs or nucleic acids with exosomes so that they can penetrate inside easily through simple diffusion mechanism. However, it may not work well for all types of cargo since sometimes loading efficiency becomes low [80].

4.4.2.2.2 Active Loading

These methods specifically been used to create modified versions of cells to increase the therapeutic efficiency. The following methods have been widely used for active loading:

(a) Electroporation: Exposing vesicles electrically creates some pores on their walls, thus allowing the entry of drug molecules. In this case, siRNA and other therapeutic agents could be used but the integrity of particles might get affected during the loading process [81].
(b) Sonication and extrusion: Ultrasound waves together with mechanical force are used to rupture membranes, temporarily enabling the incorporation of therapeutic agents which then resume encapsulation after cessation rupture [82].
(c) Genetic manipulation of donor cells: If donor cells are engineered such that they over-express specific types therapeutic proteins then those substances will be incorporated into exosomes as part natural biogenesis. For example, miRNAs loaded exosomal bodies may originate from modified cells designed to produce them [79].

4.4.2.3 Enhancing Stability and Circulation

Once injected into the bloodstream, an important factor determining how effective exosomes can be therapeutically is their ability to remain stable while circulating throughout the body; hence, various tactics have been employed including following strategies [73, 78]:

(a) PEGylation: Polyethylene glycol gets linked onto the surface of these nanovesicles which protects from quick elimination by immune system thereby extending half-life within circulation PEGylation also aids reduction opsonization followed subsequent reticuloendothelial uptake.
(b) Cryopreservation: It is critical to ensure that storage conditions used do not lead loss functional capacity during prolonged periods since very often long-term preservation is required for clinical applications where cryopreservation

techniques should therefore be optimized in order to maintain stability while preserving functionality.

4.4.2.4 Controlled Release Mechanisms

Therapeutic delivery precision can be improved if release properties are incorporated into engineering design concept. These properties are essential as they make the delivery system more advanced and targeted:

(a) Stimuli-responsive exosomes: Such vesicles would only open up and let go of the contents when triggered by particular stimuli like change in pH level or temperature rise at targeted areas, which ensures accuracy of drug administration. For example, pH sensitive lipids could cause for them to release contents within an acidic environment, thus ensuring localized chemotherapy delivery [83].
(b) Magnetic exosomes: Incorporation of magnetic nanoparticles into exosomes enables their movement and release to be controlled from outside by magnetic fields. This method can direct exosomes to particular locations within the body and manage drug discharge time [84].

Various preclinical and clinical studies have shown that designed exosomes can deliver drugs to specific areas in cancer therapy, neurologic disorders and cardiovascular diseases. They do this by targeting chemotherapeutic agents or RNA molecules into the body which are able to differentiate between healthy cells and cancerous ones [77, 80]. This means that they kill the latter while minimizing off-target effects thus providing better outcomes for patients with these conditions. A good example is their use in breast cancer treatment where modified exosomes loaded with doxorubicin were employed thereby reducing systemic toxicity levels associated with such interventions [85]. Similarly, in cases of neurological disorders like Alzheimer's or Parkinson's disease, scientists have been trying out different types of exosomes that can cross the blood–brain barrier carrying siRNA and miRNA among other neuroprotective proteins needed during the treatment stages [86]. As a matter of fact, researchers are also considering using stem cell derived exosomes packed full cardioprotective substances so as to repair damaged hearts caused by myocardial infarction events. It is clear therefore that engineering these vesicles for target drug delivery holds great potentiality toward improving accuracy and efficiency within medical care systems currently available worldwide [87].

4.4.3 Exosome-Based Therapeutics

Exosome therapeutics are an advanced approach in medicine that utilizes the characteristics of exosomes for various applications. These vesicles are biocompatible, can bear different bioactive molecules, and participate in intercellular communication naturally. Therefore, they are examined as a possible means for drug delivery systems, immune modulation agents not only within regenerative medicine area but also beyond it [77]. This part will discuss many therapeutic uses of this diversity.

4.4.3.1 Drug Delivery

(a) Targeted cancer therapy: Toxic substances used against tumor cells can be transferred directly by modified exosomes which increases their effectiveness without general damage to the body. For example, *in vivo* models showed anti-tumor activity after injection with doxorubicin- or paclitaxel-loaded exosomes. Such particles gain better ability to recognize specific cancer cells if they have targeting ligands like antibodies or peptides on their surface recognizing tumor markers [85].

(b) Neurological disorders: Due to its capability to cross the blood–brain barrier, exosome-mediated delivery becomes one of the most promising ways for brain-targeting therapy. In particular, introduction into affected regions of siRNA against neurodegeneration-related genes such as those causing Alzheimer's disease (AD) and Parkinson's disease (PD) has demonstrated significant success so far based on animal experiments using miRNAs intended for these purposes too. For instance, reduction in levels of mutant huntingtin mRNA through intracerebral infusion with siRNA-loaded exosomes led to decreased expression levels of abnormal protein aggregates followed by improvement in Huntington's disease model animals' phenotype [86].

(c) Cardiovascular diseases: Researchers have investigated whether stem cell-derived exosomes could protect heart tissue from damage caused by myocardial infarction (MI). It was found out that this is indeed possible because such particles contain growth factors required for healing processes alongside cytokines capable of reducing fibrosis there around infarcted areas, thus promoting cardiac repair overall. Moreover, treatment with them improves post-MI ventricular function, thus contributing to overall recovery [87].

4.4.3.2 *Immune Modulation*

(a) Autoimmune diseases: Exosomes have the potential to suppress immune responses, making them valuable tools in treating autoimmune conditions. For example, exosomes derived from mesenchymal stem cells (MSCs) possess immunosuppressive properties that can help decrease inflammation and autoimmunity. In such cases, these vesicles may carry anti-inflammatory cytokines or miRNAs directed toward immune cells, thereby downregulating abnormal immunological reactions seen in rheumatoid arthritis (RA) and multiple sclerosis (MS) [88].

(b) Cancer immunotherapy: Another way of utilizing exosomes for therapeutic purposes involves their use as adjuvants during cancer vaccination therapies aimed at stimulating anti-tumor immunity through activation antigen presenting cells like dendritic cells (DCs). These particles are loaded with tumor antigens before being released by DCs which then present them to T-lymphocytes resulting into robust T-cell mediated immune response against malignancies. Some ongoing clinical trials evaluate vaccines based on this principle where tumor-specific DEX activate cytotoxicity within tumors, leading to their destruction via immune mechanisms [89].

(c) Transplantation: One application area where exosome-based interventions have great promise is transplantation medicine due to ability induce long-term acceptance without graft-versus-host disease development. This means that donor exosomal MHC molecules educate recipient's immune system recognize transplanted organs leading increased tolerance toward them. Indeed donor-derived regulatory T cell (Treg) derived exosomes can promote survival times upregulated Foxp3 expression, thus preventing rejection responses while ensuring normal functionality of allografts even when self-reactive Tregs are absent [90].

4.4.3.3 *Regenerative Medicine*

(a) Wound healing: The potential role played by MSC-originating exosomes in wound healing process has attracted significant attention recent years thanks mainly to its capability to enhance tissue regeneration through delivery growth factors promoting angiogenesis alongside anti-inflammatory ones reducing inflammation. Clinical evidences demonstrated their positive impact on diabetic ulcer healing rate as well as acceleration of burn restoration [91].

(b) Bone and cartilage repair: The ability of exosomes to stimulate osteogenic and chondrogenic differentiation suggests that these particles could be used for bone fracture healing purposes. In particular, MSC-derived exosomal cargoes containing osteoinductive factors may provide an effective therapy option either alone or in combination with other biologics such as growth factors needed during cartilage repair processes following injury occurrence. Indeed preclinical models have shown increased density together with improved integrity within areas affected by osteoporosis after treatment using them, thereby making potentially useful treatments against this condition available, too [92].

(c) Regeneration of liver: Liver damage and ailments like hepatitis and cirrhosis can take advantage from exosome therapy. These regenerative molecules are carried by exosomes from stem cells or hepatocytes which induce the healing process of liver and decrease the formation of scar tissue in it. Some experiments indicate that treating liver injury models with exosomes enhances their functionality while minimizing fibrosis-induced scarring [93].

4.4.3.4 *Applications for Diagnosis*

Non-invasive diagnostics are now using exosomes which were unknown until recent times. These carry proteins, RNA, and DNA that indicate certain diseases and can be tested to diagnose or follow the course of an illness. Scientists are developing liquid biopsies based on exosomal biomarkers for early detection of cancers, heart diseases as well as brain disorders associated with aging. Exosome-based therapeutics have flexible applications in treating different ailments innovatively. Precise drug administration, immune modulation ability, and tissue regeneration promotion by these particles promise better outcomes for patients; hence, they should be embraced widely as they contribute greatly toward personalized healthcare provision. More studies into their potential through ongoing investigations alongside clinical trials will demystify them, thus creating opportunities for new treatments within various medical disciplines [94].

4.5 Reprogramming Immunity for Cancer Through CAR-T Cell Therapy

4.5.1 CAR-T Cell Design and Engineering

CAR-T therapy has shown promise as an innovative cancer treatment option, as it involves harnessing and altering the immune system to attack malignancy. The design and engineering of CAR-T cells entail a series of complex procedures aimed at ensuring that they are specific, effective, and safe [95].

4.5.1.1 CAR-T Cell Design

The core of CAR-T cell therapy is the chimeric antigen receptor (CAR), which is a synthetic receptor created to enable T cells to recognize and destroy cancerous cells. Typically, a CAR consists of four main domains:

(a) Extracellular antigen-binding domain: This section is usually comprised of a single-chain variable fragment (scFv) derived from an antibody built for specifically sticking onto antigens expressed on cancer cell surfaces. For this reason, the selection in some manner should be limited to unique or overexpressed antigens toward minimizing nontarget effects [96]. Examples include CD19 for B-cell tumors and HER2 for select solid malignancies [97].
(b) Hinge region: The extracellular domain's hinge region, derived from either immunoglobulin or other protein domains enhancing binding kinetics, serves to make it flexible enough to bind efficiently to the target antigen [98].
(c) Transmembrane domain: It provides attachment points for the CARs on the T cell membrane facilitating signal transduction upon encountering antigens; generally

obtained from naturally occurring CD3ζ domain or CD28 molecules responsible for TCR signaling [99].

(d) Intracellular signaling domains: These regions contain signal domains that trigger multiple pathways in T-cells activation. First-generation CARs only have one signaling domain—the CD3ζ chain, which initiates T cell activation—while subsequent ones carry costimulatory domains like CD28 or 4-1BB (CD137) that promote T cells proliferation, survival and cytokine secretion. The third-generation CARs have several costimulatory domains included for enhanced efficacy [100].

4.5.1.2 Genetic Engineering of CAR-T Cells

The process of engineering CAR-T cells begins with the collection of a patient's T cells, followed by their genetic modification to express the CAR construct. This involves several critical steps [101, 102]:

(a) T-cell isolation and activation: From patients' peripheral blood after leukapheresis, T cells are isolated. Then they are activated using agents such as anti-CD3/CD28 antibodies that mimic natural activation and growth signals [103].

(b) Gene transfer: Activated T cells are transfected with the CAR gene using viral vectors such as lentiviruses and retroviruses. These vectors ensure stable expression by integrating the CAR gene into the genome of T-cells. Efforts are underway to improve safety and efficiency through nonviral means such as transposon-based systems [104].

(c) Expansion and quality control: The genetically modified T-cells are grown *in vitro* until enough numbers for therapeutic administration is realized. During this expansion phase, close attention is paid to preserve functionality and desired phenotypic characteristics [105].

(d) Formulation and infusion: Engineered CAR-T cells are formulated in a suitable medium before they can be reintroduced back to patients where they home toward tumor sites recognizing target antigens on cancerous cells and killing them eventually [102].

For success in CAR-T cell therapy, it depends on appropriate design and development of a specific CAR along with careful T-cell manipulation and expansion processes coupled together with quality control measures throughout its manufacturing process leading up to formulation into product that can be administered to humans. With new developments in cancer immunotherapy, hope has been rekindled among patients with relapsed and refractory malignancies [105].

4.5.2 Clinical Applications and Efficacy of CAR-T Cell Therapy

CAR-T cell therapy has achieved remarkable clinical success, especially in curing blood cancers. It is true that there are several clinical trials that have shown the efficacy of the therapy leading to approvals of different CAR-T cell products.

4.5.2.1 Hematologic Malignancies

Hematologic malignancies are a group of diseases or cancers that instigate in blood forming-tissue and characterized by the clonal propagation of hematopoietic cells (e.g., leukemia, lymphoma, etc.):

(a) B-cell acute lymphoblastic leukemia (B-ALL): CD19-targeted CAR-T cells display unprecedented magnitude remissions for B-ALL with CR rates ranging from 70% to 90% for patients with relapsed or refractory disease. Tisagenlecleucel (Kymriah) is a good example of such treatments being approved by FDA for pediatric as well as young adult patients suffering from B-ALL [106].
(b) Diffuse large B-cell lymphoma (DLBCL): Axicabtagene ciloleucel (Yescarta) and Tisagenlecleucel are examples of CAR-T cell therapies that have shown significant clinical improvements in DLBCL which is a type of non-Hodgkin lymphoma. These therapies have improved the patient's quality life, as they have resulted in long-lasting responses and increased overall survival for those who had no other treatment options left [107].
(c) Chronic lymphocytic leukemia (CLL): Early phase trials of CD19-targeted CAR-T cells in CLL showed impressive results with major percentage of patients experiencing substantial clinical responses following extensive

pretreatment. Further studies are ongoing to optimize the therapeutic protocols and enhance efficacy [108].

4.5.2.2 *Solid Tumors*

Despite limited success compared to hematologic malignancies, current research aims at addressing specific challenges presented within solid tumor microenvironment through CAR-T cell therapy:

(a) Glioblastoma: For instance, glioblastoma can be treated using CAR-T cells against EGFRvIII, a mutant form of epidermal growth factor receptor. Problems facing this therapeutic approach include the fact that, though it has shown initial responses, EGFRvIII is not universally expressed and tumor microenvironment is immunosuppressive [109].

(b) Breast cancer: HER2-targeted CAR-T cells have been tested for treatment of HER2-positive breast cancer. However, on-target off-tumor toxicity concerns mandate strategies which improve their specificity and safety [110].

4.5.2.3 *Efficacy and Long-Term Outcomes*

The success of CAR-T cell therapy in curing hematologic cancers refractory to treatment is proved by high rates complete remissions and long survival in patients. Nevertheless, there need to be long term follow-up studies to establish the durability of any responses as well as late relapses. Combining this with standard treatments schedules as well as developing combination therapies will go a long way in improving its efficacy and making it more applicable widely [111].

4.5.3 Challenges and Future Directions of CAR-T Cell Therapy

Though CAR T-cell therapy has been transforming, it still presents some challenges which call for ongoing research and innovation.

4.5.3.1 *Toxicity*

(a) Cytokine release syndrome (CRS): This occurs when there is an infusion of excessive cytokines after the administration of CAR-T cells. High levels of circulating cytokine are detected and appropriate management includes corticosteroids as well as immunosuppressive agents like tocilizumab [112].

(b) Neurotoxicity: Severe neurological toxicities including encephalopathy or seizures have been seen in patients who received CAR T-cells. The underlying mechanisms are poorly defined and research efforts are directed at finding useful biomarkers that can be used to predict neurotoxic effects leading to more effective management strategies [113].

4.5.3.2 *Resistance and Relapse*

(a) Antigen escape: Tumor cells evade CAR-T cell therapy by downregulating or losing the target antigen. This may be addressed through development of dual-antigen-targeting CAR-T cell therapies, second-generation constructs, or targeting multiple epitopes with a single construct [114].

(b) Tumor microenvironment: CAR-T cells face obstacles in solid tumors where they encounter an immunosuppressive microenvironment. Current investigations include combining checkpoint inhibitors with CAR-T cells or engineering them to secrete pro-inflammatory cytokines within the tumor milieu [115].

4.5.3.3 *Manufacturing and Accessibility*

(a) Scalability and cost: The manufacture of such a complex product on an individual basis makes scaling up difficult because it increases costs greatly. Efforts toward standardizing manufacturing processes, automation, and exploring allogenic CAR-T cell products are geared toward reducing these barriers to scalability thereby improving access to this technology [103].

(b) Regulatory and logistical challenges: Logistics is faced with several difficulties like quality control during manufacturing process in addition addressing issues related to processing delivery including those related to transportation need serious attention if wider application is going to take place [116].

4.5.3.4 Future Directions

(a) Next-generation CAR-T cells: Research is underway to develop next-generation CAR T-cells that are more specific, less toxic, and have better efficiency. Some of the innovations include novel co-stimulatory domains or armored CAR-T cells that are resistant to immunosuppression as well as off-the shelf allogeneic CAR-T cells [117].
(b) Combination therapies: CAR-T cell therapy combined with other therapeutic modalities like checkpoint inhibitors, targeted therapies, or radiation has potential in overcoming resistance and improving outcomes. Synergistic approaches using different combinations are being explored to improve CAR-T cell-based antitumor efficacy further [118].
(c) Broadening applications: This then brings into focus the idea of utilizing CAR-T cell for treating patients with tumors other than hematological malignancies such as autoimmune diseases as well as infectious diseases. New strategies for targeting and innovative designs of the CAR-T can help achieve these aims [119].

In conclusion, CAR T-cell therapy is a leading cancer treatment giving hope where no hope existed for patients suffering from various incurable cancers. Despite the challenges at hand, ongoing research and technological advancements continue influencing the direction taken by CAR T-cell therapy toward its future applications with better patient outcomes in mind.

4.6 Integrating Nanobiomaterials with Stem Cells and CAR-T Cells

4.6.1 Nanobiomaterials for Enhanced Stem Cell Therapy

Stem cell therapy holds much promise for regenerative medicine, as it offers a way of restoring tissues and treating a wide range of diseases. However, stem cell therapies' effectiveness may be limited by poor cell survival, lack of differentiation among cells, and suboptimal migration toward specific tissues. Consequently, nanobiomaterials have been proposed to address these bottlenecks in stem cell-based therapeutic interventions.

4.6.1.1 *Proliferation Enhancement*

Nanobiomaterials can ensure supportive conditions for proliferation and viability of the stem cells. Different forms of nanomaterials such as nanoparticles, nanofibers, and hydrogels have been utilized to this end [29, 84].

4.6.1.1.1 Nanoparticles

(a) Growth factor delivery: These particles are fabricated in such a way that they can deliver some growth factors like fibroblast growth factor (FGF) and vascular endothelial growth factor (VEGF), which are essential in supporting the process of stem cell proliferation. Moreover, these factors can be steadily released from the particles [120].

(b) Antioxidant delivery: Reactive oxygen species (ROS) could lead to oxidative stress that might damage stem cells thereby causing their death. In order to protect such cells against oxidative stress-related injuries, nanoparticles containing antioxidants were used [121].

4.6.1.1.2 Nanofibers and Hydrogels

(a) Scaffold structures: A three-dimensional scaffold fashioned out of nanofibers and hydrogels is created which mimics the natural extracellular matrix environment that enhances cellular adhesion and proliferation. Also, bioactive molecules can be immobilized onto those structures to facilitate their supportive behavior [40, 60].

(b) Mechanical support: Tuning the mechanical properties like stiffness or elasticity allows them to match target tissues thus providing structural support necessary for various cellular functions [66].

4.6.1.2 *Differentiation Enhancement*

Differentiation is an important aspect in tissue regeneration through stem cells. Nanobiomaterials have been realized to possess a wide range of mechanisms that can be used to influence the differentiation process.

4.6.1.2.1 Chemical Cues

(a) Surface functionalization: Differentiation-inducing agents like peptides and small molecules, which can bind cell surface receptors and result in lineage-specific differentiation, are attached on nanomaterials [30].

(b) Controlled release: In order to ensure optimal differentiation such as osteogenic differentiation-controlled release of osteoinductive factors like bone morphogenetic proteins (BMPs) or IGF for myogenic differentiations is employed by nanoparticles [22].

4.6.1.2.2 Physical Cues

(a) Topographical features: The topography of nanomaterial surfaces affects the behavior of stem cells. Aligned nanofibers induce neural differentiation, while nanoscale patterns promote osteogenic differentiation [122].

(b) Mechanical stimulation: Mechanical properties such as stiffness and elasticity can be tailored to mimic the mechanical features of target tissues, thus providing physical signals that lead to stem cell maturation [122].

4.6.1.3 Improvement of Homing and Engraftment

Effective homing and engraftment of stem cells at site of interest are critical to their therapeutic potential. Nanobiomaterials, on the other hand, could enhance this through various strategies.

4.6.1.3.1 Homing Signals

(a) Chemokine delivery: Nanoparticles loaded with chemokines especially stromal cell-derived factor-1 (SDF-1), which improves stem cell recruitment into injured tissues. The gradient created from a slow release of chemokines will help guide stem cell homing toward desired tissues [123].

(b) Surface modification: For instance, selectins and integrins could be attached onto nanomaterial surfaces as homing ligands for adhering with other molecules in target tissues, leading to proper positioning and retention of stem cells thereon [123].

4.6.1.3.2 Immune Modulation

(a) Immune evasion: Finally, polyethylene glycol (PEG)-coated nanoparticles evade immune system attack, thereby preventing immune-mediated destruction against injected stem cells. This technique prevents immune-mediated rejection of stem cells by using nanoparticles that are coated with polyethylene glycol (PEG) [124].
(b) Immunomodulatory agents: Nanoparticles can also be used as delivery vehicles for anti-inflammatory drugs and other immunosuppressive compounds, thus creating a local immune-tolerant environment that facilitates successful engraftment of stem cells [124].

What makes nanobiomaterials unique are the multifarious enhancements they offer to stem cell therapies in the sense of promoting proliferation, directing differentiation, and enhancing homing and engraftment. By uniting nanotechnology with stem cells therapy, there is a lot of hope that regenerative medicine will be advanced making therapeutic results more efficient.

4.6.2 Nanocarriers for Delivery and Activation of CAR-T Cells

CAR-T cell therapy has been established as one of the most potent approaches toward the treatment of certain types of cancer. However, this therapy can become ineffective due to problems such as inappropriate delivery to tumor sites, insufficient activation or limited persistence. The delivery and activation of CAR-T cells can be improved through nanocarriers that provide alternative approaches for solving these challenges hence improving therapeutic outcomes [118].

4.6.2.1 Enhanced Delivery to Tumor Sites

The effectiveness of CAR-T cells at tumor sites depends on their efficient delivery. This could be made possible by employing various mechanisms using nanocarrier systems that improve targeting and homing abilities of CAR-T cells [111].

4.6.2.1.1 Targeting Ligands

(a) Surface modification: These nanocarriers may also have surface modifications including antibodies or peptides which bind target ligands on tumor cells. This targeted approach enhances the homing effect between CAR-T cells and tumor microenvironment [115].

(b) Homotypic targeting: Homotypic interactions (e.g., binding to molecules expressed by both CAR-T cells and tumor cells) can be employed in order to specifically engineer nanocarriers for selective guidance of such immunocytes into tumors [115].

4.6.2.1.2 Magnetic Targeting

(a) Magnetic nanoparticles: Furthermore, incorporation of magnetic nanoparticles within nanocarriers enables external control over location where CAR-T lymphocytes localize using magnetic fields. Thus, this strategy helps in directing CAR-T cells precisely into tumors, thereby increasing their accumulation and therapeutic action [125].

4.6.2.2 Improved Activation and Expansion

The anti-tumor activity requires activation and expansion of CAR-T cells. Nanocarriers could achieve this by different methods.

4.6.2.2.1 Activation Signals

(a) Signal delivery: The nano-carrier can deliver activation signals like interleukins such as IL-2/IL-15 or costimulatory molecules, e.g., CD28/4-1BB to CAR-T cells, thereby promoting their proliferation and activation. Controlled release of these signals from the nanocarriers ensures that there is continuous activation [126].

(b) Synthetic antigen-presenting cells (APCs): Accordingly, nanocarriers can be fabricated to mimic natural APCs, presenting antigens and co-stimulatory signals to CAR-T lymphocytes. Before infusion, these synthetic cells can efficiently activate and expand car T-cells *in vitro* [127].

4.6.2.2.2 Microenvironment Modulation

(a) Overcoming immunosuppression: There exist immunosuppressive factors within the tumor microenvironment that interfere with CAR-T cell function. These factors may be neutralized using immune checkpoint inhibitors (e.g., anti-PD-1 or anti-CTLA-4) or cytokine blockers delivered through nanocarriers, thus enhancing their activity within the tumor. In addition, nanocarriers can be programmed for local release of pro-inflammatory cytokines such as IL-12/IFN-g in the tumor microenvironment. These molecules are known to alter local immune landscape leading to enhanced CAR-T cell activation necessary for effective anti-tumor response [95].

4.6.2.3 *Enhancement of Persistence and Survival*

Persistence and survival of CAR-T cells *in vivo* are crucial for long-term therapeutic success. Several mechanisms can support these aspects using nanocarriers.

4.6.2.3.1 Cytokine Delivery [112, 113]

(a) Sustained support: Other than delivering sustained supports for CAR-T cell survival such as IL-7 and IL-15, other cytokines could also be carried by the same systems. Thus, continuous availability of supportive environment is maintained by the application of controlled release systems carrying these cytokines.
(b) Localized delivery: The delivery of cytokines to the tumor microenvironment by nanocarriers can ensure that CAR-T cells survive locally and also carry out their normal functions. This improves therapeutic outcomes while reducing systemic side effects.

4.6.2.3.2 Metabolic Support

Nanocarriers can bring indispensable food substances such as glucose or amino acids to CAR-T cells for their metabolism and persistence enhancement. This method may be especially useful in nutrient-poor environments of tumor tissues. Nanocarriers can convey agents for enhancing mitochondrial function and energy generation within CAR-T cells, which are vital to the cells' survival and fighting against tumors. The nanocarriers have proven to be adaptable and strong for the enhancement of delivery, activation and longevity of CAR-T cells. This would be achieved through making a good use of the distinctive features that nanomaterials possess which would help in addressing these constraints in current CAR-T cell therapies and also improve treatment efficacy for cancer patients [118, 126].

4.6.3 Combination Treatments for Improved Therapeutic Effectiveness

A good approach to increase therapeutic efficacy and achieve synergistic effects involves integrating nanobiomaterials, stem cells, and CAR-T cells into combination therapies. Such combination therapies can take advantage of the unique strengths of each component offering comprehensive treatment strategies for many diseases such as cancer or regenerative medicine.

4.6.3.1 Nanobiomaterials and Stem Cells

When combined with stem cell therapy, nanobiomaterials can address several problems, thereby increasing overall curative potential.

4.6.3.1.1 Augmented Homing and Engraftment

(a) Nanoparticle-coated stem cells: Coating stem cells with nanoparticles that release homing factors or chemokines could help them move more easily toward target tissues, hence improving engraftment, too. This is particularly useful in regenerative medicine where exact positioning of stem cells is vital for tissue repair [127].

(b) Hydrogel scaffolds: When stem cells are incorporated into nanofiber-based or hydrogel scaffolds they create an environment which supports their survival and differentiation better than without any support at all. It becomes possible by functionalizing these scaffolds with bioactive molecules capable of promoting both tissue regeneration as well integration thereof [29].

4.6.3.1.2 Controlled Differentiation

(a) Differentiation factors delivered via nanoparticles: Nanoparticles have the ability to guide differentiation-inducing agents directly into particular types of cells derived from stems since this approach ensures efficiency and specificity unlike others do not. In other words, it helps them differentiate faster which in turn enhances their effectiveness when used therapeutically [122].

(b) Biophysical cues: By incorporating nanomaterials having specific mechanical properties together with those having certain topographical features one may provide biophysical cues that influence behavior as well outcome following regenerative treatments involving stem cell-based therapies [122].

4.6.3.2 CAR-T Cells and Nanobiomaterials

Combining CAR-T cell therapy with nanotechnology-based materials can improve delivery systems' efficiency in terms activation levels and persistence rates besides overcoming various challenges experienced during cancer management.

4.6.3.2.1 Targeted Delivery

(a) Nanocarrier-encapsulated CAR-T cells: The localization of CAR-T cells within tumors can be improved by encapsulating them in nanocarriers which release ligands for targeting or factors facilitating homing. This method ensures that a large number of these cells are brought into direct contact with cancerous tissues, hence increasing their therapeutic effect on such sites [128].

(b) Magnetic nanoparticles: External magnetic fields may be used to guide CAR-T cell migration toward tumor regions through magnetic attraction created between particles loaded onto these cells, thus enabling controlled delivery coupled with enhanced anti-tumor activity at desired locations only [129].

4.6.3.2.2 Activation and Expansion

(a) Synthetic APC nanoparticles: CAR-T cells could be activated *ex vivo* before infusion using nanoparticles designed to act as antigen presenting cells (APCs) bearing tumor antigens and capable of delivering costimulatory signals required for efficient activation together with expansion. This approach allows the presentation of appropriate signals necessary for optimal recognition as well subsequent killing by cytotoxic T lymphocytes (CTLs) [129].

(b) Cytokine-releasing particles: Within the microenvironment surrounding neoplastic tissue, localized release of cytokines like IL-2 or IL-15 from nanoparticles would support activation plus proliferation among anti-cancer effector mechanisms such as CAR T cells. In simple terms, it means that cytokine-based therapy can work better when administered directly into tumors than systemically since higher drug concentrations become achievable there, leading to improved efficacy accordingly [113].

4.6.3.3 CAR-T Cells and Stem Cells

Treating cancer through combining stem cell-based therapy with chimeric antigen receptor modified (CAR) T cell immunotherapy offers multifaceted advantages where regenerative potentials delivered by stem cells get combined together targeted killing abilities exhibited by the latter against tumorigenic cells.

4.6.3.3.1 Tumor Microenvironment Modulation

(a) Immunomodulatory factors derived from stem cells: Immunomodulatory factors secreted by stem cells have been shown to influence CAR T cell function within the tumor microenvironment thereby alleviating immunosuppression while enhancing anti-tumor activities associated with these cells. In this regard, both treatment approaches should work together for better results in future [130].

(b) Engineered stem cells: Genetically modifying stem cells so that they express certain cytokines among other molecules capable of enhancing CAR-T cell activity could be another way forward. For instance, IL-12-secreting engineered stromal populations can create pro-inflammatory conditions within local sites leading into improved performance levels by CAR-T lymphocytes during their encounter with neoplastic tissues [130].

4.6.3.3.2 Tissue Restoration and Repair

The regenerative ability of stem cells can be used to repair injured tissues and boost patient's recovery after treatment. This twofold method looks at the cancer itself as well as the tissue destruction caused by it, thereby ensuring complete healing. Another strategy is to engineer stem cells so that they express immune-modulating factors or anti-cancer agents in addition to CAR-T cell therapy. Through this joint approach, the curative effect is improved while targeting a range of processes involved in cancer development.

To sum up, integrating nanobiomaterials with stem cells and CAR-T cells as combination therapies presents an effective method toward enhancing therapeutic efficacy. By capitalizing on the advantages peculiar to each constituent part, current treatment modalities can be addressed where

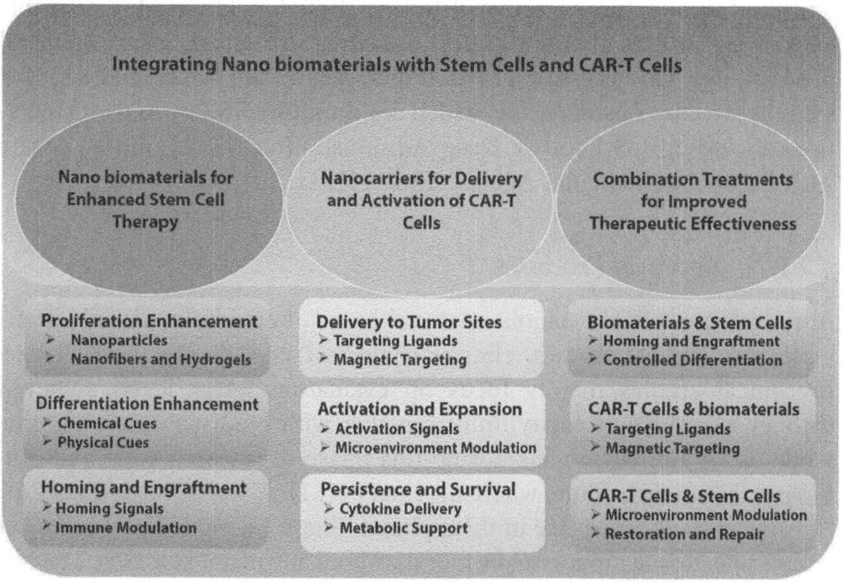

Figure 4.1 This figure illustrates the integration of nanobiomaterials with stem cells and CAR-T cells for advanced therapeutic strategies. The diagram outlines three major areas: (1) nanobiomaterials for enhanced stem cell therapy, which includes proliferation and differentiation enhancement and homing and engraftment; (2) nanocarriers for delivery and activation of CAR-T cells, focusing on targeted delivery, activation, expansion, and persistence; and (3) combination treatments that synergize biomaterials with stem cells and CAR-T cells to improve therapeutic outcomes. This multidisciplinary approach aims to optimize regenerative medicine and cancer immunotherapy by leveraging the unique properties of nanobiomaterials.

their weaknesses are known, thus leading to better outcomes both for cancer care and regenerative medicine. This holistic perspective represents a great leap forward within biomedicine, therefore paving the way for originality coupled with efficiency during treatment procedures [51]. Figure 4.1 demonstrates how nanobiomaterials optimize stem cell treatments alongside CAR-T delivery systems toward advancing combined therapeutic approaches in regenerative medicine and cancer immunotherapy.

4.7 Challenges and Future Perspectives

4.7.1 Regulatory Considerations for Nanobiomaterial-Based Therapeutics

The rules surrounding nanobiomaterial-based treatments are intricate and fluid due to the necessity of testing these breakthrough treatments extensively so as to guarantee safety measures and success rates. Approval processes for nanobiomaterial-based therapeutics involve several essential steps as well as considerations which are monitored by various organizations including US Food & Drug Administration (FDA) among others such European Medicines Agency (EMA) [131].

4.7.1.1 Approval Processes

Approval process for nanobiomaterial-based therapeutics usually comprises pre-clinical studies, clinical trials and post-market surveillance. Preclinical studies mainly focus on evaluating safety levels achieved through measuring toxicity limits together with biodistribution patterns while using animal models during this phase of research work but also *in vitro* tests may be done where necessary [131].

Clinical trials take place in three phases: Phase I designed primarily just ensure safe dosage ranges can be found without any major side effects resulting from its use which would later move into the next stage called phase II where effectiveness shall now become more visible alongside some possible adverse reactions occurring at higher dose levels than those observed earlier, i.e., now it becomes clear what works against what does not work so well under given conditions. Phase III confirms if indeed new treatment better than any other already known standard therapies showing also how they compare in terms of effectiveness between them when applied together or separately at different stages. Upon successful completion of clinical trials, a New Drug Application (NDA) or Biologics License Application (BLA) is submitted to the regulatory agency for review and approval [132].

4.7.1.2 *Safety Standards*

In order to address potential risks associated with their unique properties safety standards should be set up concerning nanobiomaterials. For instance, it has been found necessary by some regulatory bodies that there should exist comprehensive characterization methods capable of determining size distributions as well shapes alongside surface chemistries employed during synthesis processes which may introduce impurities into final products used for various biomedical purposes such drugs but not limited, too; hence, the need to evaluate these on a case by case basis might arise depending on nature toxic manifestations observed among others [133].

Moreover manufacturing practices involved in producing nanobiomaterials must comply with Good Manufacturing Practices (GMP) so as ensure repeatability in terms of purity levels achieved as well maintenance uniformity across batches produced this way making them safe during clinical applications where such materials are required, i.e., they shall always have same concentration strength irrespective whether used singly or combined together during therapeutic interventions aimed combating certain malignant growths within human body systems but sometimes scaling up how much you produce could also matter most if many patients need the same dose amounting to large quantities being made per batch without compromising quality aspects. Relevant quality control measures were taken into account here, including the determination of particle number concentrations [134].

4.7.1.3 *Challenges in Regulation*

Regulatory challenges can result from shortcomings arising due lack standard testing protocols applicable specifically toward nanomaterials thus leading to inadequate assessments regarding their bio-interactions within biological systems. Furthermore, interdisciplinary nature associated with these advanced healthcare interventions based on nanotechnology often requires experts' input drawn from different disciplines working together establish comprehensive regulatory frameworks capable of ensuring safety alongside efficient employment such products during various therapeutic procedures aimed at treating diverse ailments affecting humans globally [135].

4.7.1.4 Future Perspectives

Precision in testing methods used during future regulatory considerations for nanobiomaterial-based therapeutics is likely going become more refined over time. This can be achieved through adoption highly predictive computational models as well utilizing rapid screening approaches which take into account large numbers samples simultaneously, thereby eliminating too many animal tests. Regulatory standards harmonization among various regions, in addition to fostering worldwide development and endorsement of therapies that rely on nanobiomaterials, can also make it easier for them to be authorized. Such initiatives or collaborations like ICH (International Conference on Harmonization) are important because they help bring into line what is required by regulations while at the same time enhancing safe medical use for nanotechnology through universal involvement [134, 135].

4.7.2 Conquering Biocompatibility and Safety Issues

Biocompatibility and safety are the main worries for scientists involved in the creation of medical treatments based on nanobiomaterials. There are a number of ways to address these issues, which require that nanomaterials be safe for humans and do not produce unwanted reactions [136].

4.7.2.1 Material Selection and Surface Engineering

The choice of biocompatible materials is a key stage in designing safe nanobiomaterials. Polymers, lipids, metals and ceramics are commonly used because they have different properties that make them compatible with biological systems. For example, surface engineering techniques including PEGylation (coating with polyethylene glycol) can enhance biocompatibility by decreasing immunogenicity and extending circulation time [34].

4.7.2.2 Incorporating Biodegradability

Designing nanomaterials with the ability to biodegrade ensures their safe breakdown and elimination from the body. Biodegradable polymers like polylactic acid (PLA) or polyglycolic acid (PGA) break down into harmless products which are naturally metabolized. This approach prevents long-term accumulation that may cause toxicity [33].

4.7.2.3 *Mitigating Immune Responses*

Nanoparticles can provoke immune responses leading to inflammation or immunotoxicity. Measures taken to minimize these effects involve altering surface characteristics so as not to be recognized by immune system; stealth coatings such as PEG should be used as well as anti-inflammatory agents should be added. Also it is possible to affect size and shape optimization of nano-objects impacting their interaction with cells of immunity, thereby reducing adverse reactions provoked by this system [137].

4.7.2.4 *Evaluating Long-Term Safety*

Long-term safety assessments must account for potential chronicity in toxicity related with exposure toward nano-scale materials over extended periods of time spanned by such studies themselves being conducted upon treated animals under monitoring during which delayed response might occur thus indicating carcinogenic effect associated specifically with one particular organ among others like this. Advanced imaging methods along with various analytical techniques allow precise tracking not only distribution but also degradation kinetics associated with nanomaterials within biological systems over time [138].

4.7.2.5 *Regulatory Guidelines and Standardization*

Following regulatory requirements as well as standardizing practices will ensure that any medical treatments involving nanobiomaterials meet safety standards expected from them. Establishing uniform test methods like those detailed in International Organization for Standardization (ISO) fosters consistent appraisal of biocompatibility vis-à-vis safety across different research works done at various places or times [139].

4.7.2.6 *Interdisciplinary Collaboration*

Addressing issues related to the compatibility between biological systems and nanoparticles calls for cooperation among professionals specializing in materials science, toxicology, immunology, clinical medicine, etc., which can be achieved through interdisciplinary studies aimed at gaining comprehensive insights into these interactions so as to come up with safer therapies [140].

Therefore, it takes a multi-pronged approach toward dealing with biocompatibilities and safeties that may arise when using nano bio-manufactured drugs for treatment purposes—material selection, surface engineering, biodegradability, and immune response mitigation strategies should all be considered together against long-term evaluations in line with existing regulations between different disciplines. This is how we can make sure that they are secure and efficient especially within medical applications where they are employed mostly, thus allowing advanced therapeutic interventions [140].

4.7.3 Personalized Medicine and Patient-Specific Therapeutics

Nanobiomaterials and stem cell technology are moving personalized medicine toward treatment customization at genetic, molecular, and clinical levels that better improve the performance of therapy while at the same time minimizing its side effects.

4.7.3.1 Customized Drug Delivery

Nanobiomaterials can be used to create personalized drug delivery systems which can meet specific requirements of individual patients. For instance, one may design nanoparticles whose characteristics such as size, surface charge or targeting ligands vary so that drugs are transported directly into cancerous cells or organs. In this method, different surfaces of nanomaterials could carry various charges; meanwhile these particles might bear different sizes when necessary too. The reason behind aiming only at diseased tissues is because it reduces systemic toxicity while increasing therapeutic concentration within affected sites [26, 140].

4.7.3.2 Stem Cell Treatments for Particular Patients

By utilizing patient-specific cells in the creation of stem cell therapies, it is possible to come up with treatments that suit particular individuals. Various types of stem cells can be generated from induced pluripotent stem cells (iPSCs) which originate from a person's own body tissues before being differentiated into diverse cell categories required for regenerative medicine purposes. This means using one's own cells so as not to induce an immunological reaction against them, thereby making such treatments safer and more compatible also effective [61, 141].

4.7.3.3 *Genomic and Proteomic Profiling*

Progress made in genome-based and proteome-based profiling have facilitated identification of gene mutations; biomarkers as well molecular pathways linked to specific diseases thus aiding development of targeted treatment approaches based on individuals' biological make up against this background knowledge provided by these new technologies like nanobiomaterials science where we can design nanodevices specifically targeting certain gene mutations or use RNAi interference strategy through delivering RNAi agents using nanoscale systems such as those made from lipids or polymers carrying nucleic acids into cells to control gene expression levels [142].

4.7.3.4 *Customized Cancer Immunotherapy*

Nanobiomaterials can be integrated into personalized cancer immunotherapies like CAR-T cell therapy which will enable us improve its efficiency. For example, by attaching tumor-specific antigens onto them and using nanocarriers as vehicles for delivering these CAR-T cells we could achieve better results in terms of effectiveness and safety. Additionally, since every patient has different immune system characteristics this method could also help us design anti-tumor responses that are more potent against specific tumors while sparing normal tissues [117, 118].

4.7.3.5 *Precision Diagnostics*

The field of precision diagnostics greatly benefits from nanobiomaterials because they allow highly sensitive detection methods for disease biomarkers, thus enabling early diagnosis. For instance, diagnostic probes can be attached onto nanoparticles then used to identify very low amounts of biomolecules found in blood or tissue samples, thereby providing accurate diagnosis at an early stage when treatment is still possible. These personalized diagnostic tools make it easy to intervene promptly with appropriate therapies, hence improving outcomes among patients suffering from various illnesses [2, 7].

4.7.3.6 *Real-Time Monitoring and Feedback*

Real time monitoring systems can be achieved through combination of nanobiomaterials with wearable/implantable devices which continuously monitor changes in a patient's health status during therapy administration.

Nano-sensors are capable of detecting alterations within body fluids or tissues such as levels of certain proteins called biomarkers, physiological parameters like temperature, etc., drug concentrations so that clinicians receive continuous feedback about how well each patient responds to treatment over time. By dynamically tracking response rates and other relevant information, adjustments may then be made individually tailored toward optimizing therapeutic outcome in such cases where required [143].

4.7.3.7 Ethical and Social Considerations

Introduction of personalized medicine raises ethical issues regarding privacy rights protection against abuses: safeguarding personal data integrity while promoting responsible use thereof vis-à-vis social implications arising from unequal availability/accessibility/inclusion opportunities associated with advanced medical interventions on global scale particularly between different regions/countries or socio-economic groups [144]. Therefore, it is important that proper measures are put in place not only to protect patients' sensitive information but also ensure its accountability as well as fairness so that everyone can benefit equally from these breakthroughs regardless of their economic status [68].

Personalized medicine will be better served by integrating nanobiomaterials and stem cell technology into the system. The regenerative properties of stem cells combined with unique features found in nanomaterials could pave way for developing more targeted therapies which are effective against specific diseases at individual level. Henceforth, this approach represents a new era in healthcare where we can expect higher cure rates accompanied by lower side effects across various conditions [145].

4.8 Conclusion

4.8.1 Stem Cells, Nanotechnology, Exosomes, and CAR-T Cells: What the Future Holds

A combination of stem cells, nanotechnology, exosomes and CAR-T cells marks a front-line approach toward the development of advanced therapies that provide synergistic benefits in treating a plethora of diseases [118, 146]. The next steps for this multidisciplinary area will be aimed at increasing the combined technologies' therapeutic potential as well as safety and accuracy.

4.8.1.1 Stem Cell Treatment Improvement

The future researches will seek to develop advanced nanomaterials which promote cell proliferation, differentiation and homing thereby enhancing stem cell therapies. Revolutionary matrices and scaffolds incorporating bioactive nanomaterials will support tissue regeneration and integration thus improving outcomes in regenerative medicine. Additionally, there is going to be an investigation into finding ways through which delivery and engraftment of stem cells can be optimized so that efficient repair damaged tissues becomes possible [40].

4.8.1.2 Targeted Release Mechanisms

Nanotechnology will still have a crucial role to play in designing targeted release mechanisms for both stem cells and CAR-T cells. Efforts are going to concentrate on coming up with nanocarriers that can deliver therapeutic agents or cells precisely to specific sites within the body thereby making them more effective therapeutically while at the same time reducing off-target effects. The advancement of nanoparticle engineering allows for creation stimuli-responsive delivery systems which release cargo upon receipt certain biological signals [117, 122].

4.8.1.3 Therapeutics Based on Exosome

As natural carriers, exosomes hold great promise for drug delivery targeting only affected organs but also intercellular communication among various parts of human body including immune system regulation. In the future, scientists would like to find out how useful exosome can be when it comes delivering drugs such nucleic acid proteins etcetera into target cells where they needed most. Engineering exosomes so that their efficiency loading capacity increased could therefore serve as another breakthrough in creating new precision-based medicines against different kinds cancers [94]

4.8.1.4 CAR-T Cell Engineering

CAR-T cell technology is advancing day by day in order to overcome existing challenges like toxicity or resistance thus improving safety and efficiency of these treatment methods. In future, researchers are expected to come up with ways on how best they can use nanocarriers during delivery, activation as well as persistence when it comes targeting cancer cells using CAR-T cells. Besides that, there will be investigations carried out concerning combination therapies such immune checkpoint inhibitors together with targeted therapy among others which may produce synergistic anti-tumor effects if combined with CAR T cells [118].

4.8.1.5 *Combination Therapy*

Future researches shall pay more attention toward finding out various combination therapies involving stem cells, nanotechnology, exosomes and CAR-T cells so as to achieve better therapeutic outcomes [117, 145, 146]. The main purpose for these combinations is usually based on the fact that each technology has its own unique advantage over another hence when all put together can do much better than any single approach. For instance, a comprehensive treatment strategy against cancer other diseases could involve stem cell-based tissue regeneration coupled with CAR T cell immunotherapy plus nanoparticle mediated drug delivery [115, 119].

4.8.1.6 *Personalized Medicine*

Nanotechnology coupled with stem cell research advancements will lead to development of personalized medicine approaches for different individuals. What this means is that future researches will focus more on integration of genomic, proteomic and metabolomic data in designing personalized treatment regimens which maximize therapeutic efficacy whilst minimizing adverse effects caused by drugs used during treatments [8, 145]. Personalized diagnostic tools enabled by nanotechnology could also be used for early detection of diseases thus guiding tailored treatment plans aimed at improving patient outcome [74]

4.8.1.7 *Regulatory and Ethical Considerations*

With continuous changes witnessed within advanced therapies field; regulatory requirements become very necessary so as to guarantee safety and effectiveness whenever such products are being translated into clinical practice. This implies that future studies ought to involve collaboration between scientists, clinicians as well regulatory authorities among other stakeholders who may help in coming up comprehensive guidelines or standards which governs development, testing as well approval process for these innovative therapies [131, 139, 144].

4.8.2 The Potential of Nanobiomaterials for Precision Therapeutics

There has been a revolutionary change in the area of precision medicine with the advent of nanobiomaterials which provides unprecedented opportunities for improving the effectiveness, specificity and safety of treatment methods used in medicine. The concept is built on such properties inherent to nanoobjects as their small size, large surface area, and the ability to modify their chemical composition.

4.8.2.1 Improved Drug Delivery

Nanobiomaterials make it possible to deliver drugs directly into certain cells or tissues by establishing systems at this level thus aiding targeted drug delivery. This approach reduces side effects that may be experienced off target sites while also reducing systemic toxicity, making drugs less harmful overall. Nanocarriers are capable of releasing substances when certain conditions are met like changing pH levels or presence/absence of specific enzymes which ensures that there will always be enough amount of medication needed right where required over long periods, ensuring sustained release action [147].

4.8.2.2 More Accurate Diagnosis

Integration of nanostructures into diagnostic devices increases sensitivity thereby allowing for early detection and correct identification diseases. Assays based on nanoparticles as well as imaging agents can pick up even very low concentrations biomarkers thus giving more information about disease onset or progression useful in decision-making process concerning its management. This makes it possible to intervene timely so that each person gets treated individually at an appropriate time, leading better outcomes to being achieved [148].

4.8.2.3 Regenerative Medicine

Another important role played by nanobiomaterials is creating environment necessary for cell growth, differentiation tissue regeneration they serve as scaffolds matrices upon which these processes take place during regenerative therapy. Active biological components added into scaffold materials promote adhesion proliferation differentiation cells, thereby accelerating repair damaged tissues through natural healing mechanisms sometimes supported also artificial means like grafting skin patches etcetera. Furthermore, growth factors along other bioactive molecules can be delivered using nanosized structures stimulating new development, thus increasing the chances of a successful outcome following treatment intervention [40, 58].

4.8.2.4 Cancer Treatment

In cancer management, nanotechnologies offer a number of benefits including site-specific delivery chemotherapeutics, better visualization systems and more effective immunotherapy methods. Particles at this level are functionalized such that they recognize bind with particular tumor markers, thus ensuring that drugs act only on malignant cells. Additionally, it has been found out that these agents could also improve how immune system responds against tumors hence enhancing activity, persistence as well targeting delivery routes for CAR-T kind therapies [128].

4.8.2.5 Gene Therapy

Nanobiomaterials have different ways to enable nucleic acid delivery in gene therapy where one can use DNA, RNA or siRNA depending on what needs be done at cellular level. This is because carriers protect them from being damaged while enhancing uptake into target areas either cytoplasm or nucleus which are known control centers for all genes expression pathways. It therefore becomes possible regulating precisely when and how much each gene should produce, thus opening new possibilities toward finding cure against genetic disorders as well cancers among other diseases [149].

4.8.2.6 Individualized Approach

Combining nanobiomaterials with individualized medicine allows creation patient-specific treatments based personal genomic data obtained during diagnosis stage itself. For instance, nanostructured materials could be designed such way that they only interact mutated regions within genes or deliver customized drug cocktails directly onto these sites thereby maximizing therapeutic efficiency while minimizing side effects associated with non-selective treatments. Nanotechnology powered tests would also enable early detection tailored management strategies through which many lives may save by preventing disease progression into advanced stages before it is too late [1, 145].

4.8.2.7 Future Outlook

The prospects for using nanobiomaterials in precision medicine appear bright given ongoing studies toward creating improved carriers as well discovering innovative uses them. These include development of safer materials, enhancement transport systems within body among others so that treatment outcome is successful without causing harm elsewhere along the treatment pathway due to their smaller size and high reactivity rates compared larger counterparts like bulkier drugs. Continued collaborations between scientists' and clinicians regulators will facilitate translating these findings into practice, leading to better care provision for all patients [140].

4.9 Summary

The fusion of stem cells, nanotechnology, exosomes, and CAR-T cells represents the beginning of a new chapter in precision therapies that have the potential to change the face of treatment for complex diseases. Nanobiomaterials have completely transformed drug delivery systems through permitting targeted, regulated and efficient administration of therapy while minimizing off-target effects. The biodegradability, tunable characteristics, and biocompatibility properties of such materials make them relevant to various medical areas like cancer therapy and regenerative medicine. Stem cell technologies particularly those dealing with mesenchymal stem cells (MSCs) as well pluripotent stem cells (PSCs) are promising for tissue regeneration due to differentiation potency into numerous types of cells as well as immunomodulatory features.

These natural nanoparticles serve as essential mediators between other cells allowing intercommunication hence providing fresh approaches toward targeted pharmaceutical delivery besides immune modulations. It is a significant development in personalized treatment that has turned exosomes into therapeutic agents by modifying them so that they can be directed at specific elements in the body carrying therapeutics. CAR-T cell therapy stands out among others since it is one way where we can reprogram immunity in order to target and get rid of cancerous growths from any part of human body including brain. Moreover, improved clinical efficiency plus ensuring safety could simultaneously be addressed by employing nanotechnology for better CAR-T cell delivery mechanisms.

However, there are still several challenges such as biocompatibility issues, immune rejection concerns during transplantation, tough regulations guiding their use in treating patients over time, hence posing safety risks if not properly handled. These obstacles must therefore be surmounted through innovative engineering strategies stringent testing and interdisciplinary cooperation for effective transfer into clinical practice. A future personalized medicine depends on integrating these technologies seamlessly so that individual patients have a better chance of finding a cure that is tailored to meet their particular needs. To sum up, the integration of nanobiomaterials, stem cells, exosomes, and CAR-T cells represents a paradigm shift in precision therapeutics. With such an interdisciplinary basis, there are chances that it can transform medical approaches hence giving hope for better personalized and safe therapeutic strategies against these complicated ailments.

References

1. Marques, L., *et al.*, Advancing Precision Medicine: A Review of Innovative *In Silico* Approaches for Drug Development, Clinical Pharmacology and Personalized Healthcare. *Pharmaceutics*, 16, 3, Art. no. 3, 1–39, Mar. 2024. doi: 10.3390/pharmaceutics16030332.
2. Naithani, N., Sinha, S., Misra, P., Vasudevan, B., Sahu, R., Precision medicine: Concept and tools. *Med J. Armed Forces India*, 77, 3, 249–257, Jul. 2021. doi: 10.1016/j.mjafi.2021.06.021.
3. Hurry, R. and basit, A., *Precision Medicine: Tailoring Treatment to the Individual's Genomic Profile*, Department of Public Health, IPH, 2023, doi: 10.13140/RG.2.2.32452.19848.
4. Wang, R.C. and Wang, Z., Precision Medicine: Disease Subtyping and Tailored Treatment. *Cancers (Basel)*, 15, 15, 3837, Jul. 2023. doi: 10.3390/cancers15153837.
5. Hood, L. and Rowen, L., The Human Genome Project: big science transforms biology and medicine. *Genome Med.*, 5, 9, 79, Sep. 2013. doi: 10.1186/gm483.
6. ScienceDirect, *Cancer Genome Project - an overview | ScienceDirect Topics*, Accessed: 2008. [Online]. Available: https://www.sciencedirect.com/topics/biochemistry-genetics-and-molecular-biology/cancer-genome-project.
7. Konig, M.F., The rise of precision cellular therapies. *Nat. Rev. Rheumatol.*, 20, 2, 69–70, Feb. 2024. doi: 10.1038/s41584-023-01073-6.
8. Hasanzad, M., *et al.*, Precision medicine journey through omics approach. *J. Diabetes Metab. Disord.*, 21, 881–888, Nov. 2021. doi: 10.1007/s40200-021-00913-0.
9. Alowais, S.A., *et al.*, Revolutionizing healthcare: the role of artificial intelligence in clinical practice. *BMC Med. Educ.*, 23, 689, Sep. 2023. doi: 10.1186/s12909-023-04698-z.
10. Ahmed, Z., Mohamed, K., Zeeshan, S., Dong, X., Artificial intelligence with multi-functional machine learning platform development for better healthcare and precision medicine. *Database (Oxford)*, 2020, baaa010, Mar. 2020. doi: 10.1093/database/baaa010.

11. Javaid, M., Haleem, A., Pratap Singh, R., Suman, R., Rab, S., Significance of machine learning in healthcare: Features, pillars and applications. *Int. J. Intell. Netw.*, 3, 58–73, Jan. 2022. doi: 10.1016/j.ijin.2022.05.002.
12. Sterner, R.C. and Sterner, R.M., CAR-T cell therapy: current limitations and potential strategies. *Blood Cancer J.*, 11, 4, 1–11, Apr. 2021. doi: 10.1038/s41408-021-00459-7.
13. Janjigian, Y.Y., *et al.*, First-line pembrolizumab and trastuzumab in HER2-positive esophagogastric cancer. *Lancet Oncol.*, 21, 6, 821–831, Jun. 2020. doi: 10.1016/S1470-2045(20)30169-8.
14. Agostinetto, E., *et al.*, Immunotherapy for HER2-Positive Breast Cancer: Clinical Evidence and Future Perspectives. *Cancers (Basel)*, 14, 9, 2136, Apr. 2022. doi: 10.3390/cancers14092136.
15. PricewaterhouseCoopers, *From healthcare to life care: Transforming precision health to provide societal and financial value*, PwC, 2023, Accessed: Jul. 11, 2024. [Online]. Available: https://www.pwc.com/gx/en/issues/transformation/insights/transforming-precision-health.html.
16. Nature Education, *Pharmacogenomics and Personalized Medicine | Learn Science at Scitable*, 2008, Accessed: Jul. 11, 2024. [Online]. Available: http://www.nature.com/scitable/topicpage/pharmacogenomics-and-personalized-medicine-643.
17. Tabish, S.A., Complementary and Alternative Healthcare: Is it Evidence-based? *Int J. Health Sci. (Qassim)*, 2, 1, V–IX, Jan. 2008.
18. National Cancer Institute, *Chemotherapy to Treat Cancer - NCI*, 2022, Accessed: Jul. 11, 2024. [Online]. Available: https://www.cancer.gov/about-cancer/treatment/types/chemotherapy.
19. Andrews, G.J., Complementary and Alternative Medicine, in: *International Encyclopedia of Human Geography (Second Edition)*, A. Kobayashi (Ed.), pp. 351–355, Elsevier, Oxford, 2020, doi: 10.1016/B978-0-08-102295-5.10367-1.
20. Talib, W.H., Alsayed, A.R., Barakat, M., Abu-Taha, M., II, Mahmod, A., II, Targeting Drug Chemo-Resistance in Cancer Using Natural Products. *Biomedicines*, 9, 10, 1353, Sep. 2021. doi: 10.3390/biomedicines9101353.
21. Dong, Y., Wu, X., Chen, X., Zhou, P., Xu, F., Liang, W., Nanotechnology shaping stem cell therapy: Recent advances, application, challenges, and future outlook. *Biomed. Pharmacother.*, 137, 111236, May 2021. doi: 10.1016/j.biopha.2021.111236.
22. Abbasi, R., Shineh, G., Mobaraki, M., Doughty, S., Tayebi, L., Structural parameters of nanoparticles affecting their toxicity for biomedical applications: a review. *J. Nanopart. Res.*, 25, 3, 43, 2023. doi: 10.1007/s11051-023-05690-w.
23. Altammar, K.A., A review on nanoparticles: characteristics, synthesis, applications, and challenges. *Front. Microbiol.*, 14, 1155622, Apr. 2023. doi: 10.3389/fmicb.2023.1155622.
24. Samuel, P., Sundarraj, S., Sudarmani, D.N.P., Samuel, P., Sundarraj, S., Sudarmani, D.N.P., Nanotechnology-Based Stem Cell Therapy: Current Status and Perspectives, in: *Possibilities and Limitations in Current Translational*

Stem Cell Research, IntechOpen, 2023, doi: 10.5772/intechopen.109275. https://www.intechopen.com/chapters/86123

25. Kim, J., *et al.*, Use of Nanoparticle Contrast Agents for Cell Tracking with Computed Tomography. *Bioconjug Chem*, 28, 6, 1581–1597, Jun. 2017. doi: 10.1021/acs.bioconjchem.7b00194.
26. Patra, J.K., *et al.*, Nano based drug delivery systems: recent developments and future prospects. *J. Nanobiotechnol.*, 16, 1, 71, Sep. 2018. doi: 10.1186/s12951-018-0392-8.
27. Satchanska, G., Davidova, S., Petrov, P.D., Natural and Synthetic Polymers for Biomedical and Environmental Applications. *Polym. (Basel)*, 16, 8, 1159, Apr. 2024. doi: 10.3390/polym16081159.
28. Luo, D., Wang, X., Burda, C., Basilion, J.P., Recent Development of Gold Nanoparticles as Contrast Agents for Cancer Diagnosis. *Cancers (Basel)*, 13, 8, 1825, Apr. 2021. doi: 10.3390/cancers13081825.
29. Ahmed, E.M., Hydrogel: Preparation, characterization, and applications: A review. *J. Adv. Res.*, 6, 2, 105–121, Mar. 2015. doi: 10.1016/j.jare.2013.07.006.
30. Trucillo, P., Biomaterials for Drug Delivery and Human Applications. *Materials*, 17, 2, 1–27, Art. no. 2, Jan. 2024. doi: 10.3390/ma17020456.
31. Schubert, J. and Chanana, M., Coating Matters: Review on Colloidal Stability of Nanoparticles with Biocompatible Coatings in Biological Media, Living Cells and Organisms. *Curr. Med. Chem.*, 25, 35, 4556–4586, Oct. 2019. doi: 10.2174/0929867325666180601101859.
32. Jiang, T., Jin, K., Liu, X., Pang, Z., 8 - Nanoparticles for tumor targeting, in: *Biopolymer-Based Composites*, S. Jana, S. Maiti, S. Jana (Eds.), pp. 221–267, Woodhead Publishing, 2017, doi: 10.1016/B978-0-08-101914-6.00008-9. https://www.sciencedirect.com/science/article/abs/pii/B9780081019146000089
33. Makadia, H.K. and Siegel, S.J., Poly Lactic-co-Glycolic Acid (PLGA) as Biodegradable Controlled Drug Delivery Carrier. *Polym. (Basel)*, 3, 3, 1377–1397, Sep. 2011. doi: 10.3390/polym3031377.
34. Das, S.S., *et al.*, Stimuli-Responsive Polymeric Nanocarriers for Drug Delivery, Imaging, and Theragnosis. *Polym. (Basel)*, 12, 6, 1397, Jun. 2020. doi: 10.3390/polym12061397.
35. Mosleh-Shirazi, S., *et al.*, Nanotechnology Advances in the Detection and Treatment of Cancer: An Overview. *Nanotheranostics*, 6, 4, 400–423, Aug. 2022. doi: 10.7150/ntno.74613.
36. Subhan, M.A., Yalamarty, S.S.K., Filipczak, N., Parveen, F., Torchilin, V.P., Recent Advances in Tumor Targeting via EPR Effect for Cancer Treatment. *J. Pers Med*, 11, 6, 571, Jun. 2021. doi: 10.3390/jpm11060571.
37. Rosenblum, D., Joshi, N., Tao, W., Karp, J.M., Peer, D., Progress and challenges towards targeted delivery of cancer therapeutics. *Nat. Commun.*, 9, 1, 1410, Apr. 2018. doi: 10.1038/s41467-018-03705-y.
38. Duan, L., *et al.*, Nanoparticle Delivery of CRISPR/Cas9 for Genome Editing. *Front. Genet.*, 12, May 2021. doi: 10.3389/fgene.2021.673286.

39. Hofmann, M.-C., Stem Cells and Nanomaterials. *Adv Exp Med Biol*, 811, 255–275, 2014. doi: 10.1007/978-94-017-8739-0_13.
40. Villanueva-Flores, F., Garcia-Atutxa, I., Santos, A., Armendariz-Borunda, J., Toward a New Generation of Bio-Scaffolds for Neural Tissue Engineering: Challenges and Perspectives. *Pharmaceutics*, 15, 6, 1750, Jun. 2023. doi: 10.3390/pharmaceutics15061750.
41. Zakrzewski, J.L., van den Brink, M.R.M., Hubbell, J.A., Overcoming immunological barriers in regenerative medicine. *Nat. Biotechnol.*, 32, 8, 786–794, Aug. 2014. doi: 10.1038/nbt.2960.
42. Fu, Y., *et al.*, Applications of nanomaterial technology in biosensing. *J. Sci.: Adv. Mater. Devices*, 9, 2, 100694, Jun. 2024. doi: 10.1016/j.jsamd.2024.100694.
43. Mahla, R.S., Stem Cells Applications in Regenerative Medicine and Disease Therapeutics. *Int. J. Cell Biol.*, 2016, 6940283, 2016. doi: 10.1155/2016/6940283.
44. Miller, L.W. and Perin, E.C., Introduction and Overview of Stem Cells, in: *Stem Cell and Gene Therapy for Cardiovascular Disease*, E.C. Perin, L.W. Miller, D.A. Taylor, J.T. Willerson (Eds.), pp. 3–11, Academic Press, Boston, 2016, doi: 10.1016/B978-0-12-801888-0.00001-1.
45. Zakrzewski, W., Dobrzyński, M., Szymonowicz, M., Rybak, Z., Stem cells: past, present, and future. *Stem Cell Res. Ther.*, 10, 1, 68, Feb. 2019. doi: 10.1186/s13287-019-1165-5.
46. Nagamura-Inoue, T. and He, H., Umbilical cord-derived mesenchymal stem cells: Their advantages and potential clinical utility. *World J. Stem Cells*, 6, 2, 195–202, Apr. 2014. doi: 10.4252/wjsc.v6.i2.195.
47. Gooi, L.M. and Gopalakrishnan, J., Regulation of Stem Cells in Their Niche. *Curr. Stem Cell Rep.*, 2, 3, 282–289, Sep. 2016. doi: 10.1007/s40778-016-0048-2.
48. Cai, Y., Wang, S., Qu, J., Belmonte, J.C., II, Liu, G.-H., Rejuvenation of Tissue Stem Cells by Intrinsic and Extrinsic Factors. *Stem Cells Transl. Med.*, 11, 3, 231–238, Mar. 2022. doi: 10.1093/stcltm/szab012.
49. Oshimori, N. and Fuchs, E., The Harmonies Played by TGF-β in Stem Cell Biology. *Cell Stem Cell*, 11, 6, 751–764, Dec. 2012. doi: 10.1016/j.stem.2012.11.001.
50. Patel, D.M., Shah, J., Srivastava, A.S., Therapeutic Potential of Mesenchymal Stem Cells in Regenerative Medicine. *Stem Cells Int.*, 2013, 496218, 2013. doi: 10.1155/2013/496218.
51. Jin, Y., Li, S., Yu, Q., Chen, T., Liu, D., Application of stem cells in regeneration medicine. *MedComm (2020)*, 4, 4, e291, Jun. 2023. doi: 10.1002/mco2.291.
52. Terashvili, M. and Bosnjak, Z.J., Stem Cell Therapies in Cardiovascular Disease. *J. Cardiothorac Vasc. Anesth.*, 33, 1, 209–222, Jan. 2019. doi: 10.1053/j.jvca.2018.04.048.
53. Cecerska-Heryć, E., *et al.*, The Use of Stem Cells as a Potential Treatment Method for Selected Neurodegenerative Diseases: Review. *Cell. Mol. Neurobiol.*, 43, 6, 2643–2673, 2023. doi: 10.1007/s10571-023-01344-6.

54. Kangari, P., Talaei-Khozani, T., Razeghian-Jahromi, I., Razmkhah, M., Mesenchymal stem cells: amazing remedies for bone and cartilage defects. *Stem Cell Res. Ther.*, 11, 492, Nov. 2020. doi: 10.1186/s13287-020-02001-1.
55. Silva, I.B.B., Kimura, C.H., Colantoni, V.P., Sogayar, M.C., Stem cells differentiation into insulin-producing cells (IPCs): recent advances and current challenges. *Stem Cell Res. Ther.*, 13, 309, Jul. 2022. doi: 10.1186/s13287-022-02977-y.
56. Fiorotto, R., Amenduni, M., Mariotti, V., Fabris, L., Spirli, C., Strazzabosco, M., Liver diseases in the dish: iPSC and organoids as a new approach to modeling liver diseases. *Biochim. Biophys. Acta, Mol. Basis Dis.*, 1865, 5, 920–928, May 2019. doi: 10.1016/j.bbadis.2018.08.038.
57. Sharma, A. and Jaganathan, B.G., Stem Cell Therapy for Retinal Degeneration: The Evidence to Date. *Biologics*, 15, 299–306, Jul. 2021. doi: 10.2147/BTT.S290331.
58. Arora, D., Pant, P., Sharma, P.K., P., Trends in Functional Biomaterials in Tissue Engineering and Regenerative Medicine, in: *Biomaterials in Tissue Engineering and Regenerative Medicine: From Basic Concepts to State of the Art Approaches*, B. Bhaskar, Sreenivasa Rao, N. Kasoju, V. Nagarjuna, R.R. Baadhe (Eds.), pp. 215–269, Springer, Singapore, 2021, doi: 10.1007/978-981-16-0002-9_7.
59. Arora, D., Babakhanova, G., Simon, C.G., Jr, Tissue Engineering Measurands. *ACS Biomater. Sci. Eng.*, 6, 10, 5368–5376, Oct. 2020. doi: 10.1021/acsbiomaterials.0c00475.
60. Chouhan, D., Kaushik, S., Arora, D., Trends in Bio-Derived Biomaterials in Tissue Engineering, in: *Biomaterials in Tissue Engineering and Regenerative Medicine: From Basic Concepts to State of the Art Approaches*, B. Bhaskar, P. Sreenivasa Rao, N. Kasoju, V. Nagarjuna, R.R. Baadhe (Eds.), pp. 163–213, Springer, Singapore, 2021, doi: 10.1007/978-981-16-0002-9_6.
61. Hoang, D.M., *et al.*, Stem cell-based therapy for human diseases. *Signal Transduct. Target Ther.*, 7, 272, Aug. 2022. doi: 10.1038/s41392-022-01134-4.
62. Abdelkrim, H., Juan, D.-B., Jane, W., Mohamed, A., Bernat, S., The immune boundaries for stem cell based therapies: problems and prospective solutions. *J. Cell. Mol. Med.*, 13, 8a, 1464–1475, Aug. 2009. doi: 10.1111/j.1582-4934.2009.00837.x.
63. Rong, Z., *et al.*, An effective approach to prevent immune rejection of human ESC-derived allografts. *Cell Stem Cell*, 14, 1, 121–130, Jan. 2014. doi: 10.1016/j.stem.2013.11.014.
64. Buriak, I.A., *et al.*, Preservation and Storage of Cells for Therapy: Fundamental Aspects of Low Temperature Science, in: *Cell Engineering and Regeneration*, J.M. Gimble, D. Marolt Presen, R.O.C. Oreffo, S. Wolbank, H. Redl (Eds.), pp. 1–60, Springer International Publishing, Cham, 2020, doi: 10.1007/978-3-319-37076-7_67-2.
65. Yu, N.-H., *et al.*, Optimal Stem Cell Transporting Conditions to Maintain Cell Viability and Characteristics. *Tissue Eng. Regen Med*, 15, 5, 639–647, Jul. 2018. doi: 10.1007/s13770-018-0133-y.

66. Wu, R.-X., Xu, X.-Y., Wang, J., He, X.-T., Sun, H.-H., Chen, F.-M., Biomaterials for endogenous regenerative medicine: Coaxing stem cell homing and beyond. *Appl. Mater. Today*, 11, 144–165, Jun. 2018. doi: 10.1016/j.apmt.2018.02.004.
67. Heslop, J.A., *et al.*, Concise Review: Workshop Review: Understanding and Assessing the Risks of Stem Cell-Based Therapies. *Stem Cells Transl. Med.*, 4, 4, 389–400, Apr. 2015. doi: 10.5966/sctm.2014-0110.
68. V., V., *et al.*, Ethical and Safety Issues of Stem Cell-Based Therapy. *Int. J. Med Sci.*, 15, 1, 36–45, Jan. 2018. doi: 10.7150/ijms.21666.
69. Driscoll, D., Farnia, S., Kefalas, P., Maziarz, R.T., Concise Review: The High Cost of High Tech Medicine: Planning Ahead for Market Access. *Stem Cells Transl. Med.*, 6, 8, 1723–1729, Jul. 2017. doi: 10.1002/sctm.16-0487.
70. Kalluri, R. and LeBleu, V.S., The biology, function, and biomedical applications of exosomes. *Science*, 367, 6478, eaau6977, Feb. 2020. doi: 10.1126/science.aau6977.
71. Zhang, Y., Liu, Y., Liu, H., Tang, W.H., Exosomes: biogenesis, biologic function and clinical potential. *Cell Biosci.*, 9, 19, Feb. 2019. doi: 10.1186/s13578-019-0282-2.
72. Han, Q.-F., *et al.*, Exosome biogenesis: machinery, regulation, and therapeutic implications in cancer. *Mol. Cancer*, 21, 207, Nov. 2022. doi: 10.1186/s12943-022-01671-0.
73. Koh, H.B., Kim, H.J., Kang, S.-W., Yoo, T.-H., Exosome-Based Drug Delivery: Translation from Bench to Clinic. *Pharmaceutics*, 15, 8, 2042, Jul. 2023. doi: 10.3390/pharmaceutics15082042.
74. Crenshaw, B.J., Sims, B., Matthews, Q.L., Biological function of exosomes as diagnostic markers and therapeutic delivery vehicles in carcinogenesis and infectious diseases. *Nanomedicines*, [Preprint]. doi:10.5772/intechopen.80225.
75. Han, W., *et al.*, The emerging role of exosomes in communication between the periphery and the central nervous system. *MedComm (2020)*, 4, 6, e410, Oct. 2023. doi: 10.1002/mco2.410.
76. Nicolini, A., Ferrari, P., Biava, P.M., Exosomes and Cell Communication: From Tumour-Derived Exosomes and Their Role in Tumour Progression to the Use of Exosomal Cargo for Cancer Treatment. *Cancers (Basel)*, 13, 4, 822, Feb. 2021. doi: 10.3390/cancers13040822.
77. Liang, Y., Duan, L., Lu, J., Xia, J., Engineering exosomes for targeted drug delivery. *Theranostics*, 11, 7, 3183–3195, Jan. 2021. doi: 10.7150/thno.52570.
78. Fu, S., Wang, Y., Xia, X., Zheng, J.C., Exosome engineering: Current progress in cargo loading and targeted delivery. *NanoImpact*, 20, 100261, Oct. 2020. doi: 10.1016/j.impact.2020.100261.
79. Ren, E., Liu, C., Lv, P., Wang, J., Liu, G., Genetically Engineered Cellular Membrane Vesicles as Tailorable Shells for Therapeutics. *Adv. Sci. (Weinh)*, 8, 21, 2100460, Sep. 2021. doi: 10.1002/advs.202100460.

80. Sadeghi, S., Tehrani, F.R., Tahmasebi, S., Shafiee, A., Hashemi, S.M., Exosome engineering in cell therapy and drug delivery. *Inflammopharmacology*, 31, 1, 145–169, 2023. doi: 10.1007/s10787-022-01115-7.
81. Young, J.L. and Dean, D.A., Electroporation-Mediated Gene Delivery. *Adv. Genet.*, 89, 49–88, 2015. doi: 10.1016/bs.adgen.2014.10.003.
82. Pitt, W.G., Husseini, G.A., Staples, B.J., Ultrasonic Drug Delivery – A General Review. *Expert Opin. Drug Deliv.*, 1, 1, 37–56, Nov. 2004. doi: 10.1517/17425247.1.1.37.
83. Wang, Y., Shim, M.S., Levinson, N.S., Sung, H.-W., Xia, Y., Stimuli-Responsive Materials for Controlled Release of Theranostic Agents. *Adv. Funct. Mater.*, 24, 4206–4220, 27Jul. 2014. doi: 10.1002/adfm.201400279.
84. Silva, A.K.A., *et al.*, Combining magnetic nanoparticles with cell derived microvesicles for drug loading and targeting. *Nanomed. Nanotechnol. Biol. Med.*, 11, 3, 645–655, Apr. 2015. doi: 10.1016/j.nano.2014.11.009.
85. Wang, L., Yu, X., Zhou, J., Su, C., Extracellular Vesicles for Drug Delivery in Cancer Treatment. *Biol. Proced. Online*, 25, 28, Nov. 2023. doi: 10.1186/s12575-023-00220-3.
86. Xiao, L., Hareendran, S., Loh, Y.P., Function of exosomes in neurological disorders and brain tumors. *Extracell. Vesicles Circulating Nucleic Acids*, 2, 55, 2021. doi: 10.20517/evcna.2021.04.
87. Farahzadi, R., Fathi, E., Valipour, B., Ghaffary, S., Stem cells-derived exosomes as cardiac regenerative agents. *Int. J. Cardiol. Heart Vasc.*, 52, 101399, Mar. 2024. doi: 10.1016/j.ijcha.2024.101399.
88. Shen, Z., Huang, W., Liu, J., Tian, J., Wang, S., Rui, K., Effects of Mesenchymal Stem Cell-Derived Exosomes on Autoimmune Diseases. *Front. Immunol.*, 12, 749192, Sep. 2021. doi: 10.3389/fimmu.2021.749192.
89. Xu, Z., Zeng, S., Gong, Z., Yan, Y., Exosome-based immunotherapy: a promising approach for cancer treatment. *Mol. Cancer*, 19, 160, Nov. 2020. doi: 10.1186/s12943-020-01278-3.
90. Justiz Vaillant, A.A., Modi, P., Mohammadi, O., Graft-Versus-Host Disease, in: *StatPearls*, StatPearls Publishing, Treasure Island (FL, 2024, Accessed: Aug. 11, 2024. [Online]. Available: http://www.ncbi.nlm.nih.gov/books/NBK538235/.
91. Quiñones-Vico, M., II, *et al.*, The Role of Exosomes Derived From Mesenchymal Stromal Cells in Dermatology. *Front. Cell Dev. Biol.*, 9, 647012, Apr. 2021. doi: 10.3389/fcell.2021.647012.
92. Lu, J., Wang, Q.-Y., Sheng, J.-G., Exosomes in the Repair of Bone Defects: Next-Generation Therapeutic Tools for the Treatment of Nonunion. *Biomed Res. Int.*, 2019, 1983131, Aug. 2019. doi: 10.1155/2019/1983131.
93. Zhu, L., *et al.*, Mesenchymal Stem Cell–Derived Exosomes in Various Chronic Liver Diseases: Hype or Hope? *J. Inflamm. Res.*, 17, 171–189, Jan. 2024. doi: 10.2147/JIR.S439974.

94. Creative Biolabs Exosomes, *Diagnostic Application of Exosomes*, 2023, [Online]. Available: https://www.creative-biolabs.com/exosome/diagnostic-application-of-exosomes.htm.
95. *Why is CAR T cell therapy seen as the future of cancer treatment?*, Karkinos Healthcare, 2024, Accessed: Aug. 12, 2024. [Online]. Available: https://www.karkinos.in/why-is-car-t-cell-therapy-seen-as-the-future-of-cancer-treatment/.
96. Ahmad, Z.A., Yeap, S.K., Ali, A.M., Ho, W.Y., Alitheen, N.B.M., Hamid, M., scFv Antibody: Principles and Clinical Application. *Clin. Dev. Immunol.*, 2012, 980250, 2012. doi: 10.1155/2012/980250.
97. Ambrose, C., *et al.*, Anti-CD19 CAR T cells potently redirected to kill solid tumor cells. *PLoS One*, 16, 3, e0247701, Mar. 2021. doi: 10.1371/journal.pone.0247701.
98. Chiu, M.L., Goulet, D.R., Teplyakov, A., Gilliland, G.L., Antibody Structure and Function: The Basis for Engineering Therapeutics. *Antibodies (Basel)*, 8, 4, 55, Dec. 2019. doi: 10.3390/antib8040055.
99. Meng, X., Jing, R., Qian, L., Zhou, C., Sun, J., Engineering Cytoplasmic Signaling of CD28ζ CARs for Improved Therapeutic Functions. *Front. Immunol.*, 11, 1046, Jun. 2020. doi: 10.3389/fimmu.2020.01046.
100. Zhong, X.-S., Matsushita, M., Plotkin, J., Riviere, I., Sadelain, M., Chimeric Antigen Receptors Combining 4-1BB and CD28 Signaling Domains Augment PI3kinase/AKT/Bcl-XL Activation and CD8+ T Cell–mediated Tumor Eradication. *Mol. Ther.*, 18, 2, 413–420, Feb. 2010. doi: 10.1038/mt.2009.210.
101. Zhang, C., Liu, J., Zhong, J.F., Zhang, X., Engineering CAR-T cells. *Biomark Res.*, 5, 22, Jun. 2017. doi: 10.1186/s40364-017-0102-y.
102. National Cancer Institute, *CAR T Cells: Engineering Immune Cells to Treat Cancer - NCI*, 2024, Accessed: Aug. 12, 2024. [Online]. Available: https://www.cancer.gov/about-cancer/treatment/research/car-t-cells.
103. Abou-el-Enein, M., *et al.*, Scalable Manufacturing of CAR T Cells for Cancer Immunotherapy. *Blood Cancer Discov.*, 2, 5, 408–422, Aug. 2021. doi: 10.1158/2643-3230.BCD-21-0084.
104. Irving, M., Lanitis, E., Migliorini, D., Ivics, Z., Guedan, S., Choosing the Right Tool for Genetic Engineering: Clinical Lessons from Chimeric Antigen Receptor-T Cells. *Hum. Gene Ther.*, 32, 19–20, 1044–1058, Oct. 2021. doi: 10.1089/hum.2021.173.
105. Sharpe, M. and Mount, N., Genetically modified T cells in cancer therapy: opportunities and challenges. *Dis. Model Mech.*, 8, 4, 337–350, Apr. 2015. doi: 10.1242/dmm.018036.
106. Testa, U., Sica, S., Pelosi, E., Castelli, G., Leone, G., CAR-T Cell Therapy in B-Cell Acute Lymphoblastic Leukemia. *Mediterr. J. Hematol. Infect. Dis.*, 16, 1, e2024010, Jan. 2024. doi: 10.4084/MJHID.2024.010.

107. Denlinger, N., Bond, D., Jaglowski, S., CAR T-cell therapy for B-cell lymphoma. *Curr. Probl. Cancer*, 46, 1, 100826, Feb. 2022. doi: 10.1016/j.currproblcancer.2021.100826.
108. Testa, U., Pelosi, E., Castelli, G., Fresa, A., Laurenti, L., CAR-T Cells in Chronic Lymphocytic Leukemia. *Mediterr. J. Hematol. Infect. Dis.*, 16, 1, e2024045, May 2024. doi: 10.4084/MJHID.2024.045.
109. Abbott, R.C., *et al.*, Human EGFRvIII chimeric antigen receptor T cells demonstrate favorable safety profile and curative responses in orthotopic glioblastoma. *Clin. Transl. Immunol.*, 12, 3, e1440, Mar. 2023. doi: 10.1002/cti2.1440.
110. Flugel, C.L., *et al.*, Overcoming on-target, off-tumour toxicity of CAR T cell therapy for solid tumours. *Nat. Rev. Clin. Oncol.*, 20, 1, 49–62, Jan. 2023. doi: 10.1038/s41571-022-00704-3.
111. Cappell, K.M. and Kochenderfer, J.N., Long-term outcomes following CAR T cell therapy: what we know so far. *Nat. Rev. Clin. Oncol.*, 20, 6, 359–371, Jun. 2023. doi: 10.1038/s41571-023-00754-1.
112. Morris, E.C., Neelapu, S.S., Giavridis, T., Sadelain, M., Cytokine release syndrome and associated neurotoxicity in cancer immunotherapy. *Nat. Rev. Immunol.*, 22, 2, 85–96, Feb. 2022. doi: 10.1038/s41577-021-00547-6.
113. Gust, J., Ponce, R., Liles, W.C., Garden, G.A., Turtle, C.J., Cytokines in CAR T Cell–Associated Neurotoxicity. *Front. Immunol.*, 11, 1–23, Dec. 2020. doi: 10.3389/fimmu.2020.577027.
114. Shah, N.N. and Fry, T.J., Mechanisms of resistance to CAR T cell therapy. *Nat. Rev. Clin. Oncol.*, 16, 6, 372–385, Jun. 2019. doi: 10.1038/s41571-019-0184-6.
115. Hawkins, E.R., D'Souza, R.R., Klampatsa, A., Armored CAR T-Cells: The Next Chapter in T-Cell Cancer Immunotherapy. *Biologics*, 15, 95–105, Apr. 2021. doi: 10.2147/BTT.S291768.
116. Li, X., Operations Management of Logistics and Supply Chain: Issues and Directions. *Discrete Dyn. Nat. Soc.*, 2014, 1, 701938, 2014. doi: 10.1155/2014/701938.
117. Hong, M., Clubb, J.D., Chen, Y.Y., Engineering CAR-T Cells for Next-Generation Cancer Therapy. *Cancer Cell*, 38, 4, 473–488, Oct. 2020. doi: 10.1016/j.ccell.2020.07.005.
118. Al-Haideri, M., *et al.*, CAR-T cell combination therapy: the next revolution in cancer treatment. *Cancer Cell Int.*, 22, 365, Nov. 2022. doi: 10.1186/s12935-022-02778-6.
119. Blache, U., Tretbar, S., Koehl, U., Mougiakakos, D., Fricke, S., CAR T cells for treating autoimmune diseases. *RMD Open*, 9, 4, e002907, Nov. 2023. doi: 10.1136/rmdopen-2022-002907.
120. So, W.Y. and Leung, P.S., Fibroblast Growth Factor 21 As an Emerging Therapeutic Target for Type 2 Diabetes Mellitus. *Med. Res. Rev.*, 36, 4, 672–704, Jul. 2016. doi: 10.1002/med.21390.
121. Huang, X., He, D., Pan, Z., Luo, G., Deng, J., Reactive-oxygen-species-scavenging nanomaterials for resolving inflammation. *Mater. Today Bio*, 11, 100124, Jul. 2021. doi: 10.1016/j.mtbio.2021.100124.

122. Cao, S. and Yuan, Q., An update of nanotopographical surfaces in modulating stem cell fate: a narrative review. *Biomater. Transl.*, 3, 1, 55–64, Mar. 2022. doi: 10.12336/biomatertransl.2022.01.006.
123. Marquez-Curtis, L.A. and Janowska-Wieczorek, A., Enhancing the Migration Ability of Mesenchymal Stromal Cells by Targeting the SDF-1/CXCR4 Axis. *Biomed Res. Int.*, 2013, 561098, 2013. doi: 10.1155/2013/561098.
124. Fan, Y.-N., *et al.*, Progress in nanoparticle-based regulation of immune cells. *Med. Rev. (2021)*, 3, 2, 152–179, 2021. doi: 10.1515/mr-2022-0047.
125. Zhu, J., Wang, J., Li, Y., Recent advances in magnetic nanocarriers for tumor treatment. *Biomed. Pharmacother.*, 159, 114227, Mar. 2023. doi: 10.1016/j.biopha.2023.114227.
126. Neeser, A., Ramasubramanian, R., Wang, C., Ma, L., Engineering enhanced chimeric antigen receptor-T cell therapy for solid tumors. *Immunooncol. Technol.*, 19, 100385, May 2023. doi: 10.1016/j.iotech.2023.100385.
127. Shin, S., *et al.*, Nanoparticle-Based Chimeric Antigen Receptor Therapy for Cancer Immunotherapy. *Tissue Eng. Regen. Med.*, 20, 3, 371–387, Mar. 2023. doi: 10.1007/s13770-022-00515-8.
128. Chehelgerdi, M., *et al.*, Progressing nanotechnology to improve targeted cancer treatment: overcoming hurdles in its clinical implementation. *Mol. Cancer*, 22, 169, Oct. 2023. doi: 10.1186/s12943-023-01865-0.
129. Persano, S., Das, P., Pellegrino, T., Magnetic Nanostructures as Emerging Therapeutic Tools to Boost Anti-Tumour Immunity. *Cancers (Basel)*, 13, 11, 2735, May 2021. doi: 10.3390/cancers13112735.
130. Yang, G., *et al.*, Immunomodulatory Mechanisms and Therapeutic Potential of Mesenchymal Stem Cells. *Stem Cell Rev. Rep.*, 19, 5, 1214–1231, Jul. 2023. doi: 10.1007/s12015-023-10539-9.
131. Foulkes, R., Man, E., Thind, J., Yeung, S., Joy, A., Hoskins, C., The regulation of nanomaterials and nanomedicines for clinical application: current and future perspectives. *Biomater. Sci.*, 8, 17, 4653–4664, 2020. doi: 10.1039/D0BM00558D.
132. N. R. C. (US) P. @ on H. M. D. @ in C. Trials, Clinical Trials: Overview and Terminology, in: *The Prevention and Treatment of Missing Data in Clinical Trials*, National Academies Press US, 2010, Accessed: Aug. 12, 2024. [Online]. Available: https://www.ncbi.nlm.nih.gov/books/NBK209903/.
133. Scimeca, J.C. and Verron, E., Nano-engineered biomaterials: Safety matters and toxicity evaluation. *Mater. Today Adv.*, 15, 100260, Aug. 2022. doi: 10.1016/j.mtadv.2022.100260.
134. Souto, E.B., *et al.*, Nanopharmaceutics: Part I—Clinical Trials Legislation and Good Manufacturing Practices (GMP) of Nanotherapeutics in the EU. *Pharmaceutics*, 12, 2, 146, Feb. 2020. doi: 10.3390/pharmaceutics12020146.
135. Đorđević, S., *et al.*, Current hurdles to the translation of nanomedicines from bench to the clinic. *Drug Deliv. Transl. Res.*, 12, 3, 500–525, 2022. doi: 10.1007/s13346-021-01024-2.

136. Kyriakides, T.R., *et al.*, Biocompatibility of nanomaterials and their immunological properties. *Biomed. Mater.*, 16, 4, 10.1088/1748–605X/abe5fa, Mar. 2021. doi: 10.1088/1748-605X/abe5fa.
137. Aljabali, A.A., *et al.*, Nanomaterials and Their Impact on the Immune System. *Int. J. Mol. Sci.*, 24, 3, 2008, Jan. 2023. doi: 10.3390/ijms24032008.
138. Xuan, L., Ju, Z., Skonieczna, M., Zhou, P., Huang, R., Nanoparticles-induced potential toxicity on human health: Applications, toxicity mechanisms, and evaluation models. *MedComm (2020)*, 4, 4, e327, Jul. 2023. doi: 10.1002/mco2.327.
139. Halamoda-Kenzaoui, B., Holzwarth, U., Roebben, G., Bogni, A., Bremer-Hoffmann, S., Mapping of the available standards against the regulatory needs for nanomedicines. *Wiley Interdiscip. Rev. Nanomed. Nanobiotechnol.*, 11, 1, e1531, 2019. doi: 10.1002/wnan.1531.
140. Joseph, T.M., *et al.*, Nanoparticles: Taking a Unique Position in Medicine. *Nanomater. (Basel)*, 13, 3, 574, Jan. 2023. doi: 10.3390/nano13030574.
141. Chehelgerdi, M., *et al.*, Exploring the promising potential of induced pluripotent stem cells in cancer research and therapy. *Mol. Cancer*, 22, 189, Nov. 2023. doi: 10.1186/s12943-023-01873-0.
142. Kwon, Y.W., *et al.*, Application of Proteomics in Cancer: Recent Trends and Approaches for Biomarkers Discovery. *Front Med. (Lausanne)*, 8, 747333, Sep. 2021. doi: 10.3389/fmed.2021.747333.
143. Kumari, M., Gupta, V., Kumar, N., Arun, R.K., Microfluidics-Based Nanobiosensors for Healthcare Monitoring. *Mol. Biotechnol.*, 66, 378–401, May 2023. doi: 10.1007/s12033-023-00760-9.
144. Brothers, K.B. and Rothstein, M.A., Ethical, legal and social implications of incorporating personalized medicine into healthcare. *Per. Med.*, 12, 1, 43–51, 2015. doi: 10.2217/pme.14.65.
145. Alghamdi, M.A., Fallica, A.N., Virzì, N., Kesharwani, P., Pittalà, V., Greish, K., The Promise of Nanotechnology in Personalized Medicine. *J. Pers. Med.*, 12, 5, 673, Apr. 2022. doi: 10.3390/jpm12050673.
146. Dehghani, L., *et al.*, Stem Cell Therapy and Its Products Such as Exosomes: Modern Regenerative Medicine Approach, in: *Possibilities and Limitations in Current Translational Stem Cell Research*, IntechOpen, 2023, doi: 10.5772/intechopen.111574. https://www.intechopen.com/chapters/87046
147. Hsu, C.-Y., *et al.*, An overview of nanoparticles in drug delivery: Properties and applications. *S. Afr. J. Chem. Eng.*, 46, 233–270, Oct. 2023. doi: 10.1016/j.sajce.2023.08.009.
148. Thwala, L.N., Ndlovu, S.C., Mpofu, K.T., Lugongolo, M.Y., Mthunzi-Kufa, P., Nanotechnology-Based Diagnostics for Diseases Prevalent in Developing Countries: Current Advances in Point-of-Care Tests. *Nanomater. (Basel)*, 13, 7, 1247, Mar. 2023. doi: 10.3390/nano13071247.
149. Mendes, B.B., *et al.*, Nanodelivery of nucleic acids. *Nat. Rev. Methods Primers*, 2, 1, 1–21, Apr. 2022. doi: 10.1038/s43586-022-00104-y.

5

Exosomes from Stem Cells as Therapeutic Agents for Cardiovascular Regeneration

Maryam Taghavi Narmi[1], Nikoo Baghal Darbandi[2], Abdulwahab Teflischi Gharavi[3], Ali Babaeizad[4] and Mohsen Sheykhhasan[5]*

[1]Department of Plant, Cell and Molecular Biology, Faculty of Natural Sciences, University of Tabriz, Tabriz, Iran
[2]Department of Biotechnology, Faculty of Chemical Engineering, Tarbiat Modares University, Tehran, Iran
[3]Department of Cell and Molecular Sciences, Faculty of Biological Sciences, Kharazmi University, Tehran, Iran
[4]Faculty of Medicine, Semnan University of Medical Sciences, Semnan, Iran
[5]Cellular and Molecular Research Center, Qom University of Medical Sciences, Qom, Iran

Abstract
Exosomes secreted from stem cells (SC-Exos) have appeared to be invaluable, non-invasive, promising therapeutic agents for cardiovascular regeneration owing to their capacity to promote tissue regeneration and cell communication. These nanosized vesicles are enriched by bioactive molecules, including lipids, proteins, and RNAs. These molecules can influence important cellular processes critical for heart regeneration. Different research have reported the efficiency of exosomes secreted by different sources of stem cells, including mesenchymal stem cells from umbilical cord blood (UCB-MSCs) as well as cells from cardio-sphere (CDCs), and the exosomes secreted by mesenchymal stem cells (MSC-Exos) in promoting cardiac regeneration and angiogenesis. This chapter discusses how MSC-Exos contribute to cardiovascular regeneration, their promising applications in the field of cardiac-related disorders and diseases, and the associated clinical translation challenges. Regarding the therapeutic potential and limitations of Exos from these resources, we aim to explore the way for innovative treatments for cardiovascular conditions.

**Corresponding author*: msheykhhasan@muq.ac.ir

Ayush Madan, Sumel Ashique, Deepika Arora and Mantosh Kumar Satapathy (eds.) Stem Cell Therapeutics, (227–308) © 2025 Scrivener Publishing LLC

***Keywords*:** Exosomes, stem cells, therapeutic, cardiovascular, regeneration

5.1 Introduction

Cardiovascular diseases (CVDs) are nowadays described as high-risk health threats causing large numbers of fatalities [1, 2]. The mature cardiac tissue shows limited capability for healing procedures [3], so myocardial injury from acute myocardial infarction (AMI) mostly causes irreversible cardiomyocyte loss and heart failure (HF) in many patients. Adult stem cell transplantation has garnered the attention of researchers for cardiac repair over the past years [4]. Stem cells (SCs) are noted for their differentiative capacity into many lineages, their low immunogenicity, immunomodulatory properties [5], and various sources, making them ideal for clinical trials [6]. However, a phase II clinical trial using cardiac-derived SCs *via* intracoronary infusion showed heart function improvement in single ventricle physiology patients [7]. Evidence suggests that injected cells release paracrine factors that activate beneficial endogenous mechanisms involving several intracellular messengers such as extracellular vesicles (EVs) as well as other factors [8, 9]. Studies indicate EVs, particularly exosomes, effectively replicate the paracrine effects of SCs [10–12]. Exos are nano-sized EVs (~30–140 nm), resourcing from different cell types and enveloped by a lipid bilayer [13, 14]. When cell membrane fusion with multivesicular bodies (MVBs) occurs, Exos participate in paracrine mechanisms, carrying therapeutic payloads like proteins, nucleic acids, and signaling molecules. They aid with immune response, signal transduction, tissue repair, and antigen presentation [9]. Exos in comparison to SCs show lower immunogenicity, more stable biologically and accessible to conserve, and can be controlled in dosage and uptake by target cells [10, 15]. They have been used effectively in treating myocardial repair and regeneration, showing cardioprotective functions. Preconditioning SCs enhances their biological effects and the therapeutic impact of their Exos [12, 16], with studies demonstrating more effective therapeutic strategies in a variety of diseases [15, 17]. The following section in this chapter covers the biogenesis and application of several Exos, along with their role in cardiac regeneration as therapeutic agents.

5.1.1 Background on Cardiovascular Diseases

Despite advances nowadays, CVDs and myocardial infarction (MI) particularly remain the major reasons for mortality and morbidity in clinical settings [18]. CVDs include disorders affecting cardiac and other critical tissues, including the blood vessels in various parts of the body [19]. By 2030, nearly 24 million people will be exposed to die from CVDs annually [20]. Although high-income countries have seen a decrease in CVD mortality, low- and middle-income countries are experiencing rising rates, expected to account for over 75% of global CVD deaths due to a lack of preventive and treatment strategies. The socioeconomic costs of CVDs are significant, with a $320 billion annual burden in the United States, which is increasing [21, 22]. Thus, developing advanced techniques for global CVD management is crucial. Enhanced and timely detection can shift treatment to prevention, improving survival rates. Atherosclerosis (ASVD) causes most CVDs, where fatty substances accumulate in the arteries, resulting in restricting blood flow and circulation, which potentially leads to MI or strokes [23]. This can result in life-threatening MI or strokes. CVD can be hindered by three-stage management. The primary stage includes a balanced diet, a healthy lifestyle, avoiding smoking, and regular exercise. The secondary stage focuses on early detection to increase treatment success and reduce severe health consequences. The tertiary stage addresses chronic CVD management, aiming to enhance standards of living and control pain [24].

5.1.2 The Role of Stem Cells in Regenerative Medicine

The ability of human tissue regeneration is not as flexible as is needed in most diseases, compelling defects and necrosis to tissue/organ as significant health threats. Regenerative medicine has concentrated on SC research to tackle this issue. Multidirectional differentiation, self-replication, and homing potential are the characteristics of stem cells (Table 5.1). They can become various types of cells needed for tissue regeneration [25, 26]. They can divide indefinitely, transdifferentiate into other cell types, and are crucial for tissue and organ anomalies repair resulting from congenital defects, diseases, and aging [27]. SCs are categorized through their potential of differentiation into toti-, multi-, and uni-potent cells, with pluripotent SCs showing a high-level capacity in the subject of tissue repair [28, 29]. There exist diverse origins of stem cells, such as embryonic stem cells (ESCs), tissue-specific progenitor SCs (TSPSCs), mesenchymal SCs (MSCs), and human induced pluripotent SCs (hiPSCs). SCs might be characterized as autologous, allogeneic, or syngeneic in relation to tissue regeneration and the eradication of malignant cells or pathogens. To prevent the rejection of host-versus-graft, tissue typing for human leukocyte antigens (HLA) and immunosuppressants is optional. Major histocompatibility complex (MHC) receptors are lowly expressed in SCs and release chemokines that recruit immune and endothelial cells, authorizing the tolerance of transplanted tissue at the graft site [30].

MSCs derived from various parts of the body, such as bone marrow and adipose tissue, represent the most extensively researched and utilized cell types in the field of regenerative medicine [31]. They are self-renewing

Table 5.1 Stem cell-based therapeutic applications in regenerative medicine.

Cell type	Diseases	Mechanisms	Outcomes	Reference
ESCs	Spinal cord injuries	Secretion of angiogenic and neurogenic factors associated with tissue homing	Spinal tissue regeneration, balance, and sensation improvement	[51]
	ARMD and glaucoma	Activation of BMP, TGF-β, and Wnt by COCO and of BRN3 by CRISPER-Cas9	Macular recovery, vision restoration	[52, 53]
	Cardiovascular disease	Expression of GCaMP3, secretion of angiogenic factors, and Tbox3, ESC differentiation into SANPCs	Heart arrhythmia suppression, Integration of cardiomyocyte (CM) electro-physiologically to the heart as the pacemaker	[54, 55]
	Liver injuries	Expression of PXR, Cytp450, HNF4-α, CYPA4&29, and UGTA1 markers	Liver tissue regeneration, application as a model for drug screening	[56, 57]
	Diabetes	Differentiation of progenitors (CD24+, CD49+, and CD133+) into β-cells, insulin secretion, expression of PDX1, GCK, and GLUT2	Glucose level improvement and obesity, T1DM and T2DM treatment	[58, 59]
	Osteoarthritis	Active $SOX9^+$ and $ColII^+$ cells aggregations at the transplantation site	Cartilage tissue regeneration, particularly for the treatment of injuries concerning athletes	[60]

(Continued)

Table 5.1 Stem cell-based therapeutic applications in regenerative medicine. (*Continued*)

Cell type	Diseases	Mechanisms	Outcomes	Reference
TSPSCs	Diabetes	Developing into β-cell *via* PPCs with FGF and Notch signaling	Treatment of T1DM and T2DM through PPCs occupancy as β-cell	[58, 61]
	Neuro-dental issues	βIII-tubulin, Nestin, GFAP, and L-type Ca^{2+} channel expression	Neuro-dental abnormalities feasible treatment	[62]
	Acoustic conditions	Notch by β-catenin and shutting of Atoh1 by γ-secretase in $lrg5^{+}$IESCs, differentiation into hair cells	Cochlear regeneration, acoustic function restoration	[63]
	Intestinal damage	Signaling of Mϕ, myofibroblasts, and bacteria to IPCs; differentiation into crypt-villi organoid tissue	Goblet mucosa regeneration	[64, 65]
	Corneal diseases	Differentiation of LPSCs marked by ABCB5 into the mature cornea	Corneal tissue regeneration	[66]
	Muscular deformities	Attraction of neurogenic and vasculogenic cells to the transplant site by PDGF from MABs	Muscle fibril regeneration; treatment of skeletal muscle failure	[67, 68]
	Eye disease and retinopathy	Production of elevated vasoprotective factors by ADSCs from healthy donors	Vascularization restoration, treatment of retinopathy	[69, 70]
	Cardiac dysfunctions	Differentiation of CA-ADSCs to the epithelium	Ischemic myocardium regeneration	[71, 72]

(*Continued*)

Table 5.1 Stem cell-based therapeutic applications in regenerative medicine. (*Continued*)

Cell type	Diseases	Mechanisms	Outcomes	Reference
MSCs	Bladder deformities	Healing bladder with BDMSCs (CD34^{-}, CD45^{-}, CD73^{+}, and CD105^{+}) and SIS in 10 weeks	Bladder repair	[73]
	Dental problems	Increase in teeth unit maturation in biopolymers by EMSC-DSCs among vasculogenic factors	Oral tissue repair, application in periodontics	[74]
	Bone degeneration	Transformation of MSCs through cytochalasin-D into osteoblasts by actin modeling	Bone regeneration, less injury pain	[75, 76]
	Muscle degeneration	MSCs protection by alginate gel from immune attack	Regulation of heart damage, tissue repair	[77]
	Alopecia	GAG coating mimics the ECM, promoting the regeneration of the DPCs	Hair follicle regeneration, alopecia treatment	[78]
	Cartilage damage	IGF-1, BMP-2/-4, and TGF-β1/-β3 stimulation, applying hydrostatic pressure, and dynamic compression	Activating as well as facilitating cartilage formation, strong differentiative and proliferative capacity	[79, 80]
	Meniscus injury	Recruitment of connective tissue TGF and TGF-β3, applying mechanical stimuli	Regeneration of meniscus	[81]
	Left ventricular ejection fraction	Inducing MSC differentiation through 5-azacytidine factor	CM differentiation, high resistance to oxidative stress, immunomodulatory effects	[82]

(*Continued*)

Table 5.1 Stem cell-based therapeutic applications in regenerative medicine. (*Continued*)

Cell type	Diseases	Mechanisms	Outcomes	Reference
	Myocardial ischemia	MSCs fusion with cardiomyocytes *via* IL-7, VEGF secretion	CM differentiation, angiogenesis, cardiac repair	[83, 84]
	Tracheal damage through tracheal stenosis and tracheobronchial cancer	In combination with VEGF, BDNF, TGF-β1, and EGF, activin A, KGF, HGF, and IGF-2	Tracheal regeneration *via* MSC differentiation into epithelial cells	[85]
UCSCs	Congenital heart defects	Organogenesis promotion by VEFG to PEG-coaxed AFSCs addition	Heart tissue repair	[86, 87]
	Diabetes	Decrease IL6 and IL1β by the act of WJ-factors, WJ-SCs differentiation into β-cells by Mϕ	Diabetes treatment through β-cell function efficiency	[88, 89]
	SLE	WJ-SCs decrease SLEDAI and BILAG	Renal function efficiency through β-cell function efficiency	[90]
	LSD and neurodegenerative diseases	Neuroblasts (GFAP$^+$, Nestin$^+$, and Ki67$^+$) and SCs with factors such as OCT4$^+$, SOC2$^+$; UCSCs recover from MSE deficiency and cognition improvement	Treatment of Krabbe's disease, hurler syndrome, MLD, AD, TSD, ALD, Parkinson's, stroke, TBI, ALS, and SCI	[91, 92]
	Cartilage and tendon injuries	Promotion of regeneration of hyaline cartilage by HA gel factors	Tendons as well as cartilage repair	[93]
	Hodgkin's lymphoma	Allogenic UCSCs: improving patients' quality of life approximately by 30%	Treatment of Hodgkin's lymphoma	[94]

(*Continued*)

Table 5.1 Stem cell-based therapeutic applications in regenerative medicine. (*Continued*)

Cell type	Diseases	Mechanisms	Outcomes	Reference
	Peritoneal fibrosis	Preventing cell death programming and peritoneal wall thickness by WJ-SCs	Encapsulating peritoneal fibrosis therapy	[95]
BMMSCs	Anemia and blood cancer	Immunity reconstruction by haplo-identical BMMSCs (a significant process for minorities)	Aplastic anemia and hematological malignancy treatment	[96]
	AIDS	Expressing HIV1 anti-RNA by anti-HIV1 $CD4^+$ cells; restricting the HIV infection	AIDS treatment	[97, 98]
	Blood clotting disorders	Microtubule 3D scaffolds mimicking the bone marrow structure	Therapeutics effect on burns and blood clotting diseases	[79]
	Neurodegenerative diseases	Induction of neovascularization by LA^+BMMSCs, causing colonization from microglia	Therapeutics effect on neuronal damage disorders, cognitive restoration	[80]
	Oro dental deformities	Accelerating alveolar jawbone regeneration by $CD14^+$ and $CD90^+$ TRC	Oral skin, bone, and gum regeneration	[99]
	Diaphragm abnormalities	Similar myography and spirometry of BMMSCs niche perfused hemidiaphragm	Diaphragm regeneration *via* niched diaphragm derived from donor's body	[100]
	Tendon injury	Induction of myogenic differentiation by BMP-14 through SIRT1–JNK/SMAD1-PPARγ signaling pathway, uniaxial cyclic stretching mechanism	Tendon repair, delayed lesion progression	[101, 102]

(*Continued*)

Table 5.1 Stem cell-based therapeutic applications in regenerative medicine. (*Continued*)

Cell type	Diseases	Mechanisms	Outcomes	Reference
	Intervertebral disc (IVD) degeneration	Expression of SOX-9 and BMP-2 *via* transfection of adenovirus, Col II increase, aggrecan expression	IVD regeneration	[103]
iPSCs	Eye defects	Restoring visual acuity	ARMD treatment, as well as other age-related eye defects	[104]
	Neurodegenerative disorders	Secretion of GABA by (iGABA-INs); ASD caused by FOX1G	Therapeutic effects on ASD, Alzheimer's, seizures, and treatment-resistant epilepsy	[105, 106]
	Liver/lung diseases	A1AD mutation; drug sensitivity	COPD treatment	[107]
	Diabetes	Reprogramming skin cells into β-cells through cDE and cPF recruiting GPs	Insulin production, T1DM and T2DM treatment	[108]
	Lung degeneration	Resemblance of miniature iPSCs lung to airways and alveoli in a model drug test	Lung tissue repair	[109]
	SIDs and AIDS	Generation of iPSCs through CRISPER-Cas9 in one step; iPSCs- Mϕ resistance to HIV1	SIDs, HIV1 immunotherapy	[110]

cells and are easily isolated, highly expandable under standard conditions in laboratory settings, form colony-forming unit fibroblasts (CFU-Fs), and are able to target differentiation into neural tissue cells with low immunogenicity [32–35]. To be called MSCs, cells should express a specific cluster of differentiation (CD) markers (CD73, CD90, and CD105) as well as lack hematopoietic lineage CD markers [36]. MSCs are favored for therapies recruiting specific cells due to their privileged characteristics and their ability to secrete tissue-repair-activating molecules and modulate cells responsible for inflammation and immunity [37]. They lack HLA class II and the T cell costimulatory molecules, providing them an immune-privileged status, making them suitable for allografts in conditions like cardiomyopathy without pharmacological immunosuppression [38]. Studies have demonstrated MSCs' immunoregulatory, anti-inflammatory, and immunosuppressive capacities [39].

Bone marrow and adipose tissue MSCs retain immunomodulatory characteristics and produce extracellular matrix components. Bone marrow MSCs (BMMSCs) have higher proangiogenic activity and more excellent immunosuppressive effects than ADSCs [40, 41]. Other types of mesenchymal stem cells comprise the umbilical cord matrix MSCs (UCM-MSCs), which exhibit the ability to block T cell activation [42], and MSCs secreted by amniotic fluid (AF-MSCs) express pluripotency markers [43–45]. Circulating MSCs in peripheral blood may also express pluripotency markers but lack pluripotent differentiation capacity [45, 46]. Cardiac MSCs demonstrate a slow growth rate *in vitro* and costimulatory molecule CD40 expression, questioning their immune-privileged status [34, 47]. iPSC-derived MSCs (iPSC-MSCs) have similar phenotypes to BMMSCs such as differentiating into several cell types, inhibiting inflammation along with demonstrating potential for tissue regeneration in ischemic disease [48]. They have enhanced proliferative capacity and are mainly stable during multi-passage expansions [49]. Before clinical application, systematic preclinical studies are necessary to make sure the irreversible side effects such as polyp formation are not coming as a result [50].

Cardiomyocyte (CM), retinal ganglion cells (RGCs), sinoatrial node pacemaker cells (SANPCs), pancreatic and duodenal homeobox 1 (PDX1), glucokinase (GCK), adipose-derived stem cells (ADSCs), induced pluripotent stem cell (iPSCs), bone marrow-derived mesenchymal stem cells (BDMSCs), amniotic fluid stem cells (AFSCs), dental papilla cells (DPCs), limbal progenitor stem cells (LPSCs), tissue repair cells (TRC), neural progenitor cells (NPCs), epidermal growth factor (EGF), brain-derived neurotrophic factor (BDNF), gamma-aminobutyric acid (GABA), insulin-like growth factor (IGF), glial fibrillary acidic protein (GFAP), glycosaminoglycan (GAG), bone morphogenetic protein 2 (BMP-2), interleukin (IL), transformation growth factor (TGF), hepatocyte growth factor (HGF), keratinocyte growth factor (KGF), vascular endothelial growth factor (VEGF), extracellular matrix (ECM), hyaluronic acid (HA), post-operative pulmonary complications (PPCs), metachromatic leukodystrophy (MLD), Tay–Sachs disease (TSD), adrenoleukodystrophy (ALD), Alzheimer's disease (AD), amyotrophic lateral sclerosis (ALS), spinal cord injury (SCI), traumatic brain injury (TBI), acquired immunodeficiency syndrome (AIDS), intervertebral disc (IVD), age-related macular degeneration (ARMD), chronic obstructive pulmonary disease (COPD),

Sirtuin-1–Janus N-terminal kinase (SIRT1–JNK), SMAD1-peroxisome proliferator-activated receptor-γ (SMAD1-PPARγ), human immunodeficiency virus (HIV), sudden infant death syndrome (SIDs), autism spectrum disorder (ASD), lysosomal storage disorders (LSD), left ventricular ejection fraction (LVEF), definitive endodermal progenitors (cDE), foregut like progenitor cells (cPF).

5.1.3 Overview of Exosomes and their Biological Functions

Recent research findings have revealed that exosomes play a pivotal role in augmenting the therapeutic effects of stem cells in relation to heart function post-AMI. Investigations on Exos from different SCs, such as MSCs, cardiac SCs (CSCs), embryonic SCs, hematopoietic SCs, and CDCs, report that they promote blood vessel formation and cardiomyocyte and CPC survival and decrease scar tissue after a heart attack [111, 112] (Table 5.2). Despite their effectiveness, delivering Exos to the heart poses a challenge. Intravenous administration is suboptimal as most Exos are retained in the liver, whereas intramyocardial injection yields better outcomes [113]. Exosome-based therapies hold potential for treating cardiovascular disorders like myocardial infarction, ASVD, and dilated cardiomyopathy. Pretreatments such as hypoxia preconditioning, gene programming, or pharmacological interventions can enhance the exosomal cardioprotective effects [114, 115]. For instance, MSC-Exos improve cardiac function by reducing CD68 expression and inflammatory responses [116]. Exos pretreated with atorvastatin (MSC-ATV-Exos) improved heart function and reduced damage by decreasing interleukin 6 (IL-6) and the tumor necrosis factor (TNF)-α levels, promoting angiogenesis, and hindering apoptosis [115].

Exosomes also offer therapeutic benefits in treating ASVD by transforming M1 macrophages into M2 macrophages *via* miR-let7, reducing inflammation, and improving cardiac function post-I/R injury [117]. Lankford demonstrated that MSC-Exos treat dilated cardiomyopathy by inhibiting inflammatory cytokines and promoting M2 macrophage development [118]. Cardiomyocytes, cardiac fibroblasts, and endothelial cells contribute to heart function through complex intercellular signaling networks [119]. Exos mediate in myocardial angiogenesis, heart regeneration, reticulocyte maturation, and pathological mechanisms like ASVD, I/R damage, and cardiac remodeling [120].

Owing to the different characteristics of Exos, they exhibit significant impending as non-invasive biomarkers used for diagnosing along with predicting CVDs [121]. Exosomes are implicated in various translational

Table 5.2 Biological perspective of exosomes from various cell types to serve as therapeutic and diagnostic tools in cardiovascular diseases.

Issues	Exosome type	Biological functions	References
AVSD	Endothelial cell-derived exosomes	Prompting proliferative and migratory phenotype in VSMCs by plaque formation, intimal hyperplasia post-injury facilitated by the CD137 inflammatory signaling pathway	[124]
	Macrophage-derived exosome	VSMC migration, strong proliferative capacity *via* a PTEN-related mechanism	[125]
	Foam cell-derived exosome	Facilitating integrin transfer to VSMCs, stimulating migration and downstream signaling	[126]
	Endothelial cell-derived (exposed to ox-LDL)	Formation of neutrophil extracellular traps (NETs) by transferring lncRNA MALAT1, disease progression in mouse models	[127]
	Exosomes containing lncRNA GAS5	Apoptosis promotion in vascular endothelial cells	[128]
	Exosomes from hypoxic cardiomyocytes	Contain cytokine TNF-α and heat shock protein HSP60; may serve as biomarkers for pre-necrotic conditions	[129]

(*Continued*)

Table 5.2 Biological perspective of exosomes from various cell types to serve as therapeutic and diagnostic tools in cardiovascular diseases. (*Continued*)

Issues	Exosome type	Biological functions	References
HF	Exosomes (carrying miR-192, miR-194, miR-34a)	HF development prediction and positive correlation with left ventricular diastolic size and ejection fraction	[130]
	Exosomes (carrying miR-744, miR-21, miR-425, miR-92)	Early diagnosis of hypertrophic cardiac disorders *via* miR-92, miR-744, miR-425, and miR-21 serving as biomarkers	[131]
	Exosomes (containing miR-146a)	miR-146a; as a diagnostic biomarker for peripartum cardiomyopathy	[132]
	Cardiac intrinsic cells-derived exosome	Scar size and fibroblast reduction, cardiac function improvement	[133]
	c-Kit+ stem cell-derived exosome	Promising in cardiac repair	[134]
Acute Ischemic stroke	Serum exosomal neuronal synaptopodin	Promising biomarker in newborns, linked to short-term neurological outcomes	[135]
	Exosomes (carrying miR-223, miR-9, miR-124)	Anti-inflammatory effects, serving as diagnostic markers, immune cell activity regulation	[136]

(*Continued*)

Table 5.2 Biological perspective of exosomes from various cell types to serve as therapeutic and diagnostic tools in cardiovascular diseases. (*Continued*)

Issues	Exosome type	Biological functions	References
Cardiac fibrosis	Exosomes with TNF-α	Post-myocardial infarction improvement and remodeling	[137]
	MSC-derived exosomes enriched with let7-c		[138]
		α-SMA and type IVa1 collagen reduction, TGF-β/Smad pathway modulating, fibroblasts protection	[139]
	Exosomes (carrying miR-294, miR-19a, miR-221, miR-210, miR-146a, miR-144, miR-181b, miR-132, miR-22, miR-19a)	Enhancing anti-fibrotic properties	[140]
	High-level miR-21 exosomes	Decreasing PDCD4 expression, inhibiting scar formation post-myocardial infarction, enhancing heart function	[141]
	Exosomes containing lncRNA Mhrt	Suppressing cardiac fibrosis and cardiac myocyte enlargement	[142]
	Exosomes containing circRNA circ 000203	Lowering the expression of CTGF and Col1a2 *via* miR-26b-5p sponging, inhibition of fibrosis	[143]
			[144]

(*Continued*)

Table 5.2 Biological perspective of exosomes from various cell types to serve as therapeutic and diagnostic tools in cardiovascular diseases. (*Continued*)

Issues	Exosome type	Biological functions	References
Myocardial ischemia/ reperfusion (I/R)			[145]
	Exosomes containing circRNA circActa2	Controlling α-SMA levels by competitively inhibiting miR-548f-5p; regulating cardiac fibrosis	[146]
			[147]
			[148]
Myocardial infarction	Platelets-derived exosomes	Enriched ubiquitin/ proteasome, degrading, inflammatory chemicals, miR-320 or miR-223 suppressing ICAM-1 expression, endothelial cell absorption, improving motility along anti-inflammatory effects	[149]
Ischemic heart disease	Exosomes derived from cardio-sphere-derived cells (CDC)	Mitigates ischemia and reperfusion-induced injury; effectiveness varies with donor age	[150, 151]

medicine and biological procedures, as well as cellular homeostasis and apoptosis, and they impact inflammation across various diseases [122]. While they show therapeutic potential, Exos can also spread harmful pathology in heart disease, as seen with Exos enriched with miR-155, exacerbating inflammation and cardiac rupture [123]. Due to promising results, the clinical community is particularly interested in exosome research for heart regeneration, diagnosing diseases, and their prognosis.

Vascular smooth muscle cells (VSMCs), atherosclerotic vascular disease (ASVD), heart failure (HF), acute coronary syndrome (ACS), cardiosphere-derived cells (CDC), acute myocardial infraction (AMI), tumor necrosis factor (TNF), phosphatase and tension homologue (PTEN), intercellular adhesion molecule (ICAM), oxidized low-density lipoprotein (ox-LDL), metastasis-associated lung adenocarcinoma transcript 1

(MALAT1), α-smooth muscle actin (α-SMA), Toll-like receptor 4 (TLR4), extracellular signal-regulated kinase 1 and 2 (ERK1, ERK2), p38 mitogen-activated protein kinases (p38MAPK).

5.2 Biogenesis and Characteristics of Stem-Cell-Derived Exosomes

Exosomes have been characterized as heterogeneous, lipid bilayer-enclosed vesicles that are secreted by various cellular sources [152–158]. Several research studies on extravesical functions in cancer have emerged, classifying them according to their cargo, donor cells, and morphology [13, 159]. Their application has gained significant momentum owing to their capacity to regulate cell–cell interactions in disease and physiological pathways [160]. They contain cargo such as lipids, proteins, DNA, and RNA, namely, ncRNA and mRNA [13] as well as assist in functional molecular content exchange between cells [161]. Exos are involved in various physiologic processes, including angiogenesis, therapy resistance, cancer metastasis, immune evasion, wound healing, and neuronal communication [159, 162–165]. Notably, cancer cells modulate the quantity and composition of Exos through a variety of mechanisms for cancer progression [166]. Nonetheless, the ambiguity in exosome heterogeneity and the difficulties with harvesting EVs *in vitro* and biological systems reveal some challenges to extracellular vesicle dynamics.

Exosomes are classified in ascending size order as exomeres, ectosomes, apoptotic bodies, as well as large oncosomes, with exomeres and Exos categorized as small EVs (sEVs) and large ones [167]. Regarding exomeres' identification, non-membrane-bound particles that co-isolate with EVs have led to the introduction of the term extracellular particles (EPs) [168]. The heterogeneous set of small, non-membrane-bound particles comprises exomeres. Large oncosomes, apoptotic bodies, and ectosomes are generated from the plasma membrane, particularly exosomes of endosomal origin, and undergo various cargo loading along with regulatory processes [167].

5.2.1 Exosome Formation and Release

A multifaced process is known for an exosome being formed and released. Exos are formed through the procedure by which the inward budding of the endosomal membrane takes place, resulting in the formation of intraluminal vesicles (ILVs). ILVs subsequently change into MVBs (Figure 5.1) [169]. Multiple Rab GTPases mediate the fusion process, and the ILVs release into the endosomal lumen, which is orchestrated by the endosomal sorting complexes required for transport pathway (ESCRT), a complex consisting of TSG101, STAM1, and VPS48. The ESCRT-0 complex identifies and captures ubiquitinated proteins on the endosomal membrane,

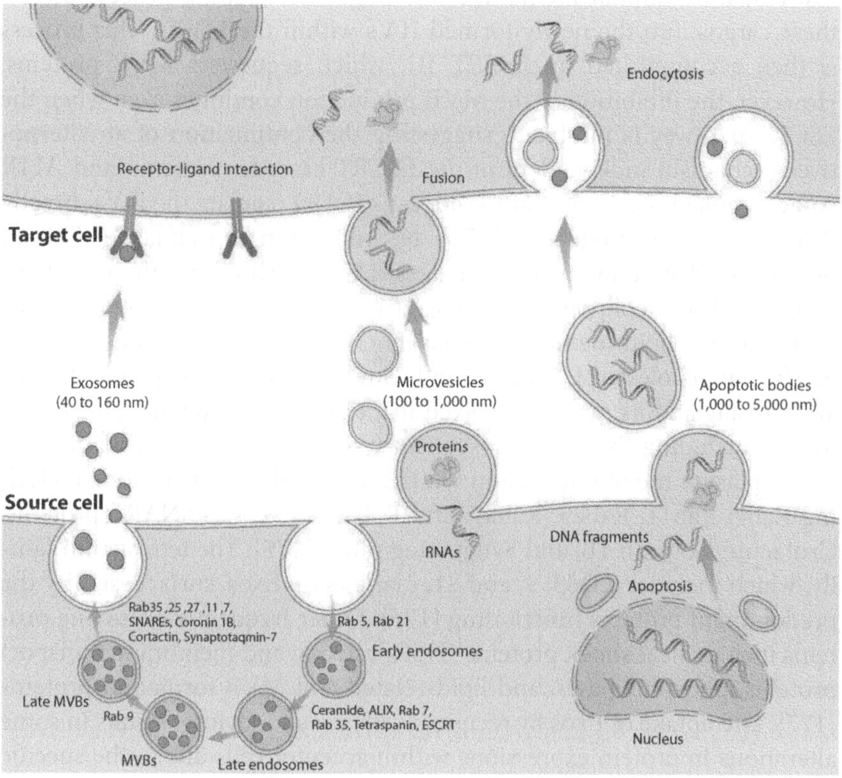

Figure 5.1 Overview of biogenesis, release, and uptake mechanism of exosomes. Microvesicles are secreted through plasma membrane inward budding. Extracellular components and cell surface proteins are internalized to form early sorting endosomes (ESEs) *via* endocytosis, accompanied by plasma membrane budding. The endoplasmic reticulum, Golgi apparatus, and mitochondria contribute to the maturation of ESEs. ESEs sequentially develop into late-sorting endosomes and MVBs. Some multivesicular bodies release their vesicles into the ECM called exosomes, while other vesicles end up in lysosomes to be degraded, with or without autophagosome fusion. Intraluminal vesicle (ILV) formation within MVBs is primarily mediated by the ESCRT processes. Rab GTPases, SNARE, and the cytoskeleton proteins control the secretion of Exos. Exos can transport particular molecules to beneficiary cells using three main procedures: endocytosis, direct membrane fusion, as well as receptor–ligand interactions [169].

subsequently recruiting ESCRT I and II. Ubiquitin guides membrane proteins or damaged cellular components to the lysosome to be disintegrated and functions as a signal molecule for exosomal cargo sorting on the endosome membrane [170, 171]. Multivesicular bodies (MVBs) can fuse with either the lysosome for degradation or the plasma membrane for exosome secretion. ESCRT I and II complexes bind to the exterior surface of the endosomal membrane close to protein cargos, which are ubiquitinated and assist in intraluminal membrane budding. The budding process directs these cargos into the newly formed ILVs within the MVBs. This process is then accomplished by ESCRT III, which sequesters MVB proteins. However, the inhibition of the MVB production continues even when the ESCRT pathway is inhibited, suggesting the continuation of an alternative mechanism independent of the ESCRT cascade. Syntenin and ALIX co-accumulate with Exos, and syndecans (CDs) regulate the ILVs directly throughout Exos biogenesis [172]. Neutral sphingomyelinase (nSMase) promotes MVB budding and enhances ILV production by enriching Exos with CD63, proteolipoproteins, TSG101, and CD81 [173]. Exos are surrounded by a lipid membrane (bilayer) consisting of ceramide, sphingomyelin, and cholesterol. Specific lipid components like phosphatidylserine and prostaglandins may be involved in exosome functionality [174]. Late MVBs, including Exos, are subsequently secreted into the ECM through MVB–plasma membrane fusion *via* the action of various factors, including Rab7, Rab11, Rab25, Rab27, and Rab35, as well as SNARE proteins, Crotactin, Coronin 1B, and Synaptotagmin-7 [175]. The tetraspanin family, which includes CD63, 9, and 81, markers on Exos' surface, signify the predominant proteins functioning [176]. Other frequently occurring proteins include heat shock proteins (HSPs), fusion and membrane transport proteins, phospholipases, and lipid-related and MVB formation proteins [177]. The uptake of Exos by recipient cells *via* endocytosis results in some alterations in protein expressions within acceptor cells due to the specific genetic information transferred by the genetic tolls (miRNAs and mRNAs) inside the Exos [178].

5.2.2 Molecular Composition and Functional Properties

SC-Exos carry crucial macromolecules and information from their origin, playing vital cell–cell interaction roles. For instance, Exos are enriched with a wide range of constitutive elements, such as ECM proteins, transcription factors, enzymes, lipids, nucleic acids, and receptors, emerging on the Exos' inner and outer surfaces, demonstrating their functional diversity and complication (Figure 5.2) [179, 180]. Exosomal proteins,

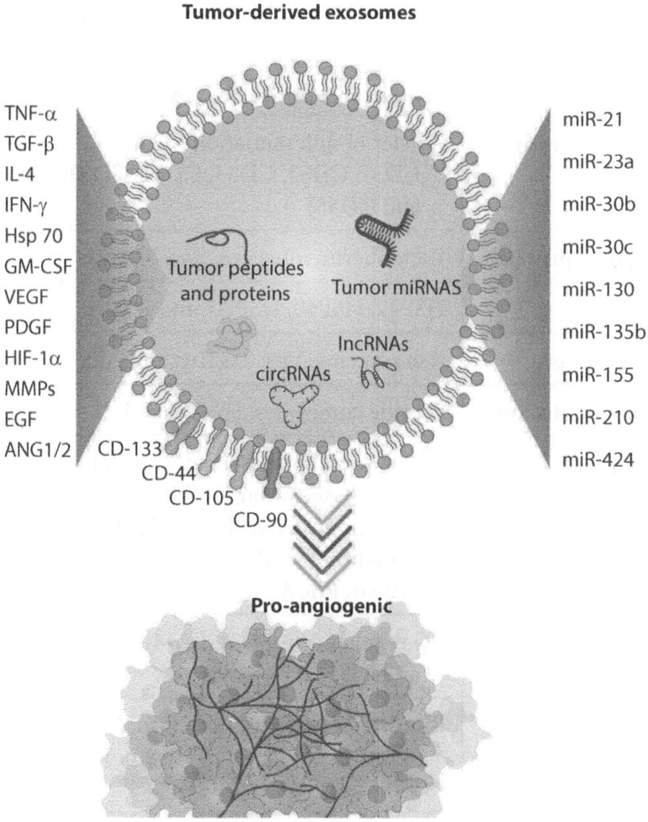

Figure 5.2 Some surface markers, microRNAs, growth factors, and cytokines are involved in tumorigenesis, angiogenesis, exocytosis, hematopoiesis, and exocytosis activities by tumor-derived exosomes.

particularly heat shock proteins (HSP70, HSP90), are involved in antigen binding and stress response. CD9, CD63, CD81, and CD82 (cluster of differentiation), as the tetraspanins family, play a role in cell penetration, invasion, and fusion processes. MVB formation proteins, such as TSG101 and Alix, are responsible for exosome secretion. Additionally, annexins and Rab proteins are accountable for membrane transport and fusion (Table 5.3) [181]. Some specific protein members, namely, Alix, flotillin, and TSG101, are engaged in the biogenesis process of Exos, which sets Exos apart from ectosomes released while the plasma membrane shedding occurs. Some proteins are particularly enriched in Exos (TSG101, HSP70, CD63, and CD81) and serve as Exos markers. The lipids in exosome-associated plasma membranes encompass cholesterol, hexosylceramides,

Table 5.3 Overview of common protein and lipid components present in exosomes.

Functional properties	Protein composition of exosome	Reference
Tetraspanins family	Cluster of differentiation (CD9, CD37, CD53, CD63, CD81, CD82, and CD53)	[185–187]
Cell adhesion proteins	Lactadherin, integrins, and ICAM-1	
HSP	HSP27, HSP60, HSP70, and HSP90	
MVB biogenesis	Rab proteins, Vps, Tsg101, and Alix	
Antigen presentation	Peptide complexes, HLA class I, II	
Signaling proteins	14-3-3 proteins, Ras-related protein, GDP dissociation inhibitor, GTPase HRas, ERK, SHP2, Syntenin-1, and transforming protein RhoA	
Membrane transport proteins	PG regulatory protein, LAMP-1/2, and CD13	
Transcription and synthesis of proteins	Ribosomal proteins, ubiquitin, MVP, complement factor 3, Histone1, 2, 3	
Components of cytoskeleton	Vimentin, actins, moesin, cofilin-1, tubulins, myosin, radixin, and ezrin	
Membrane fusion and trafficking	Annexins (I, II, IV, V, VI), SNAP, syntaxin-3, and Ras protein 5, 7	
Metabolic enzymes	ATP citrate lyase, FAS, GAPDH, PGK 1, PGAM1, PKM1/M2, ATPase, GPI, peroxiredoxin 1, aldehyde reductase, and AST	
Cytokine and growth factors	TNF-α, TGF-β, and TNF receptors	
Anti-apoptosis	Peroxidase, alix, thioredoxin	
Iron transportation	TfR	

(*Continued*)

Table 5.3 Overview of common protein and lipid components present in exosomes. (*Continued*)

Functional properties	Protein composition of exosome	Reference
Death receptor proteins	FasL, TNF-TRAIL	
Triggering polymorphonuclear leukocyte migration	LTC4, LTB4, LTA4// LTC4 synthase, LTA4 hydrolase	[188]
Inflammation	PGE2// PGE synthase	[161]
Immunosuppression, PPARγ ligand	15d-PGJ2, PGE2// COX-1/2	[189]
Accounting for the membrane curvature	AA, LPC// cPLA2, iPLA2	[189]
Increasing Exosome production	PA// PLD2, DGK	[190, 191]
Sorting cargo into MVBs	Ceramides// nSMase2	[192]
Prostaglandin biosynthesis	-// sPLA2 IIA, sPLA2 V	[189]
MVB formation followed by ILV biogenesis	BMP// -	[193, 194]
Regulating exosome secretion	Cholesterol// -	[193]
Triggering calcium influx	SM// -	[192]
Being involved in exosome fate	PS// -	[195, 196]

saturated fatty acids, phosphatidylserine, and sphingomyelin [173]. A set of proteins that have remained unchanged throughout evolution are also packaged into Exos throughout their biogenesis. These proteins include tetraspanins, integrins, MHC class II, and heat shock proteins [182]. Exos exhibit a wide range of diversity similar to the original cell types and can be crucial in determining their function. MSCs-Exos express common surface biomarkers, including CD9, CD44, CD73, CD81, and CD90, which are exclusive to MSCs [183].

Exosomes, besides selected proteins, comprise various RNA sorts that can be transferred into recipient cells, establishing an intercellular trafficking network and inducing phenotypic changes [184]. Several miRNAs involved in tumorigenesis, angiogenesis, exocytosis, hematopoiesis, and exocytosis are known to mediate exosome-based cell communication (Figure 5.2) [184].

Additionally, long RNA species, including circular RNAs (circRNAs) and long non-coding RNAs (lncRNAs), are also found in Exos, impacting diverse biological processes [185]. Exos comprise various lipids that play crucial roles in their rigidity and stability. These lipids include sphingomyelin (SM), phosphatidylserine (PS), cholesterol, phosphatidic acid, arachidonic acid, various fatty acids, as well as leukotrienes and prostaglandins. They also contain lipid-based enzymes capable of autonomously generating bioactive lipids, swallowed by target cells, and accumulate within endosomes. The interaction between exosomal lipids and lipid transfer proteins within the acceptor cells' cytosol, including fatty acid binding proteins (FABPs) or receptors like PPARγ for 15d-PGJ2, produces complexes subsequently transported to the nucleus. Consequently, 15d-PGJ2 and PGE2 derived from Exosomes can supply intracellular PGE2. Exosome- derived arachidonic acid (AA) can be transported to COXs and PGE synthase in the endoplasmic reticulum to synthesize additional PGE2. Exosome-derived unsaturated fatty acid DHA is capable of targeting the recipients' endoplasmic reticulum microsomal anti-estrogen binding site [46], thereby inhibiting cholesterol epoxide hydrolase, an enzyme responsible for converting cholesterol 5,6 epoxide into cholestanetriol [187]. Intercellular lipid exchange, mediated by exosomes, plays a significant role in developing cholesterol-related storage diseases, such as atherosclerosis. Exosome biomarkers such as lipids like cholesterol, sphingomyelin, ceramide, and phosphatidylserine, as well as RAB11B, TSG101, CHMP2A, CD63, and CD81, are crucial in cellular processes and pathophysiological conditions, including immune responses, inflammation, cell death, angiogenesis, cancer, and neurodegenerative diseases [161, 188].

Heat shock proteins (HSP), multivesicular body (MVB), cluster of differentiation (CD), intercellular adhesion molecule 1 (ICAM-1), vacuolar sorting protein (Vps), ALG-2-interacting protein X (Alix), tumor susceptibility gene 101(Tsg101), human leukocyte antigen (HLA), guanosine triphosphatases (GTPases), Harvey rat sarcoma virus (HRas), rat sarcoma virus (Ras), tumor necrosis factor (TNF), Src homology-2 domain-containing phosphatase 2 (SHP2), extracellular signal-regulated kinase (ERK), guanosine diphosphate (GDP), Ras homolog family member A (RhoA), lysosome-associated membrane proteins 1 and 2 (LAMP-1, LAMP-2), major vault protein (MVP), synaptosomal-associated protein (SNAP), transforming growth factor-β (TGF-β), transferrin receptor (TfR), TNF-related apoptosis-inducing ligand (TNF-TRAIL), Fas ligand (FasL), fatty acid synthase (FAS), glyceraldehyde 3-phosphate dehydrogenase (GAPDH), phosphoglycerate kinase 1 (PGK 1), phosphoglycerate mutase 1 (PGAM1), pyruvate kinase isozymes M1/M2 (PKM1/M2), glucose-6-phosphate isomerase (GPI), aspartate aminotransferase (AST), leukotriene A4, C4, B4 (LTA4, LTB4, LTC4), prostaglandin E2 (PGE2), 15-deoxy-Δ12,14-prostaglandin J2 (15d-PGJ2), prostaglandin (PG), cyclooxygenases 1 and 2 (COX-1, COX-2), arachidonic acid (AA), lysophosphatidylcholine (LPC), calcium-dependent phospholipases A2 (cPLA2), calcium-independent phospholipases A2 (iPLA2), phosphatidic acid (PA), phospholipases D2 (PLD2), diacylglycerol kinase (DGK), neutral sphingomyelinase 2 (nSMase2), secretory phospholipase A2 group IIA and V (sPLA2 IIA, sPLA2 V), bis(monoacylglycero)phosphate (BMP), sphingomyelin (SM), phosphatidylserine (PS).

5.3 Mechanism of Cardiovascular Regeneration Mediated by Exosomes

In cardiovascular regeneration, Exos contribute to tissue formation by facilitating the transfer of regenerative signals and genetic material, such as DNAs and RNAs, affecting cellular functionalities, including differentiation, proliferation, and survival. For example, MSC-Exos have demonstrated capabilities in enhancing angiogenesis, modulating inflammation, and reducing fibrosis [197]. The capacity of Exos to carry and deliver several valuable factors to targeted cells underscores their capability as both therapeutic agents and biomarkers in treating CVDs. Gaining a deeper understanding of how Exos influence cardiovascular regeneration could lead to innovative therapeutic strategies and improved patient outcomes in CVDs.

For instance, Exosome patches are highly potent and valuable treatment tools that can be charged with vital factors. SC-Exos patch, in conjunction with coronary artery bypass grafting (CABG), enriched by regenerative bioactive molecules, is one of the most recent therapeutic approaches restoring cardiac function in chronically ischemic myocardium by improving myocardial cell survival amongst fibrosis and angiogenesis regulation [198]. Interestingly, specified hydrogels present as supportive matrix providers, enhancing encapsulated cell survival and retention as well as their function when transplanted into ischemic cardiac tissue. These novel-engineered vehicles help MSCs release a massive variety of therapeutic agents, enhance cell functionalities more effectively, and leverage the protective and sustained delivery capabilities of these invaluable cell carriers, which offer promising therapeutic strategies for injured hearts [199]. hiPSC-Exos demonstrate significant potential in improving recovery from MI in mice. These Exos are packed with therapeutic molecules that support cardiac repair by promoting angiogenesis, reducing apoptosis, and modulating the immune response. It has been shown that administering these Exos to mice with MI results in better heart function, decreased scar tissue, and enhanced blood vessel formation in the injured areas. This approach suggests that hiPSC-Exos could be a promising non-cellular therapy for myocardial infarction, providing a novel strategy for cardiac regeneration and repair (Figure 5.3) [200, 201].

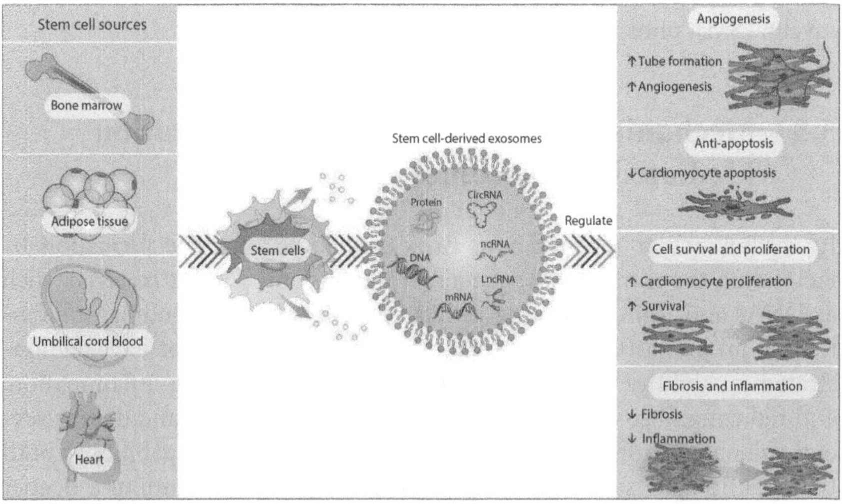

Figure 5.3 Representative configuration of exosomes, their contents, and biological roles.

5.3.1 Modulation of Inflammation

Acute MI sets off a cascade of inflammatory responses crucial for the healing and repair processes post-MI [202]. Experiments have emphasized the role of reperfusion-induced inflammation in healing mechanisms [202, 203]. First, the repair process is facilitated by infiltrating mast cells and monocytes through cytokines and growth factors complex interplay [204]. However, if inflammation persists, it exacerbates myocardial injury, resulting in detrimental left ventricular remodeling and HF [202]. Beyond localized inflammation in the injured area, AMI also initiates a systemic inflammatory response, influencing various humoral and cell-mediated response factors [205]. The activation of the complements' cascade specifically prompts immune cells' enrollment to the myocardium damage and amplifies the inflammation [202]. The mechanism leads to the neutrophils and monocytes' infiltration into the injured tissues and performs various immune functions, including degranulation, differentiation, and phagocytosis. Concurrently, a series of inflammatory cascades involving cytokines and chemokines, TNF and IL-1β, are triggers of recruited degranulated mast cells [205]. Additionally, damage-associated molecular patterns (DAMPs) are then secreted by necrotic cardiac cells, further initiating inflammation by activating Toll-like receptors (TLRs), immune cells, and inflammasome formation signaling [206–208]. ROS, oxygen species produced during AMI, exacerbate the damage to vascular cells and cardiomyocytes, perpetuating inflammatory responses in a feedback loop [209, 210]. Excessive and chronic inflammation after AMI can result in adverse left ventricular remodeling characterized by increased protease activity, ventricular dilation, cytokine expression, and excessive fibrosis due to the activation of cardiac fibroblasts [211, 212]. This adverse remodeling is closely linked with poor clinical outcomes, making the targeted post-MI inflammation management an inevitable strategy for infarct size reduction. Although the precise mechanisms of the procedure remain incompletely discovered, MSCs are well recognized for their immunomodulatory capabilities. Recent research has highlighted the increasing importance of MSCs' paracrine signaling in modulating immune responses during MI, mainly through exosome-mediated interactions. MSC-Exos exhibit immunosuppressive solid and anti-inflammatory properties (Table 5.4). Studies suggest that these Exos may shift immune response from pro-inflammatory to more tolerogenic, creating an environment that supports tissue regeneration and repair by endogenous progenitor and SCs [213–219].

A groundbreaking study in 2015 revealed that MSC-Exos mitigate inflammation during AMI by inhibiting the infiltration and proliferation

Table 5.4 A brief view of some of the components of the MSC-Exos known for their potential for immune response regulation in injured cardiac tissue.

Molecular cargo	Function	Reference
PDL-1	Acts in the PD-1/PD-L1 signaling, modulating T-cell activity and influencing immunological tolerance and immune-driven tissue injury	[243, 244]
Circular RNA circASXL1	Inhibits NLRP3 inflammasome activation	[245]
Galectin-1	Acts as a homeostatic regulator by modifying both adaptive and innate immune responses. It reduces cell proliferation and persuades cell cycle arrest and apoptosis in activated immune cells	[244, 246–249]
TGF-β	A potent cytokine that impacts various immune cells, such as T-cells and dendritic cells, can have pro-inflammatory or anti-inflammatory effects	[129, 244, 250, 251]
miR-21	A negative regulator of T-cell activation by GNAQ, CXCR4, and PLEKHA1 targeting. Also modulates anti-inflammatory responses and promotes M1 polarization to the M2 macrophages	[223, 252]
miR-24	Reduces inflammation of the aortic vascular by cooperating with CHI3L-1, regulating inflammation and tissue remodeling	[223, 225, 253]
miR-34a-5p	NF-κB regulator in T-cells and promotes differentiation toward M2 macrophage polarization	[233, 234]
miR-100-5p	FZD5/Wnt/β-catenin pathway	[254]
miR-133	Alleviates fibrosis and enhances cardiac function following MI	[242, 255]
miR-146a-5p	Promotes M1–M2 polarization by downregulating M1 marker genes	[228]

(*Continued*)

Table 5.4 A brief view of some of the components of the MSC-Exos known for their potential for immune response regulation in injured cardiac tissue. (*Continued*)

Molecular cargo	Function	Reference
miR-181a	The inflammatory response is suppressed by interacting with the c-Fos protein, a crucial immuno-activator involved in immune functions related to dendritic cells	[226]
miR-182	Involved in TLR4/NF-κB/PI3K/AKT signaling, regulating macrophage polarization	[117]

of immune cells in the rat heart's infarcted region [216]. Specifically, MSC-Exos treatment caused a notable decline in CD3+ T cells and a concurrent reduction in inflammation and cellular infiltration within the myocardial tissue. *In vitro* experiments using peripheral blood mononuclear cells (PBMCs) and MSC-Exos echoed these findings, illustrating an increase in the apoptosis of CD3+ T-cells and proliferation decline, differentiation, and immunoglobulin production in B cells under CpG stimulation [220]. Further investigations confirmed that MSC-Exos similarly restrain T-cell differentiation, activation, and proliferation in lab research [221]. Subsequent research identified several key components involved in MSC-Exos' anti-inflammatory effects. These Exos express molecules like membrane-bound TGF-β, programmed death-ligand 1 (PD-L1), and galectin-1, which promote immune tolerance [222]. A high expression of immunomodulatory miRNAs such as miR-24 and miR-29 in MSC-Exos and MSCs suggests that these Exos can partially replicate the immunosuppressive effects seen with MSC therapy [223]. miR-29 helps reduce fibrosis by repressing collagen genes, while vascular inflammation is limited by miR-24 through its interaction with chitinase-3-like protein-1 (CHI3L-1) [224, 225]. Moreover, MSC-Exos carrying miR-181a inhibit inflammation by interacting with the c-Fos protein, essential for immune functions associated with dendritic cells [226]. The miR-182 from MSC-Exos alleviates ischemia/reperfusion (I/R) injury within the myocardium through the modulation of the TLR4/NF-κB/PI3K/AKT pathway, affecting macrophage polarization [117]. Mesenchymal stem cell-derived exosomes (MSC-Exos) facilitate the transition of pro-inflammatory M1 macrophages to anti-inflammatory M2 macrophages [117, 227–232]. Unlike M1 macrophages, which act in defense and inflammation, M2 macrophages help convert inflammatory signals into reparative processes. Inflammatory environments enhance MSC-Exo synthesis, driving M2 polarization by significantly increasing miR-34a-5p, miR-146a-5, and miR-21. This M1–M2 shift has also been observed in other conditions, such as bronchopulmonary dysplasia, sepsis, and muscle damage [229–231, 233, 234]. In the aftermath of MI, extensive cardiac fibrosis is often observed as part of the repair response to ischemic cell death. While initial fibrosis helps prevent ventricular rupture, the shift from acute inflammation to fibrosis, driven *via* M2 macrophages, contributes to increased cardiac fibrosis [235]. M2 macrophages release TGF-β1, promoting myofibroblast transformation from fibroblast, producing matrix components and collagen to stabilize and scar the tissue [236, 237]. They also enhance fibrogenesis by producing arginase-activating proline and glutamate necessary for collagen synthesis [236, 238]. However, AMI can trigger an exaggerated inflammatory response beyond the injury site, potentially leading to worsening HF and cardiac function [239].

M2 macrophages, rich in IL-10, play anti-cardiac fibrosis, with a deficiency in IL-10 leading to more severe fibrosis and adverse tissue remodeling. Despite the complex role of M2 macrophages, MSC-Exos have shown promise in mitigating fibrosis post-MI [111, 219, 222, 239, 240]. Although the exact mechanisms remain unclear, some studies suggest that the miRNA components of MSC-Exos play a significant role. For example, MSC-Exos containing miR-22, preconditioned with ischemia, has been found to reduce fibrosis and improve cardiac functionality after MI [111, 241, 242].

Programmed death-ligand 1 (PDL-1), circular RNA ASXL transcriptional regulator 1 (circAXSL1), NOD-, LRR-, and pyrin domain-containing protein 3 (NLRP3), transforming growth factor beta (TGF-β), G protein subunit alpha Q (GNAQ), pleckstrin homology domain containing A1 (PLEKHA1), C-X-C motif chemokine receptor 4 (CXCR4), chitinase-3-like protein 1 (CHI3L-1), nuclear factor kappa-light-chain-enhancer of activated B cells (NF-κB), Frizzled class receptor 5 (FZD5), Wingless-related integration site (Wnt), phosphoinositide 3-kinase (PI3K), protein kinase B (AKT), Toll-like receptor 4 (TLR4).

5.3.2 Promotion of Angiogenesis

The new blood vessels' emergence, whenever and wherever that body needs them, signifies angiogenesis. Hypoxic changes are the results of the need for blood supplements inside the body. Subsequently, angiogenesis advancement appears to be the primary outcome of CVDs [256].

Exos show significant functions in promoting or inhibiting angiogenesis. Stem and mature cells release Exos, which transport macromolecules like microRNAs, nucleotide sequences, lipids, and proteins to recipient cells. These Exos can modulate angiogenesis-related RNAs and proteins, such as increasing VEGF expression, which leads to new blood vessel formation through VEGFR-2. However, excessive VEGF can activate MMP-2 and MMP-9, causing vascular leakage [257–260].

Exosomal cargos change during pathological conditions, impacting angiogenesis. For instance, some types of miRNA-212 in cardiac endothelial cells and Exos are exceptional factors in reducing angiogenesis and aiding cardiac tissue repair, whereas other types of this family enhance angiogenesis in injured cardiac tissue [261, 262]. MSC-Exos have shown benefits in cardiac infarcts by promoting angiogenesis through mechanisms due to enclosing an advanced quantity of cardioprotective miRNAs, including miR-210, miR-24, and miR-29, enhancing blood supply, and reducing apoptosis and miR-21 as well as miR-15 which are known as cardio-offensive miRNAs as compared to MSCs [121, 223, 263–266]. Additionally, miR-132 downregulates p120RasGAP and upregulates the RASA1 gene, promoting HUVEC proliferation and neovascularization in ischemic tissue [267]. Umbilical cord MSC-Exos secreting growth factor-D derived from platelets (PDGF-D) that overexpress protein kinase-B (PkB) and adipocyte MSC-Exos overexpressing SIRT1 also enhance angiogenesis by different mechanisms, including the elevated expression of CXCL12 and nuclear factor E2 related factor 2 (Nrf2) reducing infarct size and improving vascular structure (Table 5.5) [268–272].

Avian reticuloendothelial virus oncogene homolog A (RELA), nuclear factor kappa-light-chain-enhancer of activated B cells (NF-κb), matrix metalloproteinases (MMPs), wingless-type MMTV integration site family, member 5A (WNT5A), platelet-derived growth factor (PDGF), platelet-derived growth factor receptors (PDGFRs), mitogen-activated protein kinase (MAPK), protein kinase-B (PKB), phosphoinositide 3-Kinase (PI3K), mesenchymal stem cell-derived Exos (MCS-Exos), fibroblast growth factor (FGF), Harvey rat sarcoma viral oncogene homolog (H-Ras), Sirtuin 1 (SIRT1), C-X-C motif chemokine ligand 12 (CXCL12), nuclear factor erythroid 2–related factor 2 (Nrf2), endothelial progenitor cells (EPCs), GATA-binding protein 4 (GATA-4), non-receptor tyrosine kinase (NRTK), bone marrow kinase (BMX), insulin-like growth factor 1 (IGF-1), focal adhesion kinase (FAK), extracellular matrix metalloproteinase inducer (EMMPRIN), epidermal growth factor (EGF), human umbilical vein endothelial cells (HUVECs), intercellular adhesion molecule 1 (ICAM-1), long non-coding RNA urothelial carcinoma associated 1 (lncRNA UCA1), granulocyte macrophage-colony-stimulating factor

Table 5.5 Angiogenesis factors and their mechanisms in cardiac tissue.

Factor	Mechanism	Reference
VEGF-family	Increases expression, leading to decreased infarct area and initiation of new blood vessel formation through VEGFR-2. Increases the angiogenic capacity of ECs	[260]
Avian reticuloendothelial virus oncogene homolog A	In conjunction with p50, RELA forms part of the NF-κB heterodimer that drives the transactivation of NF-κB target genes, including those involved in angiogenesis	[273, 274]
IL-6, MMP-2, and MMP-9	Activated by high VEGF levels in hypoxic conditions, leading to irregular vascular spillage. Route for WNT5A signal	[257]
p195	Promotes VEGF-stimulated angiogenesis by linking VEGFR2 to VE-cadherin-containing adherence junctions	[247, 275]
Galectin-1	Enhances several stages of the angiogenesis procedure by amplifying pro-angiogenic signaling through VEGFRs and H-Ras	[247, 276]
PDGFRs and PDGF-family	Highly involved in the angiogenesis process, activation of several pro-angiogenic pathways, including MAPK and PI3K by PDGF-Rs	[129, 269, 277, 278]
FGF	A potent angiogenesis inducer, mitogenic activity on the vascular and capillary endothelial cells through driving VEGF-induced angiogenesis	[273, 279, 280]
Cadherin	VE-cadherin regulates angiogenesis and maintains the integrity of blood vessels	[273, 281]

(*Continued*)

Table 5.5 Angiogenesis factors and their mechanisms in cardiac tissue. (*Continued*)

Factor	Mechanism	Reference
SIRT1	Overexpressed in adipocyte MSC-Exos, increasing CXCL12 and Nrf2 expression in EPCs, reducing infarct size, and enhancing angiogenesis	[269–272]
GATA-4	Enhancing growth factor release	[282]
Angiopoietin-2	Tie2-unrelated route	[272, 283, 284]
NRTK	Regulates the expression of VEGFR2 and modulates angiogenesis induced by VEGF through BMX from the Tec family	[285]
IGF-1	Activation of PI3K/Akt signaling pathway	[286–288]
FAK	PI3-K activation	[289]
Annexin A2	Accelerates plasmin generation	[290]
EMMPRIN	Influences angiogenesis and cell migration by acting upstream of VEGF and MMP-9. Increases VEGF expression and promotes angiogenesis	[291–293]
EGF like repeats and Discoidin domains 3	Help in cell migration and expansion	[294]
EGF receptor III	Activates the PI3K and MAPK pathways and numerous transcription factors that adjust the transcription of VEGF, leading to VEGF signaling initiation	[295]
IL-8	Regulation *via* MAPK exerts compelling angiogenic properties on ECs through interaction with CXCR1 and CXCR2	[129, 296, 297]

(*Continued*)

Table 5.5 Angiogenesis factors and their mechanisms in cardiac tissue. (*Continued*)

Factor	Mechanism	Reference
ICAM-1	Helps with RhoA/ROCK, p38 MAPK, ERK1/2 kinase, eNOS, as well as Src kinase	[298, 299]
lncRNA UCA1	Targeting miR-184/HOXA9 axis	[300]
IL-8, PDGF	PI3K/AKT trafficking	[301]
GM-CSF, HIF1α, HIF-2α	Increases VEGF transcription	[302]
miR-9	Influences JAK/STATs besides enhancing FGF5, VEGFA, FGF12, ATG7, MMP2, MMP7, MMP10, MMP11, and MMP24 expression	[303]
miR-21	Activation of PTEN/VEGF signaling pathway in post-AMI and activation of STAT1. Induces angiogenesis through ERK and AKT activation and the expression of HIF-1α	[265, 304]
miR-21-5p	Promotes TGF-β signaling pathway, angiopoietin-1, pro-angiogenic VEGF-α, and BNP and ANP	[305]
miR-23	Interacts with Sema6A, Sema6D, and Sprouty2 in ECs, induces sprouting angiogenesis	[252]
miR-23a-3p	Along with miR-30a and miR-34a, attack prolyl promotes neovascularization and expansion of blood vessels, promotes adhesion protein molecules ZO-1 and hydroxylase, and reduces ferroptosis	[306]
miR-132	Upregulating RASA1 gene expression leads to HUVEC proliferation and neovascularization in ischemic tissue	[267]

(*Continued*)

Table 5.5 Angiogenesis factors and their mechanisms in cardiac tissue. (*Continued*)

Factor	Mechanism	Reference
miR-135b	Involved in the HIF-FIH signaling pathway; also causes overexpression of upregulated miR-210-5p, which promotes tubular structure formation in HUVECs	[307]
miR-141-3p	NF-B and JAK/STAT3 pathway activation	[308]
miR-145	Regulates VEGF signaling through N-RAS and IRS1 targeting, binding integrin-B and 8hexokinase II	[309]
miR-153-3p	Inhibition of this miRNA by means of targeting ANGPT1-mediated VEGF/PI3K/Akt/eNOS signaling promotes angiogenesis	[310]
miR-181a	MLL3 and DACT2 downregulation and VEGF upregulation	[311]
miR-182-5p	Regulates macrophage phenotype *via* the TLR4/NF-κB pathway	[312]
miR-210-3p	Promotes angiogenesis by regulating gene alterations, especially with interferon-α. Elevates angiogenic genes including VEGFR1 and VEGFR2	[262]
miRNA-212	Elevated in cardiac ECs and exosomal levels, blunting angiogenesis and aiding cardiac tissue repair	[261]
miR-373	Activates HIF downstream signaling	[313]
miR-1290/1246	Targets the SMEK1 to produce an angiogenic phenotype and induces the upregulation of SP1 and ELF5, enhancing changes in phenotype from fibroblasts to endothelial cells	[314]

(*Continued*)

Table 5.5 Angiogenesis factors and their mechanisms in cardiac tissue. (*Continued*)

Factor	Mechanism	Reference
NEU	Overexpression of MMP-9 and CXCR4	[271]
PTCH 1, SMO, SHH, Ihh	Pro-angiogenic responses in endothelium stimulated by Exos from CC cells through upregulating Hh-GLI signaling and altering upstream angiogenesis-related genes	[315]
PFN2	Promotes transforming TGF-β-mediated EMT, increases VEGF expression	[316]
Ezrin	Acts in actin-based cellular functions are essential for cell locomotion and influence angiogenesis	[247]
RAMP2-AS1 lncRNA	Depress the miR-2355-target 5p's VEGFR2 using miR-2355-5p siphoning, increasing the cell membrane receptors of the angiogenesis procedure	[317]
TGF-β	Smad and non-Smad pathways initiate, promote, or suppress ECs' migration, proliferation, permeability, and sprouting	[318]
NF-κb	Activates various pro-angiogenic-related genes, including the IL-8 and VEGF, and several MMPs	[273, 319]
Tspan8 (D6.1A)	Enhancing MMP, VEGF, and VEGFR (VEGF receptor) transcription	[320]
Vasorin	Alterations in homeostasis of calcium, levels of angiotensin II, and NO signaling, cell migration increment	[321]
SDKP-conjugated peptide	Hydrogel-based angiogenesis improvement	[322]

(GM-CSF), hypoxia-inducible factors (HIFs), Janus kinase/signal transducer and activator of transcription (JAK/STAT), phosphatase and tensin homolog (PTEN), signal transducer and activator of transcription 1 (STAT1), transforming growth factor beta (TGF- β), A serine/threonine-specific protein kinase (AKT), extracellular signal-regulated kinase (ERK), atrial natriuretic peptide (ANP), B-type natriuretic peptide (BNP), vascular endothelial-cadherin (VE-cadherin) Semaphorin (Sema), RAS p21 protein activator 1 (RASA1), hypoxia-inducible factor–factor inhibiting HIF (HIF-FIH), nitric oxide (NO), insulin receptor substrate 1 (IRS1), sonic hedgehog (SHH), Indian hedgehog (Ihh), hedgehog-GLI signaling pathway (Hh-GLI), profilin 2 (PFN2), epithelial–mesenchymal transition (ETM), zonula occludens-1 (ZO-1), receptor activity modifying protein 2–antisense 1 (RAMP2-AS1), Tetraspanin 8 (Tspan8).

5.3.3 Enhancement of Cell Survival and Proliferation

Enhancing cell survival and proliferation is crucial in cardiovascular regeneration, especially when considering the use of SC-Exos. These Exos perform a significant action in guarding cardiac cells from programmed cell death (apoptosis) and promoting their proliferation, causing the repair and regeneration of HF. The importance of these processes cannot be overstated, as the ability of cardiac cells to survive and multiply is fundamental to restoring the organization and functionality of the cardiac tissue following heart injuries. Several factors enhance cell survival and proliferation. The key point among these is the various signaling molecules and growth factors encapsulated within Exos. These include proteins, lipids, and RNAs that modulate cellular pathways, leading to increased resistance to stress and enhanced regenerative capabilities. For example, exosomal miRNAs have demonstrated the ability to suppress genes that promote apoptosis while simultaneously activating genes that facilitate cell cycle advancement and proliferation. This delicate balance of gene regulation is vital for maintaining cellular homeostasis and promoting repair mechanisms. In cardiac cells, these factors work through complex signaling networks to achieve their effects. For example, exosomal contents often activate the PI3K/Akt pathway, improving cell survival by inhibiting apoptotic processes and enhancing metabolic activity. Furthermore, the ERK1/2 pathway is another crucial signaling route that promotes cell proliferation and survival. By comprehending and leveraging these pathways, researchers strive to create targeted therapies that amplify the regenerative capabilities of stem cell-derived Exos, potentially resulting in more effective treatments for CVDs [323–325].

Exos released by BMMSCs are pivotal in mitigating myocardial injury induced through tissue reperfusion. Within these Exos, miR-486-5p is present, which inhibits myocardial apoptosis by obstructing the deactivation of PTEN [326]. The therapeutic efficacy of BMMSCs is often augmented through hypoxic pretreatment. Research indicates that hypoxia markedly elevates miR-214 levels in BMMSC-Exos, which are then transferred to cardiomyocytes to inhibit CaMKII expression. Additionally, hypoxic pretreatment of BMMSCs leads to enhancement of miR-125b-5p in their Exos, thereby enhancing cardiomyocyte resistance to apoptosis by targeting BCL2 antagonist/killer 1 (BAK1) and p53 genes [201, 327, 328]. Furthermore, Exos from hypoxic BMMSCs exhibit upregulated levels of miR-24 and miR-210, which further support cardiomyocyte survival. Although the precise mechanisms of apoptosis prevention remain undefined, miR-210 production is known to be influenced by nSMase2 levels. Modifying Exos with specific genes or microRNAs is proposed to amplify their anti-apoptotic properties. For instance, the GATA-4 modulates the miR-15 family in BMMSCs, enhancing their persistence under hypoxic conditions. Evidence suggests that Exos originating from adipose-derived SCs (ADSCs) are mighty in regulating the S1P/SK1/S1PR1 signaling pathway [329]. Elevated miR-214 levels in ADSC-derived Exos can inhibit Bcl2L11 and SLC8a1 expression in the MI region. Bcl2L11 might induce apoptosis by activating Bax or counteracting apoptosis suppressor proteins, whereas SLC8A1, encoding a sodium/calcium exchanger, can precipitate cardiomyocyte death due to calcium overload during cardiac stress. ADSC-Exos have demonstrated enhanced resistance to apoptosis in hypoxic cardiomyocytes through the overexpression of miR-146a, which utilizes its anti-apoptotic effects through inhibiting the early growth response factor 1 (EGR1). Exos overexpressing TIMP2 originating from umbilical cord MSCs bolster the anti-apoptotic properties of hypoxic cardiomyocytes by suppressing the Bax and pro-caspase levels. Similarly, iPSC-Exos take a pivotal action in avoiding miR-21 from apoptosis inhibition in injured cardiomyocytes by affecting PDCD4/AP-1 as its targets [148, 330].

MSCs overexpressing miR-375 hold promise in curbing autophagy, which enhances cell function in injured heart tissue. MiR-375's potential role in cardiac muscle involves regulating Rho-GTPase-dependent signaling pathways [331]. Exos have been shown to mitigate autophagy in the myocardium and cell death followed by MI or I/R injury. Previous studies have demonstrated that human MSC-Exos moderate I/R damage *via* cardiomyocyte autophagy prevention, even though its exact mechanisms remain elusive [332]. Delivery of miR-125b-5p carried by BMMSCs-Exos to the infarcted myocardium reduces myocardial cell autophagy and suppresses P53 production in the affected region [333]. In an I/R injury mouse model, miR-29c expression decreased, and increased myocardial cell autophagy was observed. Elevated levels of miR-29c could potentially mitigate extreme autophagy in hypoxic myocardium by downregulating PTEN expression directly, thus limiting the autophagy induced by I/R through the PTEN/AKT/mTOR pathway. Moreover, the overexpression of stromal cell-derived factor 1 (SDF1) in BMMSCs significantly raised SDF1 levels in their Exos, which correspondingly increased the Bcl-2 levels in cardiomyocytes within hypoxic conditions. In contrast, Beclin-1, Bax, LC3, and the LC3II/LC3I ratio were notably reduced, thereby preventing excessive autophagy in cardiomyocytes [325].

Given the constrained regenerative capability of adult mammalian cardiomyocytes, the human heart shows minimal ability to regenerate following injury. Thus, employing a broad spectrum of suitable cargos within engineered Exos may prove crucial for better outcomes. Jones *et al.* identified that among nearly 1,000 miRNAs screened *via* high-throughput genomic analysis, at least 40 miRNAs significantly promote cardiomyocyte proliferation by stimulating DNA synthesis, facilitating progression through the G2/S phase of the cell and supporting karyokinesis [334], some of which are presented in Table 5.6.

MYC-dependent transcriptional (MNT), vascular endothelial growth factor (VEGF), insulin-like growth factor-1 (IGF-1), hepatocyte growth factor (HGF), myelocytomatosis viral-related oncogene (Mycn), hedgehog (HH), HOX transcript antisense RNA (HOTAIR), maternally expressed 3 (MEG3), circular RNA (CircARNA), fused in sarcoma (FUS), BCL2 interacting protein 3 (BNIP3), phosphorylated protein kinase B (p-AKT), microtubule-associated proteins 1A/1B light chain 3B (LC3), programmed cell death 4 (PDCD4), phosphatase and tensin homolog (PTEN), forkhead box O3 (FoxO3), phosphoinositide 3-kinase (PI3K), Kelch-like ECH-associated protein 1 (KEPA1), nuclear factor erythroid 2-related factor 2 (Nrf2), max binding protein (MNT), suppressor of variegation 3-9 homolog 1 (Suv39H1).

5.3.4 Reduction of Fibrosis

Cardiac injury and other circumstances, such as aging, launch different pathophysiological outcomes, including fibrosis, which is also known as fibrotic scar. Fibrosis is categorized into two groups: replacement fibrosis (caused by different cardiac diseases) and reactive interstitial fibrosis

Table 5.6 Exosome cargos involved in injured cardiac cell proliferation and survival post-MI.

Cargo	Function	Reference
miR-143-3p	Mitigates I/R injury through autophagy modulation in the myocardium	[335]
Mycn mRNA	Through HH signaling pathway	[336]
miRNA-1825	Regulates expression of p16, miR-199a, Rb1, and Meis2 to be downregulated	[337]
miR-199a-3p and miR-590-3p	Downregulate both HOMER1 and CLIC5, activate p-AKT survival signaling	[338]
miR-301	Reduces LC3-II/LC3-I ratio	[339]
miR-21	Suppressed the translation PDCD4 and PTEN, inhibits apoptosis	[148, 340]
miR-132/212	Target FoxO3, elevating PI3K/Akt pathway	[341]
miR-144	Downregulate PTEN, activating Akt signaling pathway, decreasing apoptotic proteins	[342]
miR-19a	Activating Akt signaling pathway due to PTEN downregulation, thus decreasing apoptotic proteins	[343]
miR-200a-3p	Mitochondria protected by the KEAP1-Nrf2 pathway	[344]
miR-210	Regulates cell/cycle progression by targeting the MNT	[322]
HOTAIR	Targeting the miR-130a-3p/Suv39H1 axis	[345]
MEG3	Improves autophagy and apoptosis within the cardiomyocyte by regulating the miRNA-129-5p expression	[346]
HGF and IGF-1	Enhancing the incidence of GATA-4-positive cell clusters and cardiomyocyte cell cycle re-entry induction	[347]
CircFndc3b	Modulates cell proliferation *via* FUS/VEGF-A axis	[348]
CircRNA-POSTN	Regulates miR-96-5p/BNIP3 axis	[349]

(caused by cardiac tissue aging); the abnormal balance of ECM in cardiac tissue, in other words, in adult mammalian cardiac tissue often leads to a change from necrosis to cardiac fibrosis due to low regenerative capacity, that a fibrotic-reparative scar swaps cardiac tissue, and it is steadily remodeled during cell death in heart tissue. To be specific, the fibrosis procedure in cardiac tissue starts with injury/stress and goes on with processes of different pro-fibrotic factors, including angiotensin II (Ang II), aldosterone, TGF-β, and endothelin (ET), causing fibrosis initiation by getting fibroblasts in the injury area, transforming to myofibroblasts, making ECM remodeling and fibrosis as a cardiac dysfunction reason to take place. The ventricular structure, systo-diastolic function, and electrical myocardial activity might be affected by prolonged fibrosis, which finally leads to different life-threatening HF. So, it is vital to find suitable therapeutic strategies using anti-fibrotic agents to fight against the post-MI fibrotic condition. Autophagy is mighty to worsen the dysfunction of the cardiac tissue and bring about cell death once it is activated under severe ischemic conditions [350–354].

Cardiomyocytes-, ECs-, and epicardium-Exos help in maintaining cardiac structure, and Exos from sources such as fibroblasts can act in the initiation of cardiac regeneration and repair [201]. There is a massive list of therapeutic Exo-cargos used in this regard (Table 5.7). MSC-Exos significantly boost cardiac microvascular function, reducing cardiac fibrosis after I/R injury. This positive impact is facilitated through the PDGFR-β pathway modulation. By enhancing microvascular function, MSC-Exos improve blood circulation and oxygen delivery, resulting in diminished post-injury fibrotic scar formation. The role of Exos in modulating PDGFR-β underscores their ability as a therapeutic approach in treating ischemic heart diseases (IHDs) and minimizing fibrotic remodeling [355]. MSC-Exos become remarkable mighty helpers when harbored with different valuable therapeutic molecules, including the genetic materials that manipulate a wide range of cellular functions. For instance, miR-143-3p regulates autophagy effectively and reduces apoptosis, thus inhibiting myocardial injury resulting in repair promotion and helping fibrosis reduction [335]. Anti-apoptotic effects have been confirmed; MSC-Exos charged with the miR-22 targets the methyl CpG binding protein-2 (MECP2) gene, reducing fibrosis and thus promoting cardiac maintenance [111, 356]. One of the most important exosomal cargos is miR-145; its absence leads to TGF-β-associated fibrosis [357]. Exos derived from EPCs (EPC-Exos) upregulate the expression of miR-1246 and 1290 significantly, targeting ELF5 and SP1, promoting changes in phenotype from fibroblasts to ECs, enhancing angiogenesis, and proliferation in human cardiac fibroblasts (HCFs) [314].

Table 5.7 Cargos involved in fibrosis modulation of injured cardiac tissues.

Cargos	Mechanisms	References
LNA-anti-miR-21	Targets miR-21, promoting RECK, inhibiting PDCD4, along with effectors of TGF-ß signaling	[359]
miR-34a antagomirs (Ant-miR34a)	Inhibits miR-34a, attenuates the expression of the ECM protein col1, and α-SMA as well as TGF-β1 and Smad4	[360]
Antisense miR-34a antagomirs	Silence miR-34a, resulting in col1, α-SMA, TGF-β and Smad4 decrease	[361]
miR-1	Targeting cyclin D2 and CDK6	[362]
miR-15b-5p	Inhibits the TGFβ-pathway components (TGFBR1, SMAD3, SMAD7, p38, and endoglin)	[363]
miR-18a-5p	Inhibits fibroblast differentiation and TGF-β stimulation; thus, Acta2, CTGF, SMAD2, and SMAD 3 are downregulated	[364]
miR-22-3p	Downregulation of TGFβ, targeting PTAFR suppressing fibrogenesis	[365, 366]
miR-24	Regulates the secretion of TGF-β and phosphorylation of Smad2/3	[367]
miR-26a-5p	Keap-mediated activation of Nrf2; also regulates ULK1 expression and affects fibroblasts autophagy in cardiac tissue	[368, 369]
miR-27b	Targets ALK5, attenuating fibrosis	[370]
miR-29a	Targets VEGF-A/MAPK signaling pathway, thus suppresses cardiac fibroblast proliferation	[371]
miR-29b	Targets the TGF-β/Smad3 signaling pathway	[372]
miR-29b-3p	Causes silencing of circHIPK3, inhibiting α-SMA, COL1A1, and COL3A1	[373]

(Continued)

Table 5.7 Cargos involved in fibrosis modulation of injured cardiac tissues. (*Continued*)

Cargos	Mechanisms	References
miR-29c	TGFβ regulator, downregulation of COL1A1, COL1A2, COL3A1, and FBN1	[374]
miR-30b-5p	Targets PTAFR suppressing fibrogenesis	[366]
miR-101a	Targets TGFβRI on cardiac fibroblasts	[375]
miR-132	Initiated by TGF-β, downregulates pro-fibrotic Mecp2	[376]
miR-133a	Decreases the mRNA and protein expression levels of TGF-β1 and TGF-βRII, connective tissue growth factor (CTGF), col1, col3, and α-SMA	[377]
miR-145	Regulates TGFBR2 expression and matrix synthesis	[378]
miR-214-3p	Controls Ca^{2+} overload and cell death, suppression of Ctrp9 gene—resulting in TGFβ, Smad2/3, and Smad4 downregulation	[379, 380]
CPCISX-9-Exos enriched with miR-373	Significantly inhibits two key fibrogenic genes (GDF-11 and ROCK-2)	[313]
miR-384-5p	Targeting TGF-β/Wnt	[381]
miR-425	Targeting TGF-β	[131]
miR-486a-5p	Regulates Smad1	[382]
miR-590	Regulates TGF-β1 and TGF-Brii	[383]
miR-663	Regulates TGF-β1	[384]
miR-675	Smads/TGFβR1 regulation	[385]
miR-744	Targeting TGF-β	[131]

(*Continued*)

Table 5.7 Cargos involved in fibrosis modulation of injured cardiac tissues. (*Continued*)

Cargos	Mechanisms	References
B-type natriuretic peptide (BNP)	Counteracts TGF-β signaling	[386]
Sprouty 1 (Spry1)	Targets TGF-β1 in regulating fibrosis, inhibits excessive matrix production	[387]
TGF-βRIII	Inhibits miR-21 expression and weakens collagen production in fibroblasts	[388, 389]

Some miRNA families have been shown to have a fibrotic role in injured areas, which must be detected and inhibited using desired factors. One of these fibrosis antecedent miRNAs is miR-27b-3p, which function in different ways, such as targeting different genes. For example, the FGF1 gene is a target for this miRNA, increasing mitochondrial oxidative phosphorylation, and supplementation with recombinant FGF1 weakens pathological cardiac remodeling caused by Ang II. miR-27-knockdown mice encountered lower post-MI fibrosis, which is approved for its fibrotic activities [358].

Locked nucleic acid (LNA), microRNA (miR), reversion-inducing cysteine-rich protein with Kazal motifs (RECK), programmed cell death 4 (PDCD4), transforming growth factor beta (TGF-ß), miR-34a antagomirs (Ant-miR34a), extracellular matrix (ECM), collagen I (col1), alpha-smooth muscle actin (α-SMA), transforming growth factor beta 1 (TGF-β1), mothers against decapentaplegic homolog 4 (Smad4), transforming growth factor beta receptor 1 (TGFBR1), connective tissue growth factor (CTGF), actin alpha 2 (Acta2), platelet-activating factor receptor (PTAFR), Kelch-like ECH-associated protein (Keap), nuclear factor erythroid 2–related factor 2 (Nrf2), Unc-51 like autophagy activating kinase 1 (ULK1), activin receptor-like kinase 5 (ALK5), vascular endothelial growth factor A (VEGF-A), collagen type I alpha 1 chain (COL1A1), fibrillin 1 (FBN1), methyl CpG binding protein 2 (Mecp2), transforming growth factor beta receptor II (TGF-βRII), collagen type I alpha 2 chain (COL1A2), C1q/TNF-related protein 9 (Ctrp9), cardiac progenitor cells derived exosomes (CPCISX-9-Exos), growth differentiation factor 11 (GDF-11), rho-associated protein kinase 2 (ROCK-2), B-type natriuretic peptide (BNP), Sprouty 1 (Spry1), transforming growth factor beta receptor III (TGF-βRIII).

5.4 Therapeutic Applications of Exosomes in Cardiovascular Diseases

Despite advancements in the prevention and treatment of cardiovascular diseases, they continue to be the primary cause of mortality worldwide [390]. While traditional pharmacotherapy and surgery have shown effectiveness in managing symptoms and reducing mortality rates over the past decade, there is still a need for more comprehensive clinical strategies for treating CVDs [391]. Recent research in cardiovascular diseases has highlighted the potential role of exosome-mediated intercellular communication in maintaining cardiovascular system homeostasis, offering a promising target for therapeutic interventions [392]. Evidence approves that these nanovesicles can be isolated from natural or modified sources and act as cardiovascular system homeostasis protectors or utilized as therapeutics in cardiovascular disease treatment [393].

5.4.1 Myocardial Infarction

Based on the *American Journal of Cardiology* introduction in 2018, MI identifies the occurrence of acute myocardial injury as evidenced by abnormal cardiac biomarkers in conjunction with signs of acute myocardial ischemia [394]. Although common treatment methods reduce the mortality rate of MI, there remains a deficiency in clinical approaches for the restoration of impaired myocardium after this complication [395]. In this context, several studies showed that MSC-Exos have become a potential new treatment for MI. The main therapeutic impact of MSC-Exos in cardiovascular disorders is predominantly achieved through paracrine pathways facilitated by Exos. These pathways encompass the mitigation of apoptosis, suppression of ferroptosis, enhancement of pyroptosis, and attenuation of autophagic injury in myocardial cells [390, 396]. MSC-Exos that are enriched with miR-301 can inhibit myocardial autophagy. MiR-301 helps maintain cellular homeostasis and prevents myocardial damage [339]. Modified exosomes that contain curcumin and then loaded into an injectable decellularized ECM hydrogel have anti-fibrotic effects by modulating key signaling pathways involved in fibrosis regulation, thereby preventing excessive collagen deposition, preserving cardiac tissue integrity, and improving post-MI outcomes [397]. Parvaiz Ahmad Sheik *et al.* also utilized an oxygen-releasing antioxidant cardiac patch delivering exosomes using ADSC-Exos, which deliver bioactive molecules to reduce oxidative stress and facilitate angiogenesis to improve cardiac regeneration following a heart attack [398]. Additionally, ADSC-Exos with a high expression of miR-205 in hypoxic environments can decrease myocardial cell death and hinder myocardial fibrosis [399]. Furthermore, miR-196a-5p and miR-425-5p from ADSC-EXO have demonstrated efficacy in averting the apoptosis of myocardial cells, encouraging macrophage polarization toward an anti-inflammatory state and supporting angiogenesis [400]. Recent research has shown that exosomes originating from MSCs that have been modified to overexpress HIF-1α can facilitate cardiac protection and promote angiogenesis utilizing the angiogenic factors, including VEGF and PDGF [401]. Moreover, BMMSCs with increased levels of miR-129-5p have demonstrated suppressing the expression of high-mobility group box 1 (HMGB1) and inflammatory mediators. This intervention has shown promise in enhancing cardiac function in mice, offering an innovative treatment approach for treating MI [402]. Exosomes originating from mesenchymal stem cells obtained from the human umbilical cord can specifically interact with and suppress the expression of divalent metal transporter 1 (DMT1) *via* miR-23a-3p, which subsequently aids in alleviating myocardial injury by preventing ferroptosis [306]. Notably, hematopoietic stem cell (HSC)-derived exosomes have ameliorated ischemic cardiomyopathy in a rat model of MI [403]. Nevertheless, the restricted potential of MSC-Exos in therapeutic approaches underscores the importance of exploring strategies to enhance the efficacy of MSC-Exos [391, 404].

Numerous studies revealed the role of MSC pretreatment in improving the MCS-Exos therapeutic efficacy; in this regard, the application of fibronectin type III domain-containing protein 5 (FNDC5) as a pretreatment for exosomes sourced from BMMSCs has demonstrated enhanced therapeutic efficacy in the treatment of infarcted hearts when compared to exosomes derived from untreated BMMSCs [405]. In addition, research has shown that MCS-Exos prior exposure to atorvastatin results in an increase in the levels of H19 lncRNA, which leads to enhanced expression of miR-675, VEGF, and ICAM-1, thereby facilitating angiogenesis and the recovery of cardiac function [115]. Interestingly, IFN-γ has the potential to enhance the recovery from myocardial infarction by increasing the levels of miR-21 in MSC-Exos *via* the activation of STAT1 that promotes angiogenesis and mitigates the apoptosis of myocardial cells [265].

5.4.2 Heart Failure

The heart's limited ability to regenerate following damage prompted the exploration of stem cell therapies as a promising strategy for enhancing cardiac regeneration [406]. In this regard, various studies have demonstrated the importance of stem-cell-derived Exos in the case of HF and their application as drug delivery systems or regenerative therapy agents in treating individuals with HF [330, 407]. To underscore the significance of Exos roles in HF, Qiao and colleagues extracted Exos from the conditioned media of cardiac cells sourced from healthy individuals and heart failure patients. These Exos were subsequently administered *via* intra-myocardial injection into a murine model of MI induced through coronary vessel ligation. The findings of the study revealed that Exos derived from healthy volunteers that have normal levels of miR-21-5p exhibited beneficial impacts on processes such as healing, apoptosis, and tissue preservation [408]. Chen and colleagues demonstrated in their work that BMMSC-Exos can reduce cardiac hypertrophy and fibrosis in the context of pressure overload-induced remodeling, suggesting a potential therapeutic approach for heart failure [409]. In another research, Pang *et al.* revealed that Exos derived from embryonic stem cells (ESCs) have the potential to mitigate heart failure, enhance cardiac function, and stimulate myocardial angiogenesis *via* the fibroblast growth factor 2 (FGF2) signaling pathway in a model of heart failure [410]. Similarly, Kervadec and colleagues demonstrated in a study utilizing a model of HF that Exos released by ESCs have the potential to restore cardiac functionality by improving parameters such as decreased left ventricular end-systolic and end-diastolic volumes [411].

5.4.3 Peripheral Artery Disease

Peripheral artery disease (PAD) is a progressive atherosclerotic disorder that affects more than 200 million people in the world [412]. The analysis of circulating exosomes discovered their prognostic/diagnostic application in the case of PAD [413]. Moreover, ongoing studies in treating PAD focus on creating methods like utilizing stem cell-derived exosomes that promote angiogenesis. In this regard, BMMSCs-derived exosomes that contain a high concentration of VEGF protein and miR-210-3p facilitate the formation of blood vessels in ischemic limbs by elevating the expression of angiogenic genes like VEGFR1 and VEGFR2 in the endothelial cells [262]. Furthermore, the enhanced angiogenic activity observed in MSC- and CD34+ stem cell-Exos was attributed to the elevated levels of VEGF and miRNA-126-3p, which can regulate and contribute to the activation of genes associated with angiogenic processes, thereby promoting angiogenesis [272, 414]. MSC-Exos also enriched it with miR30-b and transferred it into endothelial cells causing an enhancement in MSC-mediated angiogenesis [415]. Hypoxia-treated human ADSC-Exos enhance angiogenic properties by employing the protein kinase A (PKA) signaling pathway and the upregulation of VEGF expression [416]. Furthermore, studies have shown that hydrogels can effectively prolong the release of Exos, thereby increasing their therapeutic potential, as a high concentration of Exos is required for effective treatment [417]. In this context, Zhang and colleagues revealed that chitosan hydrogel significantly improved the stability of proteins and microRNAs within Exos, resulting in the increased retention of Exos *in vivo* [418]. Similarly, Hernandez *et al.* demonstrate that decellularized ECM hydrogel enhances the delivery and potential efficacy of human cardiac progenitor cell (hCPC)-derived Exos [419].

5.5 Preclinical and Clinical Studies

Preclinical research aims to choose a top candidate from multiple options, establish the most effective method for scaling up a new therapeutics, pick the optimal dose, decide on the route, frequency, and length of exposure, and provide the necessary backing for the planned clinical trial structure [420, 421]. Animal models, which encompass a wide range of vertebrate and invertebrate species, play a crucial role in experimental studies that seek to mimic diseases in both humans and animals [421]. In order to evaluate the therapeutic potential of stem cell-derived Exos, numerous efforts have been undertaken to enhance the efficacy of CVDs treatment through the utilization of different types of SC-Exos in preclinical investigations [422]. Interestingly, preclinical outcomes revealed that Exos show significant promise in the field of therapeutics due to their capacity to circumvent immune detection and provide targeted delivery [423].

5.5.1 Animal Models of Cardiovascular Disease

Animal models have advanced our knowledge of heart disorders [424]. They continue to be essential in preclinical research, encompassing a variety of species divided into small animals such as *Drosophila*, zebrafish, *Xenopus*, mice, and rats, medium-sized animals like guinea pigs, rabbits, and cats, and large animals including dogs, pigs, sheep, and non-human primates [425]. In this context, a wide range of animal models of CVDs, including mice, rats, and pigs, was applied to assess the preclinical aspect of SC-Exos in the case of cardiovascular disease treatment [396, 422]. Wei *et al.* demonstrated that miRNA-181a-overexpressing MSC-Exos had an immunosuppressing effect that could be used as a treatment approach for myocardial ischemia-reperfusion (I/R) injury rat models [226]. Moreover, Wang and colleagues in their research showed that human umbilical MSC-Exos could potentially enhance the expression of Smad7 by blocking miR-125b-5p, leading to an increase in myocardial repair in acute myocardial infarction rat models [426]. Similarly, Liu *et al.* reported that the administration of BMMSC-Exos improved heart function in a rat model of MI, reduced heart remodeling, decreased the fragmentation of cardiomyocyte mitochondria, decreased the level of ROS, and reduced apoptosis [427]. Li *et al.* also reported that BMMSCs-derived Exos with a high expression level of miR-301-protected MI rat models by suppressing myocardial autophagy [339]. Furthermore, Du *et al.* announced that Exos released from human placenta-derived MSCs through nitric oxide stimulation showed better angiogenic properties and improved limb function in a mouse model with hind limb ischemia [414]. On the contrary, Charles *et al.* utilized pigs as CVD animal models and revealed that giving intravenous MSCs-Exos over a period of 7 days resulted in a decrease in the size of the heart attack area and maintained better heart function in pigs compared to those that did not receive this treatment [428]. In addition, Ju *et al.* showed that injecting cardiac MSC-Exos directly into the heart muscle following a heart attack improves blood vessel growth in the heart, stimulates the growth of heart muscle cells, and helps maintain heart function in mice models with MI [429]. More preclinical studies were categorized based on Exos source, animal model, and key outcomes, which are presented in Table 5.8.

Table 5.8 Animal models of CVD utilized in preclinical studies.

Study	Exosome source	Animal model	Key outcomes
Mathiyalagan *et al.* [272]	Human CD34+ stem cell	Ischemic hind limb mouse	CD34Exos demonstrating an angiogenic role.
Arslan *et al.* [430]	Human MSCs	MI mouse	Human MSC-Exos diminished MI, enhanced cardiac function, and reduced ATP loss in mouse models.

(*Continued*)

Table 5.8 Animal models of CVD utilized in preclinical studies. (*Continued*)

Study	Exosome source	Animal model	Key outcomes
Yu *et al.* [431]	Murine BMMSCs	Myocardial I/R injury rat	BMMSCs-Exos prevented CM apoptosis and enhanced cardiac function.
Bian *et al.* [432]	Human BMMSCs	MI rat	Human BMMSC-Exos enhanced cardiac performance, reduced MI size, and stimulated the formation of new blood vessels.
Feng *et al.* [111]	Murine BMMSCs	MI mouse	Exosomes derived from BMMSCs diminished MI and prevented CM apoptosis.
Ma *et al.* [269]	Human MSCs	MI rat	MCS-Exos enhanced cardiac function, prevented CM apoptosis, and stimulated new blood vessel formation.
Shao *et al.* [223]	Murine BMMSCs	MI rat	Murine BMMSCs-Exos have been shown to enhance cardiac function. cardiac function, reduce MI size, and exhibit an anti-inflammatory role.
Wang *et al.* [264]	Murine BMMSCs	MI mouse	BMMSCs-Exos stimulated the formation of new blood vessels and cardiac function.
He *et al.* [433]	Murine BMMSCs	MI mouse	BMMSCs-Exos stimulated new blood vessel formation and cardiac function and also prevented CM apoptosis.

(*Continued*)

Table 5.8 Animal models of CVD utilized in preclinical studies. (*Continued*)

Study	**Exosome source**	**Animal model**	**Key outcomes**
Wang *et al.* [327]	Murine BMMSCs	MI mouse	Exosomes stimulated angiogenesis, and the cardiac function was improved, preventing inflammation with MI size reduction.
Luther *et al.* [434]	Murine BMMSCs	MI mouse	BMMSCs-Exos had a cardio-protection role caused by the reduction of CM apoptosis.
Xiao *et al.* [332]	Human BMMSCs	MI mouse	Human stem cell-derived exosomes improved cardiac function and prevented CM apoptosis.
Xu *et al.* [435]	Human BMMSCs, ADSCs, UCMSCs	MI rat	Diverse types of SC-Exos stimulated new blood vessel formation and improved cardiac function. In addition, there was a reduction in myocardial infarction size and a decrease in inflammation.
Ni *et al.* [436]	Human UCMSCs	MI rat	Human MSC-Exos induced angiogenesis, enhanced cardiac function, improved MI size reduction, and prevented CM apoptosis.
Mao *et al.* [437]	Human MSCs	MI rat	Exosomes decreasing MI size; inflammation and CM apoptosis also stimulated new blood vessel formation.

(*Continued*)

Table 5.8 Animal models of CVD utilized in preclinical studies. (*Continued*)

Study	Exosome source	Animal model	Key outcomes
Ma *et al.* [438]	Murine BMMSCs	MI mouse	BMMSC-Exos stimulated new blood vessel formation, and cardiac function was improved.
Lv *et al.* [439]	Murine BMMSCs	MI rat	Murine BMMSCs resulted in stimulated angiogenesis with cardio-protection activity.
Zhao *et al.* [117]	Murine BMMSCs	Myocardial I/R injury mouse	Murine BMMSC-Exos with immunosuppression decreased MI size in mouse models.
Zhang *et al.* [253]	BMMSCs	MI rat	Hypoxic preconditioning BMMSCs-Exos decreased the apoptotic rate in CM of MI rat models.
Huang *et al.* [440]	ADSCs	MI mouse	ADSC-Exos increased the survival rate of CM and promoted the recovery of myocardial function, decreased MI size, induced vasculogenesis, and decreased the inflammation.
Sun *et al.* [441]	Human MSCs	MI rat	Human MSC-Exos had a cardioprotection role and prevented CM apoptosis.
Huang *et al.* [115]	Murine MSCs	MI rat	Murine MSCs had a cardio-protection role and stimulated angiogenesis.
Gao *et al.* [442]	hiPSCs	MI pig	Enhanced myocardial recovery was observed without a concomitant rise in the incidence of arrhythmogenic complications.

ADSCs, adipose-derived stem cells; BMMSCs, bone marrow mesenchymal stem cells; hiPSCs, human induced pluripotent stem cells; MI, myocardial infarction; MSCs, mesenchymal stem cells; UCMSCs, umbilical cord mesenchymal stem cells; CM, cardiomyocyte; EXO, exosome.

5.5.2 Human Clinical Trials

Stem cell therapy, a type of regenerative medicine, has attracted considerable attention for its ability to offer new treatment possibilities for people suffering from diseases that were once thought to be incurable [443]. Notably, while preclinical studies have shown promise in the therapeutic use of Exos for CVDs, clinical trials are essential to validate their therapeutic efficacy [423, 444]. In this regard, several clinical trials have been registered on ClinicalTrials.gov and encompass a wide range of CVDs, MI, stroke, hypertension, and various pathological conditions affecting peripheral arteries (Table 5.9).

5.6 Challenges and Future Directions

A substantial body of scholarly works revealed that SC-Exos exhibit cardioprotective properties in cases of heart damage resulting from diverse causes, attributed to their regenerative and reparative effects on cardiac tissue [445]. However, the utilization of Exos for treating cardiovascular ailments presents several obstacles. The first challenge lies in determining the optimal Exos dosages for patients and understanding their impacts on human subjects. For example, the promotion of angiogenesis caused by SC-Exos is beneficial for the restoration of cardiomyocytes following MI, but angiogenesis has the potential to promote the growth of tumors in various other tissues [422, 446]. Another hurdle is the limited specificity in targeting ischemic myocardium and the suboptimal retention post-transplantation, impeding the progress of Exos therapy. While conventional methods like intravenous and intra-myocardial injections exist, novel administration techniques such as cardiac patches, minimally invasive Exos-spray, and hydrogels have been introduced to address these issues partially [446]. Importantly, the challenges associated with insufficient understanding of the diversity within extracellular vesicle populations and their associated contents, limited duration of effectiveness and biological activity in the bloodstream, uncertainties surrounding the expansion of production for pharmaceutical-grade Exos, obstacles in maintaining consistent quality across different production batches, and

Table 5.9 Registered clinical trials in case of therapeutic applications of Exos for CVD treatment.

Status	Study title	Conditions	ClinicalTrials.gov identifier
Unknown	Allogenic mesenchymal stem cell-derived Exos in patients with acute ischemic stroke	Cerebrovascular disorders	NCT03384433
Recruiting	Co-transplantation of mesenchymal stem cell Derived Exos and autologous mitochondria for patients candidate for CABG surgery	Myocardial infarction Myocardial ischemia Myocardial stunning	NCT05669144
Recruiting	Effect of GD-iExo-003 in acute ischemic stroke	Acute ischemic stroke	NCT06138210
Not_Yet_Recruiting	The pilot experimental study of the neuroprotective effects of Exos in extremely low birth weight infants	Premature birth Extreme prematurity Preterm Intraventricular Hemorrhage Hypoxia–ischemia, Cerebral neurodevelopmental disorders	NCT05490173
Recruiting	The role of acupuncture-induced Exos in treating post-stroke dementia	Stroke	NCT05326724
Unknown	Role of Exos derived from epicardial fat in atrial fibrillation	Atrial fibrillation	NCT03478410
Completed	New biomarkers and difficult-to-treat hypertension	Hypertension	NCT03034265

potential impairment of Exos functionality during cryopreservation pose significant challenges in the field of therapeutic application of Exos [446, 447]. Despite facing difficulties, the Exos field is advancing steadily [448]. Enhancing the extraction methodology of Exos from SCs, comprehensive methodologies for the external loading of cargo into Exos, and targeted enrichment of Exos containing potent therapeutic cargos are technical hurdles that warrant potential future investigations [449].

5.7 Conclusion

Lately, there has been growing interest in the potential therapeutic applications of stem cell-derived Exos for cardiovascular diseases. These Exos have shown promise in preclinical studies by enhancing cardiac repair, reducing inflammation, and improving heart function post-injury. However, addressing knowledge gaps and regulatory challenges is crucial for successfully translating Exos-based therapies into clinical practice. To this end, focusing on optimizing Exos production methods, understanding their mechanisms of action, and conducting extensive clinical trials to assess their safety and efficacy in patients with cardiovascular conditions is imperative. Implementing these innovative strategies and research initiatives is crucial for the field's progression and unlocking the potential of exosome-based therapies within cardiovascular medicine.

References

1. Organization, W.H., *World Health Statistics 2016 [OP]: Monitoring Health for the Sustainable Development Goals (SDGs)*, World Health Organization, Geneva, Switzerland, 2016.
2. Adamiak, M., *et al.*, Induced pluripotent stem cell (iPSC)–derived extracellular vesicles are safer and more effective for cardiac repair than iPSCs. *Circ. Res.*, **122**, 2, 296–309, 2018.
3. Lázár, E., Sadek, H.A., Bergmann, O., Cardiomyocyte renewal in the human heart: insights from the fall-out. *European Heart J.*, **38**, 30, 2333–2342, 2017.
4. Nguyen, P.K., Rhee, J.-W., Wu, J.C., Adult stem cell therapy and heart failure, 2000 to 2016: a systematic review. *JAMA Cardiol.*, **1**, 7, 831–841, 2016.
5. Ullah, I., Subbarao, R.B., Rho, G.J., Human mesenchymal stem cells-current trends and future prospective. *Biosci. Rep.*, **35**, 2, e00191, 2015.
6. Squillaro, T., Peluso, G., Galderisi, U., Clinical trials with mesenchymal stem cells: an update. *Cell Transplant.*, **25**, 5, 829–848, 2016.

7. Ishigami, S., *et al.*, Intracoronary cardiac progenitor cells in single ventricle physiology: *the PERSEUS* (cardiac progenitor cell infusion to treat univentricular heart disease) randomized phase 2 trial. *Circ. Res.*, **120**, 7, 1162–1173, 2017.
8. Witwer, K.W., *et al.*, Standardization of sample collection, isolation and analysis methods in extracellular vesicle research. *J. Extracell. Vesicles*, **2**, 2013.
9. Johnson, K., *et al.*, A stem cell–based approach to cartilage repair. *Science*, **336**, 6082, 717–721, 2012.
10. He, N., *et al.*, Exosomes: cell-free therapy for cardiovascular diseases. *J. Cardiovasc. Transl. Res.*, **13**, 713–721, 2020.
11. Phinney, D.G. and Pittenger, M.F., Concise review: MSC-derived exosomes for cell-free therapy. *Stem Cells*, **35**, 4, 851–858, 2017.
12. Phan, J., *et al.*, Engineering mesenchymal stem cells to improve their exosome efficacy and yield for cell-free therapy. *J. Extracell. Vesicles*, **7**, 1, 1522236, 2018.
13. Kalluri, R. and LeBleu, V.S., The biology, function, and biomedical applications of exosomes. *Science*, **367**, 6478, 2020.
14. Gurunathan, S., *et al.*, Review of the isolation, characterization, biological function, and multifarious therapeutic approaches of exosomes. *Cells*, **8**, 4, 307, 2019.
15. Liang, C., *et al.*, Exosomes of human umbilical cord MSCs protect against hypoxia/reoxygenation-induced pyroptosis of cardiomyocytes *via* the miRNA-100-5p/FOXO3/NLRP3 pathway. *Front. Bioeng. Biotechnol.*, **8**, 615850, 2021.
16. Cha, J.M., *et al.*, Efficient scalable production of therapeutic microvesicles derived from human mesenchymal stem cells. *Sci. Rep.*, **8**, 1, 1171, 2018.
17. Pendse, S., Kale, V., Vaidya, A., Extracellular vesicles isolated from mesenchymal stromal cells primed with hypoxia: novel strategy in regenerative medicine. *Curr. Stem Cell Res. Ther.*, **16**, 3, 243–261, 2021.
18. Celermajer, D.S., *et al.*, Cardiovascular disease in the developing world: prevalences, patterns, and the potential of early disease detection. *J. Am Coll. Cardiol.*, **60**, 14, 1207–16, 2012.
19. Behera, S.S., Pramanik, K., Nayak, M.K., Recent advancement in the treatment of cardiovascular diseases: Conventional therapy to nanotechnology. *Curr. Pharm. Des.*, **21**, 30, 4479–4497, 2015.
20. Capewell, S., *et al.*, Cardiovascular risk factor trends and potential for reducing coronary heart disease mortality in the United States of America. *Bulletin World Health Organization*, **88**, 120–130, 2010.
21. Bloom, D.E., *et al.*, *The global economic burden of noncommunicable diseases*, Program on the Global Demography of Aging, Harvard University, Cambridge, MA, USA, 2012.
22. Leal, J., *et al.*, Economic burden of cardiovascular diseases in the enlarged European Union. *European Heart J.*, **27**, 13, 1610–1619, 2006.

23. Frostegård, J., Immunity, atherosclerosis and cardiovascular disease. *BMC Med.*, **11**, 1–13, 2013.
24. Karunathilake, S.P. and Ganegoda, G.U., Secondary prevention of cardiovascular diseases and application of technology for early diagnosis. *Biomed Res. Int.*, **2018**, 1, 5767864, 2018.
25. Alatyyat, S.M., *et al.*, Umbilical cord stem cells: Background, processing and applications. *Tissue Cell*, **65**, 101351, 2020.
26. Yamanaka, S., Pluripotent Stem Cell-Based Cell Therapy-Promise and Challenges. *Cell Stem Cell*, **27**, 4, 523–531, 2020.
27. Mason, C. and Dunnill, P., A brief definition of regenerative medicine. *Regen. Med.*, **3**, 1, 1–5, 2008.
28. Franck, C.L., *et al.*, Influence of Adipose Tissue-Derived Stem Cells on the Burn Wound Healing Process. *Stem Cells Int.*, **2019**, 2340725, 2019.
29. Tyeb, S., *et al.*, Adipose-Derived Stem Cells (ADSCs) Loaded Gelatin-Sericin-Laminin Cryogels for Tissue Regeneration in Diabetic Wounds. *Biomacromolecules*, **21**, 2, 294–304, 2020.
30. Petersdorf, E.W., *et al.*, MHC haplotype matching for unrelated hematopoietic cell transplantation. *PLoS Med.*, **4**, 1, e8, 2007.
31. Li, W., *et al.*, Bone marrow mesenchymal stem cells derived exosomal Lnc TUG1 promotes bone fracture recovery *via* miR-22-5p/Anxa8 axis. *Hum. Cell*, **36**, 3, 1041–1053, 2023.
32. Wang, X., Yu, F., Ye, L., Epigenetic control of mesenchymal stem cells orchestrates bone regeneration. *Front Endocrinol (Lausanne)*, **14**, 1126787, 2023.
33. Li, Y., *et al.*, Umbilical cord derived mesenchymal stem cell-GelMA microspheres for accelerated wound healing. *Biomed. Mater.*, **18**, 1, 2022.
34. Monsanto, M.M., *et al.*, Concurrent Isolation of 3 Distinct Cardiac Stem Cell Populations From a Single Human Heart Biopsy. *Circ. Res.*, **121**, 2, 113–124, 2017.
35. Wysoczynski, M., *et al.*, Myocardial Reparative Properties of Cardiac Mesenchymal Cells Isolated on the Basis of Adherence. *J Am Coll. Cardiol.*, **69**, 14, 1824–1838, 2017.
36. Dominici, M., *et al.*, Minimal criteria for defining multipotent mesenchymal stromal cells. The International Society for Cellular Therapy position statement. *Cytotherapy*, **8**, 4, 315–7, 2006.
37. Golpanian, S., *et al.*, Rebuilding the Damaged Heart: Mesenchymal Stem Cells, Cell-Based Therapy, and Engineered Heart Tissue. *Physiol. Rev.*, **96**, 3, 1127–68, 2016.
38. Hare, J.M., *et al.*, Comparison of allogeneic vs autologous bone marrow–derived mesenchymal stem cells delivered by transendocardial injection in patients with ischemic cardiomyopathy: the POSEIDON randomized trial. *Jama*, **308**, 22, 2369–79, 2012.
39. Bray, L.J., *et al.*, Immunosuppressive properties of mesenchymal stromal cell cultures derived from the limbus of human and rabbit corneas. *Cytotherapy*, **16**, 1, 64–73, 2014.

40. Ayatollahi, M., Talaei-Khozani, T., Razmkhah, M., Growth suppression effect of human mesenchymal stem cells from bone marrow, adipose tissue, and Wharton's jelly of umbilical cord on PBMCs. *Iran J. Basic Med. Sci.*, **19**, 2, 145–53, 2016.
41. Karaöz, E., *et al.*, Comparative Analyses of Immunosuppressive Characteristics of Bone-Marrow, Wharton's Jelly, and Adipose Tissue-Derived Human Mesenchymal Stem Cells. *Turk J. Haematol.*, **34**, 3, 213–225, 2017.
42. Ribeiro, A., *et al.*, Mesenchymal stem cells from umbilical cord matrix, adipose tissue and bone marrow exhibit different capability to suppress peripheral blood B, natural killer and T cells. *Stem Cell Res. Ther.*, **4**, 5, 125, 2013.
43. Savickiene, J., *et al.*, Human Amniotic Fluid Mesenchymal Stem Cells from Second- and Third-Trimester Amniocentesis: Differentiation Potential, Molecular Signature, and Proteome Analysis. *Stem Cells Int.*, **2015**, 319238, 2015.
44. Roubelakis, M.G., *et al.*, Molecular and proteomic characterization of human mesenchymal stem cells derived from amniotic fluid: comparison to bone marrow mesenchymal stem cells. *Stem Cells Dev.*, **16**, 6, 931–52, 2007.
45. Babaie, Y., *et al.*, Analysis of Oct4-dependent transcriptional networks regulating self-renewal and pluripotency in human embryonic stem cells. *Stem Cells*, **25**, 2, 500–10, 2007.
46. Petrini, M., *et al.*, Identification and purification of mesodermal progenitor cells from human adult bone marrow. *Stem Cells Dev.*, **18**, 6, 857–66, 2009.
47. Hatzistergos, K.E. and Vedenko, A., Cardiac Cell Therapy 3.0: The Beginning of the End or the End of the Beginning? *Circ. Res.*, **121**, 2, 95–97, 2017.
48. Sun, Y.Q., *et al.*, Insensitivity of Human iPS Cells-Derived Mesenchymal Stem Cells to Interferon-γ-induced HLA Expression Potentiates Repair Efficiency of Hind Limb Ischemia in Immune Humanized NOD Scid Gamma Mice. *Stem Cells*, **33**, 12, 3452–67, 2015.
49. Lian, Q., *et al.*, Functional mesenchymal stem cells derived from human induced pluripotent stem cells attenuate limb ischemia in mice. *Circulation*, **121**, 9, 1113–23, 2010.
50. Fukuta, M., *et al.*, Derivation of mesenchymal stromal cells from pluripotent stem cells through a neural crest lineage using small molecule compounds with defined media. *PLoS One*, **9**, 12, e112291, 2014.
51. Shroff, G. and Gupta, R., Human embryonic stem cells in the treatment of patients with spinal cord injury. *Ann. Neurosci.*, **22**, 4, 208–16, 2015.
52. Zhou, S., *et al.*, Differentiation of human embryonic stem cells into cone photoreceptors through simultaneous inhibition of BMP, TGFβ and Wnt signaling. *Development*, **142**, 19, 3294–306, 2015.
53. Sluch, V.M., *et al.*, Differentiation of human ESCs to retinal ganglion cells using a CRISPR engineered reporter cell line. *Sci. Rep.*, **5**, 16595, 2015.
54. Shiba, Y., *et al.*, Human ES-cell-derived cardiomyocytes electrically couple and suppress arrhythmias in injured hearts. *Nature*, **489**, 7415, 322–5, 2012.

55. Vedantham, V., New Approaches to Biological Pacemakers: Links to Sinoatrial Node Development. *Trends Mol. Med.*, **21**, 12, 749–761, 2015.
56. Tolosa, L., *et al.*, Transplantation of hESC-derived hepatocytes protects mice from liver injury. *Stem Cell Res. Ther.*, **6**, 246, 2015.
57. Carpentier, A., *et al.*, Hepatic differentiation of human pluripotent stem cells in miniaturized format suitable for high-throughput screen. *Stem Cell Res.*, **16**, 3, 640–50, 2016.
58. Bruin, J.E., *et al.*, Treating diet-induced diabetes and obesity with human embryonic stem cell-derived pancreatic progenitor cells and antidiabetic drugs. *Stem Cell Rep.*, **4**, 4, 605–20, 2015.
59. Salguero-Aranda, C., *et al.*, Differentiation of Mouse Embryonic Stem Cells toward Functional Pancreatic β-Cell Surrogates through Epigenetic Regulation of Pdx1 by Nitric Oxide. *Cell Transplant*, **25**, 10, 1879–1892, 2016.
60. Cheng, A., *et al.*, Cartilage repair using human embryonic stem cell-derived chondroprogenitors. *Stem Cells Transl. Med.*, **3**, 11, 1287–94, 2014.
61. Ye, L., *et al.*, An insulin signaling feedback loop regulates pancreas progenitor cell differentiation during islet development and regeneration. *Dev. Biol.*, **409**, 2, 354–69, 2016.
62. Ellis, K.M., *et al.*, Neurogenic potential of dental pulp stem cells isolated from murine incisors. *Stem Cell Res. Ther.*, **5**, 1, 30, 2014.
63. Kuo, B.R., *et al.*, *In Vivo* Cochlear Hair Cell Generation and Survival by Coactivation of β-Catenin and Atoh1. *J. Neurosci.*, **35**, 30, 10786–98, 2015.
64. Shaffiey, S.A., *et al.*, Intestinal stem cell growth and differentiation on a tubular scaffold with evaluation in small and large animals. *Regen. Med.*, **11**, 1, 45–61, 2016.
65. Yin, X., *et al.*, Niche-independent high-purity cultures of Lgr5+ intestinal stem cells and their progeny. *Nat. Methods*, **11**, 1, 106–12, 2014.
66. Ksander, B.R., *et al.*, ABCB5 is a limbal stem cell gene required for corneal development and repair. *Nature*, **511**, 7509, 353–7, 2014.
67. Fuoco, C., *et al.*, *In vivo* generation of a mature and functional artificial skeletal muscle. *EMBO Mol. Med.*, **7**, 4, 411–22, 2015.
68. Berry, S.E., Concise review: mesoangioblast and mesenchymal stem cell therapy for muscular dystrophy: progress, challenges, and future directions. *Stem Cells Transl. Med.*, **4**, 1, 91–8, 2015.
69. Mead, B., *et al.*, Stem cell treatment of degenerative eye disease. *Stem Cell Res.*, **14**, 3, 243–57, 2015.
70. Cronk, S.M., *et al.*, Adipose-derived stem cells from diabetic mice show impaired vascular stabilization in a murine model of diabetic retinopathy. *Stem Cells Transl. Med.*, **4**, 5, 459–67, 2015.
71. Nagata, H., *et al.*, Cardiac Adipose-Derived Stem Cells Exhibit High Differentiation Potential to Cardiovascular Cells in C57BL/6 Mice. *Stem Cells Transl. Med.*, **5**, 2, 141–51, 2016.
72. Chen, L., *et al.*, Application of adipose-derived stem cells in heart disease. *J. Cardiovasc. Transl. Res.*, **7**, 7, 651–63, 2014.

73. Sharma, A.K., *et al.*, A nonhuman primate model for urinary bladder regeneration using autologous sources of bone marrow-derived mesenchymal stem cells. *Stem Cells*, **29**, 2, 241–50, 2011.
74. Oshima, M. and Tsuji, T., Whole Tooth Regeneration as a Future Dental Treatment. *Adv. Exp. Med. Biol.*, **881**, 255–69, 2015.
75. Csaki, C., *et al.*, Chondrogenesis, osteogenesis and adipogenesis of canine mesenchymal stem cells: a biochemical, morphological and ultrastructural study. *Histochem. Cell Biol.*, **128**, 6, 507–20, 2007.
76. Sen, B., *et al.*, Intranuclear Actin Regulates Osteogenesis. *Stem Cells*, **33**, 10, 3065–76, 2015.
77. Levit, R.D., *et al.*, Cellular encapsulation enhances cardiac repair. *J. Am. Heart Assoc.*, **2**, 5, e000367, 2013.
78. Lin, B.J., *et al.*, Cytokine loaded layer-by-layer ultrathin matrices to deliver single dermal papilla cells for spot-by-spot hair follicle regeneration. *J. Mater. Chem. B*, **4**, 3, 489–504, 2016.
79. Di Buduo, C.A., *et al.*, Programmable 3D silk bone marrow niche for platelet generation *ex vivo* and modeling of megakaryopoiesis pathologies. *Blood*, **125**, 14, 2254–64, 2015.
80. Paradells, S., *et al.*, Lipoic acid and bone marrow derived cells therapy induce angiogenesis and cell proliferation after focal brain injury. *Brain Inj.*, **29**, 3, 380–95, 2015.
81. Tarafder, S., *et al.*, Engineered Healing of Avascular Meniscus Tears by Stem Cell Recruitment. *Sci. Rep.*, **8**, 1, 8150, 2018.
82. Butler, J., *et al.*, Intravenous Allogeneic Mesenchymal Stem Cells for Nonischemic Cardiomyopathy: Safety and Efficacy Results of a Phase II-A Randomized Trial. *Circ. Res.*, **120**, 2, 332–340, 2017.
83. Haneef, K., *et al.*, Role of interleukin-7 in fusion of rat bone marrow mesenchymal stem cells with cardiomyocytes *in vitro* and improvement of cardiac function *in vivo*. *Cardiovasc. Ther.*, **36**, 6, e12479, 2018.
84. Fujita, Y. and Kawamoto, A., Stem cell-based peripheral vascular regeneration. *Adv. Drug Deliv. Rev.*, **120**, 25–40, 2017.
85. Sun, Z., *et al.*, Secretion of rat tracheal epithelial cells induces mesenchymal stem cells to differentiate into epithelial cells. *Cell Biol. Int.*, **36**, 2, 169–75, 2012.
86. Benavides, O.M., *et al.*, *In situ* vascularization of injectable fibrin/poly(ethylene glycol) hydrogels by human amniotic fluid-derived stem cells. *J. Biomed. Mater. Res. A*, **103**p, 8, 2645–53, 2015.
87. Wolfrum, K., *et al.*, The LARGE principle of cellular reprogramming: lost, acquired and retained gene expression in foreskin and amniotic fluid-derived human iPS cells. *PLoS One*, **5**, 10, e13703, 2010.
88. Hu, J., *et al.*, Long term effects of the implantation of Wharton's jelly-derived mesenchymal stem cells from the umbilical cord for newly-onset type 1 diabetes mellitus. *Endocr. J.*, **60**, 3, 347–57, 2013.

89. Liu, X., *et al.*, A preliminary evaluation of efficacy and safety of Wharton's jelly mesenchymal stem cell transplantation in patients with type 2 diabetes mellitus. *Stem Cell Res. Ther.*, **5**, 2, 57, 2014.
90. Wang, D., *et al.*, Umbilical cord mesenchymal stem cell transplantation in active and refractory systemic lupus erythematosus: a multicenter clinical study. *Arthritis Res. Ther.*, **16**, 2, R79, 2014.
91. Jurga, M., *et al.*, Generation of functional neural artificial tissue from human umbilical cord blood stem cells. *Tissue Eng. Part C Methods*, **15**, 3, 365–72, 2009.
92. Ehrhart, J., *et al.*, Biodistribution of Infused Human Umbilical Cord Blood Cells in Alzheimer's Disease-Like Murine Model. *Cell Transplant.*, **25**, 1, 195–9, 2016.
93. Park, G.Y., Kwon, D.R., Lee, S.C., Regeneration of Full-Thickness Rotator Cuff Tendon Tear After Ultrasound-Guided Injection With Umbilical Cord Blood-Derived Mesenchymal Stem Cells in a Rabbit Model. *Stem Cells Transl. Med.*, **4**, 11, 1344–51, 2015.
94. Thompson, P.A., *et al.*, Double umbilical cord blood transplant is effective therapy for relapsed or refractory Hodgkin lymphoma. *Leuk. Lymphoma*, **57**, 7, 1607–15, 2016.
95. Fan, Y.P., *et al.*, The Therapeutic Potential of Human Umbilical Mesenchymal Stem Cells From Wharton's Jelly in the Treatment of Rat Peritoneal Dialysis-Induced Fibrosis. *Stem Cells Transl. Med.*, **5**, 2, 235–47, 2016.
96. Gaballa, S., *et al.*, A Two-Step Haploidentical Versus a Two-Step Matched Related Allogeneic Myeloablative Peripheral Blood Stem Cell Transplantation. *Biol. Blood Marrow Transplant.*, **22**, 1, 141–8, 2016.
97. Herrera-Carrillo, E. and Berkhout, B., Bone Marrow Gene Therapy for HIV/AIDS. *Viruses*, **7**, 7, 3910–36, 2015.
98. DiGiusto, D.L., *et al.*, Development of hematopoietic stem cell based gene therapy for HIV-1 infection: considerations for proof of concept studies and translation to standard medical practice. *Viruses*, **5**, 11, 2898–919, 2013.
99. Kaigler, D., *et al.*, Stem cell therapy for craniofacial bone regeneration: a randomized, controlled feasibility trial. *Cell Transplant.*, **22**, 5, 767–77, 2013.
100. Gubareva, E.A., *et al.*, Orthotopic transplantation of a tissue engineered diaphragm in rats. *Biomater.*, **77**, 320–35, 2016.
101. Wang, D., *et al.*, BMP14 induces tenogenic differentiation of bone marrow mesenchymal stem cells *in vitro*. *Exp. Ther. Med.*, **16**, 2, 1165–1174, 2018.
102. Wang, T., *et al.*, 3D uniaxial mechanical stimulation induces tenogenic differentiation of tendon-derived stem cells through a PI3K/AKT signaling pathway. *FASEB J.*, **32**, 9, 4804–4814, 2018.
103. Hou, Y., *et al.*, *In vivo* evaluation of the adenovirus-mediated Sox9 and BMP2 double gene transduction on treatment of intervertebral disc degeneration in rabbit annular puncture model. *J. Biomater. Tissue Eng.*, **8**, 8, 1091–1099, 2018.

104. Tsai, Y., *et al.*, Human iPSC-Derived Neural Progenitors Preserve Vision in an AMD-Like Model. *Stem Cells*, **33**, 8, 2537–49, 2015.
105. Paşca, A.M., *et al.*, Functional cortical neurons and astrocytes from human pluripotent stem cells in 3D culture. *Nat. Methods*, **12**, 7, 671–8, 2015.
106. Mariani, J., *et al.*, FOXG1-Dependent Dysregulation of GABA/Glutamate Neuron Differentiation in Autism Spectrum Disorders. *Cell*, **162**, 2, 375–390, 2015.
107. Wilson, A.A., *et al.*, Emergence of a stage-dependent human liver disease signature with directed differentiation of alpha-1 antitrypsin-deficient iPS cells. *Stem Cell Reports*, **4**, 5, 873–85, 2015.
108. Zhu, S., *et al.*, Human pancreatic beta-like cells converted from fibroblasts. *Nat. Commun.*, **7**, 10080, 2016.
109. Dye, B.R., *et al.*, *In vitro* generation of human pluripotent stem cell derived lung organoids. *Elife*, **4**, 2015.
110. Kambal, A., *et al.*, Generation of HIV-1 resistant and functional macrophages from hematopoietic stem cell-derived induced pluripotent stem cells. *Mol. Ther.*, **19**, 3, 584–93, 2011.
111. Feng, Y., *et al.*, Ischemic preconditioning potentiates the protective effect of stem cells through secretion of exosomes by targeting Mecp2 *via* miR-22. *PLoS One*, **9**, 2, e88685, 2014.
112. Khan, M., *et al.*, Embryonic stem cell-derived exosomes promote endogenous repair mechanisms and enhance cardiac function following myocardial infarction. *Circ. Res.*, **117**, 1, 52–64, 2015.
113. Gallet, R., *et al.*, Exosomes secreted by cardiosphere-derived cells reduce scarring, attenuate adverse remodelling, and improve function in acute and chronic porcine myocardial infarction. *Eur. Heart J.*, **38**, 3, 201–211, 2017.
114. Luo, Z., *et al.*, Hypoxia preconditioning promotes bone marrow mesenchymal stem cells survival by inducing HIF-1α in injured neuronal cells derived exosomes culture system. *Cell Death Dis.*, **10**, 2, 134, 2019.
115. Huang, P., *et al.*, Atorvastatin enhances the therapeutic efficacy of mesenchymal stem cells-derived exosomes in acute myocardial infarction *via* upregulating long non-coding RNA H19. *Cardiovasc. Res.*, **116**, 2, 353–367, 2020.
116. Pan, J., *et al.*, Exosomes derived from miR-146a-modified adipose-derived stem cells attenuate acute myocardial infarction-induced myocardial damage *via* downregulation of early growth response factor 1. *J. Cell. Biochem.*, **120**, 3, 4433–4443, 2019.
117. Zhao, J., *et al.*, Mesenchymal stromal cell-derived exosomes attenuate myocardial ischaemia-reperfusion injury through miR-182-regulated macrophage polarization. *Cardiovasc. Res.*, **115**, 7, 1205–1216, 2019.
118. Lankford, K.L., *et al.*, Intravenously delivered mesenchymal stem cell-derived exosomes target M2-type macrophages in the injured spinal cord. *PLoS One*, **13**, 1, e0190358, 2018.

119. Aghabozorgi, A.S., *et al.*, Circulating exosomal miRNAs in cardiovascular disease pathogenesis: New emerging hopes. *J. Cell Physiol.*, **234**, 12, 21796–21809, 2019.
120. Zhang, J., *et al.*, Overexpression of Exosomal Cardioprotective miRNAs Mitigates Hypoxia-Induced H9c2 Cells Apoptosis. *Int. J. Mol. Sci.*, **18**, 4, 2017.
121. Zamani, P., *et al.*, The therapeutic and diagnostic role of exosomes in cardiovascular diseases. *Trends Cardiovasc. Med.*, **29**, 6, 313–323, 2019.
122. Rezaie, J., *et al.*, Cardioprotective role of extracellular vesicles: A highlight on exosome beneficial effects in cardiovascular diseases. *J. Cell. Physiol.*, **234**, 12, 21732–21745, 2019.
123. Wang, C., *et al.*, Macrophage-Derived mir-155-Containing Exosomes Suppress Fibroblast Proliferation and Promote Fibroblast Inflammation during Cardiac Injury. *Mol. Ther.*, **25**, 1, 192–204, 2017.
124. Li, B., *et al.*, Activation of CD137 signaling promotes neointimal formation by attenuating TET2 and transferrring from endothelial cell-derived exosomes to vascular smooth muscle cells. *Biomed. Pharmacother.*, **121**, 109593, 2020.
125. Zhu, J., *et al.*, Exosomes from nicotine-stimulated macrophages accelerate atherosclerosis through miR-21-3p/PTEN-mediated VSMC migration and proliferation. *Theranostics.*, **9**, 23, 6901–6919, 2019.
126. Niu, C., *et al.*, Macrophage Foam Cell-Derived Extracellular Vesicles Promote Vascular Smooth Muscle Cell Migration and Adhesion. *J. Am. Heart Assoc.*, **5**, 10, 2016.
127. Gao, H., *et al.*, Exosomal MALAT1 derived from ox-LDL-treated endothelial cells induce neutrophil extracellular traps to aggravate atherosclerosis. *Biol. Chem.*, **401**, 3, 367–376, 2020.
128. Chen, L., *et al.*, Exosomal lncRNA GAS5 regulates the apoptosis of macrophages and vascular endothelial cells in atherosclerosis. *PLoS One*, **12**, 9, e0185406, 2017.
129. Collino, F., *et al.*, Exosome and Microvesicle-Enriched Fractions Isolated from Mesenchymal Stem Cells by Gradient Separation Showed Different Molecular Signatures and Functions on Renal Tubular Epithelial Cells. *Stem Cell Reviews Reports*, **13**, 2, 226–243, 2017.
130. Matsumoto, S., *et al.*, Circulating p53-responsive microRNAs are predictive indicators of heart failure after acute myocardial infarction. *Circ. Res.*, **113**, 3, 322–6, 2013.
131. Wang, L., *et al.*, Reduced exosome miR-425 and miR-744 in the plasma represents the progression of fibrosis and heart failure. *Kaohsiung J. Med. Sci.*, **34**, 11, 626–633, 2018.
132. Halkein, J., *et al.*, MicroRNA-146a is a therapeutic target and biomarker for peripartum cardiomyopathy. *J. Clin. Invest*, **123**, 5, 2143–54, 2013.

133. Liu, H., *et al.*, Exosomes derived from dendritic cells improve cardiac function *via* activation of CD4(+) T lymphocytes after myocardial infarction. *J. Mol. Cell Cardiol.*, **91**, 123–33, 2016.
134. Barile, L., *et al.*, Cardioprotection by cardiac progenitor cell-secreted exosomes: role of pregnancy-associated plasma protein-A. *Cardiovasc. Res.*, **114**, 7, 992–1005, 2018.
135. Goetzl, L., *et al.*, Diagnostic Potential of Neural Exosome Cargo as Biomarkers for Acute Brain Injury. *Ann Clin. Transl. Neurol.*, **5**, 1, 4–10, 2018.
136. Chen, Y., *et al.*, Increased Circulating Exosomal miRNA-223 Is Associated with Acute Ischemic Stroke. *Front. Neurol.*, **8**, 57, 2017.
137. Prathipati, P., Nandi, S.S., Mishra, P.K., Stem Cell-Derived Exosomes, Autophagy, Extracellular Matrix Turnover, and miRNAs in Cardiac Regeneration during Stem Cell Therapy. *Stem Cell Rev. Rep.*, **13**, 1, 79–91, 2017.
138. Li, J., *et al.*, MicroRNA-144 is a circulating effector of remote ischemic preconditioning. *Basic Res. Cardiol.*, **109**, 5, 423, 2014.
139. Tikhomirov, R., *et al.*, Exosomes: From Potential Culprits to New Therapeutic Promise in the Setting of Cardiac Fibrosis. *Cells*, **9**, 3, 2020.
140. Song, Y., *et al.*, Localized injection of miRNA-21-enriched extracellular vesicles effectively restores cardiac function after myocardial infarction. *Theranostics.*, **9**, 8, 2346–2360, 2019.
141. Działo, E., *et al.*, WNT3a and WNT5a Transported by Exosomes Activate WNT Signaling Pathways in Human Cardiac Fibroblasts. *Int. J. Mol. Sci.*, **20**, 6, 2019.
142. Li, Y., *et al.*, Circular RNA is enriched and stable in exosomes: a promising biomarker for cancer diagnosis. *Cell Res.*, **25**, 8, 981–4, 2015.
143. Wang, Y., *et al.*, Exosomal circRNAs: biogenesis, effect and application in human diseases. *Mol. Cancer*, **18**, 1, 116, 2019.
144. Pereira, S.S. and Alvarez-Leite, J.I., Low-Grade Inflammation, Obesity, and Diabetes. *Curr. Obes. Rep.*, **3**, 4, 422–31, 2014.
145. Sharma, S., *et al.*, A Deep Proteome Analysis Identifies the Complete Secretome as the Functional Unit of Human Cardiac Progenitor Cells. *Circ. Res.*, **120**, 5, 816–834, 2017.
146. Zampetaki, A., *et al.*, Prospective study on circulating MicroRNAs and risk of myocardial infarction. *J. Am. Coll. Cardiol.*, **60**, 4, 290–9, 2012.
147. Vicencio, J.M., *et al.*, Plasma exosomes protect the myocardium from ischemia-reperfusion injury. *J. Am Coll. Cardiol.*, **65**, 15, 1525–36, 2015.
148. Xiao, J., *et al.*, Cardiac progenitor cell-derived exosomes prevent cardiomyocytes apoptosis through exosomal miR-21 by targeting PDCD4. *Cell Death Dis.*, **7**, 6, e2277, 2016.
149. Ibrahim, A.G., Cheng, K., Marbán, E., Exosomes as critical agents of cardiac regeneration triggered by cell therapy. *Stem Cell Reports*, **2**, 5, 606–19, 2014.
150. Lu, H.Q., *et al.*, Circulating miR-214 is associated with the severity of coronary artery disease. *J. Geriatr. Cardiol.*, **10**, 1, 34–8, 2013.

151. van Balkom, B.W., *et al.*, Endothelial cells require miR-214 to secrete exosomes that suppress senescence and induce angiogenesis in human and mouse endothelial cells. *Blood*, **121**, 19, 3997–4006, 2013. s1-15.
152. Doyle, L.M. and Wang, M.Z., Overview of Extracellular Vesicles, Their Origin, Composition, Purpose, and Methods for Exosome Isolation and Analysis. *Cells*, **8**, 7, 2019.
153. Heidari, F., *et al.*, *Exosomes viral infection: Effects pathogenesis treatment strategies. Biocell*, **47**, 12, 2597–2608, 2023.
154. Ahmadieh-Yazdi, A., *et al.*, Unveiling therapeutic potential: Adipose tissue-derived mesenchymal stem cells and their exosomes in the management of diabetes mellitus, wound healing, and chronic ulcers. *Biochem. Pharmacol.*, **226**, 116399, 2024.
155. Sheykhhasan, M., *et al.*, Dual Role of Exosome in Neurodegenerative Diseases: A Review Study. *Curr. Stem Cell Res. Ther.*, **19**, 6, 852–864, 2024.
156. Davoodi Asl, F., *et al.*, Promising effects of exosomes from menstrual blood-derived mesenchymal stem cells on endometriosis. *Reprod. Biol.*, **23**, 3, 100788, 2023.
157. Sheikholeslami, A., *et al.*, Exosomes of mesenchymal stem cells and PRP restore spermatogenesis in the rat model of non-obstructive azoospermia. *Reprod.*, 23, 2024.
158. Khoei, S.G., *et al.*, The Use of Mesenchymal Stem Cells and their Derived Extracellular Vesicles in Cardiovascular Disease Treatment. *Curr. Stem Cell Res. Ther.*, **15**, 7, 623–638, 2020.
159. Rezaie, J., *et al.*, Tumor-derived extracellular vesicles: The metastatic organotropism drivers. *Life Sci.*, **289**, 120216, 2022.
160. Crescitelli, R., Lässer, C., Lötvall, J., Isolation and characterization of extracellular vesicle subpopulations from tissues. *Nat. Protoc.*, **16**, 3, 1548–1580, 2021.
161. Valadi, H., *et al.*, Exosome-mediated transfer of mRNAs and microRNAs is a novel mechanism of genetic exchange between cells. *Nat. Cell Biol.*, **9**, 6, 654–659, 2007.
162. Zhang, B., *et al.*, HucMSC-Exosome Mediated-Wnt4 Signaling Is Required for Cutaneous Wound Healing. *Stem Cells*, **33**, 7, 2158–68, 2015.
163. Marar, C., Starich, B., Wirtz, D., Extracellular vesicles in immunomodulation and tumor progression. *Nat. Immunol.*, **22**, 5, 560–570, 2021.
164. Yang, E., *et al.*, Exosome-mediated metabolic reprogramming: the emerging role in tumor microenvironment remodeling and its influence on cancer progression. *Signal Transduct. Target Ther.*, **5**, 1, 242, 2020.
165. Men, Y., *et al.*, Exosome reporter mice reveal the involvement of exosomes in mediating neuron to astroglia communication in the CNS. *Nat. Commun.*, **10**, 1, 4136, 2019.
166. Han, Q.-F., *et al.*, Exosome biogenesis: machinery, regulation, and therapeutic implications in cancer. *Mol. Cancer*, **21**, 1, 207, 2022.

167. Kang, T., Atukorala, I., Mathivanan, S., Biogenesis of Extracellular Vesicles, in: *New Frontiers: Extracellular Vesicles*, S. Mathivanan and (Eds.), pp. 19–43, Springer International Publishing, Cham, 2021.
168. Zhang, H., *et al.*, Identification of distinct nanoparticles and subsets of extracellular vesicles by asymmetric flow field-flow fractionation. *Nat. Cell Biol.*, **20**, 3, 332–343, 2018.
169. Manoochehri, H., La'ah, A.S., Babaeizad, A., Sheykhhasan, M., Rostamipoor, M., Abbaspoor, M., *et al.*, Extracellular Vesicles in Cancer Immunotherapy: Therapeutic Challenges, and Clinical Progress. *Asian J. Pharm. Sci.*, 2025.
170. Zhang, W., *et al.*, A conserved ubiquitin- and ESCRT-dependent pathway internalizes human lysosomal membrane proteins for degradation. *PLoS Biol.*, **19**, 7, e3001361, 2021.
171. Mobarak, H., *et al.*, Prokaryotic microvesicles Ortholog of eukaryotic extracellular vesicles in biomedical fields. *Cell Commun. Signaling*, **22**, 1, 80, 2024.
172. Li, Y., *et al.*, Targeted inhibition of tumor-derived exosomes as a novel therapeutic option for cancer. *Exp. Mol. Med.*, **54**, 9, 1379–1389, 2022.
173. Skotland, T., *et al.*, Exosomal lipid composition and the role of ether lipids and phosphoinositides in exosome biology. *J. Lipid Res.*, **60**, 1, 9–18, 2019.
174. Qiao, K., *et al.*, Roles of extracellular vesicles derived from immune cells in atherosclerosis. *Extracell. Vesicles*, **2**, 100028, 2023.
175. Jia, G. and Sowers, J.R., Targeting endothelial exosomes for the prevention of cardiovascular disease. *Biochim. Biophys. Acta, Mol. Basis Dis.*, **1866**, 8, 165833, 2020.
176. Zhu, J., *et al.*, Divergent sequences of tetraspanins enable plants to specifically recognize microbe-derived extracellular vesicles. *Nat. Commun.*, **14**, 1, 4877, 2023.
177. Gupta, S., *et al.*, Comparative proteomic profiling of Small Extracellular vesicles derived from iPSCs and tissue specific mesenchymal stem cells. *Exp. Cell. Res.*, **420**, 2, 113354, 2022.
178. Li, J., *et al.*, A comprehensive review on the composition, biogenesis, purification, and multifunctional role of exosome as delivery vehicles for cancer therapy. *Biomed Pharmacother*, **165**, 115087, 2023.
179. Mathivanan, S., *et al.*, ExoCarta 2012: database of exosomal proteins, RNA and lipids. *Nucleic Acids Res.*, **40**, Database issue, D1241–4, 2012.
180. Kim, D.K., *et al.*, EVpedia: an integrated database of high-throughput data for systemic analyses of extracellular vesicles. *J. Extracell Vesicles*, **2**, 2013.
181. Vlassov, A.V., *et al.*, Exosomes: current knowledge of their composition, biological functions, and diagnostic and therapeutic potentials. *Biochim. Biophys. Acta*, **1820**, 7, 940–8, 2012.
182. Kalluri, R., The biology and function of exosomes in cancer. *J. Clin. Invest.*, **126**, 4, 1208–15, 2016.
183. T, L.R., *et al.*, MSC surface markers (CD44, CD73, and CD90) can identify human MSC-derived extracellular vesicles by conventional flow cytometry. *Cell Commun. Signal*, **14**, 2, 2016.

184. Mittelbrunn, M., *et al.*, Unidirectional transfer of microRNA-loaded exosomes from T cells to antigen-presenting cells. *Nat. Commun.*, **2**, 282, 2011.
185. Yellon, D.M. and Davidson, S.M., Exosomes: nanoparticles involved in cardioprotection? *Circ. Res.*, **114**, 2, 325–32, 2014.
186. Natasha, G., *et al.*, Exosomes as immunotheranostic nanoparticles. *Clin. Ther.*, **36**, 6, 820–9, 2014.
187. Choi, D.S., *et al.*, Proteomics of extracellular vesicles: Exosomes and ectosomes. *Mass Spectrom. Rev.*, **34**, 4, 474–90, 2015.
188. Esser, J., *et al.*, Exosomes from human macrophages and dendritic cells contain enzymes for leukotriene biosynthesis and promote granulocyte migration. *J. Allergy Clin. Immunol.*, **126**, 5, 1032–40, 2010. 1040.e1-4.
189. Subra, C., *et al.*, Exosomes account for vesicle-mediated transcellular transport of activatable phospholipases and prostaglandins. *J. Lipid Res.*, **51**, 8, 2105–20, 2010.
190. Laulagnier, K., *et al.*, PLD2 is enriched on exosomes and its activity is correlated to the release of exosomes. *FEBS Lett.*, **572**, 1-3, 11–4, 2004.
191. Alonso, R., *et al.*, Diacylglycerol kinase alpha regulates the secretion of lethal exosomes bearing Fas ligand during activation-induced cell death of T lymphocytes. *J. Biol. Chem.*, **280**, 31, 28439–50, 2005.
192. Trajkovic, K., *et al.*, Ceramide triggers budding of exosome vesicles into multivesicular endosomes. *Science*, **319**, 5867, 1244–7, 2008.
192. Subra, C., *et al.*, Exosome lipidomics unravels lipid sorting at the level of multivesicular bodies. *Biochimie.*, **89**, 2, 205–12, 2007.
194. Falguières, T., Castle, D., Gruenberg, J., Regulation of the MVB pathway by SCAMP3. *Traffic*, **13**, 1, 131–42, 2012.
195. Record, M., Poirot, M., Silvente-Poirot, S., Emerging concepts on the role of exosomes in lipid metabolic diseases. *Biochimie.*, **96**, 67–74, 2014.
196. Record, M., *et al.*, Exosomes as new vesicular lipid transporters involved in cell-cell communication and various pathophysiologies. *Biochim. Biophys. Acta*, **1841**, 1, 108–20, 2014.
197. Pan, Y., *et al.*, Mesenchymal stem cell-derived exosomes in cardiovascular and cerebrovascular diseases: From mechanisms to therapy. *Biomed. Pharmacotherapy*, **163**, 114817, 2023.
198. Aggarwal, R., *et al.*, Stem cell-derived exosome patch with coronary artery bypass graft restores cardiac function in chronically ischemic porcine myocardium. *J. Thoracic Cardiovascular Surg.*, **166**, 6, e512–e530, 2023.
199. Guan, H., *et al.*, Research and application of hydrogel-encapsulated mesenchymal stem cells in the treatment of myocardial infarction. *Colloids Surf B Biointerfaces*, **239**, 113942, 2024.
200. Li, H., *et al.*, Exosomes secreted by endothelial cells derived from human induced pluripotent stem cells improve recovery from myocardial infarction in mice. *Stem Cell Res. Ther.*, **14**, 1, 278, 2023.
201. Tan, S.J.O., *et al.*, Novel Applications of Mesenchymal Stem Cell-derived Exosomes for Myocardial Infarction Therapeutics. *Biomol.*, **10**, 5, 2020.

202. Michael, L.H., *et al.*, Myocardial infarction and remodeling in mice: effect of reperfusion. *Am. J. Physiol. Heart Circ. Physiol.*, **277**, 2, H660–H668, 1999.
203. Richard, V., Murry, C.E., Reimer, K.A., Healing of Myocardial Infarcts in Dogs. *Circulation*, **92**, 7, 1891–1901, 1995.
204. Frangogiannis, N.G., Smith, C.W., Entman, M.L., The inflammatory response in myocardial infarction. *Cardiovasc. Res.*, **53**, 1, 31–47, 2002.
205. Fang, L., *et al.*, Systemic inflammatory response following acute myocardial infarction. *J. Geriatr. Cardiol.*, **12**, 3, 305–12, 2015.
206. Frangogiannis, N.G., *et al.*, Resident Cardiac Mast Cells Degranulate and Release Preformed TNF-α, Initiating the Cytokine Cascade in Experimental Canine Myocardial Ischemia/Reperfusion. *Circ.*, **98**, 7, 699–710, 1998.
207. de Haan, J.J., *et al.*, Danger Signals in the Initiation of the Inflammatory Response after Myocardial Infarction. *Mediators Inflammation*, **2013**, 1, 206039, 2013.
208. Land, W.G., The Role of Damage-Associated Molecular Patterns (DAMPs) in Human Diseases: Part II: DAMPs as diagnostics, prognostics and therapeutics in clinical medicine. *Sultan Qaboos Univ Med. J.*, **15**, 2, e157–70, 2015.
209. Mezzaroma, E., *et al.*, The inflammasome promotes adverse cardiac remodeling following acute myocardial infarction in the mouse. *Proc. Natl. Acad. Sci.*, **108**, 49, 19725–19730, 2011.
210. Sugamura, K. and Keaney, J.F., Reactive oxygen species in cardiovascular disease. *Free Radical Biol. Med.*, **51**, 5, 978–992, 2011.
211. Burchfield, J.S., Xie, M., Hill, J.A., Pathological Ventricular Remodeling. *Circ.*, **128**, 4, 388–400, 2013.
212. Ong, S.-B., *et al.*, Inflammation following acute myocardial infarction: Multiple players, dynamic roles, and novel therapeutic opportunities. *Pharmacol. Ther.*, **186**, 73–87, 2018.
213. Börger, V., *et al.*, Mesenchymal Stem/Stromal Cell-Derived Extracellular Vesicles and Their Potential as Novel Immunomodulatory Therapeutic Agents. *Int. J. Mol. Sci.*, **18**, 7, 1450, 2017.
214. Cho, K.-S., *et al.*, Dendritic cells and M2 macrophage play an important role in suppression of Th2-mediated inflammation by adipose stem cells-derived extracellular vesicles. *Stem Cell Res.*, **39**, 101500, 2019.
215. Gao, F., *et al.*, Mesenchymal stem cells and immunomodulation: current status and future prospects. *Cell Death Dis.*, **7**, 1, e2062–e2062, 2016.
216. Teng, X., *et al.*, Mesenchymal Stem Cell-Derived Exosomes Improve the Microenvironment of Infarcted Myocardium Contributing to Angiogenesis and Anti-Inflammation. *Cell. Physiol. Biochem.*, **37**, 6, 2415–2424, 2015.
217. Wang, Y., *et al.*, Plasticity of mesenchymal stem cells in immunomodulation: pathological and therapeutic implications. *Nat. Immunol.*, **15**, 11, 1009–1016, 2014.
218. Yagi, H., *et al.*, Mesenchymal Stem Cells: Mechanisms of Immunomodulation and Homing. *Cell Transplant.*, **19**, 6–7, 667–679, 2010.

219. Zhang, B., *et al.*, Mesenchymal Stem Cells Secrete Immunologically Active Exosomes. *Stem Cells Dev.*, **23**, 11, 1233–1244, 2014.
220. Di Trapani, M., *et al.*, Differential and transferable modulatory effects of mesenchymal stromal cell-derived extracellular vesicles on T, B and NK cell functions. *Sci. Rep.*, **6**, 1, 24120, 2016.
221. Ono, M., *et al.*, Exosomes from bone marrow mesenchymal stem cells contain a microRNA that promotes dormancy in metastatic breast cancer cells. *Science Signaling*, **7**, 332, ra63–ra63, 2014.
222. Zhang, Z., *et al.*, Pretreatment of Cardiac Stem Cells With Exosomes Derived From Mesenchymal Stem Cells Enhances Myocardial Repair. *J. Am. Heart Assoc.*, **5**, 1, e002856, 2016.
223. Shao, L., *et al.*, MiRNA-Sequence Indicates That Mesenchymal Stem Cells and Exosomes Have Similar Mechanism to Enhance Cardiac Repair. *BioMed Res. Int.*, **2017**, 1, 4150705, 2017.
224. Wang, B., *et al.*, Suppression of microRNA-29 Expression by TGF-β1 Promotes Collagen Expression and Renal Fibrosis. *J. Am. Soc. Nephrol.*, **23**, 2, 252–265, 2012.
225. Maegdefessel, L., *et al.*, miR-24 limits aortic vascular inflammation and murine abdominal aneurysm development. *Nat. Commun.*, **5**, 1, 5214, 2014.
226. Wei, Z., *et al.*, miRNA-181a over-expression in mesenchymal stem cell-derived exosomes influenced inflammatory response after myocardial ischemia-reperfusion injury. *Life Sci.*, **232**, 116632, 2019.
227. Ben-Mordechai, T., *et al.*, Macrophage Subpopulations Are Essential for Infarct Repair With and Without Stem Cell Therapy. *J. Am. Coll. Cardiol.*, **62**, 20, 1890–1901, 2013.
228. Domenis, R., *et al.*, Pro inflammatory stimuli enhance the immunosuppressive functions of adipose mesenchymal stem cells-derived exosomes. *Sci. Rep.*, **8**, 1, 13325, 2018.
229. Lo Sicco, C., *et al.*, Mesenchymal Stem Cell-Derived Extracellular Vesicles as Mediators of Anti-Inflammatory Effects: Endorsement of Macrophage Polarization. *Stem Cells Transl. Med.*, **6**, 3, 1018–1028, 2017.
230. Song, Y., *et al.*, Exosomal miR-146a Contributes to the Enhanced Therapeutic Efficacy of Interleukin-1β-Primed Mesenchymal Stem Cells Against Sepsis. *Stem Cells*, **35**, 5, 1208–1221, 2017.
231. Willis, G.R., *et al.*, Mesenchymal Stromal Cell Exosomes Ameliorate Experimental Bronchopulmonary Dysplasia and Restore Lung Function through Macrophage Immunomodulation. *Am. J. Respir. Crit. Care Med.*, **197**, 1, 104–116, 2018.
232. Zhang, Q.-Z., *et al.*, Human Gingiva-Derived Mesenchymal Stem Cells Elicit Polarization of M2 Macrophages and Enhance Cutaneous Wound Healing. *Stem Cells*, **28**, 10, 1856–1868, 2010.
233. Hart, M., *et al.*, miR-34a: a new player in the regulation of T cell function by modulation of NF-κB signaling. *Cell Death Dis.*, **10**, 2, 46, 2019.

234. Pan, Y., *et al.*, Adipocyte-secreted exosomal microRNA-34a inhibits M2 macrophage polarization to promote obesity-induced adipose inflammation. *J. Clin. Invest.*, **129**, 2, 834–849, 2019.
235. Talman, V. and Ruskoaho, H., Cardiac fibrosis in myocardial infarction—from repair and remodeling to regeneration. *Cell Tissue Res.*, **365**, 3, 563–581, 2016.
236. O'Rourke, S.A., Dunne, A., Monaghan, M.G., The Role of Macrophages in the Infarcted Myocardium: Orchestrators of ECM Remodeling. *Front. Cardiovasc. Med.*, **6**, 2019.
237. Tsapara, A., *et al.*, The RhoA activator GEF-H1/Lfc is a transforming growth factor-beta target gene and effector that regulates alpha-smooth muscle actin expression and cell migration. *Mol Biol Cell*, **21**, 6, 860–70, 2010.
238. Thompson, R.W., *et al.*, Cationic amino acid transporter-2 regulates immunity by modulating arginase activity. *PLoS Pathog.*, **4**, 3, e1000023, 2008.
239. Verma, S.K., *et al.*, Interleukin-10 Treatment Attenuates Pressure Overload–Induced Hypertrophic Remodeling and Improves Heart Function *via* Signal Transducers and Activators of Transcription 3–Dependent Inhibition of Nuclear Factor-κB. *Circ.*, **126**, 4, 418–429, 2012.
240. Deng, S., *et al.*, Exosomes from adipose-derived mesenchymal stem cells ameliorate cardiac damage after myocardial infarction by activating S1P/SK1/S1PR1 signaling and promoting macrophage M2 polarization. *Int. J. Biochem. Cell Biol.*, **114**, 105564, 2019.
241. Chen, Y., *et al.*, MicroRNA-133 overexpression promotes the therapeutic efficacy of mesenchymal stem cells on acute myocardial infarction. *Stem Cell Res. Ther.*, **8**, 1, 268, 2017.
242. Muraoka, N., *et al.*, MiR-133 promotes cardiac reprogramming by directly repressing Snai1 and silencing fibroblast signatures. *EMBO J.*, **33**, 14, 1565–1581, 2014.
243. Gianchecchi, E., Delfino, D.V., Fierabracci, A., Recent insights into the role of the PD-1/PD-L1 pathway in immunological tolerance and autoimmunity. *Autoimmun Rev*, **12**, 11, 1091–100, 2013.
244. Mokarizadeh, A., *et al.*, Microvesicles derived from mesenchymal stem cells: Potent organelles for induction of tolerogenic signaling. *Immunol. Lett.*, **147**, 1, 47–54, 2012.
245. Wang, Y., *et al.*, Mesenchymal stem cell-derived exosomes inhibits the activation of NLRP3 inflammasome in mice after myocardial infarction by releasing circASXL1. *J. Mol. Cell. Cardiol.*, **140**, 57, 2020.
246. Anginot, A., *et al.*, Galectin 1 Modulates Plasma Cell Homeostasis and Regulates the Humoral Immune Response. *J. Immunol.*, **190**, 11, 5526–5533, 2013.
247. Kim, H.-S., *et al.*, Proteomic Analysis of Microvesicles Derived from Human Mesenchymal Stem Cells. *J. Proteome Res.*, **11**, 2, 839–849, 2012.
248. Perillo, N.L., *et al.*, Apoptosis of T cells mediated by galectin-1. *Nature*, **378**, 6558, 736–739, 1995.

249. Rabinovich, G.A., *et al.*, Activated rat macrophages produce a galectin-1-like protein that induces apoptosis of T cells: biochemical and functional characterization. *J. Immunol.*, **160**, 10, 4831–40, 1998.
250. Burrello, J., *et al.*, Stem Cell-Derived Extracellular Vesicles and Immune-Modulation. *Front. Cell Dev. Biol.*, **4**, 2016.
251. Worthington, J.J., *et al.*, Regulation of TGFβ in the immune system: An emerging role for integrins and dendritic cells. *Immunobiology*, **217**, 12, 1259–1265, 2012.
252. Ferguson, S.W., *et al.*, The microRNA regulatory landscape of MSC-derived exosomes: a systems view. *Sci. Rep.*, **8**, 1, 1419, 2018.
253. Zhang, C.S., *et al.*, Hypoxic preconditioning BMSCs-exosomes inhibit cardiomyocyte apoptosis after acute myocardial infarction by upregulating microRNA-24. *Eur. Rev. Med. Pharmacol. Sci.*, **23**, 15, 6691–6699, 2019.
254. Gao, H., *et al.*, miR-100-5p in human umbilical cord mesenchymal stem cell-derived exosomes mediates eosinophilic inflammation to alleviate atherosclerosis *via* the FZD5/Wnt/β-catenin pathway. *Biochim. Biophys. Sin. (Shanghai)*, **53**, 9, 1166–1176, 2021.
255. Xiao, Y., *et al.*, MicroRNA-133a and Myocardial Infarction. *Cell Transplant.*, **28**, 7, 831–838, 2019.
256. Sarver, D.C. and Lusis, A.J., Linking the brain to recovery after myocardial infarction. *Nat. Cardiovasc. Res.*, **3**, 7, 780–781, 2024.
257. Bräuninger, H., *et al.*, Matrix metalloproteinases in coronary artery disease and myocardial infarction. *Basic Res. Cardiol.*, **118**, 1, 18, 2023.
258. Moeinabadi-Bidgoli, K., *et al.*, Exosomes for angiogenesis induction in ischemic disorders. *J. Cell Mol. Med.*, **27**, 6, 763–787, 2023.
259. Gutierrez-Millan, C., *et al.*, Advances in Exosomes-Based Drug Delivery Systems. *Macromol. Biosci.*, **21**, 1, e2000269, 2021.
260. Florek, K., Mendyka, D., Gomułka, K., Vascular Endothelial Growth Factor (VEGF) and Its Role in the Cardiovascular System. *Biomedicines*, **12**, 5, 1055, 2024.
261. Lei, Z., *et al.*, miR-132/212 Impairs Cardiomyocytes Contractility in the Failing Heart by Suppressing SERCA2a. *Front. Cardiovasc. Med.*, **8**, 592362, 2021.
262. Gangadaran, P., *et al.*, Extracellular vesicles from mesenchymal stem cells activates VEGF receptors and accelerates recovery of hindlimb ischemia. *J. Control. Release*, **264**, 112–126, 2017.
263. Wang, X., *et al.*, The Application Potential and Advance of Mesenchymal Stem Cell-Derived Exosomes in Myocardial Infarction. *Stem Cells Int.*, **2021**, 5579904, 2021.
264. Wang, N., *et al.*, Mesenchymal stem cells-derived extracellular vesicles, *via* miR-210, improve infarcted cardiac function by promotion of angiogenesis. *Biochim. Biophys. Acta. Mol. Basis. Dis.*, **1863**, 8, 2085–2092, 2017.

265. Zhang, J., *et al.*, IFN-γ enhances the efficacy of mesenchymal stromal cell-derived exosomes *via* miR-21 in myocardial infarction rats. *Stem Cell Res. Ther.*, **13**, 1, 333, 2022.
266. Song, R., *et al.*, MicroRNA-210 Controls Mitochondrial Metabolism and Protects Heart Function in Myocardial Infarction. *Circulation*, **145**, 15, 1140–1153, 2022.
267. Xu, K., *et al.*, Advances in miR-132-Based Biomarker and Therapeutic Potential in the Cardiovascular System. *Front. Pharmacol.*, **12**, 751487, 2021.
268. Imam, R.A.E.N., *et al.*, Role of mesenchymal stem cells-derived exosomes on inflammation, apoptosis, fibrosis and telocyte modulation in doxorubicin-induced cardiotoxicity: A closer look at the structural level. *Microsc. Res. Tech.*, **87**, 7, 1598–1614, 2024.
269. Ma, J., *et al.*, Exosomes Derived from Akt-Modified Human Umbilical Cord Mesenchymal Stem Cells Improve Cardiac Regeneration and Promote Angiogenesis *via* Activating Platelet-Derived Growth Factor D. *Stem Cells Transl. Med.*, **6**, 1, 51–59, 2017.
270. Jansen, F., *et al.*, Endothelial- and Immune Cell-Derived Extracellular Vesicles in the Regulation of Cardiovascular Health and Disease. *JACC: Basic to Translational Science*, **2**, 6, 790–807, 2017.
271. Todorova, D., *et al.*, Extracellular vesicles in angiogenesis. *Circ. Res.*, **120**, 10, 1658–1673, 2017.
272. Mathiyalagan, P., *et al.*, Angiogenic mechanisms of human CD34+ stem cell exosomes in the repair of ischemic hindlimb. *Circ. Res.*, **120**, 9, 1466–1476, 2017.
273. Anderson, J.D., *et al.*, Comprehensive Proteomic Analysis of Mesenchymal Stem Cell Exosomes Reveals Modulation of Angiogenesis *via* Nuclear Factor-KappaB Signaling. *Stem Cells*, **34**, 3, 601–613, 2016.
274. Sun, Y., Xiong, X., Wang, X., RELA promotes hypoxia-induced angiogenesis in human umbilical vascular endothelial cells *via* LINC01693/miR-302d/CXCL12 axis. *J. Cell. Biochem.*, **120**, 8, 12549–12558, 2019.
275. Yamaoka-Tojo, M., *et al.*, IQGAP1 Mediates VE-Cadherin–Based Cell–Cell Contacts and VEGF Signaling at Adherence Junctions Linked to Angiogenesis. *Arteriosclerosis, Thrombosis, Vascular Biology*, **26**, 9, 1991–1997, 2006.
276. Thijssen, V.L. and Griffioen, A.W., Galectin-1 and -9 in angiogenesis: A sweet couple. *Glycobiology*, **24**, 10, 915–920, 2014.
277. Raica, M. and Cimpean, A.M., Platelet-Derived Growth Factor (PDGF)/PDGF Receptors (PDGFR) Axis as Target for Antitumor and Antiangiogenic Therapy. *Pharmaceuticals*, **3**, 3, 572–599, 2010.
278. Wu, E., *et al.*, Comprehensive Dissection of PDGF-PDGFR Signaling Pathways in PDGFR Genetically Defined Cells. *PLOS ONE*, **3**, 11, e3794, 2008.
279. Lieu, C., *et al.*, Beyond VEGF: Inhibition of the Fibroblast Growth Factor Pathway and Antiangiogenesis. *Clin. Cancer Res.*, **17**, 19, 6130–6139, 2011.

280. Seghezzi, G., *et al.*, Fibroblast Growth Factor-2 (FGF-2) Induces Vascular Endothelial Growth Factor (VEGF) Expression in the Endothelial Cells of Forming Capillaries: An Autocrine Mechanism Contributing to Angiogenesis. *J. Cell Biol.*, **141**, 7, 1659–1673, 1998.
281. Blaschuk, O.W. and Rowlands, T.M., Cadherins as Modulators of Angiogenesis and the Structural Integrity of Blood Vessels. *Cancer Metastasis Rev.*, **19**, 1, 1–5, 2000.
282. Gong, M., *et al.*, Nano-Sized Extracellular Vesicles Secreted from GATA-4 Modified Mesenchymal Stem Cells Promote Angiogenesis by Delivering Let-7 miRNAs. *Cells*, **11**, 9, 1573, 2022.
283. Akwii, R.G., *et al.*, Role of Angiopoietin-2 in Vascular Physiology and Pathophysiology. *Cells*, **8**, 5, 2019.
284. Sack, K.D., Kellum, J.A., Parikh, S.M., The Angiopoietin-Tie2 Pathway in Critical Illness. *Crit. Care Clin.*, **36**, 2, 201–216, 2020.
285. Liu, T., *et al.*, Nuclear localization of the tyrosine kinase BMX mediates VEGFR2 expression. *J. Cell. Mol. Med.*, **24**, 1, 126–138, 2020.
286. Lin, S., *et al.*, IGF-1 promotes angiogenesis in endothelial cells/adipose-derived stem cells co-culture system with activation of PI3K/Akt signal pathway. *Cell Prolif.*, **50**, 6, 2017.
287. Higashi, Y., *et al.*, Endothelial deficiency of insulin-like growth factor-1 receptor reduces endothelial barrier function and promotes atherosclerosis in Apoe-deficient mice. *Am. J. Physiol. Heart Circ. Physiol.*, **319**, 4, H730–h743, 2020.
288. Macvanin, M., *et al.*, New insights on the cardiovascular effects of IGF-1. *Front. Endocrinol. (Lausanne)*, **14**, 1142644, 2023.
289. Casarella, S., *et al.*, Focal Adhesion's Role in Cardiomyocytes Function: From Cardiomyogenesis to Mechanotransduction. *Cells*, **13**, 8, 664, 2024.
290. Maji, S., *et al.*, Exosomal Annexin II Promotes Angiogenesis and Breast Cancer Metastasis. *Mol. Cancer Res.*, **15**, 1, 93–105, 2017.
291. Almeria, C., *et al.*, Hypoxia Conditioned Mesenchymal Stem Cell-Derived Extracellular Vesicles Induce Increased Vascular Tube Formation *in vitro*. *Front. Bioeng. Biotechnol.*, **7**, 2019.
292. Tang, Y., *et al.*, Extracellular Matrix Metalloproteinase Inducer Stimulates Tumor Angiogenesis by Elevating Vascular Endothelial Cell Growth Factor and Matrix Metalloproteinases. *Cancer Res.*, **65**, 8, 3193–3199, 2005.
293. Vrijsen, K.R., *et al.*, Exosomes from Cardiomyocyte Progenitor Cells and Mesenchymal Stem Cells Stimulate Angiogenesis *Via* EMMPRIN. *Adv. Healthcare Mater.*, **5**, 19, 2555–2565, 2016.
294. Calvisi, D.F., Pascale, R.M., Feo, F., Epidermal growth factor-like repeats and discoidin I-like domains 3: a multifaceted oncoprotein at the crossroad of MAPK and TGF-beta pathways in human hepatocellular carcinoma. *Transl. Cancer Res.*, **5**, 2, 103–109, 2016.

295. Makki, N., Thiel, K.W., Miller, Jr., F.J., The epidermal growth factor receptor and its ligands in cardiovascular disease. *Int. J. Mol. Sci.*, **14**, 10, 20597–613, 2013.
296. Fang, J., *et al.*, Exosomes and Exosomal Cargos: A Promising World for Ventricular Remodeling Following Myocardial Infarction. *Int. J. Nanomed.*, **17**, 4699–4719, 2022.
297. Heidemann, J., *et al.*, Angiogenic Effects of Interleukin 8 (CXCL8) in Human Intestinal Microvascular Endothelial Cells Are Mediated by CXCR2 *. *J. Biol. Chem.*, **278**, 10, 8508–8515, 2003.
298. Chen, D.-X., *et al.*, Endothelial progenitor cell-derived extracellular vesicles: the world of potential prospects for the treatment of cardiovascular diseases. *Cell Biosci.*, **14**, 1, 72, 2024.
299. Salvador, A.M., *et al.*, Intercellular Adhesion Molecule 1 Regulates Left Ventricular Leukocyte Infiltration, Cardiac Remodeling, and Function in Pressure Overload-Induced Heart Failure. *J. Am. Heart Assoc.*, **5**, 3, e003126, 2016.
300. Zhou, G., *et al.*, lncRNA UCA1 Is a Novel Regulator in Cardiomyocyte Hypertrophy through Targeting the miR-184/HOXA9 Axis. *Cardiorenal Med.*, **8**, 2, 130–139, 2018.
301. Kalra, K., *et al.*, Role of PDGF-A/B Ligands in Cardiac Repair After Myocardial Infarction. *Front. Cell Dev. Biol.*, **9**, 669188, 2021.
302. Anzai, A., *et al.*, The infarcted myocardium solicits GM-CSF for the detrimental oversupply of inflammatory leukocytes. *J. Exp. Med.*, **214**, 11, 3293–3310, 2017.
303. Zhu, J., *et al.*, Clarification of the Role of miR-9 in the Angiogenesis, Migration, and Autophagy of Endothelial Progenitor Cells Through RNA Sequence Analysis. *Cell Transplant.*, **29**, 963689720963936, 2020.
304. Liu, L.-Z., *et al.*, MiR-21 induced angiogenesis through AKT and ERK activation and HIF-1α expression. *PLoS One*, **6**, 4, e19139, 2011.
305. Mayourian, J., *et al.*, Exosomal microRNA-21-5p Mediates Mesenchymal Stem Cell Paracrine Effects on Human Cardiac Tissue Contractility. *Circ. Res.*, **122**, 7, 933–944, 2018.
306. Song, Y., *et al.*, Human umbilical cord blood-derived MSCs exosome attenuate myocardial injury by inhibiting ferroptosis in acute myocardial infarction mice. *Cell Biol. Toxicol.*, **37**, 1, 51–64, 2021.
307. Vieira, J.M.F., *et al.*, Overexpression of mir-135b and mir-210 in mesenchymal stromal cells for the enrichment of extracellular vesicles with angiogenic factors. *PLoS One*, **17**, 8, e0272962, 2022.
308. Masoumi-Dehghi, S., Babashah, S., Sadeghizadeh, M., microRNA-141-3p-containing small extracellular vesicles derived from epithelial ovarian cancer cells promote endothelial cell angiogenesis through activating the JAK/STAT3 and NF-κB signaling pathways. *J. Cell. Commun. Signal*, **14**, 2, 233–244, 2020.

309. Sawant, D. and Lilly, B., MicroRNA-145 targets in cancer and the cardiovascular system: evidence for common signaling pathways. *Vasc. Biol.*, **2**, 1, R115–r128, 2020.
310. Ning, W., *et al.*, Blocking exosomal miRNA-153-3p derived from bone marrow mesenchymal stem cells ameliorates hypoxia-induced myocardial and microvascular damage by targeting the ANGPT1-mediated VEGF/PI3k/Akt/eNOS pathway. *Cell. Signalling*, **77**, 109812, 2021.
311. Wang, Y., *et al.*, miR-181a, delivered by hypoxic PTC-secreted exosomes, inhibits DACT2 by downregulating MLL3, leading to YAP-VEGF-mediated angiogenesis. *Mol. Ther. Nucleic Acids*, **24**, 610–621, 2021.
312. Pei, G., *et al.*, Role of miR-182 in cardiovascular and cerebrovascular diseases. *Front. Cell Dev. Biol.*, **11**, 1181515, 2023.
313. Xuan, W., *et al.*, miRNAs in Extracellular Vesicles from iPS-Derived Cardiac Progenitor Cells Effectively Reduce Fibrosis and Promote Angiogenesis in Infarcted Heart. *Stem Cells Int.*, **2019**, 3726392, 2019.
314. Huang, Y., *et al.*, EPC-Derived Exosomal miR-1246 and miR-1290 Regulate Phenotypic Changes of Fibroblasts to Endothelial Cells to Exert Protective Effects on Myocardial Infarction by Targeting ELF5 and SP1. *Front. Cell Dev. Biol.*, **9**, 647763, 2021.
315. Thomas, N.A., *et al.*, Hedgehog signaling plays a cell-autonomous role in maximizing cardiac developmental potential. *Development*, **135**, 22, 3789–99, 2008.
316. Tang, Y.-N., *et al.*, Epigenetic regulation of Smad2 and Smad3 by profilin-2 promotes lung cancer growth and metastasis. *Nat. Commun.*, **6**, 1, 8230, 2015.
317. Cheng, C., *et al.*, Exosomal lncRNA RAMP2-AS1 Derived from Chondrosarcoma Cells Promotes Angiogenesis Through miR-2355-5p/VEGFR2 Axis. *Onco. Targets Ther.*, **13**, 3291–3301, 2020.
318. Goumans, M.J. and Ten Dijke, P., TGF-β Signaling in Control of Cardiovascular Function. *Cold Spring Harb. Perspect. Biol.*, **10**, 2, 2018.
319. Tabruyn, S.P., *et al.*, NF-κB activation in endothelial cells is critical for the activity of angiostatic agents. *Mol. Cancer Ther.*, **8**, 9, 2645–2654, 2009.
320. Bonnet, M., *et al.*, Targeting the Tetraspanins with Monoclonal Antibodies in Oncology: Focus on Tspan8/Co-029. *Cancers (Basel)*, **11**, 2, 2019.
321. Louvet, L., *et al.*, Vasorin plays a critical role in vascular smooth muscle cells and arterial functions. *J. Cell. Physiol.*, **237**, 10, 3845–3859, 2022.
322. Firoozi, S., *et al.*, A Cell-Free SDKP-Conjugated Self-Assembling Peptide Hydrogel Sufficient for Improvement of Myocardial Infarction. *Biomolecules*, **10**, 2, 2020.
323. Qu, J.L., *et al.*, Gastric cancer exosomes promote tumour cell proliferation through PI3K/Akt and MAPK/ERK activation. *Dig. Liver. Dis.*, **41**, 12, 875–80, 2009.

324. Minayi, N., *et al.*, The Effect of miR-210 Up-regulation on Proliferation and Survival of Mouse Bone Marrow Derived Mesenchymal Stem Cell. *Int. J. Hematol. Oncol. Stem Cell Res.*, **8**, 1, 15–23, 2014.
325. Li, T., *et al.*, Bone Marrow Mesenchymal Stem Cell-Derived Exosomal miRNA-29c Decreases Cardiac Ischemia/Reperfusion Injury Through Inhibition of Excessive Autophagy *via* the PTEN/Akt/mTOR Signaling Pathway. *Circ. J.*, **84**, 8, 1304–1311, 2020.
326. Sun, X.H., *et al.*, Exosomes of bone-marrow stromal cells inhibit cardiomyocyte apoptosis under ischemic and hypoxic conditions *via* miR-486-5p targeting the PTEN/PI3K/AKT signaling pathway. *Thromb. Res.*, **177**, 23–32, 2019.
327. Wang, Y., *et al.*, Exosomes Derived from miR-214-Enriched Bone Marrow-Derived Mesenchymal Stem Cells Regulate Oxidative Damage in Cardiac Stem Cells by Targeting CaMKII. *Oxid. Med. Cell. Longev.*, **2018**, 4971261, 2018.
328. Zhu, L.P., *et al.*, Hypoxia-elicited mesenchymal stem cell-derived exosomes facilitates cardiac repair through miR-125b-mediated prevention of cell death in myocardial infarction. *Theranostics*, **8**, 22, 6163–6177, 2018.
329. Shukla, S., *et al.*, BH3-only protein BIM: An emerging target in chemotherapy. *Eur. J. Cell Biol.*, **96**, 8, 728–738, 2017.
330. Farahzadi, R., *et al.*, Stem cells-derived exosomes as cardiac regenerative agents. *Int. J. Cardiol. Heart Vasc.*, **52**, 101399, 2024.
331. Osmak, G., *et al.*, From miRNA Target Gene Network to miRNA Function: miR-375 Might Regulate Apoptosis and Actin Dynamics in the Heart Muscle *via* Rho-GTPases-Dependent Pathways. *Int. J. Mol. Sci.*, **21**, 24, 9670, 2020.
332. Jiang, X., *et al.*, Human Mesenchymal Stem Cell-derived Exosomes Reduce Ischemia/Reperfusion Injury by the Inhibitions of Apoptosis and Autophagy. *Curr. Pharm. Des.*, **24**, 44, 5334–5341, 2018.
333. Xiao, C., *et al.*, Transplanted Mesenchymal Stem Cells Reduce Autophagic Flux in Infarcted Hearts *via* the Exosomal Transfer of miR-125b. *Circ. Res.*, **123**, 5, 564–578, 2018.
334. Eulalio, A., *et al.*, Functional screening identifies miRNAs inducing cardiac regeneration. *Nature*, **492**, 7429, 376–381, 2012.
335. Chen, G., *et al.*, Mesenchymal stem cell-derived exosomal miR-143-3p suppresses myocardial ischemia-reperfusion injury by regulating autophagy. *Life Sci.*, **280**, 119742, 2021.
336. Singh, B.N., *et al.*, A conserved HH-Gli1-Mycn network regulates heart regeneration from newt to human. *Nat. Commun.*, **9**, 1, 4237, 2018.
337. Pandey, R., *et al.*, MicroRNA-1825 induces proliferation of adult cardiomyocytes and promotes cardiac regeneration post ischemic injury. *Am. J. Transl. Res.*, **9**, 6, 3120–3137, 2017.
338. Lesizza, P., *et al.*, Single-Dose Intracardiac Injection of Pro-Regenerative MicroRNAs Improves Cardiac Function After Myocardial Infarction. *Circ. Res.*, **120**, 8, 1298–1304, 2017.

339. Li, Y., *et al.*, Exosomal miR-301 derived from mesenchymal stem cells protects myocardial infarction by inhibiting myocardial autophagy. *Biochem. Biophys. Res. Commun.*, **514**, 1, 323–328, 2019.
340. Yao, C., *et al.*, Self-assembly of stem cell membrane-camouflaged nanocomplex for microRNA-mediated repair of myocardial infarction injury. *Biomaterials*, **257**, 120256, 2020.
341. Ucar, A., *et al.*, The miRNA-212/132 family regulates both cardiac hypertrophy and cardiomyocyte autophagy. *Nat. Commun.*, **3**, 1, 1078, 2012.
342. Wen, Z., *et al.*, Mesenchymal stem cell-derived exosomes ameliorate cardiomyocyte apoptosis in hypoxic conditions through microRNA144 by targeting the PTEN/AKT pathway. *Stem Cell Res. Ther.*, **11**, 1, 36, 2020.
343. Yu, B., *et al.*, Exosomes secreted from GATA-4 overexpressing mesenchymal stem cells serve as a reservoir of anti-apoptotic microRNAs for cardioprotection. *Int. J. Cardiol.*, **182**, 349–60, 2015.
344. Cao, H., *et al.*, *In Vivo* Tracking of Mesenchymal Stem Cell-Derived Extracellular Vesicles Improving Mitochondrial Function in Renal Ischemia-Reperfusion Injury. *ACS Nano*, **14**, 4, 4014–4026, 2020.
345. He, W., Li, D., Zhang, X., LncRNA HOTAIR promotes the proliferation and invasion/metastasis of breast cancer cells by targeting the miR-130a-3p/Suv39H1 axis. *Biochem. Biophys. Rep.*, **30**, 101279, 2022.
346. Mi, S., *et al.*, Inhibition of MEG3 ameliorates cardiomyocyte apoptosis and autophagy by regulating the expression of miRNA-129-5p in a mouse model of heart failure. *Redox Rep.*, **28**, 1, 2224607, 2023.
347. Ruvinov, E., Leor, J., Cohen, S., The promotion of myocardial repair by the sequential delivery of IGF-1 and HGF from an injectable alginate biomaterial in a model of acute myocardial infarction. *Biomaterials*, **32**, 2, 565–78, 2011.
348. Garikipati, V.N.S., *et al.*, Circular RNA CircFndc3b modulates cardiac repair after myocardial infarction *via* FUS/VEGF-A axis. *Nat. Commun.*, **10**, 1, 4317, 2019.
349. Cheng, N., *et al.*, Circular RNA POSTN Promotes Myocardial Infarction-Induced Myocardial Injury and Cardiac Remodeling by Regulating miR-96-5p/BNIP3 Axis. *Front. Cell Dev. Biol.*, **8**, 618574, 2020.
350. Rockey, D.C., Bell, P.D., Hill, J.A., Fibrosis–a common pathway to organ injury and failure. *N. Engl. J. Med.*, **372**, 12, 1138–49, 2015.
351. Prabhu, S.D. and Frangogiannis, N.G., The biological basis for cardiac repair after myocardial infarction: from inflammation to fibrosis. *Circ. Res.*, **119**, 1, 91–112, 2016.
352. Stuart, S.D.F., *et al.*, The crossroads of inflammation, fibrosis, and arrhythmia following myocardial infarction. *J. Mol. Cell. Cardiol.*, **91**, 114–122, 2016.
353. Elnakish, M.T., Kuppusamy, P., Khan, M., Stem cell transplantation as a therapy for cardiac fibrosis. *J. Pathol.*, **229**, 2, 347–354, 2013.
354. Biernacka, A. and Frangogiannis, N.G., Aging and Cardiac Fibrosis. *Aging Dis.*, **2**, 2, 158–173, 2011.

355. Wang, X., *et al.*, Cardiac microvascular functions improved by MSC-derived exosomes attenuate cardiac fibrosis after ischemia-reperfusion *via* PDGFR-β modulation. *Int. J. Cardiol.*, **344**, 13–24, 2021.
356. Huang, Z.P. and Wang, D.Z., miR-22 in cardiac remodeling and disease. *Trends. Cardiovasc. Med.*, **24**, 7, 267–72, 2014.
357. Wang, Y.S., *et al.*, Role of miR-145 in cardiac myofibroblast differentiation. *J. Mol. Cell. Cardiol.*, **66**, 94–105, 2014.
358. Li, G., *et al.*, MicroRNA-27b-3p down-regulates FGF1 and aggravates pathological cardiac remodelling. *Cardiovasc. Res.*, **118**, 9, 2139–2151, 2021.
359. Nonaka, C.K.V., *et al.*, Therapeutic miR-21 silencing reduces cardiac fibrosis and modulates inflammatory response in chronic chagas disease. *Int. J. Mol. Sci.*, **22**, 7, 3307, 2021.
360. Huang, Y., *et al.*, MicroRNA-34a regulates cardiac fibrosis after myocardial infarction by targeting Smad4. *Expert Opin. Ther. Targets*, **18**, 12, 1355–1365, 2014.
361. Bernardo, B.C., *et al.*, Therapeutic inhibition of the miR-34 family attenuates pathological cardiac remodeling and improves heart function. *Proc. Natl. Acad. Sci. U.S.A.*, **109**, 43, 17615–20, 2012.
362. Valkov, N., *et al.*, MicroRNA-1-Mediated Inhibition of Cardiac Fibroblast Proliferation Through Targeting Cyclin D2 and CDK6. *Front. Cardiovasc. Med.*, **6**, 65, 2019.
363. Tijsen, A.J., *et al.*, The microRNA-15 family inhibits the TGFβ-pathway in the heart. *Cardiovasc. Res.*, **104**, 1, 61–71, 2014.
364. Kraus, L., *et al.*, Cortical Bone Derived Stem Cells Modulate Cardiac Fibroblast Response *via* miR-18a in the Heart After Injury. *Front. Cell Dev. Biol.*, **8**, 494, 2020.
365. Yousefi, F., *et al.*, TGF-β and WNT signaling pathways in cardiac fibrosis: non-coding RNAs come into focus. *Cell Commun Signal*, **18**, 1, 87, 2020.
366. Zhao, X.S., *et al.*, MiR-30b-5p and miR-22-3p restrain the fibrogenesis of post-myocardial infarction in mice *via* targeting PTAFR. *Eur. Rev. Med. Pharmacol. Sci.*, **24**, 7, 3993–4004, 2020.
367. Wang, J., *et al.*, MicroRNA-24 regulates cardiac fibrosis after myocardial infarction. *J. Cell. Mol. Med.*, **16**, 9, 2150–60, 2012.
368. Zheng, L., Lin, S., Lv, C., MiR-26a-5p regulates cardiac fibroblasts collagen expression by targeting ULK1. *Sci. Rep.*, **8**, 1, 2104, 2018.
369. Xiang, S., Li, J., Zhang, Z., miR-26b inhibits isoproterenol-induced cardiac fibrosis *via* the Keap1/Nrf2 signaling pathway. *Exp. Ther. Med.*, **19**, 3, 2067–2074, 2020.
370. Wang, Y., *et al.*, Atrial overexpression of microRNA-27b attenuates angiotensin II-induced atrial fibrosis and fibrillation by targeting ALK5. *Human Cell*, **31**, 3, 251–260, 2018.
371. Tao, H., *et al.*, MicroRNA-29a suppresses cardiac fibroblasts proliferation *via* targeting VEGF-A/MAPK signal pathway. *Int. J. Biol. Macromol.*, **88**, 414–23, 2016.

372. Zhang, Y., *et al.*, miR-29b as a therapeutic agent for angiotensin II-induced cardiac fibrosis by targeting TGF-β/Smad3 signaling. *Mol. Ther.*, **22**, 5, 974–85, 2014.
373. Ni, H., *et al.*, Inhibition of circHIPK3 prevents angiotensin II-induced cardiac fibrosis by sponging miR-29b-3p. *Int. J. Cardiol.*, **292**, 188–196, 2019.
374. van Rooij, E., *et al.*, Dysregulation of microRNAs after myocardial infarction reveals a role of miR-29 in cardiac fibrosis. *Proc. Natl. Acad. Sci. U.S.A.*, **105**, 35, 13027–32, 2008.
375. Zhao, X., *et al.*, MicroRNA-101a Inhibits Cardiac Fibrosis Induced by Hypoxia *via* Targeting TGFβRI on Cardiac Fibroblasts. *Cell. Physiol. Biochem.*, **35**, 1, 213–226, 2015.
376. Chistiakov, D.A., Orekhov, A.N., Bobryshev, Y.V., Cardiac Extracellular Vesicles in Normal and Infarcted Heart. *Int. J. Mol. Sci.*, **17**, 1, 63, 2016.
377. Yu, B.-T., *et al.*, Role of miR-133a in regulating TGF-β1 signaling pathway in myocardial fibrosis after acute myocardial infarction in rats. *Eur. Rev. Med. Pharmacol. Sci.*, **23**, 19, 2019.
378. Zhao, N., *et al.*, MicroRNA miR145 Regulates TGFBR2 Expression and Matrix Synthesis in Vascular Smooth Muscle Cells. *Circ. Res.*, **116**, 1, 23–34, 2015.
379. Aurora, A.B., *et al.*, MicroRNA-214 protects the mouse heart from ischemic injury by controlling Ca^{2+} overload and cell death. *J. Clin. Invest.*, **122**, 4, 1222–32, 2012.
380. Du, Y., *et al.*, miRNA-Mediated Suppression of a Cardioprotective Cardiokine as a Novel Mechanism Exacerbating Post-MI Remodeling by Sleep Breathing Disorders. *Circ. Res.*, **126**, 2, 212–228, 2020.
381. Seo, H.H., *et al.*, Multipoint targeting of TGF-β/Wnt transactivation circuit with microRNA 384-5p for cardiac fibrosis. *Cell Death Differ.*, **26**, 6, 1107–1123, 2019.
382. Zhao, H., *et al.*, Elevated IgE promotes cardiac fibrosis by suppressing miR-486a-5p. *Theranostics*, **11**, 15, 7600–7615, 2021.
383. Shan, H., *et al.*, Downregulation of miR-133 and miR-590 contributes to nicotine-induced atrial remodelling in canines. *Cardiovasc. Res.*, **83**, 3, 465–72, 2009.
384. Wu, X., *et al.*, MicroRNA663 participates in myocardial fibrosis through interaction with TGFβ1. *Exp. Ther. Med.*, **18**, 4, 3172–3176, 2019.
385. Wang, L., *et al.*, A novel mechanism of Smads/miR-675/TGFβR1 axis modulating the proliferation and remodeling of mouse cardiac fibroblasts. *J. Cell. Physiol.*, **234**, 11, 20275–20285, 2019.
386. Kapoun, A.M., *et al.*, B-type natriuretic peptide exerts broad functional opposition to transforming growth factor-beta in primary human cardiac fibroblasts: fibrosis, myofibroblast conversion, proliferation, and inflammation. *Circ. Res.*, **94**, 4, 453–61, 2004.

387. Qiao, G., *et al.*, Role of Sprouty1 (Spry1) in the pathogenesis of atrial fibrosis. *Pathol. Res. Pract.*, **214**, 2, 308–313, 2018.
388. Mehjabin, A., *et al.*, MicroRNA in Fibrotic Disorders: A Potential Target for Future Therapeutics. *FBL*, **28**, 11, 2023.
389. Liang, H., *et al.*, A novel reciprocal loop between microRNA-21 and TGFβRIII is involved in cardiac fibrosis. *Int. J. Biochem. Cell Biol.*, **44**, 12, 2152–2160, 2012.
390. Goh, R.S.J., *et al.*, The burden of cardiovascular disease in Asia from 2025 to 2050: a forecast analysis for East Asia, South Asia, South-East Asia, Central Asia, and high-income Asia Pacific regions. *Lancet Regional Health – Western Pacific*, **49**, 2024.
391. Zhang, Z., *et al.*, Advances in the study of exosomes in cardiovascular diseases. *J. Adv. Res.*, 2023, 2023.
392. Zarà, M., *et al.*, Exosomes in Cardiovascular Diseases. *Diagnostics*, **10**, 11, 943, 2020.
393. Zhang, X., *et al.*, Extracellular Vesicles in Cardiovascular Diseases: Diagnosis and Therapy. *Front. Cell Dev. Biol.*, **10**, 875376, 2022.
394. Thygesen, K., *et al.*, Fourth Universal Definition of Myocardial Infarction (2018). *J. Am. Coll. Cardiol.*, **72**, 18, 2231–2264, 2018.
395. Andersson, C. and Vasan, R.S., Epidemiology of cardiovascular disease in young individuals. *Nat. Rev. Cardiol.*, **15**, 4, 230–240, 2018.
396. Zheng, Y.L., *et al.*, Stem cell-derived exosomes in the treatment of acute myocardial infarction in preclinical animal models: a meta-analysis of randomized controlled trials. *Stem. Cell Res. Ther.*, **13**, 1, 151, 2022.
397. Wu, K., *et al.*, Injectable Decellularized Extracellular Matrix Hydrogel Containing Stromal Cell-Derived Factor 1 Promotes Transplanted Cardiomyocyte Engraftment and Functional Regeneration after Myocardial Infarction. *ACS Appl. Mater. Interfaces*, **15**, 2, 2578–2589, 2023.
398. Ahmad Shiekh, P., *et al.*, Oxygen releasing and antioxidant breathing cardiac patch delivering exosomes promotes heart repair after myocardial infarction. *Biochem. Eng. J.*, **428**, 132490, 2022.
399. Wang, T., *et al.*, ADSC-derived exosomes attenuate myocardial infarction injury by promoting miR-205-mediated cardiac angiogenesis. *Biol. Direct*, **18**, 1, 6, 2023.
400. de Almeida Oliveira, N.C., *et al.*, Multicellular regulation of miR-196a-5p and miR-425-5 from adipose stem cell-derived exosomes and cardiac repair. *Clin. Sci. (Lond)*, **136**, 17, 1281–1301, 2022.
401. Wang, Q., *et al.*, HIF-1α overexpression in mesenchymal stem cell-derived exosome-encapsulated arginine-glycine-aspartate (RGD) hydrogels boost therapeutic efficacy of cardiac repair after myocardial infarction. *Mater. Today Bio*, **12**, 100171, 2021.
402. Wang, S., *et al.*, Exosomes derived from miR-129-5p modified bone marrow mesenchymal stem cells represses ventricular remolding of mice with myocardial infarction. *J. Tissue Eng. Regen. Med.*, **16**, 2, 177–187, 2022.

403. Angulski, A.B.B., *et al.*, Systemic Infusion of Expanded CD133(+) Cells and Expanded CD133(+) Cell-Derived EVs for the Treatment of Ischemic Cardiomyopathy in a Rat Model of AMI. *Stem Cells Int.*, **2019**, 4802578, 2019.
404. Hashemi, S.M., *et al.*, A placebo controlled, dose-ranging, safety study of allogenic mesenchymal stem cells injected by endomyocardial delivery after an acute myocardial infarction. *Eur. Heart. J.*, **29**, 2, 251–9, 2008.
405. Ning, H., *et al.*, Exosomes secreted by FNDC5-BMMSCs protect myocardial infarction by anti-inflammation and macrophage polarization *via* NF-κB signaling pathway and Nrf2/HO-1 axis. *Stem Cell Res. Ther.*, **12**, 1, 519, 2021.
406. Abubakar, M., *et al.*, Unlocking the Mysteries, Bridging the Gap, and Unveiling the Multifaceted Potential of Stem Cell Therapy for Cardiac Tissue Regeneration: A Narrative Review of Current Literature, Ethical Challenges, and Future Perspectives. *Cureus*, **15**, 7, e41533, 2023.
407. Kang, I.S. and Kwon, K., Potential application of biomimetic exosomes in cardiovascular disease: focused on ischemic heart disease. *BMB Rep.*, **55**, 1, 30–38, 2022.
408. Qiao, L., *et al.*, microRNA-21-5p dysregulation in exosomes derived from heart failure patients impairs regenerative potential. *J. Clin. Invest.*, **129**, 6, 2237–2250, 2019.
409. Chen, F., *et al.*, Bone marrow mesenchymal stem cell-derived exosomes attenuate cardiac hypertrophy and fibrosis in pressure overload induced remodeling. *In Vitro Cell. Dev. Biol. Anim.*, **56**, 7, 567–576, 2020.
410. Pang, Y., *et al.*, Embryonic Stem Cell-Derived Exosomes Attenuate Transverse Aortic Constriction Induced Heart Failure by Increasing Angiogenesis. *Front. Cardiovasc. Med.*, **8**, 2021.
411. Kervadec, A., *et al.*, Cardiovascular progenitor-derived extracellular vesicles recapitulate the beneficial effects of their parent cells in the treatment of chronic heart failure. *J. Heart Lung Transplant.*, **35**, 6, 795–807, 2016.
412. Mandaglio-Collados, D., Marín, F., Rivera-Caravaca, J.M., Peripheral artery disease: Update on etiology, pathophysiology, diagnosis and treatment. *Medicina Clínica*, **161**, 8, 344–350, 2023.
413. Saenz-Pipaon, G., *et al.*, The Role of Circulating Biomarkers in Peripheral Arterial Disease. *Int. J. Mol. Sci.*, **22**, 7, 2021.
414. Du, W., *et al.*, Enhanced proangiogenic potential of mesenchymal stem cell-derived exosomes stimulated by a nitric oxide releasing polymer. *Biomaterials*, **133**, 70–81, 2017.
415. Gong, M., *et al.*, Mesenchymal stem cells release exosomes that transfer miRNAs to endothelial cells and promote angiogenesis. *Oncotarget*, **8**, 28, 45200–45212, 2017.
416. Xue, C., *et al.*, Exosomes Derived from Hypoxia-Treated Human Adipose Mesenchymal Stem Cells Enhance Angiogenesis Through the PKA Signaling Pathway. *Stem Cells Dev.*, **27**, 7, 456–465, 2018.

417. Li, C., *et al.*, Bioengineering strategies for the treatment of peripheral arterial disease. *Bioactive Materials*, **6**, 3, 684–696, 2021.
418. Zhang, K., *et al.*, Enhanced Therapeutic Effects of Mesenchymal Stem Cell-Derived Exosomes with an Injectable Hydrogel for Hindlimb Ischemia Treatment. *ACS Appl. Mater. Interfaces*, **10**, 36, 30081–30091, 2018.
419. Hernandez, M.J., *et al.*, Decellularized Extracellular Matrix Hydrogels as a Delivery Platform for MicroRNA and Extracellular Vesicle Therapeutics. *Adv. Ther. (Weinh)*, **1**, 3, 2018.
420. Reier, P.J., *et al.*, Translational spinal cord injury research: preclinical guidelines and challenges. *Handb. Clin. Neurol.*, **109**, 411–33, 2012.
421. Domínguez-Oliva, A., *et al.*, The Importance of Animal Models in Biomedical Research: Current Insights and Applications. *Animals*, **13**, 2023. DOI: 10.3390/ani13071223.
422. Suzuki, E., *et al.*, Stem cell-derived exosomes as a therapeutic tool for cardiovascular disease. *World J Stem Cells*, **8**, 9, 297–305, 2016.
423. Dykes, I.M., Exosomes in Cardiovascular Medicine. *Cardiol Ther*, **6**, 2, 225–237, 2017.
424. van der Velden, J., *et al.*, Animal models and animal-free innovations for cardiovascular research: current status and routes to be explored. Consensus document of the ESC Working Group on Myocardial Function and the ESC Working Group on Cellular Biology of the Heart. *Cardiovasc. Res.*, **118**, 15, 3016–3051, 2022.
425. van Doorn, E.C.H., *et al.*, Preclinical Models of Cardiac Disease: A Comprehensive Overview for Clinical Scientists. *Cardiovasc. Eng. Technol.*, **15**, 2, 232–249, 2024.
426. Wang, X.L., *et al.*, Exosomes derived from human umbilical cord mesenchymal stem cells improve myocardial repair *via* upregulation of Smad7. *Int. J. Mol. Med.*, **41**, 5, 3063–3072, 2018.
427. Liu, X., *et al.*, Exosomes from mesenchymal stem cells overexpressing MIF enhance myocardial repair. *J. Cell. Physiol.*, **235**, 11, 8010–8022, 2020.
428. Charles, C.J., *et al.*, Systemic Mesenchymal Stem Cell-Derived Exosomes Reduce Myocardial Infarct Size: Characterization With MRI in a Porcine Model. *Front. Cardiovasc. Med.*, **7**, 2020.
429. Ju, C., *et al.*, Transplantation of Cardiac Mesenchymal Stem Cell-Derived Exosomes Promotes Repair in Ischemic Myocardium. *J. Cardiovasc. Transl. Res.*, **11**, 5, 420–428, 2018.
430. Arslan, F., *et al.*, Mesenchymal stem cell-derived exosomes increase ATP levels, decrease oxidative stress and activate PI3K/Akt pathway to enhance myocardial viability and prevent adverse remodeling after myocardial ischemia/reperfusion injury. *Stem Cell Res.*, **10**, 3, 301–312, 2013.
431. Yu, B., *et al.*, Cardiomyocyte protection by GATA-4 gene engineered mesenchymal stem cells is partially mediated by translocation of miR-221 in microvesicles. *PLoS One*, **8**, 8, e73304, 2013.

432. Bian, S., *et al.*, Extracellular vesicles derived from human bone marrow mesenchymal stem cells promote angiogenesis in a rat myocardial infarction model. *J. Mol. Med. (Berl)*, **92**, 4, 387–97, 2014.
433. He, J.G., *et al.*, GATA-4-expressing mouse bone marrow mesenchymal stem cells improve cardiac function after myocardial infarction *via* secreted exosomes. *Sci. Rep.*, **8**, 1, 9047, 2018.
434. Luther, K.M., *et al.*, Exosomal miR-21a-5p mediates cardioprotection by mesenchymal stem cells. *J. Mol. Cell Cardiol.*, **119**, 125–137, 2018.
435. Xu, H., *et al.*, Exosomes derived from adipose tissue, bone marrow, and umbilical cord blood for cardioprotection after myocardial infarction. *J. Cell. Biochem.*, **121**, 3, 2089–2102, 2020.
436. Ni, J., *et al.*, Exosomes Derived from TIMP2-Modified Human Umbilical Cord Mesenchymal Stem Cells Enhance the Repair Effect in Rat Model with Myocardial Infarction Possibly by the Akt/Sfrp2 Pathway. *Oxid. Med. Cell. Longevity*, **2019**, 1, 1958941, 2019.
437. Mao, Q., *et al.*, LncRNA KLF3-AS1 in human mesenchymal stem cell-derived exosomes ameliorates pyroptosis of cardiomyocytes and myocardial infarction through miR-138-5p/Sirt1 axis. *Stem Cell Res. Ther.*, **10**, 1, 393, 2019.
438. Ma, T., *et al.*, MicroRNA-132, Delivered by Mesenchymal Stem Cell-Derived Exosomes, Promote Angiogenesis in Myocardial Infarction. *Stem Cells Int.*, **2018**, 3290372, 2018.
439. Lv, K., *et al.*, Incorporation of small extracellular vesicles in sodium alginate hydrogel as a novel therapeutic strategy for myocardial infarction. *Theranostics*, **9**, 24, 7403–7416, 2019.
440. Huang, H., *et al.*, Exosomes from SIRT1-Overexpressing ADSCs Restore Cardiac Function by Improving Angiogenic Function of EPCs. *Mol. Ther. Nucleic. Acids*, **21**, 737–750, 2020.
441. Sun, L., *et al.*, Long noncoding RNA UCA1 from hypoxia-conditioned hMSC-derived exosomes: a novel molecular target for cardioprotection through miR-873-5p/XIAP axis. *Cell Death Dis.*, **11**, 8, 696, 2020.
442. Gao, L., *et al.*, Exosomes secreted by hiPSC-derived cardiac cells improve recovery from myocardial infarction in swine. *Sci. Transl. Med.*, **12**, 561, 2020.
443. Tan, F., *et al.*, Clinical applications of stem cell-derived exosomes. *Signal Transduction Targeted Ther.*, **9**, 1, 17, 2024.
444. Jayaraman, S., *et al.*, Stem Cell-Derived Exosomes Potential Therapeutic Roles in Cardiovascular Diseases. *Front. Cardiovasc. Med.*, **8**, 2021.
445. Babaei, M. and Rezaie, J., Application of stem cell-derived exosomes in ischemic diseases: opportunity and limitations. *J. Transl. Med.*, **19**, 1, 196, 2021.
446. Zhang, H., *et al.*, The therapeutic efficacy and clinical translation of mesenchymal stem cell-derived exosomes in cardiovascular diseases. *Biomed. Pharmacother.*, **167**, 115551, 2023.

447. Tessier, S.N., *et al.*, Isolation of intact extracellular vesicles from cryopreserved samples. *PLoS One*, **16**, 5, e0251290, 2021.
448. Tzng, E., Bayardo, N., Yang, P.C., Current challenges surrounding exosome treatments. *Extracell. Vesicles*, **2**, 100023, 2023.
449. Li, G., *et al.*, Current challenges and future directions for engineering extracellular vesicles for heart, lung, blood and sleep diseases. *J. Extracell. Vesicles*, **12**, 2, e12305, 2023.

6

Cardiovascular Renaissance: Stem Cell Strategies for Heart Diseases

Apurva[1], Pracheta Janmeda[2], Divya Kumari[2], Narotam Sharma[3] and Priya Chaudhary[1]*

[1]Department of Biotechnology, School of Applied and Life Sciences, Uttaranchal University, Dehrdaun, Uttarakhand, India
[2]Department of Bioscience and Biotechnology, Banasthali Vidyapith, Banasthali, Rajasthan, India
[3]Department of Biotechnology, DNA Labs-A Centre for Applied Sciences, Dehrdaun, Uttarakhand, India

Abstract

Cardiovascular illness is continuing on the rise despite significant medical advancements, with ischemic heart disease being the main reason for mortality and disability globally. Therefore, significant efforts are still being made to develop methods of therapy that would enhance an individual's quality of life and chances of survival. Current research is looking into novel treatments that can both repair and regenerate the heart while also shielding the myocardium from ischemia-reperfusion damage. Additionally, to address the molecular foundation of heart conditioning, stem cell treatment which may include the application of human mesenchymal stem cells, induced pluripotent stem cells, and exosomes will enable the discovery of new treatments for ischemic cardiac disease. Clinical research had mixed outcomes, and there has been little evidence of stem cell-based cardiac remodeling and performance enhancement. Clarifying the underlying processes driving the positive effects of stem cell transplantation is essential to optimizing stem cell efficiency. Several ways to improve stem cell repair efficacy and create the "next generation" of stem cell therapies have been devised by researchers based on these principles. Furthermore, it could be challenging to offer a consistent therapeutic approach to all patient categories because cardiovascular illnesses are complicated disorders with multiple disease patterns and pathologic causes. Future tactics should therefore focus on more individualized stem cell (SC) therapy,

Corresponding author: priyachaudhary@uumail.in

Ayush Madan, Sumel Ashique, Deepika Arora and Mantosh Kumar Satapathy (eds.) Stem Cell Therapeutics, (309–350) © 2025 Scrivener Publishing LLC

where the best cell type, dosage, and delivery method are chosen based on specific illness factors.

***Keywords*:** Cardiovascular disease, stem cell therapy, ischemia–reperfusion injury, pluripotent stem cell, mesenchymal stem cell

6.1 Introduction

Globally, heart disorders are now one of the major reasons of fatality. Idiopathic dilated cardiomyopathy, coronary artery disease, hypertension, and ischemia are all prevalent reasons for cardiac failure. While left ventricular (LV) systolic function is primarily unchanged, diastolic deterioration affects up to 40% of individuals with cardiac disease. For those with cardiac disorders, there are proven medical and surgical methods accessible, but there has not been a great response to these conventional therapy techniques [1]. Since cardiac disorders are also linked to various other issues, the proportion of patients seeking a successful course of therapy that can maintain their regular health has been steadily rising. Alternative remedies are desperately needed to meet the currently enormous demand. Cardiomyocytes are harmed by cardiac illnesses, which impact the heart muscles. Scar development may result from these injuries, which can ultimately lead to an excess of blood flow and extend the cardiac cells. Heart attack and frequent death result from this in the end [2]. Therefore, the present scenario of affairs is generally beneficial for alternative therapies that may significantly enhance the disease, at least in terms of longevity or comfort of life [3].

Primitive cells with the capacity for multifaceted development and self-replication are known as stem cells. Stem cell therapy, sometimes referred to as stem cell transplantation, involves injecting stem cells locally or systemically into a particular area of the human body in order to heal injured or deteriorating tissue [4]. According to Fernández-Lázaro *et al.* [5], aberrant lesions and tissue death within certain organs are the root cause of many human disorders.

The health of a cell can be restored by the introduction of stem cells into the areas that are injured which will enhance the functioning of organs and ultimately result in the cure of illnesses. There is hope that alternative treatments have the potential to treat some of the serious ailments such as cardiovascular conditions, autoimmune disorders, neurological system diseases, blood system diseases, and metabolic disorders [6].

Thus, this chapter aims to give a brief summary of the latest information available about the potential applications of stem cells as a strategy for the cure of cardiac disease. The stem cells are also capable to tackle basic scientific issues that are associated to human health, comprising their difficulties, constraints, and abilities along with their hopes for the future.

6.2 Stem Cell Types Considered for the Treatment of Heart Disease

6.2.1 Skeletal Myoblast

A novel option for therapy that aims at boosting the dismal prognosis of those with heart failure is the transplantation of cells. The idea behind it is that post-infarction wounds could be effectively rejuvenated by implanting contractile cells into them. Autologous skeletal myoblasts have been the first to be investigated for a clinical usage, mostly because to practical considerations. Strong myoblast engraftment, their in-scar conversion into myotubes, and the corresponding enhancement of left ventricular function have all been repeatedly shown in a wide number of experimental studies [7].

Nonetheless, it is currently evident that the myoblast transfer result could probably be enhanced by maximizing the cell yield's purity (by choosing muscle-derived progenitors that are less lineage-committed than the myoblasts), the delivery method (by improving the precision of cell injections while reducing their intrusion), and the established cells' survival (by simultaneously vascularizing the graft and incorporating the cells into three-dimensional matrices) [8].

Skeletal myoblast transplant has now made its way into the clinical setting following 10 years of research as a possible novel way to enhance the deteriorating heart's function. It is more probable that the functional advantages of myogenic cell transplantation are associated with limiting detrimental post-infarction remodeling and/or paracrine impacts on recipient tissue rather than a synchronous role of the graft to systolic pump operating because of the apparent absence of relationships between the engrafted myoblasts and the host cardiomyocytes [7].

It is therefore crucial to determine if the physiological advantages seen in the research context transfer into significant increases in cardiac contractility and final patient results since the early clinical investigations mainly confirmed the procedure's viability. Randomized studies are the only way to adequately examine efficacy. The potentially pro-arrhythmic risk of myoblast engraftment, which may be connected to the different electrical characteristics of donor and recipient cells, should also be made clear by these investigations [9]. Meanwhile, more research is still required to tackle a few important issues, which include how to best optimize myoblast survival after transplantation, how to create simpler cell transfer transportation methods, how myoblast transplanting functions in non-ischemic cardiac failure settings, and how skeletal myoblasts contrast to other possible treatment options [10].

Stem cells offer a different therapeutic approach for the infarcted heart by making up for the deprivation of cardiomyocytes that occurs after myocardial damage. Emerging novel breakthroughs in stem-cell-based regenerative therapies include the existence of progenitor and resident stem cell types in the heart and genetic reprogramming of resident cells with genomic activation of pluripotency markers. However, bone marrow cells and myoblasts collected from skeletal muscle continue to be the most thoroughly researched donor cell categories for repair and regeneration of myocardial until the viability and safety of these cells are determined by comprehensive testing in *in vitro* and in living models [7].

6.2.2 Bone-Marrow-Derived Mononuclear Cells

Preclinical and clinical research have been conducted for more than 20 years to evaluate the management of cardiovascular illnesses utilizing bone marrow mononuclear cells (BM-MNC) [11]. The potential of BMMCs to enhance cardiac healing and help individuals with severe ischemic heart disease and cardiac failure has been studied. The investigation which was done on a specific area within 2006–2011 reported the efficacy and safety of intramyocardial injections of BM-MNCs in combination with the bypass of coronary artery [12]. Over the time period of 10 years of treatment, 18 individuals passed away, and about 21 individuals were called for participation. Among the initial and after the 10 years of follow-up, the cardiac magnetic resonance imaging revealed a 2.1-fold higher decrease in the recovery rate including low fibrosis or scarring which followed after the BM-MNC treatment as compared to the control [13].

6.2.3 BM-Derived Endothelial Progenitor Cells (EPCs)

The generation of endothelial progenitor cells (EPCs) from the blood of fetus has been found to restore the reconstitution of hematopoietic cells after the transplantation. One of the well-distinguished sources for progenitor/stem cell therapy in cardiac ailment is the bone marrow (BM) [14]. According to different pre-clinical and clinical research, EPCs have been involved in the process of angiogenesis and the healing mechanism of vascular endothelium. Any alteration in the micro-environment and aging process adversely affects the EPC and results in the dysfunctioning of LV [15]. Similarly, Rurali *et al.* [16] reported that the positive effect of EPC administration in young animals was observed to be minimized in case of elderly individuals who were treated for ischemic cardiac ailment. Thus, growing older is determined to be a major factor that limits the efficacy of EPC-derived treatment in various investigations.

6.2.4 BM-Derived Mesenchymal Stem Cells (MSCs)

Among the several cell types employed in these investigations, bone marrow (BM)-derived mesenchymal stem cells (MSCs) are the best-characterized stem cells during the past 20 years and have demonstrated great potential for cell-based heart therapy. As part of their paracrine function, MSCs are strong secretors of bioactive substances that are both soluble and insoluble [11]. MSCs are almost perfect for cell-based therapy because of their accessibility, capacity for multilineage distinction, favorable immunogenicity because they are not able to express main histocompatibility complex II (MHC II), and immunomodulatory qualities [17].

Numerous studies conducted for both cardiac and non-cardiac illnesses have demonstrated both the immediate and long-term security of autologous BM stem cell (BMSC)-based treatment for usage by humans since the ground-breaking work that reported the beneficial effects of BMSC-based treatment in five individuals as a supplement to coronary artery bypass graft [14]. The therapeutic effectiveness of the BM-MSCs-based intervention, however, has been inconsistently reported in published data from clinical trials; the results range from negligible effects to markedly enhanced cardiac function. The quality of the cell preparations utilized in the trials and their diversity in terms of end-points and assessment techniques are blamed for this inconsistency [17].

6.2.5 Mobilized Stem and Progenitor Cells

In steady individuals with angiographic indications of CAD, EPCs have been reliably linked to cardiovascular prognosis and death. Overall, there is a roughly twofold rise in the risk of significant cardiovascular events when baseline EPC numbers are aberrant. Lower numbers of $CD34^{+}KDR^{+}$ cells were linked to a shorter-term chance of heart disease. Conversely, in later studies, individuals with stable angina who received percutaneous coronary intervention (PCI) and had greater baseline EPC counts were linked to a worse long-term prognosis. Therefore, an extended release of circulating EPCs could be seen as the arteries' defensive mechanism trying to make up for stronger pathogenesis factors causing atherosclerosis [18].

In addition to their biological significance, EPCs have been thoroughly investigated as a potential cardiovascular risk biomarker prototype. There is growing evidence that endogenous progenitor cells (APCs) may considerably enhance the long-term stratification of patients with heart disease, despite the fact that the exact methods of action of these cells are yet unknown [19].

A patient-level review of information from five longitudinal investigations discovered that when baseline circulating progenitor cell count, comprising EPCs, was incorporated to a standard risk evaluation model, it increased the differentiation between individuals with a future cardiovascular event. In recent years, a combination of numerous plasma indicators based on inflammation, tissue remodeling, renal failure, and liver fibrosis was found to be predictive of cardiovascular prognosis. More studies are required to determine whether a multimarker strategy using EPC count and machine learning could be an intriguing strategy for improving risk categorization in CAD [20].

6.2.6 Adipose-Derived Stem and Progenitor Cells (ASPCs)

White adipose tissue consists of a number of anatomic categories over the entire body (such as visceral versus subcutaneous) that are made up of loosely connected tissue and have adipocytes as the primary unit [21]. ASPCs are characterized as the SVF devoid of red blood ($TER119^+$, $CD235a^+$), endothelial ($CD31^+$), and immunological ($CD45^+$) cells. As a result, resolving the cell hierarchy inside ASPCs has long been a field-wide difficulty. According to studies, a particular group of ASPCs is responsible for adipogenesis, serving as the "real" adipose stem cells, while another type of ASPCs is in a temporary phase, already associated to becoming adult adipocytes [22].

To tackle this idea, the first thing to determine is the degree to which ASPCs are cellularly diverse, as well as whether discrete immunophenotypic and molecular ASPC cell subtypes exist. Other concerns are how the ASPC cell landscape varies across subcutaneous and visceral adipose deposits and how much these variances represent the depots physiological features. Finally, addressing ASPC diversity is important in the setting of unrestrained fat build-up, considering the benefits of adipocyte hyperplasia in comparison to hypertrophy in case of obesity along with the disadvantages of visceral over fat expansion of the subcutaneous region [23]. However, one significant disadvantage of this sequencing research is that they are restricted in scope, as they only investigate very less gene markers at a time, as will be explained later. Due to this genome-wide tests, for instance, scRNA-seq provides a new viewpoint on the heterogeneity of particular biologic processes, namely, ASPCs, yet scRNA-seq is not a simple procedure, particularly in respect to interpreting enormous information sets. This is why, for instance, Internet programs were created to make accessible scRNA-seq analysis [24].

6.2.7 Cardiac Stem and Progenitor Cells

As the study and comprehension of the cardiovascular adverse reactions of anticancer therapy advances and a patient population develops, particularly those who are bearer of childhood malignancies, it is critical to address the complete profile of impacted myocardial cell types. While the major effects of the toxins on cardiac myocytes cause serious acute harm over time, the myocardial capability to adapt or heal from this loss becomes an increasingly important parameter of the illness phenotype. One component is the possibility of spontaneous repair and regeneration of myocardia, which can be hindered by cardiotoxins that deplete the stem cells that are crucial to this mechanism [25].

Strong proof appears for the production of new cardiomyocytes in adult human myocardium, as well as the recognition of progenitor cell/endogenous stem populations with pro-regeneration capabilities. Cardiotoxins that affect any of these mechanisms will have a negative effect on the diagnosis in the long run. While the function of progenitor/cardiac stem cells in cardiomyocyte regeneration seems to be restricted (although there is more powerful proof of this mechanism in the context of loss of diffused cardiomyocyte), there are substantial signs of a pro-regenerative function *via* their encouragement of damaged cell survival [26].

A variety of recent investigations have found that anticancer medicines have a deleterious influence on progenitor cells/cardiac stem, with unfavorable impact observed for both widespread chemotherapeutic drugs including doxorubicin and newer, less obviously cardiotoxic compounds like inhibitors of tyrosine kinase. Defile effects are found not only on cell number and sustainability but also on the capacity of cells to sustain the myocardium by producing pro-survival differentiated cells and secretome. Managing anticancer medication adverse effects has been a vital constituent of oncology care since its inception, with cardiotoxicity being one of the most common [27].

As chemotherapy for cancer regimens grow more successful and individuals' long-term survival rates rise, avoiding cardiotoxicity and managing it in the long run become increasingly important. Two changes in health maintenance remain to drive this: one favorable and the other bad. The initial goal is to improve carcinoma-related viability and another fact is that cardiovascular ailment is still very common in the general community, and survivors of cancer are no exception. In actuality, because the mechanism of cardiotoxicity caused by various anticancer medications includes cardiac ischemia or coronary artery harm, as well as the fact that cardiotoxicity can appear in survivors of cancer only in the long run, the risk is expected to be significantly higher [28].

Microvascular damage is an initial indicator of various cardiac disorder procedures, and research into its role in cardiotoxicity has led to a greater awareness of the fundamental pathology, which extends outside the causes of cardiomyocyte injury. Doxorubicin *ex vivo* treatment of vessel specimens of human cardiac microvascular results in a substantial decrease in flow-directed narrowing replies in both pediatric and adult vessels, though insignificantly in pediatric samples, as well as loss of acetylcholine-directed vessel compression responses. Tyrosine kinase inhibitors (TKIs) can potentially cause microvascular damage, as demonstrated by one TKI with clinically recognized cardiotoxicity [29].

6.2.8 Embryonic Stem Cells

Subsequently, the development of hESC lines into distinct progenitor cells demonstrated its potential for treating various terminal diseases. However, this intriguing topic has continuously demonstrated significant moral and scientific constraints. This study provides an overview of clinical trial research involving hESC, including their benefits, constraints, and other unique considerations. Some of the drawbacks of hESC have been addressed, and various research investigations are currently underway, so that current studies have aimed to improve the therapeutic uses for hESC, particularly in macular degeneration and neurodegenerative illnesses [30]. However, there are some major concerns that require further investigation and discussion in relation to hESC therapy. Despite extensive research, hESC-based treatment is not yet ready for typical clinical applications, and additional research and information are required to address present clinical and ethical restrictions. When all of the drawbacks of ESC are solved entirely, hESC may become better to other stem cell sources [31].

6.2.9 Induced Pluripotent Stem Cells

Induced pluripotent stem cells (iPSCs) provide a new chance to investigate human health and illness at the cellular level. They also have a chance to be used in precision medicine practices, such as tailored drug testing [32]. iPSCs allow researchers to investigate cells that are biologically matched to specific individuals, while genome-editing technologies enable the creation or repair of genetic abnormalities [33].

Human pluripotent stem cells can be considered as a viable option for therapeutic purposes as it include embryonic stem cells and induced pluripotent stem cells as the self-regenerating progenitors that are capable in giving rise to endoderm, mesoderm, and ectoderm lineages. Various studies reported the differentiating ability of human ESCs into oxygen carrying functioning erythrocytes. However, *ex vivo* culturing of RBC from the human ESC raised an ethical concern [34].

Human iPSCs may be an appropriate treatment option for the *ex vivo/in vitro* generation of red blood cells. Reprogramming the human somatic cells *via* ectopic production of some of the transcription factors (NANOG, LIN28, c-MYC, KLF4, SOX2, and OCT4) has opened up novel possibilities for modeling diseases and regenerative therapy. Several strategies have been established for producing enucleated RBCs by the help of human iPSCs. *In vitro* synthesis of human iPSC-derived RBCs may provide an alternate therapy choice for individuals with blood diseases [35].

6.2.10 Adult Stem Cell

Pharmacological interventions are still the most common interventional technique for slowing development and controlling symptoms, although cellular therapies offer an opportunity to cure or even induce remission of cardiovascular ailment [36]. The attempt to repair compromised work and recover damaged tissue in the failing heart has resulted in research studies using adult stem cells over the last 15 years, building on the effective usage of stem cells since the finding of incredible potential for regeneration in the latter half of the 20th century. Numerous clinical studies utilizing diverse adult stem cells with varying trial designs underscore the critical requirement for innovative approaches to both stem cell treatment and clinical trial design in order to progress cardiac regenerative medicine [37].

6.3 Stem Cell Therapy Against Different Heart Diseases

Infusion of embryonic stem cells into a damaged heart starts myocardial restoration through a variety of direct and indirect processes, such as distinction into cardiac and vascular cells, cell nuclear fusion, and paracrine

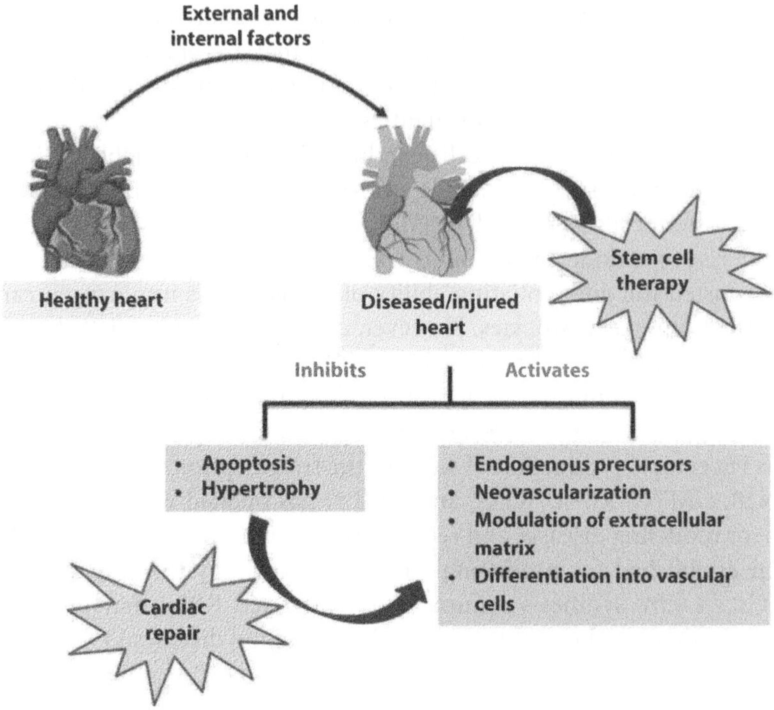

Figure 6.1 Potential therapeutic mechanism of stem cells.

consequences that include endogenous precursor stimulation, neovascularization advancement, extracellular matrix modification, apoptosis restriction, and hypertrophy repression. Collectively, these occurrences increase LV function, perfusion, and cardiac output, resulting in an improvement in medical condition [38] as shown in Figure 6.1 and Table 6.1.

6.3.1 Acute and Chronic Coronary Syndromes

Uric acid (UA) is the final result of the breakdown of indigenous and foreign purine nucleotides. While the connection with articular gout and kidney illness has long been known, fresh research has shown that UA is also linked to cardiovascular (CV) problems. UA has been discovered as a major predictor of a variety of results, including all-cause and cardiovascular mortality, as well as CV events (mostly acute coronary syndromes (ACS) and strokes). In addition, UA has been linked to a risk of coronary artery disease, increased mortality in degenerative individuals, and the emergence of atrial fibrillation [39].

Table 6.1 Pre-clinical and clinical trials of stem cell therapy in heart failure.

Stem cell type	Disease model	Delivery route	Outcome	Side effects	References
Pre-clinical studies					
Human pleuripotent stem cell (hPSC)	Mice, rats, pigs, and monkey	hPSC injection into the myocardium	Improved the function of heart	Post-transplantation arrhythmias and immature rejection	[58]
Induced pleuripotent stem cell-derived cardiomyocytes	Ishemic heart disease animal models	Intra-coronary injection	Cell-based treatment reduced the fibrosis area of left ventricle which ultimately reduces the left ventricular end-diastolic and end-systolic volumes	No major alterations in the fatality and chances of arrhythmia	[59]
Adipose-derived stem cells (ADSC)	Rat	Injection into the intraperitoneal cavity	Injected ADScs penetrates and migrates to the myocardium which has a therapeutic impact on the heart	No such major mortality observed	[60]
Adipose-derived stromal cells (ASCs)	Rat	Alginate hydrogel	Resulted in improvement in function after the myocardial infarction	Retention problem of the alginate hydrogel	[61]

(*Continued*)

Table 6.1 Pre-clinical and clinical trials of stem cell therapy in heart failure. (*Continued*)

Stem cell type	Disease model	Delivery route	Outcome	Side effects	References
Induced pleuripotent stem cells	Mice	Injection into the intramyocardial	Improves the function of left ventricular and result in neovascularization along with amelioration of hypertrophy and apoptosis	No such major mortality observed	[62]
Clinical studies					
Human embryonic stem cell-derived cardiovascular progenitors	ESCORT	Fibrin patch	Increases the systolic motion of the treated area in all patients	Silent alloimmunization	[63]
Intracoronary cardiosphere-derived cells	CADUCEUS	Infused through an over-the-wire angioplasty catheter	Reduction in the scar mass along with the increment in regional contractility, heart mass, and regional systolic wall thickening	Chest pain and acute myocardial infarction	[64]
Mesenchymal precursor cells	Randomized study	Injection into the intramyocardial region	Results in cardiac recovery	Results in weaning	[65]

(*Continued*)

Table 6.1 Pre-clinical and clinical trials of stem cell therapy in heart failure. (*Continued*)

Stem cell type	Disease model	Delivery route	Outcome	Side effects	References
Adipose tissue-derived mesenchymal stromal cells	A European multicenter, double-blind study (SCIENCE trial)	Injection into the intramyocardial region	The study demonstrated the safety of cell-based therapy in individuals with cardiac problems	No major alterations were observed within the groups	[66]
$CD133^+$ cells	Experimental study; parallel, open-label, single-blinded design	Injected stem cell	Observed to have a major improvement in scar size, wall motion, and ejection fraction along with a remarkable rise in the level of endothelial growth factor such as VEGF	During treatment, one individual died due to hemodynamic instability after the fasciotomy	[67]
Pluripotent stem cell-derived engineered human myocardium (EHM)	BioVAT-DZHK20 trial	Mini-thoracotomy	Engraftment of EHM was found to be safe and feasible for the cure of cardiac failure	Resulted in the death of an individual due to progressive cardio-renal failure	[68]
Intracoronary administered autologous BM derived MSC	MSC-HF trial phase II	Injection into the intramyocardial region	Resulted in improvement of the mass and myocardial function in individuals with heart failure	No such mortality was observed	[69]

UA has long been associated with articular gout and kidney failure, but new research has shown that it is also linked to cardiovascular (CV) diseases. In fact, UA has been discovered as a substantial predictor of several different cardiovascular outcomes, including all-cause and CV mortality as well as CV events (mostly acute coronary syndrome, ACS) and stroke [40].

In addition, UA is related with a raised risk of heart failure (HF), a greater mortality rate in this category of individuals, and the occurrence of atrial fibrillation. All of these noteworthy results prompted the most recent European Guidelines on Arterial Hypertension to include UA among the CV risk variables that should be evaluated to divide a patient's risk. Hyperuricemia is an epidemiologic concern, especially when cardiovascular conditions are present. The incidence of it varies from 6% in healthy people to 14% in hypertensives, with an alarming rise to 23% among those suffering from ACS and chronic coronary syndrome (CCS) [41].

6.3.2 Myocardial Infarction

Periodontal illness is usually characterized as an infection impacting the tissues that protect and sustain the teeth (i.e., the periodontium and gum). According to latest data, the frequency of this issue is steadily increasing across the globe, creating serious medical worries not just about local issues caused by bad oral health but also about the possibility of discovering systemic disorders. Taken jointly, the currently published data tends to show a strong relationship within CHD and periodontitis, whereas the likelihood of acute ischemic heart attacks can be seen to be increased in individuals with pre-existing coronary artery disease [42].

This epidemiological connect is backed by accurate biological proof, which shows that periodontal conditions may unfavorably regulate the risk of cardiovascular disease. Individuals with periodontitis are more likely to be overweight, have hypertension, dysfunction of the endothelial system, dyslipidemia, and platelet hyperreactivity, and may be prothrombotic. Aside from these key atherogenic variables, the transfer of periodontal bacteria into the blood circulation and subsequent aggregation inside atherosclerotic plaques could lead to increased plaque destabilization and the likelihood of acute ischemic coronary events [43].

6.3.3 Ischemia–Reperfusion Injury

Acute myocardial infarction is determined by an unexpected coronary artery obstruction, which causes ischemia in the affected myocardial region and, in most cases, myocardial necrosis. Without recovery of coronary perfusion, myocardial scarring will result in unfavorable cardiac remodeling and cardiac failure, yet reperfusion by itself causes an increased and expedited damage in the myocardium referred as ischemia–reperfusion (I/R) injury [44]. Inflammation contributes to this process *via* a number of interconnected mechanisms. Several pharmaceutical and percutaneous treatment techniques (ischemic conditioning) have shown promise during I/R in preclinical animals but have been infamously unsuccessful in human trials [45].

As reperfusion therapy became more frequently employed among individuals with ischemia-induced myocardial, acute myocardial infarction damage was significantly decreased, but reperfusion-induced cardiac injury became more apparent. Cardiac ischemia–reperfusion (I/R) injury is distinguished by microvascular perfusion abnormalities, activation of platelets, and mortality of sequential cardiomyocyte caused by further ischemic episodes during the reperfusion phase. Microvascular blockage, described as a no-reflow condition, influences the infarct area, myocardial operation, and perioperative death [46].

Cardiac microvascular endothelial damage can happen significantly early and with more serious consequences than cardiomyocyte damage. Endothelial cells have very less mitochondria in comparison to other heart cells, and various pathogenic changes occur following cardiac microvascular I/R injury. Although mROS are essential biological second messengers, elevated amounts cause oxidative damage, endothelial aging, and apoptosis [47].

Mitochondrial factors, such as mitophagy, fusion, and fission, influence the form, dispersion, size, and activity of mitochondria. This process to adaptation alter extracellular signals while orchestrating intracellular activities like cell growth, permeability transition, movement, metabolic processes, angiogenesis, sticky molecule communication, endothelial barrier functioning, and anticoagulation [48].

6.3.4 Ischemic Cardiomyopathy

Ischemic cardiomyopathy is defined as systolic LV dysfunction in the presence of obstructive coronary artery disease (CAD) and is the leading reason for heart failure globally. This has important pharmacological and prognostic significance since damage of elasticity in hibernating myocardium may be reversible once ischemia is resolved. Individuals with ischemic cardiomyopathy may experience some level of heart failure features, regardless of the underlying cause [49].

Enhancements in acute myocardial infarction therapy, revascularization methods, and guideline-directed medical treatments have resulted in a reduction in the frequency and extent of systolic malfunction and detrimental remodelling after AMI, but the chance of HF stays high, implying that other processes besides LV systolic dysfunction and reconstruction may also be involved in HF symptoms following AMI [50].

The extent of the infarct after an AMI is the best indicator of heart failure and modification. Greater size of infarct because of late presentation or reperfusion, no or little collateral passage, or anterior placement is linked to increased dysfunctioning of LV, unfavorable cardiac transformation, and HF with time. In an individual-level meta-analysis study which includes individuals receiving primary percutaneous coronary intervention (PCI), size of infarct was highly related with later death and hospitalization for heart failure [51].

6.3.5 Ischemic Heart Failure

Cardiac failure is one of the intricate problems that deal with mortality or diseases. Ischemic heart disease is one of the reasons of heart failure and it is also associated with various other coronary diseases. This disease is characterized by the formation of one or more than that obstructive plaque that restricts the flow of blood to the coronary artery and ultimately results to myocardial ischemia or cardiac failure. So, coronary blockage is just one factor of the complicated pathological process that could be responsible for myocardial ischemia [52].

Coronary dysfunction is defined as the process in which there is a lack of regulation to the coronary blood flow like ion channels for the fulfilment of energy need of the myocardium which ultimately results to tissue death, fibrosis, and hypoxia. Thus, these ion channels may consider being the linkage between the ischemic heart disease, microvascular problem, and heart failure [53].

Coronary artery disease (CAD) results in a decrease in supply of oxygen to the myocardium, impairing contractions of the heart and relaxation. In this regard, "modern" cardiology concentrates heavily on the investigation of epicardial atheromatous plaque, including its origin, mitigation, and diagnosis and therapeutic evaluation [52]. A decrease in a coronary artery diameter of ≥50% restricts its greatest vasodilatory ability, while a decrease of ≥85% results in a drop in flow even at rest. The pathophysiological link between heart failure, myocardial ischemia, and obstructive atherosclerosis of the epicardial coronary arteries is now well understood.

Angiographic results corroborate the link between the seriousness and severity of coronary atherosclerosis and mortality [54].

A number of studies, however, imply that coronary blockage is only one component of an intricate multifactorial pathologic process that results in myocardial ischemia. In addition, it is recognized that anomalies in both the functioning and the makeup of the coronary microcirculation are associated with a variety of medical disorders. There is inadequate fascination with coronary microcirculation and its pathophysiological significance for medical care [55].

6.3.6 Modeling Heart Diseases on a Chip

Heart disease places a tremendous strain on worldwide healthcare systems and is the major cause of fatality each year. For a deeper comprehension of cardiac conditions, we need superior disease models. Historically, researchers used 2D monolayer frameworks or animal specimens of heart failure to study pathophysiology and medication responses [56].

Heart-on-a-chip (HOC) technology is a growing area in which cardiomyocytes and other heart cell types can be utilized to create useful, beating cardiac microtissues that mimic many aspects of the heart of a person. HOC models are demonstrating considerable potential as disease simulation platforms and have the potential to be crucial tools in the process of developing drugs. Diseased HOCs are strongly tuneable thanks to developments in human pluripotent stem cell-derived cardiomyocyte biology and microfabrication technological advances. They can be produced in a variety of ways, including *via* cells with defined genetic mutations (patient-derived cells), adding tiny substances, modifying the cells' surroundings and changing the cell ratio/composition of tiny tissues [57].

6.4 Mechanism of Stem-Cell Based Therapy for Heart Disease

6.4.1 Differentiation of Transplanted Stem Cells into Cardiac Cells

As of now, mesenchymal stem cells (MSCs) are the most broadly investigated stem cells utilized in clinical settings and as a test subject for cellular treatment, with a focus on cardiac regeneration and repair. Numerous methods for differentiation have been widely employed in preclinical studies ever since the finding of cardiomyogenesis activation in MSCs. The lack of research studies involving pre-differentiated MSC-derived cardiomyocytes, however, highlights the limitations and contradictions in their application as an alternative source of derived cardiomyocytes for implantation to restore impaired heart function [70].

The methods for getting cells to the injured heart were then covered, along with the importance of cell survival for the therapy's overall success. Researchers have also discussed the range of MSC distribution methods, both local and systemic, as well as the application of biomaterials to improve the overall effectiveness and performance of delivery. Cardiovascular diseases (CVDs) are recognized worldwide as major contributors to death and disability. Due to the ineffectiveness of therapies for myocardial necrosis, typical therapy options for CVDs, such as pharmaceutical medicines, impose major hurdles [71].

On the other hand, medicines based on stem cells (SCs) are thought to be potential methods of treating CVDs. Cardiomyocytes are distinguished from SCs in these methods. The technique should be used to produce cardiomyocytes with a highly developed shape and efficient functions in order to realize the full potential of SCs. A large-scale and medical-grade cardiomyocyte production is required for heart-repairing technologies. However, there are several difficulties, including the possibility of graft cell death, cancer, arrhythmogenesis, and immunological rejection [72].

6.4.2 Formation of New Blood Vessels from Transplanted Stem Cells

The procedure by which MSCs stimulate neovascularization requires several phases. First, when ischemia develops as a result of stress, MSCs can detect the alterations and are recruited from their niches to relocate and attach to ischemic tissue, allowing them to multiply and specialize. Notably, MSCs create a variety of substances, comprising chemokines and growth factors, which perform a paracrine activity throughout the neovascularization process. All biological operations rely on the cooperation of several types of cells, and neovascularization necessitates the cooperation of ECs, endothelial progenitor cells (EPCs), and pericytes. Furthermore, MSC-derived exosomes serve as messengers for cell-to-cell communication [73].

6.4.3 Paracrine Effect

Cytoprotective endothelial progenitor cells (EPCs) are crucial for neovascularization and vascular restoration. It is still unknown yet exactly how

EPCs encourage vascular healing. There currently exist two schools of thought about how endothelial cells (ECs) can be directly differentiated and integrated into wounded vessels by EPCs or how EPCs can influence on blood vessels and cells by producing paracrine chemicals. However, more data currently point to the latter. Thus, more research on the paracrine processes of EPCs is worthwhile in order to back up the theory that EPCs restore blood vessels by means of paracrine effects instead of incorporating right away into impaired vessels [74].

6.4.4 Cell Fusion

The main cause of death worldwide is cardiovascular illnesses, among which aging is a significant risk element. Abdominal aortic aneurysm, hypertension, heart regeneration, hypertrophy, pressure overload-induced cardiac aging, atherosclerosis, and acute myocardial infarction are among the cardiovascular diseases that cell senescence has drawn possible consideration as an ideal target for avoiding recently. The processes of cell death are examined here, along with the function of senescent heart cells in age-associated cardiovascular disorder [75]. Next, the targets that control cellular senescence in cardiovascular disorders are emphasized: SIRT1 and mTOR. Researchers also go over the data showing that, depending on the circumstances, senescent cells may have advantageous or harmful effects on cardiovascular illnesses [76].

6.4.5 Treatment by Extracellular Vesicle-Derived Stem Cells

MSC-derived extracellular vessels (EVs), which include microvesicles and exosomes, act as important carriers of bioactive compounds such as microRNAs, proteins, and lipids, which can have a major impact on the tissues and cells of an individual. These EVs loaded with various types of molecules such as microRNAs are responsible for the expression of different genes which results in the modification in the fate of cells by improving angiogenic capacity, decreasing apoptosis, and survival. This ability of gene expression regulation is also crucial for the process of tissue repair, mainly damaged and ischemia heart by the regeneration of cellular function and vascular supply [77].

EV cells generated from the human $CD34^+$ stem cells were reported to have proangiogenic activity which indicates its ability to form new blood vessels at the site of injured tissue [78]. Luo *et al.* [79] conducted the research on the complicated composition of vesicles and their research revealed the variety of substances such as nucleic acid, lipids, and proteins found in the EVs responsible for the regeneration capacity.

6.5 Gene-Modified Stem Cells

Another method for increasing transplanted cell viability and integration is to genetically modify stem cells to excessively produce certain growth factors, anti-apoptotic proteins, or pro-survival genes as shown in Figure 6.2.

6.5.1 Increase the Capacity of Antiapoptotic

MSCs have been extensively utilized in the area of stem cell treatment in over a thousand clinical studies as well as numerous *in vivo* and *in vitro* investigations. These are multipotent stem cells that need to fulfill some minimal requirements, like adhesion to plastic, surface marker expression,

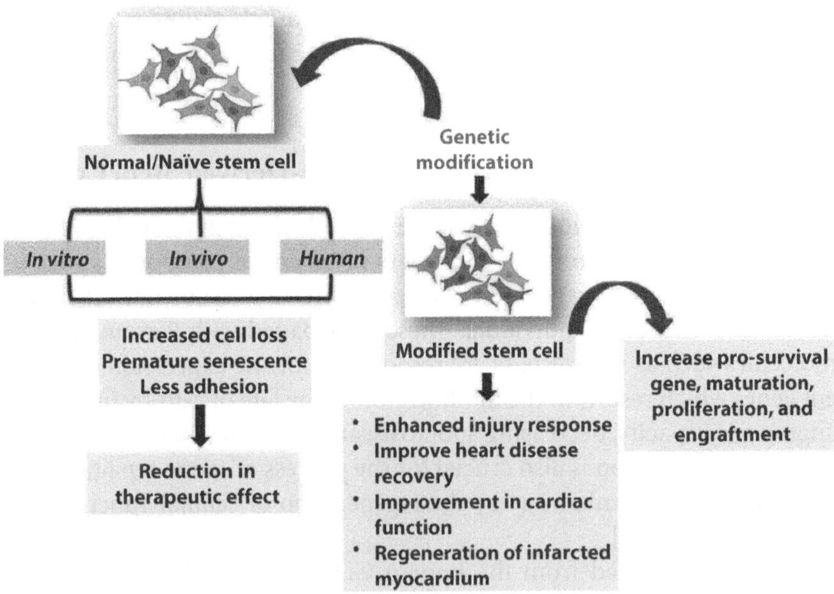

Figure 6.2 Gene-modified stem cells for the treatment of cardiovascular diseases.

and the capacity to develop into chondrocytes, osteocytes, and adipocytes [80].

One benefit of using MSCs over other stem cell types like induced pluripotent and embryonic stem cells is that their application is thought to be safer in terms of the potential for development of tumors. Moreover, it is very easy to extract these cells from the various allogeneic and autologous sources [81]. These MSCs are capable in releasing a variety of proteins associated with regeneration and revascularization of tissues along with immune-modulation and anti-apoptotic activity. MSCs demonstrated to have beneficial properties, yet their transplantation was not able to produce the desired outcome as reported by some studies [82].

The undesirable micro-environment of injured tissue may be responsible for the low survival and multiplication of transplanted MSCs. Additionally, the expression of some variables may contributes in achieving desired outcomes in the targeted disorder with some favourable effects. Thus, strategies to increase the MSCs transplantation associated effectiveness are required to develop for getting the desired outcomes [83]. These MSCs need to be genetically modified further in order to refine the therapeutic efficacy of MSCs and the outcome of following transplantation. Depending on the intended outcome such as increased rate of survival, proliferation, and pro-regenerative activity, the MSCs are modified using the viral or non-viral vectors with the induction of expression of various factors [84].

6.5.2 Promote Migration

MSCs which are generated by the help of bone marrow are referred as adult multipotent stem cells of mesodermal origin. Due to the multilineage property, MSCs are utilized for various therapeutic applications. These MSCs are utilized as cell drivers for different systemic and local treatments by being separated from the bone marrow and maintain in a culture. Different molecular factors direct MSCs movement to the targeted site and transformed them into neuronal cells, myocytes, adipocytes, chondrocytes, and osteoblasts when given the right stimuli [85]. MSCs are perfect for regenerative therapeutics, graft-versus-host illness, and autoimmune disorders due to their pluripotent nature. The extended destiny of the implanted MSC *in vivo* after *ex vivo* alterations and the *in vitro* multiplication of these cells without the loss of their stem cell qualities are key concerns that need to be severely addressed despite their potential for therapy in a range of disorders [80].

6.5.3 Anti-Inflammatory

Apart from their potential for mesodermal differentiation, MSCs can improve the processes involved in repair and alter the way that innate and adaptive immune system cells act as effectors. Based on these characteristics, MSCs have undergone large-scale efficacy testing in various animal models of degenerative and inflammatory illnesses, including EAE. Their incredible therapeutic value is supported by multiple mechanisms of action, such as inducing T-cell anergy, polarizing Th2 responses, inhibiting T helper (Th)17 cells by the help of CCL2 antagonist, and enhancing neuroprotective mechanisms through the paracrine release of antioxidant substances and growth factors. Significantly, systemically delivered MSCs from mice or humans have been discovered at distant locations in secondary lymphoid tissues and wounded tissue areas, indicating that MSCs may locate inflammatory areas [86]. Pleiotropic cytokine IL-4 plays a role in both adaptive and innate immunity. IL-4 has numerous uses, including stimulating T-cell proliferation and helping naïve T helper cells (Th0 cells) differentiate into Th2 cells [87]. According to Payne *et al.* [88], IL-4 activates transcription (STAT)6 molecules, signal transducer, and Janus kinase 1 and through a heterodimeric receptor.

6.5.4 Multigene-Modified MSC Exploration

Myocardial reperfusion and the continued existence of transplanted cells should be the minimum requirements for the perfect intervention therapies for heart injury. A single gene has a restricted effect. By applying to a range of gene combinations, several researchers were investigating a cooperative method to satisfy various therapy requests, and they have made some progress. Mingliang *et al.* [89] treated myocardial infarction with MSCs transfected with VEGF and IGF-1 plasmid. After 1 week, they discovered that MSCs with coupled genes had the greatest rate of survival, and after 3 weeks, their improved cardiac function also demonstrated the better condition.

All point to the potential therapeutic role that multigene combination therapy may have for the syndrome. In order to address myocardial infarction and satisfy the requirements of angiogenesis and antiapoptosis, some other research have jointly modified MSC transplantation through the use of both Ang-1 and Akt. They also got the anticipated outcomes. Rats' heart function has been further enhanced, and the effects of the therapy have been demonstrated to endure for 3 months [90]. While some scientists believe that HGF can promote cell migration, angiogenesis, and anti-apoptosis, the optimum gene modification scheme is still unknown at this time. There is now renewed hope for heart injury healing thanks to tissue engineering investigations on genetic improvement using MSCs as the carrier [91].

6.6 Limitations of Stem Cell Strategies for Cardiovascular Diseases

6.6.1 Heterogeneity of Stem Cells

First of all, MSCs are extremely difficult to describe since they exhibit surface characteristics that are shared by other kinds of cells and do not express any unique markers [83]. As per Togarrati *et al.* [92], MSCs have a notable expression profile of CD29 (integrin ß1), a protein that is likewise generated by myoepithelial and epithelial cells.

Similarly, keratinocytic and hematopoietic stem cells co-express CD90 (Thy-1). Furthermore, endothelial cells and some hematopoietic stem cells express CD105 or endoglin [93]. In addition to MSCs, various degrees of heterogeneity have also been identified in pluripotent stem cells. Controlling the differentiated cell population produced by ESCs and iPSCs is difficult because of their recognized pluripotent and infinitely multiplying properties. Differentiation capacity which favors various kinds of cells is exhibited by each of ESC and iPSC cell line [94]. The time period and genetic characteristics are two major drivers of interculture variability among various aspects. The complexities of stem cell heterogeneity have increased recently as examination of single cell culture has shown the various types of cells in a single population of pluripotent stem cells [95].

Exome sequencing revealed 24 types of iPSC clones obtained from mouse fibroblast comprises of nearly 400 different types of iPSCs. In another study, time-dependent single-cell RNA sequencing is used to uncover the diversity of human PSC-derived stem cell. Moreover, various research labs have generated different types of pluripotent stem cells due to a lack of standard procedure, but low repeatability and increased variability of ESC and iPSC are greatly affected by all of these factors [96].

6.6.2 Survival of Transplanted Stem Cell *In Vivo*

In relation to the challenges related with the features of stem cells, much more efforts are required to be done to assure the MSCs' survival and integration into the ischemic myocardium. Decreased flow of blood, elevated level of free radicals, and increased myocardial directed inflammation are some of the main factors contributing in the mortality of cell in case of ischemic myocardium. However, boosting up of cell integration and upregulation of cell survival proteins in wounded heart cells promote the survival of stem cells after overcoming the obstacles in the clinical findings [97].

Another way to enhance the survival of stem cells in the heart is done through genetic manipulation by the release of pro-survival genes. MSCs overexpressing Akt have been observed to result in a significant reduction in the process of apoptosis along with the rise in the release of growth factor (VEGF) which will ultimately enhance the functioning of cardiac cells following MI [98].

By focusing on VEGF signaling, a different study demonstrated how greater stretching can enhance MSCs' angiogenic capacity and reduce apoptosis. Hypoxic pre-conditioning of stem cells before transplantation has been demonstrated to have beneficial benefits in a number of previous studies [99]. Enhancing stem cell survival *in vivo* was impacted similarly by chemical or physical hypoxia induction. Additionally, it has been demonstrated that pre-conditioned MSCs with inhibitors of caspase in the existence of hyperoxia enhanced *in vivo* cell existence. Shi *et al.* [100] examined the difference in the rate of apoptosis and improvement in cell survival between hypoxia-preconditioned MSCs and normally grown MSCs in a monkey model of MI. Pro-angiogenic factors including angiopoietin-I, erythropoietin, and HGF were secreted, and after 3 months, there was a noticeable advancement in LV function and scar size. Many researchers suggest that 2%–5% O_2 levels are beneficial for improving the curative properties of stem cells and inhibiting the spontaneous maturation of ESCs because of the body's hypoxic milieu [101].

Additionally, after transplantation, inadequate vascularization and a high incidence of necrosis have been linked to poor iPSC survival because of their incapacity to adapt to their surroundings. The co-delivery of fibrin patch comprising IGF-enclosed microsphere has been considered to be a solution that promotes the release of growth factor and brings improvement in the functioning and survival of cardiac cell. It has been discovered that the addition of iPSCs to fibrin patch improved their ability of engraftment by up to 20-fold higher [102].

6.6.3 Migration and Homing of Stem Cell *In Vivo*

The migration of cells to the damaged area and their decreased aggregation capacity at the site of transplantation are considered as an issue for the treatment associated with stem cells. According to some studies, the integration and migration of stem cells to the targeted site are greatly affected by the expression of surface markers, size of cell, and culture condition [103].

According to De Becker and Riet [104], stem cells with overexpressed CXCR4 in the ischemic myocardium were able to move and engraft more easily and ultimately encouraged angiogenesis. Moreover, pre-treated stem cells with specific factors and chemokines, such as integrin β1 and TNFα, have the capacity to improve the movement of administered cells to the targeted organ site [105]. However, the translocation of MSCs to the targeted site is greatly affected by the restriction of adhesion molecules such as VCAM on the outer surface. The NF-kB pathways and high temperature of culture results were reported to have a deleterious effect on the migration of stem cells [106].

It is important to take into account the size of the stem cell prior to injecting it into the heart. Larger sizes of stem cells such as 7–10 μm need to be avoided in order to prevent chances of microinfarctions brought on by microvascular clogging [72]. In a different study, the authors discovered that smaller MSCs could cross the barrier of pulmonary microvascular more easily than the larger MSCs. Another study reported similar results, with patients exhibiting pulmonary embolism following an IV injection of AD-MSCs [80].

6.6.4 Immunogenicity of Stem Cells After Transplantation

Since MSCs are thought to be protected, they ought to be capable to elude the host immune mechanism and prevent immunological dismissal. Nonetheless, a number of investigations have revealed that MSCs in the ischemic heart develop immunity upon transplantation [107].

Preserving the stem cells' immune privilege following transplantation is thus a significant obstacle in the clinical application of MSC-based treatments. According to Gill *et al.* [108] (2024), an evaluation of the heart environment at 6 months revealed that allogeneic MSCs transition from an immune-tolerant to an immune-reactive condition. Numerous mechanisms have been linked to immune state alteration, such as altered expression profile of cell surface immunological antigen, MHC-II component on MSC surfaces, or decreased secretion of soluble immunosuppressive substances as PGE2 [109].

Wu *et al.* [110] demonstrated the remarkable impact of donor age in the control of stem cell survival by demonstrating that both ADSCs and BM-MSCs from older animals exhibit a decreased expression of $CD8^+$, $CD4^+$ T lymphocytes in comparison to the samples procured from younger animals. The immunosuppressive capability of MSCs derived from the neonatal and adult tissues was also compared by Luckheeram *et al.* [111], who discovered that these cells differed in their potential to inhibit $CD8^+$ and $CD4^+$ T cell expression. Thus, this finding highlights the significance of stem cell source for the mechanism of transplantation.

6.6.5 Uncontrolled Differentiation of Stem Cells *In Vivo* Resulting in Undesired Phenotype

Concerns in respect to the development of stem cells into specific cells with undesirable phenotypes have been raised by a few of the research studies. Cells which are exposed to the host milieu after transplantation may react with the existing stimuli and differentiate spontaneously. Furthermore, the CMs produced during the differentiation of iPSC or ESC may have an immature type of phenotype which results variations in sarcomere structure and calcium handling [112]. This could lead to a rise in arrhythmias once these cells are transplanted into the heart. Moreover, unfractionated bone marrow cells have been demonstrated in past research to undergo osteoblast differentiation, which can lead to the ossification of cardiac tissue. As a result, it is advised to incorporate enough MSCs into the perfusate while also making sure that they are pure [113].

Another strategy to think about could be transplantation of cells infused with factors for guided differentiation. Conversely, it may be advantageous to completely disregard the utilization of cells and give consideration to cell-derived products. Systems without cells are becoming more and more common because they offer the best of both worlds. Since they do not express antigen-presenting cell surface proteins, they can evade the immunological response. Furthermore, they show benefits that are comparable to those of whole cell transplantation [114].

6.6.6 Acquisition of Adequate Stem Cells

However, because bone marrow-derived stem cells (BMCs) may be extracted from either a bone marrow barrow or peripheral blood, their extraction is comparatively simpler. Examples of BMCs include hematopoietic, mesenchymal, and endothelial progenitor stem cells. Additionally, due to their unique site and small quantity of stem cells, organ-originated stem cells like cardiac stem cells (CSCs) may be challenging to harvest [115].

6.6.7 *Ex Vivo* Expansion Efficiency

Furthermore, *in vitro* pre-differentiation for embryonic stem cells (ESCs) can lessen the possibility that teratomas would form. Different cell types require different stimulation strategies for best growth and targeted selection. For example, exogenous cytokine injection is frequently employed to promote ESC differentiation into target cells. New cardiomyocytes are positively selected using flow cytometry. According to Ellison *et al.* [116], ANP, β-MHC, α-MHC, Nkx2.5, and GATA4 are popular cardiac markers for fluorescence labeling. Targeted gene encoding with increased green fluorescent protein (GFP) is another widely used technique to monitor the production of cardiomyocytes. After the GFP-positive cells are sorted, a highly pure myocyte fraction is produced. Before transplantation, the chosen cells need to be tested to make sure the cardiomyocytes are functional. Action potentials and automaticity should be present in the myocytes. Early myocytes may have immature AP characteristics, but after a longer time in culture, functional development is anticipated [117].

6.7 Strategies for Improving Stem Cell Therapy

6.7.1 Stem Cell Delivery

Stem cell delivery is an important aspect of regenerative healthcare since it includes injecting stem cells into particular bodily locations to substitute or restore injured organs and tissues. Once the stem cells get to their destination, effective transport methods are critical to their existence, integration, and functionality. Commonly used delivery methods include an intravenous injection, direct tissue injection, and scaffold-based implantation. Each approach has advantages and disadvantages, such as ensuring precise targeting, minimizing cell loss, and limiting immunological rejection. Advancements in biomaterials, nanotechnology, and imaging methods constantly improves the efficiency and efficacy of stem cell, their distribution which paves the way of efficient therapeutic treatment for a broad varieties of ailments and wounds [118].

6.7.2 Homing

The movement and integration of stem cells to the targeted location is referred as homing, and it is a crucial stage in stem cell directed therapeutic treatments. This homing method is critical to determine the effectiveness of stem cell therapies because it is the only way to ensures that the cells are transported to the injured site and begin working. Different substances such as adhesion molecules, chemokines, and cytokines are capable in influencing the homing which ultimately guides stem cells to specific locations throughout the body. Understanding the strengthening of homing mechanisms can make stem cell therapies somewhat safer and more effective in treating a wide range of disorders, including tissue regeneration, neurological difficulties, and cardiovascular diseases [119].

6.7.3 Survival/Engraftment

Transplantation and survival are the most critical factors in determining the stem cell treatment effectiveness. The capacity of transplanted stem cells to remain viable and functional after being administered into the patient's body is known as survival. By contrast, engraftment refers to the capability of stem cells to settle in the host tissue, then expand, and differentiate into the specific type of the cell. Successful engraftment can be done by overcoming a variety of challenges, including immune rejection, inflammatory processes, and creating an atmosphere in which stem cells can flourish. These characteristics must be addressed in order to maximize the medicinal value and improve the outcomes of patients. Immunosuppression, genetic alteration, and the use of scaffolds or supportive biomaterials are commonly utilized procedures that enhance the lifespan and engraftment of donated organs. Artificial scaffolds or supporting biomaterials are also some of the commonly employed strategies to promote the transplanted stem cell engraftment and survival [120].

6.8 Bioengineering Strategies: New Era

6.8.1 Cell-Free Therapy

As studies previously indicated, tumorigenicity or immunogenicity is an inevitable risk associated with stem cell transplants. Because of this, a lot of researchers in the area have focused on studying the bioactive substances that emerge from stem cells. These compounds have demonstrated similar therapeutic benefits, raising the prospect of a viable substitute for stem cell therapies. EVs, which include microvesicles and exosomes, are the most significant bioactive agents under research at the moment. They are composed of the physiologically active substances present in stem cells [121].

These have been demonstrated to have positive impacts on cardiac healing following injury, much like cell treatments. Exosomes, as opposed to other EVs, are more commonly recognized as potential candidates for cell-free therapy. For instance, studies have discovered that exosomes derived from MSCs that have been pretreated with atorvastatin greatly improve the effectiveness of therapy [122]. However, a lot of preclinical or clinical procedures involving the application of exosomes including their extraction and purification have not been standardized. Therefore, there is still a long way to go before exosomes prove beneficial in clinical studies.

6.8.2 Improving Target Ability

The efficient distribution of stem cell treatment is the initial obstacle it faces. More specifically, it is critical to deliver cells to the damaged site in a focused fashion, much like with standard drug-based therapy. Among the objectives of bioengineering is to do this. In the past, our group has attempted magnetic localization in the body using FDA-approved ferumoxytol nanoparticles [6].

An artificially generated magnetic field was placed close to the site of the heart damage as a method for improving this field's techniques is constantly being refined. The magnetic field draws iron-labeled (ferumoxytol) stem cells straight to the injured heart tissue during administration. On the other side, applying a high magnetic field to the patient during surgery could have unanticipated effects on the tools. It was necessary to design an additional biosafe approach to targeting. So we turned our attention to the platelet, a special part of the blood that has the ability to clump together and adhere directly to damaged endothelial cells on blood channels. In the past, our team has created a novel technique to use platelet nanovesicles (PNVs) to adorn the membrane on the exterior of CSCs [123].

6.8.3 Overcoming Low Cell Retention

Once more effective targeted administration of the cellular treatment has been achieved, the next hurdle is to enhance the cell integration in the area of injury. Over the last 10 years, a number of novel biomaterials have been established to encapsulate and safeguard transplanted stem cells, which have decreased retention and rates of survival. A variety of materials have been used to create injectable form of hydrogels, which is a combination with specific stem cells inside. Research studies have previously shown that encapsulation of CSCs in temperature-sensitive poly(N-isopropylacrylamide-co-acrylic acid) or P(NIPAM-AA) nanogels is secure and efficient when used in the pig and mouse models with myocardial infarction (MI) [124].

In the latest study, researchers have utilized poly(N-isopropylacrylamide-co-itaconic acid) for the generation of a water-friendly and negative-charged microcosm, which is proved to be beneficial in the maintenance of high survival of CSCs. This therapeutic mechanism boosted the amount of cell retention while inhibiting apoptosis and activating angiogenesis. Moreover, several hydrogels were considered to be the best option of treatment due to their ability to prevent both localized T cell infiltration and widespread inflammation in studies of immunocompetent mice [125].

Apart from the hydrogels, the cardiac patch is one of the innovative biomaterial which was utilized in the area of cardiac restoration. In addition to increasing cell retention, the distribution of curative cells in a cardiac patch serves other functionalization objectives [126].

6.8.4 Biomimetic Strategies in Stem-Cell-Based Therapy

The functionalization of stem cells has been the primary objective of the therapeutic approaches previously discussed, Li *et al.* [127] report another groundbreaking discovery centered on the production of "super stem cells," or synthetic stem cells. Researchers have just made two attempts at treating heart problems using this concept. We have presented a novel therapeutic microparticle (MP) design that resembles stem cell biointerfacing during regeneration: the "core–shell" design [128]. In the animal model of myocardial infarction, CMMP injection led to a similar enhancement of cardiac output comparable to CSC therapy. More importantly, CMMPs did not induce infiltration of T cells in immunized mice, indicating a high possibility of clinical studies. Following that, researchers aimed to develop a more complicated stem cell-mimicking composite [129]. These medicinal particles, known as "synthetic MSC" (or synMSC), displayed regeneration ability in mice with AMI.

6.9 Conclusion

Stem cell therapy has a chance to provide lasting advantages for individuals suffering from cardiac conditions. The stem cells have paracrine actions that inhibit immune system activity in the ischemic heart, promoting reverse rebuilding. Numerous investigations have found that they help improve the functioning of the heart, reduce cardiac fibrosis, reduce CM mortality, and promote vascular regrowth of the myocardial after ischemia, yet the research community of stem cells lacks consistency in regard to cell culture procedures and cell evaluation, which may be leading to the inconsistent results observed in clinics. These issues can be solved by establishing established processes for all labs to follow. Stem cells can confront challenges such as host immune system rejection and translocation to undesirable regions after implantation, yet using artificial biomaterials for injection directly into the EVs and myocardium as carriers for paracrine variables is a potential alternative to cell treatment. Though these techniques are possible, more research is needed to determine their safety and effectiveness for translational application in patients with cardiac disease.

References

1. Elgendy, I.Y., Mahtta, D., Pepinr, C.J., Medical therapy for heart failure caused by ischemic heart disease. *Circ. Res.*, 124, 11, 1520–1535, 2019. https://doi.org/10.1161/CIRCRESAHA.118.313568.
2. Upadhyay, R.K., Emerging risk biomarkers in cardiovascular diseases and disorders. *J. Lipids*, 2015, 971453, 2015. doi: 10.1155/2015/971453.
3. Bilgimol, J.C., Ragupathi, S., Vengadassalapathy, L., Senthil, N.S., Selvakumar, K., Ganesan, M., Manjunath, S.R., Stem cells: An eventual treatment option for heart diseases. *World J. Stem Cells*, 7, 8, 1118–26, 2015. doi: 10.4252/wjsc.v7.i8.1118.
4. Rafi, M.A., Gene and stem cell therapy: alone or in combination? *Bioimpacts*, 1, 4, 213–8, 2011. doi: 10.5681/bi.2011.030.
5. Fernández-Lázaro, D., Sanz, B., Seco-Calvo, J., The Mechanisms of Regulated Cell Death: Structural and Functional Proteomic Pathways Induced or Inhibited by a Specific Protein-A Narrative Review. *Proteomes*, 12, 1, 3, 2024. doi: 10.3390/proteomes12010003.
6. Hoang, D.M., Pham, P.T., Bach, T.Q., Ngo, A.T.L., Nguyen, Q.T., Phan, T.T.K., Nguyen, G.H., Le, P.T.T., Hoang, V.T., Forsyth, N.R., Heke, M., Nguyen, L.T., Stem cell-based therapy for human diseases. *Signal Transduct. Target. Ther.*, 7, 1, 272, 2022. doi: 10.1038/s41392-022-01134-4.
7. Durrani, S., Konoplyannikov, M., Ashraf, M., Haider, K.H., Skeletal myoblasts for cardiac repair. *Regener. Med.*, 5, 6, 919–32, 2010. doi: 10.2217/rme.10.65.
8. Menasché, P., Skeletal myoblasts as a therapeutic agent. *Prog. Cardiovasc. Dis.*, 50, 1, 7–17, 2007. doi: 10.1016/j.pcad.2007.02.002.
9. Löffler, A.I. and Kramer, C.M., Myocardial Viability Testing to Guide Coronary Revascularization. *Interventional Cardiol Clin.*, 7, 3, 355–365, 2018. doi: 10.1016/j.iccl.2018.03.005.
10. Menasché, P., Skeletal myoblast for cell therapy. *Coronary Artery Dis.*, 16, 2, 105–110, 2005. https://doi.org/10.1097/00019501-200503000-00005.

11. Attar, A., Hosseinpour, A., Hosseinpour, H., Kazemi, A., Major cardiovascular events after bone marrow mononuclear cell transplantation following acute myocardial infarction: an updated post-BAMI meta-analysis of randomized controlled trials. *BMC Cardiovasc. Disord.*, 22, 1, 259, 2022. https://doi.org/10.1186/s12872-022-02701-x.
12. Correia, C.D., Ferreira, A., Fernandes, M.T., Silva, B.M., Esteves, F., Leitão, H.S., Bragança, J., Calado, S.M., Human Stem Cells for Cardiac Disease Modeling and Preclinical and Clinical Applications—Are We on the Road to Success? *Cells*, 12, 13, 1727, 2023. https://doi.org/10.3390/cells12131727.
13. Mulari, S., Kesävuori, R., Stewart, J.A., Karjalainen, P., Holmström, M., Lehtinen, M., Peltonen, J., Laine, M., Sinisalo, J., Juvonen, T., Kupari, M., Harjula, A., Pätilä, T., Kivistö, S., Kankuri, E., Vento, A., Helsinki BMMC Collaboration. Follow-up of intramyocardial bone marrow mononuclear cell transplantation beyond 10 years. *Sci. Rep.*, 14, 1, 3747, 2024. https://doi.org/10.1038/s41598-024-53776-9.
14. Yan, Z., Zeng, L., Li, Z., Zhang, H., Chen, W., Jia, L., Chen, C., Cheng, H., Xu, K., Bone Marrow-derived Endothelial Progenitor Cells Promote Hematopoietic Reconstitution After Hematopoietic Stem Cell Transplantation. *Transplant. Proc.*, 45, 1, 427–433, 2013. https://doi.org/10.1016/j.transproceed.2012.03.064.
15. Ungvari, Z., Tarantini, S., Kiss, T., Wren, J.D., Giles, C.B., Griffin, C.T., Murfee, W.L., Pacher, P., Csiszar, A., Endothelial dysfunction and angiogenesis impairment in the ageing vasculature. *Nat. Rev. Cardiol.*, 15, 9, 555–565, 2018.
16. Rurali, E., Bassetti, B., Perrucci, G.L., Zanobini, M., Malafronte, C., Achilli, F., Gambini, E., BM ageing: Implication for cell therapy with EPCs. *Mech. Ageing Dev.*, 159, 4–13, 2016. https://doi.org/10.1016/j.mad.2016.04.002.
17. Kalou, Y., Al-Khani, A.M., Haider, K.H., Bone marrow mesenchymal stem cells for heart failure treatment: a systematic review and meta-analysis. *Heart Lung Circ.*, 32, 7, 870–880, 2023. https://doi.org/10.1016/j.hlc.2023.01.012.
18. Pelliccia, F., Zimarino, M., De Luca, G., Viceconte, N., Tanzilli, G., De Caterina, R., Endothelial Progenitor Cells in Coronary Artery Disease: From Bench to Bedside. *Stem Cells Transl. Med.*, 11, 5, 451–460, 2022. doi: 10.1093/stcltm/szac010.
19. Harshfield, E.L., Pennells, L., Schwartz, J.E., Willeit, P., Kaptoge, S., Bell, S., Shaffer, J.A., Bolten, T., Spackman, S., Wassertheil-Smoller, S., Kee, F., Amouyel, P., Shea, S.J., Kuller, L.H., Kauhanen, J., van Zutphen, E.M., Blazer, D.G., Krumholz, H., Nietert, P.J., Kromhout, D., Laughlin, G., Berkman, L., Wallace, R.B., Simons, L.A., Dennison, E.M., Barr, E.L.M., Meyer, H.E., Wood, A.M., Danesh, J., Angelantonio, E.D., Davidson, K.W., Emerging Risk Factors Collaborations. Association Between Depressive Symptoms and Incident Cardiovascular Diseases. *JAMA*, 324, 23, 2396–2405, 2020. doi:10.1001/jama.2020.23068.

20. Bouloukaki, I., Fanaridis, M., Stathakis, G., Ermidou, C., Kallergis, E., Moniaki, V., Mauroudi, E., Schiza, S.E., Characteristics of Patients with Obstructive Sleep Apnea at High Risk for Cardiovascular Disease. *Medicina*, 57, 11, 1265, 2021. https://doi.org/10.3390/medicina57111265.
21. Ferrero, R., Rainer, P., Deplancke, B., Toward a Consensus View of Mammalian Adipocyte Stem and Progenitor Cell Heterogeneity. *Trends Cell Biol.*, 30, 12, 937–950, 2020. https://doi.org/10.1016/j.tcb.2020.09.007.
22. Yang, Z., Wang, Z., Wu, L., Wang, Y., Xu, Z., Liu, Y., Wang, F., Yu, D., B lymphocytes transdifferentiate into immunosuppressive erythroblast-like cells. *Front. Immunol.*, 14, 1202943, 2023. doi: 10.3389/fimmu.2023.1202943.
23. Jovic, D., Liang, X., Zeng, H., Lin, L., Xu, F., Luo, Y., Single-cell RNA sequencing technologies and applications: A brief overview. *Clin. Transl. Med.*, 12, 3, e694, 2022. doi: 10.1002/ctm2.694.
24. Griffiths, J.A., Richard, A.C., Bach, K., Lun, A.T.L., Marioni, J.C., Detection and removal of barcode swapping in single-cell RNA-seq data. *Nat. Commun.*, 9, 2667, 2018. https://doi.org/10.1038/s41467-018-05083-x.
25. Smith, A.J., Effects of Cardiotoxins on Cardiac Stem and Progenitor Cell Populations. *Front. Cardiovasc. Med.*, 8, 624028, 2021. doi: 10.3389/fcvm.2021.624028.
26. Mehanna, R.A., Essawy, M.M., Barkat, M.A., Awaad, A.K., Thabet, E.H., Hamed, H.A., Elkafrawy, H., Khalil, N.A., Sallam, A., Kholief, M.A., Ibrahim, S.S., Mourad, G.M., Cardiac stem cells: Current knowledge and future prospects. *World J. Stem Cells*, 14, 1, 1–40, 2022. doi: 10.4252/wjsc.v14.i1.1.
27. Albini, A., Pennesi, G., Donatelli, F., Cammarota, R., De Flora, S., Noonan, D.M., Cardiotoxicity of anticancer drugs: the need for cardio-oncology and cardio-oncological prevention. *J. Natl. Cancer Inst.*, 102, 1, 14–25, 2010. doi: 10.1093/jnci/djp440.
28. Gillman, A.S., Vo, J.B., Nohria, A., Ferrer, R.A., Decision Science Can Inform Clinical Trade-Offs Regarding Cardiotoxic Cancer Treatments. *JNCI Cancer Spectr.*, 5, 4, pkab053, 2021. doi: 10.1093/jncics/pkab053.
29. Sayegh, N., Yirerong, J., Agarwal, N., Addison, D., Fradley, M., Cortes, J., Weintraub, N.L., Sayed, N., Raval, G., Guha, A., Cardiovascular Toxicities Associated with Tyrosine Kinase Inhibitors. *Curr. Cardiol Rep.*, 25, 4, 269–280, 2023. doi: 10.1007/s11886-023-01845-2.
30. Eguizabal, C., Aran, B., Chuva de Sousa Lopes, S.M., Geens, M., Heindryckx, B., Panula, S., Popovic, M., Vassena, R., Veiga, A., Two decades of embryonic stem cells: a historical overview. *Hum. Reprod. Open*, 2019, 1, hoy024, 2019. doi: 10.1093/hropen/hoy024.
31. Volarevic, V., Markovic, B.S., Gazdic, M., Volarevic, A., Jovicic, N., Arsenijevic, N., Armstrong, L., Djonov, V., Lako, M., Stojkovic, M., Ethical and Safety Issues of Stem Cell-Based Therapy. *Int. J. Med. Sci.*, 15, 1, 36–45, 2018. doi: 10.7150/ijms.21666.
32. Musunuru, K., Sheikh, F., Gupta, R.M., Houser, S.R., Maher, K.O., Milan, D.J., Terzic, A., Wu, J.C., American Heart Association Council on Functional

Genomics and Translational Biology; Council on Cardiovascular Disease in the Young; and Council on Cardiovascular and Stroke Nursing. Induced Pluripotent Stem Cells for Cardiovascular Disease Modeling and Precision Medicine: A Scientific Statement From the American Heart Association. *Circ.: Genomic Precis. Med.*, 11, 1, e000043, 2018. doi: 10.1161/HCG.0000000000000043.

33. Hockemeyer, D. and Jaenisch, R., Induced Pluripotent Stem Cells Meet Genome Editing. *Cell Stem Cell*, 18, 5, 573–86, 2016. doi: 10.1016/j.stem.2016.04.013.
34. Ebrahimi, M., Forouzesh, M., Raoufi, S., Ramazii, M., Ghaedrahmati, F., Farzaneh, M., Differentiation of human induced pluripotent stem cells into erythroid cells. *Stem Cell Res. Therpy*, 11, 1, 483, 2020. doi: 10.1186/s13287-020-01998-9.
35. Al Abbar, A., Ngai, S.C., Nograles, N., Alhaji, S.Y., Abdullah, S., Induced Pluripotent Stem Cells: Reprogramming Platforms and Applications in Cell Replacement Therapy. *Biores. Open Access*, 9, 1, 121–136, 2020. doi: 10.1089/biores.2019.0046.
36. Broughton, K.M. and Sussman, M.A., Enhancement Strategies for Cardiac Regenerative Cell Therapy: Focus on Adult Stem Cells. *Circ. Res.*, 123, 2, 177–187, 2018. doi: 10.1161/CIRCRESAHA.118.311207.
37. Rheault-Henry, M., White, I., Grover, D., Atoui, R., Stem cell therapy for heart failure: Medical breakthrough, or dead end? *World J. Stem Cells*, 13, 4, 236–259, 2021. doi: 10.4252/wjsc.v13.i4.236.
38. Sun, R., Li, X., Liu, M., Zeng, Y., Chen, S., Zhang, P., Advances in stem cell therapy for cardiovascular disease (Review). *Int. J. Mol. Med.*, 38, 1, 23–9, 2016. doi: 10.3892/ijmm.2016.2607.
39. Maloberti, A., Biolcati, M., Ruzzenenti, G., Giani, V., Leidi, F., Monticelli, M., Algeri, M., Scarpellini, S., Nava, S., Soriano, F., Oreglia, J., Sacco, A., Morici, N., Oliva, F., Piani, F., Borghi, C., Giannattasio, C., The Role of Uric Acid in Acute and Chronic Coronary Syndromes. *J. Clin. Med.*, 10, 20, 4750, 2021. doi: 10.3390/jcm10204750.
40. Pignochino, Y., Grignani, G., Cavalloni, G., Motta, M., Tapparo, M., Bruno, S., Bottos, A., Gammaitoni, L., Migliardi, G., Camussi, G., Alberghini, M., Torchio, B., Ferrari, S., Bussolino, F., Fagioli, F., Picci, P., Aglietta, M., Sorafenib blocks tumour growth, angiogenesis and metastatic potential in preclinical models of osteosarcoma through a mechanism potentially involving the inhibition of ERK1/2, MCL-1 and ezrin pathways. *Mol. Cancer*, 8, 118, 2009. doi: 10.1186/1476-4598-8-118.
41. Verhaert, D.V.M., Brunner-La Rocca, H.P., van Veldhuisen, D.J., Vernooy, K., The bidirectional interaction between atrial fibrillation and heart failure: consequences for the management of both diseases. *Europace*, 23, 23 Suppl 2, ii40–ii45, 2021. doi: 10.1093/europace/euaa368.
42. Nocini, R., Lippi, G., Mattiuzzi, C., Periodontal disease: the portrait of an epidemic. *J. Public Health Emergency*, 4, 10, 2020. doi: 10.21037/jphe.2020.03.01.

43. Sanz, M., Marco Del Castillo, A., Jepsen, S., Gonzalez-Juanatey, J.R., D'Aiuto, F., Bouchard, P., Chapple, I., Dietrich, T., Gotsman, I., Graziani, F., Herrera, D., Loos, B., Madianos, P., Michel, J.B., Perel, P., Pieske, B., Shapira, L., Shechter, M., Tonetti, M., Vlachopoulos, C., Wimmer, G., Periodontitis and cardiovascular diseases: Consensus report. *J. Clin. Periodontology*, 47, 3, 268–288, 2020. doi: 10.1111/jcpe.13189.
44. Algoet, M., Janssens, S., Himmelreich, U., Gsell, W., Pusovnik, M., Van den Eynde, J., Oosterlinck, W., Myocardial ischemia-reperfusion injury and the influence of inflammation. *Trends Cardiovasc. Med.*, 33, 6, 357–366, 2023. doi: 10.1016/j.tcm.2022.02.005.
45. Xu, Y., Wang, Y., Ji, X., Immune and inflammatory mechanism of remote ischemic conditioning: A narrative review. *Brain Circ.*, 9, 2, 77–87, 2023. doi: 10.4103/bc.bc_57_22.
46. Basalay, M.V., Yellon, D.M., Davidson, S.M., Targeting myocardial ischaemic injury in the absence of reperfusion. *Basic Res. Cardiol.*, 115, 6, 63, 2020. doi: 10.1007/s00395-020-00825-9.
47. Juni, R.P., Kuster, D.W.D., Goebel, M., Helmes, M., Musters, R.J.P., van der Velden, J., Koolwijk, P., Paulus, W.J., van Hinsbergh, V.W.M., Cardiac Microvascular Endothelial Enhancement of Cardiomyocyte Function Is Impaired by Inflammation and Restored by Empagliflozin. *JACC Basic to Transl. Sci.*, 4, 5, 575–591, 2019. doi: 10.1016/j.jacbts.2019.04.003.
48. Madan, S., Uttekar, B., Chowdhary, S., Rikhy, R., Mitochondria Lead the Way: Mitochondrial Dynamics and Function in Cellular Movements in Development and Disease. *Front. Cell Dev. Biol.*, 9, 781933, 2022. doi: 10.3389/fcell.2021.781933.
49. Del Buono, M.G., Moroni, F., Montone, R.A., Azzalini, L., Sanna, T., Abbate, A., Ischemic Cardiomyopathy and Heart Failure After Acute Myocardial Infarction. *Curr. Cardiol. Rep.*, 24, 10, 1505–1515, 2022. doi: 10.1007/s11886-022-01766-6.
50. Reddy, K., Khaliq, A., Henning, R.J., Recent advances in the diagnosis and treatment of acute myocardial infarction. *World J. Cardiol.*, 7, 5, 243–76, 2015. doi: 10.4330/wjc.v7.i5.243.
51. Jenča, D., Melenovský, V., Stehlik, J., Staněk, V., Kettner, J., Kautzner, J., Adámková, V., Wohlfahrt, P., Heart failure after myocardial infarction: incidence and predictors. *ESC Heart Fail.*, 8, 1, 222–237, 2021. doi: 10.1002/ehf2.13144.
52. Severino, P., D'Amato, A., Pucci, M., Infusino, F., Birtolo, L.I., Mariani, M.V., Lavalle, C., Maestrini, V., Mancone, M., Fedele, F., Ischemic Heart Disease and Heart Failure: Role of Coronary Ion Channels. *Int. J. Mol. Sci.*, 21, 9, 3167, 2020. doi: 10.3390/ijms21093167.
53. Heusch, G., Myocardial ischemia: lack of coronary blood flow, myocardial oxygen supply-demand imbalance, or what? *Am. J. Physiol. Heart Circ. Physiol.*, 316, 6, H1439–H1446, 2019. doi: 10.1152/ajpheart.00139.2019.

54. Stegehuis, V.E., Wijntjens, G.W., Murai, T., Piek, J.J., van de Hoef, T.P., Assessing the Haemodynamic Impact of Coronary Artery Stenoses: Intracoronary Flow Versus Pressure Measurements. *Eur. Cardiol. Rev.*, 13, 1, 46–53, 2018. doi: 10.15420/ecr.2018:7:2.
55. Fathallah-Shaykh, S.A. and Cramer, M.T., Uric acid and the kidney. *Pediatr. Nephrol.*, 29, 6, 999–1008, 2014. https://doi.org/10.1007/s00467-013-2549-x.
56. Mourad, O., Yee, R., Li, M., Nunes, S.S., Modeling Heart Diseases on a Chip: Advantages and Future Opportunities. *Circ. Res.*, 132, 4, 483–497, 2023. https://doi.org/10.1161/CIRCRESAHA.122.321670.
57. Deir, S., Mofrad, Y.M., Mashayekhan, S., Shamloo, A., Mansoori-Kermani, A., Step-by-step fabrication of heart-on-chip systems as models for cardiac disease modeling and drug screening. *Talanta*, 266, Part 1, 124901, 2024. https://doi.org/10.1016/j.talanta.2023.124901.
58. Soma, Y., Tani, H., Morita-Umei, Y., Kishino, Y., Fukuda, K., Tohyama, S., Pluripotent stem cell-based cardiac regenerative therapy for heart failure. *J. Mol. Cell. Cardiol.*, 187, 90–100, 2024. https://doi.org/10.1016/j.yjmcc.2023.12.001.
59. Vo, Q.D., Saito, Y., Ida, T., Nakamura, K., Yuasa, S., The use of artificial intelligence in induced pluripotent stem cell-based technology over 10-year period: A systematic scoping review. *PLoS One*, 19, 5, e0302537, 2024. https://doi.org/10.1371/journal.pone.0302537.
60. Guo, W., Xu, Y., Liu, X., Dou, J., Guo, Z., Therapeutic effect of adipose-derived stem cells injected into pericardial cavity in rat heart failure. *ESC Heart Fail.*, 11, 1, 492–502, 2024. doi: 10.1002/ehf2.14606.
61. Follin, B., Ghotbi, A.A., Clemmensen, A.E., Bentsen, S., Juhl, M., Søndergaard, R.H., Lund, L.D., Haack-Sørensen, M., Hasbak, P., Cohen, S., Ripa, R.S., Kastrup, J., Ekblond, A., Kjaer, A., Retention and Functional Effect of Adipose-Derived Stromal Cells Administered in Alginate Hydrogel in a Rat Model of Acute Myocardial Infarction. *Stem Cells Int.*, 2018, 1, 7821461, 2018. https://doi.org/10.1155/2018/7821461.
62. Adamiak, M., Cheng, G., Bobis-Wozowicz, S., Zhao, L., Kedracka-Krok, S., Samanta, A., Karnas, E., Xuan, Y.T., Skupien-Rabian, B., Chen, X., Jankowska, U., Girgis, M., Sekula, M., Davani, A., Lasota, S., Vincent, R.J., Sarna, M., Newell, K.L., Wang, O.L., Dudley, N., Madeja, Z., Dawn, B., Zuba-Surma, E.K., Induced Pluripotent Stem Cell (iPSC)-Derived Extracellular Vesicles Are Safer and More Effective for Cardiac Repair Than iPSCs. *Circ. Res.*, 122, 2, 296–309, 2018. doi: 10.1161/CIRCRESAHA.117.311769.
63. Menasché, P., Vanneaux, V., Hahège, A., Bel, A., Cholley, B., Parouchev, A., Cacciapuoti, I., Al-Daccak, R., Benhamouda, N., Blons, H., Agbulut, O., Tosca, L., Trouvin, J.-H., Fabreguettes, J.-R., Bellamy, V., Charron, D., Tartour, E., Tachdjian, G., Desnos, M., Larghero, J., Transplantation of Human Embryonic Stem Cell–Derived Cardiovascular Progenitors for Severe Ischemic Left Ventricular Dysfunction. *J. Am. Coll. Cardiol.*, 71, 4, 429–438, 2018. https://doi.org/10.1016/j.jacc.2017.11.047.

64. Makkar, R.R., Smith, R.R., Cheng, K., Malliaras, K., Thomson, L.E., Berman, D., Czer, L.S., Marbán, L., Mendizabal, A., Johnston, P.V., Russell, S.D., Schuleri, K.H., Lardo, A.C., Gerstenblith, G., Marbán, E., Intracoronary cardiosphere-derived cells for heart regeneration after myocardial infarction (CADUCEUS): a prospective, randomised phase 1 trial. *Lancet*, 379, 9819, 895–904, 2012. doi: 10.1016/S0140-6736(12)60195-0.
65. Yau, T.M., Pagani, F.D., Mancini, D.M., Chang, H.L., Lala, A., Woo, Y.J., Acker, M.A., Selzman, C.H., Soltesz, E.G., Kern, J.A., Maltais, S., Charbonneau, E., Pan, S., Marks, M.E., Moquete, E.G., O'Sullivan, K.L., Taddei-Peters, W.C., McGowan, L.K., Green, C., Rose, E.A., Jeffries, N., Parides, M.K., Weisel, R.D., Miller, M.A., Hung, J., O'Gara, P.T., Moskowitz, A.J., Gelijns, A.C., Bagiella, E., Milano, C.A., Cardiothoracic Surgical Trials Network. Intramyocardial Injection of Mesenchymal Precursor Cells and Successful Temporary Weaning From Left Ventricular Assist Device Support in Patients With Advanced Heart Failure: A Randomized Clinical Trial. *J. Am. Med. Assoc.*, 321, 12, 1176–1186, 2019. doi: 10.1001/jama.2019.2341.
66. Qayyum, A.A., Mouridsen, M., Nilsson, B., Gustafsson, I., Schou, M., Nielsen, O.W., Hove, J.D., Mathiasen, A.B., Jørgensen, E., Helqvist, S., Joshi, F.R., Johansen, E.M., Follin, B., Juhl, M., Højgaard, L.D., Haack-Sørensen, M., Ekblond, A., Kastrup, J., Danish phase II trial using adipose tissue derived mesenchymal stromal cells for patients with ischaemic heart failure. *ESC Heart Fail.*, 10, 2, 1170–1183, 2023. doi: 10.1002/ehf2.14281.
67. Soetisna, T.W., Sukmawan, R., Setianto, B., Mansyur, M., Murni, T.W., Listiyaningsih, E., Santoso, A., Combined transepicardial and transseptal implantation of autologous CD 133+ bone marrow cells during bypass grafting improves cardiac function in patients with low ejection fraction. *J. Card. Surg.*, 35, 4, 740–746, 2020. doi: 10.1111/jocs.14454.
68. Tiburcy, M., Meyer, T., Liaw, N.Y., Zimmermann, W.H., Generation of Engineered Human Myocardium in a Multi-well Format. *STAR Protoc.*, 1, 1, 100032, 2020. doi: 10.1016/j.xpro.2020.100032.
69. Mazine, A., Rushani, D., Yau, T.M., Clinical mesenchymal stem cell therapy in ischemic cardiomyopathy. *JTCVS Open*, 8, 135–141, 2021. doi: 10.1016/j.xjon.2021.06.030.
70. Nitkin, C.R. and Bonfield, T.L., Concise Review: Mesenchymal Stem Cell Therapy for Pediatric Disease: Perspectives on Success and Potential Improvements. *Stem Cells Transl. Med.*, 6, 2, 539–565, 2017. doi: 10.5966/sctm.2015-0427.
71. Liu, Z., Mikrani, R., Zubair, H.M., Taleb, A., Neveed, M., Baig, M.M.F.A., Zhang, Q., Li, C., Habib, M., Cui, X., Sembatya, K.R., Lei, H., Zhou, X., Systemic and local delivery of mesenchymal stem cells for heart renovation: Challenges and innovations. *Eur. J. Pharmacol.*, 876, 173049, 2020. https://doi.org/10.1016/j.ejphar.2020.173049.

72. Terashvili, M. and Bosnjak, Z.J., Stem Cell Therapies in Cardiovascular Disease. *J. Cardiothorac. Vasc. Anesthesia*, 33, 1, 209–222, 2019. doi: 10.1053/j.jvca.2018.04.048.
73. Shi, W., Xin, Q., Yuan, R., Yuan, Y., Cong, W., Chen, K., Neovascularization: The Main Mechanism of MSCs in Ischemic Heart Disease Therapy. *Front. Cardiovasc. Med.*, 8, 633300, 2021. doi: 10.3389/fcvm.2021.633300.
74. Kou, F., Zhu, C., Wan, H., Xue, F., Wang, J., Xiang, L., Li, J., Endothelial progenitor cells as the target for cardiovascular disease prediction, personalized prevention, and treatments: progressing beyond the state-of-the-art. *EPMA J.*, 11, 4, 629–643, 2020. doi: 10.1007/s13167-020-00223-0.
75. Hu, C., Zhang, X., Teng, T., Ma, Z.G., Tang, Q.Z., Cellular Senescence in Cardiovascular Diseases: A Systematic Review. *Aging Dis.*, 13, 1, 103–128, 2022. doi: 10.14336/AD.2021.0927.
76. Baum, J. and Duffy, H.S., Fibroblasts and myofibroblasts: what are we talking about? *J. Cardiovasc. Pharmacol.*, 7, 4, 376–9, 2011. doi: 10.1097/FJC.0b013e3182116e39.
77. Yuan, Q.L., Zhang, Y.G., Chen, Q., Mesenchymal Stem Cell (MSC)-Derived Extracellular Vesicles: Potential Therapeutics as MSC Trophic Mediators in Regenerative Medicine. *Anat. Rec. (Hoboken)*, 03, 6, 1735–1742, 2020. doi: 10.1002/ar.24186.
78. Sahoo, S., Klychko, E., Thorne, T., Misener, S., Schultz, K.M., Millay, M., Ito, A., Liu, T., Kamide, C., Agrawal, H., Perlman, H., Qin, G., Kishore, R., Losordo, D.W., Exosomes from human CD34(+) stem cells mediate their proangiogenic paracrine activity. *Circ. Res.*, 109, 7, 724–8, 2011. doi: 10.1161/CIRCRESAHA.111.253286.
79. Luo, H., Birjandi, A.A., Ren, F., Sun, T., Sharpe, P.T., Sun, H., An, Z., Advances in oral mesenchymal stem cell-derived extracellular vesicles in health and disease. *Genes Discovery*, 11, 1, 346–357, 2023. doi: 10.1016/j.gendis.2023.03.015.
80. Zhou, T., Yuan, Z., Weng, J., Pei, D., Du, X., He, C., Lai, P., Challenges and advances in clinical applications of mesenchymal stromal cells. *J. Hematol. Oncol.*, 14, 1, 24, 2021. doi: 10.1186/s13045-021-01037-x.
81. Chehelgerdi, M., Behdarvand Dehkordi, F., Chehelgerdi, M., Kabiri, H., Salehian-Dehkordi, H., Abdolvand, M., Salmanizadeh, S., Rashidi, M., Niazmand, A., Ahmadi, S., Feizbakhshan, S., Kabiri, S., Vatandoost, N., Ranjbarnejad, T., Exploring the promising potential of induced pluripotent stem cells in cancer research and therapy. *Mol. Cancer*, 22, 1, 189, 2023. doi: 10.1186/s12943-023-01873-0.
82. Huang, N.F. and Li, S., Mesenchymal stem cells for vascular regeneration. *Regener. Med.*, 3, 6, 877–92, 2008. doi: 10.2217/17460751.3.6.877.
83. Mastrolia, I., Foppiani, E.M., Murgia, A., Candini, O., Samarelli, A.V., Grisendi, G., Veronesi, E., Horwitz, E.M., Dominici, M., Challenges in Clinical Development of Mesenchymal Stromal/Stem Cells: Concise Review. *Stem Cells Transl. Med.*, 8, 11, 1135–1148, 2019. doi: 10.1002/sctm.19-0044.

84. Damasceno, P.K.F., de Santana, T.A., Santos, G.C., Orge, I.D., Silva, D.N., Albuquerque, J.F., Golinelli, G., Grisendi, G., Pinelli, M., dos Santos, R.R., Dominici, M., Soares, M.B.P., Genetic Engineering as a Strategy to Improve the Therapeutic Efficacy of Mesenchymal Stem/Stromal Cells in Regenerative Medicine. *Front. Cell Dev. Biol.*, 8, 737, 20202020. https://doi.org/10.3389/fcell.2020.00737.
85. Salem, H.K. and Thiemermann, C., Mesenchymal stromal cells: current understanding and clinical status. *Stem Cells*, 28, 3, 585–96, 2010. doi: 10.1002/stem.269.
86. Zappia, E., Casazza, S., Pedemonte, E., Benvenuto, F., Bonanni, I., Gerdoni, E., Giunti, D., Ceravolo, A., Cazzanti, F., Frassoni, F., Mancardi, G., Uccelli, A., Mesenchymal stem cells ameliorate experimental autoimmune encephalomyelitis inducing T-cell anergy. *Blood*, 106, 5, 1755–61, 2005. doi: 10.1182/blood-2005-04-1496.
87. Dardalhon, V., Korn, T., Kuchroo, V.K., Anderson, A.C., Role of Th1 and Th17 cells in organ-specific autoimmunity. *J. Autoimmunity*, 31, 3, 252–6, 2008. doi: 10.1016/j.jaut.2008.04.017.
88. Payne, N.L., Dantanarayana, A., Sun, G., Moussa., L., Caine, S., McDonald, C., Herszfeld, D., Bernard, C.C.A., Siatskas, C., Early intervention with gene-modified mesenchymal stem cells overexpressing interleukin-4 enhances anti-inflammatory responses and functional recovery in experimental autoimmune demyelination. *Cell Adhes. Migr.*, 6, 3, 179–189, 2012. https://doi.org/10.4161/cam.20341.
89. Mingliang, R., Bo, Z., Zhengguo, W., Stem cells for cardiac repair: status, mechanisms, and new strategies. *Stem Cells Int.*, 2011, 310928, 2011. doi: 10.4061/2011/310928.
90. Hou, L., Kim, J.J., Woo, Y.J., Huang, N.F., Stem cell-based therapies to promote angiogenesis in ischemic cardiovascular disease. *Am. J. Physiol. Heart Circ. Physiol.*, 310, 4, H455–65, 2016. doi: 10.1152/ajpheart.00726.2015.
91. Jin, Y., Li, S., Yu, Q., Chen, T., Liu, D., Application of stem cells in regeneration medicine. *MedComm*, 4, 4, e291, 2023. doi: 10.1002/mco2.291.
92. Togarrati, P.P., Dinglasan, N., Desai, S., Ryan, W.R., Muench, M.O., CD29 is highly expressed on epithelial, myoepithelial, and mesenchymal stromal cells of human salivary glands. *Oral Dis.*, 24, 4, 561–572, 2018. doi: 10.1111/odi.12812.
93. Tepekoy, F., Ozturk, S., Sozen, B., Ozay, R.S., Akkoyunlu, G., Demir, N., CD90 and CD105 expression in the mouse ovary and testis at different stages of postnatal development. *Reprod. Biol.*, 15, 4, 195–204, 2015. doi: 10.1016/j.repbio.2015.10.004.
94. Romito, A. and Cobellis, G., Pluripotent Stem Cells: Current Understanding and Future Directions. *Stem Cells Int.*, 2016, 9451492, 2016. doi: 10.1155/2016/9451492.
95. Mason, E.A., Mar, J.C., Laslett, A.L., Pera, M.F., Quackenbush, J., Wolvetang, E., Wells, C.A., Gene expression variability as a unifying element of the

pluripotency network. *Stem Cell Rep.*, 3, 2, 365–77, 2014. doi: 10.1016/j.stemcr.2014.06.008.

96. Chehelgerdi, M., Behdarvand Dehkordi, F., Chehelgerdi, M., Kabiri, H., Salehian-Dehkordi, H., Abdolvand, M., Salmanizadeh, S., Rashidi, M., Niazmand, A., Ahmadi, S., Feizbakhshan, S., Kabiri, S., Vatandoost, N., Ranjbarnejad, T., Exploring the promising potential of induced pluripotent stem cells in cancer research and therapy. *Mol. Cancer*, 22, 1, 189, 2023. doi: 10.1186/s12943-023-01873-0.
97. Abdelwahid, E., Kalvelyte, A., Stulpinas, A., de Carvalho, K.A., Guarita-Souza, L.C., Foldes, G., Stem cell death and survival in heart regeneration and repair. *Apoptosis*, 21, 3, 252–68, 2016. doi: 10.1007/s10495-015-1203-4.
98. Mirotsou, M., Jayawardena, T.M., Schmeckpeper, J., Gnecchi, M., Dzau, V.J., Paracrine mechanisms of stem cell reparative and regenerative actions in the heart. *J. Mol. Cell. Cardiol.*, 50, 2, 280–9, 2011. doi: 10.1016/j.yjmcc.2010.08.005.
99. Hu, C. and Li, L., Preconditioning influences mesenchymal stem cell properties *in vitro* and *in vivo*. *J. Cell. Mol. Med.*, 22, 3, 1428–1442, 2018.
100. Shi, W., Xin, Q., Yuan, R., Yuan, Y., Cong, W., Chen, K., Neovascularization: The Main Mechanism of MSCs in Ischemic Heart Disease Therapy. *Front. Cardiovasc. Med.*, 8, 633300, 2021. doi: 10.3389/fcvm.2021.633300.
101. Mas-Bargues, C., Sanz-Ros, J., Román-Domínguez, A., Inglés, M., Gimeno-Mallench, L., El Alami, M., Viña-Almunia, J., Gambini, J., Viña, J., Borrás, C., Relevance of Oxygen Concentration in Stem Cell Culture for Regenerative Medicine. *Int. J. Mol. Sci.*, 20, 5, 1195, 2019. doi: 10.3390/ijms20051195.
102. Gorecka, J., Kostiuk, V., Fereydooni, A., Gonzalez, L., Luo, J., Dash, B., Isaji, T., Ono, S., Liu, S., Lee, S.R., Xu, J., Liu, J., Taniguchi, R., Yastula, B., Hsia, H.C., Qyang, Y., Dardik, A., The potential and limitations of induced pluripotent stem cells to achieve wound healing. *Stem Cell Res. Ther.*, 10, 1, 87, 2019. doi: 10.1186/s13287-019-1185-1.
103. Jin, Y., Li, S., Yu, Q., Chen, T., Liu, D., Application of stem cells in regeneration medicine. *MedComm*, 4, 4, e291, 2023. doi: 10.1002/mco2.291.
104. De Becker, A. and Riet, I.V., Homing and migration of mesenchymal stromal cells: How to improve the efficacy of cell therapy? *World J. Stem Cell*, 8, 3, 73–87, 2016. doi: 10.4252/wjsc.v8.i3.73.
105. Smith, J.A., Das, A., Ray, S.K., Banik, N.L., Role of pro-inflammatory cytokines released from microglia in neurodegenerative diseases. *Brain Res. Bull.*, 87, 1, 10–20, 2012. doi: 10.1016/j.brainresbull.2011.10.004.
106. Sen, A. and Ta, M., Altered Adhesion and Migration of Human Mesenchymal Stromal Cells under Febrile Temperature Stress Involves NF-κβ Pathway. *Sci. Rep.*, 10, 1, 4473, 2020. doi: 10.1038/s41598-020-61361-z. PMID: 32161303; PMCID: PMC7066177.
107. Jiang, W. and Xu, J., Immune modulation by mesenchymal stem cells. *Cell Proliferation*, 53, 1, e12712, 2020. doi: 10.1111/cpr.12712.

108. Gill, J.K., Rehsia, S.K., Verma, E., Sareen, N., Dhingra, S., Stem cell therapy for cardiac regeneration: past, present, and future. *Can. J. Physiol. Pharmacol.*, 102, 3, 161–179, 2024. https://doi.org/10.1139/cjpp-2023-0202.
109. Han, Y., Yang, J., Fang, J., Zhou, Y., Candi, E., Wang, J., Hua, D., Shao, C., Shi, Y., The secretion profile of mesenchymal stem cells and potential applications in treating human diseases. *Signal Transduct. Target. Ther.*, 7, 1, 92, 2022. doi: 10.1038/s41392-022-00932-0.
110. Wu, L.W., Wang, Y.L., Christensen, J.M., Khalifian, S., Schneeberger, S., Raimondi, G., Cooney, D.S., Lee, W.P., Brandacher, G., Donor age negatively affects the immunoregulatory properties of both adipose and bone marrow derived mesenchymal stem cells. *Transplant Immunol.*, 30, 4, 122–7, 2014. doi: 10.1016/j.trim.2014.03.001.
111. Luckheeram, R.V., Zhou, R., Verma, A.D., Xia, B., $CD4^{+}T$ cells: differentiation and functions. *Clin. Dev. Immunol.*, 2012, 925135, 2012. doi: 10.1155/2012/925135.
112. Campostrini, G., Kosmidis, G., Ward-van Oostwaard, D., Davis, R.P., Yiangou, L., Ottaviani, D., Veerman, C.C., Mei, H., Orlova, V.V., Wilde, A.A.M., Bezzina, C.R., Verkerk, A.O., Mummery, C.L., Bellin, M., Maturation of hiPSC-derived cardiomyocytes promotes adult alternative splicing of SCN5A and reveals changes in sodium current associated with cardiac arrhythmia. *Cardiovasc. Res.*, 119, 1, 167–182, 2023. doi: 10.1093/cvr/cvac059.
113. Hernández, R., Jiménez-Luna, C., Perales-Adán, J., Perazzoli, G., Melguizo, C., Prados, J., Differentiation of Human Mesenchymal Stem Cells towards Neuronal Lineage: Clinical Trials in Nervous System Disorders. *Biomol. Ther. (Seoul)*, 28, 1, 34–44, 2020. doi: 10.4062/biomolther.2019.065.
114. Fox, I.J., Daley, G.Q., Goldman, S.A., Huard, J., Kamp, T.J., Trucco, M., Stem cell therapy. Use of differentiated pluripotent stem cells as replacement therapy for treating disease. *Science*, 345, 6199, 1247391, 2014. doi: 10.1126/science.1247391.
115. Wu, Y., Zhao, R.C., Tredget, E.E., Concise review: bone marrow-derived stem/progenitor cells in cutaneous repair and regeneration. *Stem*, 28, 5, 905–15, 2010. doi: 10.1002/stem.420.
116. Ellison, G.M., Vicinanza, C., Smith, A.J., Aquila, I., Leone, A., Waring, C.D., Henning, B.J., Stirparo, G.G., Papait, R., Scarfo, M., Agosti, V., Viglietto, G., Condorelli, G., Indolfi, C., Ottolenghi, S., Torella, D., Nadal-Ginard, B., Adult c-kitpos Cardiac Stem Cells Are Necessary and Sufficient for Functional Cardiac Regeneration and Repair. *Cell*, 154, 4, 827–842, 2013. https://doi.org/10.1016/j.cell.2013.07.039.
117. Floy, M.E., Shabnam, F., Simmons, A.D., Bhute, V.J., Jin, G., Friedrich, W.A., Steinberg, A.B., Palecek, S.P., Advances in Manufacturing Cardiomyocytes from Human Pluripotent Stem Cells. *Annu. Rev. Chem. Bimolecular Eng.*, 13, 255–278, 2022. doi: 10.1146/annurev-chembioeng-092120-033922.

118. Labusca, L., Herea, D.D., Mashayekhi, K., Stem cells as delivery vehicles for regenerative medicine-challenges and perspectives. *World J. Stem Cells*, 10, 5, 43–56, 2018. doi: 10.4252/wjsc.v10.i5.43.
119. Kang, S.K., Shin, I.S., Ko, M.S., Jo, J.Y., Ra, J.C., Journey of mesenchymal stem cells for homing: strategies to enhance efficacy and safety of stem cell therapy. *Stem Cells Int.*, 2012, 342968, 2012. doi: 10.1155/2012/342968.
120. Kharbikar, B.N., Mohindra, P., Desai, T.A., Biomaterials to enhance stem cell transplantation. *Cell Stem Cell*, 29, 5, 692–721, 2022. doi: 10.1016/j.stem.2022.04.002.
121. Musiał-Wysocka, A., Kot, M., Majka, M., The Pros and Cons of Mesenchymal Stem Cell-Based Therapies. *Cell Transplant.*, 28, 7, 801–812, 2019. doi: 10.1177/0963689719837897.
122. Farahzadi, R., Fathi, E., Valipour, B., Ghaffary, S., Stem cells-derived exosomes as cardiac regenerative agents. *Int. J. Cardiology. Heart Vasculature*, 52, 101399, 2024. doi: 10.1016/j.ijcha.2024.101399.
123. Tang, J., Su, T., Huang, K., Dinh, P.U., Wang, Z., Vandergriff, A., Hensley, M.T., Cores, J., Allen, T., Li, T., Sproul, E., Mihalko, E., Lobo, L.J., Ruterbories, L., Lynch, A., Brown, A., Caranasos, T.G., Shen, D., Stouffer, G.A., Gu, Z., Zhang, J., Cheng, K., Targeted repair of heart injury by stem cells fused with platelet nanovesicles. *Nat. Biomed. Eng.*, 2, 17–26, 2018. doi: 10.1038/s41551-017-0182-x.
124. Mohd Satar, A., Othman, F.A., Tan, S.C., Biomaterial application strategies to enhance stem cell-based therapy for ischemic stroke. *World J. Stem Cells*, 14, 12, 851–867, 2022. doi: 10.4252/wjsc.v14.i12.851.
125. Taşdelen, B., Kayaman-Apohan, N., Güven, O., Baysal, B.M., Preparation of poly(N-isopropylacrylamide/itaconic acid) copolymeric hydrogels and their drug release behavior. *Int. J. Pharm.*, 278, 2, 343–51, 2004. doi: 10.1016/j.ijpharm.2004.03.017.
126. Vasu, S., Zhou, J., Chen, J., Johnston, P.V., Kim, D.H., Biomaterials-based Approaches for Cardiac Regeneration. *Korean Circ. J.*, 51, 12, 943–960, 2021. doi: 10.4070/kcj.2021.0291.
127. Li, J., Hu, S., Cheng, K., Engineering better stem cell therapies for treating heart diseases. *Ann. Transl. Med.*, 8, 8, 569, 2020. doi: 10.21037/atm.2020.03.44.
128. Tang, J., Shen, D., Caranasos, T.G., wang, Z., Vandergriff, A.C., Allen, T.A., Hensley, M.T., Dinh, P.-U., Cores, J., Li, T.-S., Zhang, J., Kan, Q., Cheng, K., Therapeutic microparticles functionalized with biomimetic cardiac stem cell membranes and secretome. *Nat. Commun.*, 8, 13724, 2017. https://doi.org/10.1038/ncomms13724.
129. Yan, H., Shao, D., Lao, Y.-H., Li, M., Hu, H., Leong, K.W., Engineering Cell Membrane-Based Nanotherapeutics to Target Inflammation. *Adv. Sci.*, 6, 15, 1900605, 2019. https://doi.org/10.1002/advs.201900605.

7

Neurological Disorders: Advancements in Stem-Cell-Based Treatments

Dheeraj Sharma[1] **and Shriyansh Srivastava**[1,2*]

[1]*Department of Pharmacy, School of Medical and Allied Sciences, Galgotias University, Greater Noida, India*
[2]*Department of Pharmacology, Delhi Pharmaceutical Sciences and Research University (DPSRU), Sector 3 Pushp Vihar, New Delhi, India*

Abstract
A well-functioning nervous system comprises myriad elements that involve the central and peripheral nervous systems, which cause patients to experience physiological, cognitive, and emotional problems that are very often debilitating. The inability of traditional methods of treatment to sort out, cure the ailment, or produce complete healing and restoration is still obtainable despite the enormous progress achieved in medical science. Stem-cell-based therapies have been transformed into a new paradigm that presents the ability to repair, regenerate, and restore damaged brain tissues. This chapter focuses on the most recent developments in stem cell research and the treatment of diseases of the nervous system by this technology. We analyze the different stem cells, their working mechanisms, and the high-end appliances pushing this area ahead. Case studies and clinical trials are practical and provide visibility as to the significance of these treatments. Moreover, ethical aspects, regulatory issues, and directions into the future are described in greater detail to give a complete picture of this unnoticed opportunity in neurological treatment.

***Keywords*:** Neuro-regeneration, pluripotent, neuroprotection, clinical trials, gene editing, immunomodulation

**Corresponding author*: shriyanshsrivastav@gmail.com; shriyanshsrivastav@galgotiasuniversit.edu.in

Ayush Madan, Sumel Ashique, Deepika Arora and Mantosh Kumar Satapathy (eds.) Stem Cell Therapeutics, (351–380) © 2025 Scrivener Publishing LLC

7.1 Introduction

Neurological disorders refer to a cluster of conditions that manifest in the central nervous system (CNS) and peripheral nervous system (PNS), causing severe physical, cognitive, and emotional disabilities. In the wake of medical science advancements, most of the traditional medical therapies for these disorders are just addressing the symptoms, with no reasonable cure. Globally, neurological disorders are a huge cause of disability and death. They are putting an increasing load on the healthcare sector. It is estimated by the WHO that about 6.3% of disability-adjusted life years (DALYs) are due to neurological disorders, with hundreds of millions suffering worldwide [1]. For people with neurological disorders, these difficulties often take the form of challenges in the physical and mental aspects of their lives. They are usually treated in the long term, and though there have been breakthroughs in the treatment of neurological disorders, it is still a daily practice. Conventional therapeutic strategies merely tackle the symptoms; thus, they are mainly preventive rather than curative. The introduction of advanced therapies nowadays may open a new way of treating disorders that can change the disease state, with stem cell therapy being the most promising one [2]. When it comes to traditionally treating neurological disorders, it often involves drug procedures, operations, as well as rehabilitative therapies. However, these methods are only successful in the effort to reduce the symptoms and enhance the quality of life; they often do not deal with the real causes of the diseases. To illustrate, in Parkinson's disease, dopamine drugs are capable of relieving the symptoms but not stopping the loss of neurons [3]. Correspondingly, in Alzheimer's disease, the current medications mainly work on the neurotransmitter systems to temporarily lessen cognitive problems, but they do not thwart the course of neurodegeneration [4]. Additionally, there are many neurological disorders, such as amyotrophic lateral sclerosis (ALS) and spinal cord injuries, that have very few effective treatments, and hence patients are left with hardly any options. The central nervous system in adults is unable to regenerate damaged neurons, and the lack of treatments available to replace lost neural cells is an insurmountable problem in neurology [5]. These constraints signify that there is a necessity for innovations in drug therapy that could mend or restore damaged neural tissue, and stem cell therapy remains the most hopeful for its application [6]. Stem cell therapy involves the use of revolutionary technologies for the treatment of neurological disorders by repairing, regenerating, and restoring damaged neural tissues. Stem cells can turn into various kinds of cells, such as

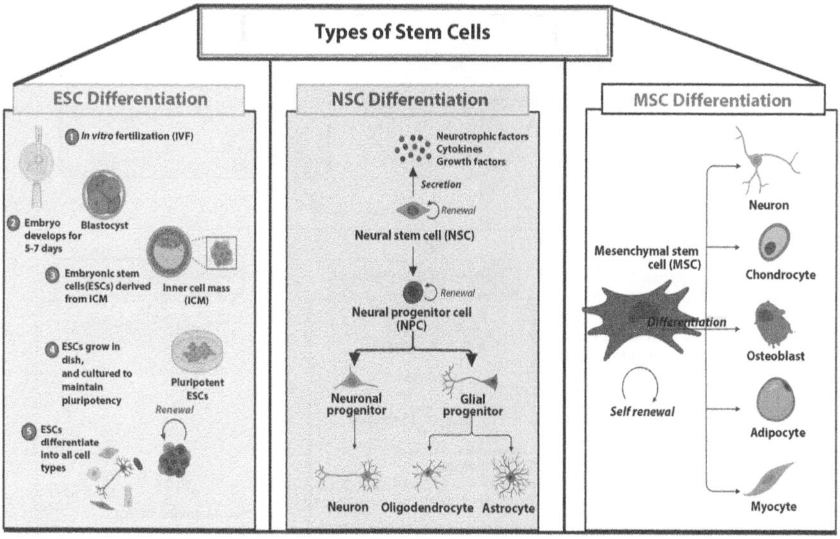

Figure 7.1 Types of stem cells and their differentiation.

neurons, oligodendrocytes, and astrocytes, all crucial for the proper functioning of the nervous system. In the last two decades, stem cell treatment has experienced explosive growth, with studies showing the possibility of stem cells not only indeed substituting worn-out or dead cells but also toughening the disease environment, hence promoting healing and recovery [6]. Various kinds of stem cells were used and are under research in neurological research and therapy, like embryonic stem cells, adult stem cells, induced pluripotent stem cells (iPSCs), and neural stem cells. Each type of stem cell presents both pros and cons, mainly due to its ability to be advanced through differentiation potential, the matter of ethics, and ideas about how the treatment works [7]. Currently, the issue of stem cell use in practices is becoming closer to reality in aspects of research that are currently assessing the treatments in clinical trials for various neurological disorders [8] (Figure 7.1).

7.2 Types of Stem Cells

Stem cells work through their ability to self-renew and differentiate into several cell types. They are invaluable in regenerative medicine and therapeutic applications for neurological disorders. The most significant types of stem cells used in neurological research are embryonic stem cells (ESCs)

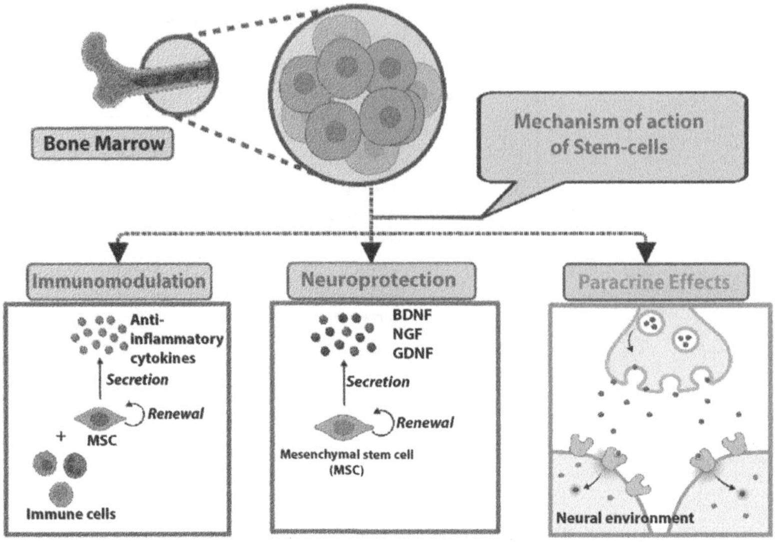

Figure 7.2 Mechanism of action of stem cells.

[9], adult stem cells [10], induced pluripotent stem cells (iPSCs) [11], and neural stem cells (NSCs) [12]. It is worth noting that each type of stem cell presents different characteristics, advantages, and drawbacks that directly influence its therapeutic potential (Figure 7.2).

7.2.1 Embryonic Stem Cells

Embryonic stem cells come from the inner cell mass of the blastocyst; a stage-early embryo that is just being developed. They are pluripotent, and this means that they can be turned into any cell type in the body to make, among others, neurons, oligodendrocytes, and astrocytes, which are required for the treatment of neurodegenerative disorders [13]. The pluripotent feature of ESCs is the main factor that enables their application in regenerative medicine, besides the metamorphosis it duplicates. Anyway, their benefits are accompanied by fundamental problems because of the complicated ethical issues and the risk of tumorigenesis [14]. ESC has given impressive results in preclinical animal models of neural disorders. An example would be that some studies have found that ESC-derived neural progenitor cells are a promising treatment for not only human but also animal Parkinson's disease models and can enhance activity, motor function, and neurogenesis "in animal models of Parkinson's disease" [15].

Moreover, the ability of ESCs to correct spinal cord injuries and cure diseases such as Huntington's disease and amyotrophic lateral sclerosis has also been the subject of research [16]. Despite these advancements, compulsory ethical matters and the risk of immune rejection are the major problems hindering their application in humans [17].

7.2.2 Adult Stem Cells

Adult stem cells and somatic stem cells are the other names for these cells, which are found in various body tissues and form a network of intercellular communication and self-repair of the tissue in which they reside. As opposed to those, adult stem cells are multipotent but not pluripotent, that is, they are only able to differentiate into a limited range of cell types that are closely related to their tissue of origin. The context of neurological disorders makes the neural stem cells which are NSCs and the mesenchymal stem cells the most important ones [18]. The NSCs are located mainly in the subventricular zone of the brain and the hippocampus. They can become not only neurons but also astrocytes and oligodendrocytes. Therefore, they are good candidates for treating neurodegenerative diseases [19]. The study was carried out based on which NSCs were proven to be able to migrate to the injury location in the brain, differentiate into neural cells, and then eventually become an already-existent neural circuit, which offered the potential for recovery in situations such as stroke and traumatic brain injury [2, 20]. MSCs previously taken from bone marrow, adipose tissue, and other sources are now under the spotlight for their neuronal protective and immune modulation abilities. The cells can produce many trophic factors that contribute to neuron survival and inflammation reduction, the latter being very important, especially for the treatment of multiple sclerosis and spinal cord injuries [21]. In addition, MSCs have been investigated in clinical trials that have the potential to improve the results of strokes by impacting the brain repair process and achieving functional recovery [22].

7.2.3 Induced Pluripotent Stem Cells (iPSCs)

Pluripotent-induced stem cells belong to the stem cell family. They can be made by changing the genes of mature somatic cells to make them act embryonic through transcription factors such as Oct4, Sox2, Klf4, and c-Myc [23]. Induced pluripotent stem cells have the same property as ESCs, i.e., pluripotency, which enables them to give rise to any cell type, even neurons and glial cells, which are essential for neurological applications [24]. iPSCs have the upper hand because of their derivation capability from the patient's cells, which in turn reduces the risks of immune rejection as well as the moral issues that come along with ESCs. Modeling of different neurogenic disorders via patient-derived iPSCs *in vitro* has made it possible to examine disease mechanisms and drug screening [25]. For instance, the use of iPSC-derived neurons has led to the study of the pathophysiology of Alzheimer's disease, Parkinson's disease, and amyotrophic lateral sclerosis, which in turn has resulted in the accelerated introduction of target therapies [26, 27]. Moreover, the iPSCs have been experimented with as a potential source for regenerative medicine. Pluripotent stem cell-product applications fall under the category of regenerative medicine. PSC sheets and other fragments are the first that should be restored to make human beings healthy again. To say that cells derived from iPS could restore motor function in animal models of spinal cord injury and promote recovery in ischemic stroke models was first done in preclinical trials [28, 29]. Notwithstanding these favorable findings, iPS cells must first solve the issues of genetic instability and tumorigenesis before they get into clinical settings [30].

7.2.4 Neural Stem Cells

Neural stem cells are a specific type of stem cell that is found in the central nervous system. They are a type of stem cell that is capable of differentiating into the main cell types of the nervous system, such as neurons, astrocytes, and oligodendrocytes [18]. Given their homing property in damaged or degenerated parts, neural stem cells are particularly exciting for the treatment of neurological disorders in the brain and spinal cord [31]. Studies about NSCs are pointing to the fact that they can be a definite source of treatment for many neurological disorders. To cite a particular example, NSCs have been used in experimenting with Parkinson's, in which, using their differentiation, they can produce dopamine neurons to theoretically replace the lost ones of the disease [32]. Furthermore, the potential afforded by transplanted NSCs to contribute to the repair of spinal cord injury models has been reported [33]. Attempts to utilize NSCs in clinics have not stopped with the first experiment. Various studies aimed at finding out if NSCs can be injected into patients with multiple sclerosis, stroke, and spinal cord injuries and their safety and efficacy were also evaluated [34]. Despite these advances, several problems still exist, including transplanted neural stem cell survival, integration, and functional contribution, that need ongoing research for development [35].

7.3 Mechanisms of Action

Stem-cell-based treatments provide allopathic solutions in the management of neurologic diseases. These are examples such as differences and integration into the existing neurological network, neuroprotection, immunomodulation, and paracrine effects. All of these methods are included in the therapeutic benefits of stem cells, however, in different ways.

7.3.1 Differentiation and Integration

The therapeutic effects of stem cells on neurological disorders mainly result from their potential to evolve into a variety of types of neural cells, especially neurons, astrocytes, and oligodendrocytes. Upon maturity, these cells can adopt the status of the actual neural sequences; in this way, they specialize in the lost or destroyed cells, and thereby, they regain their performance. A case in point is Parkinson's disease, where the removal of dopaminergic neurons in the substantia nigra is an important health problem. The cells induced by pluripotent stem cells (iPSCs) and embryonic stem cells (ESCs) have been effective in converting to dopaminergic neurons, which can then be shifted into the brain, where they will be integrated into current networks and thus facilitate motor function recovery [36]. In addition, human beings with injuries to the spinal cord are capable of differentiating into demyelinated axons that will improve motor function recovery [37].

7.3.2 Neuroprotection

Benefits of stem cells include neuroprotection, particularly through the blocking of neuron damage and the promotion of the long life of already existing neurons. There is the fact that the stem cells produce and ultimately release trophic factors, which are the main providers of the neurons' health and functions [38]. Quite typically, mesenchymal stem cells (MSCs) are notable for their ability to cause BDNF, NGF, and GDNF production [39]. All of these substances are needed for the maintenance of synaptic plasticity, the prevention of brain damage caused by inflammation, and the stem cell repair of neuron debris. They can be used in treatments involving neurological diseases and damage. For instance, they can help treat such diseases as stroke and traumatic brain injury because the mentioned substances have cytoprotective effects on the neurons [40].

7.3.3 Immunomodulation

One more critical modality of the action of stem cells is their impact on the immune system, especially MSCs that are connected to immunity. One of the neurological disorders is closely related to the inflammatory response it raises, which in turn can only intensify the neural damage. MSCs practice their function in the immune system by joining forces with immune cells such as T cells, B cells, and macrophages, and then the activation of pro-inflammatory cytokines is inhibited and the production of anti-inflammatory cytokines is stimulated [41]. This change in the immune system is the main factor in some autoimmune diseases, such as multiple sclerosis, that have demonstrated the capacity of MSC to reduce immune cell movement to the central nervous system, which, in turn, is used to minimize the disease patient's sufferings [21, 42]. Also, in neurological diseases like Alzheimer's, these cells have another substantial function, which is to reduce chronic inflammation by participating in the growth of the nerve tissue [43].

7.3.4 Paracrine Effects

Not only do stem cells differentiate and integrate directly but they also follow ordinary therapy mechanisms through paracrine signalling. That is, bioactive molecules such as cytokines, growth factors, and exosomes are secreted, which can affect the nearby neural environment and regenerate injury processes [44]. For example, MSCs not only produce a set of trophic factors that help neurons survive but also induce the repair mechanisms that are endogens, such as the activation of resident neural cells and the enhancement of synaptic plasticity [45]. In stroke models, the therapeutic effects of transplanted stem cells included the reduction of infarct size and the enhancement of functional recovery. These have been the different aspects that underpinned the regeneration of the nervous system [46].

7.4 Technological Advancements in Stem Cell Research

The advances in technology have resulted in stem cells being used for the treatment of neurodegenerative diseases with better success than ever before. They imply innovations in CRISPR and gene editing, making organoids and 3D bioprinting, the use of biomaterials and scaffolding, and the application of single-cell sequencing [47].

7.4.1 CRISPR and Gene Editing

The CRISPR-Cas9 technology has considerably influenced gene editing. It offers an opportunity to make precise and effective modifications to the genome. In stem cell research, CRISPR is used to rectify the genetic mutations that are causing the occurrence of neurological disorders, thus rendering it so far a useful tool in personalized therapy development [48]. Thus far, CRISPR has been utilized to rectify the mutation in the genes related to Huntington's disease and amyotrophic lateral sclerosis (ALS) in iPSCs, which has been considered the hope for future treatments [49, 50]. Through the editing of patient iPSCs' DNA, scientists can create these cells by incorporating the corrected gene and transferring them back into the individual, thus potentially leading to the regression of the disease pathology [51]. Moreover, the CRISPR technology eradicates the process of disease model creation by inserting specific mutations into the stem cells. Later, these cells can be converted to neural cells through differentiation. These models are very helpful for the study of the mechanisms of neurological diseases as well as the testing of the proposed treatments [52].

7.4.2 Organoids and 3D Bioprinting

Brain organoids and 3D bioprinting technologies developed by scientific analysis have paved a new path to examine neurological illnesses and assess new remedies. Organoids are minimally re-created, more natural bodies that are produced from stem cells *in vitro*. Brain organoids, made from iPSCs or ESCs, reproduce the structure and function of the human brain and are increasingly used in the studies of complex diseases or neurological disorders [53]. On the other hand, by using a disease-in-a-dish model, stem-cell-derived mini brains that mimic the human brain and disease progression are also becoming more popular. Moreover, it is also being employed for drug testing in a completely authentic human environment in a far more precise and faster way [54]. Moreover, organoids can be manipulated by following the genetic codes using CRISPR to examine their influence on brain development and function [55]. These are also technologies used apart from bioprinting, but they still contribute to increasing our ability to conduct research. An indispensable development that facilitates this kind of study is 3D printing technology, which can be used to electronically place stem cells and other biomaterials to make intricate tissue structures [56].

7.4.3 Biomaterials and Scaffolding

Biomaterials are very vital in the aid of stem cells that are targeted by neural treatments, they also support the growth and differentiation of brain cells. These materials are the scaffold that resembles the natural extracellular matrix and, thus, allow cell adhesion, proliferation, and differentiation into neural lineages [57]. A familiar one at this point is hydrogels, which are materials that are often employed as racks with high water contents and biocompatibility. Biomaterials have the additional capability to work in coordination with 3D bioprinting to fabricate more exact models of neural tissue, which can be used for research and therapeutic applications [58]. You can mold them to release growth factors that guide stem cells to differentiate into specific neural cells, thus increasing the success of stem cell therapies for diseases like spinal cord injury and stroke [59].

7.4.4 Single-Cell Sequencing

Single-cell sequencing is a cutting-edge technology that gives scientists the possibility to measure the gene expression profiles of the cells separately. This research procedure is especially important for stem cell research as it shows the heterogeneity of stem cell populations and how the cells specialize to form tissues [61]. In the realm of neurological disorders, single-cell sequencing has been waiting in the wings to solve the obscurity of the differentiation of iPSCs to neural cells ever since it was used for the first time. It has opened up the individual pathways in the neural development of the cell and given insight into the most possible targets for therapy [62]. This technology is also used to uncover the ways that stem cells interact with a microenvironment and how they are altered by certain treatments, which is why new and positive therapies are increasingly being found [63]. Moreover, single-cell RNA sequencing has been implemented as part of a study concerning the cellular composition of brain organoids, which stands for creating maps of cell types and states; hence, this model is very important for the understanding of disease roots and treatments [64] (Table 7.1).

Table 7.1 Technological advancements in stem cell research.

Technology	**Description**	**Applications in neurological disorders**	**References**
CRISPR-Cas9 gene editing	CRISPR-Cas9 is a precise gene editing tool that allows for effective modifications of the genome.	Used to correct genetic mutations in iPSCs related to Huntington's disease and ALS	[48–52]
Organoids and 3D bioprinting	Brain organoids and 3D bioprinting technologies create miniaturized, *in vitro* models of the human brain.	Employed in disease modeling, drug testing, and studying neurological disorders like Alzheimer's and Parkinson's	[53–56]
Biomaterials and scaffolding	Biomaterials serve as scaffolds, mimicking the natural extracellular matrix to support cell growth.	Aid in the differentiation of stem cells into neural lineages, particularly useful in spinal cord injury and stroke therapies	[57–60]
Single-cell sequencing	A cutting-edge technology that measures gene expression profiles at the single-cell level.	Helps in understanding the heterogeneity of stem cell populations, crucial for developing targeted therapies	[61–64]

7.5 Stem-Cell-Based Treatments for Specific Neurological Disorders

7.5.1 Parkinson's Disease

Parkinson's disease (PD) is a gradual, degenerative brain disorder characterized by the steady loss of dopamine-producing neurons that gets progressively worse with time in the substantia nigra and causes a wide range of motor deficits such as tremors, rigidity, and bradykinesia. Stem cell therapy is a relatively new long-term curative approach for PD, which is unique because it looks to replace the missing neurons and restore the production of dopamine in the brain [3]. Using iPSCs and ESCs to prepare dopaminergic neurons for transplantation has been found to provide breakthrough results in preclinical models and also in the case of initial clinical trials. For example, the very real situation in Japan saw successful therapies in which iPSC-derived dopaminergic neurons engendered *in situ* brain tissues. The treatment, in this case, resulted in the restoration of motor function as well as further effects on the regeneration of cells [65]. However, the success of transplanted "iPSC"-derived cells remains a mystery for patients with Parkinson's disease. ESCs have even been proven to be the means through which the restoration of motor function in animal models of PD is being attained and are currently being tested in clinical trials [66]. Regulatory budgets concerning the case of patients with PD will be higher since we will need to monitor the progress of the patient after the stem cell treatment in a more intense way. Tumor formation and immune rejection are the major safety issues that should be properly addressed through intensive clinical trials [14].

7.5.2 Alzheimer's Disease

Alzheimer's disease (AD) is the most common form of dementia and is characterized by the accumulation of plaques that are made of amyloid-beta and tangles that are made of tau in the brain, causing neuronal death and cognitive decline. Using stem cells that are made in the lab can provide the body with the sufficient amount of cells required to fix the damage that has been done to the neurons and, at the same time, allow the remaining cells in the brain to generate new ones [67]. The demonstration of MSCs and NSCs' potential to produce a significant number of neurotrophins that can prolong the lives of neurons as well as reduce the amounts of amyloid-beta accumulation, which, in turn, slows down the progression of the disease, was done through research [68]. NSC transplantation has resulted in the improvement of cognitive function and a decrease in neuropathology in animals. Moreover, the development of clinical trials has also been revealed [69]. Clinical trials that will assess the safety and efficacy of stem cell therapy in AD patients are now ongoing. Initial results indicate that stem cells can regulate the immune response and curb inflammation in the brain, which in turn may shield neuronal death from happening [70]. However, more research is required to determine the long-term advantages and possible risks of this method.

7.5.3 Multiple Sclerosis

The cause of multiple sclerosis (MS) is the breakdown of myelin, an insulating layer that guards the nerve fibres, and hence the impaired transmission of signals and neurological symptoms. The application of stem cell technology is being studied as a future method to fix damaged myelin and retrieve neurological function in cases of MS [71]. Hematopoietic stem cell transplantation (HSCT) has helped in many cases and in stopping the progression of the disease in severe forms by hitting the reset switch of the immune system. Auto-HSCT, a technique using the patient's stem cells, has resulted in much improvement in the number of relapse cases and thus in the level of neurological function [72]. Besides HSCT, the MSCs have undergone research for remineralization and degradation protection. It has been observed that the MSCs mediate the immune response, which results in the alleviation of inflammation, tissue repair, and consequent improvement in clinical outcomes in patients with MS [73]. However, the long-term efficiency and safety of these treatments remain to be properly established.

7.5.4 Stroke

Stroke is a leading cause of disability in the world and arises because of blood flow blockage to the brain, followed by neuron death. Researchers have demonstrated that stem cells can be a potential platform for attaining neurodegeneration in the case of stroke sufferers and also as a means to increase recovery [74]. Recent research has found that transplantation of NSCs and MSCs can promote neurogenesis (neuron production), angiogenesis (the formation of new blood capillaries), and synaptic plasticity (the strengthening of connections occurring in the nervous system between nerve cells) in the brains of animal models. Moreover, this caused the animals to achieve even higher functional outcomes [75]. Also, through the clinical application of stem cells, it was revealed that, in addition to being safe in stroke therapy, stem cells may even improve brain function and quality of life [76]. One of the most potent methods employed with iPSCs was to set them to the corresponding neurons of the patients for transplantation. To see how it helps to treat liability or not, the availability of exchange of iPSCs was verified with the help of surveys [77].

7.5.5 Spinal Cord Injuries

Spinal cord injuries (SCIs) often cause permanent paralysis due to the incapacity of damaged neurons to regenerate. The treatment of the stem cell is aimed at growing cells that will replace the lost ones, carry out axonal re-growth, and repair affected areas in patients with SCI [78]. The testing of NSCs and MSCs in animals has led to a great development of the technique through processes of cell regeneration, myelination, and behavior recovery [33]. Trial tests have also shown stem cell therapy to be safe, as well as its potential to provide a cure for patients with chronic SCI. Some studies have even reported improvements in their functional recovery and quality of life [79]. Discoveries made in stem cell transplantation and their use of biomaterials and scaffolds make it possible to talk about cell therapy for SCI in this way. These applications facilitate axonal growth and create suitable conditions for neural regeneration, thus increasing the probability of a successful recovery [80].

7.5.6 Amyotrophic Lateral Sclerosis (ALS)

Amyotrophic lateral sclerosis (ALS) is a progressive neurodegenerative disease that affects the motor neurons and leads to weakness of the muscles and eventually paralysis. Stem cell treatment is being explored as a hopeful strategy to decelerate illness, as shown by the stem cells that protect the motoneurons [81]. Previous research has indicated that MSCs and NSCs can release nerve growth factors that support motor neuron survival and lower spinal cord inflammation, which consequently could be ALS [82]. The trials have shown that the transplant of stem cells might be used safely with ALS patients, as some research studies have reported a disease pace that is slower and has better chances of survival [83]. Nevertheless, the efficiency of stem cell therapy in ALS is still unknown, and more examination has to be done to get an expert opinion on the best treatment for this highly life-threatening disorder [84].

7.6 Clinical Trials and Case Studies

7.6.1 Overview of Current Clinical Trials

In the field of stem cell clinical trials, covering neurological disorders, such research provides a vast and rapidly changing landscape. This enormous number of clinical trials has been started in different parts of the world to investigate the safety and efficacy of diverse kinds of stem cells, including ESCs, iPSCs, MSCs, and NSCs, in the treatment of diseases such as Parkinson's disease, Alzheimer's disease, multiple sclerosis, stroke, and spinal cord injuries. For example, the NCT02452723 study, which was carried out by the University of California, San Francisco, is for the transplantation of iPSC-derived dopaminergic progenitor cells into patients with Parkinson's disease. The information that we have so far from this trial data is promising: the transplanted cells can get established and survive in the brain of the host and consequently cause motor function benefits. The success of stem cell transplantation is in a race with the actual surgery [85]. In the same way, the NCT01217008 investigation is a clinical trial that aims at using NSCs in patients with chronic spinal cord injuries. The first results seem to be extremely promising since the transplanted cells not only promote the healing of the nerve tissue but also improve the motor functions of the patients, even those who have irreversible injuries [79]. As for Alzheimer's disease, the NCT03117738 trial is a research project to uncover how MSCs could be factors that control the immune response and stop amyloid-beta from choosing the brain as a target. A strict precaution to ensure safety first among the hovered MSCs has been reached. In total, 40% of the participants have stunningly agreed to continuous cognitive reassessment [70]. In general, these trials represent a very important phase in the development of stem cell therapies for neurological disorders, providing information about their potential research and guiding future activities in this domain.

7.6.2 Success Stories and Outcomes

The success of experiments with stem cell therapies in clinical procedures has been exhibited in several high-profile cases, thereby underscoring the possibility of the complete transformation of the handling of neurological disorders with these treatments. An example of a very successful experiment is the case of a patient who had Parkinson's disease and was part of a clinical trial in which iPSC-derived neuron dopaminergic cells were going to be transplanted. The patient showed significant improvement in motor skills and quality of life, with no disadvantages being noted [85]. This case has been highly referenced as proof of the use of iPSCs regarding neurodegenerative diseases and has, therefore, led to more research being carried out on this treatment. Yet another touchstone for success is the clinical trial, where the NSCs approach was employed and included patients with chronic spinal cord injuries. After receiving the stem cell treatment, one paralysis patient, who had been so for years and years, recovered the capability to move body parts and sensory functions [79]. This superb outcome has shifted the focus of SCI patient treatment and has also brought to light stem cell therapies' power to induce neural regeneration. During the clinical trial of a patient with multiple sclerosis (MS) treated with HSCT in the NCT02674217 trial, a patient was fortunate enough to experience a complete cessation of disease activity and a substantial improvement in the function of the nervous system with the use of autologous stem cell therapy [72]. This is the case that has made it clear that HSCT can be used to reset the immune system and stop disease progression in severe MS cases. These cases of success provide solid evidence of stem cell therapy's ability to produce guided neural regeneration, thereby inadvertently predicting a future that is burgeoning with other newly developed breakthrough innovations in this field.

7.6.3 Challenges and Setbacks

Stem cell treatments for neurological disorders have not been a bed of roses, despite the beautiful results reported in some clinical trials. The barriers to this field have a lot of implications for its future as well as for the necessity of further research and innovation. One of the biggest pitfalls of stem cell therapy is the risk of tumour formation, especially with ESCs and iPSCs. Cell transplants forming tumours are the main factor behind the studies being protective, and in some cases, the discarding of trials that go wrong [30]. The way to resolve this problem is by learning more about cells and also developing methods that will prevent the generation of side effects and the proper insertion of the cells into healthy tissues. On the other hand, a common challenge that patients come across on their way to therapy with stem cells is the different responses that these human beings have. These are the causing factors for major improvements for some patients, the impossibility of progress simultaneously with others, and, on the contrary, the generation of side effects. The problem of the same patient's condition, the kind and quality of the cells, as well as the administration style, can be nominated as the major one [86]. Personalized technologies designed for stem cell treatment, such as the application of patients's own induced pluripotent stem cells, will benefit the solution to this issue. More studies need to be performed to make these therapies more effective. Additionally, one more issue that the patients might come across is the unevenness in their reaction to stem cell therapy. The immune attack can also be a cause of some patients undergoing graft failure and not working, even though immunosuppressive therapies are given [14]. Besides stem cell therapy, other alternative methods of immune rejection also exist. The introduction of a screening regimen to create "universal" cells that will be accepted by immune cells will also lead to a very active area of scientific research, with the cell cells that are created thus being capable of being used in any host [14, 17].

7.7 Ethical Considerations and Regulatory Challenges

A milestone in the growth and exploitation of stem-cell-based therapies in treating neurological disorders is the emergence of an entire spectrum of ethical and regulatory issues. The issue of these problems lies in the fact that stem cells have a unique nature, and embryonic stem cells (ESCs) in particular, differentiate into any cell type in the human body. The impetus behind these therapies is to reach clinical reality; therefore, it is of utmost importance to figure out all the ethical and regulatory bottlenecks that will define the way ahead of this so-considered promising field.

7.7.1 Ethical Issues in Stem Cell Research

An important ethical issue in stem cell research is the use of embryonic stem cells (ESCs), which are derived from early-stage embryos. These cells are usually collected by destroying the embryo and thus pose serious moral and ethical dilemmas, especially for those who claim that life will commence at the moment of fertilization. The ethical debate faces a further complication with the availability of ESCs to treat a wide range of diseases because of the paradox of potential benefits to patients versus the moral status of the embryo [17]. Some of the solutions to this issue have been found by using other types of pluripotent stem cells. These include how iPSCs are made that are reprogrammed from adult cells to pluripotent cells and fixing genetic defects in the human body without the use of embryos. Nevertheless, while iPSCs evade some of the ethical issues of ESCs, they create new problems, like concerns about genetic modifications and the potential occurrence of unintended consequences [87]. The other ethical question that arises is about the commercialization of stem cell therapies. High costs associated with the development and administration of these treatments have caused a lot of queries about the fairness and accessibility of the treatments to patients in low- or middle-income countries. Arguments about the over-commercialisation of unproven stem cell therapies have been the kernel of "stem cell tourism," a practice where people go to countries with less strict regulations to get treatments that may be unsafe or ineffective [88]. There is a moral issue with stem cell research that relates to the donor's consent and the use of donor cells. In one case, the use of donor stem cells such as umbilical cord blood or bone marrow raised the issue of informed consent, especially in cases where the cells have been received from minors or people who do not completely realize the effect of their donation [89].

7.7.2 Regulatory Landscape

The implementation of stem cell therapies is strikingly different from one state to another and reflects the ethics, culture, and law of the area. In the majority of countries, the framework for the regulation of stem cell research and therapy still has gaps and needs to be updated as governments deal with new obstacles due to these technologies. In the United States, the Food and Drug Administration (FDA) takes care of the regulation of stem cell therapies under the Public Health Service Act and the Federal Food, Drug, and Cosmetic Act. The FDA has categorized stem cell products as "biological products," which undergo strict examination in a clinical trial to prove their safety and efficiency before being approved for sale. The FDA has also issued guidance documents to help clarify the regulatory requirements for stem cell therapies, such as the distinction between minimally manipulated and more substantially altered cells, which are regulated under different standards [90]. Europe has its own set of guidelines for the regulation of stem cell therapies through the European Medicines Agency (EMA), which has introduced a centralized approval process for advanced therapy medicinal products (ATMPs), such as stem-cell-based treatments. The EMA emphasizes the regulation of stem cell therapies in terms of quality, safety, and effectiveness while enforcing quality control in how stem cell therapies are made and used. This is the quest to establish the best scientific as well as ethical standards [91]. However, some, particularly in the Asian region, have adapted the regulatory approach, which is so lenient that it has even led to the fast development of stem cell technologies, but some questions that relate to quality and oversight have also been raised. For example, in Japan, the government has created a new drug approval system that allows the clinical testing of some stem cell therapies on people, after which the products can be commercialized. As the method has decreased the delivery lead time, it still has to face criticism for its potential risk to human safety [92]. One of the most complex issues for world stem cell research today is the fact that the regulatory framework for stem cell therapies is different from one country to another, and it is indeed a labyrinth. This means that the standards for stem cell research and therapy differ enormously, leading to the difficulties of international collaboration and the global development of these treatments [93].

7.7.3 Patient Consent and Safety

Introducing patient consent and safety is a crucial element in the clinical application of stem cell therapies. Because many of these treatments are still experimental, patients have to have a complete grasp of everything before they will agree to be in the clinical trials, or they can also decide to take treatment. Informed consent is a fundamental part of ethical medical practice, which involves the patients being given the exact nature of the treatment and any potential risks, side effects, and uncertainties. This is the one thing that is different from the other kinds of medical technologies: stem cell technologies offer long-term results, which are, in most cases, unknown. This is why the situation is so critical in stem cell therapies, where there is so much uncertainty. The potential for the development of adverse effects, like immune rejection or tumour formation, remains a concern. Besides the informed consent, measures have to be taken against safety incidents so patients safely go through stem cell treatment. To that end, physicians need to carefully select patients, watch out for negative interactions, and adhere to the principles of the standard procedures that secure the quality and consistency of the stem cells used. Regulatory agencies are essential for promoting the oversight of these security measures and the enforcement of stem cell therapies that are made and administered in such a way as to put the well-being of the patient above everything else [94]. The ethical and regulatory issues stemming from stem cell therapy emphasize the need for constant dialogue amongst researchers, clinicians, ethicists, regulators, and patient advocacy groups. To achieve this objective, these actors must come together to find possible solutions to complex research topics and improve the quality of human life using ethical research and the application of bioscience.

7.8 Future Directions and Potential Innovations

Frontiers in stem cell biology and applications of regenerative neurology are being painted by trailblazing research. The use of gene therapy tools such as CRISPR-Cas9 to facilitate stem cell therapy is the most promising area of investigation. This is because gene therapy helps pinpoint the mutation in the gene that is directly responsible for a brain disease or the production of novel therapeutic factors that encourage the repair of neurons [95]. Brain organoids are a newly discovered frontier in stem cell research that allows the development of 3D human brain models. They are miniatures developed from stem cells. New insights into these organoids have shown their usage as a platform for the study of the development of neurological diseases and testing potential therapies in a controlled laboratory environment. Breakthroughs in organoid technology have allowed for the making of more intricate and physiologically close models, which are now seen in the use of Alzheimer's disease, Parkinson's disease, and epilepsy studies [96]. Furthermore, gene editing and organoids are not the only two areas of application that AI and machine learning technologies could incorporate in stem cell research, setting the stage for completely new ways of drug discovery and personalized treatments. AI algorithms can analyze vast amounts of data from stem cell experiments to identify patterns and predict the outcomes of different treatments, accelerating the development of new therapies [97]. As these technologies become more and more advanced, they should become key players in the acquisition of information on neurological disorders and the development of better treatments. Personalized medicine, which tailors treatments to the individual characteristics of each patient, is going to change the face of brain stem cell therapy. The emergence of technologies for induced pluripotent stem cells (iPSC) is a great revolution in this context since iPSCs can be derived from patients' cells and used to make customized models of diseases. With this, researchers are now able to learn more about the individual genetic and molecular characteristics of the patient's condition and to experiment with different therapeutic strategies in a personalized context [98]. Personalized stem cell therapies are believed to be powerful in defeating various traditional treatment' problems, which are mainly immune rejection and the patient's variability of reactions. The transplantation of a patient's cells can evade immune rejection. The therapy is precisely customized based on the patient's case. For instance, in the case of Parkinson's, iPSCs from the patient's cells can be differentiated into dopaminergic neurons and then put back into the brain to restore the function lost [85]. A lot of progress is being made in research in this field, so it is expected that the development of stem cell therapies that are tailored to the individual will be more targeted, efficient, and safe not only for neurological disorders but also for a variety of other diseases. Nevertheless, we still have to overcome a few major issues so that we can enjoy the full benefits of personalized medicine. The most important ones are the high costs of personalized medicine, the need for sophisticated infrastructure, and the difficulty in manufacturing and delivering patient-specific stem cells. Solving the above-mentioned problems is a prerequisite for the full realization of personalized medicine applications in the treatment of neurological diseases [99].

7.8.1 Potential for Preventive Therapies

New methods of stem cell therapy are currently being tried out, mostly to deal with neurological disorders that are already in existence, but one can also observe keen curiosity in the application of stem cells in preventive care. The premise of this potential medical technique is to employ stem cells at the initial stage of the illness, at a time when the destruction of some cells is at a very small level or completely absent (invisible), to stop or delay the disease symptoms. This therapeutic strategy would be very useful for neurodegenerative diseases, such as Alzheimer's disease, that usually gradually develop over a long period before they can be identified [91]. One of the potential real-life applications of stem cell therapy for prevention could be the care of some chronic conditions linked to the increased risk of neurodegeneration, such as hypertension. Stem cells must be allowed to construct new tissue or blood vessels whose integrity is already lost. It shall be accompanied by lower body inflammation and preservation of the brain's well-being, and, therefore, the possibility of stroke, dementia, and the health of the nervous system will be considerably decreased [100, 101]. Though the idea of stem cell techniques for the prevention of diseases is newly introduced, ongoing research in this field promises great opportunities. If these therapies are successful, new healthcare methods will emerge, shifting the focus from managing problems by treating them to keeping the patients' brains healthy and functioning over time.

7.8.2 The Future of Stem-Cell-Based Treatments in Neurology

The perspectives pointing to the horizon are full of percentages, and it looks like neurology will be one of the first to benefit from them. The inherence of technology and a deeper understanding of stem cell biology with progress towards developing significant alternatives that can modify the landscape of neurological medicine is quite clear. The cell therapy techniques (stem cells) that are personalized for each patient according to their genetic and molecular profiles are very likely to be realized, thus offering even a greater number of effective rather than dangerous ways of handling an array of neurological disorders. Moreover, the power to use stem cells in preventive medicine can be opened wide, which affords the possibility of retarding or stopping the onset of neurodegenerative diseases. In this way, the quality of life of millions of people would be significantly improved. Nevertheless, the implementation of such ideas will first of all depend on continuing investments in research, the development of strong regulatory frameworks, and the willingness to face the ethical challenges that arise. In conclusion, even though there are still very serious challenges to be faced, the stormy road to stem-cell therapy in neurology is open. These cures are announced to have the chance of not only treating but also completely healing some of the most severe neurological disorders. Consequently, they favor patients and their families all over the world.

7.9 Conclusion

The outlined section of this book chapter has given a tour of the latest stem cell therapies for neurological disorders, where we put forward a full-fledged review of where this section of medicine reaches and the possible benefits of it in the future. Our journey started with the exploration of the categories of stem cells applied in the treatment of disorders, especially those related to the specific features of embryonic stem cells (ESCs) and induced pluripotent stem cells (iPSCs). We further got into the inner mechanisms by which stem cells are therapeutically functional, which include differentiation, neuroprotection, immunomodulation, and paracrine signalling. The new technological skills that were noted as essential facilitators of stem cell therapies are CRISPR, a gene editing tool, 3D bioprinting and biomaterials, organoid models, and single-cell sequencing. These inventions are speeding up the process of disease treatment. Besides this, we also explored the use of stem cell therapies on different specific neurological disorders, such as Parkinson's, Alzheimer's, MS, stroke, spinal cord injuries, and ALS, which illustrated the advancement achieved in clinical trials and real-world case studies. Furthermore, we saw the ethical considerations as well as regulatory challenges in stem cell research that were discussed, along with the need for a fair scientific progress enhancer that is not in conflict with moral and ethical issues. Finally, in the honed-up chapter, we explored future routes to the beginning of the field, such as the newest scientific research, methylation affecting them, the introduction of personalized treatment, and the detection of cumulative preventive drugs that can reduce the burden of neurodegenerative diseases.

References

1. Organization, W.H., *Neurological disorders: public health challenges: World Health Organization*, 2006.
2. Lindvall, O. and Kokaia, Z., Stem cells for the treatment of neurological disorders. *Nature*, 441, 7097, 1094–6, 2006.
3. Kalia, L.V. and Lang, A.E., Parkinson's disease. *Lancet*, 386, 9996, 896–912, 2015.
4. Cummings, J., Lee, G., Mortsdorf, T., Ritter, A., Zhong, K., Alzheimer's disease drug development pipeline: 2017. *Alzheimer's Dementia Transl. Res. Clin. Interv.*, 3, 3, 367–84, 2017.
5. Rowland, L.P. and Shneider, N.A., Amyotrophic lateral sclerosis. *N. Engl. J. Med.*, 344, 22, 1688–700, 2001.
6. Glenn, J.D. and Whartenby, K.A., Mesenchymal stem cells: emerging mechanisms of immunomodulation and therapy. *World J. Stem Cells*, 6, 5, 526, 2014.
7. Masoudi Asil, S., Ahlawat, J., Guillama Barroso, G., Narayan, M., Application of nanotechnology in stem-cell-based therapy of neurodegenerative diseases. *Appl. Sci.*, 10, 14, 4852, 2020.
8. Salado-Manzano, C., Perpiña, U., Straccia, M., Molina-Ruiz, F.J., Cozzi, E., Rosser, A.E., *et al.*, Is the immunological response a bottleneck for cell therapy in neurodegenerative diseases? *Front. Cell. Neurosci.*, 14, 250, 2020.
9. Rippon, H. and Bishop, A., Embryonic stem cells. *Cell Proliferation*, 37, 1, 23–34, 2004.
10. Young, H.E. and Black Jr., A.C., Adult stem cells. *Anat. Rec. Part A Discoveries Molecular Cellular Evol. Biology An Off. Publ. Am. Assoc. Anatomists.*, 276, 1, 75–102, 2004.
11. Cerneckis, J., Cai, H., Shi, Y., Induced pluripotent stem cells (iPSCs): molecular mechanisms of induction and applications. *Signal Transduct. Target. Ther.*, 9, 1, 112, 2024.
12. De Gioia, R., Biella, F., Citterio, G., Rizzo, F., Abati, E., Nizzardo, M., *et al.*, Neural stem cell transplantation for neurodegenerative diseases. *Int. J. Mol. Sci.*, 21, 9, 3103, 2020.
13. Thomson, J.A., Itskovitz-Eldor, J., Shapiro, S.S., Waknitz, M.A., Swiergiel, J.J., Marshall, V.S., *et al.*, Embryonic stem cell lines derived from human blastocysts. *Science*, 282, 5391, 1145–7, 1998.

14. Lewandowski, J. and Kurpisz, M., Techniques of human embryonic stem cell and induced pluripotent stem cell derivation. *Arch. Immunol. Ther. Exp.*, 64, 349–70, 2016.
15. Kim, J.-H., Auerbach, J.M., Rodríguez-Gómez, J.A., Velasco, I., Gavin, D., Lumelsky, N., *et al.*, Dopamine neurons derived from embryonic stem cells function in an animal model of Parkinson's disease. *Nature*, 418, 6893, 50–6, 2002.
16. Tao, Y. and Zhang, S.-C., Neural subtype specification from human pluripotent stem cells. *Cell Stem Cell*, 19, 5, 573–86, 2016.
17. Lo, B. and Parham, L., Ethical issues in stem cell research. *Endocr. Rev.*, 30, 3, 204–13, 2009.
18. Gage, F.H., Mammalian neural stem cells. *Science*, 287, 5457, 1433–8, 2000.
19. Matta, R. and Gonzalez, A.L., Stroke repair via biomimicry of the subventricular zone. *Front. Mater.*, 5, 15, 2018.
20. Garzón-Muvdi, T. and Quiñones-Hinojosa, A., Neural stem cell niches and homing: recruitment and integration into functional tissues. *Ilar J.*, 51, 1, 3–23, 2010.
21. Uccelli, A., Laroni, A., Freedman, M.S., Mesenchymal stem cells as a treatment for MS–progress to date. *Multiple Sclerosis J.*, 19, 5, 515–9, 2013.
22. Lukomska, B., Stanaszek, L., Zuba-Surma, E., Legosz, P., Sarzynska, S., Drela, K., Challenges and controversies in human mesenchymal stem cell therapy. *Stem Cells Int.*, 2019, 1, 9628536, 2019.
23. Maali, A., Maroufi, F., Sadeghi, F., Atashi, A., Kouchaki, R., Moghadami, M., *et al.*, Induced pluripotent stem cell technology: trends in molecular biology, from genetics to epigenetics. *Epigenomics*, 13, 08, 631–47, 2021.
24. Selvaraj, V., Jiang, P., Chechneva, O., Lo, U.-G., Deng, W., Differentiating human stem cells into neurons and glial cells for neural repair. *Front Biosci.*, 17, 65–89, 2012.
25. Yamanaka, S., Pluripotent stem cell-based cell therapy—promise and challenges. *Cell Stem Cell*, 27, 4, 523–31, 2020.
26. Zhang, X., Hu, D., Shang, Y., Qi, X., Using induced pluripotent stem cell neuronal models to study neurodegenerative diseases. *Biochim. Biophys. Acta, Mol. Basis Dis.*, 1866, 4, 165431, 2020.
27. Pasteuning-Vuhman, S., de Jongh, R., Timmers, A., Pasterkamp, R.J., Towards advanced iPSC-based drug development for neurodegenerative disease. *Trends Mol. Med.*, 27, 3, 263–79, 2021.
28. Lu, P., Woodruff, G., Wang, Y., Graham, L., Hunt, M., Wu, D., *et al.*, Long-distance axonal growth from human induced pluripotent stem cells after spinal cord injury. *Neuron*, 83, 4, 789–96, 2014.
29. Tornero, D., Wattananit, S., Grønning Madsen, M., Koch, P., Wood, J., Tatarishvili, J., *et al.*, Human induced pluripotent stem cell-derived cortical neurons integrate in stroke-injured cortex and improve functional recovery. *Brain*, 136, 12, 3561–77, 2013.

30. Yoshihara, M., Hayashizaki, Y., Murakawa, Y., Genomic instability of iPSCs: challenges towards their clinical applications. *Stem Cell Rev. Rep.*, 13, 7–16, 2017.
31. Martino, G. and Pluchino, S., The therapeutic potential of neural stem cells. *Nat. Rev. Neurosci.*, 7, 5, 395–406, 2006.
32. Wernig, M., Zhao, J.-P., Pruszak, J., Hedlund, E., Fu, D., Soldner, F., *et al.*, Neurons derived from reprogrammed fibroblasts functionally integrate into the fetal brain and improve symptoms of rats with Parkinson's disease. *Proc. Natl. Acad. Sci.*, 105, 15, 5856–61, 2008.
33. Cummings, B.J., Uchida, N., Tamaki, S.J., Salazar, D.L., Hooshmand, M., Summers, R., *et al.*, Human neural stem cells differentiate and promote locomotor recovery in spinal cord-injured mice. *Proc. Natl. Acad. Sci.*, 102, 39, 14069–74, 2005.
34. Vishwakarma, S.K., Bardia, A., Tiwari, S.K., Paspala, S.A., Khan, A.A., Current concept in neural regeneration research: NSCs isolation, characterization and transplantation in various neurodegenerative diseases and stroke: A review. *J. Adv. Res.*, 5, 3, 277–94, 2014.
35. Grade, S. and Götz, M., Neuronal replacement therapy: previous achievements and challenges ahead. *npj Regener. Med.*, 2, 1, 29, 2017.
36. Hallett, P.J., Deleidi, M., Astradsson, A., Smith, G.A., Cooper, O., Osborn, T.M., *et al.*, Successful function of autologous iPSC-derived dopamine neurons following transplantation in a non-human primate model of Parkinson's disease. *Cell Stem Cell*, 16, 3, 269–74, 2015.
37. Keirstead, H.S., Nistor, G., Bernal, G., Totoiu, M., Cloutier, F., Sharp, K., *et al.*, Human embryonic stem cell-derived oligodendrocyte progenitor cell transplants remyelinate and restore locomotion after spinal cord injury. *J. Neurosci.*, 25, 19, 4694–705, 2005.
38. Wang, J., Hu, W.-W., Jiang, Z., Feng, M.-J., Advances in treatment of neurodegenerative diseases: perspectives for combination of stem cells with neurotrophic factors. *World J. Stem Cells*, 12, 5, 323, 2020.
39. Rahbaran, M., Zekiy, A.O., Bahramali, M., Jahangir, M., Mardasi, M., Sakhaei, D., *et al.*, Therapeutic utility of mesenchymal stromal cell (MSC)-based approaches in chronic neurodegeneration: a glimpse into underlying mechanisms, current status, and prospects. *Cell. Mol. Biol. Lett.*, 27, 1, 56, 2022.
40. Aboutaleb, N., Faezi, M., Maleki, S.N., Nazarinia, D., Tousi, S.M.T.R., Hashemirad, N., Conditioned medium obtained from mesenchymal stem cells attenuates focal cerebral ischemia reperfusion injury through activation of ERK1/ERK2-BDNF signaling pathway. *J. Chem. Neuroanat.*, 97, 87–98, 2019.
41. Jiang, W. and Xu, J., Immune modulation by mesenchymal stem cells. *Cell Proliferation*, 53, 1, e12712, 2020.
42. Gordon, D., Pavlovska, G., Glover, C.P., Uney, J.B., Wraith, D., Scolding, N.J., Human mesenchymal stem cells abrogate experimental allergic

encephalomyelitis after intraperitoneal injection, and with sparse CNS infiltration. *Neurosci. Lett.*, 448, 1, 71–3, 2008.

43. Lee, J.K., Jin, H.K., Endo, S., Schuchman, E.H., Carter, J.E., Bae, J.-., Intracerebral transplantation of bone marrow-derived mesenchymal stem cells reduces amyloid-beta deposition and rescues memory deficits in Alzheimer's disease mice by modulation of immune responses. *Stem Cells*, 28, 2, 329–43, 2010.
44. Kusuma, G.D., Carthew, J., Lim, R., Frith, J.E., Effect of the microenvironment on mesenchymal stem cell paracrine signaling: opportunities to engineer the therapeutic effect. *Stem Cells Dev.*, 26, 9, 617–31, 2017.
45. Andres, R.H., Horie, N., Slikker, W., Keren-Gill, H., Zhan, K., Sun, G., *et al.*, Human neural stem cells enhance structural plasticity and axonal transport in the ischaemic brain. *Brain*, 134, 6, 1777–89, 2011.
46. Tatarishvili, J., Oki, K., Monni, E., Koch, P., Memanishvili, T., Buga, A.-M., *et al.*, Human induced pluripotent stem cells improve recovery in stroke-injured aged rats. *Restor. Neurol. Neurosci.*, 32, 4, 547–58, 2014.
47. Rauth, S., Karmakar, S., Batra, S.K., Ponnusamy, M.P., Recent advances in organoid development and applications in disease modeling. *Biochim. Biophys. Acta, Rev. Cancer*, 1875, 2, 188527, 2021.
48. Liang, N., Trujillo, C.A., Negraes, P.D., Muotri, A.R., Lameu, C., Ulrich, H., Stem cell contributions to neurological disease modeling and personalized medicine. *Prog. Neuro-Psychopharmacol. Biol. Psychiatry*, 80, 54–62, 2018.
49. Yang, S., Chang, R., Yang, H., Zhao, T., Hong, Y., Kong, H.E., *et al.*, CRISPR/Cas9-mediated gene editing ameliorates neurotoxicity in mouse model of Huntington's disease. *J. Clin. Invest.*, 127, 7, 2719–24, 2017.
50. Wang, L., Yi, F., Fu, L., Yang, J., Wang, S., Wang, Z., *et al.*, CRISPR/Cas9-mediated targeted gene correction in amyotrophic lateral sclerosis patient iPSCs. *Protein Cell*, 8, 5, 365–78, 2017.
51. Christensen, C.L. and Choy, F.Y., A prospective treatment option for lysosomal storage diseases: CRISPR/Cas9 gene editing technology for mutation correction in induced pluripotent stem cells. *Diseases*, 5, 1, 6, 2017.
52. Bhinge, A., Namboori, S.C., Zhang, X., VanDongen, A.M., Stanton, L.W., Genetic correction of SOD1 mutant iPSCs reveals ERK and JNK activated AP1 as a driver of neurodegeneration in amyotrophic lateral sclerosis. *Stem Cell Rep.*, 8, 4, 856–69, 2017.
53. Villa, C., Combi, R., Conconi, D., Lavitrano, M., Patient-derived induced pluripotent stem cells (iPSCs) and cerebral organoids for drug screening and development in autism spectrum disorder: opportunities and challenges. *Pharmaceutics*, 13, 2, 280, 2021.
54. Lancaster, M.A. and Knoblich, J.A., Generation of cerebral organoids from human pluripotent stem cells. *Nat. Protoc.*, 9, 10, 2329–40, 2014.
55. Qian, X., Song, H., Ming, G.-., Brain organoids: advances, applications and challenges. *Development*, 146, 8, dev166074, 2019.

56. Murphy, S.V. and Atala, A., 3D bioprinting of tissues and organs. *Nat. Biotechnol.*, 32, 8, 773–85, 2014.
57. Yaldiz, B., Saglam-Metiner, P., Yesil-Celiktas, O., Decellularised extracellular matrix-based biomaterials for repair and regeneration of central nervous system. *Expert Rev. Mol. Med.*, 23, e25, 2021.
58. Bedir, T., Ulag, S., Ustundag, C.B., Gunduz, O., 3D bioprinting applications in neural tissue engineering for spinal cord injury repair. *Mater. Sci. Engineering C.*, 110, 110741, 2020.
59. Gupta, D., Tator, C.H., Shoichet, M.S., Fast-gelling injectable blend of hyaluronan and methylcellulose for intrathecal, localized delivery to the injured spinal cord. *Biomaterials*, 27, 11, 2370–9, 2006.
60. Qu, M., Jiang, X., Zhou, X., Wang, C., Wu, Q., Ren, L., *et al.*, Stimuli-responsive delivery of growth factors for tissue engineering. *Adv. Healthcare Mater.*, 9, 7, 1901714, 2020.
61. Zheng, H., Pomyen, Y., Hernandez, M.O., Li, C., Livak, F., Tang, W., *et al.*, Single-cell analysis reveals cancer stem cell heterogeneity in hepatocellular carcinoma. *Hepatology*, 68, 1, 127–40, 2018.
62. Lang, C., Campbell, K.R., Ryan, B.J., Carling, P., Attar, M., Vowles, J., *et al.*, Single-cell sequencing of iPSC-dopamine neurons reconstructs disease progression and identifies HDAC4 as a regulator of Parkinson cell phenotypes. *Cell Stem Cell*, 24, 1, 93–106.e6, 2019.
63. Shang, Z., Chen, D., Wang, Q., Wang, S., Deng, Q., Wu, L., *et al.*, Single-cell RNA-seq reveals dynamic transcriptome profiling in human early neural differentiation. *Gigascience*, 7, 11, giy117, 2018.
64. Camp, J.G., Badsha, F., Florio, M., Kanton, S., Gerber, T., Wilsch-Bräuninger, M., *et al.*, Human cerebral organoids recapitulate gene expression programs of fetal neocortex development. *Proc. Natl. Acad. Sci.*, 112, 51, 15672–7, 2015.
65. Takahashi, J., Strategies for bringing stem cell-derived dopamine neurons to the clinic: the Kyoto trial. *Prog. Brain Res.*, 230, 213–26, 2017.
66. Kirkeby, A., Parmar, M., Barker, R.A., Strategies for bringing stem cell-derived dopamine neurons to the clinic: a European approach (STEM-PD). *Prog. Brain Res.*, 230, 165–90, 2017.
67. Liu, X.-Y., Yang, L.-P., Zhao, L., Stem cell therapy for Alzheimer's disease. *World J. Stem Cells*, 12, 8, 787, 2020.
68. Blurton-Jones, M., Kitazawa, M., Martinez-Coria, H., Castello, N.A., Müller, F.-J., Loring, J.F., *et al.*, Neural stem cells improve cognition via BDNF in a transgenic model of Alzheimer disease. *Proc. Natl. Acad. Sci.*, 106, 32, 13594–9, 2009.
69. Teng, Y.D., Benn, S.C., Kalkanis, S.N., Shefner, J.M., Onario, R.C., Cheng, B., *et al.*, Multimodal actions of neural stem cells in a mouse model of ALS: a meta-analysis. *Sci. Transl. Med.*, 4, 165, 165ra4–ra4, 2012.
70. Kang, J.M., Yeon, B.K., Cho, S.-J., Suh, Y.-H., Stem cell therapy for Alzheimer's disease: a review of recent clinical trials. *J. Alzheimers Dis.*, 54, 3, 879–89, 2016.

71. Genc, B., Bozan, H.R., Genc, S., Genc, K., Stem cell therapy for multiple sclerosis. *Tissue Eng. Regener. Med.*, 1084, 145–74, 2019.
72. Muraro, P.A., Pasquini, M., Atkins, H.L., Bowen, J.D., Farge, D., Fassas, A., *et al.*, Long-term outcomes after autologous hematopoietic stem cell transplantation for multiple sclerosis. *JAMA Neurol.*, 74, 4, 459–69, 2017.
73. Alanazi, A., Alassiri, M., Jawdat, D., Almalik, Y., Mesenchymal stem cell therapy: a review of clinical trials for multiple sclerosis. *Regener. Therapy.*, 21, 201–9, 2022.
74. Locatelli, F., Bersano, A., Ballabio, E., Lanfranconi, S., Papadimitriou, D., Strazzer, S., *et al.*, Stem cell therapy in stroke. *Cell. Mol. Life Sci.*, 66, 757–72, 2009.
75. Zhang, G.-L., Zhu, Z.-H., Wang, Y.-Z., Neural stem cell transplantation therapy for brain ischemic stroke: Review and perspectives. *World J. Stem Cells.*, 11, 10, 817, 2019.
76. Kalladka, D. and Muir, K.W., Stem cell therapy in stroke: designing clinical trials. *Neurochem. Int.*, 59, 3, 367–70, 2011.
77. Pandey, S., Jirásko, M., Lochman, J., Chvátal, A., Chottova Dvorakova, M., Kučera, R., iPSCs in neurodegenerative disorders: A unique platform for clinical research and personalized medicine. *J. Personalized Med.*, 12, 9, 1485, 2022.
78. Gao, L., Peng, Y., Xu, W., He, P., Li, T., Lu, X., *et al.*, Progress in stem cell therapy for spinal cord injury. *Stem Cells Int.*, 2020, 1, 2853650, 2020.
79. Curtis, E., Martin, J.R., Gabel, B., Sidhu, N., Rzesiewicz, T.K., Mandeville, R., *et al.*, A first-in-human, phase I study of neural stem cell transplantation for chronic spinal cord injury. *Cell Stem Cell*, 22, 6, 941–50, 2018.
80. Katoh, H., Yokota, K., Fehlings, M.G., Regeneration of spinal cord connectivity through stem cell transplantation and biomaterial scaffolds. *Front. Cell. Neurosci.*, 13, 248, 2019.
81. Mazzini, L., Ferrari, D., Andjus, P.R., Buzanska, L., Cantello, R., De Marchi, F., *et al.*, Advances in stem cell therapy for amyotrophic lateral sclerosis. *Expert Opin. Biol. Ther.*, 18, 8, 865–81, 2018.
82. Mazzini, L., Vercelli, A., Ferrero, I., Mareschi, K., Boido, M., Servo, S., *et al.*, Stem cells in amyotrophic lateral sclerosis: state of the art. *Expert Opin. Biol. Ther.*, 9, 10, 1245–58, 2009.
83. Glass, J.D., Hertzberg, V.S., Boulis, N.M., Riley, J., Federici, T., Polak, M., *et al.*, Transplantation of spinal cord–derived neural stem cells for ALS: analysis of phase 1 and 2 trials. *Neurology*, 87, 4, 392–400, 2016.
84. Petrou, P., Gothelf, Y., Argov, Z., Gotkine, M., Levy, Y.S., Kassis, I., *et al.*, Safety and clinical effects of mesenchymal stem cells secreting neurotrophic factor transplantation in patients with amyotrophic lateral sclerosis: results of phase 1/2 and 2a clinical trials. *JAMA Neurol.*, 73, 3, 337–44, 2016.

85. Schweitzer, J.S., Song, B., Herrington, T.M., Park, T.-Y., Lee, N., Ko, S., *et al.*, Personalized iPSC-derived dopamine progenitor cells for Parkinson's disease. *N. Engl. J. Med.*, 382, 20, 1926–32, 2020.
86. Cohen, J.A., Baldassari, L.E., Atkins, H.L., Bowen, J.D., Bredeson, C., Carpenter, P.A., *et al.*, Autologous hematopoietic cell transplantation for treatment-refractory relapsing multiple sclerosis: position statement from the American Society for Blood and Marrow Transplantation. *Biol. Blood Marrow Transplant.*, 25, 5, 845–54, 2019.
87. Hyun, I., The bioethics of stem cell research and therapy. *J. Clin. Invest.*, 120, 1, 71–5, 2010.
88. Lau, D., Ogbogu, U., Taylor, B., Stafinski, T., Menon, D., Caulfield, T., Stem cell clinics online: the direct-to-consumer portrayal of stem cell medicine. *Cell Stem Cell*, 3, 6, 591–4, 2008.
89. Sugarman, J., Kaalund, V., Kodish, E., Marshall, M.F., Reisner, E.G., Wilfond, B.S., *et al.*, Ethical issues in umbilical cord blood banking. *Jama*, 278, 11, 938–43, 1997.
90. USFDA, *Current Good Tissue Practice (CGTP) and Additional Requirements for Manufacturers of Human Cells, Tissues, and Cellular and Tissue-Based Products (HCT/Ps)*, 2011.
91. Dwenger, A., Strassburger, J., Schwerdtfeger, W., Regulation (EC) No. 1394/2007 on advanced therapy medicinal products: Incorporation into national law. *Bundesgesundheitsblatt-Gesundheitsforschung-Gesundheitsschutz.*, 53, 14–9, 2010.
92. Sipp, D. and Turner, L., US regulation of stem cells as medical products. *Science.*, 338, 6112, 1296–7, 2012.
93. Isasi, R.M. and Knoppers, B.M., Governing stem cell banks and registries: emerging issues. *Stem Cell Res.*, 3, 2-3, 96–105, 2009.
94. Sugarman, J., Barker, R.A., Charo, R.A., A professional standard for informed consent for stem cell therapies. *JAMA.*, 322, 17, 1651–2, 2019.
95. Li, M., Suzuki, K., Kim, N.Y., Liu, G.-H., Belmonte, J.C.I., A cut above the rest: targeted genome editing technologies in human pluripotent stem cells. *J. Biol. Chem.*, 289, 8, 4594–9, 2014.
96. Quadrato, G., Nguyen, T., Macosko, E.Z., Sherwood, J.L., Min Yang, S., Berger, D.R., *et al.*, Cell diversity and network dynamics in photosensitive human brain organoids. *Nature*, 545, 7652, 48–53, 2017.
97. Kim, M. and Hong, S., Integrating Artificial Intelligence to Biomedical Science: New Applications for Innovative Stem Cell Research and Drug Development. *Technologies*, 12, 7, 95, 2024.
98. Shi, Y., Inoue, H., Wu, J.C., Yamanaka, S., Induced pluripotent stem cell technology: a decade of progress. *Nat. Rev. Drug Discov.*, 16, 2, 115–30, 2017.
99. Okita, K. and Yamanaka, S., Induced pluripotent stem cells: opportunities and challenges. *Philos. Trans. R. Soc. B Biol. Sciences.*, 366, 1575, 2198–207, 2011.

100. Sart, S., Ma, T., Li, Y., Preconditioning stem cells for *in vivo* delivery. *Biores. Open Access*, 3, 4, 137–49, 2014.
101. Mirotsou, M., Jayawardena, T.M., Schmeckpeper, J., Gnecchi, M., Dzau, V.J., Paracrine mechanisms of stem cell reparative and regenerative actions in the heart. *J. Mol. Cell. Cardiol.*, 50, 2, 280–9, 2011.

8

Stem Cell Therapy for Metabolic Disorders: Harnessing the Power of Stem Cells to Treat Diabetes, Obesity, and Other Metabolic Disorders

Sindhu D. Bali[1,2], Anand Kumar Shukla[1,2], Rosaleen Sahoo[1,2], Varsha A. Mahadik[1,2] and Narendra Kadoo[1,2*]

[1]Biochemical Sciences Division, CSIR-National Chemical Laboratory, Pune, India
[2]Academy of Scientific and Innovative Research (AcSIR), Ghaziabad, India

Abstract

Diabetes and obesity are metabolic disorders that have become major public health concerns worldwide. Stem cell therapy is emerging as a promising approach to treat these disorders due to the ability of stem cells to repair or replace damaged tissues. This chapter provides an in-depth review of the present state of stem cell therapy to treat metabolic disorders, highlighting the latest advances and future directions in this field. The chapter begins by discussing the fundamental principles of stem cell therapy, including different types of stem cells and their unique properties. It then explores current clinical applications of stem cell therapy for metabolic disorders, which includes the use of stem cells like mesenchymal stem cells, induced pluripotent stem cells, and hematopoietic stem cells. The chapter also delves into the mechanisms that stem cells adopt to regulate the immune response and aid in tissue repair in metabolic disorders, as well as the potential challenges and limitations associated with this therapeutic approach. Finally, the chapter concludes by discussing the future directions for stem cell therapy in combating metabolic disorders, including the integration of stem cell therapy with other therapeutic approaches and the development of new stem cell sources.

***Keywords*:** Metabolic disorder, stem cell therapy, stem cell homing, immune regulation, regenerative medicine

**Corresponding author*: ny.kadoo@ncl.res.in

Ayush Madan, Sumel Ashique, Deepika Arora and Mantosh Kumar Satapathy (eds.) Stem Cell Therapeutics, (381–408) © 2025 Scrivener Publishing LLC

8.1 Introduction

8.1.1 Overview of Metabolic Disorders and Their Current Treatment Options

A metabolic disorder is defined as a condition where the metabolic processes of a healthy person are abruptly altered due to varied chemical reactions. Metabolic disorders can result when the normal metabolic processes, such as glycolysis, Kreb's cycle, protein synthesis, hormone metabolism, etc., get altered due to abnormal chemical reactions. Some of the metabolic disorders include diabetes, obesity, thyroid problems, and inherited conditions like phenylketonuria and Gaucher's disease [1]. Elevated blood sugar levels resulting either due to insulin resistance or reduced insulin production lead to diabetes [2]. Obesity involves excessive fat accumulation that poses health risks and is often linked to other metabolic disorders [3]. Currently, available treatment options broadly include medication, surgery, and lifestyle changes. However, many patients still have difficulty managing these disorders over a long term or lifelong and achieving remission in spite of these treatments [4].

Diabetes is a long-term metabolic disorder characterized by high blood glucose levels due to defects in insulin secretion, insulin resistance, or both. The condition is broadly categorized into two main types: type 1 diabetes, the condition of autoimmunity where the body attacks its own insulin-producing beta cells in the pancreas, and type 2 diabetes, which is more common and results due to conditions like insulin resistance or its relative deficiency [5]. Diabetes symptoms include excessive thirst, increased frequency of urination, sudden weight loss, and tiredness. Continued and long-term complications can damage various organs, leading to cardiovascular diseases, neuropathy, nephropathy, and retinopathy. While type 1 diabetes demands life-long insulin therapy, type 2 diabetes can be managed with oral medications, insulin therapy, or other injectable medications. Monitoring the levels of blood sugar, maintaining a balanced diet, practicing regular physical exercises, and regular medical checkups are crucial to managing diabetes and preventing complications like heart-related diseases, nerve damage, and issues related to the kidney [6].

Similarly, obesity is a complex health condition resulting from the accumulation of excessive body fat that increases the risks of several diseases and disorders, like heart disease, diabetes, and certain cancers. It results from genetic, behavioral, metabolic, and environmental factors [7]. Treatments for obesity typically involve lifestyle changes, including a balanced diet, increased physical activity, and behavioral therapy. Medical interventions include medications such as orlistat, liraglutide, and phentermine-topiramate, which can increase feelings of fullness, help reduce appetite, or inhibit fat absorption [8]. For individuals with severe obesity or those who do not respond to other treatments, bariatric surgery options, like gastric bypass and sleeve gastrectomy, can be considered [9]. A comprehensive approach that includes psychological support and continuous monitoring is essential for effective and long-term weight management.

The deficiency of the enzyme phenylalanine hydroxylase can lead to phenylketonuria (PKU). This rare genetic disorder is characterized by the body's inability to metabolize the amino acid phenylalanine into tyrosine, leading to its over-accumulation in the blood. This can lead to developmental delays, intellectual disabilities, seizures, and other neurological problems if not managed properly [10]. Early diagnosis through newborn screening is crucial for effective treatment. The primary treatment for PKU involves a strict and balanced lifetime diet low in phenylalanine and proteins. However, adequate nutrition should be ensured with unique medical formulae and low-protein foods. Additionally, some patients may benefit from the medication sapropterin dihydrochloride, which can enhance enzyme activity and help reduce phenylalanine levels in mild cases of PKU. Gene therapy and enzyme substitution therapies are also being explored as potential treatments [11]. Regular monitoring of phenylalanine by healthcare professionals is essential to manage PKU effectively and prevent complications.

Table 8.1 shows the most common metabolic disorders and currently available treatments. However, despite the available treatments, patients often take a long time to get cured. Some metabolic disorders cannot be cured, requiring lifelong treatments to manage the disorder. Figure 8.1 shows the correlation between blood glucose levels and HbA1c scores, indicating normal, prediabetic, and diabetic ranges. The prediabetic and diabetic ranges indicate the severity of altered blood glucose levels from the normal range.

The goals of the current metabolic disorder therapy options are managing symptoms and avoiding complications. The primary therapies include lifestyle modifications such as consuming a nutritious diet, controlling weight, and regular exercise to increase metabolism. Drugs that help in lowering blood sugar levels, improving insulin sensitivity, and helping lose weight include insulin, metformin, and GLP-1 receptor agonists for diabetes, as well as orlistat and liraglutide for obesity [19]. Patients with severe obesity who do not respond to these treatments need to undergo bariatric surgery or other surgical procedures to remove the excess body fat. Significant weight loss and the improvement or resolution of metabolic

Table 8.1 Common metabolic disorders, their descriptions, and common treatments available

Metabolic disorder	Description	Common treatments	References
Obesity	Excessive fat accumulation in the body leading to various health issues, including heart disease and diabetes	Lifestyle changes (diet and exercise) Medications (e.g., semaglutide, liraglutide) Bariatric surgery in severe cases Psychological management for emotional support	[12]
Diabetes mellitus	A metabolic disorder resulting from inappropriately elevated blood glucose levels	Lifestyle changes (diet, exercise) Insulin therapy Oral hypoglycemics (e.g., metformin)	[13]
Phenylketonuria (PKU)	A genetic disorder resulting in the inability to metabolize the amino acid phenylalanine	Special low-phenylalanine or protein diet Sapropterin dihydrochloride (Kuvan) in some cases	[14]
Cystic fibrosis	A genetic disorder that affects the lungs and digestive system due to thick mucus production	Enzyme replacement therapy CFTR modulators (e.g., ivacaftor) Nutritional support	[15]

(*Continued*)

Table 8.1 Common metabolic disorders, their descriptions, and common treatments available (*Continued*)

Metabolic disorder	Description	Common treatments	References
Metabolic syndrome (insulin resistance, atherogenic dyslipidemia, obesity, and hypertension)	A cluster of conditions increasing the chance of heart-related diseases, stroke, and type 2 diabetes	Lifestyle modifications (diet, exercise) Medications for hypertension, dyslipidemia, and hyperglycemia Weight management	[16]
Urea cycle disorders	Genetic disorders that affect the ability of the body to remove ammonia from the bloodstream	Dietary protein restriction Medications (e.g., arginine, sodium benzoate) Liver transplant in severe cases	[17]

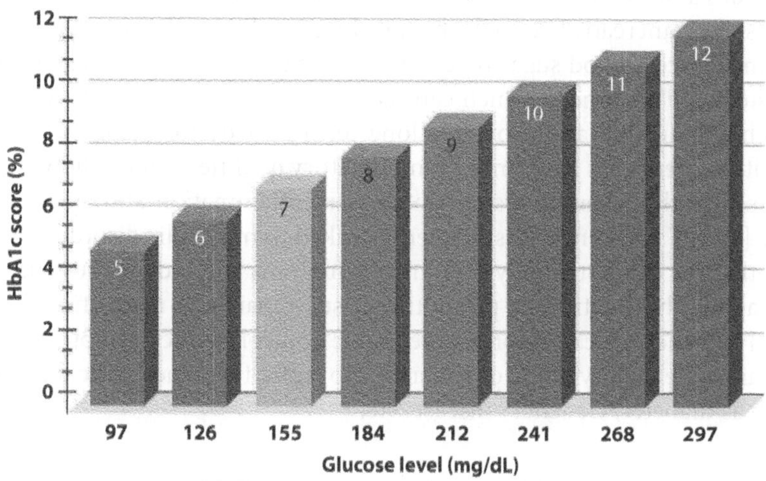

Figure 8.1 Blood glucose levels in mg/dL and HbA1c score (%), categorized into normal (green), prediabetic (yellow), and diabetic ranges (red) [18].

disorders, including type 2 diabetes, can result after bariatric surgery [20]. However, despite these therapeutic possibilities, patients face challenges with long-term maintenance and remission. This may be because of socio-economic hurdles, genetic predisposition, and inability to adhere to the treatment plan [21]. New and improved therapies that provide more long-term and effective solutions are therefore desperately needed. In view of this, stem cell therapy is becoming a more viable option with potential reparative and regenerative effects that can completely change the treatment of metabolic disorders.

8.1.2 Significance of Stem Cell Therapy in Metabolic Disorders

Metabolic disorders, which are often caused by mutations in genetic material, leading to enzyme deficiencies or dysfunctional metabolic pathways, can be treated with various therapies, including stem cell therapy. Stem cells, mainly induced pluripotent stem cells (iPSCs) and mesenchymal stem cells (MSCs), have the unique property of differentiating into various cell types and replacing or repairing damaged tissues [22]. This therapeutic approach can mainly focus on the root cause of metabolic disorders by providing a source of healthy, functional cells capable of restoring normal metabolic functions. Hence, with these remarkable qualities, employing stem cells provides a more potent treatment method for metabolic disorders. In diabetes, the pancreatic beta cells that produce insulin get damaged, resulting in unbalanced blood sugar regulation. These beta cells can be regenerated by stem cell treatment, which can reestablish the capacity of the body to synthesize insulin and provide a long-term remedy for diabetes [23]. In addition, stem cells have immunomodulatory qualities, which allows them to control the immune system and lessen inflammation and autoimmunity [4], the two conditions frequently linked to metabolic disorders. Stem cell therapy can potentially both treat symptoms and reduce the causes of metabolic dysfunction by promoting tissue repair and restoration. Thus, stem cell therapy can improve metabolic control, reduce symptoms, and enhance the quality of life of patients who suffer from metabolic disorders such as diabetes, Gaucher's disease, etc.

8.1.3 Fundamental Principles of Stem Cell Therapy

8.1.3.1 Stem Cell Types

Stem cells have the unique ability to differentiate into various specialized cell types. The stem cells are categorized into different types based on their origin and potency. Embryonic stem cells (ESCs) are pluripotent and derived from the inner cell mass of blastocysts (early-stage embryos). Hence, they can differentiate into nearly all cell types of an adult organism (Figure 8.2). They can differentiate to form ectoderm, endoderm, and mesoderm cell lineages. Somatic stem cells, found in or derived from various tissues like bone marrow (hematopoietic stem cells) and brain (neural stem cells), are generally multipotent or unipotent, having a limited differentiation capacity, which is usually restricted to the cell types of their tissue of origin. These are also known as progenitor cells and can differentiate into a few closely related cell types [24]. Induced pluripotent stem cells (iPSCs) are adult somatic cells that can be reprogrammed back into a pluripotent stem cell-like state, offering the versatility of embryonic stem cells without the associated ethical concerns for regenerative medicine. Lastly, the perinatal stem cells, found in amniotic fluid and umbilical cord blood, possess capabilities between embryonic and adult stem cells. They have

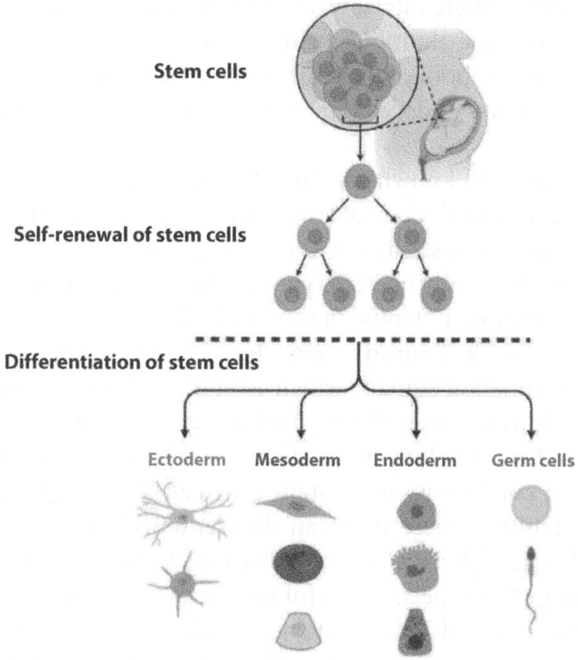

Figure 8.2 Differentiation of pluripotent stem cells in the ectoderm (neural and pigment cells), mesoderm (muscle, blood, and kidney cells), endoderm (beta, lung, and pancreatic cells), and germ cells (egg and sperm cells). The image was rendered using Biorender (https://www.biorender.com/).

potential applications in regenerative medicine. Thus, every type of stem cells has distinct applications and potential in therapeutic interventions and medical research.

8.1.3.2 Unique Characteristics of Stem Cells

Stem cells are unique cells due to several fundamental properties that distinguish them from other types of cells in the body. Self-renewal, potency, differentiation, homing and migration, immunomodulation, and paracrine signaling are a few of the important characteristics of stem cells [25]. These characteristics are essential for their use in scientific research and regenerative medicine. Understanding and harnessing these properties are essential to unlock the full therapeutic potential of stem cells. One of the most distinguishing features of stem cells is their ability to proliferate, meaning they can undergo several cycles of cell division while maintaining an undifferentiated state, contributing to their longevity throughout an organism's lifespan. Another critical property is their potency, that is, the potential to differentiate into several specialized types of cells. This ranges from totipotency (the ability to form a complete organism with different cell types, including both embryonic and extraembryonic tissues) to pluripotency (the ability to develop into nearly all cell types, typically seen in ESCs) [26].

Stem cells are also known for their multipotency, where they can differentiate into a limited range of related cell types, as seen with adult stem cells like hematopoietic stem cells. Unipotency is the property of some stem cells, which produce only one cell type but still retain self-renewal capability. The differentiation potential of stem cells allows them to specialize into distinct cell types, like neurons, myocytes, or erythrocytes, in response to specific environmental signals, which is vital for tissue homeostasis and regeneration. Furthermore, their clonality signifies that a single stem cell proliferates to form a colony of genetically and functionally identical cells. Under certain conditions, some stem cells can also exhibit plasticity, or transdifferentiation, where they can differentiate into cell types outside their usual lineage. They can successfully target and heal damaged tissues due to homing. It is especially useful for therapeutic purposes, like managing autoimmune and inflammatory disorders [27]. These unique properties make stem cells invaluable in biomedical research and therapeutic applications, particularly in tissue engineering, regenerative medicine, and treating various pathologies.

8.2 Clinical Applications of Stem Cell Therapy for Metabolic Disorders

Conventional methods of treatment, like changing the lifestyle (inculcating a healthy routine), timely intake of medications, etc., can help in reducing the effects of metabolic disorders but often fail to cure them or provide long-term solutions. Hence, stem cell therapies are emerging as promising approaches to treat metabolic disorders. Stem cells can replace damaged cells and restore normal metabolic functions, improving patient outcomes. Various metabolic disorders, including type 1 and type 2 diabetes, are among the most actively researched areas for stem cell therapy. In diabetes treatment, stem cell therapy aims to replace or regenerate the insulin-producing beta cells that have been destroyed [28].

8.2.1 Mesenchymal Stem Cells (MSCs)

Several studies and clinical trials have used mesenchymal stem cells (MSCs) that are derived from bone marrow, adipose tissue, or umbilical cord to treat diabetes. Several studies on stem cell therapy have proven them to be safe with no side effects [29]. However, the MSCs' tissue origin, the infusion route, dosage and administration period, or preconditioning approaches need to be optimized for maximum effect. Some examples of promising studies include the following:

1. In patients with bone marrow-derived MSC therapy, elevated levels of C-peptide were observed, indicating improved levels of insulin production [30].
2. Allogeneic adipose-derived MSCs improved glucose homeostasis in animal models of type 2 diabetes [31].
3. Umbilical cord-derived MSCs are the most favored cell source for clinical trials, with 28 trials currently underway. While these results are encouraging, further longer-term trials are needed to fully assess the efficacy and long-term safety of stem cell therapy for diabetes [32].

8.2.2 Induced Pluripotent Stem Cells (iPSCs)

Induced pluripotent stem cells (iPSCs) from individuals with metabolic illnesses can simulate diseases *in vitro*, providing a platform for pathophysiology evaluation and therapeutic testing. Cardiomyocytes derived from iPSCs, for example, have been used to model several cardiomyopathies associated with metabolic disorders [33]. Diabetes patient-specific iPSCs can be genetically altered and developed into the appropriate cell type to create pancreatic beta cells or hepatocytes for autologous transplantation. By doing this, immunosuppressive measures and immunological rejection can be avoided. Creating iPSCs from several donors allows for high throughput drug library screening to discover new molecules to treat metabolic disorders. Genetic heterogeneity in the population can also be modeled using iPSCs [34–36].

8.2.3 Hematopoietic Stem Cells (HSCs)

The qualities of hematopoietic stem cells (HSCs) of differentiating into multiple blood cell lineages and self-renewing are essential to treat metabolic disorders. Lysosomal storage illnesses are metabolic disorders caused by accumulated harmful chemicals due to enzyme deficiencies. Numerous studies have investigated using HSCs to treat various conditions. Clinical studies have shown that the recipients of HSC transplantation frequently show significant improvements in metabolic functions and quality of life, which highlights their therapeutic promise [37]. Investigations are being done to explore HSCs further to treat various metabolic disorders. Hematopoietic progenitor cells that are healthy and functional can be obtained through HSC transplantation, which can produce the missing enzymes, stop cellular damage, and reduce symptoms.

According to several studies, the functionality of the HSCs can be influenced by their metabolic status. Anaerobic glycolysis is one of the metabolic processes that HSCs depend on to keep themselves quiescent and capable of self-renewal. Understanding how metabolic programming affects HSC functions will help create new strategies that enhance the HSC engraftment and function in the transplantation settings, mainly improving the outcomes for patients with metabolic disorders [38].

HSCs can be used in gene therapy to treat metabolic disorders by correcting genetic defects in stem cells. For example, functional alleles of mutated genes that cause enzymatic deficiencies in metabolic disorders can be expressed in HSCs. After the transplantation, the modified HSCs would express the functional enzymes in the patients. Furthermore, these stem cells can be differentiated into different blood cells, thereby propagating the cells and producing the necessary enzymes to restore the metabolic balance [39].

8.2.4 Embryonic Stem Cells (ESCs)

Embryonic stem cells (ESCs) possess the pluripotency that allows them to differentiate into many types of cells in the body, including beta cells of the pancreas, which secrete insulin. Restoring insulin production is crucial in diabetes treatment. ESCs can differentiate into beta cells and be transplanted into diabetic patients for treatment. Gene editing tools like CRISPR/Cas9 can be applied to ESCs to fix the abnormalities that can cause metabolic disorders, wherein they result in the production of genetically altered cells for use in transplantation [40].

8.2.5 Neural Stem Cells (NSCs)

The neural stem cells (NSCs) develop into various neurons like astrocytes and oligodendrocytes, which may prove advantageous in treating metabolic disorders that cause illnesses that impact the nervous system. For example, the NSCs can repair the injured nerve tissue in diseases such as diabetic neuropathy. This can enhance brain activity and general metabolic control. According to recent research, these NSCs control the body's metabolism [41]. The therapeutic medications can be directly delivered into the affected tissues by transplanting genetically engineered NSCs. This focused strategy can improve the effectiveness of therapies for metabolic illnesses by promising reduced systemic side effects by functioning precisely at the site of tissue damage.

8.3 Molecular Mechanisms Underlying the Action of Stem Cells in Metabolic Disorders

The distinct features and capacities of various types of stem cells utilized in stem cell therapy present a promising approach to treating metabolic disorders. Stem cells can mitigate metabolic disorders through direct and indirect processes. The property of stem cells to differentiate into various cell types impacted by metabolic disorders is one of the main processes. For example, in diabetes, stem cells can be reprogrammed to differentiate into pancreatic beta cells that make insulin. These cells can replace the damaged or malfunctioning cells and help the body produce insulin again. Stem cells can be modified similarly to generate the missing enzymes in lysosomal storage diseases, preventing the buildup of harmful compounds [42]. Directly replacing damaged cells with functional and healthy cells can significantly enhance metabolic processes and general well-being. An additional instance concerns the inherent metabolic state of stem cells, which plays a vital role in defining their function and fate. By altering their metabolism, stem cells can meet the metabolic demands necessary for self-renewal, differentiation, and tissue regeneration. These cells can adjust to external environmental fluctuations and preserve metabolic equilibrium by carefully regulating their metabolic pathways. By understanding metabolic programming, researchers can devise strategies to enhance stem cell therapy and explore its potential for treating metabolic disorders. The integration of these mechanisms offers a significant potential for developing effective stem cell-based therapies for a wide range of metabolic disorders [43]. Some of the important properties of stem cells are discussed below.

8.3.1 Homing to Target Tissues

An essential feature of stem cell therapy is the ability of the stem cells to home in on the target tissues, allowing them to migrate to specific areas of injury or disease. This mechanism is particularly crucial to treat metabolic disorders, where the normal cellular functions in the affected tissues need to be restored on priority. Stem cell homing is governed by various biological processes and signaling pathways [44].

Chemotaxis and chemokine signaling are the pivotal mechanisms driving stem cell migration toward damaged tissues. These cells respond to chemical signals released by the injured tissues, which create a chemical gradient detectable by stem cells. This gradient consists of chemokines and cytokines. For instance, in conditions such as obesity or diabetes, the inflammatory environment produces specific signals that guide stem cells to reach the affected areas [45]. Stem cell surface receptors mediate the interaction with these signaling molecules, allowing stem cells to navigate the local environment effectively [46]. Thus, inflammation plays a significant role in stem cell homing. It enhances the expression of adhesion molecules and chemokines that attract stem cells. In diseases like non-alcoholic fatty liver disease (NAFLD), the inflammatory environment may facilitate the recruitment of stem cells to the liver, where they can contribute to tissue repair [47].

8.3.2 Tissue Repair and Regeneration

Different growth factors, cytokines, and extracellular vesicles secreted by stem cells can influence the surrounding microenvironment. These secreted substances can have a higher ability to improve metabolic functions, lower inflammation, and promote tissue regeneration. Stem cells have the property to produce substances that enhance energy expenditure and improve metabolic homeostasis by encouraging the browning of white fat tissue, particularly in conditions such as obesity and metabolic syndrome [48]. Furthermore, stem cells can change the immunological function, lowering chronic inflammation linked to metabolic disorders. For example, stem cells can be differentiated into pancreatic beta cells, which produce insulin, restoring the body's ability to control blood sugar levels. Thus, stem cell therapy can effectively help in the repair and regeneration of damaged tissue and restore its function.

8.3.3 Modulation of Immune Response

Stem cells can release numerous cytokines, growth factors, and extracellular vesicles to regulate the immune response. They have emerged as a promising tool for modulating immune responses in the treatment of metabolic disorders, such as diabetes and obesity. These disorders are often associated with chronic inflammation and immune system dysregulation. Mesenchymal stem cells (MSCs) have shown significant potential in regulating immune responses due to their ability to secrete anti-inflammatory cytokines and modulate the activity of various immune cells, including T cells, B cells, and macrophages [49]. MSCs can promote an anti-inflammatory environment by suppressing pro-inflammatory cytokines and enhancing the production of regulatory T cells (Tregs), which are crucial for maintaining immune tolerance. This immunomodulatory effect can reduce inflammation-related damage in tissues, improve insulin sensitivity, and ultimately contribute to the restoration of metabolic balance. Additionally, stem cell therapy may help repair damaged tissues and regenerate pancreatic beta cells in diabetes [50], offering a multifaceted approach to treating metabolic disorders. In fact, studies have demonstrated that stem cell transplantation can improve insulin sensitivity and glucose metabolism, indicating a multifaceted role in metabolic regulation [51].

Different kinds of immune cells have varied differentiation properties that are greatly influenced by stem cells. All the blood cell lineages involved in innate and adaptive immunological responses originate from HSCs. Stem cells can aid in reestablishing immunological balance by regulating the generation of immune cells, encouraging a change from the pro-inflammatory to the anti-inflammatory condition. It is important because hyperactive immune responses in metabolic disorders can harm tissue and exacerbate existing metabolic problems [52]. Stem cells can facilitate recruiting additional regenerative cells, including immune cells, that can aid in healing by generating a favorable milieu through the substances they produce. This interaction is essential in metabolic illnesses, such as fatty liver disease or diabetes, where tissue integrity is compromised [53]. Further research is needed to fully understand the molecular mechanisms underlying stem cell-mediated immune modulation and to optimize therapeutic strategies for clinical use.

8.3.4 Signaling Pathways Involved in Metabolic Disorders

Signaling pathways significantly impact physiological functions, including energy metabolism, nutrition sensing, and inflammation, which are essential in the onset and development of metabolic disorders. Some of the critical signaling pathways that are linked to metabolic disorders are discussed in the following subsections.

8.3.4.1 The mTOR Signaling Pathway

The mTOR (mechanistic target of rapamycin) signaling pathway is crucial for regulating cellular growth, proliferation, and metabolism, making it a key regulator in metabolic disorders. mTOR functions as a central hub that integrates signals from nutrients, growth factors, and energy status to modulate anabolic and catabolic processes. Thus, dysregulation of mTOR signaling can result in various metabolic disorders such as obesity, type 2 diabetes, and non-alcoholic fatty liver disease (NAFLD). Hyperactivation of mTOR complex 1 (mTORC1) promotes lipogenesis and protein synthesis and inhibits autophagy, contributing to insulin resistance and adiposity. Conversely, mTOR complex 2 (mTORC2) regulates glucose and lipid metabolism by modulating Akt signaling [54]. Therapeutic targeting of the mTOR pathway using inhibitors such as rapamycin and its analogs (rapalogs) shows potential in ameliorating metabolic dysfunctions. Systemic inhibition of mTOR, while potentially beneficial, can also lead to adverse side effects. Therefore, employing precise modulation strategies to maximize therapeutic benefits while minimizing unfavorable outcomes is essential. A thorough understanding of the complex dynamics of mTOR signaling within metabolic pathways is critical for developing effective interventions against metabolic disorders [55].

8.3.4.2 The JNK Signaling Pathway

Stress signals cause activation of the c-Jun N-terminal kinase (JNK) pathway, which plays a crucial role in insulin signaling and inflammation. JNK isoforms have been associated with regulating insulin receptor substrate 2 (IRS2), a key component of insulin signaling in pancreatic beta cells. JNK signaling dysregulation impacts the emergence of metabolic syndrome and insulin resistance. The JNK inhibition has shown a promising effect as a therapeutic target for enhancing insulin sensitivity and lowering problems associated with obesity [56].

8.3.4.3 The AMPK Pathway

AMP-activated protein kinase (AMPK), functioning as a sensor, can control metabolism in response to variations in energy levels. AMPK activation inhibits anabolic processes that use ATP and encourages catabolic processes that produce ATP when cellular energy levels are low. Obesity, type 2 diabetes, and other metabolic disorders can be addressed by improving fatty acid oxidation and glucose absorption through this pathway. AMPK signaling dysregulation results in impaired metabolic homeostasis and an elevated risk of metabolic disorders [57].

8.3.4.4 Inflammatory Signaling Pathways

Chronic inflammation is commonly linked with metabolic disorders and is driven by various inflammatory signaling pathways. For instance, pro-inflammatory cytokines can activate the nuclear factor kappa B (NF-κB) pathway, which regulates inflammatory responses. Adipose tissue with elevated NF-κB activity is associated with metabolic dysfunction and insulin resistance. Furthermore, stress-induced inflammation is influenced by the hypothalamic–pituitary–adrenal (HPA) axis. This inflammation can worsen insulin resistance and dyslipidemia [58].

8.3.4.5 Signaling by Leptin and Ghrelin

The leptin and ghrelin hormones control the metabolism and energy balance. Adipose tissue secretes the hormone leptin, which controls calorie consumption and indicates fullness. Insulin resistance and obesity may result from dysregulation of leptin signaling. In contrast, the stomach secretes ghrelin, which increases hunger and encourages fat storage. Metabolic disorders can arise from modifications in these hormones' signaling pathways, underscoring the significance of neuroendocrine regulation in maintaining energy homeostasis [59].

8.4 Challenges and Limitations of Stem Cell Therapy for Metabolic Disorders

Stem cell therapy is a potentially effective treatment for several metabolic conditions, such as diabetes, obesity, and lysosomal storage diseases. Despite its apparent advantages, stem cell therapy is hindered by several obstacles and constraints. The main challenges that scientists and medical professionals have while bringing stem cell treatments from the bench to the patient's side include concerns related to safety, ethical and regulatory issues, variability in stem cell sources, limited understanding of the molecular mechanisms, scaling up of production of stem cells, patient-specific factors, and many more [60]. Some of these challenges and limitations are described below.

8.4.1 Off-Target Effects

A major challenge in stem cell therapy is ensuring the safety of the treatment. The unintended differentiation of the stem cells at the off-target sites can lead to severe complications. In particular, the concern is highly regarded with pluripotent stem cells like ESCs and iPSCs, which have a high risk of tumor formation because of their potential to differentiate into any cell type. The stem cells can be redirected from the intended target of the tissue, which can reduce their therapeutic efficiency. For example, the transplanted stem cells may move to another place and diffuse in the non-target areas, thereby preventing the desired therapeutic benefits [61]. The stem cells that get differentiated in the off-target sites can produce changed surface markers that set off an immune reaction and cause rejection of the transplanted cells. This may negate the therapeutic benefits and need for immunosuppressive medications. The off-target effects make it difficult to identify the accurate mechanisms by which stem cell therapies work their therapeutic effects. The wrong target hinders the development of biomarkers for predicting treatment responses and optimizing therapies.

8.4.2 Immune Responses

Immune responses following a stem cell transplant are critical determinants of both the success and potential complications of the procedure. Transplanted stem cells, typically sourced from bone marrow or peripheral blood, can trigger graft-versus-host disease (GVHD) if the donor and recipient are not immunologically compatible. This mainly occurs in the case of an allogeneic stem cell transplant, wherein the donor-originated healthy stem cells are transplanted instead of the patient's own stem cells. GVHD arises when the donor's immune cells identify the recipient's tissues as foreign, leading to an immune-mediated attack on various organs. Conversely, the recipient's immune system can mount a response against the transplanted cells, known as graft rejection, compromising the engraftment process [62]. However, immune rejection can sometimes result, even if the patient's own stem cells are utilized. This is because the developed or differentiated cells sometimes exhibit modified surface markers that elicit an immunological response after transplant. For example, in the case of diabetes patients, the patient's immune system may reject the insulin-producing beta cells produced from the patient's own stem cells. Such immunological reactions against the transplanted stem cells may cause inflammation. This inflammatory condition can impair the function of the transplanted cells and exacerbate metabolic dysfunction. For instance, the inflammation resulting from stem cell rejection could worsen the condition in cases of obesity and metabolic syndrome [63, 64]. To manage these responses, immunosuppressive therapies are commonly used to promote tolerance and improve the chances of successful engraftment. Alternatively, in the graft-versus-tumor effect, the donor's immune cells can target the patient's remaining malignant cells, which can be advantageous in lowering the risk of recurrence of cancer. Therefore, achieving a careful balance between suppressing harmful immune reactions and leveraging beneficial immune responses is crucial in stem cell transplantation.

8.4.3 Tissue Heterogeneity

Stem cells derived from the same tissue source can exhibit variability in their biological functions and therapeutic effectiveness when incorporated into treating metabolic disorders [65]. This variability can complicate the determination of dose formulation, regimen, and timing for treatment. Therefore, further research is required to address these challenges. Likewise, stem cells transplanted into heterogeneous tissue environments might not integrate or perform optimally. The extracellular matrix and signaling pathways of surrounding cells can influence the fate of these transplanted stem cells [66]. A mismatch between stem cells and the target tissue site can significantly impact their effectiveness.

Factors such as age, genetic background, and coexisting conditions further complicate the development of predictive models for stem cell therapy [67]. Due to the variability in both target tissues and stem cells, predicting therapy outcomes is difficult. Personalized stem cell therapies are emerging to address individual patient needs and overcome the challenges of tissue heterogeneity. This approach involves isolating, analyzing, and selecting the patient's stem cells with desirable traits before transplantation. Although promising, this method adds complexity and cost to the treatment process [68].

To overcome the tissue heterogeneity in stem cell treatments, several strategies can be employed to ensure therapeutic efficacy and safety. Single-cell RNA sequencing can provide detailed profiles of diverse cell populations within a tissue, identifying specific cell types that may contribute to disease or therapeutic resistance [69]. Advanced bioengineering techniques, such as organoid cultures and tissue scaffolding, can recreate more uniform tissue environments, improving the consistency and functionality of stem cell-derived tissues [70]. Additionally, targeted genetic modifications using CRISPR-Cas9 can precisely correct mutations in specific cell populations, reducing variability. High-throughput screening can optimize culture conditions to minimize heterogeneity. These strategies can help to refine stem cell therapy, addressing the challenges of tissue heterogeneity and enhancing clinical outcomes.

8.4.4 Stem Cell Exhaustion

Stem cell exhaustion refers to the gradual decline in the functional abilities of stem cells, marked by diminished self-renewal, proliferation, and differentiation capabilities, eventually leading to a loss of their regenerative potential. This issue significantly affects the effectiveness of stem cell therapies, especially for metabolic disorders. The success and longevity of stem cell-based treatments can be enhanced by understanding the mechanisms behind stem cell exhaustion and developing targeted approaches to address it. Addressing stem cell exhaustion is essential for advancing stem cell technologies and improving patient outcomes across various disorders [71].

Several approaches can be employed to combat stem cell exhaustion. For example, modulation of signaling pathways, such as the Wnt, Notch, and mTOR, can help maintain stem cell self-renewal and differentiation potential.

Pharmacological agents like small molecule inhibitors or activators can be used to enhance stem cell functionality. Similarly, telomerase activators can be used to maintain telomere length, thereby improving cellular longevity. Genetic and epigenetic reprogramming techniques, including CRISPR-Cas9 and histone deacetylase inhibitors, can also rejuvenate exhausted stem cells by resetting their molecular clocks. However, periodic transplantation of fresh, healthy stem cells or using iPSCs is the most straightforward approach to replenishing the stem cell pool, counteracting the effects of exhaustion, and promoting sustained tissue regeneration.

8.5 Future Directions for Stem Cell Therapy in Metabolic Disorders

Stem cell therapy offers significant potential for treating metabolic disorders like diabetes, obesity, and lysosomal storage diseases. Despite substantial progress, there are still many challenges and opportunities for future research and clinical applications. This overview highlights potential future directions based on current research and clinical insights.

8.5.1 Advancements in Sources of Stem Cells

Future research may explore diverse stem cell sources. Currently, mesenchymal stem cells (MSCs) originating from adipose tissue, umbilical cord, and bone marrow are commonly utilized. The advent of induced pluripotent stem cells (iPSCs) presents a new opportunity for creating patient-specific stem cells. iPSCs can be derived from somatic cells, offering an abundant source of pluripotent cells that can be differentiated into various cell types, such as insulin-producing beta cells for diabetes treatment [72]. Somatic cells from patient-specific iPSCs can be genetically modified using CRISPR-Cas9 to correct metabolic defects before being differentiated into desired cell types, like hepatocytes or pancreatic beta cells. Refining iPSC generation and differentiation methods can significantly improve their therapeutic potential.

8.5.2 Enhancing Treatment Safety and Efficacy

Safety concerns in stem cell therapy are paramount due to risks such as tumor formation and off-target effects. Future research should focus on developing advanced safety measures for stem cell transplants, including genetic modifications like suicide genes that can be triggered in case of adverse reactions [73]. Improving differentiation protocols to ensure that stem cells accurately develop into the intended cell types with minimal risk of abnormal differentiation is also vital for enhancing therapeutic outcomes [74].

8.5.3 Personalized Medicine Approaches

The variability in patient responses to stem cell therapy highlights the importance of personalized medicine. Research should explore the genetic and epigenetic factors influencing individual responses to these treatments. Customizing therapies based on a patient's unique genetic and metabolic profiles can optimize efficacy and reduce adverse effects [75]. Identifying biomarkers that predict treatment responses will be crucial for guiding personalized therapeutic approaches.

8.5.4 Combination Therapies

Combining stem cell therapy with other treatments might improve outcomes for metabolic disorders. For instance, integrating stem cell therapy with pharmacological agents, lifestyle changes, or gene therapy could offer synergistic benefits [76]. Developing effective combinations that address the complex nature of metabolic disorders, such as obesity and type 2 diabetes, is crucial for achieving better treatment results.

8.5.5 Longitudinal Studies and Clinical Trials

Extensive longitudinal studies and clinical trials are essential to fully grasp the long-lasting effects and effectiveness of stem cell therapies. Although early clinical trials have demonstrated promising outcomes, further and more detailed research is required to evaluate the longevity of treatment benefits and identify any potential late-onset complications. These studies should also assess the quality of life and metabolic health of patients receiving stem cell treatments, providing valuable insights into the overall impact of these treatments [77].

8.5.6 Regulatory and Ethical Considerations

Addressing regulatory and ethical considerations will be crucial as stem cell therapies advance. Regulatory frameworks need to be evolved to accommodate the unique challenges posed by stem cell treatments, prioritizing patient safety and fostering innovation. The use of stem cells, particularly iPSCs and ESCs, is expected to remain vital as the field progresses [78]. Engaging stakeholders, including patients, healthcare providers, and policymakers, is essential during the development of responsible and equitable stem cell therapies.

Other approaches involve optimization of cell delivery methods, including encapsulation to protect the transplanted cells from immune rejection while ensuring their sustained functionality. Metabolic reprogramming of stem cells through the modulation of critical metabolic pathways can enhance their therapeutic potential. Understanding the stem cell microenvironment can help design strategies to improve stem cell integration and function post-transplantation. Combining stem cell therapy with gene therapy, small molecule drugs, or personalized medicine can yield synergistic effects for effectively treating metabolic disorders.

8.6 Conclusions

Stem cell therapy offers a transformative approach to treating metabolic disorders such as diabetes and obesity by leveraging the regenerative potential of stem cells. These cells can replace the damaged or malfunctioning cells, restoring their normal metabolic functions and enhancing patient outcomes. Significant advancements have been made in understanding how stem cells achieve their therapeutic effects, including tissue regeneration, immune system modulation, and metabolic adjustments. Despite these advances, challenges such as safety issues, variability in patient responses, and the complexities of tissue heterogeneity and stem cell exhaustion remain. Overcoming these obstacles is crucial for advancing stem cell therapies from preclinical studies to practical clinical applications. Future research should aim at optimizing stem cell sources, improving treatment safety and effectiveness, and developing personalized treatment plans tailored to individual patient needs. At the same time, collaboration between researchers, clinicians, and regulatory bodies will be vital in addressing the ethical and logistic challenges associated with stem cell therapy. By addressing current limitations and fully utilizing the potential of stem cells, it would be possible to create innovative treatments that significantly enhance the quality of life for those with metabolic disorders. Integrating stem cell therapy into clinical practice holds the promise of revolutionizing the management of complex metabolic conditions, leading to improved health outcomes and patient care.

Acknowledgments

VAM acknowledges the University Grants Commission (UGC), New Delhi, India, for junior and senior research fellowships. SDB, AKS, VAM, and RS acknowledge the Academy of Scientific and Innovative Research (AcSIR), Ghaziabad, Uttar Pradesh, India, for the M.Tech. and Ph.D. programs. All authors thank the CSIR-National Chemical Laboratory, Pune, India, for providing the necessary facilities to undertake this study.

References

1. Ferreira, E.A., Buijs, M.J.N., Wijngaard, R., Daams, J.G., Datema, M.R., Engelen, M., van Karnebeek, C.D.M., Oud, M.M., Vaz, F.M., Wamelink, M.M.C., van der Crabben, S.N., Langeveld, M., Inherited metabolic disorders in adults: systematic review on patient characteristics and diagnostic yield of broad sequencing techniques (exome and genome sequencing). *Front. Neurol.,*, 14, 1206106, 2023.
2. Sapra, A. and Bhandari, P., Diabetes Mellitus. [Updated 2022 Jun 26], in: *StatPearls [Internet]*, StatPearls Publishing, Treasure Island FL, 2022.
3. Panuganti, K.K., Nguyen, M., Kshirsagar, R.K., Obesity. [Updated 2022 Aug 8], in: *StatPearls [Internet]*, StatPearls Publishing, Treasure Island FL, 2023.
4. Ruban, A., Stoenchev, K., Ashrafian, H., Teare, J., Current treatments for obesity. *Clin. Med.*, 19, 3, 205–212, 2019.
5. Diagnosis and Classification of Diabetes Mellitus. *Diabetes Care*, 37, Supplement_1, S81–S90, Dec. 2013. doi: 10.2337/dc14-S081.
6. Nathan, D.M., Diabetes: Advances in Diagnosis and Treatment. *JAMA*, 314, 10, 1052–1062, Sep. 2015. doi: 10.1001/jama.2015.9536.
7. Hruby, A. and Hu, F.B., The Epidemiology of Obesity: A Big Picture. *Pharmacoeconomics*, 33, 7, 673–689, Jul. 2015. doi: 10.1007/s40273-014-0243-x.
8. Apovian, C.M., *et al.*, Pharmacological Management of Obesity: An Endocrine Society Clinical Practice Guideline. *J. Clin. Endocrinol. Metab.*, 100, 2, 342–362, Feb. 2015. doi: 10.1210/jc.2014-3415.
9. Courcoulas, A.P., *et al.*, Long-term outcomes of bariatric surgery: a National Institutes of Health symposium. *JAMA Surg.*, 149, 12, 1323–1329, Dec. 2014. doi: 10.1001/jamasurg.2014.2440.
10. van Spronsen, F.J., Phenylketonuria: a 21st century perspective. *Nat. Rev. Endocrinol.*, 6, 9, 509–514, Sep. 2010. doi: 10.1038/nrendo.2010.125.
11. van Spronsen, F.J., Blau, N., Harding, C., Burlina, A., Longo, N., Bosch, A.M., Phenylketonuria. *Nat. Rev. Dis. Primers*, 7, 1, 36, May 2021. doi: 10.1038/s41572-021-00267-0.

12. Yang, M., Liu, S., Zhang, C., The Related Metabolic Diseases and Treatments of Obesity. *Healthcare (Basel)*, 10, 9, 1616, Aug. 2022. doi: 10.3390/healthcare10091616.
13. Patel, D., Kumar, R., Laloo, D., Hemalatha, S., Diabetes mellitus: An overview on its pharmacological aspects and reported medicinal plants having antidiabetic activity. *Asian Pac. J. Trop. Biomed.*, 2, 5, 411–420, May 2012. doi: 10.1016/S2221-1691(12)60067-7.
14. Vinueza, A.M.Z., Recent Advances in Phenylketonuria: A Review. *Cureus*, 15, 6, e40459, Jun. 2023. doi: 10.7759/cureus.40459.
15. Jaques, R., Shakeel, A., Hoyle, C., Novel therapeutic approaches for the management of cystic fibrosis. *Multidiscip. Respir. Med.*, 15, 1, 690, Nov. 2020. doi: 10.4081/mrm.2020.690.
16. Fahed, G., *et al.*, Metabolic syndrome: updates on pathophysiology and management in 2021. *Int. J. Mol. Sci.*, 23, 2, 786, 2022.
17. Summar, M.L., Dobbelaere, D., Brusilow, S., Lee, B., Diagnosis, symptoms, frequency and mortality of 260 patients with urea cycle disorders from a 21-year, multicentre study of acute hyperammonaemic episodes. *Acta Paediatrica*, 97, 10, 1420–1425, Oct. 2008. doi: 10.1111/j.1651-2227.2008.00952.x.
18. Sherwani, S.I., Khan, H.A., Ekhzaimy, A., Masood, A., Sakharkar, M.K., Significance of HbA1c test in diagnosis and prognosis of diabetic patients. *Biomark. Insights*, 11, 95–104, 2016, doi: 10.4137/BMI.S38440.
19. Fujiwara, Y., *et al.*, Relationship between diet/exercise and pharmacotherapy to enhance the GLP-1 levels in type 2 diabetes. *Endocrinol. Diabetes Metab.*, 2, 3, e00068, May 2019. doi: 10.1002/edm2.68.
20. Kashyap, S.R., Gatmaitan, P., Brethauer, S., Schauer, P., Bariatric surgery for type 2 diabetes: Weighing the impact for obese patients. *CCJM*, 77, 7, 468–476, Jul. 2010. doi: 10.3949/ccjm.77a.09135.
21. Carratalá-Munuera, C., *et al.*, Barriers and Solutions to Improve Therapeutic Adherence from the Perspective of Primary Care and Hospital-Based Physicians. *Patient Prefer Adherence*, 16, 697–707, Mar. 2022. doi: 10.2147/PPA.S319084.
22. Hoang, D.M., *et al.*, Stem cell-based therapy for human diseases. *Signal Transduct. Target Ther.*, 7, 272, Aug. 2022. doi: 10.1038/s41392-022-01134-4.
23. Mukherjee, P.K., Harwansh, R.K., Bahadur, S., Chanda, J., Biswas, S., Banerjee, S., Enzyme inhibition assay for metabolic disorders—exploring leads from medicinal plants, in: *Animal Biotechnology*, Second, A.S. Verma and A. Singh (Eds.), pp. 631–653, Academic Press, Boston, 2020, doi: 10.1016/B978-0-12-811710-1.00033-1.
24. Ghazimoradi, M.H., Khalafizadeh, A., Babashah, S., A critical review on induced totipotent stem cells: Types and methods. *Stem Cell Res.*, 63, 102857, Aug. 2022. doi: 10.1016/j.scr.2022.102857.

25. Mahla, R.S., Stem Cells Applications in Regenerative Medicine and Disease Therapeutics. *Int. J. Cell Biol.*, 2016, 6940283, 2016. doi: 10.1155/2016/6940283.
26. Shi, Y., Inoue, H., Wu, J.C., Yamanaka, S., Induced pluripotent stem cell technology: a decade of progress. *Nat. Rev. Drug Discov.*, 16, 2, 115–130, Feb. 2017. doi: 10.1038/nrd.2016.245.
27. Regmi, S., Pathak, S., Kim, J.O., Yong, C.S., Jeong, J.-H., Mesenchymal stem cell therapy for the treatment of inflammatory diseases: Challenges, opportunities, and future perspectives. *Eur. J. Cell Biol.*, 98, 5, 151041, Dec. 2019. doi: 10.1016/j.ejcb.2019.04.002.
28. Basile, G., *et al.*, Emerging diabetes therapies: Bringing back the β-cells. *Mol. Metab.*, 60, 101477, Mar. 2022. doi: 10.1016/j.molmet.2022.101477.
29. Galderisi, U., Peluso, G., Di Bernardo, G., Clinical Trials Based on Mesenchymal Stromal Cells are Exponentially Increasing: Where are We in Recent Years? *Stem Cell Rev. Rep.*, 18, 1, 23–36, 2022. doi: 10.1007/s12015-021-10231-w.
30. Habiba, U.E., Khan, N., Greene, D.L., Ahmad, K., Shamim, S., Umer, A., Meta-analysis shows that mesenchymal stem cell therapy can be a possible treatment for diabetes. *Front. Endocrinol.*, 15, 1380443, May 2024. doi: 10.3389/fendo.2024.1380443.
31. Cao, M., *et al.*, Adipose-derived mesenchymal stem cells improve glucose homeostasis in high-fat diet-induced obese mice. *Stem Cell Res. Ther.*, 6, 208, Oct. 2015. doi: 10.1186/s13287-015-0201-3.
32. Liau, L.L., Looi, Q.H., Chia, W.C., Subramaniam, T., Ng, M.H., Law, J.X., Treatment of spinal cord injury with mesenchymal stem cells. *Cell Biosci.*, 10, 112, Sep. 2020. doi: 10.1186/s13578-020-00475-3.
33. Paik, D.T., Chandy, M., Wu, J.C., Patient and Disease–Specific Induced Pluripotent Stem Cells for Discovery of Personalized Cardiovascular Drugs and Therapeutics. *Pharmacol. Rev.*, 72, 1, 320–342, Jan. 2020. doi: 10.1124/pr.116.013003.
34. Du, S., *et al.*, Engineering Islets From Stem Cells: The Optimal Solution for the Treatment of Diabetes? *Front. Immunol.*, 13, 869514, 2022. doi: 10.3389/fimmu.2022.869514.
35. Chehelgerdi, M., *et al.*, Exploring the promising potential of induced pluripotent stem cells in cancer research and therapy. *Mol. Cancer*, 22, 189, Nov. 2023. doi: 10.1186/s12943-023-01873-0.
36. Pellegrini, S., Zamarian, V., Sordi, V., Strategies to Improve the Safety of iPSC-Derived β Cells for β Cell Replacement in Diabetes. *Transpl. Int.*, 35, 10575, 2022. doi: 10.3389/ti.2022.10575.
37. Bastani, S., Staal, F.J.T., Canté-Barrett, K., The quest for the holy grail: overcoming challenges in expanding human hematopoietic stem cells for clinical use. *Stem Cell Invest.*, 10, 15, 2023. doi: 10.21037/sci-2023-016.
38. Ito, K. and Ito, K., Hematopoietic stem cell fate through metabolic control. *Exp. Hematol.*, 64, 1–11, 2018, doi: 10.1016/j.exphem.2018.05.005.

39. Morgan, R.A., Gray, D., Lomova, A., Kohn, D.B., Hematopoietic Stem Cell Gene Therapy –Progress and Lessons Learned. *Cell Stem Cell*, 21, 5, 574–590, Nov. 2017. doi: 10.1016/j.stem.2017.10.010.
40. Redman, M., King, A., Watson, C., King, D., What is CRISPR/Cas9? *Arch. Dis. Child Educ. Pract. Ed.*, 101, 4, 213–215, Aug. 2016. doi: 10.1136/archdischild-2016-310459.
41. Vishwakarma, S.K., Bardia, A., Tiwari, S.K., Paspala, S.A.B., Khan, A.A., Current concept in neural regeneration research: NSCs isolation, characterization and transplantation in various neurodegenerative diseases and stroke: A review. *J. Adv. Res.*, 5, 3, 277–294, May 2014. doi: 10.1016/j.jare.2013.04.005.
42. Bonam, S.R., Wang, F., Muller, S., Lysosomes as a therapeutic target. *Nat. Rev. Drug Discov.*, 18, 12, 923–948, 2019. doi: 10.1038/s41573-019-0036-1.
43. Ito, K. and Suda, T., Metabolic requirements for the maintenance of self-renewing stem cells. *Nat. Rev. Mol. Cell Biol.*, 15, 4, 243–256, Apr. 2014. doi: 10.1038/nrm3772.
44. Liesveld, J.L., Sharma, N., Aljitawi, O.S., Stem cell homing: From physiology to therapeutics. *Stem Cells*, 38, 10, 1241–1253, 2020. doi: 10.1002/stem.3242.
45. Moore, J.E., Brook, B.S., Nibbs, R.J.B., Chemokine Transport Dynamics and Emerging Recognition of Their Role in Immune Function. *Curr. Opin. Biomed. Eng.*, 5, 90–95, Mar. 2018. doi: 10.1016/j.cobme.2018.03.001.
46. Sun, Y., Chen, C.S., Fu, J., Forcing stem cells to behave: a biophysical perspective of the cellular microenvironment. *Annu. Rev. Biophys.*, 41, 1, 519–542, 2012, doi: 10.1146/annurev-biophys-042910-155306.
47. Lepperdinger, G., Inflammation and mesenchymal stem cell aging. *Curr. Opin. Immunol.*, 23, 4, 518, Aug. 2011. doi: 10.1016/j.coi.2011.05.007.
48. Shook, B., Gonzalez, G.R., Ebmeier, S., Grisotti, G., Zwick, R., Horsley, V., The Role of Adipocytes in Tissue Regeneration and Stem Cell Niches. *Annu. Rev. Cell Dev. Biol.*, 32, 609, Oct. 2016. doi: 10.1146/annurev-cellbio-111315-125426.
49. Tolar, J., Le Blanc, K., Keating, A., Blazar, B.R., Concise Review: Hitting the Right Spot with Mesenchymal Stromal Cells. *Stem Cells*, 28, 8, 1446–1455, 2010. doi: 10.1002/stem.459.
50. Bhartiya, D., Stem cells to replace or regenerate the diabetic pancreas: Huge potential & existing hurdles. *Indian J. Med. Res.*, 143, 3, 267–274, Mar. 2016. doi: 10.4103/0971-5916.182615.
51. Wan, X.-X., *et al.*, Stem Cell Transplantation in the Treatment of Type 1 Diabetes Mellitus: From Insulin Replacement to Beta-Cell Replacement. *Front. Endocrinol (Lausanne)*, 13, 859638, Mar. 2022. doi: 10.3389/fendo.2022.859638.
52. Blanco, L.P. and Kaplan, M.J., Metabolic alterations of the immune system in the pathogenesis of autoimmune diseases. *PLoS Biol.*, 21, 4, e3002084, Apr. 2023. doi: 10.1371/journal.pbio.3002084.

53. Bhatt, H.B. and Smith, R.J., Fatty liver disease in diabetes mellitus. *Hepatobiliary Surg. Nutr.*, 4, 2, 101–108, 2015. doi: 10.3978/j.issn.2304-3881.2015.01.03
54. Linke, M., Fritsch, S.D., Sukhbaatar, N., Hengstschläger, M., Weichhart, T., mTORC1 and mTORC2 as regulators of cell metabolism in immunity. *FEBS Lett.*, 591, 19, 3089–3103, Oct. 2017. doi: 10.1002/1873-3468.12711.
55. Marafie, S.K., Al-Mulla, F., Abubaker, J., mTOR: Its Critical Role in Metabolic Diseases, Cancer, and the Aging Process. *Int. J. Mol. Sci.*, 25, 11, 6141, Jan. 2024. doi: 10.3390/ijms25116141.
56. Solinas, G. and Becattini, B., JNK at the crossroad of obesity, insulin resistance, and cell stress response. *Mol. Metab.*, 6, 2, 174–184, Dec. 2016. doi: 10.1016/j.molmet.2016.12.001.
57. Layne, A.S., *et al.*, Impaired Muscle AMPK Activation in the Metabolic Syndrome May Attenuate Improved Insulin Action after Exercise Training. *J. Clin. Endocrinol. Metab.*, 96, 6, 1815–1826, Jun. 2011. doi: 10.1210/jc.2010-2532.
58. Baker, R.G., Hayden, M.S., Ghosh, S., NF-κB, inflammation and metabolic disease. *Cell Metab.*, 13, 1, 11–22, Jan. 2011. doi: 10.1016/j.cmet.2010.12.008.
59. Clemente-Suárez, V.J., Beltrán-Velasco, A., II, Redondo-Flórez, L., Martín-Rodríguez, A., Yáñez-Sepúlveda, R., Tornero-Aguilera, J.F., Neuro-Vulnerability in Energy Metabolism Regulation: A Comprehensive Narrative Review,. *Nutrients*, 15, 14, 3106, Jul. 2023. doi: 10.3390/nu15143106.
60. Volarevic, V., *et al.*, Ethical and Safety Issues of Stem Cell-Based Therapy. *Int. J. Med. Sci.*, 15, 1, 36–45, Jan. 2018. doi: 10.7150/ijms.21666.
61. Romito, A. and Cobellis, G., Pluripotent Stem Cells: Current Understanding and Future Directions. *Stem Cells Int.*, 2016, 9451492, 2016. doi: 10.1155/2016/9451492.
62. Charles A Janeway, J., Travers, P., Walport, M., Shlomchik, M.J., Responses to alloantigens and transplant rejection, in: *Immunobiology: The Immune System in Health and DiseaseGarland Science*, 5th, 2001, Accessed: Aug. 01, 2024. [Online]. Available: https://www.ncbi.nlm.nih.gov/books/NBK27163/.
63. Michniacki, T.F., Choi, S.W., Peltier, D.C., Immune Suppression in Allogeneic Hematopoietic Stem Cell Transplantation. *Handb. Exp. Pharmacol.*, 272, 209–243, 2022. doi: 10.1007/164_2021_544.
64. Nguyen, H., *et al.*, Stem Cell Therapy For Neurological Disorders: A Focus On Aging. *Neurobiol. Dis.*, 126, 85–104, Jun. 2019. doi: 10.1016/j.nbd.2018.09.011.
65. Costa, L.A., *et al.*, Functional heterogeneity of mesenchymal stem cells from natural niches to culture conditions: implications for further clinical uses. *Cell. Mol. Life Sci.*, 78, 2, 447–467, Jan. 2021. doi: 10.1007/s00018-020-03600-0.
66. Stoltz, J.-F., *et al.*, Stem Cells and Regenerative Medicine: Myth or Reality of the 21th Century. *Stem Cells Int.*, 2015, 734731, 2015. doi: 10.1155/2015/734731.
67. Patel, S.A., *et al.*, Personalizing Stem Cell Research and Therapy: The Arduous Road Ahead or Missed Opportunity? *Curr. Pharmacogenomics Person Med.*, 8, 1, 25–36, Mar. 2010.

68. Coelho, M.B., Cabral, J.M.S., Karp, J.M., Intraoperative Stem Cell Therapy. *Annu. Rev. Biomed. Eng.*, 14, 325–349, 2012. doi: 10.1146/annurev-bioeng-071811-150041.
69. Paul, S., *et al.*, Application of Single-Cell Sequencing on Stem Cell Research, in: *Handbook of Stem Cell Applications*, K.H. Haider (Ed.), pp. 1–26, Springer Nature, Singapore, 2023, doi: 10.1007/978-981-99-0846-2_43-1.
70. Marchini, A. and Gelain, F., Synthetic scaffolds for 3D cell cultures and organoids: applications in regenerative medicine. *Crit. Rev. Biotechnol.*, 42, 3, 468–486, Apr. 2022. doi: 10.1080/07388551.2021.1932716.
71. Aly, R.M., Current state of stem cell-based therapies: an overview. *Stem Cell Investig.*, 7, 8, May 2020. doi: 10.21037/sci-2020-001.
72. Liu, G., David, B.T., Trawczynski, M., Fessler, R.G., Advances in Pluripotent Stem Cells: History, Mechanisms, Technologies, and Applications. *Stem Cell Rev. Rep.*, 16, 1, 3–32, Feb. 2020. doi: 10.1007/s12015-019-09935-x.
73. Schuldiner, M., Itskovitz-Eldor, J., Benvenisty, N., Selective Ablation of Human Embryonic Stem Cells Expressing a 'Suicide' Gene. *Stem Cells*, 21, 3, 257–265, May 2003. doi: 10.1634/stemcells.21-3-257.
74. Muneekaew, S., Wang, M.J., Chen, S.Y., Control of stem cell differentiation by using extrinsic photobiomodulation in conjunction with cell adhesion pattern. *Sci. Rep.*, 12, 1, 1812, 2022, doi: 10.1038/s41598-022-05888-3.
75. Puccetti, M., Pariano, M., Schoubben, A., Giovagnoli, S., Ricci, M., Biologics, theranostics, and personalized medicine in drug delivery systems. *Pharmacol. Res.*, 201, 107086, Mar. 2024. doi: 10.1016/j.phrs.2024.107086.
76. Rafi, M.A., Gene and Stem Cell Therapy: Alone or in Combination? *Bioimpacts*, 1, 4, 213–218, 2011. doi: 10.5681/bi.2011.030.
77. Fogel, D.B., Factors associated with clinical trials that fail and opportunities for improving the likelihood of success: A review. *Contemp. Clin. Trials Commun.*, 11, 156–164, Aug. 2018. doi: 10.1016/j.conctc.2018.08.001.
78. Board on Health Sciences Policy, Board on Life Sciences, Division on Earth and Life Studies, Institute of Medicine, & National Academy of Sciences. in: *Stem Cell Therapies: Opportunities for Ensuring the Quality and Safety of Clinical Offerings: Summary of a Joint Workshop*. National Academies Press (US), 2014. PMID: 25057697. doi: 10.17226/18746.

9

Immunomodulation in Autoimmune Disorders: Harnessing the Power of Stem Cell

Neha Kamboj, Rahul Kumar* and Debasis Mitra†

Department of Microbiology, Graphic Era (Deemed to be University), Dehradun, Uttarakhand, India

Abstract

Autoimmune disorders, characterized by an aberrant immune response against self-antigens, present significant therapeutic challenges due to their chronic nature and the complexity of immune system dysregulation. Conventional treatments, including immunosuppressive drugs and biological agents, often fall short in providing lasting remission and are associated with substantial side effects. This chapter explores the innovative potential of SC therapy as a novel approach to immunomodulation in autoimmune diseases. We provide an in-depth overview of SC types pluripotent, adult, and embryonic (iPSCs), highlighting their unique characteristics and immunomodulatory properties. SC mechanisms interacting with the immune system, including anti-inflammatory effects and the regulation of immune cell populations, are examined to understand their therapeutic potential. Clinical applications of SCs, particularly hematopoietic SC transplantation (HSCT) and mesenchymal SCs (MSCs), are discussed with a focus on their efficacy and outcomes in treating autoimmune conditions such as rheumatoid arthritis and multiple sclerosis. We delve into the current research on iPSCs, evaluating their promise and the challenges that need to be addressed for their clinical use. The chapter also addresses critical challenges and considerations such as immunogenicity, ethical issue, and safety concern included in SC therapy and the need for standardization in SC production. Future directions in SC research, including advances in genetic engineering, personalized medicine, and combination therapies, are explored to provide a comprehensive outlook on the evolving landscape

**Corresponding author*: rkpma.dpmi@gmail.com
† *Corresponding author*: debasismitra.mb@geu.ac.in

Ayush Madan, Sumel Ashique, Deepika Arora and Mantosh Kumar Satapathy (eds.) Stem Cell Therapeutics, (409–462) © 2025 Scrivener Publishing LLC

of autoimmune disease treatment. Through detailed case studies and clinical experiences, we illustrate the practical applications and benefits of SC therapy, drawing insights for future research and clinical practice. This chapter aims to present a thorough understanding of the potential of SCs to revolutionize the treatment of autoimmune disorders, offering hope for more effective and sustainable therapeutic strategies.

***Keywords*:** Autoimmune disorders, immunomodulation, SC therapy, anti-inflammatory effects, immune cell regulation

9.1 Introduction

SC therapy shows promise for treating many conditions, from autoimmune to degenerative diseases and injuries. Yet, the interaction between transplanted cells and the recipient's immune system poses challenges. To boost treatment effectiveness, scientists are exploring ways to modulate immune responses, aiming to improve SC therapy outcomes (Tetila *et al.*, 2020). Immunomodulation, a crucial concept in immunology and medical research, involves deliberately regulating the immune response to achieve the desired therapeutic outcomes. The defensive system, comprising a complex system of cells and tissues, plays a vital role in defending the body against harmful pathogens while maintaining tolerance to its own components. By developing techniques to modify immune responses, scientists aim to improve the safety and efficacy of SC treatment. This approach could potentially address issues such as graft rejection and optimize the regenerative capabilities of displaced SCs. While research in this field progresses, immunomodulation may become an essential tool in maximizing the therapeutic importance of SC treatments across some medical situation (Kamilaris & Prenafeta-Boldú, 2018). However, the immune system's functions can fall out of balance in several disease situations, including cancer, organ transplantation, and autoimmune disorders, which can result in tissue damage or the inability to get rid of aberrant cells (Mamdouh, & Khattab, 2021). By adjusting the components of the immune system, immunomodulation aims to correct or reroute immune responses. To prevent rejection during transplantation, this may entail controlling immune reactions, reducing hyperactive reactions as observed in autoimmune illnesses, or boosting the immune response against infectious pathogens or cancer cells. Researchers and medical professionals are developing a variety of immunomodulation techniques, like biologics, pharmaceuticals, vaccinations, and cutting-edge innovations such as gene editing

(Brunelli, & Benini, 2020). Recent developments in immunomodulation have completely changed how medicine is practiced.

Immune checkpoint-targeting therapies, for example, have shown impressive results in the treatment of several malignancies. Furthermore, because SCs and their descendants may control immune responses and promote tissue regeneration, they present interesting pathways for immunomodulation (Suto, 2022). The field of immunomodulation continues to offer enormous potential for influencing the future of medicine by utilizing the immune system to successfully battle diseases and reduce unwanted responses as our understanding of immunology develops and new techniques arise (Jellish, & Edelstein, 2015).

9.1.1 Importance of Immunomodulation in Managing Autoimmune Diseases

Immunomodulation is crucial for managing autoimmune diseases, which occur due to immune system dysregulation causing self-tissue attacks. It aims to restore immune balance and suppress overactive responses, thus preventing tissue damage in conditions like rheumatoid arthritis, lupus, and multiple sclerosis. By reducing chronic inflammation through targeted therapies, it helps preserve tissue function and integrity. Immunomodulation also alleviates symptoms such as pain, swelling, and fatigue, improving the patients' quality of life. Additionally, it can modify disease progression, potentially leading to long-term remission. Advances in this field enable personalized treatments tailored to individual immune profiles. Often combined with other therapies like anti-inflammatory drugs or physical therapy, immunomodulation offers a comprehensive approach to managing symptoms and addressing the root causes of autoimmune diseases, making it a critical component of multidisciplinary treatment strategies.

9.1.2 Role of SCs in Medical Research and Treatment

SCs are crucial in medicine because they can multiply indefinitely and transform into different cell types, making them key players in research and therapeutic applications. Their unique properties make them invaluable in the field of regenerative medicine, where they show promise for organs and tissue damage for replacing or repairing. This potential offers new avenues for addressing challenging conditions such as heart disease, diabetes, and spinal cord injuries, bringing hope to patients and researchers alike.

They are also crucial in drug testing and development, where they help create disease models to test new drugs' safety and efficacy before human trials. Understanding disease mechanisms is another critical area, as SC research helps unravel the fundamental processes of development and disease, leading to new therapeutic targets and strategies. Personalized medicine benefits from SCs derived from a patient's tissues, minimizing immune rejection and enabling treatments tailored to individual genetic profiles. In gene therapy, genetically modified SCs can correct defective genes, potentially curing genetic disorders like cystic fibrosis and muscular dystrophy. Research into cancer SCs aims to target these cells specifically, addressing tumor growth and recurrence. Additionally, SCs play a significant act in biomedical engineering, contributing to the creation of bioengineered organs and tissues for transplantation, which can alleviate the shortage of donor organs. Their versatility extends to basic research, providing insights into cellular differentiation and development that inform various medical and biological fields. SCs represent a groundbreaking area of medical science, offering the potential to transform healthcare and treatment approaches for a broad range of injuries and diseases. Their versatility and regenerative capabilities open up new possibilities in medicine, potentially reshaping how we address complex health challenges in the future.

9.2 SCs: Types and Characteristics

SCs are rare yet powerful components in multicellular organisms. These unique cells can self-replicate, divide, and transform into various specialized cell types. Their versatility, or plasticity, differs among SC types. Scientists classify SCs based on their origin and adaptability level, reflecting their diverse potential in biological processes and medical applications (Figure 9.1). The classification of SCs range from unipotent cells, which can only become one specific cell type, to multipotent cells that can develop into multiple cell types within a particular tissue. Pluripotent SCs have even greater potential, capable of becoming almost any cell in the body. At the top of the hierarchy are totipotent SCs, which can form any cell type and even a complete organism. Scientists have also developed induced pluripotent SCs by reprogramming adult cells. Additionally, cancer SCs have been identified in various tumors, potentially contributing to cancer progression. This diverse array of SC types offers a wide spectrum of possibilities for research and medical applications, each with its own unique characteristics and potential uses.

Gastric cancer; CD_{44}^{+}	Melanoma; CD_{4}^{+} / CD_{25}^{+} / Fox P_{3}^{+}
Prostate cancer; CD_{133}^{+} / CD_{44}^{+}	Pancreatic cancer; CD_{44}^{+} / CD_{24}^{+}
Lung cancer; CD_{133}^{+}	Ovary cancer; CD_{44}^{+} / CD_{117}^{+}
Liver cancer; CD_{133}^{+}	Colon cancer; CD_{133}^{+}
Multiple myeloma; CD_{138}^{+}	Breast cancer; CD_{44}^{+} / CD_{24}^{+}
Brain tumor; CD_{133}^{+}	Acute myeloid leukemia; CD_{34}^{+} / CD_{38}^{+}

SCs are broadly categorized into adult and embryonic types, each with distinct origins and potentials. Adult SCs, primarily of mesodermal origin, include various subtypes found in different tissues. These range from muscle satellite cells to bone marrow and hematopoietic SCs, each playing crucial roles in tissue maintenance and repair. Embryonic SCs, on the other hand, are derived from early-stage embryos and possess pluripotent capabilities. Additionally, several SC types originate from the endoderm, including those found in reproductive organs, mammary glands, liver,

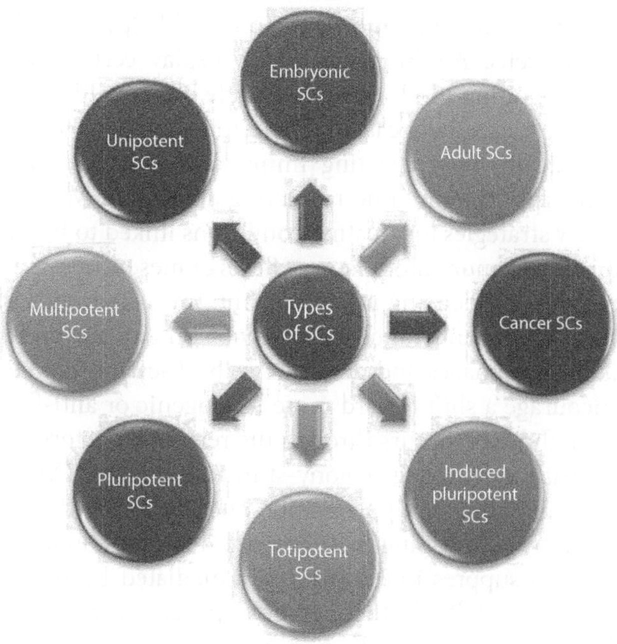

Figure 9.1 SCs.

pancreas, and lungs. This diverse array of SCs contributes to the body's ability to regenerate and maintain various tissues and organs throughout life, offering significant potential for medical research and therapeutic applications.

9.3 Mechanisms of SC-Based Immunomodulation

Mesenchymal SCs (MSCs) are leading the way in innovative immunomodulation strategies for regenerative medicine and immunotherapy. This approach shows significant potential for addressing various immune-related conditions. The process involves complex, interlinked mechanisms that combine direct cellular interactions with the release of diverse soluble factors. MSCs exert their immunomodulatory effects primarily through secreting a rich blend of bioactive substances. These include various cytokines, chemokines, growth factors, and extracellular vesicles. Together, these components create an environment that dampens immune responses such as PGE2, IDO, TGF-β, HGF and IL-10 (Matsuzaka and Yashiro, 2024). The bioactive substances released by MSCs work together to regulate immune responses in several ways. They reduce T-cell multiplication, encourage the development of regulatory T cells (Tregs), and curb the activity of inflammation-promoting T helper 17 (Th17) and T helper 1 (Th1) cells. Additionally, MSCs display certain molecules on their surface, including PD-L1. This molecule binds to PD-1 receptors on activated T cells, potentially causing them to become inactive or undergo programmed death, thus affecting immune responses. Together, these processes underpin the immune-regulating abilities of MSCs, suggesting promising new strategies for treating conditions linked to immune system dysfunction. The immunomodulatory effects of mesenchymal SCs (MSCs) are wide-ranging, influencing numerous immune cell groups beyond just T cells. MSCs significantly alter the behavior on antigen-presenting cells (APCs), like macrophages and dendritic cells. They inhibit APC maturation and encourage a shift toward more tolerogenic or anti-inflammatory states. This modulation occurs through the release of factors like interleukin-6 (IL-6) with macrophage colony-stimulating factor (M-CSF), which affect the differentiation and polarization of these cells. MSCs also impact B cell function by suppressing their growth, differentiation, and antibody production. This suppression is primarily mediated by the secretion of chemokine ligand 2 (CCL2) that has been processed by metalloproteinases. MSCs also regulate NK cells, curbing their growth and reducing their cell-killing ability. This regulation involves substances like PGE2 and IDO.

Notably, MSCs' immune-modulating effects are flexible, changing based on environmental cues rather than being static. This dynamic nature allows MSCs to fine-tune their immunoregulatory functions based on the specific conditions of their surroundings (Jiang and Xu, 2020). The immunomodulatory capabilities of MSCs are enhanced in the presence of pro-inflammatory cytokines like tumor necrosis factor-α (TNF-α) and interferon-γ (IFN-γ). This process, known as "licensing" or activation, allows MSCs to adapt their immunosuppressive functions to specific inflammatory conditions. Current studies have shown the importance of extracellular vesicles (EVs) secreted by MSCs in mediating their immunomodulatory effects. These EVs, which include exosomes and microvesicles, contain proteins, lipids, and nucleic acids like microRNAs that can be transferred to target cells, altering their function and gene expression. The ability of MSC-derived EVs to mimic many of the immunomodulatory effects of whole MSCs has opened out prospect for cell-free therapeutic approaches. The mixture of these various mechanisms results in a powerful immunosuppressive effect, making SC-based therapies promising for treating conditions characterized by excessive or dysregulated immune responses. These include autoimmune disorders, graft-versus-host disease (GVHD), and chronic inflammatory position (Chen *et al.*, 2024). In-progress research aims to elucidate the other complex molecular pathways involved in SC-mediated immunomodulation. This work is critical for optimizing the therapeutic potential of MSCs and developing more targeted interventions. As our understanding grows, it may lead to the creation of next-generation cell therapies as well as novel approaches to manipulate the immune system for therapeutic benefit. These advancements have the power to transform the therapy of immune-mediated disorder and contribute into the advancement to personalized medicine.

9.3.1 SC Interactions with the Immune System

SC-based immunomodulation, particularly using mesenchymal SCs (MSCs), is a cutting-edge approach in regenerative medicine and immunotherapy. This method shows great promise for treating various immune-mediated disorders. Mesenchymal SCs (MSCs) exert complex immunomodulatory effects through various interconnected mechanisms. These involve both direct cellular interactions and the release of diverse soluble factors. MSCs produce a rich array of bioactive molecules, including various cytokines, chemokines, growth factors, and extracellular vesicles. This intricate mixture works in concert to create an environment that suppresses immune responses. Several key factors secreted by MSCs play crucial roles in this process, such as PGE2, IDO, TGF-β, HGF, and IL-10. These molecules collaborate to inhibit T-cell proliferation, promote the development of regulatory T cells, and dampen the activity of certain pro-inflammatory T cell subsets. Furthermore, MSCs express surface proteins like PD-L1 that interact with receptors on T cells, potentially leading to their inactivation or programmed cell death. This multifaceted approach allows MSCs to modulate the immune system in a sophisticated and nuanced manner. This multifaceted approach to immunomodulation makes MSCs a powerful tool in treating immune-related disorders, offering new possibilities in personalized medicine and targeted therapies (Ou *et al.*, 2023).

MSCs have wide-ranging immunomodulatory outcome that impact various immune cell types beyond T cells. They influence antigen-presenting cells (APCs) like dendritic cells and macrophages by inhibiting their maturation and promoting a more tolerogenic or anti-inflammatory state. This modulation occurs through factors like IL-6 as well as macrophage colony-stimulating factor (M-CSF), which affect the differentiation and polarization of these cells. MSCs also affect B cells by suppressing their growth, differentiation, and antibody production. This suppression primarily involves the secretion of metalloproteinase-processed CC-chemokine ligand 2 (CCL2). The connection between MSCs and natural killer (NK) cells is another important aspect of their immunomodulatory capacity. MSCs prohibit NK cell increase and reduce their cytotoxic activity through mechanisms involving indoleamine 2,3-dioxygenase (IDO) and prostaglandin E2 (PGE2). Notably, the immunomodulatory properties of MSCs are dynamic and respond to environmental signals. In the presence of pro-inflammatory cytokines such as interferon-γ (IFN-γ) and tumor necrosis factor-α (TNF-α), MSCs become licensed or activated, enhancing their immunosuppressive abilities. This adaptive response allows MSCs to effectively address inflammatory conditions and tailor their immunomodulatory effects to specific microenvironments (Zaripova *et al.*, 2023).

Recent studies have shed light on the importance of extracellular vesicles (EVs) produced by MSCs in mediating their immunomodulatory effects. These EVs, which include exosomes and microvesicles, contain proteins, lipids, and nucleic acids such as microRNAs that can be transferred to target cells, altering their function and gene expression. The ability of MSC-derived EVs to mimic many of the immunomodulatory effects of whole MSCs has created opportunities for cell-free therapeutic approaches. The intricate interplay of these various mechanisms results in a powerful immunosuppressive effect, making SC-based therapies promising for treating conditions characterized by excessive or dysregulated immune responses. These include autoimmune disorders, GVHD, and chronic inflammatory conditions. Ongoing research aims to further elucidate the complex molecular pathways involved in SC-mediated immunomodulation. This work is critical for optimizing the beneficial ability of MSCs and developing more targeted interventions. As our understanding grows, it may lead to the creation of next-generation cell therapies and novel approaches to manipulate the immune system for therapeutic benefit. These advancements have the potential to transform the treatment of immune-mediated diseases and contribute to the advancement of personalized medicine (Table 9.1) (Rozmus, 2021).

Table 9.1 Anti-inflammatory effects of SCs.

S. no.	SC type	Anti-inflammatory effect	References
1.	Mesenchymal SCs (MSCs)	Reduction of pro-inflammatory cytokine production	Matsuzaka and Yashiro, 2024
2.	Neural SCs (NSCs)	Modulation of microglial activation in neuroinflammation	Trunk *et al.*, 2024
3.	Hematopoietic SCs (HSCs)	Regulation of immune cell differentiation and function	Banche-*Niclot et al.*, 2024
4.	Adipose-derived SCs (ADSCs)	Secretion of anti-inflammatory exosomes	Trunk *et al.*, 2024
5.	Dental pulp SCs (DPSCs)	Inhibition of T-cell proliferation	Min *et al.*, 2023
6.	SCs of umbilical cord	Suppression of inflammatory mediator release	Fu *et al.*, 2022
7.	Amniotic fluid SCs	Promotion of M2 macrophage polarization	Dong *et al.*, 2023
8.	Embryonic SCs (ESCs)	Reduction of oxidative stress in inflamed tissues	Yamanaka, 2020
9.	Induced pluripotent SCs (iPSCs)	Modulation of innate immune responses	Cerneckis *et al.*, 2024

(*Continued*)

Table 9.1 Anti-inflammatory effects of SCs. (*Continued*)

S. no.	SC type	Anti-inflammatory effect	References
10.	Liver SCs	Attenuation of hepatic inflammation	Sun *et al.*, 2022
11.	Cardiac SCs	Reduction of myocardial inflammation post-injury	Shao *et al.*, 2022
12.	Endothelial progenitor cells	Promotion of vascular repair and reduced inflammation	Yan *et al.*, 2022
13.	Muscle satellite cells	Mitigation of inflammatory muscle damage	Kaczmarek *et al.*, 2021
14.	Olfactory ensheathing cells	Reduction of neuroinflammation in spinal cord injuries	Zhang *et al.*, 2021
15.	Placenta-derived SCs	Suppression of pro-inflammatory T helper 17 (Th17) cells	Zaripova *et al.*, 2023
16.	Bone marrow mononuclear cells	Modulation of systemic inflammatory responses	Eldaly *et al.*, 2022
17.	Periodontal ligament SCs	Inhibition of inflammatory bone resorption	Wang *et al.*, 2022
18.	Urine-derived SCs	Reduction of renal inflammation in kidney diseases	Li *et al.*, 2023
19.	Menstrual blood-derived SCs	Suppression of inflammatory cytokine production	Shokri *et al.*, 2019
20.	Intestinal SCs	Promotion of mucosal healing and reduced gut inflammation	Zhang *et al.*, 2022
21.	Epidermal SCs	Acceleration of wound healing and reduction of skin inflammation	Lei *et al.*, 2022

(*Continued*)

Table 9.1 Anti-inflammatory effects of SCs. (*Continued*)

S. no.	SC type	Anti-inflammatory effect	References
22.	Corneal limbal SCs	Mitigation of ocular surface inflammation	Elhusseiny *et al.*, 2022
23.	Synovial mesenchymal SCs	Reduction of joint inflammation in arthritis	Li *et al.*, 2024
24.	Pancreatic SCs	Modulation of islet inflammation in diabetes	Basile *et al.*, 2022
25.	Lung SCs	Attenuation of pulmonary inflammation in respiratory diseases	Guo and Wang, 2024
26.	Testicular SCs	Reduction of inflammatory responses in testicular tissue	Sharifian *et al.*, 2022
27.	Mammary SCs	Modulation of breast tissue inflammation	Liu *et al.*, 2024
28.	Gingival mesenchymal SCs	Suppression of periodontal inflammation	Balaban *et al.*, 2024
29.	Hair follicle SCs	Reduction of scalp inflammation in alopecia	Sadeghi *et al.*, 2024
30.	Spermatogonial SCs	Mitigation of inflammatory damage in male reproductive system	Di Persio and Neuhaus, 2023

9.3.2 Anti-Inflammatory Effects of SCs

SC-based immunomodulation, especially applying MSCs, is an innovative approach in regenerative medicine and immunotherapy. This method shows great potential for treating various immune-mediated disorders. The mechanisms behind MSC-mediated immunomodulation are complex and interconnected, involving both cell-to-cell release and reaction of the diverse soluble factors. A key aspect of MSC-mediated immunomodulation is their ability to secrete a complex mixture of bioactive molecules. These include extracellular vesicles, chemokines, cytokines, and growth factors which work together to create an immunosuppressive environment. Among the important secreted factors are prostaglandin E2 (PGE2), indoleamine 2,3-dioxygenase (IDO), transforming growth factor-β (TGF-β), hepatocyte growth factor (HGF), and interleukin-10 (IL-10). This multifaceted approach to immunomodulation makes MSCs a powerful tool in treating immune-related disorders, offering new possibilities in personalized medicine and targeted therapies (Matsuzaka and Yashiro, 2024).

The bioactive molecules secreted by mesenchymal SCs (MSCs) work together to regulate immune responses in several ways. They inhibit T-cell proliferation, stimulate the growth of regulatory T cells (Tregs), and suppress the activity of pro-inflammatory T helper 1 (Th1) as well as T helper 17 (Th17) cells. MSCs also express surface molecules like programmed death-ligand 1 (PD-L1), which also interconnected PD-1 on T cells, leading to apoptosis in activated T cells.

They modulate the function of APCs, including macrophages and dendritic cells, by inhibiting their maturation and promoting a shift on a more tolerogenic or anti-inflammatory state. This modulation occurs through factors like IL-6 and M-CSF, which affect the differentiation and polarization of these cells. The bioactive molecules secreted by mesenchymal SCs (MSCs) work together to regulate immune responses in several ways. They inhibit T-cell proliferation, encourage the growth of managing T cells (Tregs), and suppress the activity of pro-inflammatory T helper 1 (Th1) and T helper 17 (Th17) cells. MSCs also express surface molecules like programmed death-ligand 1 (PD-L1), which interacts with PD-1 on T cells, leading to anergy or apoptosis in activated T cells. The antitumor effectiveness of MSCs extend beyond T cells to influence other immune cell types. They modulate the role of APCs, including macrophages and dendritic cells, by inhibiting their maturation and promoting a shift about a more tolerogenic or anti-inflammatory state. This modulation occurs through factors including IL-6 and M-CSF, which affect the differentiation and polarization of these cells (Ou *et al.*, 2023).

Moreover, recent research has highlighted the role of extracellular vesicles (EVs) produced by MSCs as key mediators of their immunomodulatory effects. These EVs, including microvesicles and exosomes, carry a cargo of lipids, nucleic acids, and proteins (such as microRNAs) that can be transferred to target cells, modulating their function and gene expression (Manzoor *et al.*, 2023). The ability of MSC-derived EVs to recapitulate many of the immunomodulatory effects of their parent cells has spread out avenues for cell-free therapies. The complex interplay of these diverse mechanisms results in a potent immunosuppressive effect, making

SC-based therapies particularly promising for treating conditions characterized by excessive or dysregulated immune responses, such as chronic inflammatory conditions, autoimmune disorders, and GVHD. Ongoing research continues to unravel the intricate molecular pathways involved in SC-mediated immunomodulation, aiming to optimize their therapeutic potential and develop more targeted interventions. As our understanding of these mechanisms deepens, it paves the way for the development of next-generation cryotherapies and novel approaches to manipulate the immune system for therapeutic benefit, potentially revolutionizing the therapy of immune-mediated infection and advancing the field of personalized medicine (Puccetti *et al.*, 2024).

9.3.3 Paracrine Signaling and Secretion of Immunomodulatory Factors

SC-based immunomodulation, particularly using MSCs, is an innovative approach in regenerative medicine and immunotherapy with significant potential for treating various immune-related disorders. This process involves a complex structure, as well as direct cell interactions and the release of soluble factors. MSCs exert their immunomodulatory effects primarily through secreting a diverse mix of bioactive molecules. These include chemokines, extracellular vesicles, cytokines, and growth factors, which together create an environment that suppresses immune responses. Some of the key factors secreted by MSCs are interleukin-10 (IL-10), hepatocyte growth factor (HGF), prostaglandin E2 (PGE2), transforming growth factor-β (TGF-β), and indoleamine 2,3-dioxygenase (IDO). These molecules work together to achieve several immunomodulatory effects, inhibiting T-cell proliferation, promoting the development of regulatory T cells (Tregs), and suppressing the activity of pro-inflammatory T helper 17 (Th17) cells and T helper 1 (Th1) (Han *et al.*, 2022).

Furthermore, MSCs express surface molecules such as PD-L1, which directly engage with PD-1 on T cells, inducing anergy or apoptosis in activated T cells. The immunomodulatory effects of MSCs extend beyond T cells, significantly impacting various other immune cell populations. For instance, MSCs modulate the role of antigen-presenting cells (APCs), as well as dendritic cells and macrophages, by inhibiting their maturation and promoting a shift toward a more tolerogenic or anti-inflammatory phenotype. This is achieved through the secretion of factors like IL-6 and macrophage colony-stimulating factor (M-CSF), which alter the differentiation and polarization of these cells. Additionally, MSCs influence B-cell functions by suppressing their proliferation, differentiation, and antibody production, primarily through the secretion of metalloproteinase-processed CC-chemokine ligand 2 (CCL2) (Wang *et al.*, 2022).

The relationship between MSCs and NK cells is another crucial aspect of their immunomodulatory capacity, with MSCs inhibiting cytotoxicity and NK cell proliferation through mechanisms involving PGE2 and IDO. Importantly, the immunomodulatory properties of MSCs are not static but rather responsive to environmental cues. In the presence of IFN-γ and TNF-α, MSCs become licensed or activated, enhancing their immunosuppressive capabilities. This adaptive response allows MSCs to effectively respond to inflammatory conditions and tailor their immunomodulatory effects to the specific microenvironment. Moreover, recent research has highlighted the role of extracellular vesicles (EVs) produced by MSCs as key mediators of their immunomodulatory effects. These EVs, including microvesicles and exosomes, carry a cargo of lipids, nucleic acids, and proteins (such as microRNAs) that can be transferred to target cells, modulating their function and gene expression.

The ability of MSC-derived EVs to recapitulate many of the immunomodulatory effects of their parent cells has opened new avenues for cell-free therapies. The complex interplay of these diverse mechanisms results in a potent immunosuppressive effect, making SC-based therapies particularly promising for treating conditions characterized by excessive or dysregulated immune responses, such as chronic inflammatory, and autoimmune disorders, GVHD conditions. Ongoing research continues to unravel the intricate molecular pathways involved in SC-mediated immunomodulation, aiming to optimize their therapeutic potential and develop more targeted interventions. As our understanding of these mechanisms deepens, it paves the way for the transformation of next-generation cell therapies and novel approaches to manipulate the immune system for therapeutic benefit, potentially revolutionizing the therapy of immune-mediated diseases and advancing the field of personalized medicine (Bashor *et al.*, 2022).

9.3.4 Regulation of Immune Cell Populations (e.g., T Cells, B Cells, Dendritic Cells)

The regulation of immune cell populations, particularly, dendritic cells, B cells, and T cells, is a complex and multifaceted process that plays a crucial role in maintaining immune homeostasis and orchestrating effective immune responses against pathogens while preventing autoimmunity.

This intricate regulation involves a myriad of molecular mechanisms, including cytokine signaling, transcription factor activation, epigenetic modifications, and metabolic reprogramming, all of which work in concert to fine-tune the development, activation, differentiation, and function of these immune cell subsets. T cell regulation, for instance, begins in the thymus with positive and negative selection processes that shape the T cell receptor (TCR) repertoire, ensuring self-tolerance while maintaining the ability to recognize foreign antigens.

T cell regulation in the periphery involves multiple mechanisms, with regulatory T cells (Tregs) playing a crucial role. Tregs suppress excessive immune responses by producing inhibitory molecules and through direct cellular interactions. The development of different T cell subsets is influenced by the surrounding environment and specific genetic factors. B cell regulation is also intricate, involving processes that depend on T cells and others that do not. In the bone marrow, developing B cells go through checks to remove those that might react against the body's own tissues. These regulatory processes help maintain a balanced immune system that can effectively fight pathogens while minimizing the risk of attacking the body's own cells.

In peripheral tissues, B cells undergo activation and differentiation based on various signals. These include inputs from their B cell receptors (BCRs), co-stimulatory molecules, and cytokines. The germinal center reaction, vital for producing high-affinity antibodies and memory B cells, is carefully regulated by two types of T cells: follicular helper T cells (Tfh) that promote B cell responses and follicular regulatory T cells (Tfr) that suppress them. Dendritic cells (DCs) are crucial in linking innate and adaptive immunity. Their regulation is essential for maintaining immune balance. Environmental cues control DC maturation and activation, including PAMPs and DAMPs. These signals activate pattern recognition receptors (PRRs) such as Toll-like receptors (TLRs) on DCs. The activation state of DCs influences their ability to prime T cell responses, mature DCs encourage effector T cell responses, and immature or tolerogenic DCs promote T cell anergy or regulatory T cell development. This intricate regulatory system helps maintain immune homeostasis while allowing effective responses to pathogens (Goldmann *et al.*, 2024).

The plasticity of immune cell populations adds another layer of complexity to their regulation. For example, the Th17/Treg axis demonstrates remarkable plasticity, with the balance between these subsets being influenced by environmental cues such as the presence of IL-6 (promoting Th17 differentiation) or retinoic acid (favoring Treg development). This plasticity allows for rapid adaptation to changing immunological contexts. Metabolic regulation has emerged as a critical factor in shaping immune cell fate and function. For instance, the switch from oxidative phosphorylation to aerobic glycolysis (known as the Warburg effect) is essential for T cell activation and effector function, while fatty acid oxidation supports the development and function of memory T cells and Tregs.

Similarly, the metabolic state of DCs influences their immunogenic versus tolerogenic potential, with tolerogenic DCs relying more on oxidative phosphorylation. The intricate regulation of these immune cell populations is further modulated by the microbiome, hormones, and neural inputs, highlighting the complex interplay connecting the immune system and other physiological systems. Understanding these regulatory mechanisms is crucial for developing targeted immunotherapies and interventions for various immune-mediated disorders, including autoimmune diseases, cancer, and chronic infections (Table 9.2) (Brescia *et al.*, 2024).

Table 9.2 Regulation of immune cell populations.

S. no.	Regulatory mechanism	Effect on immune cell population	References
1.	Thymic selection	Elimination of self-reactive T cells	Jiang and Xu, 2020
2.	Bone marrow hematopoiesis	Production of diverse immune cell precursors	Schirrmacher, 2023
3.	Cytokine-mediated differentiation	Directed development of specific immune cell types	Cui *et al.*, 2024
4.	Apoptosis	Removal of excess or dysfunctional immune cells	Cao *et al.*, 2019
5.	Regulatory T cell suppression	Inhibition of effector T cell proliferation	So *et al.*, 2023
6.	Clonal expansion	Rapid increase in antigen-specific lymphocytes	Mansourabadi *et al.*, 2023
7.	Myeloid-derived suppressor cells	Suppression of T cell responses	Veglia *et al.*, 2021

(Continued)

Table 9.2 Regulation of immune cell populations. (*Continued*)

S. no.	Regulatory mechanism	Effect on immune cell population	References
8.	Feedback inhibition	Self-limiting immune responses	Majdoul and Compton, 2022
9.	Antigenic stimulation	Activation and proliferation of specific lymphocytes	Waldman *et al.*, 2020
10.	Metabolic regulation	Control of immune cell activation and function	Wu *et al.*, 2023
11.	Epigenetic modifications	Long-term alterations in immune cell gene expression	Scott-Browne and Shih, 2022
12.	Circadian rhythms	Temporal regulation of immune cell activity	Ding *et al.*, 2024
13.	Gut microbiome interactions	Modulation of mucosal immune populations	Lin *et al.*, 2021
14.	Hormonal influences	Endocrine system effects on immune cell development	Santana-Sánchez *et al.*, 2024
15.	Chemokine gradients	Directed migration and positioning of immune cells	Moore *et al.*, 2018
16.	Antigen presentation	Activation of T cells by professional APCs	Eiz-Vesper and Schmetzer, 2020
17.	Neutrophil extracellular traps (NETs)	Regulation of neutrophil turnover	Liana *et al.*, 2022
18.	B cell receptor editing	Elimination of autoreactive B cells	Bonasia *et al.*, 2021
19.	Complement system activation	Recruitment and activation of phagocytes	Vandendriessche *et al.*, 2021

(*Continued*)

Table 9.2 Regulation of immune cell populations. (*Continued*)

S. no.	Regulatory mechanism	Effect on immune cell population	References
20.	Interferon signaling	Modulation of innate and adaptive immune responses	Shuo *et al.*, 2023
21.	Natural killer cell education	Tuning of NK cell responsiveness	Rascle *et al.*, 2023
22.	Dendritic cell maturation	Regulation of T cell priming capacity	Blumenthal *et al.*, 2020
23.	Macrophage polarization	Shifts in pro- and anti-inflammatory macrophage populations	Merz *et al.*, 2021
24.	Plasma cell differentiation	Control of antibody-producing cell numbers	Saito *et al.*, 2024
25.	Memory cell formation	Maintenance of long-lived antigen-specific populations	Syeda *et al.*, 2024
26.	Granulopoiesis	Regulation of granulocyte production and release	Paudel *et al.*, 2022
27.	Lymph node trafficking	Control of lymphocyte recirculation	Najibi *et al.*, 2024
28.	Thymic involution	Age-related decline in naive T cell production	Liang *et al.*, 2022
29.	Extramedullary hematopoiesis	Emergency immune cell production in peripheral tissues	Cenariu *et al.*, 2021
30.	Immune checkpoint inhibition	Modulation of T cell activation thresholds	Elliot *et al.*, 2021

9.4 Clinical Applications of SCs in Autoimmune Disorders

The work of SCs in treating autoimmune disorders represents a groundbreaking area in regenerative medicine, introducing new therapeutic possibilities for situation that have been difficult to manage with traditional treatments. Mesenchymal SCs (MSCs), in particular, have shown significant potential due to their immunomodulatory and tissue-repair capabilities. These SCs can differentiate into various cell types, release bioactive molecules, and regulate immune responses, making them well-suited for addressing the complex mechanisms underlying autoimmune diseases. In clinical practice, SC therapies are being developed to reset the immune system, support tissue repair, and promote immune tolerance.

The administration routes vary, including intravenous infusion, local injection, and even encapsulated delivery systems, each tailored to the specific disease and target tissue. One of the most extensively studied applications is the treatment of multiple sclerosis (MS), where mesenchymal stem cells (MSCs) and hematopoietic stem cells (HSCs) have shown promising results. In MS, SC therapies aim to halt disease progression, reduce inflammation, and potentially promote remyelination (Banche-Niclot *et al.*, 2024). Clinical trials have demonstrated improvements in disability scores, reduction in brain lesions, and, in some cases, long-term remission following autologous HSC transplantation. Another significant application is in the management of a multi-system autoimmune disease, systemic lupus erythematosus (SLE). MSC therapy in SLE patients has shown the potential to modulate the aberrant immune response, reduce disease activity, and improve organ function. Clinical studies have reported decreased levels of autoantibodies, improved renal function in cases of lupus nephritis, and a reduction in the need for immunosuppressive medications following MSC treatment.

In the realm of rheumatological disorders, rheumatoid arthritis (RA) has been a key target for SC therapies. MSCs have demonstrated the ability to suppress the inflammatory processes driving joint destruction in RA. Clinical trials have shown improvements in disease activity scores, reduced joint inflammation, and even evidence of cartilage repair in some patients (Tsiapalis *et al.*, 2023). The immunomodulatory effects of MSCs have been particularly beneficial in mitigating the chronic inflammatory state characteristic of RA. SC approaches have also shown promise in treating type 1 diabetes mellitus (T1DM), the autoimmune condition affecting the insulin-producing beta cells of the pancreas. Clinical studies have explored the use of MSCs to modulate the autoimmune attack on beta cells and potentially stimulate the regeneration of insulin-producing cells. Some trials have reported improved glycemic control, reduced insulin requirements, and even temporary insulin independence in a subset of patients following SC therapy.

This approach offers hope for preserving beta cell function and potentially altering the course of the disease, especially when intervention occurs early in the disease process. In the field of gastroenterology, Crohn's disease has emerged as another important application for SC therapies. Local injection of MSCs has shown efficacy in healing recalcitrant fistulas, a common and debilitating complication of Crohn's disease. Systemic administration of MSCs has also been explored for managing the broader inflammatory aspects of the disease, with some studies reporting improvements in disease activity indices and quality of life measures.

The potential of MSCs to promote mucosal healing and modulate the dysregulated immune response in the gut makes them a promising option for patients with refractory Crohn's disease (Qiao *et al.*, 2024). Despite these encouraging results, it is important to note that SC therapies for autoimmune disorders are still largely experimental, with ongoing research aimed at optimizing treatment protocols, understanding long-term outcomes, and identifying the most responsive patient populations. Challenges remain, including standardizing cell preparation methods, determining optimal dosing regimens, and addressing potential long-term safety concerns. Moreover, the heterogeneity of autoimmune disorders necessitates a personalized approach, tailoring SC therapies to individual disease characteristics and patient factors. As research progresses, combination therapies incorporating SCs with other immunomodulatory agents are being explored to enhance therapeutic efficacy. The evolving landscape of SC applications in autoimmune disorders holds great promise for transforming the management of these complex diseases, potentially offering new hope for patients who have exhausted conventional treatment options (Ren *et al.*, 2024).

9.4.1 Hematopoietic SC Transplantation (HSCT)

HSCT is a groundbreaking therapy approach in hematology and oncology, offering potential cures for various malignant and non-malignant blood and immune system disorders. This procedure involves infusing multipotent hematopoietic SCs, which are usually sourced from peripheral blood and bone marrow to restore hematopoietic function in patients with damaged or defective bone marrow or immune systems. The HSCT process is intricate, consisting of several key stages: patient evaluation, donor selection, administration of a conditioning regimen, SC collection, infusion, and post-transplant care. The conditioning regimen, often involving high-dose chemotherapy or radiation, aims to eliminate diseased cells, create space for the donor SCs, and suppress the recipient's immune system to prevent graft rejection. Hematopoietic SC transplantation (HSCT) encompasses two primary approaches, each defined by the SC source. Autologous HSCT uses the patient's own cells, while allogeneic HSCT employs donor cells. In autologous transplants, the patient's SCs are collected and later reinfused, avoiding rejection issues but potentially carrying disease risks. Allogeneic transplants introduce healthy donor cells, offering potential curative effects but with increased complications like graft-versus-host disease. The choice between these methods depends on the specific condition, patient factors, and donor availability (Lachance *et al.*, 2021).

The choice between autologous and allogeneic HSCT depends on factors like the underlying disease, patient characteristics, and availability of suitable donors. One of the most significant uses of HSCT is in treating acute leukemias, particularly acute myeloid leukemia (AML) and acute lymphoblastic leukemia (ALL). For these aggressive blood cancers, HSCT offers the potential for long-term remission and even a cure, particularly for high-risk patients or those who have relapsed after initial chemotherapy. In allogeneic HSCT, the graft-versus-leukemia effect provides an additional immune mechanism to target any remaining leukemic cells, leading to improved outcomes. Another critical application of HSCT is in the treatment of lymphomas, including both non-Hodgkin and Hodgkin types. Autologous HSCT has become the standard of care for relapsed or refractory aggressive lymphomas, offering a chance for durable remission in patients who have not responded to conventional therapies. The use of high-dose chemotherapy, supported by autologous SCs, allows for a more intensive treatment of chemotherapy-sensitive diseases. In the context of non-malignant disorders, HSCT has shown significant success in treating severe aplastic anemia, a life-threatening condition involving bone marrow failure. For young patients with severe aplastic anemia, allogeneic HSCT from a matched sibling donor is often the preferred treatment, offering the possibility of complete restoration of blood cell production and long-term cure.

The procedure has dramatically improved the prognosis of this once-fatal condition, with long-term survival rates exceeding 90% in well-matched transplants. HSCT has also emerged as a groundbreaking therapy for certain inherited metabolic disorders, such as adrenoleukodystrophy and Hurler syndrome. In these conditions, HSCT can provide a source of enzymatically normal cells, potentially halting disease progression and preventing further neurological deterioration. The ability of donor-derived cells to cross the blood-brain barrier along with provide cross-correction of enzymatic deficiencies has made HSCT a vital intervention, particularly when performed early in the disease course (Merli *et al.*, 2022). In recent years, HSCT has gained attention as a potential treatment for severe autoimmune diseases refractory to conventional therapies. Conditions such as multiple sclerosis, systemic sclerosis, and Crohn's disease have been targets of this approach.

The rationale involves using high-dose immunosuppression to ablate the autoreactive immune system, followed by reconstitution with autologous hematopoietic SCs, essentially "resetting" the immune system. While still considered experimental in many autoimmune indications, this approach has shown promising results in carefully selected patients, with some experiencing long-term remission and improved quality of life. Despite its potential, HSCT is associated with significant risks, including graft-versus-host disease in allogeneic transplants, infections due to prolonged immunosuppression, and organ toxicities from conditioning regimens. Ongoing research focuses on refining conditioning protocols, improving graft-versus-host disease prophylaxis, and expanding the donor pool through advances in haploidentical and cord blood transplantation. The field is also witnessing exciting developments in gene therapy and gene editing techniques, which may further expand the applications of HSCT by allowing for correction of genetic defects in autologous SCs. As our understanding of SC biology and immunology continues to grow, HSCT remains a dynamic and evolving field, holding the promise of new cures and improved outcomes for patients with a diverse array of challenging diseases (Haltalli *et al.*, 2022).

9.4.2 Mechanism of Action and Therapeutic Effects

Hematopoietic SC transplantation (HSCT) stands as a comprehensive therapeutic approach in the field of hematology and oncology, offering potentially curative treatments for a wide array of malignant and non-malignant disorders of the blood and immune system. This complex procedure involves the infusion of multipotent hematopoietic SCs, usually derived from peripheral blood and bone marrow to rebuild hematopoietic function in patients whose immune system or bone marrow is damaged or defective. The process of HSCT is multifaceted, encompassing several crucial stages including patient evaluation, donor selection, conditioning regimen administration, SC collection, infusion, and post-transplant care.

The conditioning regimen, which may involve high-dose chemotherapy or radiation therapy, serves to eradicate infected cells and create space for donor SCs, while also suppressing the recipient's immune system to prevent graft rejection. HSCT can be categorized into two primary types: autologous, where the patients have SCs are utilized, and allogeneic, where the SCs are sourced out from a donor (Lachance *et al.*, 2021). The choice between these types depends on the underlying disease, patient characteristics, and availability of suitable donors. One of the most significant applications of HSCT is in the treatment of acute leukemias, particularly acute lymphoblastic leukemia (ALL) and acute myeloid leukemia (AML). In these destructive blood cancers, HSCT offers the potential for long-term remission and even cure, especially for high-risk patients or those who have relapsed after initial chemotherapy.

The graft-versus-leukemia reaction in allogeneic HSCT provides an additional immunological mechanism to target residual leukemic cells, contributing to improved outcomes. Another crucial application is in the management of lymphomas, both Hodgkin and non-Hodgkin lymphomas. Autologous HSCT has become a level of care for relapsed or refractory aggressive lymphomas, offering a chance for durable remissions in patients who have failed conventional therapies. The high-dosage chemotherapy enabled by autologous SC support allows for a more intensive treatment of chemo-sensitive diseases (Chen *et al.*, 2023). In the realm of non-malignant disorders, HSCT has shown remarkable efficacy in treating severe aplastic anemia, a life-threatening bone marrow failure syndrome. Allogeneic HSCT from a matched sibling donor is considered the treatment of choice for young patients with severe aplastic anemia, offering the potential for complete restoration of hematopoiesis and long-term cure. The procedure has dramatically improved the prognosis of this once-fatal condition, with long-term survival rates exceeding 90% in well-matched transplants. HSCT has also emerged as a groundbreaking therapy for certain inherited metabolic disorders, such as adrenoleukodystrophy and Hurler syndrome.

In these conditions, HSCT can provide a source of enzymatically normal cells, potentially halting disease progression and preventing further neurological deterioration. The ability of donor-derived cells to cross the blood–brain barrier and provide a cross-correction of enzymatic deficiencies has made HSCT a vital intervention, particularly when performed early in the disease course (Alexander *et al.*, 2024). HSCT has gained attention as a potential treatment for severe autoimmune diseases refractory to conventional therapies. Conditions such as Crohn's disease, multiple sclerosis, and systemic sclerosis have been targets of this approach. The rationale involves using high-dose immunosuppression to ablate the autoreactive immune system, followed by reconstitution with autologous hematopoietic SCs, essentially resetting the immune system. While still considered experimental in many autoimmune indications, this approach has shown promising results in carefully selected patients, with some experiencing long-term remission and improved quality of life.

Despite its potential, HSCT is associated with significant risks, including graft-versus-host disease in allogeneic transplants, infections due to prolonged immunosuppression, and organ toxicities from conditioning regimens. Ongoing research focuses on refining conditioning protocols, improving graft-versus-host disease prophylaxis, and expanding the donor pool through advances in haploidentical and cord blood transplantation. The field is also witnessing exciting developments in gene therapy and gene editing techniques, which may further expand the applications of HSCT by allowing for correction of genetic defects in autologous SCs. As our understanding of SC biology and immunology continues to grow, HSCT remains a dynamic and evolving field, holding the promise of new cures and improved outcomes for patients with a diverse array of challenging diseases (Fraint *et al.*, 2021).

9.4.3 Clinical Outcomes and Case Studies

The field of SC research and therapy has witnessed remarkable advancements in recent years, with numerous clinical trials and case studies demonstrating the potential of these versatile cells to revolutionize medical treatments across a wide spectrum of diseases. Clinical outcomes have varied widely depending on the specific condition, type of SCs used, and method of administration, but several notable successes have emerged, offering hope for patients with previously intractable conditions. In the realm of hematological disorders, a groundbreaking case study involved a patient with sickle cell disease who underwent autologous hematopoietic SC transplantation (HSCT) using genetically modified SCs. The patients who have SCs were edited using CRISPR-Cas9 technology to produce fetal hemoglobin, effectively alleviating the symptoms of sickle cell disease. Follow-up studies showed a sustained production of fetal hemoglobin and a marked reduction in vaso-occlusive crises, demonstrating the development of combining SC therapy with gene editing techniques.

Moving to neurological disorders, a series of clinical trials focusing on amyotrophic lateral sclerosis (ALS) has shown promising results using mesenchymal SCs (MSCs). In one notable study, intrathecal and intravenous administration of MSCs in ALS patients led to a slowing of disease progression in some participants, with improvements in respiratory function and quality of life measures. While not a cure, these outcomes suggest that SC therapy could potentially extend survival and improve the quality of life for ALS patients (Park *et al.*, 2024). In the field of ophthalmology, SC therapy has offered new hope for patients with certain forms of blindness. A landmark case study involved the treatment of age-related macular degeneration using embryonic SC-derived retinal pigment epithelium cells. The patient, who had severe vision loss, experienced significant improvement in visual acuity following the transplantation, with the effects persisting for several years post-treatment.

This case paved the way for larger clinical trials in retinal diseases, showcasing the potential of SCs to restore vision in previously untreatable conditions. Cardiovascular medicine has also seen promising applications of SC therapy. In a large-scale clinical trial for chronic heart failure, transendocardial injection of mesenchymal precursor cells demonstrated significant improvements in left ventricular function and reduction in major adverse cardiovascular events. Patients receiving the SC treatment showed enhanced exercise capacity and reduced hospitalization rates balanced to the control group, suggesting that SC therapy could become a valuable adjunct in managing chronic heart failure. The treatment of spinal cord injuries has been another area of intense focus for SC research.

A case study involving a patient with complete cervical spinal cord injury reported remarkable outcomes following the transplantation of autologous olfactory ensheathing cells and olfactory nerve fibroblasts. The patient, who was initially paralyzed from the chest down, regained some motor function and sensory perception below the level of injury, challenging long-held beliefs about the irreversibility of complete spinal cord injuries. While the results are still considered preliminary, they offer hope for developing more effective treatments for spinal cord injury patients (Perin *et al.*, 2026). In the realm of autoimmune diseases, a notable clinical trial focused on the use of hematopoietic SC transplantation for severe, treatment-resistant multiple sclerosis.

The trial, which involved intense immunosuppression observe by autologous HSCT, reported that a significant proportion of patients remained free from disease activity for several years post-treatment, with some even showing improvements in disability scores. These results suggest that SC therapy could potentially halt or even reverse the progression of aggressive forms of multiple sclerosis, offering a new avenue for patients who have exhausted conventional treatment options. Lastly, in the field of regenerative medicine, a groundbreaking case study involved the use of induced pluripotent SCs (iPSCs) to treat macular degeneration. Researchers generated retinal pigment epithelium cells from the patient's own skin cells, reprogrammed into iPSCs. The transplanted cells integrated successfully into the retina, with the patient experiencing improved visual acuity and no signs of rejection or tumor formation over a 5-year follow-up period.

This case not only demonstrated the feasibility of iPSC-based therapies but also highlighted their potential to provide personalized treatments while minimizing the risk of immune rejection. These varied examples highlight the wide-ranging potential of SC therapies in different medical fields. However, it is important to recognize that many of these studies are still in the early phases, and long-term monitoring is essential to thoroughly evaluate their safety and effectiveness. Several challenges persist, including the need to optimize cell delivery techniques, ensure long-term survival and integration of the cells, and address potential risks such as tumor development or uncontrolled cell differentiation. Despite these obstacles, the encouraging results seen in clinical studies and case reports lay a strong groundwork for future research and hold promise for developing new treatments for various serious diseases (Skidmore *et al.*, 2023).

9.4.4 Mesenchymal SCs (MSCs)

MSCs have appeared as a cornerstone in cell-based therapies, owing to their remarkable plasticity, immunomodulatory properties, and therapeutic potential across a wide spectrum of diseases. These multipotent stromal cells, initially identified in the bone marrow but subsequently isolated from some adult tissues including dental pulp, adipose tissue, umbilical cord, and placenta possess the range to convert into multiple cell lineages of mesodermal origin, including osteoblasts, chondrocytes, and adipocytes (Yang *et al.*, 2023). The defining characteristics of MSCs include their adherence to plastic surfaces in culture, expression of specific surface antigens (CD90, CD105, and CD73), and lack of hematogenic markers (CD34, CD14, CD45, HLA-DR, and CD19 or CD79α).

Mesenchymal SCs (MSCs) are recognized not only for their ability to differentiate into various cell types but also for their therapeutic impact through paracrine signaling. They release a wide range of bioactive compound, such as, cytokines, extracellular vesicles, and growth factors, which help shape the tissue microenvironment. These secretions play a critical role in promoting blood vessel formation, preventing cell death, and modulating immune responses. The immunomodulatory capabilities of MSCs are especially important in clinical settings. They have been shown to suppress the proliferation of T cells, influence B-cell activity, hinder the maturation of dendritic cells, and encourage the development of regulatory T cells. These multifaceted capabilities have led to the exploration of MSCs in treating a diverse array of pathological conditions (Summer *et al.*, 2024).

In orthopedics, mesenchymal SCs (MSCs) have demonstrated significant potential for cartilage repair and the treatment of osteoarthritis. Clinical studies involving intra-articular injections of MSCs in individuals with knee osteoarthritis have shown notable improvements in pain levels, functional abilities, and cartilage quality, as observed through magnetic resonance imaging (MRI).

The capacity of MSCs to transform into chondrocytes, coupled with their anti-inflammatory effects, positions them as a potential game-changer in managing degenerative joint diseases. The application of MSCs in treating GvHD represents another significant advancement. In a landmark study, patients and steroid-resistant, severe acute GvHD received intravenous infusions of allogeneic MSCs, resulting in complete responses in a substantial proportion of participants. This success has led to the approval of MSC-based products for GvHD treatment in several countries, highlighting the potent immunomodulatory effects of these cells. In the realm of cardiovascular medicine, MSCs have been explored for their potential in treating myocardial infarction and chronic heart failure. Clinical trials involving intramyocardial or intracoronary administration of MSCs have demonstrated improvements in increased left ventricular role and decreased infarct size and enhanced exercise capacity in some patients. The mechanisms underlying these benefits are thought to involve paracrine effects promoting angiogenesis and reducing fibrosis rather than direct differentiation of MSCs into cardiomyocytes (Kadri *et al.*, 2023). The neurological applications of MSCs have garnered significant attention, particularly in the treatment of multiple sclerosis (MS). Clinical studies exploring the intravenous or intrathecal administration of MSCs in MS patients have reported reductions in inflammatory lesions, improvements in visual acuity, and stabilization of disease progression in some cases.

The neuroprotective and immunomodulatory properties of MSCs are believed to contribute to these beneficial effects, offering hope for patients with this debilitating condition. In the field of wound healing and tissue repair, MSCs have demonstrated remarkable efficacy, especially in the treatment of chronic wounds like diabetic foot ulcers. Clinical trials utilizing MSC-seeded scaffolds or direct application of MSCs to chronic wounds have shown accelerated wound closure, improved vascularization, and reduced scarring. The ability of MSCs to secrete growth factors that promote angiogenesis as well as modulate the inflammatory response makes them particularly suited for enhancing the wound healing process. Lastly, the potential of MSCs in treating liver cirrhosis has been an area of active investigation. Clinical studies involving the infusion of MSCs along the hepatic artery or portal vein in patients with liver cirrhosis have reported improvements in liver function parameters, reduction in ascites, and enhanced quality of life in some participants.

The capacity of MSCs to modulate hepatic stellate cell activation and promote hepatocyte regeneration underlies their therapeutic potential in liver diseases. Despite these promising outcomes, challenges remain in fully harnessing the therapeutic potential of MSCs. Issues such as optimal cell dosing, timing of administration, and delivery routes are still being refined. Additionally, the heterogeneity of MSC populations and the variability in their preparation methods pose challenges for standardization and regulatory approval. Ongoing research is focused on enhancing the efficacy of MSC therapies through genetic modification, preconditioning strategies, and combination with biomaterials or other therapeutic agents. As our understanding of MSC biology continues to evolve and with advances in cell manufacturing and delivery technologies, the future holds great promise for expanding the clinical applications of these versatile cells, potentially revolutionizing the therapy of a wide range of infection (Lu *et al.*, 2024).

9.4.5 Mechanism of Action and Therapeutic Effects

The mechanism of action describes the particular biochemical process by which a drug exerts its therapeutic effects. Understanding this process is crucial for developing effective medications and predicting their therapeutic outcomes. The mechanism of action is intricately linked to a drug's therapeutic effects, which are the beneficial changes in physiological or biochemical processes resulting from the drug's interaction with the body. Drugs generally function by binding to particular molecular attack within the human body, like receptors, enzymes. These reactions can either enhance or inhibit natural biological processes, leading to the desired therapeutic outcome.

The specificity of a drug's mechanism of action often determines its efficacy and potential side effects. Some examples are aspirin (acetylsalicylic acid), penicillin, metformin, selective serotonin reuptake inhibitors (SSRIs), and statins (HMG-CoA reductase inhibitors) (Ialongo *et al.*, 2023). The mechanism of action of aspirin is such that it works primarily by inhibiting the enzyme cyclooxygenase (COX). This enzyme is responsible for producing prostaglandins, which are involved in inflammation, pain, and fever. By irreversibly binding to COX, aspirin prevents the production of prostaglandins. In terms of therapeutic effects, the inhibition of prostaglandin synthesis leads to several beneficial outcomes. Aspirin reduces inflammation, alleviates pain, and lowers fever. Additionally, it inhibits platelet aggregation, which gives it anticoagulant properties. This makes aspirin useful for preventing heart attacks and strokes in high-risk individuals.

Penicillin is part of the β-lactam antibiotic class and works by targeting and inhibiting penicillin-binding proteins (PBPs), which are key enzymes involved in the synthesis of bacterial cell walls. This inhibition interferes the final movement of peptidoglycan production, an essential module of bacterial cell walls. As a result, penicillin weakens the structural integrity of growing bacteria, ultimately leading to cell lysis and death. This mechanism makes penicillin highly effective against a broad spectrum of bacterial infections, especially those caused by gram-positive bacteria. Its ability to selectively target bacterial cell walls while leaving human cells unaffected contributes to its relatively low toxicity in humans. Metformin, a widely used antidiabetic medication, works through multiple mechanisms. The drug mainly works by lowering liver sugar production. It blocks the liver's ability to make new glucose, helping control blood sugar levels. It also enhances insulin sensitivity in peripheral tissues and increases glucose uptake in skeletal muscle (Rouzer and Marnett, 2020). By reducing glucose production and improving insulin sensitivity, metformin helps lower blood glucose levels in individuals with type 2 diabetes. This leads to better glycemic control, reducing the risk of diabetic complications. Additionally, metformin has been associated with weight loss and improved lipid profiles in some patients, providing additional cardiovascular benefits. Selective serotonin reuptake inhibitors (SSRIs), like sertraline and fluoxetine, work by blocking the consumption again of serotonin in the synaptic cleft between neurons.

This action increases the availability of serotonin, a neurotransmitter involved in mood regulation, in synaptic space. SSRIs (selective serotonin reuptake inhibitors) work by boosting serotonergic neurotransmission, which can help relieve symptoms of depression, anxiety, and other mood disorders. The increased serotonin levels are believed to enhance mood, alleviate anxiety, and regulate sleep patterns. Due to their effectiveness and relatively better side effect profile compared to older antidepressants, SSRIs are commonly prescribed for various mental health conditions. Statins or HMG-CoA reductase inhibitors function by blocking the enzyme HMG-CoA reductase, which is essential for cholesterol production in the liver. This enzyme controls the rate-limiting step in the mevalonate pathway, responsible for generating cholesterol and other isoprenoids. By inhibiting cholesterol synthesis, statins effectively lower the blood levels of low-density lipoprotein (LDL) cholesterol, commonly known as bad cholesterol. Lowering the LDL cholesterol levels helps to prevent the development and progression of atherosclerotic plaques in the blood vessels, thereby significantly decreasing the risk of cardiovascular events like heart attack and stroke. Statins also have pleiotropic effects, including anti-inflammatory and antioxidant properties, which may contribute to their cardiovascular benefits beyond lipid lowering. The mechanism of action and therapeutic effects of medications is crucial for healthcare providers and patients alike. It allows for more informed decision-making in treatment plans, helps predict potential drug interactions and side effects, and guides the development of new and improved medications. As medical research advances, our knowledge of drug mechanisms continues to expand, covering the way for extra targeted and potent therapies across various medical fields (Moncrieff *et al.*, 2023).

9.4.6 Clinical Trials and Outcomes in Autoimmune Diseases (e.g., Rheumatoid Arthritis, Multiple Sclerosis)

Clinical trials are essential for improving our knowledge and therapy of autoimmune diseases like multiple sclerosis (MS) and rheumatoid arthritis (RA). These trials evaluate the safety and efficacy of new therapies, compare different treatment strategies, and help refine management approaches for patients. Typically, they involve randomized, double-blind, placebo-controlled studies to minimize bias and provide reliable evidence. Outcomes are measured using disease-specific scales, biomarkers, imaging studies, and patient-reported outcomes. Long-term research is often necessary to assess the durability of treatment effects and monitor for potential late-onset side effects. One significant example is the REFLEX trial, which evaluated rituximab in patients and active RA who had an inadequate response to anti-TNF therapies.

In this phase III study, 520 patients were treated with either rituximab or a placebo alongside methotrexate. The main objective was to measure the proportion of patients who achieved an ACR20 response after 24 weeks (Jung and Kim, 2022). The results showed that 51% of patients in the rituximab class achieved an ACR20 response compared to 18% in the placebo group, with significant improvements also observed in secondary endpoints. This trial established rituximab as an effective treatment option for RA patients who had failed anti-TNF therapy, leading to its approval and changing treatment paradigms in refractory RA. Another notable example is the OPERA I and OPERA II trials, which evaluated ocrelizumab in relapsing-remitting multiple sclerosis (RRMS). These twin phase III studies compared ocrelizumab to interferon beta-1a in 1,656 patients over 96 weeks. The primary end terminal was the annual relapse rate, with secondary endpoints including confirmed disability progression and MRI measures.

Ocrelizumab decreased the annualized relapse rate by 46%–47% in comparison to interferon beta-1a and led to significant improvements in MRI outcomes. These trials led to the approval of ocrelizumab for RRMS and primary progressive MS, marking a significant advance in MS treatment by offering a highly effective B-cell-depleting therapy (Brummer *et al.*, 2021). The CARE-MS I and CARE-MS II trials investigated the advantage of alemtuzumab in patients and relapsing-remitting multiple sclerosis (RRMS), comparing it to interferon beta-1a. CARE-MS I focused on treatment-naive patients, while CARE-MS II involved those who had an inadequate response to previous therapy. These randomized, rater-blinded, active-controlled trials evaluated relapse rate and time to confirmed disability progression as co-primary endpoints. Alemtuzumab significantly reduced relapse rates in both studies—by 49% in CARE-MS I and 55% in CARE-MS II—and also demonstrated positive effects on MRI outcomes. However, these trials also revealed important safety considerations, including autoimmune adverse events, which influenced subsequent risk management strategies. The CARE-MS trials established alemtuzumab as a highly effective treatment for RRMS, particularly in patients with highly active disease, while also highlighting the need for careful monitoring of potential side effects. These clinical trials demonstrate the significant progress made in treating autoimmune diseases like RA and MS, leading to the approval of new therapies, refined treatment strategies, and improved outcomes for patients. They also underscore the complexity of these diseases and the need for ongoing research to address unmet needs, such as predictive biomarkers for treatment response and strategies to achieve long-term remission or cure (Jacob *et al.*, 2024).

9.4.7 Induced Pluripotent SCs (iPSCs)

Induced pluripotent SCs (iPSCs) represent a significant breakthrough in SC research and regenerative medicine. First developed by Shinya Yamanaka in 2006, iPSCs are adult cells reprogrammed to a state alike to embryonic SCs, allowing them to differentiate into any cell species in the body. This technology addresses ethical issues related to the usage of embryonic SCs and offers new opportunities in personalized medicine, disease modeling, and drug development. The creation of iPSCs involves introducing specific transcript features, known as Yamanaka factors (Oct4, Sox2, Klf4, and c-Myc), into adult cells. These factors reset the cell's epigenetic state, reverting it to a pluripotent condition. Since their discovery, various methods have been developed to generate iPSCs, including viral vectors, non-viral vectors, and small molecules, each with their own advantages and limitations.

One important application of iPSCs is in disease modeling. For example, researchers have used iPSCs derived from patients with Parkinson's disease to create dopaminergic neurons, the cell type affected in this condition. These neurons exhibit characteristics of the disease, allowing scientists to study its progression and test potential treatments in a dish. This approach has been particularly valuable for neurological disorders, where access to affected tissue is limited. In the field of drug discovery, iPSCs offer a powerful tool for screening potential therapeutic compounds. A notable example is the use of iPSC-derived cardiomyocytes to test drug cardiotoxicity (Liu *et al.*, 2020). These heart cells can be used to assess a drug's effect on cardiac function before clinical trials, potentially reducing the risk of late-stage drug failures due to unforeseen cardiac side effects. This application has become increasingly important in the pharmaceutical industry, improving the efficiency and safety of drug development processes. Regenerative medicine is another area where iPSCs show great promise. In Japan, researchers have successfully treated macular degeneration using iPSC-derived retinal pigment epithelium cells. This clinical trial marked a significant milestone as one of the first applications of iPSC technology in human patients. The treatment demonstrated the potential of iPSCs to restore function in damaged tissues, surface the way for similar approaches in other degenerative diseases. Lastly, iPSCs have revolutionized our understanding of cellular reprogramming and development. For instance, studies using iPSCs have provided insights into the mechanisms of X chromosome inactivation, a process crucial for female mammalian development.

By reprogramming adult cells and observing the reactivation and subsequent inactivation of the X chromosome, researchers have gained new understanding of this fundamental biological process. These examples highlight the versatility and potential of iPSC technology. As research progresses, iPSCs are expected to play an increasingly significant role in personalized medicine, offering tailored treatments established on a patient's own cells. However, challenges remain, including improving the efficiency of iPSC generation, ensuring their genetic stability, and developing standardized protocols for their use in clinical applications. Despite these challenges, the field of iPSC research continues to advance rapidly, holding promise for transformative developments in biomedical science and healthcare (Mao *et al.*, 2022).

9.4.8 Challenges and Future Directions

SC-based immunomodulation is a promptly advancing area of research focused on how SCs can modulate immune responses, presenting potential therapies for autoimmune diseases, inflammatory conditions, and transplant rejection. This approach harnesses the distinctive properties of SCs, particularly mesenchymal SCs (MSCs) and hematopoietic SCs (HSCs), to impact various aspects of the immune system. The mechanisms involved in SC-based immunomodulation are complex. A primary mechanism includes the release of soluble factors by SCs, such as cytokines and growth factors, which can reduce inflammation and support tissue healing. For instance, MSCs release substances like indoleamine 2,3-dioxygenase (IDO), prostaglandin E2 (PGE2), and transforming growth factor-β (TGF-β), which inhibit T cell proliferation and encourage the formation of regulatory T cells. This process has been explored in treating graft-versus-host disease (GVHD), where MSC infusions have shown potential in mitigating immune responses in transplant patients. Another crucial mechanism involves direct cell-to-cell connections between SCs and immune cells (Han *et al.*, 2022).

Through direct interactions, SCs can modulate the activities of various immune cell types, containing T cells, B cells, and natural killer cells. For instance, in multiple sclerosis models, neural SCs have been shown to directly interact with T cells, inducing their apoptosis and reducing inflammatory responses in the central nervous system. This direct cellular interaction complements the effects of secreted factors, providing a more comprehensive immunomodulatory effect. SCs also exert their immunomodulatory effects by influencing the polarization of immune cells. In the case of macrophages, MSCs can promote the shift from a pro-inflammatory M1 phenotype to an anti-inflammatory M2 phenotype. This mechanism has been studied in the context of wound healing and tissue repair, where MSC treatment has been shown to accelerate healing by promoting a more favorable immune environment.

Similarly, in models of inflammatory bowel disease, MSCs have demonstrated the ability to reduce inflammation by modulating the balance between pro-inflammatory and regulatory T cell subsets. The field of SC-based immunomodulation faces several challenges and exciting future directions. One major challenge is the variability in SC preparations and the need for standardized protocols to ensure consistent therapeutic effects. Researchers are working on developing more uniform cell populations and defining optimal dosing regimens. Another challenge is the limited engraftment and survival of transplanted SCs, which can reduce their long-term efficacy. To address this, scientists are exploring various strategies, including genetic modification of SCs to enhance their survival and immunomodulatory properties (Yan *et al.*, 2024).

Looking to the future, one promising direction is the development of "off-the-shelf" SC products that can be readily available for clinical use. This approach aims to overcome the logistical challenges associated with autologous SC therapies. Additionally, researchers are investigating the potential of combining SC therapy with other immunomodulatory approaches, such as checkpoint inhibitors or CAR-T cell therapy, to enhance overall therapeutic efficacy. Another exciting avenue is the use of extracellular vesicles (EVs) derived from SCs. These EVs, which carry various bioactive molecules, have shown immunomodulatory effects like their parent cells but with potentially fewer safety concerns. Research is ongoing to harness the power of these EVs for therapeutic applications. Lastly, advances in gene editing technologies like CRISPR-Cas9 open new possibilities for enhancing the immunomodulatory properties of SCs. By precisely modifying genes involved in immune regulation, researchers aim to create super SCs with amplified therapeutic potential (Lotfi *et al.*, 2023).

9.5 Challenges and Considerations in SC Therapy

SC therapy offers significant potential for treating various diseases and injuries, yet it also presents several challenges and issues that researchers and clinicians need to address. These challenges span scientific, ethical, regulatory, and practical domains, each requiring careful attention to ensure the safe and effective development of SC-based treatments. One of the primary challenges in SC therapy is ensuring the safety and long-term efficacy of treatments. For example, in the field of induced pluripotent SCs (iPSCs), there are concerns about the potential for tumor formation due to residual undifferentiated cells or genetic instability introduced during the reprogramming process.

This was highlighted in a clinical trial for age-related macular degeneration in Japan, where the discovery of genetic mutations in a patient's iPSCs led to a temporary halt of the trial. This case underscores the need for severe safety testing and long-term follow-up studies to monitor for potential adverse effects. Another significant challenge is achieving precise control over SC differentiation and integration into host tissues. This is particularly crucial in neurological disorders, where the complex architecture of the brain poses unique challenges (Doss and Sachinidis, 2019). In Parkinson's disease trials using fetal SCs, some patients experienced graft-induced dyskinesias, likely due to the presence of serotonergic neurons in the transplanted tissue.

This outcome emphasizes the importance of developing methods to generate pure populations of the desired cell types and to ensure their proper addition and function within the host tissue. Immune rejection remains a considerable hurdle in allogeneic SC therapies. While strategies like immunosuppression or the use of autologous cells can mitigate this risk, they come with their own set of challenges. For instance, in hematopoietic SC transplantation for leukemia, graft-versus-host disease (GVHD) remains a significant complication. Researchers are exploring various approaches to reduce GVHD, such as T-cell depletion or the use of mesenchymal SCs as adjunct therapy but finding the right balance between preventing rejection and maintaining the graft-versus-leukemia effect remains challenging. Regulatory and ethical considerations also play a crucial role in the development of SC therapies. The use of embryonic SCs, while scientifically promising, continues to face ethical debates and regulatory restrictions in many countries. This has led to a surge in research on alternative sources like adult SCs and iPSCs. However, these alternatives come with their own set of challenges. For example, the FDA approval process for SC therapies is complex and time-consuming, as evidenced by the lengthy development of CAR-T cell therapies for cancer treatment. Balancing the need for thorough safety evaluation with the urgency of bringing potentially life-saving treatments to patients remains an ongoing challenge. Looking forward, addressing these challenges will require a multifaceted approach. Advances in gene editing tools like CRISPR-Cas9 offer potential keys for enhancing SC safety and efficacy (Liu *et al.*, 2020).

Improved imaging techniques and biomarkers will be crucial for tracking the fate of transplanted cells and monitoring their long-term effects. Standardization of protocols for SC derivation, increase, and difference will be essential for ensuring reproducibility and facilitating clinical translation. Furthermore, international collaboration and harmonization of regulatory frameworks will be key to accelerating the development of SC therapies while maintaining high safety standards. This includes addressing issues of SC tourism and unproven treatments, which pose risks to patients and undermine public trust in legitimate SC research. As the field progresses, it will be crucial to maintain a balance between scientific innovation and rigorous safety standards. Continued investment in basic research, alongside well-designed clinical trials, will be necessary to unlock the full potential of SC therapy. By addressing these encounters and considerations, the field of SC therapy can move closer to realizing its promise of revolutionizing the treatment of numerous diseases and injuries (Zeng, 2023).

9.5.1 Safety Concerns and Risk of Tumorigenesis

Safety issues and the possibility of tumor development are paramount considerations in SC therapy, presenting significant challenges that researchers and clinicians must address to safeguard the safe and active application of these promising cures. The pluripotent nature of SCs, while key to their therapeutic potential, also raises concerns about uncontrolled growth and tumor formation. A major safety concern is the potential for teratoma development, especially with pluripotent SCs like embryonic SCs (ESCs) and induced pluripotent SCs (iPSCs). Teratomas are tumors containing tissues from all three germ layers, which can develop if undifferentiated pluripotent cells are transplanted. A notable example of this risk was observed in a study where undifferentiated human ESCs were injected into immunodeficient mice, resulting in teratoma formation. This highlighted the crucial need for complete differentiation of SCs before transplantation and rigorous purification techniques to remove any remaining pluripotent cells from the final cell product. Another significant concern is the potential for genetic instability in cultured SCs (Yamanaka, 2020).

Long-term culture and growth of SCs can lead to the accumulation of genetic and epigenetic changes, potentially increasing the risk of tumorigenesis. This was exemplified in a clinical trial for age-related macular deterioration in Japan, where the discovery of genetic mutations in a patient's iPSC line led to the suspension of the trial. The mutations, which were not present in the original fibroblasts, were detected in genes linked to increased cancer risk. This case underscored the importance of comprehensive genetic screening of SC lines before clinical application and the need for improved culture techniques to maintain genetic stability. The threat of malignant transformation is also a worry, particularly with adult SCs. While generally considered safer than pluripotent SCs, adult SCs can still pose risks if they acquire mutations during expansion or manipulation. For instance, in a case report involving a patient who received autologous mesenchymal SC (MSC) therapy for lupus nephritis, the patient developed a renal tumor at the injection site.

Although rare, this case highlighted the potential risks associated with *ex vivo* expansion of adult SCs and the need for careful monitoring of patient's post-treatment. The immunogenicity of transplanted SCs presents another safety challenge. While autologous SC therapies mitigate the risk of immune rejection, allogeneic SC treatments can potentially trigger immune responses. This was observed in early clinical trials of fetal neural SC transplantation for Huntington's disease, where some patients developed antibodies against the transplanted cells. Such immune reactions not only reduce the efficacy of the treatment but can also pose safety risks to the patient, emphasizing the need for strategies to modulate immune responses or develop universal donor cell lines (Derks and van Boxtel, 2023). Addressing these safety concerns requires a multifaceted approach. Improved differentiation protocols and purification techniques are being developed to ensure the complete elimination of undifferentiated cells from therapeutic products. Advanced genetic screening methods, including whole-genome sequencing and karyotyping, are being employed to detect genetic abnormalities in SC lines. Researchers are also exploring genetic engineering approaches to introduce suicide genes into SCs, allowing for their selective elimination if uncontrolled growth occurs post-transplantation. Furthermore, the development of *in vitro* tumor formation assays and more predictive animal models is essential for assessing the tumorigenic potential of SC-based products before clinical application. Long-term follow-up studies of patients receiving SC therapies are essential to monitor for delayed onset of tumors or other adverse effects. Regulatory bodies play a critical role in confirming the safety of SC therapies. The FDA and other international controlling agencies have established guidelines for the advancement and testing of SC-based products, emphasizing the need for robust preclinical data and carefully designed clinical trials. As the field advances, balancing the therapeutic potential of SCs with rigorous safety measures remains a key challenge. Continued research into the mechanisms of tumorigenesis in SCs, paired with advances in cell engineering and monitoring technologies, will be crucial in mitigating these risks and realizing the full ability of SC treatments (Madrid *et al.*, 2024).

9.5.2 Immunogenicity and Rejection Issues

Immunogenicity and rejection issues are critical concerns in transplantation and the development of biological therapies. These phenomena occur when the immune system recognizes foreign substances or tissues as potential threats, leading to an immune response that can compromise the effectiveness of treatments or cause rejection of transplanted organs. The process begins with the detection of non-self-antigens by the recipient's immune system, triggering a cascade of cellular and molecular events aimed at eliminating the perceived threat. This immune response can manifest in various forms, including acute rejection, chronic rejection, and graft-versus-host disease in the context of organ transplantation. In the realm of biological therapies, immunogenicity can lead to the production of anti-drug antibodies, reducing the efficacy of treatments and potentially causing adverse reactions. To mitigate these issues, researchers and clinicians employ various strategies, such as immunosuppression, tissue typing, and the development of less immunogenic therapeutic proteins. For example, in solid organ transplantation, careful HLA matching between donor and recipient can reduce the risk of rejection. In the case of biologic drugs, humanization of monoclonal antibodies has been used to decrease their immunogenicity. For cell therapies, genetic modification of cells to express immunomodulatory factors can help evade immune recognition. Lastly, in the context of gene therapy, the use of specific viral vectors or non-viral delivery systems can minimize immune responses to the therapeutic transgene. Despite these advances, immunogenicity and rejection remain significant challenges in the field of medicine, necessitating ongoing research and innovation to improve patient outcomes and expand the applications of transplantation and biological therapies (Li and Lan, 2023).

9.5.3 Standardization and Quality Control in SC Production

Regularization and excellence control in SC production are crucial aspects of regenerative medicine and cell-based therapies. These processes ensure the consistency, care, and efficacy of SC products, which is essential for their successful clinical application. The field of SC research and manufacture is rapidly evolving, and with it comes the need for robust standardization and quality control measures to meet regulatory requirements and maintain public trust. One of the primary challenges in SC production is the standardization of cell culture conditions. This includes the optimization of growth media, supplements, and environmental factors such as temperature, pH, and oxygen levels. For example, researchers at a prominent SC institute developed a chemically defined, xeno-free culture medium that supports the long-term expansion of human pluripotent SCs while maintaining their differentiation potential.

This advancement allows for more consistent and reproducible cell production across different laboratories and manufacturing facilities. Quality control in SC production involves rigorous testing and characterization of the cell products (Cao *et al.*, 2021). This includes assessing cell identity, purity, potency, and genetic stability. For instance, a biotechnology company specializing in SC therapies implemented a comprehensive quality control pipeline that utilizes flow cytometry, karyotyping, and next-generation sequencing to ensure the integrity of their cell lines. These measures help detect any genetic abnormalities or contamination that could compromise the security and efficacy of the ending product. Researchers at a prominent university developed a reliable protocol for producing functional cardiomyocytes from human induced pluripotent SCs (iPSCs).

This protocol includes precise timing of growth factor additions and carefully controlled culture conditions, resulting in a consistently high yield of cardiomyocytes with well-defined functional properties. Automation plays an increasingly important role in standardization and quality control of SC production (Cerneckis *et al.*, 2024). A cutting-edge SC manufacturing facility implemented a fully automated cell culture system that minimizes human intervention and reduces variability in cell handling. This system incorporates real-time monitoring of cell growth and metabolic parameters, allowing for data-driven optimization of culture conditions and early detection of potential issues. Ensuring the consistency of SC-derived products across different manufacturing batches is a significant challenge.

To address this, a consortium of SC researchers and industry partners developed a set of standardized reference materials and assays for characterizing iPSCs and their derivatives. These tools enable better comparison of results between laboratories and facilitate the establishment of industry-wide quality standards. Lastly, the implementation of good manufacturing practices (GMP) is necessary for the clinical-grade production of SCs A prominent cell therapy company built a state-of-the-art GMP facility specifically designed for large-scale SC manufacturing. This facility incorporates cleanroom technology, stringent environmental controls, and comprehensive documentation systems to safeguard compliance with supervisory requirements and preserve the highest values of quality control throughout the production process (Seet *et al.*, 2023).

9.5.4 Regulatory and Clinical Trial Design Challenges

Regulatory and clinical trial design challenges in the field of regenerative medicine and advanced therapies present significant hurdles for researchers, clinicians, and pharmaceutical companies. These challenges stem from the complex nature of cell-based healings, gene therapies, and tissue-engineered effects, which often do not fit neatly into traditional regulatory frameworks or clinical trial designs. Addressing these challenges is crucial for advancing innovative treatments while ensuring patient safety and treatment efficacy. One major regulatory challenge is the classification of advanced therapy medicinal products (ATMPs). These novel therapies often blur the lines between traditional drug categories, making it difficult to apply existing regulatory pathways. For instance, a groundbreaking gene therapy for a rare genetic disorder faced regulatory uncertainty as it combined elements of both a biological product and a medical device. The regulatory agency had to develop a new framework to assess its safety and efficacy, considering both the viral vector used for gene delivery and the administration device. This case highlighted the need for more flexible regulatory approaches that can adapt to rapidly evolving technologies. Clinical trial design for cell remedies presents unique dares due to the adaptability inherent in living cell products (Devi *et al.*, 2024). A notable example is a trial for an autologous CAR-T cell therapy for leukemia. The researchers had to grapple with how to account for differences in cell quality and quantity between patients as well as how to define appropriate endpoints for a treatment that may have delayed effects.

They ultimately developed a novel adaptive trial design that allowed for dose adjustments based on individual patient responses, setting a precedent for future cell therapy trials. The long-term follow-up requirements for gene therapies pose another significant challenge. Regulatory agencies often require extended monitoring periods to assess potential delayed adverse effects, which can be logistically complex and financially burdensome. A prominent gene therapy company faced this issue when designing a trial for a treatment targeting a neurodegenerative disorder. They had to implement a 15-year follow-up plan, which included innovative strategies for patient retention and long-term data collection, such as wearable devices and remote monitoring technologies. Designing appropriate control groups for trials of regenerative medicine products is often problematic, especially for rare diseases or conditions with no standard treatment. A research team developing a tissue-engineered product for severe burns encountered this challenge.

Ethical considerations made it difficult to justify a placebo control, while the uniqueness of each patient's injury made standardized comparisons challenging. The team ultimately opted for a within-subject control design, using different areas of the same patient's burns for treatment and control, demonstrating the need for creative solutions in trial design for these novel therapies. The scalability and manufacturing consistency of cell-based products present both regulatory and clinical trial design challenges. A company developing an allogeneic SC therapy for heart failure had to address concerns about batch-to-batch variability in their product. They implemented a comprehensive characterization protocol and in-process controls to ensure product consistency (Kaizer *et al.*, 2023). However, this raised questions about how to design clinical trials that could account for potential variability between manufacturing runs while still providing meaningful efficacy data. The company worked closely with regulators to develop a trial design that incorporated manufacturing data into the analysis of clinical outcomes. Lastly, the accelerated approval pathways for regenerative medicine advanced therapies (RMATs) in some jurisdictions, while beneficial for speeding development, create challenges in balancing the need for robust efficacy data with the desire for rapid patient access. A small biotech company developing a novel gene therapy for a rare pediatric condition utilized the RMAT designation to expedite their development process. However, they faced the challenge of designing a trial that could provide sufficient evidence for accelerated approval while also planning for the collection of long-term data to support full approval. This required careful negotiation with regulatory agencies and innovative approaches to post-market surveillance (Devi *et al.*, 2024).

9.5.5 Advances in SC Engineering and Gene Editing

Improvements in SC engineering and gene editing have reformed the field of regenerative medicine and opened new options for treating previously incurable diseases. These cutting-edge technologies allow researchers to manipulate SCs and genetic material with unprecedented precision, paving the way for personalized therapies and novel approaches to understanding human development and disease. One of the most significant breakthroughs in SC engineering has been the development of induced pluripotent SCs (iPSCs). This technology, pioneered by Shinya Yamanaka and colleagues, allows researchers to reprogram adult somatic cells into a pluripotent state, capable of differentiating into any cell type in the body. Building on this foundation, a team of researchers recently developed a highly efficient method for generating iPSCs from blood cells using a novel combination of reprogramming factors and small molecules.

This advancement has made it possible to create patient-specific SC lines from a simple blood draw, greatly expanding the potential for personalized medicine and disease modeling (Seet *et al.*, 2023). In the realm of gene editing, the CRISPR-Cas9 system has emerged as a powerful tool for precise genetic modifications. A groundbreaking application of this technology was demonstrated by a group of scientists who successfully corrected a disease-causing mutation in human embryos. The researchers targeted a gene responsible for hypertrophic cardiomyopathy, a heart condition that affects millions worldwide. By editing the mutated gene in embryos at the single-cell stage, they were able to produce embryos free of the disease-causing mutation, showcasing the potential of gene editing for preventing genetic disorders before birth. Combining SC engineering with gene editing has led to remarkable advancements in cell therapy. A team of researchers utilized CRISPR-Cas9 to engineer CAR-T cells derived from iPSCs, creating a potentially unlimited source of off-the-shelf immunotherapy products for cancer treatment. This approach addresses many of the challenges associated with traditional CAR-T cell therapies, such as manufacturing complexity and variability between patient-derived cells. The engineered iPSC-derived CAR-T cells demonstrated potent antitumor activity in preclinical models, offering hope for more accessible and effective cancer immunotherapies. Advances in 3D bioprinting have opened new avenues for tissue engineering using SCs. A group of bioengineers developed a novel 3D bioprinting technique that allows for the precise deposition of multiple cell types and biomaterials to create complex tissue structures.

Using this method, they successfully bioprinted a functional mini-liver organoid using a combination of hepatocytes, endothelial cells, and supporting cells derived from iPSCs. The bioprinted organoids demonstrated liver-specific functions and maintained their structure for extended periods in culture, representing a significant step towards creating transplantable organs and advanced *in vitro* models for drug testing (Ansori *et al.*, 2023). The field of epigenetic engineering has also seen remarkable progress, offering new ways to control SC fate and function. Researchers developed a programmable epigenome editing system that allows for targeted modification of DNA methylation and histone modifications. Using this system, they were able to precisely control the differentiation of neural SCs into specific subtypes of neurons. This technology holds great promise for generating pure populations of desired cell types for therapeutic applications and for studying the role of epigenetic modifications in development and disease.

Lastly, the development of *in vivo* reprogramming techniques has opened up exciting possibilities for regenerative medicine. A team of scientists demonstrated the feasibility of converting astrocytes into functional neurons directly in the brain using a combination of transcription factors delivered by adeno-associated viruses. This approach successfully generated new neurons in a mouse model of Parkinson's disease, leading to significant improvements in motor function. The ability to reprogram cells *in situ* represents a paradigm shift in regenerative medicine, potentially eliminating the need for cell transplantation in some applications. These advancements in SC engineering and gene editing are rapidly transforming the landscape of biomedical research and therapeutic development. As these technologies continue to evolve and mature, we can expect to see increasingly sophisticated applications in disease modeling, drug discovery, and regenerative medicine. However, it is crucial to address the ethical and safety considerations associated with these powerful tools to ensure their responsible development and application. The coming years will likely bring even more exciting breakthroughs, bringing us closer to realizing the full potential of SC and gene therapies in treating a wide range of diseases and injuries (Kharbikar *et al.*, 2022).

9.5.6 Combination Therapies with SCs and Other Immunomodulatory Agents

Combination therapies that integrate SCs with other immunomodulatory agents are an innovative approach in regenerative medicine and immunotherapy. This strategy seeks to boost the effectiveness of SCs by modifying the immune environment, which could lead to better results in treating a range of diseases and conditions. The synergistic effects of combining SCs with immunomodulatory agents can address multiple aspects of disease pathology, offering more comprehensive and effective treatments. One promising example of this approach is the combination of mesenchymal SCs (MSCs) with interleukin-10 (IL-10) use for inflammatory bowel disease (IBD) treatment. Researchers developed a novel delivery system where MSCs were genetically modified to overexpress IL-10, a potent anti-inflammatory cytokine. In a preclinical model of colitis, these engineered MSCs demonstrated superior therapeutic efficacy compared to either MSCs or IL-10 alone. The combined therapy not only reduced inflammation but also promoted tissue regeneration in the damaged intestinal mucosa.

This approach showcases how the immunomodulatory properties of MSCs can be enhanced through genetic engineering to deliver targeted anti-inflammatory factors, potentially offering a more effective treatment for chronic inflammatory conditions (Aderinto *et al.*, 2023). Another innovative combination therapy involves the use of neural SCs (NSCs) with a checkpoint inhibitor for the treatment of glioblastoma, an aggressive brain cancer. In this approach, NSCs were engineered to express a tumor-targeting protein and were administered in conjunction with an anti-PD-1 checkpoint inhibitor. The NSCs served as a delivery vehicle for the anti-tumor agent, while the checkpoint inhibitor helped to activate the immune system against the cancer cells. In preclinical studies, this combination therapy showed remarkable efficacy in reducing tumor size and improving survival rates compared to either treatment alone. This example highlights how SC-based therapies can be combined with immunotherapy to create multi-pronged approaches for tackling challenging cancers. In transplantation medicine, researchers have explored the combination of hematopoietic SC transplantation (HSCT) with regulatory T cell (Treg) therapy to prevent graft-versus-host disease (GVHD). In this approach, patients undergoing allogeneic HSCT for leukemia treatment also received infusions of *ex vivo* expanded Tregs. The Tregs helped to modulate the immune response, reducing the threat of GVHD without compromising the graft-versus-leukemia cause. Clinical trials using this combination therapy have shown promising results, with reduced incidence and severity of GVHD compared to standard protocols. This strategy demonstrates how combining cellular therapies with different immunomodulatory properties can improve outcomes in complex medical procedures.

The mixture of adipose-derived SCs (ADSCs) with platelet-rich plasma (PRP) has shown remarkable potential in the field of recovering medicine, particularly for tissue restore and wound healing. In a novel approach to treat chronic diabetic ulcers, researchers developed a bioengineered skin substitute incorporating ADSCs and PRP (Trunk *et al.*, 2024). The ADSCs provided regenerative and anti-inflammatory effects, while the growth factors in PRP promoted angiogenesis and tissue repair. Clinical studies using this combination therapy demonstrated significantly faster wound closure and improved healing quality compared to standard treatments. This example illustrates how combining SCs with naturally derived immunomodulatory agents can enhance tissue regeneration and wound healing processes. In the realm of autoimmune diseases, a groundbreaking approach combines encouraged pluripotent SC (iPSC)-derived thymic epithelial progenitor cells with a short course of immunosuppressive drugs treatment of type 1 diabetes.

The thymic progenitors were used to generate a functional thymus organoid, which was transplanted to help re-educate the immune system and induce tolerance to pancreatic beta cells. Transient immunosuppression allowed for the engraftment and maturation of the thymus organoid. In preclinical studies, this combination therapy showed promising results in halting the autoimmune destruction of beta cells and even promoting some degree of beta cell regeneration. This innovative approach demonstrates how SC-based therapies can be combined with traditional immunomodulatory strategies to address the root causes of autoimmune diseases. Lastly, an emerging area of research involves combining neural SCs with extracellular vesicles (EVs) derived from mesenchymal SCs for the treatment of traumatic brain injury (TBI). In this approach, neural SCs are administered to upgrade the redevelopment and repair of injured neural tissue, while MSC-derived EVs provide immunomodulatory and neuroprotective effects (Ramos *et al.*, 2023). The EVs, loaded with anti-inflammatory cytokines and growth factors, help create a more favorable environment for neural SC survival and integration. Preclinical studies have shown that this combination therapy leads to improved cognitive and motor function recovery compared to either treatment alone. This example highlights the potential of combining different types of SCs and their secreted factors to create more effective therapies for complex neurological conditions. These examples illustrate the diverse and innovative ways in which SCs are being combined with immunomodulatory agents to create more effective therapies. As our understanding of SC biology and immunology continues to grow, we can expect to see even more sophisticated combination therapies emerging. These approaches connect great suggest for treating a wide range of illnesses and conditions, potentially offering hope to patients who have not responded to traditional treatments. Conversely, it is important to note that many of these combination therapies are still in experimental stages and require additional research and clinical experiments to fully establish their care and efficacy. The field of regenerative medicine and immunotherapy is rapidly evolving, and these combination approaches represent an exciting frontier in the development of next-generation treatments (Yuan *et al.*, 2022).

9.6 Conclusion

Immunomodulation in autoimmune disorders through the application of SC therapies represents a groundbreaking frontier in medical science. This approach offers new hope for patients suffering from a wide range of autoimmune conditions that have traditionally been challenging to treat. The ability of SCs to modulate the immune system stems from their capacity to secrete anti-inflammatory factors, interact with immune cells, and even differentiate into specialized cell types that can replace damaged tissues. This multifaceted approach allows for a more comprehensive treatment strategy that goes beyond simply suppressing the immune system, as is often the case with conventional therapies. Instead, SC-based immunomodulation aims to restore balance to the dysregulated immune response characteristic of autoimmune disorders. As research in this field progresses, we are witnessing the development of increasingly sophisticated and targeted approaches. These include the use of genetically modified SCs designed to enhance their immunomodulatory properties, combination therapies that synergize the effects of SCs with other immunomodulatory agents, and novel delivery methods to improve the efficacy and specificity of treatment. While significant challenges remain, including optimizing cell dosing, addressing potential long-term effects, and overcoming controlling hurdles, the promise of SC-based immunomodulation in autoimmune disorders is undeniable. As our understanding of SC biology and autoimmune disease mechanisms continues to grow, we can anticipate further refinements and innovations in this field. Ultimately, the goal of harnessing the power of SCs for immunomodulation in autoimmune disorders is to develop treatments that not only alleviate symptoms but also address the root causes of these conditions. This approximate has the probable to revolutionize the management of autoimmune diseases and disorders, offering hope for long-term remission or even cure in conditions that were once considered chronic and incurable. As we move forward, continued research, clinical trials, and collaboration between scientists, clinicians, and patients will be crucial in achieving the full capability of this agreeing therapeutic strategy.

References

1. Alexander, T., Tassy, N., Domenech, A., Kramer, E., Jessop, H., Kenyon, M., Greco, R., , Patient-reported outcomes in HSCT for autoimmune diseases: Considerations on behalf of PAC, and Nurses Group. *J. Allergy Clin. Immunology Global, 3*, 3, 100283, 2024.
2. Ansori, A.N., Antonius, Y., Susilo, R.J., Hayaza, S., Kharisma, V.D., Parikesit, A.A., Zainul, R., Jakhmola, V., Saklani, T., Rebezov, M., Ullah, M. E., Maksimiuk, N., Derkho, M., Burkov, P., Application of CRISPR-Cas9 genome editing technology in various fields: A review. *Narra J.*, 3, 2, e184, 2023, https://doi.org/10.52225/narra.v3i2.184.
3. Aderinto, N., Abdulbasit, M.O., Olatunji, D., SC-based combinatorial therapies for spinal cord injury: a narrative review of current research and future directions. *Ann. Med. Surg., 85*, 8, 3943–3954, 2023.
4. Bashor, C.J., Hilton, I.B., Bandukwala, H., Smith, D.M., Veiseh, O., Engineering the next generation of cell-based therapeutics. *Nat. Rev. Drug Discov., 21*, 9, 655–675, 2022.
5. Brunelli, D., Polonelli, T., Benini, L., Ultra-low energy pest detection for smart agriculture, in: *2020 IEEE Sensors*, pp. 1–4, Rotterdam, Netherlands, 2020, doi: 10.1109/SENSORS47125.2020.9278587.
6. Bonasia, C.G., Abdulahad, W.H., Rutgers, A., Heeringa, P., Bos, N.A., B cell activation and escape of tolerance checkpoints: recent insights from studying autoreactive B cells. *Cells, 10*, 5, 1190, 2021.
7. Basile, G., Qadir, M.M.F., Mauvais-Jarvis, F., Vetere, A., Shoba, V., Modell, A.E., Dominguez-Bendala, J., Emerging diabetes therapies: Bringing back the β-cells. *Mol. Metab., 60*, 101477, 2022.
8. Brummer, T., Ruck, T., Meuth, S.G., Zipp, F., Bittner, S., Treatment approaches to patients with multiple sclerosis and coexisting autoimmune disorders. *Ther. Adv. Neurological Disord., 14*, 17562864211035542, 2021.
9. Balaban, Y.E., Akbaba, S., Bozkurt, S.B., Buyuksungur, A., Akgun, E.E., Gonen, Z.B., Hakki, S.S., Local application of gingiva-derived mesenchymal

SCs on experimental periodontitis in rats. *J. Periodontology, 95*, 5, 456–468, 2024.

10. Banche-Niclot, F., Lim, J., McCulloch, P., Corradetti, B., Taraballi, F., Mesenchymal Stromal Cell Immunomodulatory Potential for Orthopedic Applications can be fine-tuned *via* 3D nano-engineered Scaffolds. *Curr. Stem Cell Rep.*, 10, 4, 65–76, 2024.
11. Brescia, C., Audia, S., Pugliano, A., Scaglione, F., Iuliano, R., Trapasso, F., Perrotti, N., Chiarella, E., Amato, R., Metabolic drives affecting Th17/Treg gene expression changes and differentiation: impact on immune-microenvironment regulation. *APMIS: acta pathologica, microbiologica, et immunologica Scandinavica*, 132, 12, 1026–1045, 2024, https://doi.org/10.1111/apm.13378.
12. Blumenthal, D., Chandra, V., Avery, L., Burkhardt, J.K., Mouse T cell priming is enhanced by maturation-dependent stiffening of the dendritic cell cortex. *Elife, 9*, e55995, 2020.
13. Cenariu, D., Iluta, S., Zimta, A.A., Petrushev, B., Qian, L., Dirzu, N., Zaharie, F., Extramedullary hematopoiesis of the liver and spleen. *J. Clin. Med., 10*, 24, 5831, 2021.
14. Cui, A., Huang, T., Li, S., Ma, A., Pérez, J.L., Sander, C., Hacohen, N., Dictionary of immune responses to cytokines at single-cell resolution. *Nature, 625*, 7994, 377–384, 2024.
15. Cao, C., Yu, M., Chai, Y., Pathological alteration and therapeutic implications of sepsis-induced immune cell apoptosis. *Cell Death Dis., 10*, 10, 782, 2019.
16. Chen, Z., Yao, M.W., Ao, X., Gong, Q.J., Yang, Y., Liu, J.X., Zuo, L.J., The expression mechanism of programmed cell death 1 ligand 1 and its role in immunomodulatory ability of mesenchymal SCs. *Chin. J. Traumatol., 27*, 01, 1–10, 2024.
17. Chen, Y.F., Li, J., Xu, L.L., Găman, M.A., Zou, Z.Y., Allogeneic SC transplantation in the treatment of acute myeloid leukemia: An overview of obstacles and opportunities. *World J. Clin. Cases, 11*, 2, 268, 2023.
18. Cao, J., Hao, J., Wang, L., Tan, Y., Tian, Y., Li, S., Zhao, T., Developing standards to support the clinical translation of SCs. *SCs Trans. Med., 10*, S2, S85–S95, 2021.
19. Cerneckis, J., Cai, H., Shi, Y., Induced pluripotent SCs (iPSCs): molecular mechanisms of induction and applications. *Signal Transduction Targeted Ther., 9*, 1, 112, 2024.
20. Doss, M.X. and Sachinidis, A., Current challenges of iPSC-based disease modeling and therapeutic implications. *Cells, 8*, 5, 403, 2019.
21. Devi, S.K., Joga, R., Srivastava, S., Nagpal, K., Dhamija, I., Grover, P., Kumar, S., Regulatory landscape and challenges in CAR-T cell therapy development in the US, EU, Japan, and India. *Eur. J. Pharm. Biopharm.: official journal of Arbeitsgemeinschaft fur Pharmazeutische Verfahrenstechnik e.V*, 201, 114361, 2024. https://doi.org/10.1016/j.ejpb.2024.114361

22. Dong, J., Wu, B., Tian, W., Human adipose tissue-derived small extracellular vesicles promote soft tissue repair through modulating M1-to-M2 polarization of macrophages. *SC Res. Ther., 14*, 1, 67, 2023.
23. Derks, L.L.M., and van Boxtel, R., Stem cell mutations, associated cancer risk, and consequences for regenerative medicine. *Cell stem cell.*, 30, 11, 1421–1433, 2023, https://doi.org/10.1016/j.stem.2023.09.008.
24. Di Persio, S. and Neuhaus, N., Human spermatogonial SCs and their niche in male (in) fertility: novel concepts from single-cell RNA-sequencing. *Hum. Reprod., 38*, 1, 1–13, 2023.
25. Dejas, L., Santoni, K., Meunier, E., Lamkanfi, M., Regulated cell death in neutrophils: From apoptosis to NETosis and pyroptosis, in: *Seminars in Immunology*, vol. 70, p. 101849, Academic Press, USA, 2023, November.
26. Ding, J., Chen, P., Qi, C., Circadian rhythm regulation in the immune system. *Immunology, 171*, 4, 525–533, 2024.
27. Eldaly, A.S., Mashaly, S.M., Fouda, E., Emam, O.S., Aglan, A., Abuasbeh, J., Fath, A.R., Systemic anti-inflammatory effects of mesenchymal SCs in burn: A systematic review of animal studies. *J. Clin. Transl. Res., 8*, 4, 276, 2022.
28. Elhusseiny, A.M., Soleimani, M., Eleiwa, T.K., ElSheikh, R.H., Frank, C.R., Naderan, M., Djalilian, A.R., Current and emerging therapies for limbal SC deficiency. *SCs Trans. Med., 11*, 3, 259–268, 2022.
29. Eiz-Vesper, B. and Schmetzer, H.M., Antigen-presenting cells: potential of proven und new players in immune therapies. *Transfus. Med. Hemotherapy, 47*, 6, 429–431, 2020.
30. Elliot, T.A., Jennings, E.K., Lecky, D.A., Thawait, N., Flores-Langarica, A., Copland, A., Bending, D., Antigen and checkpoint receptor engagement recalibrates T cell receptor signal strength. *Immunity, 54*, 11, 2481–2496, 2021.
31. Fraint, E., Ulloa, B.A., Feliz Norberto, M., Potts, K.S., Bowman, T.V., Advances in preclinical hematopoietic SC models and possible implications for improving therapeutic transplantation. *SCs Transl. Med., 10*, 3, 337–345, 2021.
32. Fu, Y., Li, J., Li, M., Xu, J., Rong, Z., Ren, F., Chang, Z., Umbilical Cord Mesenchymal SCs Ameliorate Inflammation-Related Tumorigenesis *via* Modulating Macrophages. *SCs Int., 2022*, 1, 1617229, 2022.
33. Goldmann, O., Nwofor, O.V., Chen, Q., Medina, E., Mechanisms underlying immunosuppression by regulatory cells. *Front. Immunol., 15*, 1328193, 2024.
34. Guo, S. and Wang, D., Novel insights into the potential applications of SCs in pulmonary hypertension therapy. *Respir. Res., 25*, 1, 237, 2024.
35. Han, Y., Yang, J., Fang, J., Zhou, Y., Candi, E., Wang, J., Shi, Y., The secretion profile of mesenchymal SCs and potential applications in treating human diseases. *Signal Transduction Targeted Ther., 7*, 1, 92, 2022.
36. Haltalli, M.L., Wilkinson, A.C., Rodriguez-Fraticelli, A., Porteus, M., Hematopoietic SC gene editing and expansion: State-of-the-art technologies and recent applications. *Exp. Hematol., 107*, 9–13, 2022.

37. Jiang, W. and Xu, J., Immune modulation by mesenchymal SCs. *Cell Proliferation*, *53*, 1, e12712, 2020.
38. Jellish, W.S. and Edelstein, S.B., General anesthesia versus conscious sedation for the endovascular treatment of acute ischemic stroke. *J. Stroke Cerebrovascular Dis.*, *24*, 9, 1957–1960, 2015.
39. Jung, S.M., Kim, W.U., Targeted Immunotherapy for Autoimmune Disease. *Immune Netw.*, 22, 1, e9, 2022, https://doi.org/10.4110/in.2022.22.e9.
40. Jacob, A., Shatila, A.O., Inshasi, J., Massouh, J., Mir, R., Noori, S., Yamout, B., UAE National Multiple Sclerosis Society. Disease modifying treatment guidelines for multiple sclerosis in the United Arab Emirates. *Mult. Scler. Relat. Disord.*, 88, 105703, 2024. https://doi.org/10.1016/j.msard.2024.105703.
41. Kadri, N., Amu, S., Iacobaeus, E., Boberg, E., Le Blanc, K., Current perspectives on mesenchymal stromal cell therapy for graft versus host disease. *Cell. Mol. Immunol.*, *20*, 6, 613–625, 2023.
42. Kamilaris, A. and Prenafeta-Boldú, F.X., Deep learning in agriculture: A survey. *Comput. Electron. Agric.*, *147*, 70–90, 2018.
43. Kharbikar, B.N., Mohindra, P., Desai, T.A., Biomaterials to enhance SC transplantation. *Cell SC*, *29*, 5, 692–721, 2022.
44. Kaizer, A.M., Belli, H.M., Ma, Z., Nicklawsky, A.G., Roberts, S.C., Wild, J., Sabo, R.T., Recent innovations in adaptive trial designs: a review of design opportunities in translational research. *J. Clin. Transl. Sci.*, *7*, 1, e125, 2023.
45. Kaczmarek, A., Kaczmarek, M., Ciałowicz, M., Clemente, F.M., Wolański, P., Badicu, G., Murawska-Ciałowicz, E., The role of satellite cells in skeletal muscle regeneration—the effect of exercise and age. *Biology*, *10*, 10, 1056, 2021.
46. Lachance, S., Gomes, M.M., Bambace, N.M., Bittencourt, H., Lepic, K., Shafey, M., Guilcher, G.M., Recognition of Hematopoietic SC Transplantation and Cellular Therapy Expertise to Promote Care Accessibility: A Formally Credentialed Area of Focused Competence in Canada. *Transplant. Cell. Ther.*, *27*, 9, 702–706, 2021.
47. Liana, P., Liberty, I.A., Murti, K., Hafy, Z., Salim, E.M., Zulkarnain, M., Umar, T.P., A systematic review on neutrophil extracellular traps and its prognostication role in COVID-19 patients. *Immunol. Res.*, *70*, 4, 449–460, 2022.
48. Lu, P., Ruan, D., Huang, M., Tian, M., Zhu, K., Gan, Z., Xiao, Z., Harnessing the potential of hydrogels for advanced therapeutic applications: current achievements and future directions. *Signal Transduct. Target. Ther.*, *9*, 1, 166, 2024.
49. Lin, S., Mukherjee, S., Li, J., Hou, W., Pan, C., Liu, J., Mucosal immunity–mediated modulation of the gut microbiome by oral delivery of probiotics into Peyer's patches. *Sci. Adv.*, *7*, 20, eabf0677, 2021.
50. Liu, G., David, B.T., Trawczynski, M., Fessler, R.G., Advances in pluripotent SCs: history, mechanisms, technologies, and applications. *SC Rev. Rep.*, *16*, 1, 3–32, 2020.

51. Li, F., Zhao, B., Zhang, L., Chen, G.Q., Zhu, L., Feng, X.L., Liu, Y.Q., Therapeutic potential of urine-derived SCs in renal regeneration following acute kidney injury: A comparative analysis with mesenchymal SCs. *World J. SCs*, *16*, 5, 525, 2024.
52. Li, C., Sun, Y., Xu, W., Chang, F., Wang, Y., Ding, J., Mesenchymal Stem Cells-Involved Strategies for Rheumatoid Arthritis Therapy. *Adv. Sci., (Weinheim, Baden-Wurttemberg, Germany)*, 11, 24, e2305116, 2024. https://doi.org/10.1002/advs.202305116.
53. Li, Q. and Lan, P., Activation of immune signals during organ transplantation. *Signal Transduct. Target. Ther.*, *8*, 1, 110, 2023.
54. Lei, M., Lien, W.H., Li, J., Inflammation, SCs and wound healing in skin aging. *Front. Cell Dev. Biol.*, *10*, 1046022, 2022.
55. Liu, C., Xu, Y., Yang, G., Tao, Y., Chang, J., Wang, S., Zeng, Y.A., Niche inflammatory signals control oscillating mammary regeneration and protect SCs from cytotoxic stress. *Cell SC*, *31*, 1, 89–105, 2024.
56. Liang, Z., Dong, X., Zhang, Z., Zhang, Q., Zhao, Y., Age-related thymic involution: Mechanisms and functional impact. *Aging Cell*, *21*, 8, e13671, 2022.
57. Lotfi, M., Morshedi Rad, D., Mashhadi, S.S., Ashouri, A., Mojarrad, M., Mozaffari-Jovin, S., Abbaszadegan, M.R., Recent Advances in CRISPR/Cas9 Delivery Approaches for Therapeutic Gene Editing of SCs. *SC Rev. Rep.*, *19*, 8, 2576–2596, 2023.
58. Ialongo, D., Tudino, V., Arpacioglu, M., Messore, A., Patacchini, E., Costi, R., Madia, V.N., Synergistic effects of caffeine in combination with conventional drugs: perspectives of a drug that never ages. *Pharmaceuticals*, *16*, 5, 730, 2023.
59. Merz, J., Nettesheim, A., Von Garlen, S., Albrecht, P., Saller, B.S., Engelmann, J., Stachon, P., Pro-and anti-inflammatory macrophages express a sub-type specific purinergic receptor profile. *Purinergic Signal.*, *17*, 3, 481–492, 2021.
60. Min, Q., Yang, L., Tian, H., Tang, L., Xiao, Z., Shen, J., Immunomodulatory mechanism and potential application of dental pulp-derived SCs in immune-mediated diseases. *Int. J. Mol. Sci.*, *24*, 9, 8068, 2023.
61. Madrid, M., Lakshmipathy, U., Zhang, X., Bharti, K., Wall, D.M., Sato, Y., Muschler, G., Ting, A., Smith, N., Deguchi, S., Kawamata, S., Moore, J.C., Makovoz, B., Sullivan, S., Falco, V., Al-Riyami, A.Z., Considerations for the development of iPSC-derived cell therapies: a review of key challenges by the JSRM-ISCT iPSC Committee. *Cytotherapy*, 26, 11, 1382–1399, 2024. https://doi.org/10.1016/j.jcyt.2024.05.022.
62. Mamdouh, N. and Khattab, A., YOLO-based deep learning framework for olive fruit fly detection and counting. *IEEE Access*, *9*, 84252–84262, 2021.
63. Merli, P., Pagliara, D., Mina, T., Bertaina, V., Pira, G.L., Lazzaro, S., Locatelli, F., αβT-and B-cell-depleted HLA-haploidentical hematopoietic SC transplantation in children with myelodysplastic syndromes. *Haematologica*, *107*, 12, 2966, 2022.

64. Manzoor, T., Saleem, A., Farooq, N., Dar, L.A., Nazir, J., Saleem, S., Ahmad, S.M., Extracellular vesicles derived from mesenchymal SCs—a novel therapeutic tool in infectious diseases. *Inflamm. Regener.*, *43*, 1, 17, 2023.
65. Majdoul, S. and Compton, A.A., Lessons in self-defence: inhibition of virus entry by intrinsic immunity. *Nat. Rev. Immunol.*, *22*, 6, 339–352, 2022.
66. Matsuzaka, Y. and Yashiro, R., Current Strategies and Therapeutic Applications of Mesenchymal SC-Based Drug Delivery. *Pharmaceuticals*, *17*, 6, 707, 2024.
67. Moore Jr., J.E., Brook, B.S., Nibbs, R.J., Chemokine transport dynamics and emerging recognition of their role in immune function. *Curr. Opin. Biomed. Eng.*, *5*, 90–95, 2018.
68. Mao, J., Saiding, Q., Qian, S., Liu, Z., Zhao, B., Zhao, Q., Cui, W., Reprogramming SCs in regenerative medicine. *Smart Med.*, *1*, 1, e20220005, 2022.
69. Mansourabadi, A.H., Aghamajidi, A., Dorfaki, M., Keshavarz, F., Shafeghat, Z., Moazzeni, A., Jafari, R., B lymphocytes in COVID-19: a tale of harmony and discordance. *Arch. Virol*, *168*, 5, 148, 2023.
70. Moncrieff, J., Cooper, R.E., Stockmann, T., Amendola, S., Hengartner, M.P., Horowitz, M.A., The serotonin theory of depression: a systematic umbrella review of the evidence. *Mol. Psychiatry*, *28*, 8, 3243–3256, 2023.
71. Najibi, A.J., Lane, R.S., Sobral, M.C., Bovone, G., Kang, S., Freedman, B.R., Gutierrez Estupinan, J., Elosegui-Artola, A., Tringides, C.M., Dellacherie, M.O., Williams, K., Ijaz, H., Müller, S., Turley, S.J., Mooney, D.J., Durable lymph-node expansion is associated with the efficacy of therapeutic vaccination. *Nat. Biomed. Eng.*, 8, 10, 1226–1242, 2024. https://doi.org/10.1038/s41551-024-01209-3.
72. Puccetti, M., Pariano, M., Schoubben, A., Giovagnoli, S., Ricci, M., Biologics, theranostics, and personalized medicine in drug delivery systems. *Pharmacol. Res.*, 201, 107086, 2024. https://doi.org/10.1016/j.phrs.2024.107086.
73. Perin, E.C., Borow, K.M., Henry, T.D., Mendelsohn, F.O., Miller, L.W., Swiggum, E., Greenberg, B., Randomized trial of targeted transendocardial mesenchymal precursor cell therapy in patients with heart failure. *J. Am. Coll. Cardiol.*, *81*, 9, 849–863, 2023.
74. Paudel, S., Ghimire, L., Jin, L., Jeansonne, D., Jeyaseelan, S., Regulation of emergency granulopoiesis during infection. *Front. Immunol.*, *13*, 961601, 2022.
75. Ou, Q., Power, R., Griffin, M.D., Revisiting regulatory T cells as modulators of innate immune response and inflammatory diseases. *Front. Immunol.*, *14*, 1287465, 2023.
76. Qiao, Y., Tang, X., Liu, Z., Ocansey, D.K.W., Zhou, M., Shang, A., Mao, F., Therapeutic Prospects of Mesenchymal SC and Their Derived Exosomes in the Regulation of the Gut Microbiota in Inflammatory Bowel Disease. *Pharmaceuticals*, *17*, 5, 607, 2024.

77. Rozmus, J., Monogenic immune diseases provide insights into the mechanisms and treatment of chronic graft-versus-host disease. *Front. Immunol.*, *11*, 574569, 2021.
78. Ramos, S.A., Armitage, L.H., Morton, J.J., Alzofon, N., Handler, D., Kelly, G., Russ, H.A., Generation of functional thymic organoids from human pluripotent SCs. *SC Rep.*, *18*, 4, 829–840, 2023.
79. Rouzer, C.A. and Marnett, L.J., Structural and chemical biology of the interaction of cyclooxygenase with substrates and non-steroidal anti-inflammatory drugs. *Chem. Rev.*, *120*, 15, 7592–7641, 2020.
80. Ren, R., Jiang, J., Li, X., Zhang, G., Research progress of autoimmune diseases based on induced pluripotent SCs. *Front. Immunol.*, *15*, 1349138, 2024.
81. Rascle, P., Woolley, G., Jost, S., Manickam, C., Reeves, R.K., NK cell education: Physiological and pathological influences. *Front. Immunol.*, *14*, 1087155, 2023.
82. Sadeghi, S., Ghane, Y., Hajizadeh, N., Goodarzi, A., Autologous adipose tissue injection in the treatment of alopecia: A mini-review. *J. Cosmet. Dermatol.*, *23*, 3, 758–765, 2024.
83. Skidmore, S. and Barker, R.A., Challenges in the clinical advancement of cell therapies for Parkinson's disease. *Nat. Biomed. Eng.*, *7*, 4, 370–386, 2023.
84. Syeda, M.Z., Hong, T., Huang, C., Huang, W., Mu, Q., B cell memory: from generation to reactivation: a multipronged defense wall against pathogens. *Cell Death Discov.*, *10*, 1, 117, 2024.
85. Saito, Y., Harada, A., Ushijima, M., Tanaka, K., Higuchi, R., Baba, A., Baba, Y., Plasma cell differentiation is regulated by the expression of histone variant H3. 3. *Nat. Commun.*, *15*, 1, 5004, 2024.
86. Suto, J., Codling moth monitoring with camera-equipped automated traps: A review. *Agriculture*, *12*, 10, 1721, 2022.
87. Santana-Sánchez, P., Vaquero-García, R., Legorreta-Haquet, M.V., Chávez-Sánchez, L., Chávez-Rueda, A.K., Hormones and B-cell development in health and autoimmunity. *Front. Immunol.*, *15*, 1385501, 2024.
88. So, L., Obata-Ninomiya, K., Hu, A., Muir, V.S., Takamori, A., Song, J., Ziegler, S.F., Regulatory T cells suppress CD4+ effector T cell activation by controlling protein synthesis. *J. Exp. Med.*, *220*, 3, e20221676, 2023.
89. Sharifian, P., Yari, S., Hasanein, P., Manteghi Nezhad, Y., Conditioned medium of bone marrow mesenchymal SCs improves sperm parameters and reduces histological alteration in rat testicular ischaemia/reperfusion model. *Andrologia*, *54*, 11, e14624, 2022.
90. Seet, W.T., Mat Afandi, M.A., Ishak, M.F., Hassan, M.N.F., Ahmat, N., Ng, M.H., Maarof, M., Quality management overview for the production of a tissue-engineered human skin substitute in Malaysia. *SC Res. Ther.*, *14*, 1, 298, 2023.
91. Sun, H., Shi, C., Ye, Z., Yao, B., Li, C., Wang, X., Qian, Q., The role of mesenchymal SCs in liver injury. *Cell Biol. Int.*, *46*, 4, 501–511, 2022.

92. Summers, B.S., Thomas Broome, S., Pang, T.W.R., Mundell, H.D., Koh Belic, N., Tom, N.C., Brew, B.J., A Review of the Evidence for Tryptophan and the Kynurenine Pathway as a Regulator of SC Niches in Health and Disease. *Int. J. Tryptophan Res.*, *17*, 11786469241248287, 2024.
93. Shao, L., Shen, Y., Ren, C., Kobayashi, S., Asahara, T., Yang, J., Inflammation in myocardial infarction: roles of mesenchymal SCs and their secretome. *Cell Death Discov.*, *8*, 1, 452, 2022.
94. Shuo, L., II, Guifang, F.A.N., Xiaojiaoyang, L., II, Yajie, C.A., II, Runping, L., II, Modulation of type I interferon signaling by natural products in the treatment of immune-related diseases. *Chin. J. Nat. Med.*, *21*, 1, 3–18, 2023.
95. Scott-Browne, J. and Shih, H.Y., Stability and change in epigenetic regulation of immune cells. *Immunol. Rev.*, *305*, 1, 5–8, 2022.
96. Shokri, M.R., Bozorgmehr, M., Ghanavatinejad, A., Falak, R., Aleahmad, M., Kazemnejad, S., Zarnani, A.H., Human menstrual blood-derived stromal/SCs modulate functional features of natural killer cells. *Sci. Rep.*, *9*, 1, 10007, 2019.
97. Schirrmacher, V., Bone marrow: The central immune system. *Immuno*, *3*, 3, 289–329, 2023.
98. Tsiapalis, D., Floudas, A., Tertel, T., Boerger, V., Giebel, B., Veale, D.J., O'Driscoll, L., Therapeutic effects of mesenchymal/stromal SCs and their derived extracellular vesicles in rheumatoid arthritis. *SCs Transl. Med.*, *12*, 12, 849–862, 2023.
99. Tetila, E.C., Machado, B.B., Astolfi, G., de Souza Belete, N.A., Amorim, W.P., Roel, A.R., Pistori, H., Detection and classification of soybean pests using deep learning with UAV images. *Comput. Electron. Agric.*, *179*, 105836, 2020.
100. Trunk, A.D., Guo, M., Budvytyte, L., Islam, N.S., Khera, N., Hamilton, B.K., Amonoo, H.L., Hematopoietic Stem-Cell Transplantation: Exploring the Latest Advances and Gaps in Disparities, Psychosocial and Symptom Management Interventions, and Chronic Graft-Versus-Host Disease Care. *Am. Soc. Clin. Oncol. Educ. Book*, *44*, 3, e432186, 2024.
101. Veglia, F., Sanseviero, E., Gabrilovich, D., II, Myeloid-derived suppressor cells in the era of increasing myeloid cell diversity. *Nat. Rev. Immunol.*, *21*, 8, 485–498, 2021.
102. Vandendriessche, S., Cambier, S., Proost, P., Marques, P.E., Complement receptors and their role in leukocyte recruitment and phagocytosis. *Front. Cell Dev. Biol.*, *9*, 624025, 2021.
103. Wang, Y., Fang, J., Liu, B., Shao, C., Shi, Y., Reciprocal regulation of mesenchymal SCs and immune responses. *Cell SC*, *29*, 11, 1515–1530, 2022.
104. Wu, L., Yan, Z., Jiang, Y., Chen, Y., Du, J., Guo, L., Liu, Y., Metabolic regulation of dendritic cell activation and immune function during inflammation. *Front. Immunol.*, *14*, 1140749, 2023.
105. Waldman, A.D., Fritz, J.M., Lenardo, M.J., A guide to cancer immunotherapy: from T cell basic science to clinical practice. *Nat. Rev. Immunol.*, *20*, 11, 651–668, 2020.

106. Wang, M., Xie, J., Wang, C., Zhong, D., Xie, L., Fang, H., Immunomodulatory properties of SCs in periodontitis: current status and future prospective. *SCs Int., 2020,* 1, 9836518, 2020.
107. Yang, G., Fan, X., Liu, Y., Jie, P., Mazhar, M., Liu, Y., Wang, L., Immunomodulatory mechanisms and therapeutic potential of mesenchymal SCs. *SC Rev. Rep., 19,* 5, 1214–1231, 2023.
108. Yan, W., Xia, Y., Zhao, H., Xu, X., Ma, X., Tao, L., SC-based therapy in cardiac repair after myocardial infarction: Promise, challenges, and future directions. *J. Mol. Cell. Cardiol, 188,* 1–14, 2024.
109. Yan, F., Liu, X., Ding, H., Zhang, W., Paracrine mechanisms of endothelial progenitor cells in vascular repair. *Acta Histochem., 124,* 1, 151833, 2022.
110. Yuan, X., Sun, L., Jeske, R., Nkosi, D., York, S.B., Liu, Y., Li, Y., Engineering extracellular vesicles by three-dimensional dynamic culture of human mesenchymal SCs. *J. Extracell. Vesicles, 11,* 6, e12235, 2022.
111. Yamanaka, S., Pluripotent SC-based cell therapy—promise and challenges. *Cell SC, 27,* 4, 523–531, 2020.
112. Zaripova, L.N., Midgley, A., Christmas, S.E., Beresford, M.W., Pain, C., Baildam, E.M., Oldershaw, R.A., Mesenchymal SCs in the pathogenesis and therapy of autoimmune and autoinflammatory diseases. *Int. J. Mol. Sci., 24,* 22, 16040, 2023.
113. Zeng, C.W., Advancing spinal cord injury treatment through SC therapy: A comprehensive review of cell types, challenges, and emerging technologies in regenerative medicine. *Int. J. Mol. Sci., 24,* 18, 14349, 2023.
114. Zhang, L., Zhuang, X., Kotitalo, P., Keller, T., Krzyczmonik, A., Haaparanta-Solin, M., Xia, H., Intravenous transplantation of olfactory ensheathing cells reduces neuroinflammation after spinal cord injury *via* interleukin-1 receptor antagonist. *Theranostics, 11,* 3, 1147, 2021.
115. Zaripova, L.N., Midgley, A., Christmas, S.E., Beresford, M.W., Pain, C., Baildam, E.M., Oldershaw, R.A., Mesenchymal SCs in the pathogenesis and therapy of autoimmune and autoinflammatory diseases. *Int. J. Mol. Sci., 24,* 22, 16040, 2023.
116. Zhang, H.M., Yuan, S., Meng, H., Hou, X.T., Li, J., Xue, J.C., Zhang, Q.G., SC-based therapies for inflammatory bowel disease. *Int. J. Mol. Sci., 23,* 15, 8494, 2022.

10

The Microbiome Connection: Exploring Interactions with Stem Cell Therapies

Rizwan Ahamad[1], Asad Ali[2], Priya Manna[3], Neha Quadri[4], Zuber Khan[5], Radheshyam Pal[6], Mohammad Muztaba[7], Shakira Ghazanfar[8], Anas Islam[9], Asif Iqbal[10] and Sumel Ashique[6]*

[1]Department of Pharmacology, School of Pharmaceutical Education and Research (SPER), Jamia Hamdard, New Delhi, India
[2]Department of Pharmaceutics, School of Pharmaceutical Education and Research (SPER), Jamia Hamdard, New Delhi, India
[3]Department of Pharmaceutical Technology, Brainware University, Barasat, Kolkata, WB, India
[4]Jamuna Prasad Memorial College Pharmacy, Baheri, Uttar Pradesh, India
[5]Division of Neuroscience, Department of Pharmacology, ISF College of Pharmacy (Affiliated to IK Gujral Punjab Technical University, Jalandhar, Punjab, India), Moga, Punjab, India
[6]Department of Pharmaceutical Technology, Bharat Technology, Uluberia, West Bengal, India
[7]Department of Pharmacology, Praduman Singh Pharmacy College Phutahiya, Sansarpur, Basti, Uttar Pradesh, India
[8]National Institute for Genomics and Advanced Biotechnology (NIGAB), National Agricultural Research Centre (NARC), Islamabad, Pakistan
[9]Faculty of Pharmacy, Integral University, Lucknow, Uttar Pradesh, India
[10]Department of Pharmacology, Dr. B.C. Roy College of Pharmacy and Allied Health Sciences, Durgapur, West Bengal, India

Abstract

The intestinal tract, home to over 100 trillion microorganisms, functions crucially within the framework of health and illness. Human microbiome, often called second genome, contributes to genetic diversity, modulates immunity, influences metabolic processes, modifies disease pathways, and interacts with drugs.

**Corresponding author*: ashiquesumel007@gmail.com; ORCID: https://orcid.org/0000-0003-4362-2830

Ayush Madan, Sumel Ashique, Deepika Arora and Mantosh Kumar Satapathy (eds.) Stem Cell Therapeutics, (463–504) © 2025 Scrivener Publishing LLC

The microbiota collaborates with intestinal stem cells within the gut to maintain homeostasis. These multipotent stem cells are the progenitors of various intestinal cell types. The bidirectional interaction between the microbiota and stem cells significantly impacts the treatment of neurodegenerative conditions, gastrointestinal diseases, and cancer. This chapter explores an analysis of the correlation between the microbiome and stem cells within the framework of regenerative medicine and discusses the potential for developing personalized therapeutic strategies.

Keywords: Stem cell, microbiome, gut health, health benefits, therapeutic approach

10.1 Introduction

Metabolic diseases that are prevalent worldwide includes obesity, nonalcoholic fatty liver disease (NAFLD), insulin resistance, type 2 diabetes mellitus (T2DM), and atherosclerosis, is widespread. The prevalence of androgen excess (AS) and polycystic ovarian syndrome (PCOS) has increased significantly [1]. The last several decades have witnessed an increasing tendency of raised utilization of calorie-dense meals and not engaging in physical activities during leisure time. The combination of physical inactivity with activities that involve sitting for long periods has ultimately led to a positive energy balance that occurs when energy consumption surpasses energy expended. Energy expenditure, which denotes to the amount of energy used by the body, has emerged as the primary factor contributing to risk. Contributing factors to obesity and diseases are associated with obesity [2, 3].

10.2 The Function of Microbiota in Gut Health

The excessive build-up of fat in adipocytes initiates the heightened manufacturing and release of pro-inflammatory substances. Adipokines have a contribution to the pathogenesis of resistance to insulin, associated with the onset of T2DM and NAFLD [4]. In terms of genetics, the microbial component accounts for around 99% of human genes [5]. It was shown that microbial cells are present in equal or greater numbers compared to human cells. Somatic cells were studied by Sender *et al.* Gastrointestinal microbiota is the whole population of microorganisms that reside in a standard environment. The gut microbiota comprises various microbes, including bacteria, viruses, fungus, and archaea phages, and protozoa, as described by Whitman *et al.*, which possess the mentioned characteristics. Having the ability to engage with the host through various means, some gut bacteria are considered pathogens by certain individuals. The host immune system identifies and eradicates those individuals. However, the greater part of the gut microbiota is nonpathogenic and mutually beneficially associated with intestinal epithelial cells. Intestinal microbiota exerts a substantial impact on food metabolism. The processes involved in the body's utilization of nutrients from food, the breakdown of foreign substances and medications, and the preservation of the gut form and function. The study focuses on the function of the barrier and GIT in preventing the movement of intestinal infections [6].

10.3 The Intersection of Microbiome and Stem Cell Research

Intestinal stem cells (ISCs) are a restricted group of specialized cells contained in the crypts of the intestines. A healthy gut microbiota is crucial for maintaining the balance and health of the intestinal tissue, as well as for restoring damage resulting from injury or inflammation. Intestinal stem cells (ISCs) function vitally in preserving the structural integrity of the mucosal barrier by undergoing controlled and continual multiplication, leading to the ongoing renewal of the intestinal epithelium. This process serves to preserve the integrity of the intestinal barrier and inhibit the infiltration of infections and physical harm. This helps to prevent the development of chronic or acute inflammation in both pathological and physiological situations [7].

The modifications of mechanisms that regulate the proliferation of ISCs play a vital role in the advancement of intestinal illnesses, including intestinal bowel disease (IBD) and colorectal cancer (CRC) [8]. Specifically, numerous studies have concentrated on the mechanisms that cause the transformation in cancer stem cells (CSCs) by ISCs, sometimes referred to as tumor-initiating cells (TICs) [9]. This research has provided a new opportunity to discover targeted therapeutics for CRC that specifically target CSCs. Nevertheless, studying the enteric crypt before the formation of a CSC niche is a captivating field of research that holds great potential for further discoveries. The microenvironment plays a crucial role in transformation ISC niches, as these niches are situated at the boundary separating the host organism from the microbiota. There is an intricate relationship interposed among the epithelial obstacles and its microbiological biota and the indigenous immune system [10]. Typically, the lining of the intestines engages with a diverse assortment of microorganisms, encompassing bacteria, fungi, and viruses, referred to as the intestinal microbiota [11]. The gut microbiota exhibits significant diversity, which is carefully regulated to safeguard the integrity of the enteric mucosal membrane. Epithelial cells establish recognition and communication with the indigenous microbiota through pattern recognition receptors (PRRs), such as Toll-like receptors (TLRs). The receptors are situated on each of the intestine-differentiated epithelial cells and intestinal stem cells (ISCs) [12]. Contrary to other types of stem cells, intestinal stem cells (ISCs) do in fact exist alongside the intestinal flora. Certain species belonging to the genus *Lactobacillus* possess the capacity to directly or indirectly impact the growth and specialization of intestinal stem cells (ISCs). In contrast, ISCs defend themselves against the presence of butyrate, which is generated by beneficial microorganisms in the gut lumen [13]. According to recent studies, if the natural balance between the host's mucosal cells and the gut microbiota is disturbed, it can lead to atypical immune reactions against the resident companions. This can lead to persistent inflammation and ultimately increase the individual's susceptibility to colorectal cancer (CRC) [14]. During this state of inflammation, intestinal stem cells (ISCs) detect and react to the presence of microbiota [15]. External factors, such as alterations in the gut microbiota, that exert effects on the ISC niche are a fascinating but intricate and debated area of study that centers on stem cell biology. In this summary, we present the most recent facts regarding molecular pathways and metabolic events that are involved in the intricate relevance of the intestinal microbiota to intestinal stem cells. This connection is essential for maintaining the normal behavior of ISCs and understanding how they become reprogrammed in the onset of CRC. In addition, the review examines the progress made in findings on OMICS technologies used in pre-clinical and clinical examination to evaluate molecular underpinnings of the inequality in the intestinal microbiota–ISCs interaction connected to colorectal cancer.

10.4 Microbiota Metabolites and Stem Cell Differentiation

Every surface of the human body harbors microorganisms, but the GI system and gut are home to a large quantity of them. The human gut microbiome refers to an intricate ecological assemblage composed of more than a thousand distinct bacterial species [16]. Proteobacteria and bacteria support the immune system, aid in the breakdown of carbohydrates, regulate the gut microbiota, and safeguard the body from infiltration of pathogens [17]. Microbes in the intestine mostly rely on food substrates that have not been fully digested in the upper digestive tract to survive. When there are not enough carbs available for fermentation, the bacteria will switch to another energy source, which might result in the synthesis of additional metabolites that may have greater detrimental impacts on people's health. Saccharolytic bacterial fermentation normally produces beneficial metabolites [18, 19]. Anaerobic and fermentation-based dietary fiber, protein, and peptide breakdown are associated to the human gut microbiota [20]. Investigation indicates that short-chain fatty acids (SCFAs) impact the maturation of ISCs [21]. Within an organism, stem cells are a class of cells that have not yet developed into fully functional adult cells that have the capacity to experience self-renewal and adapt into any cellular form [22]. The body's capacity to create and reproduce particular cells depends on the stem cells' capacity for self-renewal and differentiation into any type of cell. Based on their highest capacity for differentiation, stem cells are classified as totipotent, pluripotent, multipotent, oligopotent, and ultimately unipotent [23], which is equivalent to having the capacity to differentiate into just one lineage of cells from a complete organism and its extraembryonic tissues. Although commensal gut microbiota and pathogenic microorganisms compete the same finite resources—nutrients, habitat, and energy—in the host GIT to proliferate, the gut microbiota provides the host with antibiotic resistance to pathogenic microorganisms. To prevent the growth of invading microorganisms *via* direct and indirect channels, it has been discovered that the native gut microbiota has evolved and employs specific strategies [24]. The term "direct pathway" describes how the natural intestinal microbes in the GI tract directly prevent the progress and colonization of pathogenic bacteria by changing the pH of the GI tract [25, 26], synthesizing noxious compounds that are lethal to pathogens [27], and contending for the restricted supply of minerals [28]. This is accomplished by the secretion of certain bacterial metabolites that promote the proliferation of the indigenous gut microbiota and suppress the growth of potentially detrimental strains and species. By recognizing and reacting to pathogen recognition receptors, the gut microbiota is thought to indirectly maintain the enteric epithelium layer [29] by enlisting the assistance pertaining to the host immune system's defense against invading infections and the metabolism of bile acids [30]. Figure 10.1 depicts the correlation between gut-microbiota and stem cell therapy [31].

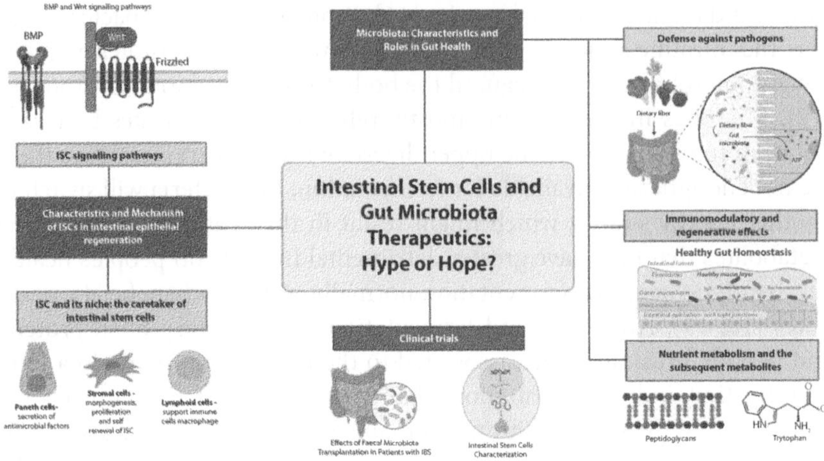

Figure 10.1 Gut microbiota and stem cell connection [31].

10.5 Modulation of Mucosal Immunity and Intestinal Inflammation

Live microorganisms that benefit gut health are called probiotics. The World Health Organization (WHO) defines probiotics as advantageous microorganisms that enhance health when ingested in optimal amounts [32]. They strongly induce and regulate both systemic and mucosal adaptive immune responses [33]. Recent progress in synthetic biology has facilitated the manipulation of probiotics for the purpose of disease treatment, including cancer, infections, and metabolic disorders that conventional medicine struggles to cure [34]. Engineered probiotic vaccines have induced protective immune responses at mucosal sites, including the urogenital, intestinal, and respiratory tracts [35]. Evidence suggests that probiotics mitigate the intensity and length of viral respiratory infections [36]. Studies report that probiotics used preventively can induce effective antiviral responses [37]. *E. coli*, *Bacteroides*, and LAB are the primary bacterial chassis for engineered live Therapeutics [38]. Rapidly spreading SARS-CoV-2 necessitates urgent vaccine development. Bioengineered probiotics, offering feasibility and cost-effectiveness, have emerged as a promising platform. The gut–lung nexus is crucial in COVID-19 [39]. The gut microbiome significantly impacts COVID-19 immune responses [40, 41]. To effectively combat COVID-19, including preventing viral shedding, a vaccine targeting both respiratory and intestinal infections is essential. Mucosal immunity, especially in the gut, is crucial for this [42].

A microbiome is a heterogeneous assemblage of microorganisms inhabiting the body's exposed surfaces, primarily the gut [43]. The fungal microbiome (mycobiome) is a (Iliev ID, 2017) vital part of the body's microbial community, developing alongside bacteria and viruses from birth [44]. Recent advances have revealed a diverse fungal community, yet most research focus on *Candida albicans*, providing deep insights into fungal behavior [45]. Research is expanding to include other fungi, exploring their impact on overall microbiome health and immune function. Advanced studies are uncovering how fungal imbalances contribute to various diseases.

10.5.1 Fungal Colonization in the Gut

Fungi are far less common than bacteria in the gut, making up only 1%–3% of gut microbes [46]. Despite their low abundance, gut fungi play a crucial role in early immune development [47].

10.5.2 Fungi's Role in Gut Immune Balance

Intestinal immune cells detect fungi through specialized receptors that recognize fungal components [48]. C-type lectin receptors (CLRs) are key to fungal recognition due to their ability to bind fungal cell wall sugars [49]. Dectin-1 deficiency leads to increased fungal infections and spread due to weakened immune response [50]. Dectin-3 and MBL are additional CLRs crucial for fungal recognition and gut health [51]. CX3CR1+ mononuclear phagocytes in the gut initiate immune responses against fungal colonizers [52]. Gut CX3CR1+ MNP express fungal recognition receptors and effectively engulf fungi.

10.5.3 Fungal–Immune Interactions Extend Beyond the Gut: Gut Fungi Influence Lung Health

Mycobacteria in the gut modulate immunological responses in several body regions, particularly the lungs. Early exposure to specific fungi can impact lung inflammation [53], and antibiotics can disrupt the gut microbiome, leading to fungal overgrowth. This imbalance can trigger a specific type of immune response (M2 macrophage polarization) in the lungs, contributing to inflammation from allergies. A specific yeast linked to early childhood asthma worsened allergic lung inflammation in mice [54]. Specific fungi in the gut can elicit immune responses that promote allergic reactions in the lungs. This involves stimulating both type 2 and type 17 immune responses, which are associated with inflammation. These immune reactions can also increase sensitivity to other allergens, contributing to chronic lung conditions.

Certain gut fungi, such as *P. kudriavzevii*, *A. amstelodami*, *W. mellicola*, and *E. nigrum*, are linked to increased allergic airway inflammation. These fungi can trigger an excessive immune response (Th2) in the lungs, contributing to conditions like asthma [55]. Gut *Candida* and lung inflammation is the Th17 cell involvement: Candida albicans in the gut can trigger an increase in Th17 immune cells, which can also react to other fungi like *Aspergillus fumigatus*, leading to lung inflammation [46]. Monocyte influence: Early exposure to *Candida albicans* can increase the lung monocyte numbers, contributing to heightened inflammation in conditions like asthma.

10.5.4 Gut Fungi Influence Overall Immune Defense

Gut fungi can boost the overall immunity. For instance, *Candida albicans* in the gut can strengthen defenses against invasive fungal infections by increasing specific immune cells (Th17) and their supporting factors (IL-17) [56]. Gut fungi can influence the protection against fungal infections. Mixed results: While *Candida albicans* in the gut can boost the immunity against certain fungal infections, its protective effect against other fungi is less clear. Organ-specific protection: The immune response triggered by gut *C. albicans* can protect against *Candida auris* infection in the brain but not the kidneys [57].

10.5.5 Impact of Different Fungi on Immune Response

Not all fungi are created equal when it comes to stimulating the immune system. Limited immunogenicity: Common fungi like *Cladosporium cladosporioides*, *M. restricta*, and *S. cerevisiae* do not effectively trigger a strong immune response (IgG antibodies) compared to *Candida albicans*. Phylogenetic relationship: It seems that protection against related fungal species is more likely. In other words, fungi with a closer genetic relationship are more prone to having comparable components that the immune system can identify. Innate immunity is key: Analogous to the function of innate immunity in mucosal responses, the CARD9 signaling route in specific immune cells (CX3CR1+ MNPs) is crucial for producing antibodies against *Candida albicans* [58].

10.6 Influences on Stem Cell Engraftment

Hematopoietic stem cells (HSC) use a process called engraftment to navigate (homing) to vacant bone marrow (BM) niches where they are able to settle in the best environment for survival and proliferation. HSCs must multiply once they enter the BM milieu to produce all hematopoietic cell subsets [59]. The ability of the transplanted HSC to maintain long-term functional hematopoiesis—the creation of red blood cells (RBCs), white blood cells (WBCs), and platelets—as well as their liberation into the peripheral circulation, is a crucial requirement for successful engraftment [60]. The most crucial factor for improved overall survival following engraftment is the process of stem cell replacement [61]. The intricate procedure of stem cell engraftment permits transplanted stem cells to successfully reside in the BM and initiate the production of fresh blood cells. The field of hematopoietic stem cell transplantation (HSCT) is constantly developing, with continual attempts being made to maximize results. Techniques such as high-resolution human leukocyte antigen (HLA) and programs of reduced-intensity conditioning (RIC) have greatly enhanced patient outcomes [62]. Refer to Table 10.1 for the list of authorized applications of hematopoietic stem cell transplantation (HSCT). Transplanted stem cells must successfully establish themselves in the bone marrow to initiate generating new blood cells, a process known as stem cell engraftment [63]. Hematopoietic stem cells (HSCs) in the BM surroundings compete with host cells after infusion. Donor HSCs can successfully engraft when they pass through the endothelium barrier, find a niche, and start proliferating [64]. Following hemodialysis, rapid engraftment can result in fewer blood component transfusion needs, lower rates of infection and bleeding, shorter hospital stays, cheaper expenses, and eventually better survival results. Determining the factors impacting outcomes is critical because transplant centers have different patient populations and treatment methods [65].

10.6.1 Factors Influencing Stem Cell Engraftment

Even with so many variables at play, this study offers a targeted summary of the major parameters that determine HSCT outcomes and investigates possible approaches to improve patient outcomes, such as altered conditioning schedules, graft manipulation, and predictive indicators [66].

10.6.2 Graft-Versus-Host Disease (GVHD)

In allogeneic stem cell transplantation, HLA compatibility is the main criterion adopted for donor selection [67]. The short arm of chromosome 3 contains both HLA types, which are expressed on most cells, especially hematopoietic cells, where they are essential for identifying foreign antigens [68]. While improving disease-free survival (DFS), HLA compatibility dramatically lowers the risk of GVHD, graft rejection, and fatality [10]. Successful HSCT is hampered by the high degree of HLA polymorphism, which ranges from 13 alleles for HLA-DRB4 to 699 for HLA-B [69]. An elevated occurrence of mild GVHD has been linked to minor HLA incompatibility in allogeneic HSCT [70]. Less than 30% of patients have the opportunity to receive sibling donors who match their HLA [71]. HLA-matched separate donors, haploidentical corresponding donors, and incompatible unrelated donors are the next most suitable alternatives for umbilical cord blood stem cells [72].

10.6.3 Additional Factors

Body mass index (BMI) is a key determinant of GVHD. Weight loss before and after HSCT is linked to a higher susceptibility of GVHD and increased toxicity from conditioning regimens [73]. Another contributing factor is malnutrition that lowers patient survival and raises the likelihood of GVHD, which harms the allogeneic HSCT outcomes [74].

10.6.4 A Novel Approach to Improving Outcomes

In the BM, mesenchymal stem/stromal cells (MSCs), which frequently resemble fibroblasts, form a unique microenvironment. Hematopoietic stem cell (HSC) function depends on this habitat. MSCs secrete a wide range of chemicals that actively participate in the control of HSCs. These consist of extracellular vesicles (EVs) that transport cargo, such as proteins and nucleic acids, to affect HSC behavior, extracellular matrix proteins that offer structural support and biochemical cues, and cytokines and growth factors that function as signaling molecules. MSCs are essential for sustaining HSC quiescence, encouraging proliferation, and directing differentiation into distinct blood cell lineages through these processes. MScs essentially play the role of hematopoiesis orchestrators, providing an environment that is conducive to HSC viability and the creation of blood cells [75]. Through a variety of mechanisms, MSCs are essential for the regulation of hematopoietic stem cells (HSCs). Secretion of stem cell factor (SCF), sometimes referred to as CXCL12, is one important function. This cytokine is necessary to keep the HSC pool healthy and protect it from harm. MSCs also express leptin receptors (LepR), and a decrease in quiescent HSCs in the BM has been linked with the downregulation of these receptors. This implies that leptin signaling plays a crucial function in maintaining the HSC pool in a resting condition [76]. Extracellular vesicles (EVs) have become important players in the communication between cells. The cargo of these membrane-bound, nanoscale structures, which include proteins, mRNA, DNA, and microRNA, is diverse and is released by cells, including MSCs. EVs use membrane fusion to transfer their contents to destination cells, acting as cellular messengers. Through this procedure, genetic material and functional proteins can be transferred, changing the phenotype and behavior of target cells. Essentially, EVs are sophisticated messengers for biological data, enabling nuanced cellular reactions and interactions [77]. Bone marrow (BM) extracellular vesicles (EVs) have become important mediators in both normal and pathological processes. Essential cellular processes in the normal BM, including survival, proliferation, differentiation, and angiogenesis, are greatly aided by EVs. On the other hand, when it comes to cancer, EVs are essential for promoting the growth of tumors, avoiding medical treatments, and reducing immunological responses.

These results highlight the intricate and varied function that EVs play in sculpting the BM microenvironment and affecting cellular activity [78]. Through graft modification, MSC-derived extracellular vesicles (MSC-EVs) have become a viable approach to optimize allogeneic hematopoietic stem cell transplantation (allo-HSCT). It has been demonstrated that pretreating CD34+ hematopoietic stem cells (HSCs) generated by umbilical cord blood (UCB) with MSC-EVs increases CXCR4 expression, suppresses premature differentiation, and improves cell survival. A key receptor that mediates HSC homing to the bone marrow surroundings is CXCR4. All of these results point to the possibility that MSC-EVs can enhance HSC engraftment and lead to better outcomes for allo-HSCT users [79]. Furthermore, it has been demonstrated that extracellular vesicles (BM-MSC-EVs) generated from mesenchymal stem cells obtained from BM (miRNAs) and Piwi-interacting RNAs (piRNAs) enhance the survival and inhibit the differentiation of UCB-HSCs originating from human umbilical cord blood. These results have important ramifications for improving transplant outcomes. MSC-EVs provide a number of benefits over applying soluble factors or MSCs directly. They are a desirable treatment alternative because of the stability of their encapsulated contents and the lower possibility of negative side effects [80].

10.7 Role in Neurological Diseases

10.7.1 Gut Microbiota–Brain Axis in Neuro-Inflammation

Modern lifestyles drive a surge in neurological diseases. The gut microbiome likely influences brain function through microbial metabolites [81]. Recent research has unveiled a complex interplay between the brain and gut microbes. These interactions, involving immune, neural, and chemical factors, are crucial for brain health and offer new insights into neurological diseases [82]. The gut microbiota profoundly impacts overall health through the gut–brain nexus. The bidirectional interactions encompass neuronal, hormonal, and immunological transmission routes, regulate brain function, and foster a symbiotic host–microbe relationship [83]. A healthy gut microbiome produces molecules that influence brain function. Gut imbalances are linked to various neurological disorders. The complex brain controls behavior through intricate neural networks [84]. The gut microbiota pertains to the species of bacteria that inhabit the GIT [85]. Ongoing technological developments are enhancing our comprehension of the gut–brain interface. This explores intricate interaction between our gut bacteria and neurological disorders, including neuropsychiatric, mood, and neurodegenerative conditions [86].

10.7.2 Gut–Brain Axis

The gut microbiome is a complex interconnected community of bacteria, archaea, viruses, and fungus that inhabit the gastrointestinal tract [87]. It comprises 100–150 times the number of genes found in the human genome and comprises thousands of bacterial species [88], dominated by Firmicutes and Bacteroidetes [89]. Dysbiosis is a disparity in nutritional combination of the gut microbiota. This imbalance is linked with a diverse array of health concerns, spanning from gastrointestinal disorders to long-term illnesses [90].

10.7.3 Gut Dysbiosis, an Imbalance of Gut Bacteria, is Linked to Brain Problems

Changes in gut microbes have been tied to brain inflammation, protein clumps, weakened immune function, and reduced brain cell activity [81]. Gut microbes influence brain function [91]. Microbial metabolites impact the immune system, affecting cognition, mood, and movement [92]. The gut–brain nexus regulates digestion, behavior, and stress. Recent research emphasizes the complex correlation between gut microbiota and brain health [93].

10.7.4 The Gut Microbiota Influences Brain Function

Information exchange between the brain and gut takes place *via* neuronal, immunological, and hormonal relays [94]. This is a complicated network connecting the gut and brain referred to as the gut-brain interface. This bidirectional messaging system encompasses the neurological, immune, and endocrine systems, influencing digestion, mood, and overall health [95]. Gut bacteria significantly impact brain function. These microbes influence brain development, aging, and overall health by altering brain chemistry [96].

10.7.5 The Gut–Brain Axis: An Intricate Connection

The gut microbiome profoundly influences brain health. Gut bacteria communicate with the brain *via* the vagus nerve and immune system, affecting mood, behavior, as well as cognitive function. While research has linked gut imbalances to various neurological disorders, further investigation is required to comprehensively grasp these complex interactions and develop effective treatments.

10.8 Emerging Significance of Gut Microbiota in Neurogenesis

10.8.1 Gut Bacteria Significantly Impact Brain Health

Gut bacteria impact brain health. Changes in gut microbes are associated to neurological and psychiatric states [97]. Imbalances in gut microbes (dysbiosis) are associated with a range of neurological diseases, including Parkinson's disease, Alzheimer's disease, and depression [98]. Vagus nerve stimulation modulates the monoaminergic system, impacting mood and anxiety. It triggers neurotransmitter release (GABA, serotonin) and SCFA production, exerting anti-stress and anti-anxiety effects, like gut–brain axis [99].

10.8.2 Neurological Pathway in the Gut–Brain Axis (GBA)

Gut neurotransmitters, *via* ENS and vagus nerve, stimulate hormone and neurotransmitter release, influencing the gut–brain nexus and interacting with endocrine and immune systems [100]. Inflammasome activation, triggered by MAMPs, induces the transformation and synthesis of pro-inflammatory cytokines IL-1β and IL-18 facilitated by caspase-1, contributing to neurological disorders [101].

10.8.3 Neurological Disorders and the Gut–Brain Axis (GBA)

Parkinson's disorder involves α-synuclein deposition in the substantia nigra, resulting in the death of dopaminergic neurons. α-Synuclein can spread from the gut to the brain through the vagus nerve [102]. Antibiotic therapy improves PD symptoms, suggesting gut microbiota influence. Microbial tyrosine decarboxylase reduces levodopa efficacy, highlighting implications of gut microbiota in Parkinson's development and therapy [103]. Microbial balance promotes brain health through SCFAs, neurotrophic factors, antioxidants, and anti-inflammatory cytokines, while dysbiosis drives PD pathogenesis *via* inflammation, oxidative stress, and α-synuclein accumulation. Alzheimer's disease (AD) is a neurodegenerative condition marked by memory loss and cognitive deterioration caused by synaptic degeneration [104, 105]. Amyloid-beta (Aβ) plaques and tau protein tangles are pathologenic hallmarks of AD [106]. Gut microbiota dysbiosis, marked by reduced SCFAs and altered metabolic pathways, is implicated in AD pathogenesis. Probiotics and ketogenic diets may mitigate disease progression, highlighting the potential of microbiome-based AD therapies [107]. Early-life microbiome influences brain development, with disruptions linked to ASD. Maternal factors like valproate exposure and obesity alter gut microbiota composition. Probiotic and prebiotic interventions show promise in modulating social behavior, suggesting potential microbiota-based ASD therapies [108]. Neurological disorders are rising, demanding novel biomarkers and treatments. The gut–brain axis (GBA) offers a promising research avenue. The microbiome, a complex ecosystem, influences health and disease. Its imbalance is linked to various physiological abnormalities, emphasizing the GBA's critical role in human health. Gut microbiota (GM) dysbiosis underlies multiple neurological disorders by disrupting the gut–brain axis. Various factors modulate associated pathways, influencing diseases like PD, AD, MS, ALS, ASD, stroke, epilepsy, and HD. Microbial therapies (probiotics, prebiotics, etc.) target these pathways to ameliorate symptoms [109]. Gut microbiota (GM) imbalance underlies numerous neurological disorders. Therapeutic interventions like probiotics, prebiotics, and fecal transplants can restore GM balance and potentially mitigate disease progression [110, 111].

10.8.4 The Gut–Brain Axis and Neurological Disorders

An essential function of the gut microbiota is to coordinate brain activity and behavior *via* gut–brain axis (GBA). Microbiome disturbances, known as dysbiosis, have been associated with several neurological diseases such as Alzheimer's, Parkinson's, autism, and depression. The gut microbiota plays a crucial role in brain health maintenance, and understanding its role in neurological disorders holds a significant potential for developing novel therapeutic strategies.

10.9 Clinical Trials

Clinical trials investigating the direct influence of gut microbiota modulation on neurological disease are still in their early stages. However, several studies are exploring the utilization of probiotics, prebiotics, and fecal microbiota transplantation as prospective therapeutic actions. A comprehensive knowledge of the gut–brain nexus is expected to result in innovative microbiota-based therapies for neurological disorders.

10.10 Role in Gastrointestinal Diseases

Gastrointestinal (GI) diseases affect the digestive tract. They are classified as functional (normal appearance but impaired function) or structural (abnormal appearance). Examples include IBS, colitis, and food poisoning. COVID-19 significantly impacted IBD patients. Initial uncertainty about SARS-CoV-2 risks for immune-compromised IBD patients quickly evolved into comprehensive clinical guidelines and vaccination recommendations.

10.10.1 Inflammatory Bowel Disease (IBD)

Inflammatory bowel disease (IBD) impacts 0.8% of Canadians (over 320,000 people) in 2023 [112, 113]. IBD patients were initially deemed to be at a high risk of severe COVID-19 due to their immunosuppressed states [114]. Due to limited COVID-19 data, Crohn's and Colitis Canada formed a task squad in March 2020 to address IBD patient concerns [115].

A multidisciplinary task force was formed to provide guidance for IBD patients during the COVID-19 pandemic. Early in the pandemic, IBD patients were considered to be at a high risk of COVID-19 infection. To address data gaps, the SECURE-IBD registry enrolled over 6,000 IBD patients with COVID-19 [116]. Early SECURE-IBD data suggested no increased COVID-19 risk for IBD patients compared to the general population [117]. IBD patients had similar COVID-19 infection rates as the general population [110]. Prednisone use increases the risk of severe COVID-19 outcomes in IBD individuals [119]. The effect of public health measures on IBD patients' COVID-19 risk needs further study. IBD medications like anti-TNFs, vedolizumab, ustekinumab, and tofacitinib were linked to lower COVID-19 hospitalization and death risks [120]. Prednisone use increases serious COVID-19 risk, especially flaring in younger IBD patients [121]. IBD medications like anti-TNFs lowered this risk. Continuing IBD therapies was recommended to prevent flares, while high-dose prednisone users should be isolated [122]. Prior IBD surgery did not increase COVID-19 severity. Paxlovid decreased the hospitalization risk in high-risk IBD patients with COVID-19 [123].

10.10.2 COVID-19 Vaccines in IBD Patients

mRNA vaccines showed superior antibody responses compared to adenovirus vector vaccines in IBD patients [124]. However, immunosuppressants like infliximab and tofacitinib reduced the vaccine efficacy [125]. Adolescents with IBD had strong immune responses to mRNA vaccines. Antibody levels waned over time, highlighting the need for potential booster doses [126].

10.10.3 Bivalent Vaccines are Suggested for IBD Patients Due to Higher Breakthrough Infections

Before Omicron, breakthrough infections were rare in vaccinated IBD patients. The antibody levels were lower in those with breakthrough infections. Studies showed a similar break through infection rates in a comparison between IBD patients and the general population [127].

10.10.4 SARS-CoV-2 Vaccines are Safe and Effective in IBD Patients, with No Increased Risk of Disease Flares

Adverse effects of COVID-19 vaccines in IBD individuals are corresponding to the general population, primarily mild and short-lived. The gastrointestinal symptoms slightly increased after the third dose. No increased risk of IBD flares was observed after vaccination [128].

10.10.5 COVID-19 Complications in IBD Patients Include Mental Health Issues and Potential Long-Term Effects (Long COVID)

Long COVID is a potential complication for IBD patients [129]. While overall risk may be comparable to that of the general population, the convergence of symptoms might highly affect the quality of life [130]. Further study is required to comprehend the precise risks and challenges for IBD patients. COVID-19 exacerbated the mental health challenges in the IBD community (Meringer and Saurabh *et al.*, 2022). Increased stress, anxiety, and depression impacted the quality of life and worsened the IBD symptoms [131].

10.10.6 COVID-19 and IBD: Key Findings

IBD patients had similar COVID-19 risks as the general population but faced unique challenges. Prednisone use increased the severity risk. Vaccines were effective but breakthrough infections occurred, especially with Omicron. Mental health significantly deteriorated for many IBD patients [132].

10.10.7 Patients' Perspectives on COVID-19 and IBD

Individuals with IBD have comparable COVID-19 risks and outcomes to those in the general population. Preventing flares and high-dose steroid use was crucial. Continuing IBD medications was essential. Vaccines were deemed safe and effective, with boosters recommended. Mental health support was highlighted due to increased stress and isolation. Clinically vulnerable patients (e.g., those on high-dose prednisone) required ongoing focus [133].

10.10.8 Significance of the Microbiota to Liver Pathophysiology

A major worldwide health burden is caused by liver disease, in which the gut microbiota plays an important role in its advancement. Malnutrition, common in advanced liver disease, exacerbates gut dysbiosis. This asymmetry results in heightened intestinal permeability, therefore enabling bacterial compounds to access the liver, triggering inflammation and disease progression [134]. Intestinal microbiota links with the hepatic immunological system, amplifying inflammation and contributing to liver injury. Addressing gut health holds promise for managing chronic liver diseases. The gut–liver axis plays a vital role in liver health. Dysbiosis, an imbalance in gut bacteria, contributes several liver diseases [135]. Increased intestinal permeability, bacterial translocation, and inflammation are key mechanisms linking gut dysbiosis to liver damage. While specific microbiota alterations have been identified in different liver diseases, the intricate interplay between the gut and liver requires further exploration. Targeting the gut microbiota offers promising therapeutic potential for managing chronic liver diseases [136].

10.10.9 Gut–Liver Axis: A Bidirectional Relationship

Intestinal penetration and microbiota have a strong association with chronic liver disease due to the liver's intimate connection with the gut. Intestinal barrier integrity, composed of physical, immune, and microbial components, is crucial for liver health [137]. Mucus, rich in antimicrobial peptides and immunoglobulins, forms a protective layer influenced by gut bacteria. Dysbiosis, characterized by shifts in bacterial populations like Bacteroidetes and Firmicutes, can compromise this barrier [138]. Impaired intestinal permeability is linked to liver disease progression [139]. Lifestyle factors like diet and alcohol consumption contribute to gut dysbiosis and elevated intestinal permeability. This allows gut-derived endotoxins to enter the bloodstream, triggering liver inflammation and fibrosis. Endotoxemia is of paramount importance in the development of liver cirrhosis and hepatocellular carcinoma (HCC) [140]. Targeting the gut microbiota presents a promising therapeutic avenue for managing chronic liver diseases [141].

10.10.10 Microbiota and Hemopoietic Stem Cell Transplantation

The microbiota refers to a complex network of bacteria that colonize the human body [142]. Bacteria dominate, but fungi (mycobiome), viruses (virome), and other microbes also contribute. The gut microbiota is extensively researched with its varied microbial communities existing in various body sites [143]. Analyzing these microbiomes independently and collectively provides valuable insights into health and disease. Advancements in metagenomics have unveiled the pivotal significance of the human microbiome in well-being and pathology [144]. Dysbiosis, an imbalance in microbial communities, is associated with various pathologies [145]. The immune system intricately interacts with the microbiota, shaping both innate and adaptive immunity [146]. This interplay is crucial for transplant success and preventing graft-versus-host disease (GVHD). Subsequent sections will examine the influence of various microbiomes on stem cell and organ transplantation therapies [147].

10.10.11 Hematopoietic Stem Cell Transplantation (HSCT)

In medicine, hematopoietic stem cell transplantation (HSCT) is the replacement of injured or destroyed bone marrow with viable stem cells. These progenitor cells are capable of undergoing differentiation into erythrocytes, leukocytes, and thrombocytes. HSCT is a critical therapy for various hematological malignancies [148]. It entails substituting impaired bone marrow with viable stem cells. While it offers a chance for cure, it is associated with significant risks like GVHD, infections, and graft failure [149]. Recent advancements have improved outcomes, but challenges persist. The impact of the gut microbiota on HSCT is increasingly recognized [150]. Its impact on GVHD, immune reconstitution, and overall survival is being studied. Gaining insight into the complex dynamics between the microbiota and the host will be essential for devising innovative therapeutic approaches to improve the HSCT results [151].

10.10.12 Chronic Kidney Disease (CKD) and Renal Transplantation

CKD is a worldwide health challenge that advances to end-stage renal disease (ESRD), necessitating renal replacement therapy (RRT). Renal transplantation is the recommended therapeutic intervention, offering improved quality of life compared to dialysis. However, it necessitates lifelong immunosuppression to prevent rejection. Immunosuppression has evolved with drugs like cyclosporine, tacrolimus, and monoclonal antibodies. Despite advancements, long-term graft survival remains a task. The impact of the microbiota in kidney transplantation is emerging as a critical factor. Unraveling the intricate interaction between the immune system and the microbiome holds potential for improving transplant outcomes and developing novel therapeutic strategies [152].

10.11 Microbiome Influence on Tumor Microenvironment

In the present scenario, the lifestyle of the population is affected by several factors, among which food habits and different social habits significantly alter the microbial flora of the gut. Microbiota dysbiosis leads to different categories of cancer like breast cancer, colon cancer, gastric cancer, etc. [153]. The tumor microenvironment (TME) is the cellular environment where tumors or cancer stem cells reside. Cancer stem cells may have the ability to drive the genesis of a tumor [154]. In a previous study, it was stated that the extracellular matrix (ECM), blood arteries, lymphocytes, fibroblasts, bone marrow-derived inflammatory cells, and signaling particles are all included in the tumor micro environment (TME) [155]. Basically, when a non-malignant cell interacts with a malignant cell, they create a TME that affects cancer development as well as its progression [156]. Numerous studies have shown that few bacteria can promote the growth of tumors under unfavorable conditions [157]. For example, *Helicobacter pylori* can trigger immunological reactions and persistent inflammation that may cause gastric cancer [158]. Globally, stomach cancer is the most prevalent kind of cancer. *H. pylori* can cause gastritis, and the gastritis may lead to gastric cancer [154]. According to 16S rRNA sequencing studies of GIT and fecal samples, dysbiosis of the gut microbiota may contribute to the progression of cancer [159]. Colon cancer is the third most common cancer globally; over 1.9 million new cases are introduced, with 930,000 deaths that occurred in 2020. The chances of colon cancer may increase due to the tumor environment factors [160]. In a previous study, it was stated that the microbiota, like *B. fragilis*, *C. septum*, *E. coli*, *E. faecalis*, *Fusobacterium* spp., and *H. pylori*, causes colon carcinoma [161]. Breast cancer is the second most prevalent cancer globally, predominantly affecting women. Aged women are more affected with breast cancer than middle-aged ones [162]. In a previous study, it was mentioned that there are some microbiome promoting breast carcinoma—for example, breast

Table 10.1 Micro-organisms and enzymes involved in promoting a tumor microenvironment.

Microbiota	Enzyme	Promoting tumor	Types of tumor	Reference
H. pylori	Cytotoxin-linked gene-A	Interfere with AKT pathways of the host cell causing gastric cancer or Cag+ strain infection elevates inflammatory factors like IL-1, IL-6, and IL-8	Gastric cancer or adenocarcinoma of the stomach	[166, 167]
Clostridium leptum and *Clostridium coccoides*	β-glucuronidase	Stimulates the target cells' expression of estrogen receptors; causes breast and endometrial, activation of estrogen receptors which stimulates cell growth	Breast cancer	[168]
E. coli	Colibactin and cytolethal distending toxin (CDT)	Breaks the double strand break in DNA which induces cell arrest	Tumorigenesis	[169]
Fusobacterium nucleatum	Natural killer (NK) cell	Inhibits NK cell in myeloid suppressor cell	Tumorigenesis	[168]
Fusobacterium nucleatum	FadA (E-cadherin)	Activates β-catenin signaling which induces inflammation in colon tissue	Colorectal cancer	[170]

tumor tissues contain high concentrations of *Methylobacterium radiotoler* [163]. Overall, cervical cancer ranks as the fourth most prevalent form of cancer worldwide. Previous research suggested that a chronic HPV infection followed by cervical cancer is caused by immune avoidance and a host-dependent immune response generated by the human papillomavirus (HPV) [164]. Women who tested positive for HPV had more diverse vaginal microbiomes, with a higher abundance of some bacteria, such as *L. gasseri* and *G. vaginalis* [165] (Table 10.1).

10.11.1 Microbiome-Supported Stem Cell Treatment

Moreover, the tumorigenesis effect of the microbiome has several effects as anti-cancer agent in chemotherapy as well as immunotherapy [170]. Just as we have previously discussed, *H. pylori* stimulates the β-catenin signaling pathway, which promotes the growth of tumor. On the other hand, eliminating *H. pylori* mitigates the likelihood of inflammation in gastric cells (Table 10.2).

10.11.2 Microbiome Diversity

The human gut microbiome encompasses the entire population of microorganisms, including bacteria and other microbes, inhabiting the GIT [176]. The human body hosts around 500 to 1,000 different species of gut microbes, which collectively possess a genetic repertoire that is 100 times exceeding than that of the human genome [177]. An extensive variety of gut microbiota persisted largely uncleared until the advent of culture-independent high-throughput sequencing technologies as well as advanced

Table 10.2 Microbiome with a tumor-suppressing effect.

Microbiome	Effect	Type of tumor	Reference
Lactobacillus johnsonii, Enterococcus hirae, Barnesiella intestinihominis	Enhances the efficacy of cyclophosphamide (CTX) in cancer therapy by T helper 1 and 17 cell	Anti-tumor effect	[171]
Escherichia coli and *Caenorhabditis elegans*	Interacts the complex interaction among host cell and fluoropyrimidine	Anti-cancer effect	[172]
Bifidobacterium longum and *B. breve*	Causes cytotoxic T-cell infiltration into cancerous areas		[173]
B. rodentium	Decreases the upregulation signaling mechanism	Anti-cancer effect against melanoma	[174]
Lactobacillus coryniformis and *Lacticaseibacillus casei*	Inhibits the JNK signaling pathway, inhibits colon cancer cell progression	Anti-cancer effect against colon cancer	[175]

analytical and bioinformatics tools [178]. Substantial data indicates that the vast number of bacteria and archaea in the human gut is essential for preserving the well-being of the host [179]. Microbial species participate in diverse metabolic activities such as biomolecule fermentation and digestion, xenobiotic degradation, heavy metal biotransformation, immune regulator and specialized metabolite creation, energy generation, and epithelial homeostasis [180]. Besides its metabolism-related role, it also contributes in the management of brain health and inflammatory processes. Extensive research on the gut microbiome has demonstrated the secondary metabolites generated by bacterial flora to have the potential to impact the functioning of geographically distant tissues, including the liver, brain, muscles, and adipose tissues [181]. An optimal and harmonious microbiota not only nurtures gastrointestinal processes but also governs the diverse activities of the central nervous system [182]. Disruptions in this balance may lead to neuropsychiatric effects, including anxiety, depression, schizophrenia, and autism [182]. In the infant gut microbiota, the dominant phyla are Actinobacteria, Proteobacteria, Firmicutes, and Bacteroidetes, with their abundance reflecting the order of their appearance [183, 184].

10.12 Challenges of Stem Cell Therapy

A defining characteristic of stem cells (SCs) is their capacity to undergo self-renewal and differentiation into specific cell types, distinguished by their source and strength [185]. Pluripotent SCs have two sources: one is embryonic stem cells (ESCs) and another is mesenchymal stem/stromal cells (MSCs) and they have self-renewal potency [186]. As per the previous study, stem cell has a significance in understanding fundamental biochemical and developmental process. Additionally, stem cells hold a remarkable promise for creating *in vitro* disease models, advancing tissue engineering, regenerative drug, cell therapy, and pharmaceutical application [187], but there are several challenges in the use of stem cell therapy.

10.12.1 Inadequate Source of Stem Cells

Limited availability of stem cells is a result of the reduction in both number and role of bone marrow mesenchymal stem cells (MSCs) in older individuals as well as the loss of stem cell characteristics during culture expansion [188].

10.12.2 Presence of Optimal Time Window

The ideal temporal range for cell treatment is crucial, as stem cell tropism toward the brain diminishes over time. Consequently, the mechanistic targets for effective cell treatment vary depending on the temporal window [189].

10.12.3 Inherited Limitation of Stem Cells

An inherent limitation of stem cells, specifically mesenchymal stem cells (MSCs), is their heterogeneity and diminished neurorestorative capabilities, particularly in older patients with chronic conditions [190].

10.12.4 Adverse Effect

To achieve the therapeutic benefits of stem cell treatments, adequate drug administration is necessary. However, this approach may carry potential adverse effects, such as tumor formation, entrapment of cells within lung or brain vessels, coagulation activation, and immune rejection [191]. The optimal dosage and administration route for stem cell therapy of cardiovascular disorders remain undetermined, ensuring an adequate cell dose is crucial for achieving therapeutic benefits [192]. Unlike hematogenous stem cells, the set of mesenchymal stem cells (MSCs) in the bone marrow significantly decreases with age, necessitating culture expansion. Reports indicate that MSCs can undergo senescence and lose their stem cell characteristics, such as plasticity and homing ability, during *ex vivo* culture [193]. Adult stem cells may possess intrinsic constraints. Mesenchymal stem cells (MSCs) are heterogeneous, comprising various progenitor and stem cell types with differing capacities for progress, trophic support, and differentiation. In the elderly, the neurorestorative potential of MSCs could be constrained due to the reduced number of neural stem/progenitor cells (NSCs) [194]. A prior study demonstrated the use of bone marrow MSCs in the treatment of type I diabetes rats that led to elevated morbidity and blood–brain barrier (BBB) leakage, causing brain hemorrhage. This finding highlights the potential risks of stem cell therapy in diabetic subjects [195].

10.13 Microbiome Engineering and Integration of Multi-Omics for Personalized Medicine

The microbiota is a heterogeneous and ever-changing assemblage of microorganisms—for example, bacteria and viruses, eukaryotic microbes and archaea—that live in and on host organism (humans, animals, and plants) as well as the environment [196]. Furthermore, from a technical perspective, the constraints of traditional microbiological techniques were mostly surmounted by the swift growth of next-generation sequencing platforms, which have considerably accelerated and reduced the cost of sequencing. This, alongside continuing developments in bioinformatics for the analysis and integration of extensive datasets, has led to revolutionary progress in microbiome research and its prospective functional uses [197]. According to a previous study, amplicon sequencing techniques specifically designed to target the bacterial 16S ribosomal RNA gene have been extensively employed. An early finding of microbiome in sputum samples from 28 patients with severe asthma revealed that the presence of the *Haemophilus* and *Streptococcus* genera or *Moraxella catarrhalis* was more common in those samples when there was neutrophilic airway inflammation.

References

1. Virani, S.S., Alonso, A., Benjamin, E.J., Bittencourt, M.S., Callaway, C.W., Carson, A.P., Chamberlain, A.M., Chang, A.R., Cheng, S., Delling, F.N., Djousse, L., Heart disease and stroke statistics—2020 update: a report from the American Heart Association. *Circulation*, 141, 9, e139–596, 2020 Mar 3.
2. Heymsfield, S.B. and Wadden, T.A., Mechanisms, pathophysiology, and management of obesity. *N. Engl. J. Med.*, 376, 3, 254–66, 2017 Jan 19.
3. Corpeleijn, E., Saris, W.H., Blaak, E.E., Metabolic flexibility in the development of insulin resistance and type 2 diabetes: effects of lifestyle. *Obes. Rev.*, 10, 2, 178–93, 2009 Mar.
4. Reilly, S.M. and Saltiel, A.R., Adapting to obesity with adipose tissue inflammation. *Nat. Rev. Endocr.*, 13, 11, 633–43, 2017 Nov.
5. Gilbert, J.A., Blaser, M.J., Caporaso, J.G., Jansson, J.K., Lynch, S.V., Knight, R., Current understanding of the human microbiome. *Nat. Med.*, 24, 4, 392–400, 2018 Apr 1.
6. Sender, R., Fuchs, S., Milo, R., Are we really vastly outnumbered? Revisiting the ratio of bacterial to host cells in humans. *Cell*, 164, 3, 337–40, 2016 Jan 28.

7. Whitman, W.B., Coleman, D.C., Wiebe, W.J., Prokaryotes: the unseen majority. *Proc. Natl. Acad. Sci.*, 95, 12, 6578–83, 1998 Jun 9.
8. Jandhyala, S.M., Talukdar, R., Subramanyam, C., Vuyyuru, H., Sasikala, M., Reddy, D.N., Role of the normal gut microbiota. *World J. Gastroenterol. WJG.*, 21, 29, 8787, 2015 Aug 8.
9. Thoo, L., Noti, M., Krebs, P., Keep calm: the intestinal barrier at the interface of peace and war. *Cell Death Disease*, 10, 11, 849, 2019 Nov 7.
10. Moossavi, S., Zhang, H., Sun, J., Rezaei, N., Host–microbiota interaction and intestinal stem cells in chronic inflammation and colorectal cancer. *Expert Rev. Clin. Immunol.*, 9, 5, 409–22, 2013 May 1.
11. Kreso, A. and Dick, J.E., Evolution of the cancer stem cell model. *Cell Stem Cell*, 14, 3, 275–91, 2014 Mar 6.
12. Fidelle, M., Yonekura, S., Picard, M., Cogdill, A., Hollebecque, A., Roberti, M.P., Zitvogel, L., Resolving the paradox of colon cancer through the integration of genetics, immunology, and the microbiota. *Front. Immunol.*, 11, 600886, 2020 Dec 14.
13. Grice, E.A. and Segre, J.A., The human microbiome: our second genome. *Annu. Rev. Genomics Hum. Genet.*, 13, 1, 151–70, 2012 Sep 22.
14. Kaiko, G.E., Ryu, S.H., Koues, O.I., Collins, P.L., Solnica-Krezel, L., Pearce, E.J., Pearce, E.L., Oltz, E.M., Stappenbeck, T.S., The colonic crypt protects stem cells from microbiota-derived metabolites. *Cell*, 165, 7, 1708–20, 2016 Jun 16.
15. O'keefe, S.J., Diet, microorganisms and their metabolites, and colon cancer. *Nat. Rev. Gastroenterol. Hepatol.*, 13, 12, 691–706, 2016 Dec.
16. Saffarian, A., Mulet, C., Regnault, B., Amiot, A., Tran-Van-Nhieu, J., Ravel, J., Sobhani, I., Sansonetti, P.J., Pédron, T., Crypt-and mucosa-associated core microbiotas in humans and their alteration in colon cancer patients. *MBio.*, 10, 4, 10–128, 2019 Aug 27.
17. Lagier, J.C., Khelaifia, S., Alou, M.T., Ndongo, S., Dione, N., Hugon, P., Caputo, A., Cadoret, F., Traore, S.I., Seck, E.H., Dubourg, G., Culture of previously uncultured members of the human gut microbiota by culturomics. *Nat. Microbiol.*, 1, 12, 2016 Dec.
18. Rosser, E.C. and Mauri, C., A clinical update on the significance of the gut microbiota in systemic autoimmunity. *J. Autoimmun.*, 74, 85–93, 2016 Nov 1.
19. Rowland, I., Gibson, G., Heinken, A., Scott, K., Swann, J., Thiele, I., Tuohy, K., Gut microbiota functions: metabolism of nutrients and other food components. *Eur. J. Nutr.*, 57, 1–24, 2018 Feb.
20. Cardona, D. and Roman, P., New perspectives in health: gut microbiota. *Int. J. Environ. Res. Public Health*, 19, 10, 5828, 2022 May 10.
21. Yadav, M., Verma, M.K., Chauhan, N.S., A review of metabolic potential of human gut microbiome in human nutrition. *Arch. Microbiol.*, 200, 203–17, 2018 Mar.

22. Markandey, M., Bajaj, A., Ilott, N.E., Kedia, S., Travis, S., Powrie, F., Ahuja, V., Gut microbiota: sculptors of the intestinal stem cell niche in health and inflammatory bowel disease. *Gut Microbes*, 13, 1, 1990827, 2021 Jan 1.
23. Zakrzewski, W., Dobrzyński, M., Szymonowicz, M., Rybak, Z., Stem cells: past, present, and future. *Stem Cell Res. Ther.*, 10, 1, 1–22, 2019 Dec.
24. Thanaskody, K., Jusop, A.S., Tye, G.J., Wan Kamarul Zaman, W.S., Dass, S.A., Nordin, F., MSCs vs. iPSCs: Potential in therapeutic applications. *Front. Cell Dev. Biol.*, 10, 1005926, 2022 Nov 2.
25. Khan, I., Bai, Y., Zha, L., Ullah, N., Ullah, H., Shah, S.R., Sun, H., Zhang, C., Mechanism of the gut microbiota colonization resistance and enteric pathogen infection. *Front. Cell. Infect. Microbiol.*, 11, 716299, 2021 Dec 23.
26. Fukuda, S., Toh, H., Hase, K., Oshima, K., Nakanishi, Y., Yoshimura, K., Tobe, T., Clarke, J.M., Topping, D.L., Suzuki, T., Taylor, T.D., Bifidobacteria can protect from enteropathogenic infection through production of acetate. *Nature*, 469, 7331, 543–7, 2011 Jan 27.
27. Shin, R., Suzuki, M., Morishita, Y., Influence of intestinal anaerobes and organic acids on the growth of enterohaemorrhagic Escherichia coli O157: H7. *J. Med. Microbiol.*, 51, 3, 201–6, 2002 Mar.
28. Ducarmon, Q.R., Zwittink, R.D., Hornung, B.V., Van Schaik, W., Young, V.B., Kuijper, E.J., Gut microbiota and colonization resistance against bacterial enteric infection. *Microbiol. Mol. Biol. Rev.*, 83, 3, 10–128, 2019 Aug 21.
29. Fabich, A.J., Jones, S.A., Chowdhury, F.Z., Cernosek, A., Anderson, A., Smalley, D., McHargue, J.W., Hightower, G.A., Smith, J.T., Autieri, S.M., Leatham, M.P., Comparison of carbon nutrition for pathogenic and commensal Escherichia coli strains in the mouse intestine. *Infect. Immun.*, 76, 3, 1143–52, 2008 Mar.
30. Kim, S., Covington, A., Pamer, E.G., The intestinal microbiota: antibiotics, colonization resistance, and enteric pathogens. *Immunol. Rev.*, 279, 1, 90–105, 2017 Sep.
31. Ahmad Sophien, A.N., Jusop, A.S., Tye, G.J., Tan, Y.F., Wan Kamarul Zaman, W.S., Nordin, F., Intestinal stem cells and gut microbiota therapeutics: hype or hope? *Front. Med.*, 10, 1195374, 2023 Jul 21.
32. Jayawardena, R., Sooriyaarachchi, P., Chourdakis, M., Jeewandara, C., Ranasinghe, P., Enhancing immunity in viral infections, with special emphasis on COVID-19: A review. *Diabetes Metab. Syndrome Clin. Res. Rev.*, 14, 4, 367–82, 2020 Jul 1.
33. Wyszyńska, A., Kobierecka, P., Bardowski, J., Jagusztyn-Krynicka, E.K., Lactic acid bacteria—20 years exploring their potential as live vectors for mucosal vaccination. *Appl. Microbiol. Biotechnol.*, 99, 2967–77, 2015 Apr.
34. Kanauchi, O., Andoh, A., AbuBakar, S., Yamamoto, N., Probiotics and para probiotics in viral infection: clinical application and effects on the innate and acquired immune systems. *Curr. Pharm. Des.*, 24, 6, 710–7, 2018 Feb 1.

35. Pedrolli, D.B., Ribeiro, N.V., Squizato, P.N., Cozetto, D.A., De Jesus, V.N., Freire, P.J., Lins, M.R., Correa, G.G., Engineering microbial living therapeutics, in: *Precision Medicine for Investigators, Practitioners and Providers*, pp. 71–82, Academic Press, 2020 Jan 1.
36. Guardiola, F.A., Bahi, A., Jiménez-Monreal, A.M., Martínez-Tomé, M., Murcia, M.A., Esteban, M.A., Dietary administration effects of fenugreek seeds on skin mucosal antioxidant and immunity status of gilthead seabream (Sparusaurata L.). *Fish Shellfish Immunol.*, 75, 357–64, 2018 Apr 1.
37. Berggren, A., LazouAhrén, I., Larsson, N., Önning, G., Randomised, double-blind and placebo-controlled study using new probiotic lactobacilli for strengthening the body immune defence against viral infections. *Eur. J. Nutr.*, 50, 203–10, 2011 Apr.
38. Tiwari, S.K., Dicks, L.M., Popov, I.V., Karaseva, A., Ermakov, A.M., Suvorov, A., Tagg, J.R., Weeks, R., Chikindas, M.L., Probiotics at war against viruses: what is missing from the picture? *Front. Microbiol.*, 11, 1877, 2020 Aug 20.
39. Claesen, J. and Fischbach, M.A., Synthetic microbes as drug delivery systems. *ACS Synth. Biol.*, 4, 4, 358–64, 2015 Apr 17.
40. Ahlawat, S. and Sharma, K.K., Immunological co-ordination between gut and lungs in SARS-CoV-2 infection. *Virus Res.*, 286, 198103, 2020 Sep 1.
41. Fu, Y., Cheng, Y., Wu, Y., Understanding SARS-CoV-2-mediated inflammatory responses: from mechanisms to potential therapeutic tools. *VirologicaSinica*, 35, 3, 266–71, 2020 Jun.
42. Sender, R., Fuchs, S., Milo, R., Revised estimates for the number of human and bacteria cells in the body. *PLoS Biol.*, 14, 8, e1002533, 2016 Aug 19.
43. Iliev, I.D. and Leonardi, I., Fungal dysbiosis: immunity and interactions at mucosal barriers. *Nat. Rev. Immunol.*, 17, 10, 635–46, 2017 Oct.
44. Raimondi, S., Amaretti, A., Gozzoli, C., Simone, M., Righini, L., Candeliere, F., Brun, P., Ardizzoni, A., Colombari, B., Paulone, S., Castagliuolo, I., Longitudinal survey of fungi in the human gut: ITS profiling, phenotyping, and colonization. *Front. Microbiol.*, 10, 1575, 2019 Jul 10.
45. Bacher, P., Hohnstein, T., Beerbaum, E., Röcker, M., Blango, M.G., Kaufmann, S., Röhmel, J., Eschenhagen, P., Grehn, C., Seidel, K., Rickerts, V., Human anti-fungal Th17 immunity and pathology rely on cross-reactivity against Candida albicans. *Cell*, 176, 6, 1340–55, 2019 Mar 7.
46. Doron, I., Leonardi, I., Li, X.V., Fiers, W.D., Semon, A., Bialt-DeCelie, M., Migaud, M., Gao, I.H., Lin, W.Y., Kusakabe, T., Puel, A., Human gut mycobiota tune immunity *via* CARD9-dependent induction of anti-fungal IgG antibodies. *Cell*, 184, 4, 1017–31, 2021 Feb 18.
47. van Tilburg Bernardes, E., Pettersen, V.K., Gutierrez, M.W., Laforest-Lapointe, I., Jendzjowsky, N.G., Cavin, J.B., Vicentini, F.A., Keenan, C.M., Ramay, H.R., Samara, J., MacNaughton, W.K., Intestinal fungi are causally implicated in microbiome assembly and immune development in mice. *Nat. Commun.*, 11, 1, 2577, 2020 May 22.

48. Sancho, D. and Reis e Sousa, C., Signaling by myeloid C-type lectin receptors in immunity and homeostasis. *Annu. Rev. Immunol.*, 30, 1, 491–529, 2012 Apr 23.
49. Plato, A., Hardison, S.E., Brown, G.D., Pattern recognition receptors in antifungal immunity, in: *Seminars in immunopathology*, vol. 37, pp. 97–106, Springer Berlin Heidelberg, 2015 Mar.
50. Cohen-Kedar, S., Baram, L., Elad, H., Brazowski, E., Guzner-Gur, H., Dotan, I., Human intestinal epithelial cells respond to β-glucans *via* Dectin-1 and Syk. *Eur. J. Immunol.*, 44, 12, 3729–40, 2014 Dec.
51. Wang, T., Pan, D., Zhou, Z., You, Y., Jiang, C., Zhao, X., Lin, X., Dectin-3 deficiency promotes colitis development due to impaired antifungal innate immune responses in the gut. *PLoS Pathog.*, 12, 6, e1005662, 2016 Jun 9.
52. Leonardi, I., Li, X., Semon, A., Li, D., Doron, I., Putzel, G., Bar, A., Prieto, D., Rescigno, M., McGovern, D.P., Pla, J., CX3CR1+ mononuclear phagocytes control immunity to intestinal fungi. *Science*, 359, 6372, 232–6, 2018 Jan 12.
53. Arrieta, M.C., Arévalo, A., Stiemsma, L., Dimitriu, P., Chico, M.E., Loor, S., Vaca, M., Boutin, R.C., Morien, E., Jin, M., Turvey, S.E., Associations between infant fungal and bacterial dysbiosis and childhood atopic wheeze in a nonindustrialized setting. *J. Allergy Clin. Immunol.*, 142, 2, 424–34, 2018 Aug 1.
54. Skalski, J.H., Limon, J.J., Sharma, P., Gargus, M.D., Nguyen, C., Tang, J., Coelho, A.L., Hogaboam, C.M., Crother, T.R., Underhill, D.M., Expansion of commensal fungus Wallemiamellicola in the gastrointestinal mycobiota enhances the severity of allergic airway disease in mice. *PLoS Pathog.*, 14, 9, e1007260, 2018 Sep 20.
55. Boutin, R.C., Petersen, C., Woodward, S.E., Serapio-Palacios, A., Bozorgmehr, T., Loo, R., Chalanuchpong, A., Cirstea, M., Lo, B., Huus, K.E., Barcik, W., Bacterial–fungal interactions in the neonatal gut influence asthma outcomes later in life. *Elife*, 10, e67740, 2021 Apr 20.
56. Servais, S., Beguin, Y., Baron, F., Emerging drugs for prevention of graft failure after allogeneic hematopoietic stem cell transplantation. *Expert Opin. Emerg. Drugs*, 18, 2, 173–92, 2013 Jun 1.
57. Locatelli, F., Lucarelli, B., Merli, P., Current and future approaches to treat graft failure after allogeneic hematopoietic stem cell transplantation. *Expert Opin. Pharmacother.*, 15, 1, 23–36, 2014 Jan 1.
58. Cluzeau, T., Lambert, J., Raus, N., Dessaux, K., Absi, L., Delbos, F., Devys, A., De Matteis, M., Dubois, V., Filloux, M., Fort, M., Risk factors and outcome of graft failure after HLA matched and mismatched unrelated donor hematopoietic stem cell transplantation: a study on behalf of SFGM-TC and SFHI. *Bone Marrow Transplant*, 51, 5, 687–91, 2016 May.
59. Fürst, D., Müller, C., Vucinic, V., Bunjes, D., Herr, W., Gramatzki, M., Schwerdtfeger, R., Arnold, R., Einsele, H., Wulf, G., Pfreundschuh, M., High-resolution HLA matching in hematopoietic stem cell transplantation:

a retrospective collaborative analysis. *Blood J. Am. Soc. Hematology*, 122, 18, 3220–9, 2013 Oct 31.
60. Hatzimichael, E. and Tuthill, M., Hematopoietic stem cell transplantation. *Stem Cells Cloning Adv. Applications*, 26, 105–17, 2010 Aug.
61. Caocci, G., Greco, M., La Nasa, G., Bone marrow homing and engraftment defects of human hematopoietic stem and progenitor cells. *Mediterr. J. Hematol. Infect. Dis.*, 9, 1, 2017.
62. Hassan, M.N., Fauzi, H.M., Husin, A., Mustaffa, R., Hassan, R., Ibrahim, M.I., Noor, N.H., Autologous peripheral blood stem cell transplantation among lymphoproliferative disease patients: factors influencing engraftment. *Oman Med. J.*, 34, 1, 34, 2019 Jan.
63. Fürst, D., Neuchel, C., Tsamadou, C., Schrezenmeier, H., Mytilineos, J., HLA matching in unrelated stem cell transplantation up to date. *Transfus. Med. Hemotherapy*, 46, 5, 326–36, 2019 Sep 3.
64. Bertaina, A. and Andreani, M., Major histocompatibility complex and hematopoietic stem cell transplantation: Beyond the classical HLA polymorphism. *Int. J. Mol. Sci.*, 19, 2, 621, 2018 Feb 22.
65. Park, M. and Seo, J.J., Role of HLA in hematopoietic stem cell transplantation. *Bone Marrow Res.*, 2012, 1, 680841, 2012.
66. Stussi, G., Muntwyler, J., Passweg, J.R., Seebach, L., Schanz, U., Gmür, J., Gratwohl, A., Seebach, J.D., Consequences of ABO incompatibility in allogeneic hematopoietic stem cell transplantation. *Bone Marrow Transplant*, 30, 2, 87–93, 2002 Jul.
67. Kekre, N. and Antin, J.H., Hematopoietic stem cell transplantation donor sources in the 21st century: choosing the ideal donor when a perfect match does not exist. *Blood J. Am. Soc. Hematol.*, 124, 3, 334–43, 2014 Jul 17.
68. Shouval, R., Fein, J.A., Labopin, M., Kröger, N., Duarte, R.F., Bader, P., Chabannon, C., Kuball, J., Basak, G.W., Dufour, C., Galimard, J.E., Outcomes of allogeneic haematopoietic stem cell transplantation from HLA-matched and alternative donors: a European Society for Blood and Marrow Transplantation registry retrospective analysis. *Lancet Haematol.*, 6, 11, e573–84, 2019 Nov 1.
69. Ueda, M., Body composition assessment in hematopoietic stem cell transplantation. *Revistabrasileira de Hematologia e Hemoterapia*, 39, 4, 297–8, 2017.
70. Hirose, E.Y., de Molla, V.C., Gonçalves, M.V., Pereira, A.D., Szor, R.S., da Fonseca, A.R., Fatobene, G., Serpa, M.G., Xavier, E.M., Tucunduva, L., Rocha, V., The impact of pretransplant malnutrition on allogeneic hematopoietic stem cell transplantation outcomes. *Clin. Nutr. ESPEN*, 33, 213–9, 2019 Oct 1.
71. Cominal, J.G., da Costa Cacemiro, M., Pinto-Simões, B., Kolb, H.J., Malmegrim, K.C., Castro, F.A., Emerging role of mesenchymal stromal cell-derived extracellular vesicles in pathogenesis of haematological malignancies. *Stem Cells Int.*, 2019, 1, 6854080, 2019.

72. Crippa, S. and Bernardo, M.E., Mesenchymal stromal cells: role in the BM niche and in the support of hematopoietic stem cell transplantation. *Hemasphere*, 2, 6, e151, 2018 Dec 1.
73. Qiu, G., Zheng, G., Ge, M., Wang, J., Huang, R., Shu, Q., Xu, J., Mesenchymal stem cell-derived extracellular vesicles affect disease outcomes *via* transfer of microRNAs. *Stem Cell Res. Ther.*, 9, 1–9, 2018 Dec.
74. Laurenzana, I., Lamorte, D., Trino, S., De Luca, L., Ambrosino, C., Zoppoli, P., Ruggieri, V., Del Vecchio, L., Musto, P., Caivano, A., Falco, G., Extracellular vesicles: a new prospective in crosstalk between microenvironment and stem cells in hematological malignancies. *Stem Cells Int.*, 2018, 1, 9863194, 2018.
75. De Luca, L., Trino, S., Laurenzana, I., Lamorte, D., Caivano, A., Del Vecchio, L., Musto, P., Mesenchymal stem cell derived extracellular vesicles: a role in hematopoietic transplantation? *Int. J. Mol. Sci.*, 18, 5, 1022, 2017 May 9.
76. Grochowska, M., Laskus, T., Radkowski, M., Gut microbiota in neurological disorders. *Archivumimmunologiae Therapiaeexperimentalis.*, 67, 375–83, 2019 Dec.
77. Morais, L.H., Schreiber IV, H.L., Mazmanian, S.K., The gut microbiota–brain axis in behaviour and brain disorders. *Nat. Rev. Microbiol.*, 19, 4, 241–55, 2021 Apr.
78. Collins, S.M., Surette, M., Bercik, P., The interplay between the intestinal microbiota and the brain. *Nat. Rev. Microbiol.*, 10, 11, 735–42, 2012 Nov.
79. Deidda, G. and Biazzo, M., Gut and brain: investigating physiological and pathological interactions between microbiota and brain to gain new therapeutic avenues for brain diseases. *Front. Neurosci.*, 15, 753915, 2021 Oct 12.
80. Ma, Q., Xing, C., Long, W., Wang, H.Y., Liu, Q., Wang, R.F., Impact of microbiota on central nervous system and neurological diseases: the gut-brain axis. *J. Neuroinflammation*, 16, 1–4, 2019 Dec.
81. Martin, C.R., Osadchiy, V., Kalani, A., Mayer, E.A., The brain-gut-microbiome axis. *Cell. Mol. Gastroenterol. Hepatol.*, 6, 2, 133–48, 2018 Jan 1.
82. Flowers, S.A. and Ellingrod, V.L., The microbiome in mental health: potential contribution of gut microbiota in disease and pharmacotherapy management. *Pharmacotherapy J. Hum. Pharmacol. Drug Ther.*, 35, 10, 910–6, 2015 Oct.
83. Askarova, S., Umbayev, B., Masoud, A.R., Kaiyrlykyzy, A., Safarova, Y., Tsoy, A., Olzhayev, F., Kushugulova, A., The links between the gut microbiome, aging, modern lifestyle and Alzheimer's disease. *Front. Cell. Infect. Microbiol.*, 10, 104, 2020 Mar 18.
84. Blaser, M.J., The theory of disappearing microbiota and the epidemics of chronic diseases. *Nat. Rev. Immunol.*, 17, 8, 461–3, 2017 Aug.
85. Perry, R.J., Peng, L., Barry, N.A., Cline, G.W., Zhang, D., Cardone, R.L., Petersen, K.F., Kibbey, R.G., Goodman, A.L., Shulman, G.I., Acetate mediates a microbiome–brain–β-cell axis to promote metabolic syndrome. *Nature*, 534, 7606, 213–7, 2016 Jun 9.

86. Cryan, J.F., O'Riordan, K.J., Sandhu, K., Peterson, V., Dinan, T.G., The gut microbiome in neurological disorders. *Lancet Neurol.*, 19, 2, 179–94, 2020 Feb 1.
87. Fond, G., Boukouaci, W., Chevalier, G., Regnault, A., Eberl, G., Hamdani, N., Dickerson, F., Macgregor, A., Boyer, L., Dargel, A., Oliveira, J., The "psychomicrobiotic": Targeting microbiota in major psychiatric disorders: A systematic review. *PathologieBiologie*, 63, 1, 35–42, 2015 Feb 1.
88. Kennedy, P.J., Cryan, J.F., Dinan, T.G., Clarke, G., Kynurenine pathway metabolism and the microbiota-gut-brain axis. *Neuropharmacology*, 112, 399–412, 2017 Jan 1.
89. Knight, R., Vrbanac, A., Taylor, B.C., Aksenov, A., Callewaert, C., Debelius, J., Gonzalez, A., Kosciolek, T., McCall, L.I., McDonald, D., Melnik, A.V., Best practices for analysing microbiomes. *Nat. Rev. Microbiol.*, 16, 7, 410–22, 2018 Jul.
90. Yoo, B.B. and Mazmanian, S.K., The enteric network: interactions between the immune and nervous systems of the gut. *Immunity*, 46, 6, 910–26, 2017 Jun 20.
91. Al Omran, Y. and Aziz, Q., The brain-gut axis in health and disease. *Microb. Endocrinology Microbiota-gut-brain Axis Health Disease*, 9, 135–53, 2014 Jun.
92. Tian, Y., Ullah, H., Gu, J., Li, K., Immune-metabolic mechanisms of post-traumatic stress disorder and atherosclerosis. *Front. Physiol.*, 14, 1123692, 2023 Feb 8.
93. Claesson, M.J., Jeffery, I.B., Conde, S., Power, S.E., O'connor, E.M., Cusack, S., Harris, H., Coakley, M., Lakshminarayanan, B., O'sullivan, O., Fitzgerald, G.F., Gut microbiota composition correlates with diet and health in the elderly. *Nature*, 488, 7410, 178–84, 2012 Aug.
94. Tiwari, P., Dwivedi, R., Bansal, M., Tripathi, M., Dada, R., Role of gut microbiota in neurological disorders and its therapeutic significance. *J. Clin. Medicine*, 12, 4, 1650, 2023 Feb 19.
95. Azzam, I., Gilad, S., Limor, R., Stern, N., Greenman, Y., Ghrelin stimulation by hypothalamic–pituitary–adrenal axis activation depends on increasing cortisol levels. *Endocr. Connect.*, 6, 8, 847–55, 2017 Nov 1.
96. Kelley, N., Jeltema, D., Duan, Y., He, Y., The NLRP3 inflammasome: an overview of mechanisms of activation and regulation. *Int. J. Mol. Sci.*, 20, 13, 3328, 2019 Jul 6.
97. Holmqvist, S., Chutna, O., Bousset, L., Aldrin-Kirk, P., Li, W., Björklund, T., Wang, Z.Y., Roybon, L., Melki, R., Li, J.Y., Direct evidence of Parkinson pathology spread from the gastrointestinal tract to the brain in rats. *Actaneuropathologica*, 128, 805–20, 2014 Dec.
98. Scheperjans, F., Aho, V., Pereira, P.A., Koskinen, K., Paulin, L., Pekkonen, E., Haapaniemi, E., Kaakkola, S., Eerola-Rautio, J., Pohja, M., Kinnunen, E., Gut

microbiota are related to Parkinson's disease and clinical phenotype. *Mov. Disord.*, 30, 3, 350–8, 2015 Mar.
99. Chandra, S., Sisodia, S.S., Vassar, R.J., The gut microbiome in Alzheimer's disease: what we know and what remains to be explored. *Mol. Neurodegener.*, 18, 1, 9, 2023 Feb 1.
100. Balan, Y., Gaur, A., Sakthivadivel, V., Kamble, B., Sundaramurthy, R., Is the gut microbiota a neglected aspect of gut and brain disorders? *Cureus*, 13, 11, 2021 Nov.
101. Jin, J., Xu, Z., Zhang, L., Zhang, C., Zhao, X., Mao, Y., Zhang, H., Liang, X., Wu, J., Yang, Y., Zhang, J., Gut-derived β-amyloid: Likely a centerpiece of the gut–brain axis contributing to Alzheimer's pathogenesis. *Gut Microbes*, 15, 1, 2167172, 2023 Dec 31.
102. Naomi, R., Embong, H., Othman, F., Ghazi, H.F., Maruthey, N., Bahari, H., Probiotics for Alzheimer's disease: a systematic review. *Nutrients*, 14, 1, 20, 2021 Dec 22.
103. Burokas, A., Arboleya, S., Moloney, R.D., Peterson, V.L., Murphy, K., Clarke, G., Stanton, C., Dinan, T.G., Cryan, J.F., Targeting the microbiota-gut-brain axis: prebiotics have anxiolytic and antidepressant-like effects and reverse the impact of chronic stress in mice. *Biol. Psychiatry*, 82, 7, 472–87, 2017 Oct 1.
104. Zhu, S., Jiang, Y., Xu, K., Cui, M., Ye, W., Zhao, G., Jin, L., Chen, X., The progress of gut microbiome research related to brain disorders. *J. Neuroinflammation*, 17, 1–20, 2020 Dec.
105. Nandwana, V., Nandwana, N.K., Das, Y., Saito, M., Panda, T., Das, S., Almaguel, F., Hosmane, N.S., Das, B.C., The role of microbiome in brain development and neurodegenerative diseases. *Molecules*, 27, 11, 3402, 2022 May 25.
106. Kandpal, M., Indari, O., Baral, B., Jakhmola, S., Tiwari, D., Bhandari, V., Pandey, R.K., Bala, K., Sonawane, A., Jha, H.C., Dysbiosis of Gut Microbiota from the Perspective of the Gut–Brain Axis: Role in the Provocation of Neurological Disorders. *Metabolites*, 12, 11, 1064, 2022 Nov 3.
107. Kaplan, G.G. and Windsor, J.W., The four epidemiological stages in the global evolution of inflammatory bowel disease. *Nat. Rev. Gastroenterol. Hepatol.*, 18, 1, 56–66, 2021 Jan.
108. Coward, S., Clement, F., Benchimol, E.I., Bernstein, C.N., Avina-Zubieta, J.A., Bitton, A., Carroll, M.W., Hazlewood, G., Jacobson, K., Jelinski, S., Deardon, R., Past and future burden of inflammatory bowel diseases based on modeling of population-based data. *Gastroenterology*, 156, 5, 1345–53, 2019 Apr 1.
109. Magro, F., Rahier, J.F., Abreu, C., MacMahon, E., Hart, A., van der Woude, C.J., Gordon, H., Adamina, M., Viget, N., Vavricka, S., Kucharzik, T., Inflammatory bowel disease management during the COVID-19 outbreak:

the ten do's and don'ts from the ECCO-COVID taskforce. *J. Crohn's Colitis*, 14, Supplement_3, S798–806, 2020 Oct.

110. Kaplan, G.G., Windsor, J.W., Crain, J., Barrett, L., Bernstein, C.N., Bitton, A., Chauhan, U., Coward, S., Fowler, S., Ghia, J.E., Gibson, D.L., Crohn's and Colitis Canada's 2021 Impact of COVID-19 & inflammatory bowel disease in canada: a knowledge translation strategy. *J. Can. Assoc. Gastroenterol.*, 4, Supplement_2, S10–9, 2021 Dec 1.
111. Windsor, J.W., Underwood, F.E., Brenner, E., Colombel, J.F., Kappelman, M.D., Ungaro, R., Zhang, X., Kaplan, G.G., Data visualization in the era of COVID-19: an interactive map of the SECURE-IBD registry. *Off. J. Am. Coll. Gastroenterol.| ACG*, 115, 11, 1923–4, 2020 Nov 1.
112. Kaplan, G.G., Underwood, F.E., Coward, S., Agrawal, M., Ungaro, R.C., Brenner, E.J., Gearry, R.B., Kissous-Hunt, M., Lewis, J.D., Ng, S.C., Rahier, J.F., The multiple waves of COVID-19 in patients with inflammatory bowel disease: a temporal trend analysis. *Inflamm. Bowel Dis.*, 28, 11, 1687–95, 2022 Nov 1.
113. Singh, A.K., Jena, A., Kumar, M.P., Sharma, V., Sebastian, S., Risk and outcomes of coronavirus disease in patients with inflammatory bowel disease: a systematic review and meta-analysis. *U. Eur. Gastroenterol. J.*, 9, 2, 159–76, 2021 Mar.
114. Ludvigsson, J.F., Axelrad, J., Halfvarson, J., Khalili, H., Larsson, E., Lochhead, P., Roelstraete, B., Simon, T.G., Söderling, J., Olen, O., Inflammatory bowel disease and risk of severe COVID-19: a nationwide population-based cohort study in Sweden. *U. Eur. Gastroenterol. J.*, 9, 2, 177–92, 2021 Mar.
115. Ungaro, R.C., Brenner, E.J., Agrawal, M., Zhang, X., Kappelman, M.D., Colombel, J.F., Gearry, R.B., Kaplan, G.G., Kissous-Hunt, M., Lewis, J.D., Ng, S.C., Impact of medications on COVID-19 outcomes in inflammatory bowel disease: analysis of more than 6000 patients from an international registry. *Gastroenterology*, 162, 1, 316–9, 2022 Jan 1.
116. Ricciuto, A., Lamb, C.A., Benchimol, E.I., Walker, G.J., Kennedy, N.A., Kuenzig, M.E., Kaplan, G.G., Kappelman, M.D., Ungaro, R.C., Colombel, J.F., Brenner, E.J., Inflammatory bowel disease clinical activity is associated with COVID-19 severity especially in younger patients. *J. Crohn's Colitis*, 16, 4, 591–600, 2022 Apr 1.
117. Remzi, F.H., Panis, Y., Spinelli, A., Kotze, P.G., Mantzaris, G., Söderholm, J.D., d'Hoore, A., Bemelman, W.A., Yamamoto, T., Pemberton, J.H., Tiret, E., International organization for the study of IBD recommendations for surgery in patients with IBD during the coronavirus disease 2019 pandemic. *Dis. Colon Rectum.*, 63, 7, 870–3, 2020 Jul 1.
118. Hashash, J.G., Desai, A., Kochhar, G.S., Farraye, F.A., Efficacy of paxlovid and lagevrio for COVID-19 infection in patients with inflammatory bowel disease: a propensity-matched study. *Clin. Gastroenterol. Hepatol.*, 21, 3, 841, 2023 Mar.

119. Kennedy, N.A., Lin, S., Goodhand, J.R., Chanchlani, N., Hamilton, B., Bewshea, C., Nice, R., Chee, D., Cummings, J.F., Fraser, A., Irving, P.M., Infliximab is associated with attenuated immunogenicity to BNT162b2 and ChAdOx1 nCoV-19 SARS-CoV-2 vaccines in patients with IBD. *Gut*, 70, 10, 1884–93, 2021 Oct 1.
120. Alexander, J.L., Kennedy, N.A., Ibraheim, H., Anandabaskaran, S., Saifuddin, A., Seoane, R.C., Liu, Z., Nice, R., Bewshea, C., D'Mello, A., Constable, L., COVID-19 vaccine-induced antibody responses in immunosuppressed patients with inflammatory bowel disease (VIP): a multicentre, prospective, case-control study. *Lancet Gastroenterol. Hepatol.*, 7, 4, 342–52, 2022 Apr 1.
121. Bronsky, J., Copova, I., Durilova, M., Kazeka, D., Kubat, M., Lerchova, T., Vlckova, E., Mitrova, K., Rataj, M., Klocperk, A., Sediva, A., Postvaccination immunogenicity of BNT162b2 SARS-CoV-2 vaccine and its predictors in pediatric inflammatory bowel disease. *J. Pediatr. Gastroenterol. Nutrition*, 76, 2, e36–44, 2023 Feb 1.
122. Weaver, K.N., Zhang, X., Dai, X., Chen, W., Watkins, R., Adler, J., Dubinsky, M.C., Kastl, A., Bousvaros, A., Strople, J.A., Cross, R.K., Low rates of breakthrough COVID-19 infection after SARS-CoV-2 vaccination in patients with inflammatory bowel disease. *Inflamm. Bowel Dis.*, 29, 3, 483–6, 2023 Mar 1.
123. Li, D., Debbas, P., Mujukian, A., Cheng, S., Braun, J., McGovern, D.P., Melmed, G.Y., Postvaccination symptoms after a third dose of mRNA SARS-CoV-2 vaccination in patients with inflammatory bowel disease: results from CORALE-IBD. *Inflamm. Bowel Dis.*, 29, 6, 883–7, 2023 Jun 1.
124. Zabana, Y., Marin-Jimenez, I., Rodriguez-Lago, I., Vera, I., Martin-Arranz, M.D., Guerra, I., Gisbert, J.P., Mesonero, F., Benitez, O., Taxonera, C., Ponferrada-Diaz, A., Nationwide COVID-19-EII study: incidence, environmental risk factors and long-term follow-up of patients with inflammatory bowel disease and COVID-19 of the ENEIDA registry. *J. Clin. Med.*, 11, 2, 421, 2022 Jan 14.
125. McNaughton, C.D., Austin, P.C., Sivaswamy, A., Fang, J., Abdel-Qadir, H., Daneman, N., Udell, J.A., Wodchis, W.P., Mostarac, I., Lee, D.S., Atzema, C.L., Post-acute health care burden after SARS-CoV-2 infection: a retrospective cohort study. *Cmaj.*, 194, 40, E1368–76, 2022 Oct 17.
126. Meringer, H. and Mehandru, S., Gastrointestinal post-acute COVID-19 syndrome. *Nat. Rev. Gastroenterol. Hepatol.*, 19, 6, 345–6, 2022 Jun.
127. Goodday, S.M., Travis, S., Walsh, A., Friend, S.H., Stress-related consequences of the coronavirus disease 2019 pandemic on symptoms of Crohn's disease. *Eur. J. Gastroenterol. Hepatol.*, 33, 12, 1511–6, 2021 Dec 1.
128. De Bock, E., Filipe, M.D., Meij, V., Oldenburg, B., Van Schaik, F.D., Bastian, O.W., Fidder, H.F., Vriens, M.R., Richir, M.C., Quality of life in patients with IBD during the COVID-19 pandemic in the Netherlands. *BMJ Open Gastroenterol.*, 8, 1, e000670, 2021 Jul 1.

129. Fedele, F., Martinelli, M., Strisciuglio, C., Dolce, P., Giugliano, F.P., Scarpato, E., Staiano, A., Miele, E., Health-Related Quality of Life in Pediatric Inflammatory Bowel Disease During COVID-19 Pandemic: A Prospective Study. *J. Pediatr. Gastroenterol. Nutr.*, 75, 5, 595–600, 2022 Nov 1.
130. Younossi, Z.M., Stepanova, M., Younossi, Y., Golabi, P., Mishra, A., Rafiq, N., Henry, L., Epidemiology of chronic liver diseases in the USA in the past three decades. *Gut*, 69, 3, 564–8, 2020 Mar 1.
131. Sarin, S.K., Kumar, M., Eslam, M., George, J., Al Mahtab, M., Akbar, S.M., Jia, J., Tian, Q., Aggarwal, R., Muljono, D.H., Omata, M., Liver diseases in the Asia-Pacific region: a lancet gastroenterology & hepatology commission. *Lancet Gastroenterol. Hepatol.*, 5, 2, 167–228, 2020 Feb 1.
132. Rinella, M.E., *Nonalcoholic fatty liver disease: a systematic review*, Jama, 2015.
133. Yang, X., Guo, Y., Chen, C., Shao, B., Zhao, L., Zhou, Q., Liu, J., Wang, G., Yuan, W., Sun, Z., Interaction between intestinal microbiota and tumour immunity in the tumour microenvironment. *Immunology*, 164, 3, 476–93, 2021 Nov.
134. Chopyk, D.M. and Grakoui, A., Contribution of the intestinal microbiome and gut barrier to hepatic disorders. *Gastroenterology*, 159, 3, 849–63, 2020 Sep 1.
135. Birchenough, G.M. and Johansson, M.E., Forming a mucus barrier along the colon. *Science.*, 370, 6515, 402–3, 2020 Oct 23.
136. Ren, Z., Li, A., Jiang, J., Zhou, L., Yu, Z., Lu, H., Xie, H., Chen, X., Shao, L., Zhang, R., Xu, S., Gut microbiome analysis as a tool towards targeted non-invasive biomarkers for early hepatocellular carcinoma. *Gut*, 68, 6, 1014–23, 2019 Jun 1.
137. Adolph, T.E., Grander, C., Moschen, A.R., Tilg, H., Liver–microbiome axis in health and disease. *Trends Immunol.*, 39, 9, 712–23, 2018 Sep 1.
138. Long-Smith, C., O'Riordan, K.J., Clarke, G., Stanton, C., Dinan, T.G., Cryan, J.F., Microbiota-gut-brain axis: new therapeutic opportunities. *Annu. Rev. Pharmacol. Toxicol.*, 60, 1, 477–502, 2020 Jan 6.
139. Nagu, P., Parashar, A., Behl, T., Mehta, V., Gut microbiota composition and epigenetic molecular changes connected to the pathogenesis of Alzheimer's disease. *J. Mol. Neurosci.*, 71, 7, 1436–55, 2021 Jul.
140. Badia, P., Andersen, H., Haslam, D., Nelson, A.S., Pate, A.R., Golkari, S., Teusink-Cross, A., Flesch, L., Bedel, A., Hickey, V., Kramer, K., Improving oral health and modulating the oral microbiome to reduce bloodstream infections from oral organisms in pediatric and young adult hematopoietic stem cell transplantation recipients: a randomized controlled trial. *Biol. Blood Marrow Transplant, 26*, 9, 1704–1710, 2020.
141. Chen, X., Lu, Y., Chen, T., Li, R., The female vaginal microbiome in health and bacterial vaginosis. *Front. Cell. Infect. Microbiol.*, 11, 631972, 2021 Apr 7.

142. Matysiak, A., Kabza, M., Karolak, J.A., Jaworska, M.M., Rydzanicz, M., Ploski, R., Szaflik, J.P., Gajecka, M., Characterization of ocular surface microbial profiles revealed discrepancies between conjunctival and corneal microbiota. *Pathogens*, 10, 4, 405, 2021 Mar 30.
143. Robinson, S., Peterson, C.B., Sahasrabhojane, P., Ajami, N.J., Shelburne, S.A., Kontoyiannis, D.P., Galloway-Peña, J.R., Observational cohort study of oral mycobiome and interkingdom interactions over the course of induction therapy for leukemia. *MSphere*, 5, 2, 10–128, 2020 Apr 29.
144. Zheng, D., Liwinski, T., Elinav, E., Interaction between microbiota and immunity in health and disease. *Cell Res.*, 30, 6, 492–506, 2020 Jun.
145. Al-Anazi, K.A., Hematopoietic stem cell transplantation in multiple myeloma in the era of novel therapies, in: *Update on Multiple Myeloma*, IntechOpen, 2018 Nov 5.
146. Mosaad, Y.M., Immunology of hematopoietic stem cell transplant. *Immunol. Invest.*, 43, 8, 858–87, 2014 Nov 1.
147. El Fakih, R., Greinix, H., Koh, M., Shaw, B., Mohty, M., Al Nahedh, M., Saber, W., Kharfan-Dabaja, M.A., Perales, M.A., Savani, B.N., Majhail, N.S., Worldwide Network for Blood and Marrow Transplantation (WBMT) recommendations regarding essential medications required to establish an early stage hematopoietic cell transplantation program. *Transplant. Cell. Ther.*, 27, 3, 267–e1, 2021 Mar 1.
148. Sepulveda, M., Pirozzolo, I., Alegre, M.L., Impact of the microbiota on solid organ transplant rejection. *Curr. Opin. Organ Transplant.*, 24, 6, 679–86, 2019 Dec 1.
149. Lai, X., Zheng, X., Mathew, J.M., Gallon, L., Leventhal, J.R., Zhang, Z.J., Tackling chronic kidney transplant rejection: challenges and promises. *Front. Immunol.*, 12, 661643, 2021 May 20.
150. Ahmad, S. and Bromberg, J.S., Current status of the microbiome in renal transplantation. *Curr. Opin. Nephrol. Hypertension*, 25, 6, 570–6, 2016 Nov 1.
151. Salvadori, M., The Microbiota and Kidney Transplantation: Influence on the Graft. *Urology.*, 9, 95–105, 2021 Aug.
152. Ramirez, J., Guarner, F., Bustos Fernandez, L., Maruy, A., Sdepanian, V.L., Cohen, H., Antibiotics as major disruptors of gut microbiota. *Front. Cell. Infect. Microbiol.*, 10, 572912, 2020 Nov 24.
153. Dickson, R.P., Erb-Downward, J.R., Falkowski, N.R., Hunter, E.M., Ashley, S.L., Huffnagle, G.B., The lung microbiota of healthy mice are highly variable, cluster by environment, and reflect variation in baseline lung innate immunity. *Am. J. Respiratory Crit. Care Med.*, 198, 4, 497–508, 2018 Aug 15.
154. Meng, C., Bai, C., Brown, T.D., Hood, L.E., Tian, Q., Human Gut Microbiota and Gastrointestinal Cancer. *Genom. Proteom. Bioinf.*, 16, 1, 33–49, 2018 Feb 1.
155. Grivennikov, S.I., Greten, F.R., Karin, M., Immunity, Inflammation, and Cancer. *Cell*, 140, 6, 883–99, 2010 Mar.

156. Spill, F., Reynolds, D.S., Kamm, R.D., Zaman, M.H., Impact of the physical microenvironment on tumor progression and metastasis. *CurrOpinBiotechnol.*, 40, 41–8, 2016 Aug.
157. Balkwill, F.R., Capasso, M., Hagemann, T., The tumor microenvironment at a glance. *J. Cell Sci.*, 125, 23, 5591–6, 2012 Dec 1.
158. Doorakkers, E., Lagergren, J., Engstrand, L., Brusselaers, N., Eradication of Helicobacter pylori and Gastric Cancer: A Systematic Review and Meta-analysis of Cohort Studies. *J. Natl. Cancer Inst.*, 108, 9, djw132, 2016 Sep 14.
159. Watanabe, T., Tada, M., Nagai, H., Sasaki, S., Nakao, M., Helicobacter pylori infection induces gastric cancer in Mongolian gerbils. *Gastroenterology*, 115, 3, 642–8, 1998 Sep.
160. Chen, W., Liu, F., Ling, Z., Tong, X., Xiang, C., Human Intestinal Lumen and Mucosa-Associated Microbiota in Patients with Colorectal Cancer. *PLoS One*, 7, 6, e39743, 2012 Jun 28.
161. Ahmed, M., Colon Cancer: A Clinician's Perspective in 2019. *Gastroenterol. Res.*, 13, 1, 1–10, 2020.
162. Abdulamir, A.S., Hafidh, R.R., Bakar, F.A., The association of Streptococcus bovis/gallolyticus with colorectal tumors: The nature and the underlying mechanisms of its etiological role. *J. Exp. Clin. Cancer Res.*, 30, 1, 11, 2011 Dec 20.
163. Bray, F., Ferlay, J., Soerjomataram, I., Siegel, R.L., Torre, L.A., Jemal, A., Global cancer statistics 2018: GLOBOCAN estimates of incidence and mortality worldwide for 36 cancers in 185 countries. *CA Cancer J. Clin.*, 68, 6, 394–424, 2018 Nov 12.
164. Xuan, C., Shamonki, J.M., Chung, A., DiNome, M.L., Chung, M., Sieling, P.A., *et al.*, Microbial Dysbiosis Is Associated with Human Breast Cancer. *PLoS One*, 9, 1, e83744, 2014 Jan 8.
165. Conesa-Zamora, P., Immune responses against virus and tumor in cervical carcinogenesis: Treatment strategies for avoiding the HPV-induced immune escape. *GynecolOncol.*, 131, 2, 480–8, 2013 Nov.
166. Khatoon, J., Role of Helicobacter pylori in gastric cancer: Updates. *World J. GastrointestOncol*, 8, 2, 147, 2016.
167. Odenbreit, S., Püls, J., Sedlmaier, B., Gerland, E., Fischer, W., Haas, R., Translocation of Helicobacter pylori CagA into Gastric Epithelial Cells by Type IV Secretion. *Sci. (1979)*, 287, 5457, 1497–500, 2000 Feb 25.
168. Vivarelli, S., Salemi, R., Candido, S., Falzone, L., Santagati, M., Stefani, S., *et al.*, Gut Microbiota and Cancer: From Pathogenesis to Therapy. *Cancers (Basel).*, 11, 1, 38, 2019 Jan 3.
169. Lara-Tejero, M. and Galán, J.E., A Bacterial Toxin That Controls Cell Cycle Progression as a Deoxyribonuclease I-Like Protein. *Sci. (1979).*, 290, 5490, 354–7, 2000 Oct 13.
170. Rubinstein, M.R., Wang, X., Liu, W., Hao, Y., Cai, G., Han, Y.W., Fusobacterium nucleatum promotes colorectal carcinogenesis by modulating

E-cadherin/β-catenin signaling *via* its FadAadhesin. *Cell Host Microbe*, 14, 2, 195–206, 2013 Aug 14.

171. Viaud, S., Saccheri, F., Mignot, G., Yamazaki, T., Daillère, R., Hannani, D., *et al.*, The Intestinal Microbiota Modulates the Anticancer Immune Effects of Cyclophosphamide. *Sci. (1979)*, 342, 6161, 971–6, 2013 Nov 22.
172. Iida, N., Dzutsev, A., Stewart, C.A., Smith, L., Bouladoux, N., Weingarten, R.A., *et al.*, Commensal Bacteria Control Cancer Response to Therapy by Modulating the Tumor Microenvironment. *Sci. (1979)*, 342, 6161, 967–70, 2013 Nov 22.
173. Akbar, N., Khan, N.A., Muhammad, J.S., Siddiqui, R., The role of gut microbiome in cancer genesis and cancer prevention. *Health Sci. Rev.*, 2, 100010, 2022 Mar.
174. Konishi, H., Fujiya, M., Tanaka, H., Ueno, N., Moriichi, K., Sasajima, J., *et al.*, Probiotic-derived ferrichrome inhibits colon cancer progression *via* JNK-mediated apoptosis. *Nat. Commun.*, 7, 1, 12365, 2016 Aug 10.
175. Eckburg, P.B., Bik, E.M., Bernstein, C.N., Purdom, E., Dethlefsen, L., Sargent, M., *et al.*, Diversity of the Human Intestinal Microbial Flora. *Sci. (1979)*, 308, 5728, 1635–8, 2005 Jun 10.
176. Takahashi, K. and Yamanaka, S., Induction of Pluripotent Stem Cells from Mouse Embryonic and Adult Fibroblast Cultures by Defined Factors. *Cell*, 126, 4, 663–76, 2006 Aug.
177. Li, X., Brejnrod, A.D., Ernst, M., Rykær, M., Herschend, J., Olsen, N.M.C., *et al.*, Heavy metal exposure causes changes in the metabolic health-associated gut microbiome and metabolites. *Environ. Int.*, 126, 454–67, 2019 May.
178. Lladó, S. and Baldrian, P., Community-level physiological profiling analyses show potential to identify the copiotrophic bacteria present in soil environments. *PLoS One*, 12, 2, e0171638, 2017 Feb 7.
179. Mosca, A., Leclerc, M., Hugot, J.P., Gut Microbiota Diversity and Human Diseases: Should We Reintroduce Key Predators in Our Ecosystem? *Front. Microbiol.*, 7, 2016 Mar 31.
180. Nkamga, V.D., Henrissat, B., Drancourt, M., Archaea: Essential inhabitants of the human digestive microbiota. *Hum. Microb. J.*, 3, 1–8, 2017 Mar.
181. Oliphant, K. and Allen-Vercoe, E., Macronutrient metabolism by the human gut microbiome: major fermentation by-products and their impact on host health. *Microbiome*, 7, 1, 91, 2019 Dec 13.
182. Zahran, S.A., Ali-Tammam, M., Hashem, A.M., Aziz, R.K., Ali, A.E., Azoreductase activity of dye-decolorizing bacteria isolated from the human gut microbiota. *Sci. Rep.*, 9, 1, 5508, 2019 Apr 2.
183. Turnbaugh, P.J., Ley, R.E., Hamady, M., Fraser-Liggett, C.M., Knight, R., Gordon, J.I., The Human Microbiome Project. *Nature*, 449, 7164, 804–10, 2007 Oct 17.
184. Huang, Y.J., Nariya, S., Harris, J.M., Lynch, S.V., Choy, D.F., Arron, J.R., *et al.*, The airway microbiome in patients with severe asthma: Associations with

disease features and severity. *J. Allergy Clin. Immunol.*, 136, 4, 874–84, 2015 Oct.

185. Kumar, P., Tan, Y., Cahan, P., Understanding development and stem cells using single cell-based analyses of gene expression. *Development*, 144, 1, 17–32, 2017 Jan 1.
186. Hung, S.S.C., Khan, S., Lo, C.Y., Hewitt, A.W., Wong, R.C.B., Drug discovery using induced pluripotent stem cell models of neurodegenerative and ocular diseases. *Pharmacol. Ther.*, 177, 32–43, 2017 Sep.
187. Prasad, K., Sharma, A., Garg, A., Mohanty, S., Bhatnagar, S., Johri, S., *et al.*, Intravenous Autologous Bone Marrow Mononuclear Stem Cell Therapy for Ischemic Stroke. *Stroke*, 45, 12, 3618–24, 2014 Dec.
188. Shen, L.H., Li, Y., Chen, J., Zacharek, A., Gao, Q., Kapke, A., *et al.*, Therapeutic Benefit of Bone Marrow Stromal Cells Administered 1 Month after Stroke. *J. Cereb. Blood Flow Metab.*, 27, 1, 6–13, 2007 Jan 5.
189. Paik, M.J., Li, W.Y., Ahn, Y.H., Lee, P.H., Choi, S., Kim, K.R., *et al.*, The free fatty acid metabolome in cerebral ischemia following human mesenchymal stem cell transplantation in rats. *ClinicaChimicaActa.*, 402, 1–2, 25–30, 2009 Apr.
190. Lee, J.S., Hong, J.M., Moon, G.J., Lee, P.H., Ahn, Y.H., Bang, O.Y., A Long-Term Follow-Up Study of Intravenous Autologous Mesenchymal Stem Cell Transplantation in Patients With Ischemic Stroke. *Stem Cells*, 28, 6, 1099–106, 2010 Jun 1.
191. Caplan, A., Why are MSCs therapeutic? New data: new insight. *J. Pathol.*, 217, 2, 318–24, 2009 Jan 20.
192. Golpanian, S., Schulman, I.H., Ebert, R.F., Heldman, A.W., DiFede, D.L., Yang, P.C., *et al.*, Concise Review: Review and Perspective of Cell Dosage and Routes of Administration From Preclinical and Clinical Studies of Stem Cell Therapy for Heart Disease. *Stem Cells Transl. Med.*, 5, 2, 186–91, 2016 Feb 1.
193. Manganas, L.N., Zhang, X., Li, Y., Hazel, R.D., Smith, S.D., Wagshul, M.E., *et al.*, Magnetic Resonance Spectroscopy Identifies Neural Progenitor Cells in the Live Human Brain. *Sci. (1979).*, 318, 5852, 980–5, 2007 Nov 9.
194. Chen, J., Ye, X., Yan, T., Zhang, C., Yang, X.P., Cui, X., *et al.*, Adverse Effects of Bone Marrow Stromal Cell Treatment of Stroke in Diabetic Rats. *Stroke.*, 42, 12, 3551–8, 2011 Dec.
195. Berg, G., Rybakova, D., Fischer, D., Cernava, T., Vergès, M.C.C., Charles, T., *et al.*, Microbiome definition re-visited: old concepts and new challenges. *Microbiome.*, 8, 1, 103, 2020 Dec 30.
196. Logotheti, M., Agioutantis, P., Katsaounou, P., Loutrari, H., Microbiome Research and Multi-Omics Integration for Personalized Medicine in Asthma. *J. Pers. Med.*, 11, 12, 1299, 2021 Dec 5.

197. Breitwieser, F.P., Lu, J., Salzberg, S.L., A review of methods and databases for metagenomic classification and assembly. *Brief Bioinform.*, 20, 4, 1125–36, 2019 Jul 19.
198. Green, B.J., Wiriyachaiporn, S., Grainge, C., Rogers, G.B., Kehagia, V., Lau, L., *et al.*, Potentially Pathogenic Airway Bacteria and Neutrophilic Inflammation in Treatment Resistant Severe Asthma. *PLoS One*, 9, 6, e100645, 2014 Jun 23.

11

Cellular Synchrony: Coordinated Strategies for Effective Tissue Regeneration Using Stem Cells

Amit Anand[1], Santhepete Nanjundiah Manjula[2] and Mruthunjaya Kenganora[1]*

[1]Department of Pharmacognosy, JSS College of Pharmacy, Mysuru, JSS Academy of Higher Education and Research, Mysuru, Karnataka, India
[2]Department of Pharmacology, JSS College of Pharmacy, Mysuru, JSS Academy of Higher Education and Research, Mysuru, Karnataka, India

Abstract

Cell synchrony has a vital function in tissue regeneration, a process vital for repairing and replacing damaged tissue. This chapter offers a comprehensive exploration of the interplay between cell synchrony and tissue regeneration, emphasizing the fundamental role of stem cells. The chapter begins by providing an insight on the importance of cellular synchrony and regenerative capabilities of the stem cells focusing on various types and potency levels. Key properties such as differentiation, self-renewal, and clonogenicity are examined to highlight the versatility of stem cells in regenerative medicine. The mechanism underlying cellular synchrony are detailed, including molecular mechanism and signaling pathways as well as methods to increase synchrony. Investigation techniques encompassing computational, biological, and biochemical methods were comprehensively reviewed. Strategies to enhance cellular synchrony such as genetic engineering, biomaterials scaffolds, etc., are also explored, providing innovative approaches to improve tissue regeneration outcomes. Applications of stem-cell-based tissue regeneration are diverse with significant advancements in regenerating bone, cartilage, cardiac, neural tissues, and sensory organs. Case studies and clinical studies illustrate the practical implications and success of these approaches. The chapter concludes by addressing the challenges and future direction, emphasizing the need for continued research.

**Corresponding author*: Kmruthunjaya@jssuni.edu.in

Ayush Madan, Sumel Ashique, Deepika Arora and Mantosh Kumar Satapathy (eds.) Stem Cell Therapeutics, (505–540) © 2025 Scrivener Publishing LLC

***Keywords*:** Cellular synchrony, stem cells, tissue regeneration, computational modeling, regenerative medicine, clonogenicity

Abbreviations

APC	adenomatous polyposis coli
CK	casein kinase
DNA	deoxyribonucleic acid
ECM	extracellular matrix
ESCs	embryonic stem cells
FDA	food and drug administration
GPCR	G-protein-coupled receptors
HBCs	horizontal basal cells
HSCs	haematopoietic stem cells
MSCs	mesenchymal stem cells
RTKs	receptor tyrosine kinase
SC	stem cells
TSCs	totipotent stem cells

11.1 Introduction to Cellular Synchrony and Tissue Regeneration

11.1.1 Overview of Tissue Regeneration

Tissue regeneration is an organized process of regrowth and renewal used to replace or mend damaged tissue. Complex biological processes including signaling, cell migration, proliferation, extracellular matrix (ECM) production, and fine tuning are all included into this process. In humans, tissue regeneration is restricted to repair; in echinoderms, amphibians, and many reptiles, including newts and salamanders, body parts and tissues may regenerate completely [1]. The aging process, pathological conditions, energy allocation, metabolic rates, microarchitectural characteristics, and tissue-specific functional characteristics all contribute to the heterogeneity of tissue regeneration in the human body [2]. During regeneration, a number of cells are used, which includes dedifferentiated cells, lineage-restricted tissue-specific stem cells, and pluripotent stem cells. Tissues are made up of numerous cells with varying transcriptional programs, division capabilities, and activities [3]. In essence, coordinated damage stimuli are needed to control the degree and timing of the healing process, a process known as cellular plasticity, which is critical for flawless tissue regeneration. In addition to cells, the tissue or cellular microenvironment where cells cooperate to produce a functioning organ is also necessary for regeneration [4]. Delivering certain cell types or cell products to damaged tissues or organs in order to restore tissue and organ function is known as tissue regeneration. Since autologous tissue transplantation of patients can overcome some limits, stem cell treatment has garnered a lot of attention. Nevertheless, it is not a perfect means of treating illnesses. The integration of stem cell technology with tissue engineering has opened new pathways for creating tissue replacements. Numerous studies have shown that these combined approaches enhance the therapeutic efficacy, differentiation, and survival of transplanted stem cells [5].

11.1.2 Importance of Cellular Synchrony

A special method for examining the structural and chemical changes that occur in cells throughout their advancement *via* the cell cycle is the synchronization of cell populations. It facilitates the precise investigation of distinct stages within the cell cycle, the identification of regulatory mechanisms which govern cell cycle control at the level of posttranscriptional protein modifications and gene expression, and aids in the search for new drugs. Cells from various cell cycle phases are chosen and brought to the same phase during the synchronization procedure. The S phase of genetic material replication and the M phase, which is the progressive distribution of genetic material and other cell components onto two daughter cells, make up the cell cycle. These two processes progress at different times, with G1 and G2 acting as gaps in between. This movement is known as the cell cycle [6]. Cell synchronization is necessary to investigate phenomena particular to the cell cycle. In cases where a gene is necessary for a certain stage of the cell cycle, synchronization is also helpful. For example, if a gene is required for mitosis, we may analyze its function in interphase and track the synchronized development of the cell cycle to M phase by synchronizing the cells in G1 phase and simultaneously inactivating the gene [7]. Contrary to medication synchronization, which might have undesirable side effects, centrifugal elutriation can be accomplished in a shorter time and with relatively minimal physiological damage to the cells. Examples of such medications are nocodazole and mimocine [8]. Below Figure 11.1 represents the step by step process in tissue regeneration.

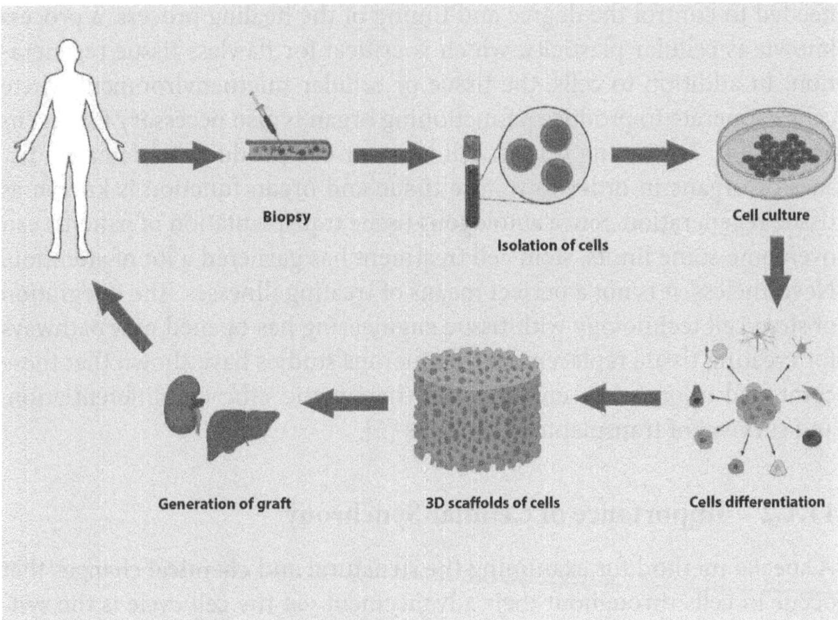

Figure 11.1 Process involved in regeneration of tissues.

11.1.3 Function of Stem Cells in Tissue Repair and Regeneration

The deterioration of tissues and the reduction of organ functioning as a result of ageing, chronic illnesses, infections, or traumas can seriously impair an individual's quality of life. Over the past few decades, the use of stem cell transplant and tissue regeneration techniques to treat a wide range of disorders has increased dramatically [9]. While both stem cell exosomes and stem cells share similar biological characteristics, stem cell exosomes are considered safer and more efficient in promoting tissue regeneration and repair and accelerating wound healing. The body's vital organs, the skin, and bone tissues are necessary for maintaining daily functions [10]. An individual's quality of life is significantly impacted by their poor ability to mend, a problem that stem cell exosome therapy may be able to resolve. Nevertheless, there are barriers that stem cells and stem cell exosomes must overcome in order to be more bioavailable *via* the skin [8, 9]. Additionally, by their immunomodulatory activities, stem cells can effectively reduce the inflammatory response and aid in wound healing for skin regeneration Stem cells have emerged as a promising alternative for bone tissue repair. They possess the capacity to move to the site of an injury, differentiate into local tissue components, and facilitate tissue regeneration [11]. For instance, MSCs were extracted from rat femur bone marrow and injected into a rat model that had cranial abnormalities according to the literature source. These findings imply that MSC transplantation could be a novel cranioplasty technique. MSCs have the ability to produce new cranial bone and form suture-like gaps in addition to repairing bone tissue. It has been shown that distinct Chronic illnesses, cytokines, and growth factors secreted by recruited stem cells during the healing process are known as paracrine factors and are required to support tissue regeneration and repair or tissue differentiation [12].

11.2 Fundamentals of Stem Cells

Every organ, tissue, and cell in the body is built on stem cells. They resemble an empty microchip that can be eventually configured to carry out certain functions. Undifferentiated or "blank" cells are those that have not yet reached the point of complete specialization. Stem cells can differentiate into specific tissues and organs when the right circumstances are met. Furthermore, stem cells possess the ability to self-renew or divide and proliferate into new stem cells [13]. Cells that possess the capacity to differentiate into mature cells and also replicate themselves through self-renewal are preferred as stem cells. Because of this, it is necessary to precisely identify and purify stem cells before studying their characteristics. While the idea that each tissue develops from a stem cell that is particular to that tissue sounds plausible, the somatic stem cells have only been rigorously identified and isolated in a few cases [14]. For instance, it has been demonstrated that hematopoietic stem cells (HSCs), which have been identified from mice and humans are in charge of the development and regeneration of the immunological and blood-forming systems. Although stem cells derived from many organs may one day be employed in therapeutic contexts, hematopoietic stem cells (HSCs), which are essential components of bone marrow transplantation, have already seen widespread application in therapeutic contexts [15]. Stem cells are unspecialized cells in the human body that can self-renew and differentiate into various cell types. They are present in both embryonic and adult cells. Stem cells undergo various stages of specialization, with each stage progressively limiting their developmental potential. For example, pluripotent stem cells can differentiate into a wider range of cell types compared to unipotent stem cells, which have a more restricted differentiation capacity [16].

11.2.1 Types of Stem Cells

11.2.1.1 Embryonic Stem Cells

Mammalian stem cells have the ability to multiply and grow unrestrictedly *in vitro* without differentiation. It has been demonstrated that in *in vivo* settings, these cells may develop into endoderm, the progenitor of hepatocytes. While embryonic stem cells have demonstrated potential in animal models, ethical issues surrounding the termination of human embryos prevent them from being widely used in human models. The inner cell mass of human blastocysts, which represent an early development stage of the embryo occurring between the fourth and seventh day after conception, contains embryonic stem cells (ESCs) [17]. Embryonic stem cells in their undifferentiated state have the potential to differentiate into cells originating from all three embryonic germ layers. The ethical debate surrounding the use of stem cells in regenerative medicine primarily focuses on research involving human ESCs as the extraction of these cells necessitates the destruction of embryos [18]. Embryonic stem cells which are pluripotent originate from the inner cell mass of the blastocysts and possess the ability to differentiate into nearly all types of cell except those of the trophoblast lineage [19]. Phase I clinical studies including individuals with macular degeneration and spinal cord injury (SCI) have examined ESCs. In one research, five patients with SCI received transplants of hESC-derived oligodendrocyte precursors, and four of the patients showed a decrease in spinal cord cavitation. Adverse effects were not reported by any of the patients [20]. In a different research, individuals with Stargardt disease and age-related degeneration received hESC-derived retinal pigment epithelium, and some showed clinical improvement without any sign of tumorigenicity. They can trigger a host of immunological responses since they are allogenic cells to start with. Second, because they are undifferentiated cells, there is a chance that they will cause teratomas to develop [21]. Embryonic stem cells possess special qualities that allow them to self-renew and develop into any form of cell, including germ cells. Because of these characteristics, mouse embryonic stem cells are now a useful tool for genetically modifying the mouse genome [19]. However, understanding the processes governing lineage choices during embryogenesis appears to be a prerequisite for further advancement in this area. At the blastocyst stage, embryonic stem cells are extracted from the embryo and possess the potential to differentiate into fully developed body cells such as cardiomyocytes and endothelial cells [22]. Furthermore, it has been demonstrated that pre-conditioning ES cells increases the lifespan of transplanted cells and offers a straightforward yet efficient method of enhancing cell therapy. Embryonic stem cells (ESCs) are pluripotent stem cells that may develop into any type of cell and are derived from the blastocyst stage of early mammalian embryogenesis. Because of this, ESCs are able to differentiate and proliferate without undergoing reprogramming-induced genetic changes or mutations. Embryonic stem cells possess the capacity for unlimited self-renewal, making them an excellent option for regenerative medicine therapies [23]. Multipotent cells called embryonic stem cells (ESC) have the ability to produce progenitors of oligodendrocytes. Mice with neonatal MLD were transplanted with engineered ESC that overexpress ARSA. As a result, the amount of sulfatide stored was reduced by over 50%, but the ESC's capacity to differentiate into oligodendrocytes and astrocytes *in vivo* was preserved [24].

11.2.1.2 Adult Stem Cells (ASCs)

Adult stem cells, also referred to as somatic stem cells, are undifferentiated cells present in various organs across nearly all species. These cells are found in various organs including skin, heart, liver, skin, blood, and bone marrow, which possess the ability to differentiate into any cell type specific to their tissue or organ. In contrast, embryonic stem cells are pluripotent. ASCs, which can persist in tissues for decade, are important in replenishing the damaged tissue or replacing lost cells [25]. In stem cell biology, adult stem cells, particularly hematopoietic stem cells, represent the most well-understood cell type. Nonetheless, since adult stem cells have the potential to treat a wide range of degenerative diseases, there is still much research being done in this field. In the past 10 years, adult stem cell populations have been found in the gastrointestinal tract, bone marrow, and a number of adult tissues, including the brain, skin, and muscle. Numerous different subtypes of adult stem cells have been found in various organs throughout the body, and it is believed that these cells act as the main sources of healing for the organs in question [26]. The MSC, often referred to as the multipotent adult progenitor cell, is a noteworthy exception to the tissue specificity of adult stem cells. The source of this cell type is the stroma of the bone marrow. These cells have the ability to grow into many distinct tissue types *in vitro* and can also do so if they are implanted into a blastocyst. Adult progenitor cells with multiple developmental potential may differentiate into many tissues such as neurons, adipose, muscle, liver, lungs, spleen, and gut tissue, but they cannot differentiate into bone marrow or gonads [24]. When needed, these adult stem cells which have been present in the tissue for decades replace lost tissue cells, as shown in the daily development of new skin in humans [22]. Adult stem cells are thought to hold the key to developing treatments for a wide range of conditions, from neurological disease (regenerating lost neurons in the brain or spinal cord) to type 1 diabetes (providing insulin-producing cells) to heart attacks (repairing cardiac muscle) [27]. In an attempt to repair damaged tissues, these adult stem cells are being stimulated to grow new cells. With this strategy, adult stem cells will be guided and stimulated to properly regenerate just the appropriate cell types by making use of the current tissue organization and chemicals. An alternative would be to separate the adult stem cells from the tissue and cultivate them in cultures outside of the body [28]. This would make it possible to control the cells with ease, even though they are frequently scarce and challenging to cultivate in culture. "Adult stem cells" have the ability to support the establishment of homeostasis throughout the life cycle of an organism. Nearly all of the body's specialized tissues include adult stem cells in a metabolic quiescent state [29]. Many researchers believe that adult stem cells are the best option available for stem cell therapy [30, 31]. In hematopoietic transplantation, adult stem cells have a therapeutic potential. Adult stem cells have the potential to be therapeutically successful, especially in hematopoietic transplantation. Unlike ESCs, the origin of adult stem cells is not up for debate. ASCs can give rise to a wide variety of specialized cells found in tissues and organs [32].

11.2.2 Potency of Stem Cells

11.2.2.1 Totipotent

Cells known as totipotent stem cells possess the ability to develop into a whole embryo. Numerous attempts have been undertaken to create "induced totipotent stem cells," which are stem cells that can be produced from other types of cells. These cells are special in several ways, including their transcriptional and epigenetic networks. Effective techniques to produce totipotent stem cells have been developed by using these features. Even while this is a huge accomplishment, there are still many unanswered questions about induction, including the fundamental process [33]. Totipotent stem cells have the ability to differentiate into all adult cell types, can infiltrate the germ line, and have demonstrated self-replication capabilities. The fertilized eggs from *in vitro* fertilization (IVF), a process in which sperm and eggs (ova) obtained by laparoscopic operations are put into culture dishes and allowed to undergo fertilization, can be used to create totipotent stem cells. To create a pregnancy, the fertilized eggs that are produced can be placed within the uterus (*in utero*). Fertilized eggs that are not needed can be kept in liquid nitrogen cryopreserved for an endless period of time [34]. Only when totipotent stem cells are implanted *in utero*, the location where innumerable signals and circumstances occur at the proper time and in the correct quantity, can they progress through all of the phases of development normally to become an animal (or human). Implanting totipotent stem cells in ectopic sites, locations other than *in utero*, allows the cells to develop into a wide variety of tissue types [35]. All adult cell types can be produced by totipotent stem cells, which may also contribute genetic material to future generations by entering the germ line and reproducing themselves into daughter cells that are exact replicas of their parents. Only early embryonic tissues include totipotent stem cells, which are well characterized and often originate from the first few cell divisions following fertilization. Because they may grow into whole creatures, totipotent stem cells, or TSCs, are employed in the biological sciences in areas including conservation, mammalian breeding, and regenerative medicine [36]. One major obstacle to TSC research, nevertheless, is the challenge of preserving the developmental totipotency and self-renewal potential of cells cultivated from early-stage embryos. Research on induced pluripotent stem cells that revert to their totipotent state has advanced to a new level, allowing for the creation of numerous TSCs and sparking a new direction in stem cell studies. Moreover, established TSCs have the ability to produce blastocyst-like structures, which paves the way for *in vitro* synthetic embryogenesis [37].

11.2.2.2 Pluripotent

Pluripotent stem cells possess the remarkable ability to both self-renew and differentiate into any cell type within the body's tissues. These cells are considered the most versatile form of stem cells and are typically found in the early stages of embryonic development [38]. In fact, a number of clinical trials using tissues and cells derived from pluripotent stem cells are currently being conducted to see if they may treat or completely eradicate a range of diseases. Their extraordinary potential allows them to form all three germ layers- ectoderm, mesoderm and endoderm as well as germ cells. In essence, pluripotent stem cells can generate any cell or tissue required by the body for repair and regeneration [39].

11.2.2.3 Unipotent

An even more specialized form of stem cell than multipotent stem cells are called unipotent stem cells. Unlike pluripotent and multipotent stem cells, which can develop into a variety of cell types, unipotent stem cells have very limited differentiation potential and can only grow into one specialized cell type. Usually, they are connected to the particular regrowth of a certain kind of tissue or organ [40]. For instance, unipotent germline stem cells in the ovaries are used to create oocytes, or eggs, in the female reproductive system. This is a very specialized and unipotent differentiation, as these cells can only develop into oocytes [41]. Unipotent stem cells have a more specialized role in the regeneration and renewal of certain tissues and are generally less numerous than multipotent or pluripotent stem cells. Although they lack the versatility of other stem cell types, their ability to regenerate vital specialized cells is vital to the preservation and maintenance of body processes. Given that unipotent stem cells may differentiate into just one kind of cell, certain biomarkers can be employed to recognize and describe these cells [40]. Unipotent stem cells vary from other stem cells in that they are not as flexible as pluripotent and totipotent stem cells; yet, because of their capacity for self-renewal, researchers are very interested in studying them. Their ability to produce healthy cells for transplantation and this characteristic may prove to be very helpful in the treatment of illnesses [42].

11.2.2.4 Multipotent

In bone tissue engineering, mesenchymal stem cells (MSC), a kind of multipotent stem cell, are the focus of an extensive study. MSC may be obtained from an increasing variety of sources, such as bone marrow, peripheral blood, adipose tissue, umbilical cord blood, and others [43]. Since MSCs are easier to get and their growth paths are better understood, most bone tissue engineering researchers favor MSCs in many situations. Although the use of MSC is not morally dubious, the process of obtaining these cells for allogeneic use presents certain difficulties because two people will be transferring these cells. However, unlike complete blood or organ donation, when donors donate their blood or organs to a receiver who receives the minimally changed blood or organ, this transfer will not be a direct transfer [44]. Interest in using mesenchymal stem cells for possible applications is developing, also known as multipotent stem cells, in cell treatments. According to Oh and Choo's, MSCs are a broad class of fibroblast-like, anchorage-dependent cells obtained from various sources, including adipose tissue, periosteum, skeletal muscle connective tissue, human trabecular bones, fetal blood and liver, and umbilical cord blood. Human embryonic stem cells (hESCs) can also be used to generate homogeneous MSCs. According to Dominici *et al.*, MSCs express the primitive Stro-1 marker as well as CD29, CD44, CD73, CD90, and CD105 [45].

11.2.2.5 Oligopotent

The ocular surface of pigs, particularly the cornea, contains oligopotent stem cells capable of forming distinct colonies of corneal and conjunctival cells. These stem cells have the ability of self-renew and generate two or more cell lineages within a single tissue. Hemopoietic stem cells, which can differentiate into both lymphoid and myeloid lineages, are a typical example of oligopotent stem cells [46]. Studies suggest that alveolar and bronchiolar epithelium in the lungs may originate from cells located at the bronchioalveolar duct junction. These stem cells have limited cell differentiation potential; among them are the myeloblast stem cells, which give rise to neutrophils, basophils, and eosinophils, three separate kinds of white blood cells [47].

11.2.3 Properties of Stem Cells

11.2.3.1 Differentiation of Stem Cells

The potency and adaptability of stem cells are two of their many distinctive qualities. When transferred to a novel stem cell environment, they possess the capacity to generate several cell lineages and undergo trans differentiation into entirely distinct cell types. Stem cells are involved in the self-renewal of most tissues [48]. Corneo-scleral limbus is thought to provide corneal epithelial stem cells with a habitat at the ocular surface. It has been possible to detect epithelial cell populations near the limbus that may contain potential stem cells by using markers for cytokeratin, vimentin, epidermal growth factor receptors, p63, and other molecules. Conversely, neither CD34 nor CD133, two established markers of hematopoietic stem cells, show up as discrete subsets of corneal or limbal epithelial cells. The ability of SC to produce unique cell lineages is one of its key traits. But not every SC has the same efficacy, and there is a possible hierarchy among them [49]. Multipotent, totipotent, pluripotent, or unipotent SC are possible. Many investigations have concentrated on pinpointing the molecular and

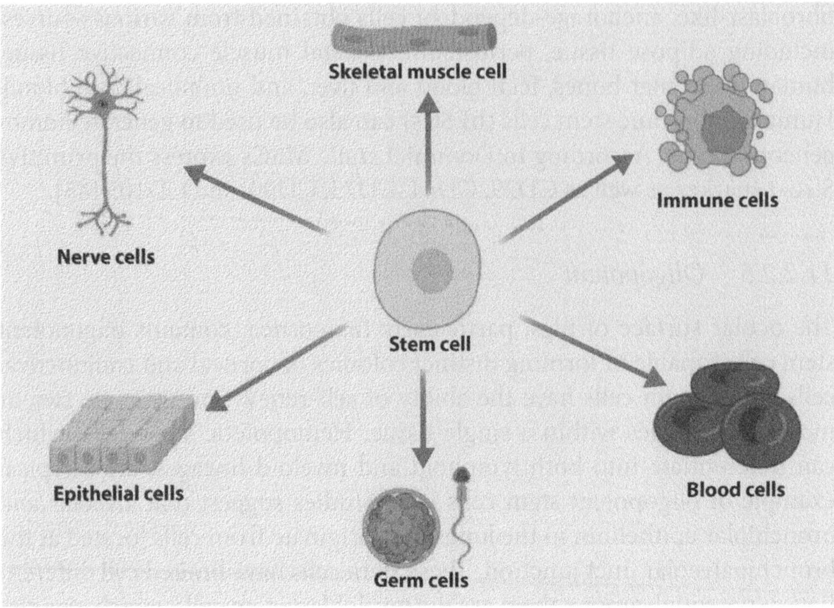

Figure 11.2 Differentiation of stem cells into various animal tissues.

biological elements that control stem cell differentiation; nonetheless, more recent findings indicate that mechanical cues could be just as significant [48]. Figure 11.2 indicates the tissues differentiation into various animal tissues.

11.2.3.2 Self-Renewal of Stem Cells

Self-renewal is a distinct cellular process including both proliferation and maintenance of multipotency and tissue regenerative potential. For a cell to achieve self-renewal, two conditions must be fulfilled: it must divide and progress through the cell cycle, and at least one of it must divide and progress through the cell cycle and at least one of its progenies must remain in an undifferentiated state. Cell depletion and ultimately tissue dysfunction result from failure in either one of these two areas [50]. The topic of self-renewal in specialized adult stem cells, such as those of the gut and hematopoietic system, has been the subject of several outstanding reviews. Nonetheless, progenitor cells may also self-renew, therefore stem cells are not the only ones capable of doing so [51]. The main distinction between progenitor cells and stem cells lies in their capacity for self-renewal potential. The process by which stem cells divide to generate new stem cells, called self-renewal, maintains the stem cell reservoir throughout an individual's lifespan. This process involves division while preserving the cells in an undifferentiated state, ensuring continuous self-renewal [52]. To achieve this, it is imperative to regulate the cell cycle while preserving pluripotency or multipotency depending on the type of stem cell. Self-renewal mechanism involves intricate networks that balance proto-oncogenes, which promotes self-renewal, gatekeeper, tumor suppressor, which inhibits it, and caretaker suppressors which uphold genomic integrity. As stem cells adapt their developmental potential and cell cycle status in response to evolving tissue needs, different self-renewal strategies are necessary for various life stages [53].

11.2.3.3 Clonogenicity of Stem Cells

Clonogenicity refers to the ability of cells to self-replicate and form a colony of genetically identical cells. It is commonly known that stem cells have the ability to divide and become many kinds of cells. The capacity of cells to divide and form clones in response to cytotoxic chemicals is determined by a clonogenic test, also known as a colony formation assay, which is frequently used to determine a cell's clonogenicity. This test is frequently used to examine how chemotherapy and radiation affect various cell lines and helps gauge how much each treatment affects the biochemistry and clonogenicity of cells [54]. Clonogenic tests are helpful in the development of novel cancer treatments. In a ground-breaking study, Barrandon and Green demonstrated the connection between primary keratinocytes' ability to form colonies and their stem cell capability. Barrandon and Green discovered that single cells could generate three different forms of colonies dubbed holoclones, meroclones, and paraclones derived from cells with varying capacity for proliferation using primary cultures of human keratinocytes. While meroclones have a limited capacity for proliferation and are unable to self-renew, paraclones are incapable of further proliferation, and only holoclones are capable of substantial proliferation and self-renewal [55]. Colony morphology is often used as surrogate test to identify and characterize the stem cells from skin follicular and limbal tissues, as it reflects the hierarchy of colony-forming cells. This test is so applied to evaluate the stem cells intended for use in tissue engineering. Holoclones express genes related to stem cell capability, such as BMI-1 and p63, which are involved in survival and self-renewal. In these investigations, the terms "holoclones," "meroclones," and "paraclones" refer to the colonies rather than the single cell from which they are produced [56].

11.3 Mechanisms of Cellular Synchrony

11.3.1 Signaling Pathways Involved in Synchrony

11.3.1.1 MAPK Signaling Pathways

Receptor tyrosine kinases (RTKs), triggered by the interaction of ligand receptor, help in the phosphorylation of tyrosine residues on the cytoplasmic receptor. Signal molecules such as growth factor receptor bound protein 2 get attached to the phosphotyrosine binding proteins and SHP2 [57]. These molecules work together to bind to a range of proteins and activate the Ras and Raf GTPases. This, in turn, initiates a cascade of events that culminate in the activation of three different families of MAPKs: p38MAPKs, c-Jun N-terminal kinases (JNKs), and extracellular signal-regulated kinases (ERKs) [58].

11.3.1.2 PI3K/AKT Signaling Pathway

AKT/PI3K signaling is connected to a number of biological processes, such as cell division, proliferation, migration, and survival. The p85 and p110 subunits of PI3K are drawn to the membrane in response to the activation of RTKs or GPCRs. Phosphatidylinositol-4, 5-bis-phosphate (PIP2) is then further converted to PIP3 by PI3K phosphorylating it at position 30 of its inositol ring. AKT is activated by elevated PIP3 through phosphoinositide-dependent kinase-1 (PDK1) [59]. AKT also triggers the activation of many downstream proteins, including mTOR, which impact various cellular processes. Additionally, mTOR contributes to improving translation initiation [60]. In mice with homozygous deletion of AKT1, the number of proliferating trophoblast cells is reduced. EGF activates the PI3K and MAPK pathways in HTR-8/SVneo cells, increasing the levels of MMP-9 and TIMP-1; blocking of either route decreases their release. It has been shown that AKT1 and AKT3 are essential for both basal and EGF-induced migration of SGHPL-5 trophoblast cells by suppressing their expression. MAPK and PI3K signaling are both necessary for SGHPL-4 cell motility caused by HGF. These cells' ability to migrate is inhibited by blocking the PI3K pathway with LY294002 and the MAPK pathway with PD98059 inhibitor [61]. Research on SGHPL-5 and primary EVTs isolated from first-trimester placentas demonstrates that the presence of the Wnt ligand activates the canonical Wnt and AKT signaling, promoting migration *via* increasing the expression of MMP-2 [62].

11.3.1.3 JAK-STAT Signaling Pathways

Janus kinases, also known as receptor-associated tyrosine kinases, are brought together during receptor aggregation by cytokine binding, allowing for the phosphorylation of both JAKs and the receptor's cytoplasmic domains. These phosphorylated domains have the ability to bind and phosphorylate certain tyrosine and serine residues on PTB proteins that include SH2-like signal transducer and activator of transcription (STATs). After activation, STATs separate from their corresponding ligands and move as homo- or heterodimers into the nucleus, where they attach to certain promoter regions of the genes they are targeting to control transcription [63]. In mice, STAT3 is activated post-implantation and is crucial for development; embryos lacking STAT3 deteriorate and ultimately perish, but they can be saved by replacing STAT3 with an alternative version [64]. While STAT3 is absent from trophoblast cells separated from term placenta, it can be found in choriocarcinoma cells and primordial trophoblast cells obtained from first-trimester placenta. When JEG-3 cells are treated with LIF, STAT3 and ERK½ are activated [11]. LIF-mediated JEG-3 trophoblastic cell invasion is mostly dependent on both ERK½ and JAK-STAT-mediated STAT3 (Ser727) activation. This process is also linked to invasion-associated molecules including mucin1, Fos, and Jun [65]. It has also been shown that STAT3 activation is relevant to IL-11-stimulated human EVT migration. The tyrosine and serine phosphorylation of STAT3 also triggers the invasion-promoting action of IL-11 in JEG-3 cells. Apart from STAT3, activation of STAT1 has also been seen in HTR-8/SVneo and JEG-3 after LIF treatment; suppression of STAT1 resulted in a minor reduction in invasion. STAT5 is a member of another STAT protein family that promotes trophoblast invasion; human placental growth hormone (hPGH) induces invasion in primary EVT by activating STAT5 *via* JAK2 [66].

11.3.1.4 Wnt Signaling Pathway

A family of secreted glycoproteins is known as Wnts signals when they bind to many coreceptors, including lipoprotein-receptor-related protein (LRP-5/6), receptor-like tyrosine kinase, and receptor tyrosine kinase-like orphan receptor, as well as the GPCR Frizzled (Fz). By means of a canonical or non-canonical signaling cascade, wnt-receptor binding operates [67]. Dishevelled protein (Dvl) is activated by Wnt ligand, which also attracts axin to the plasma membrane and breaks apart the APC/axin/GSK-3b/

CK1a complex. B-catenin is then stabilized and recruited to the nucleus as a result. B-catenin binds to the T-cell factor (TCF)/lymphoid enhancer factor family of transcription factors in the nucleus, where it stimulates the transcription of Wnt target genes, including BCL9L and BCL9. In the placental tissue of the first trimester, 14 out of 19 Wnt ligands and eight out of 10 Frizzled receptors have been identified [68]. Wnt3a induces the production of the target gene, cyclin D1, and active b-catenin/TCF transcription factor complexes in first-trimester trophoblast cells, hence stimulating trophoblast growth. The placentae with full hydatidiform mole had higher numbers of b-catenin-positive EVT nuclei, which may indicate that aberrant Wnt signaling plays a role in the abnormal invasion observed in this pregnancy disease [69].

11.4 Methods of Investigating Cell Synchrony

11.4.1 Gene Regulation Model

A core oscillator from a previously reported gene regulatory model was adapted for the computational model to enable the integration of communication across numerous cells. Each of the 16 ordinary differential equations in the original model defined a negative feedback loop in which dimers made of the transcription factors CLOCK and BMAL1 promote the transcription of the Period (Per) gene [70]. A protein complex called PER-CRY suppresses transcriptional activity. Over the course of a day, these two proteins accumulate and then degrade, resulting in circadian rhythmicity. Conditions equivalent to either constant darkness or 24-h cycles of light and dark can entrain people into sustained oscillations [71].

11.4.2 Simulation Studies

The final signaling model included five algebraic equations representing VPAC2 saturation levels and intracellular calcium concentration. A full cell model consisting of five algebraic equations and 17 ordinary differential equations were created by adding intercellular signaling to the core oscillator [72]. Based on 400 cells arranged on a 20320 grid, a multicellular model comprising 6,800 ordinary differential equations was generated through simulations. The parameter values related to VIP signaling were selected within physiologically conceivable limits to replicate experimentally observed synchronization and desynchronization behavior [73].

11.4.3 Biological Methods

11.4.3.1 Serum Starvation

To understand the molecular mechanism behind cellular growth and division, it is essential to study the regulatory processes that generally occur in a phase-dependent manner within the cell cycle. Synchronizing cell population is a crucial step for a research focused on cell cycle regulation and progression. Consequently, various methods have been devised to synchronize cells at specific stages of the cell cycle [74]. One typical approach used in this context is biological methods. With the use of human pluripotent stem cells (hPSCs), organogenesis research and disease simulation are made possible. Notably, genetic modification methods can be used to maximize these cells' potential. The ability of hPSCs to differentiate into three germ layers and their capacity for infinite self-renewal make them extremely promising for study on human diseases and regenerative therapy. Genetic modification remains a potent tool for researching hPSC characteristics [75]. However, in many hPSC lines, regulated differentiation also produced an inconsistent and variable efficiency. According to recent research, serum starvation-induced cell cycle synchronization enhanced transient gene transport into mouse embryonic stem cells and dramatically boosted the effectiveness of reprogramming in human fibroblasts. However, Knock Out Serum Replacement (KO-SR), a serum-free formulation that replaces serum in culture directly, is regularly used to cultivate human papillomastis cells [76]. Accordingly, we postulated that KOSR decrease has a significant impact on the transfection and differentiation efficiency of hPSCs. By starving the cells for 2 days with 5% KOSR, the synchronizing effect of serum starvation on the cell cycle was attained. More than 48 h of prolonged serum deprivation did not considerably raise the percentage of G0/G1 cells. When compared to non-starved cells, the highly synchronized higher cells' nucleofection demonstrated a differentiation efficiency [77].

11.4.3.2 Double Thymidine Block

Treatment with double thymidine block effectively stops EO771 cells in the S-phase without affecting the cells' ability to proliferate or survive. Using 5(6)-carboxy fluorescein diacetate N-succinimidyl ester (CFSE) labeling, a long-term examination of cell behavior reveals that synchronization is robust and constant across time. Murine tumor implantation models may be created using the EO771 cell line, a medullary breast-adenocarcinoma cell line obtained from a spontaneous mammary tumour in mice. Cell synchronization is achieved by a variety of physical (elutriation, mitotic shake-off), biological, and pharmacological (colchicine, nocodazole, thymidine) therapies [78]. When studying functional and biochemical activities happening at certain phases of the cell cycle development, a large amount of fully synchronized cells is necessary. Of the several approaches evaluated, the double thymidine block is the most effective for synchronizing murine EO771 cells. Thymidine prevents DNA replication in cells by preventing the conversion of cytidine 5′-diphosphate into deoxycytidine 5′-diphosphate [79]. It is important to use this medication cautiously since too much thymidine might harm chromosomes. To increase the effectiveness of cell line synchronization, double thymidine blockage (db) is utilized. This method is predicated on treating a S obstruction sequentially with thymidine and leaving a gap between the two treatment sessions. However, using a lower thymidine concentration for blocking is made possible by this twofold treatment, which is very effective on a variety of cell types. Thymidine was effective in synchronizing EO771, as evidenced by the significant number of cells in the same cycle phase [78, 80]. As anticipated, a twofold thymidine block yielded the highest effectiveness, especially when considering the highest cell viability. By generating nitroxide-free radicals, hydroxyurea deactivates ribonucleotide reductase and prevents the conversion of ribonucleotides into deoxy ribonucleotides. Under our circumstances, the combination of hydroxyurea and thymidine produced a moderate synchronization efficiency with good cell survival. It is possible to target the mitotic spindle with many medications, including nocodazole, colchicine, and colcemid [81]. This technique can be used to cure or synchronize cancer cells by blocking them when they are in the G2/M phase. However, it may produce a range of abnormalities in cells, including apoptotic activation, multiple mitotic spindle formation, or a return to interphase, depending on the length of therapy and the kind of cell [82].

11.5 Strategies for Enhancing Cellular Synchrony in Tissue Regeneration

11.5.1 Overexpression of Cell Cycle Regulators

11.5.1.1 Cyclins and Cyclin-Dependent Kinase

Cyclin-dependent kinase (CDK) are a group of protein kinase that requires cyclin, a unique regulatory subunit, for their enzymatic activity. CDKs

Table 11.1 Various methods of cell synchronization, including their advantages, mechanisms, and disadvantages.

Methods	Mechanisms	Advantages	Disadvantages	References
Serum starvation	Halts cell cycle in G_0/G_1 phase	This method is effective for many cell types	Prolonged starvation can cause cell death	[83]
Thymidine block	Thymidine helps in inhibition of DNA synthesis	Effective synchronization in S phase	Prolonged exposure can cause DNA damage	[84]
Nocodazole treatment	Causes the arrest of cells in mitosis, disrupts the formation of microtubules	Effective in mitotic synchronization	Toxic at higher concentration that can cause mitotic slippage of cells	[85]
Mitotic shake off	Causes physical dislodgement of mitotic cells from the cell surface	Effective in collecting mitotic cells with less chemical intervention	Not suitable for all types of instruments	[73]
Colcemid treatment	Arrests cells in mitotic phase	Effective in synchronization of mitosis	Cells may undergo mitotic slippage	[86]
Double thymidine block	Arrest cell cycle in S phase	Precise synchronization	Possibility of incomplete synchronization	[81]
Centrifugal elutriation	Causes physical separation of cells based on their size	Non-toxic	Less precise, requires specialized instruments	[87]
Lovastatin treatment	Inhibits HMG Co A reductase that results in decrease of cholesterol level causing cell arrest in G1 phase	Effective G1 synchronization, less toxic	Prolonged exposure may affect the cell viability	[88]
Hydroxyurea treatment	Inhibition of ribonucleotide reductase that causes cell arrest in S phase	Effective for S phase synchronization	Prolonged treatment can lead to cell death	[89]

play a crucial role in controlling cell division by modulating transcription in response to various extracellular and intracellular signals [90]. Many cyclin-dependent kinase interact with one or more cyclins, unlike the model Cdc28 kinase in budding yeast indicating functional specialization throughout evolution. CDKs, a family of serine/threonine kinases, require the regulatory protein cyclin for their activity. The CMGC group of kinases, named after some of its members, includes CDKs, mitogen activated protein kinases (MAPKs), glycogen synthase kinase- beta (Gsk3), members of the dual specificity tyrosine regulated kinase (DYRK) family, and CDK-like kinases [91]. The above-mentioned Table 11.1 represents the different methods of cell synchrony along with their advantages and disadvantages.

11.5.1.1.1 CDK Inhibitors

At the G2/M transition, cyclin-dependent kinase 1 (CDK1), a key cell cycle driver, provides an excellent target for cell cycle control. Both the completion of mitosis and the cell's shift from the G2 to the M phase depend on CDK1 activity. While CDK1/cyclin B complexes build up during the G2 phase, inhibitory phosphorylation keeps them dormant. The inhibitory phosphates from CDK1 are removed by CDC25 phosphatases at the G2/M boundary, accurately signaling the start of mitosis [92]. Activated CDK1 phosphorylates several protein substrates to organize the intricate biochemical and morphological processes that occur during cell division. According to this mode of cell cycle control, cells with decreased CDK1 activity should arrest at the G2/M boundary rather than begin mitosis. An efficient G2 synchronization drug might be a reversible inhibitor of CDK1 catalytic activity, as CDK1 has not been demonstrated to be involved in cell cycle progression other than mitosis. Nevertheless, the majority of known CDK inhibitors target the ATP binding site, which is where the great degree of homology between CDK1 and CDK2 makes it difficult to produce selective CDK1 inhibitors. Every CDK1 inhibitor is a dual CDK1/CDK2 pan-CDK inhibitor, or off-target actions have probably impacted their cell cycle patterns [93].

11.5.2 Controlled Release of Growth Factors

Growth factors are soluble signaling molecules that specifically bind to transmembrane receptors on target cells influencing a wide variety of physiological processes. The biological effect of growth factor is determined by the identity. Therefore, selecting the important growth factor, either individually or in combination, is crucial for developing a controlled drug delivery system that optimizes tissue healing [94]. Growth factors are essential in tissue engineering because they control cell fate and facilitate the development of tissues. Gaining knowledge of the essential conditions for growth factor delivery can lead to the development of innovative tissue engineering platforms. Growth factors have a propensity to become less bioactive in response to physical and environmental changes [95]. Other recent studies are investigating novel approaches to delivering growth factors, such using bioreactors and delivering them by molecular recognition. Compared to traditional methods, technologies that enable the administration of growth factors in three dimensions show promise. They are now thought of as a subgroup of cytokines. In order to support osteoprogenitor cells' infiltration and proliferation for integrative tissue healing, the carrier can also function as a substrate that mimics extracellular matrix [96]. The article outlines the fundamental needs for protein, cell, and gene therapy and elaborates on how the choice of materials, arrangements, and procedures impacts the distribution of growth factors and cells as well as the effectiveness of regenerative medicine, both *in vitro* and *in vivo* [97].

11.5.3 Electrical and Mechanical Stimulation

One can achieve mechanical stimulation by varying the stiffness. Silently, there has been a recent upsurge in research on the processes behind the cellular response to electrical stimulation (ES) with an eye toward possible clinical uses. However, this methodology's unconventionality has given rise to a wide range of approaches that challenge the interpretation and comparability of experimental results. The three main types of electrode-based electro-stimulation (ES), conductive substrate-mediated ES, and non-invasive stimulation are applied in biomedical research and clinical settings [98]. It has been discovered that ES increases stem cell differentiation, migration, growth, and proliferation of cells, indicating a significant potential for regulating cellular activities in both healthy and pathological settings. Inappropriate setup or specifications, however, might backfire. The manner and format in which the electric signals are created determines their level of complexity [99]. Additionally, it is challenging to compare one set of criteria to another. According to Wolff's Law, mechanical stimulation causes the force line direction to reshape the morphology of bone [100]. In high-stress areas, bone mass rises, whereas in low-stress areas, it falls. According to Wolff's Law, bone can perceive mechanical stress from the outside world, react to it, and adjust its metabolism accordingly. Bedridden patients experience bone loss because of disuse osteoporosis in the clinic,

Table 11.2 Strategies for enhancing cellular synchrony.

Strategy	Description	Mechanism of action	Applications	References
Regulation of cell cycle through small molecules	Small molecules ensure the cell division in a synchronized manner	Small molecules modulate key proteins in cell cycle which promotes synchronized cell proliferation	Treatment of cancer, regenerative medicine	[103]
Genetic manipulation	Alteration of gene regulation to enhance cell synchronization	Modulation in genes related to cell cycle occurs which ensures coordinated cell growth	Stem cell therapy, muscle regeneration	[104]
Bioelectrical stimulation	Stimulation to tissues with electrical stimulation to synchronize the cellular activities	Electrical stimulation promotes cell alignment and accelerates healing process	Generation of nerve, cardiac repair	[105]
Tissue engineering scaffolds	Designing of 3D structures that mimics the extracellular matrix to guide cell growth	Scaffolds provide a framework for cells to attach, proliferate, and differentiate in a coordinated manner	Bone, cartilage, and organ regeneration	[106]
Pre-conditioning stem cell therapy	Using preconditioning stem cells to enhance the cell synchrony	Preconditioned cells are most likely to integrate, survive, and function	Heart repair, spinal cord injury, organ regeneration	[107]
Microenvironment engineering	Modify the cellular microenvironment to promote synchronized cellular behavior	Factors like pH, substrate, oxygen levels help in optimization of cell functions	Tissue regeneration, organ transplantation	[108]

which is caused by inadequate loading [101]. Osteocytes and bone marrow MSCs are two of the bone cell types that are capable of sensing mechanical stimulation. These cells respond to mechanical inputs from the outside and perform a variety of physiological functions [102]. Table 11.2 illustrates the different kinds of techniques which can be employed for enhancing synchrony of the cells.

11.6 Applications of Stem Cell-Based Tissue Regeneration

11.6.1 Bone and Cartilage Tissue Regeneration

Either an indirect (secondary) or direct (primary) procedure can heal a bone. The remodeling machinery starts the mending process during direct healing, moving from the cortex to the other side of the cortex along the fracture line. In order to promote primary healing, the broken section must be tightly fastened with the use of external supports like cords, plaster, or wooden pieces and returned to its original anatomical position. Indirect or secondary healing techniques based on the overlapping processes of remodeling, regeneration, and inflammation are used to mend the fractured bone [109]. The availability of progenitors at the fracture site that later develop into osteoblasts is essential for the bone healing process [110]. Osteoclasts modify newly formed bone as a result of intramembranous and endochondral ossification. The extracellular matrix (ECM), which is made up of glycosaminoglycans, proteoglycans, glycoproteins, and fibrillar proteins, is where newly formed intramembranous bone is placed [111].

11.6.2 Cardiac Tissue Regeneration

Adult bone marrow cells have been linked in some cases to both functional improvement in infarcted hearts and myocardial regeneration. Based on available evidence, bone marrow cells enhance heart performance through a paracrine process that relies on soluble factor release rather than trans-differentiation. However, before these cells can be safely employed for cardiac treatment, concerns about the potential production of bone and cartilage from these cells as well as the inconsistent outcomes of bone marrow transplantation in patients with myocardial infarction must to be resolved [112]. Resident cardiac stem cells represent another potential supply of cells for heart engineering. Stem cell antigen-1 expressing cardiac progenitor cells have been found in the adult mouse myocardium; these cells are associated with cardiac homing and differentiation after myocardial infarction. According to other reports, there are cells called Lin (–) c-kit (+) that may be injected into an ischemic heart to restore well-differentiated myocardium and resident Isl1+ cardiac progenitors, which have demonstrated the ability to develop into cardiomyocytes. Yet, because these cells are present in such little quantities, it is yet uncertain what their true potential for cardiovascular tissue engineering is [113]. Numerous studies have been conducted utilizing various cell types to illustrate the efficacy and safety of stem cells treatment for MI since the first human use of stem cells occurred in 2001. Nevertheless, the majority of clinical trials indicate only modest improvements in cardiac function, and none of the existing cell treatment techniques have been demonstrated to successfully rebuild the damaged heart following MI. For cell-based heart regeneration, there are still many unresolved questions that needs further investigations [114]. A well-structured cardiac tissue must be developed to replace the damage area after a myocardial infarction, in order to restore normal heart function. The optimal cell treatment should produce new muscle fibers, a new vascular network, and coordinate with the host heart's contractions. While pluripotent stem cells are used to create cardiomyocytes, smooth muscle cells, and endothelial cells, no cell transplant has been able to properly restore a completely functioning and ordered cardiac tissue because the transplanted cells have not aligned and integrated with the host cells. Following MI, cardiac fibroblasts take the place of the infarction area [112]. Collagen is excreted by cardiac fibroblasts to create new extracellular matrix (ECM). Collagen deposition causes scar fibrosis, which skews the alignment of the cardiomyocytes. Consequently, collagen prevents the transplanted cells from aligning correctly. Connexin43 is a crucial protein for cell–cell connection and electrical signal transduction in the host heart, which is another need for the mechanical and electrical integration of the transplanted cells [115]. Restoring gap junction coupling is crucial because overexpression of Connexin43 in transplanted cells might lessen post-infarct arrhythmia and enhance intercellular mechanical and electrical connection. Nevertheless, during MI, cardiac fibroblasts increase the production of Connexin, which hinders the incorporation of transplanted cells with host cells, and it also inhibits the formation of strong gap junction coupling [116].

11.6.3 Regeneration of Neural Tissue

Regeneration of nerves is an intricate biological process. If the damage is minor, nerves in the peripheral nervous system can repair themselves.

Larger wounds require surgical intervention, usually including the harvesting of nerve grafts from other parts of the body. Because the body has components that prevent healing, spinal cord injuries are more difficult to treat. Regretfully, there is currently no known way to fully heal spinal cord injuries. Therefore, whereas attempts to treat spinal cord injuries concentrate on fostering an environment that is favorable for regeneration, bioengineering initiatives for the peripheral nervous system are concentrated on alternatives to the nerve transplant [117]. Fortunately, there is hope for novel therapies for nerve damage due to recent advancements in neurology, cell culture, genetics, and biomaterials. Neural stem cells (NSCs) are pluripotent cells which enable remyelination, differentiate into particular neuronal or glial cells, and provide nourishment [118]. Furthermore, because NSCs can develop into neurons and constitutively release brain-derived neurotropic substances, they are appropriate for fundamental spinal cord tissue engineering studies. According to Zhong D *et al.*, by delivering vascular endothelial growth factor A to spinal cord microvascular endothelial cells, NSCs-exosomes demonstrated a proangiogenic impact, hence encouraging microvascular regeneration and tissue repair [119].

11.6.4 Regeneration of Sensory Organ

Receptors found in certain epithelial organs mediate the sensory reception of vision, smell, hearing, and balance. A rising number of people have sensorineural hearing loss and macular degeneration, which cause age-related degeneration of the cochlear hair cells and retinal photoreceptors, respectively [120]. Certain characteristics of the cellular and molecular processes that permit the continuous genesis of receptor cells in diverse specialized sensory epithelia in different animals offer insights into what elements may be essential for regeneration. A reasonably steady population of olfactory receptor cells is maintained by an equilibrium between the rate of synthesis and disappearance of these cells. Instead of being produced to replace dead cells, rod photoreceptor cells are produced as the retina expands. These epithelia need a large number of mitotic cells which can function similarly to the stem cells in non-neural epithelia in order to continue producing sensory receptor cells [121]. In the healthy epithelium, the GBCs are mitotically active and serve as a multipotent progenitor for all other olfactory cell types, including the sensory receptors. In non-mammalian vestibular systems, the situation is slightly different, as there does not seem to be a dedicated hair cell progenitor [121, 122]. Instead, it seems that as the epithelium expands, some or all of the support cells retain the ability to divide mitotically and do so slowly to create new hair cells as well as support cells. A population of cells known as rod precursors in the retina of fish is the source of new rods. Under normal circumstances, these cells only produce rod photoreceptors and are most likely descended from Müller glia. The rod precursor in the retina expresses NeuroD1, indicating that these cells are committed to developing as rod photoreceptors and that their expression of other transcription factors specific to photoreceptors indicates that they are at a slightly "later" stage of development [123]. In all the studied species, regeneration is possible for all cell types, including sensory receptor neurons. Within the basal cell populations, the number of mitotic figures grows significantly, and new receptor neurons become visible in a matter of days [124]. By using traceable thymidine analogues to label the epithelium, researchers were able to observe that the proliferating cells produced new sustentacular cells and receptor neurons, resulting in full epithelial restoration 4 weeks following the damage. The newly formed receptor neurons carry out their regular functions by extending their axons back to the olfactory bulb. The poison methyl bromide (MeBr) damages the sensory epithelium, causing a considerably greater degeneration of the receptor neurons, sustentacular supporting cells, and many GBC [125].

11.7 Challenges and Future Directions

Because of their intrinsic heterogeneity, stem cell populations might react differently to stimuli related to differentiation. This heterogeneity might result in incomplete or uneven tissue creation because it makes it difficult to achieve coordinated cellular behavior during tissue regeneration. One of the biggest challenges is translating synchronized stem cell methods from small-scale laboratory settings to therapeutically relevant proportions. One major challenge is maintaining coordinated cellular behavior as the tissue regeneration process gets larger and incorporating these approaches into the current therapeutic frameworks [126]. Advanced synchronization procedures notwithstanding, transplanting stem-cell-derived tissues into patients may elicit immunological responses, resulting in tissue rejection or diminished efficacy. For stem-cell-based regenerative treatments to be successful, this immunological barrier must be overcome. Proper development of functional tissues depends on achieving temporal synchronization in stem cell differentiation and maturation [127]. However, it is difficult and yet in its early stages to coordinate the timing of these events over a large population of cells. Furthermore, the regulatory environment around stem cell treatments is continually changing, which may impede the creation and application of fresh synchronization techniques. It will be essential to develop sophisticated biomaterials that more closely resemble the natural stem cell niche. By offering both temporal and geographical control over the microenvironment, these materials may enhance the results of tissue regeneration by encouraging synchronized stem cell behavior [128]. CRISPR and other gene editing technologies may be utilized to produce stem cells with improved behavior synchronization capabilities. It could be able to create stem cells with more consistent and predictable responses by altering certain genes that control cellular timing and differentiation. Using 3D bioprinting, it is possible to precisely regulate the orientation and positioning of cells, resulting in highly ordered tissue architectures. With the use of this technique, stem cells might be printed in precise patterns that encourage coordinated behavior and improve tissue regeneration [129]. Immune rejection may be decreased by creating stem cells with immune-modulatory capabilities or by combining immune-suppressive medications with stem cell treatments. This would increase the rate of tissue regeneration based on stem cells being successful. Customizing stem cell treatments for each patient according to their immune system and genetic composition may increase the efficacy and synchronization of these interventions. Using customized strategies might guarantee that the stem cells behave consistently and more easily blend in with the patient's natural tissues [130].

Conclusion

The concept of cellular synchrony in tissue regeneration using stem cells points to a promising area of exploration in regenerative medicine. The ability to achieve coordinated behavior among stem cells is essential for effective tissue repair and regeneration as it mirrors the natural processes of embryonic development and wound healing. Cellular synchrony ensures that stem cells differentiate, proliferate, and integrate into damaged tissues in a controlled and organized manner, minimizing risks such as tumor formation or improper tissue development. Recent advances in understanding the mechanisms of cellular communication, such as signaling pathways, extracellular matrix interactions, and epigenetic regulation, have paved the way for innovative strategies to induce and maintain synchrony among stem cells. Additionally, the development of biomaterials and scaffolds that can support synchronized cell behavior further augments the effectiveness of stem cell-based therapies. However, despite these advances, challenges remain in translating these strategies from the laboratory to clinical settings. The complexity of human tissues and the requirement for precise regulation of stem cell behavior in living organisms call for continued research and refinement of these method. Moreover, understanding the long-term effects of induced cellular synchrony on tissue function and overall organismal health is crucial to understand the safety and efficacy of these therapies. In conclusion, cellular synchrony plays a crucial role in the effectiveness of stem-cell-based tissue regeneration. Ongoing research and innovation in this field have the potential to transform regenerative medicine to provide renewed hope for patients suffering from various degenerative diseases and injuries. The integration of interdisciplinary approaches will be essential in overcoming current challenges and unlocking the full therapeutic range of stem cells.

Acknowledgments

The authors express their heartfelt gratitude toward the leadership and management of JSS College of Pharmacy, JSS Academy of Higher Education and Research (JSS AHER), Mysuru, Karnataka, India.

References

1. George, J., Sreekala, M.S., Thomas, S.A., A review on interface modification and characterization of natural fiber reinforced plastic composites. *Polym. Eng. Sci.*, 41, 1471, 2001.
2. Seetharaman, R., Mahmood, A., Kshatriya, P., Patel, D., Srivastava, A., An Overview on Stem Cells in Tissue Regeneration. *Curr. Pharm. Des.*, 25, 18, 2086–2098, 2019.
3. Astarita, C. and Arora, C.L., and Trovato, L.,Tissue regeneration: an overview from stem cells to micrografts. *J. Int. Med. Res.*, 48, 6, 030006052091479, 2020.
4. Jazayeri, H.E., Tahriri, M., Razavi, M., Khoshroo, K., Fahimipour, F., Dashtimoghadam, E., Almeida, L., Tayebi, L., A current overview of materials and strategies for potential use in maxillofacial tissue regeneration. *Mater. Sci. Eng. C*, 70, 913–929, 2017.
5. Yi, B., Xu, Q., Liu, W., An overview of substrate stiffness guided cellular response and its applications in tissue regeneration. *Bioact. Mater.*, 15, 82–102, 2022.
6. St. John, J.C., Genomic Balance: Two Genomes Establishing Synchrony to Modulate Cellular Fate and Function. *Cells*, 8, 11, 1306, 2019.
7. Padilla, G.M., Cameron, I.L., Whitson, G.L., Cell Synchrony. *Science*, 147, 3654, 175–177, 1979.

8. Mohawk, J.A. and Takahashi, J.S., Cell autonomy and synchrony of suprachiasmatic nucleus circadian oscillators. *Trends Neurosci.*, 34, 7, 349–358, 2011.
9. Stappenbeck, T.S. and Miyoshi, H., The Role of Stromal Stem Cells in Tissue Regeneration and Wound Repair. *Science*, 324, 5935, 1666–1669, 2009.
10. Mimeault, M. and Batra, S.K., Concise Review: Recent Advances on the Significance of Stem Cells in Tissue Regeneration and Cancer Therapies. *Stem Cells*, 24, 11, 2319–2345, 2006.
11. Lee, J.Y. and Hong, S.-H., Hematopoietic Stem Cells and Their Roles in Tissue Regeneration. *Int. J. Stem Cells*, 13, 1, 1–12, 2020.
12. Mizuno, H., Adipose-derived stem cells for tissue repair and regeneration: ten years of research and a literature review. *J. Nippon Med. Sch.*, 76, 2, 56–66, 2009.
13. Terskikh, A.V., Bryant, P.J., Schwartz, P.H., Mammalian stem cells. *Pediatr. Res.*, 59, 4, 13–20, 2006.
14. Sun, C., Yue, J., He, N., Liu, Y., Zhang, X., Zhang, Y., *Fundamental Principles of Stem Cell Banking*, Springer Nature, Beijing, China, pp. 31–45, 2016.
15. Tian, Z., Yu, T., Liu, J., Wang, T., Higuchi, A., *Introduction to stem cells*, Elsevier, Amsterdam, Netherlands, pp. 3–32, 2023.
16. Levenberg, S., Khademhosseini, A., Langer, R., Embryonic Stem Cells in Tissue Engineering, in: *Essentials of Stem Cell Biology*, pp. 571–581, Elsevier, Amsterdam, Netherlands, 2009.
17. Rippon, H.J. and Bishop, A.E., Embryonic stem cells. *Cell Prolif.*, 37, 1, 23–34, 2004.
18. Young, R.A., Control of the Embryonic Stem Cell State. *Cell*, 144, 6, 940–954, 2011.
19. Martello, G. and Smith, A., The Nature of Embryonic Stem Cells. *Annu. Rev. Cell Dev. Biol.*, 30, 1, 647–675, 2014.
20. Wobus, A.M., Potential of embryonic stem cells. *Mol. Aspects Med.*, 22, 3, 149–164, 2011.
21. Pera, M.F., Reubinoff, B., Trounson, A., Human embryonic stem cells. *J. Cell Sci.*, 113, 1, 5–10, 2000.
22. Nichols, J. and Smith, A., The origin and identity of embryonic stem cells. *Development*, 138, 1, 3–8, 2011.
23. Solter, D., From teratocarcinomas to embryonic stem cells and beyond: a history of embryonic stem cell research. *Nat. Rev. Genet.*, 7, 4, 319–327, 2006.
24. Lerou, P. and Daley, G., Therapeutic potential of embryonic stem cells. *Blood Rev.*, 19, 6, 321–331, 2005.
25. Young, H.E. and Black, A.C., Adult stem cells. *Anat. Rec. A Discov. Mol. Cell Evol. Biol.*, 276A, 75–102, 1, 2004.
26. Alison, M. and Islam, S., Attributes of adult stem cells. *J. Pathol.*, 217, 2, 144–160, 2009.
27. Clevers, H., What is an adult stem cell? *Sci. (1979)*, 350, 6266, 1319–1320, 2015.

28. Snippert, H.J. and Clevers, H., Tracking adult stem cells. *EMBO Rep.*, 12, 2, 113–122, 2011.
29. Clevers, H. and Watt, F.M., Defining Adult Stem Cells by Function, not by Phenotype. *Annu. Rev. Biochem.*, 87, 1, 1015–1027, 2018.
30. Prentice, D.A., Adult Stem Cells. *Circ. Res.*, 124, 6, 837–839, 2019.
31. Andrzejewska, A., Lukomska, B., Janowski, M., Concise Review: Mesenchymal Stem Cells: From Roots to Boost. *Stem Cells*, 37, 7, 855–864, 2019.
32. Liu, S., Zhou, J., Zhang, X., Liu, Y., Chen, J., Hu, B., Song, J., Zhang, Y., Strategies to Optimize Adult Stem Cell Therapy for Tissue Regeneration. *Int. J. Mol. Sci.*, 17, 6, 982, 2016.
33. Baker, C.L. and Pera, M.F., Capturing Totipotent Stem Cells. *Cell Stem Cell*, 22, 1, 25–34, 2018.
34. Condic, M.L., Totipotency: What It Is and What It Is Not. *Stem Cells Dev.*, 23, 8, 796–812, 2014.
35. Ghazimoradi, M.H., Khalafizadeh, A., Babashah, S., A critical review on induced totipotent stem cells: Types and methods. *Stem Cell Res.*, 63, 102857, 2024.
36. Morgani, S.M., Canham, M.A., Nichols, J., Sharov, A.A., Migueles, R.P., Ko, M.S.H., Brickman, J.M., Totipotent Embryonic Stem Cells Arise in Ground-State Culture Conditions. *Cell Rep.*, 3, 6, 1945–1957, 2013.
37. Hu, Y., Yang, Y., Tan, P., Zhang, Y., Han, M., Yu, J., Zhang, X., Jia, Z., Wang, D., Yao, K., Pang, H., Hu, Z., Li, Y., Ma, T., Liu, K., Ding, S., Induction of mouse totipotent stem cells by a defined chemical cocktail. *Nature*, 617, 7962, 792–797, 2023.
38. Martí, M., Mulero, L., Pardo, C., Morera, C., Carrió, M., Laricchia-Robbio, L., Esteban, C.R., Belmonte, J.C.I., Characterization of pluripotent stem cells. *Nat. Protoc.*, 8, 2, 223–253, 2013.
39. Okita, K. and Yamanaka, S., Induced pluripotent stem cells: opportunities and challenges. *Philos. Trans. R. Soc. B Biol. Sci.*, 366, 1575, 2198–2207, 2011.
40. Visvader, J.E. and Lindeman, G.J., The unmasking of novel unipotent stem cells in the mammary gland. *EMBO J.*, 30, 24, 4858–4859, 2011.
41. Ko, K., Tapia, N., Wu, G., Kim, J.B., Bravo, M.J.A., Sasse, P., Glaser, T., Ruau, D., Han, D.W., Greber, B., Hausdörfer, K., Sebastiano, V., Stehling, M., Fleischmann, B.K., Brüstle, O., Zenke, M., Schöler, H.R., Induction of Pluripotency in Adult Unipotent Germline Stem Cells. *Cell Stem Cell*, 5, 1, 87–96, 2009.
42. Ko, K., Araúzo-Bravo, M.J., Kim, J., Stehling, M., Schöler, H.R., Conversion of adult mouse unipotent germline stem cells into pluripotent stem cells. *Nat. Protoc.*, 5, 5, 921–928, 2010.
43. Krampera, M., Pasini, A., Pizzolo, G., Cosmi, L., Romagnani, S., Annunziato, F., Regenerative and immunomodulatory potential of mesenchymal stem cells. *Curr. Opin. Pharmacol.*, 6, 4, 435–441, 2006.

44. Krampera, M., Marconi, S., Pasini, A., Galiè, M., Rigotti, G., Mosna, F., Tinelli, M., Lovato, L., Anghileri, E., Andreini, A., Pizzolo, G., Sbarbati, A., Bonetti, B., Induction of neural-like differentiation in human mesenchymal stem cells derived from bone marrow, fat, spleen and thymus. *Bone*, 40, 2, 382–390, 2007.
45. Dominici, M., Paolucci, P., Conte, P., Horwitz, E.M., Heterogeneity of Multipotent Mesenchymal Stromal Cells: From Stromal Cells to Stem Cells and Vice Versa. *Transplantation*, 87, 9S, S36–S42, 2009.
46. Majo, F., Rochat, A., Nicolas, M., Jaoudé, G.A., Barrandon, Y., Oligopotent stem cells are distributed throughout the mammalian ocular surface. *Nature*, 456, 7219, 250–254, 2008.
47. Kolios, G. and Moodley, Y., Introduction to Stem Cells and Regenerative Medicine. *Respiration*, 85, 1, 3–10, 2013.
48. Till, J.E. and McCulloch, E.A., Hemopoietic stem cell differentiation. *Biochim. Biophys. Acta (BBA) - Rev. Cancer*, 605, 4, 431–459, 1980.
49. Keller, G., Embryonic stem cell differentiation: emergence of a new era in biology and medicine. *Genes Dev.*, 19, 10, 1129–1155, 2005.
50. Fuchs, E. and Chen, T., A matter of life and death: self-renewal in stem cells. *EMBO Rep.*, 14, 1, 39–48, 2013.
51. Shenghui, H., Nakada, D., Morrison, S.J., Mechanisms of Stem Cell Self-Renewal. *Annu. Rev. Cell Dev. Biol.*, 25, 1, 377–406, 2009.
52. Ivanova, N., Dobrin, R., Lu, R., Kotenko, I., Levorse, J., DeCoste, C., Schafer, X., Lun, Y., Lemischka, I.R., Dissecting self-renewal in stem cells with RNA interference. *Nature*, 442, 7102, 533–538, 2006.
53. Molofsky, A.V., Pardal, R., Morrison, S.J., Diverse mechanisms regulate stem cell self-renewal. *Curr. Opin. Cell Biol.*, 16, 6, 700–707, 2004.
54. Beaver, C.M., Ahmed, A., Masters, J.R., Clonogenicity: Holoclones and Meroclones Contain Stem Cells. *PLoS One*, 9, 2, e89834, 2014.
55. Matsui, W., Wang, Q., Barber, J.P., Brennan, S., Smith, B.D., Borrello, I., McNiece, I., Lin, L., Ambinder, R.F., Peacock, C., Watkins, D.N., Huff, C.A., Jones, R.J., Clonogenic Multiple Myeloma Progenitors, Stem Cell Properties, and Drug Resistance. *Cancer Res.*, 68, 1, 190–197, 2008.
56. Potten, C.S. and Hendry, J.H., Clonogenic Cells and Stem Cells in Epidermis. *Int. J. Radiat. Biol. Relat. Stud. Phys. Chem. Med.*, 24, 5, 537–540, 1973.
57. Goldsmith, C.S. and Bell-Pedersen, D., *Diverse Roles for MAPK Signaling in Circadian Clocks*, Elsevier, Amsterdam, Netherland, pp. 1–39, 2013.
58. Roberts, E.C., Shapiro, P.S., Nahreini, T.S., Pages, G., Pouyssegur, J., Ahn, N.G., Distinct Cell Cycle Timing Requirements for Extracellular Signal-Regulated Kinase and Phosphoinositide 3-Kinase Signaling Pathways in Somatic Cell Mitosis. *Mol. Cell. Biol.*, 22, 20, 7226–7241, 2002.
59. Ganesan, H., Balasubramanian, V., Iyer, M., Venugopal, A., Subramaniam, M.D., Cho, S.-G., Vellingiri, B., mTOR signalling pathway - A root cause for idiopathic autism? *BMB Rep.*, 52, 7, 424–433, 2019.

60. MA, X. and BAI, Y., IGF-1 activates the P13K/AKT signalling pathway *via* upregulation of secretory clusterin. *Mol. Med. Rep.*, 6, 6, 1433–1437, 2012.
61. Xu, J., Li, Z., Su, Q., Zhao, J., Ma, J., TRIM29 promotes progression of thyroid carcinoma *via* activating P13K/AKT signalling pathway. *Oncol. Rep.*, 37, 3, 1555–1564, 2017.
62. Xu, W., Yang, Z., Lu, N., A new role for the PI3K/Akt signalling pathway in the epithelial-mesenchymal transition. *Cell Adh. Migr.*, 9, 4, 317–324, 2105.
63. Lenhart, K.F., Capozzoli, B., Warrick, G.S.D., DiNardo, S., Diminished Jak/STAT Signaling Causes Early-Onset Aging Defects in Stem Cell Cytokinesis. *Curr. Biol.*, 29, 2, 256–267.e3, 2019.
64. Sun, H., Ma, D., Cheng, Y., Li, J., Zhang, W., Jiang, T., Li, Z., Li, X., Meng, H., The JAK-STAT Signalling Pathway in Epilepsy. *Curr. Neuropharmacol.*, 21, 10, 2049–2069, 2023.
65. Pansky, A., Hildebrand, P., Fasler-Kan, E., Baselgia, L., Ketterer, S., Beglinger, C., Heim, M.H., Defective Jak-STAT signal transduction pathway in melanoma cells resistant to growth inhibition by interferon-? *Int. J. Cancer*, 85, 5, 720–725, 2000.
66. Goody, R.J., Beckham, J.D., Rubtsova, K., Tyler, K.L., JAK-STAT signaling pathways are activated in the brain following reovirus infection. *J. Neurovirol.*, 13, 4, 373–383, 2007.
67. Matsu-ura, T., Moore, S.R., Hong, C.I., WNT Takes Two to Tango: Molecular Links between the Circadian Clock and the Cell Cycle in Adult Stem Cells. *J. Biol. Rhythms*, 33, 1, 5–14, 2018.
68. Waterman, M.L., Lymphoid enhancer factor/T cell factor expression in colorectal cancer. *Cancer Metastasis Rev.*, 23, 41–52, 2004.
69. Saga, Y., The Synchrony and Cyclicity of Developmental Events. *Cold Spring Harb. Perspect. Biol.*, 4, 4, a008201–a008201, 2012.
70. Klevecz, R.R., Li, C.M., Marcus, I., Frankel, P.H., Collective behavior in gene regulation: The cell is an oscillator, the cell cycle a developmental process. *FEBS J.*, 275, 10, 2372–2384, 2008.
71. Aguiar, M.A.D., Dias, A.P.S., Ruan, H., Synchrony patterns in gene regulatory networks. *Physica D*, 429, 133065, 2022.
72. Orlando, D., Lin, C.Y., Bernard, A., Iversen, E.S., Hartemink, A.J., Haase, S.B., A Probabilistic Model for Cell Cycle Distributions in Synchrony Experiments. *Cell Cycle*, 6, 4, 478–488, 2007.
73. Grdina, D.J., Meistrich, M.L., Meyn, R.E., Johnson, T.S., White, R.A., Cell Synchrony Techniques, in: *Techniques in Cell Cycle Analysis*, pp. 367–402, Humana Press, Totowa, NJ, 1987.
74. Liliensiek, S.J., Schell, K., Howard, E., Nealey, P., Murphy, C.J., Cell sorting but not serum starvation is effective for SV40 human corneal epithelial cell cycle synchronization. *Exp. Eye Res.*, 83, 1, 61–68, 2006.
75. Cooper, S. and Gonzalez-Hernandez, M., Experimental reconsideration of the utility of serum starvation as a method for synchronizing mammalian cells. *Cell Biol. Int.*, 33, 1, 71–77, 2009.

76. Griffin, M.J., Synchronization of some human cell strains by serum and calcium starvation. *In Vitro*, 12, 5, 393–398, 1976.
77. Sun, X., Wang, S., Zhang, Y., Wang, H., Wang, L., Ying, L., Li, R., Li, N., Cell-cycle synchronization of fibroblasts derived from transgenic cloned cattle ear skin: effects of serum starvation, roscovitine and contact inhibition. *Zygote*, 16, 2, 111–116, 2008.
78. Wang, R.C. and Wang, Z., *Synchronization of Cultured Cells to G1, S, G2, and M Phases by Double Thymidine Block*, pp. 61–71, Springer Nature, Beijing, China, 2022.
79. Goepp, M., Le Guennec, D., Rossary, A., Vasson, M., Cell Cycle Synchronization of the Murine EO771 Cell Line Using Double Thymidine Block Treatment. *BioEssays*, 42, 9, 1900116, 2020.
80. Carrasco, E., Álvarez, P.J., Melguizo, C., Prados, J., Alvarez-Manzaneda, E., Chahboun, R., Messouri, I., Vázquez-Vázquez, M.I., Aranega, A., Rodriguez-Serrano, F., Novel merosesquiterpene exerts a potent antitumor activity against breast cancer cells *in vitro* and *in vivo*. *Eur. J. Med. Chem.*, 22, 79, 1–2, 2014.
81. BOSTOCK, C., An evaluation of the double thymidine block for synchronizing mammalian cells at the G1-S border*1. *Exp. Cell. Res.*, 68, 1, 163–168, 1971.
82. Shedden, K. and Cooper, S., Analysis of cell-cycle-specific gene expression in human cells as determined by microarrays and double-thymidine block synchronization. *Proc. Natl. Acad. Sci.*, 99, 7, 4379–4384, 2002.
83. Dalman, A., Eftekhari-Yazdi, P., Valojerdi, M., Shahverdi, A., Gourabi, H., Janzamin, E., Fakheri, R., Sadeghian, F., Hasani, F., Synchronizing Cell Cycle of Goat Fibroblasts by Serum Starvation Causes Apoptosis. *Reprod. Domest. Anim.*, 45, 5, e46–53, 2002.
84. Thomas, D.B. and Lingwood, C.A., A model of cell cycle control: Effects of thymidine on synchronous cell cultures. *Cell*, 5, 1, 37–44, 1975.
85. Cooper, S., Iyer, G., Tarquini, M., Bissett, P., Nocodazole does not synchronize cells: implications for cell-cycle control and whole-culture synchronization. *Cell Tissue Res.*, 324, 2, 237–242, 2006.
86. Ohara, H. and Terasima, T., Colcemid Effects on Mitotic Synchrony and Synchronized Cells. *Cell Struct. Funct.*, 1, 2, 187–195, 1976.
87. Banfalvi, G., Cell cycle synchronization of animal cells and nuclei by centrifugal elutriation. *Nat. Protoc.*, 3, 4, 663–673, 20082008.
88. JavanMoghadam-Kamrani, S. and Keyomarsi, K., Synchronization of the cell cycle using Lovastatin. *Cell Cycle*, 7, 15, 2434–2440, 2008.
89. Miyamoto, T., Watanabe, M., Takabe, Y., Terasima, T., Synchrony of Burkitt Lymphoma Cells Induced by Hydroxyurea. *Cell Struct. Funct.*, 1, 2, 177–185, 1976.
90. Wang, Z., *Cell Cycle Progression and Synchronization: An Overview*, pp. 3–23, Springer Cham, Beijing, China, 2022.

91. Jackman, J. and O'Connor, P.M., Methods for Synchronizing Cells at Specific Stages of the Cell Cycle. *Curr. Protoc. Cell Biol.*, 00, 1, 8–3, 1998.
92. Pérez-Benavente, B. and Farràs, R., *Cell Synchronization Techniques to Study the Action of CDK Inhibitors*, pp. 85–93, Springer Nature, Valencia, Spain, 2016.
93. Vassilev, L.T., Cell Cycle Synchronization at the G 2 /M Phase Border by Reversible Inhibition of CDK1. *Cell Cycle*, 5, 22, 2555–2556, 2006.
94. Caballero Aguilar, L.M., Silva, S.M., Moulton, S.E., Growth factor delivery: Defining the next generation platforms for tissue engineering. *J. Control. Release*, 306, 40–58, 2019.
95. Zisch, A.H., Lutolf, M.P., Ehrbar, M., Raeber, G.P., Rizzi, S.C., Davies, N., Schmökel, H., Bezuidenhout, D., Djonov, V., Zilla, P., Hubbell, J.A., Cell-demanded release of VEGF from synthetic, biointeractive cell-ingrowth matrices for vascularized tissue growth. *FASEB J.*, 17, 15, 2260–2262, 2003.
96. Hom, D.B., Medhi, K., Juhn, S.K., Assefa, G., Johnston, T.P., Vascular Effects of Sustained-Release Fibroblast Growth Factors. *Ann. Otology Rhinology Laryngology*, 105, 2, 109–116, 1996.
97. Al-Rubeai, M. and Emery, A.N., Mechanisms and kinetics of monoclonal antibody synthesis and secretion in synchronous and asynchronous hybridoma cell cultures. *J. Biotechnol.*, 16, 1–2, 67–85, 1990.
98. Stoppel, W.L., Kaplan, D.L., Black, L.D., Electrical and mechanical stimulation of cardiac cells and tissue constructs. *Adv. Drug Deliv. Rev.*, 96, 135–155, 2016.
99. Dang, S.M., Gerecht-Nir, S., Chen, J., Itskovitz-Eldor, J., Zandstra, P.W., Controlled, scable embryonic stem cell differentiation culture. *Stem Cells*, 22, 3, 275–282, 2004.
100. Urbani, L., Sherwood, S.W., Schimke, R.T., Dissociation of Nuclear and Cytoplasmic Cell Cycle Progression by Drugs Employed in Cell Synchronization. *Exp. Cell. Res.*, 219, 1, 159–168, 1995.
101. Xuan, Q., Zhang, Z.-Y., Fu, C., Hu, H.-X., Filkov, V., Social Synchrony on Complex Networks. *IEEE Trans. Cybern.*, 48, 5, 1420–1431, 2018.
102. Thompson, W.R., Rubin, C.T., Rubin, J., Mechanical regulation of signaling pathways in bone. *Gene*, 503, 2, 179–193, 2012.
103. Tayar, A.M., Karzbrun, E., Noireaux, V., Bar-Ziv, R.H., Synchrony and pattern formation of coupled genetic oscillators on a chip of artificial cells. *Proc. Natl. Acad. Sci.*, 114, 44, 11609–11614, 2017.
104. McMillen, D., Kopell, N., Hasty, J., Collins, J.J., Synchronizing genetic relaxation oscillators by intercell signaling. *Proc. Natl. Acad. Sci.*, 99, 2, 679–684, 2002.
105. Medeiros, D., de, C., Moraes, M.F.D., Focus on desynchronization rather than excitability: A new strategy for intraencephalic electrical stimulation. *Epilepsy Behav.*, 38, 32–36, 2014.

106. Freed, L.E., Engelmayr, G.C., Borenstein, J.T., Moutos, F.T., Guilak, F., Advanced Material Strategies for Tissue Engineering Scaffolds. *Adv. Mater.*, 21, 32–33, 3410–3418, 2009.
107. Liska, G.M., Lippert, T., Russo, E., Nieves, N., Borlongan, C.V., A Dual Role for Hyperbaric Oxygen in Stroke Neuroprotection: Preconditioning of the Brain and Stem Cells. *Cond. Med.*, 1, 4, 151–166, 2018.
108. Keung, A.J., Healy, K.E., Kumar, S., Schaffer, D.V., Biophysics and dynamics of natural and engineered stem cell microenvironments. *WIREs Syst. Biol. Med.*, 2, 1, 49–64, 2010.
109. Dimitriou, R., Jones, E., McGonagle, D., Giannoudis, P.V., Bone regeneration: current concepts and future directions. *BMC Med.*, 9, 1, 66, 2011.
110. Bigham-Sadegh, A. and Oryan, A., Selection of animal models for pre-clinical strategies in evaluating the fracture healing, bone graft substitutes and bone tissue regeneration and engineering. *Connect Tissue Res.*, 56, 3, 175–194, 2015.
111. Lin, X., Patil, S., Gao, Y.-G., Qian, A., The Bone Extracellular Matrix in Bone Formation and Regeneration. *Front. Pharmacol.*, 11, 757, 2020.
112. Mancuso, A., Barone, A., Cristiano, M.C., Cianflone, E., Fresta, M., Paolino, D., Cardiac Stem Cell-Loaded Delivery Systems: A New Challenge for Myocardial Tissue Regeneration. *Int. J. Mol. Sci.*, 21, 20, 7701, 2020.
113. Shi, C., Li, Q., Zhao, Y., Chen, W., Chen, B., Xiao, Z., Lin, H., Nie, L., Wang, D., Dai, Stem-cell-capturing collagen scaffold promotes cardiac tissue regeneration. *Biomaterials*, 32, 10, 2508–2515, 2011.
114. Nunes, S.S., Song, H., Chiang, C.K., Radisic, M., Stem Cell-Based Cardiac Tissue Engineering. *J. Cardiovasc. Transl. Res.*, 4, 5, 592–602, 2011.
115. Tian, S., Ma, X., Wang, Z., Liu, Q., Gnatovskiy, L., Heart Regeneration with Embryonic Cardiac Progenitor Cells and Cardiac Tissue Engineering. *J. Stem Cell Transplant. Biol.*, 01, 01, 104, 2015.
116. Martinez, E.C. and Kofidis, T., Adult stem cells for cardiac tissue engineering. *J. Mol. Cell Cardiol.*, 50, 2, 312–319, 2011.
117. Li, H., Ye, A.Q., Su, M., Application of Stem Cells and Advanced Materials in Nerve Tissue Regeneration. *Stem Cells Int.*, 2018, 1–2, 2018.
118. Binan, L., Ajji, A., De Crescenzo, G., Jolicoeur, M., Approaches for Neural Tissue Regeneration. *Stem Cell Rev. Rep.*, 10, 1, 44–59, 2014.
119. Zhong, D., Cao, Y., Li, C.-J., Li, M., Rong, Z.-J., Jiang, L., Guo, Z., Lu, H.-B., Hu, J.-Z., Neural stem cell-derived exosomes facilitate spinal cord functional recovery after injury by promoting angiogenesis. *Exp. Biol. Med.*, 245, 1, 54–65, 2020.
120. Vlastarakos, P.V., Nikolopoulos, T.P., Tavoulari, E., Papacharalambous, G., Tzagaroulakis, A., Dazert, S., Sensory Cell Regeneration and Stem Cells. *Otol. Neurotol.*, 29, 6, 758–768, 2008.
121. Bermingham-McDonogh, O. and Reh, T.A., Regulated Reprogramming in the Regeneration of Sensory Receptor Cells. *Neuron*, 71, 3, 389–405, 2011.

122. Denans, N., Baek, S., Piotrowski, T., Comparing Sensory Organs to Define the Path for Hair Cell Regeneration. *Annu. Rev. Cell Dev. Biol.*, 35, 1, 567–589, 2019.
123. Baddour, J.A., Sousounis, K., Tsonis, P.A., Organ repair and regeneration: An overview. *Birth Defects Res. C Embryo Today*, 96, 1, 1–29, 2012.
124. Child, K.M., Herrick, D.B., Schwob, J.E., Holbrook, E.H., Jang, W., The Neuroregenerative Capacity of Olfactory Stem Cells Is Not Limitless: Implications for Aging. *J. Neurosci.*, 38, 31, 6806–6824, 2018.
125. Choi, R. and Goldstein, B.J., Olfactory epithelium: Cells, clinical disorders, and insights from an adult stem cell niche. *Laryngoscope Investig. Otolaryngol*, 3, 1, 35–42, 2018.
126. Hoshino, A., Horvath, S., Sridhar, A., Chitsazan, A., Reh, T.A., Synchrony and asynchrony between an epigenetic clock and developmental timing. *Sci. Rep.*, 9, 1, 3770, 2019.
127. Davidesco, I., Brain-to-Brain Synchrony in the STEM Classroom. *CBE—Life Sci. Educ.*, 19, 3, es8, 2020.
128. D'Amato, M., Morra, F., Di Meo, I., Tiranti, V., Mitochondrial Transplantation in Mitochondrial Medicine: Current Challenges and Future Perspectives. *Int. J. Mol. Sci.*, 24, 3, 1969, 2003.
129. Gansel, K.S., Neural synchrony in cortical networks: mechanisms and implications for neural information processing and coding. *Front. Integr. Neurosci.*, 16, 900715, 2022.
130. Duchesne de Lamotte, J., Perrier, A., Martinat, C., Nicoleau, C., Emerging Opportunities in Human Pluripotent Stem-Cells Based Assays to Explore the Diversity of Botulinum Neurotoxins as Future Therapeutics. *Int. J. Mol. Sci.*, 22, 14, 7524, 2021.

12

Tissue Engineering Paradigms: Creating Functional Replacements with Stem Cells

Karan Goel[1], Isha Chawla[1], Garima[2], Sarita Sharma[1] and Sumeet Gupta[1]*

[1]M.M. College of Pharmacy, Maharishi Markandeshwar (Deemed to be University), Mullana-Ambala, Haryana, India
[2]Chitkara College of Pharmacy, Chitkara University, Rajpura, Punjab, India

Abstract

Tissue engineering is an interdisciplinary field that integrates knowledge from stem cell and cellular biology, biochemistry, materials science and engineering, and bio-mechanics to develop biological substitutes suitable for restoring, maintaining, or improving tissue function in traumatic or disease-degenerated conditions. The issue of stem cell-based solutions to anatomic replacement considers the paradigms that stem cells can be used to create functional replacements and reviewing the history, properties, and clinical applications of stem cell-based replacement. It starts with an introduction to stem cells: embryonic, adult, and induced pluripotent stem cells, highlighting their properties and applications in regenerative medicine. They explore biomaterials, specifically natural, synthetic, and composite (i.e., natural/synthetic) biomaterials, along with uses for such materials, properties, and revealing properties that must be considered in designing extracellular matrix (ECM) scaffolds. The efficacies of advanced cell seeding practices, such as scaffold-based methods and bioprinting technologies, in tissue fabrication are reviewed. Here we explore how different types of bioreactors can mimic *in vivo* conditions better suited for cell proliferation and differentiation. It concludes with the clinical translation of tissue engineering products and highlights the areas in which there has been some success and where current limitations, especially vascularization and the immune response, lay. Guidelines, patient safety, and informed consent are reviewed from a regulatory and ethical point of view. An engaging overview of the future of tissue engineering discusses exciting

**Corresponding author*: sumeetgupta25@gmail.com

Ayush Madan, Sumel Ashique, Deepika Arora and Mantosh Kumar Satapathy (eds.) Stem Cell Therapeutics, (541–574) © 2025 Scrivener Publishing LLC

new technologies such as gene editing, smart biomaterials, and artificial intelligence. Their potential in curing autoimmune diseases is in talks but yet not much is done and it is even further far from clinical trials. They encompass comprehensive tissue engineering, motivational paradigms, integration of the latest technological and strategic developments and aspects to further the field to clinical and therapeutic efficacy.

Keywords: Tissue engineering, stem cells, biomaterials, bioprinting, regenerative medicine

12.1 Introduction to Tissue Engineering

Tissue engineering is a multidisciplinary area of study that integrates concepts from biology, engineering, and materials science to create biological replacements that can heal, sustain, or enhance tissue function. Tissue engineering encompasses the creation of basic tissue structures as well as the regeneration of intricate organs. It aims to overcome the constraints of conventional organ transplantation and offers novel approaches for the treatment of many illnesses (de Isla *et al.*, 2010). The inception of tissue engineering may be traced back to the early 1980s when scientists began investigations into the possibility of merging cells with biomaterials to restore or substitute damaged tissues. Initial advancements included the creation of skin grafts and cartilage repair methods, establishing the groundwork for future intricate implementations. Advancements in stem cell biology, biomaterials, and bioreactor technologies have greatly transformed the area, resulting in the development of intricate and highly functional tissue constructions (Berthiaume *et al.*, 2011; de Isla *et al.*, 2010).

12.2 Fundamentals of Stem Cells

12.2.1 Types of Stem Cells

Stem cells possess the exceptional capacity to differentiate into many specialized cell types and possess the potential for self-regeneration. They are categorized into many sorts according to their source and capacity. Table 12.1 represents the list of different types of stems cells.

Table 12.1 Types of stem cells.

Type of stem cell	Source	Characteristics	Applications	Reference
Embryonic stem cells (ESCs)	Inner cell mass of blastocysts	• Pluripotent • Indefinite self-renewal	• Tissue engineering • Disease modeling • Drug screening	(Esmaeili *et al.*, 2023; Pérez *et al.*, 2018)
Adult stem cells (ASCs)	Various tissues (e.g., bone marrow, brain)	• Multipotent • Tissue-specific maintenance and repair	• Regenerative therapies • Tissue repair (e.g., bone marrow transplants, grafts for skin, bone, cartilage)	(Baptista, 2024; Gheitasi *et al.*, 2024)
Induced pluripotent stem cells (iPSCs)	Reprogrammed somatic cells	• Pluripotent • Indefinite self-renewal • Ethical advantage	• Personalized medicine • Disease modeling • Drug discovery	(Kwon *et al.*, 2018)
Perinatal stem cells	Umbilical cord blood, placenta, amniotic fluid	• Pluripotent or multipotent • Less ethical controversy	• Banking for future use • Regenerative medicine	(Tian *et al.*, 2023)
Cancer stem cells (CSCs)	Found within tumors	• Self-renew and differentiate into various tumor cell types	• Targeting cancer treatments • Understanding cancer progression • Potential for more effective therapies	(Kharkar, 2020)

12.2.2 Characteristics and Properties

Stem cells have distinct qualities and features that are crucial for a diverse array of uses in regenerative medicine (Goetzl *et al.*, 2024).

12.2.2.1 Pluripotency and Multipotency

The concept of pluripotency involves the capacity of stem cells to develop into types of cells found in the body. This characteristic is observed in both stem cells (ESCs). Induced pluripotent stem cells (iPSCs). Conversely hematopoietic stem cells have the capability to mature into blood cells and other specific cell varieties. In the case of adult stem cells (ASCs), their multipotency pertains to their capacity to transform into an limited range of cell types (Hovland *et al.*, 2022; Pan *et al.*, 2022).

12.2.2.2 Self-Renewal and Differentiation Potential

One important quality is the ability of stem cells to renew themselves going through divisions while keeping their original form. Keeping a supply of stem cells over time is vital for maintaining tissue equilibrium and aiding in tissue healing. Furthermore, stem cells can transform into cell types when prompted by signals that direct their growth. This aspect plays a role in biology and medical treatments as it allows for the creation of specific cell types essential for tissue repair (C. Li *et al.*, 2023).

12.2.2.3 Regenerative Medicine Applications

In the field of regenerative medicine, stem cells are essential. They are used to substitute impaired or afflicted tissues in the pancreas, heart, and neurological system, among several other locations. A further significant benefit of iPSCs for illness modeling is their capacity to examine the etiology of diseases and evaluate potential therapies in a controlled laboratory setting. By facilitating the analysis of a medication's impact on many types of human cells, they also enhance drug development protocols. Furthermore, altered stem cells exhibit significant promise in the treatment of many conditions, such as injuries, degenerative illnesses, and particular blood disorders, showcasing their groundbreaking potential in modern medicine (Goetzl *et al.*, 2024; Y.-R. Li *et al.*, 2021).

12.3 Biomaterials and Components in Tissue Engineering

12.3.1 Types of Biomaterials

Figure 12.1 shows the various types of bio materials in tissue engineering.

12.3.2 Components in Tissue Engineering

Figure 12.2 represents the component of tissue engineering.

Biomaterials in Tissue Engineering			
Types of Biomaterials	**Natural Biomaterials**	**Synthetic Biomaterials**	**Composite Biomaterials**
Description	Derived from natural sources [e.g., collagen, chitosan, alginate]	Man-made materials [e.g., poly(lactic acid), polycaprolactone]	Combinations of natural and synthetic materials to enhance properties
Characteristics	Biocompatible Biodegradable Promote Cell Adhesion Proliferation	Controlled Mechanical Properties Tailored Degradation	Enhanced Mechanical Strength Improved Biocompatibility Tailored Degradation Rates
Applications	Skin Cartilage Soft Tissue Engineering	Bone Regeneration Cartilage Engineering Drug Delivery Systems	Bone Regeneration Soft Tissue Applications Tissue Scaffolding

Figure 12.1 Types of biomaterials.

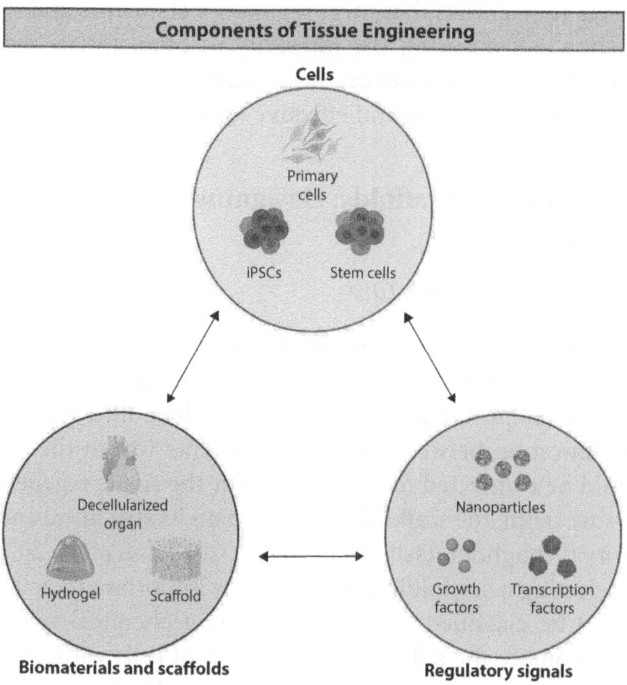

Figure 12.2 Components of tissue engineering.

12.3.3 Properties and Selection Criteria

Choosing biomaterials for tissue engineering involves careful consideration of many crucial characteristics and prerequisites. Biocompatibility is crucial to prevent detrimental consequences while engaging with biological systems. Moreover, biodegradability is essential; the pace at which it breaks down must align with the tissue regeneration process to provide enough support throughout the healing process (Akter, 2016; Doblado *et al.*, 2021). To function properly, biomaterials must possess mechanical properties that closely resemble those of actual tissue as well as an adequate level of porosity to facilitate the movement of nutrients, migration of cells, and development of blood vessels. Surface features, such as chemical composition and roughness, significantly influence cell adhesion, proliferation, and differentiation. The hydrophilicity or hydrophobicity of a substance also impacts its capacity to adsorb proteins and interact with tissues (Tsiapalis & O'Driscoll, 2020).

The selection of biomaterials is influenced by the specific requirements of different tissues, making the intended usage an important factor to consider. Key factors for achieving outcomes in tissue engineering involve having access to materials that meet regulatory standards and thus ensure cost efficiency and ensuring the longevity of biomaterials regardless of their intended lifespan. In essence, these qualities and criteria aid in selecting biomaterials that are safe and effective for tissue regeneration.

12.3.4 Biomaterial Scaffolds: Designing the Extracellular Matrix

12.3.4.1 Scaffold Architecture

Tissue engineering is a critical area that heavily depends on scaffold architecture. The porosity and pore size of a material are very important for cell migration, cell proliferation, and waste product diffusion. For effective intercommunication between cells and nutrients within the scaffold, the pores should be connected to each other. For the tissue regeneration process to be successful, the scaffold must maintain its structural and mechanical integrity throughout itself (Goetzl *et al.*, 2024; Liu *et al.*, 2024).

When selecting a scaffolding material, it is important to consider other variables. To be classified as biocompatible, a chemical must have the capacity to promote the adhesion and growth of cells without causing an immune reaction. For a material to be classified as biodegradable, its rate of decomposition must correspond to the pace at which tissues regenerate. To provide sufficient support, it is necessary for the mechanical characteristics of the material to closely mimic those of the specific tissue being targeted. For material processability to achieve the intended scaffold architecture, it is necessary to use a material that is compatible with electrospinning, 3D printing, or freeze-drying methods (Y.-R. Li *et al.*, 2021; Papantoniou *et al.*, 2021; Xu *et al.*, 2019).

Functionalization refers to the process of altering the scaffold to improve its performance and its ability to interact with the biological surroundings. One approach is to use a surface coating to improve cell adhesion, development, and specialization by applying bioactive chemicals onto the surface of the scaffold. These molecules have the potential to be peptides, proteins, or growth factors. Chemical modification involves changing the chemical makeup of a scaffold by adding functional groups that may attach to cells or bioactive substances. The incorporation of nanoparticles into the scaffold enhances its mechanical strength, controls the distribution of medicine, and imparts antimicrobial characteristics (Papantoniou *et al.*, 2021; Ude *et al.*, 2018).

12.4 Cell-Seeding Techniques

12.4.1 Scaffold-Based Methods

12.4.1.1 Scaffold Fabrication Techniques

Scaffolding is essential for promoting cellular connection and growth. Several often-used strategies include the following: electrospinning is a method used to fabricate fibrous scaffolds that have a strong resemblance to the extracellular matrix; 3D printing allows for the precise manipulation of scaffold architecture, including the size and distribution of pores; cryodesiccation is a method that entails extracting ice from a cryogenic polymer solution to produce scaffolds with a substantial degree of porosity; and solvent casting/particle leaching is a technique that involves pouring a polymer solution and then eliminating particle porogens to create porous structures (Doblado *et al.*, 2021; Liu *et al.*, 2024).

12.4.1.2 Seeding Efficiency

The seeding efficiency of a scaffold pertains to the rate at which cells are deposited onto or inside it. The effectiveness of seeding is determined by critical factors such as cell density, which is the measure of the number of cells inside a certain volume of seeding fluid, ensuring uniform dispersion of the cells across the scaffold. To improve the adherence of cells to the scaffold material, extracellular matrix proteins are added or modified on its surface during the attachment process (Brown *et al.*, 2019; Y.-R. Li *et al.*, 2021).

12.4.1.3 Dynamic Seeding Methods

Dynamic seeding methods using fluidic or mechanical forces may enhance cell–scaffold interactions. By using shear and gravity, cells are uniformly disseminated via scaffolds in bioreactors that rotate. A perfusion bioreactor is a device that improves nutrition delivery and waste removal by allowing cell culture media to flow through a scaffold. By stirring the contents of a spinner flask, one may improve the diffusion and contact of cell scaffolds (Z. Li *et al.*, 2023; Liu *et al.*, 2024).

12.4.2 Bioprinting Technologies

12.4.2.1 Bioprinting Principles

Bioprinters can construct complex tissue structures by layering bio-inks, which consist of cells, biomaterials, and growth factors. The concept of exactness is encompassed by several ideas: cells and biomaterials are accurately arranged; the capacity to construct large and intricate structures (scalability); and cell viability is the measure of the extent to which printed cells remain alive and operational (Vanaei *et al.*, 2021).

12.4.2.2 Types of Bioprinting

Several bioprinting techniques have been developed, including achieving the precise deposition of bioink droplets; inkjet bioprinting utilizes either piezoelectric or thermal forces. Materials are often extruded via microextrusion, which involves working on a millimeter scale or smaller. Bioprinting is the process of employing a nozzle to expel uninterrupted strands of bioink. Using a laser, bioprinting is the process of depositing cells and biomaterials onto a substrate using laser light (Doblado *et al.*, 2021; Vanaei *et al.*, 2021).

Bioprinting is often used in tissue engineering and regenerative medicine to create *in vitro* models of organs and tissues. These models are specifically designed for investigating illnesses and evaluating the effectiveness of drugs. Regenerative implantology refers to the specialized production of implants used to repair or replace injured tissues. Cancer researchers examine tumor growth to evaluate the effectiveness of potential therapies.

Utilizing distinct biological data to create tissues tailored to individual patients is a crucial component of customized therapy (Akter, 2016; Xu *et al.*, 2019).

12.5 Bioreactor Systems for Tissue Culture

Bioreactor systems are necessary for the precise and consistent culture of tissues. Cells can grow, differentiate, and form tissues because they control several factors such as temperature, pH, oxygen levels, and nutrition availability.

12.5.1 Types of Bioreactors

12.5.1.1 Static Bioreactors

Simple systems known as static bioreactors allow for the cultivation of cells or tissues in a fixed environment, free from the need for active mixing or movement. Their ability to transport oxygen and nutrients to cells and eliminate waste is dependent on diffusion. Academic research often makes use of static bioreactors for small-scale tissue growth because of their simplicity and inexpensive cost (Gustin & Prasad, 2024; Muenst *et al.*, 2015).

12.5.1.2 Dynamic Bioreactors

Nutrient and oxygen supply, as well as waste collection, are both improved in dynamic bioreactors by the use of mechanical agitation or fluid flow. Many methods exist for accomplishing this kind of motion, including perfusion systems, shaking, and stirring. Cultures on a grand scale and more intricate tissue engineering projects often make use of dynamic bioreactors (Z. Li *et al.*, 2023; Muenst *et al.*, 2015; Niu *et al.*, 2023).

12.5.1.3 Microfluidic Bioreactors

A microfluidic bioreactor is a miniature bioreactor that can precisely regulate the biological microenvironment via the use of microfabrication techniques. Cells are cultivated in these bioreactors using chambers or channels with regulated fluid flow. Microfluidic bioreactors shine when it comes to developing complicated tissue models, high-throughput screening, and single-cell analysis (Gustin & Prasad, 2024; Muenst *et al.*, 2015; Niu *et al.*, 2023).

12.5.2 Applications in Tissue Engineering

Table 12.2 shows the list of the applications in various areas.

Table 12.2 Applications in tissue engineering.

Application area	Description	Uses	Examples	Reference
Skin regeneration	Promotes healing, reduces scarring, and restores skin function in burn victims, chronic wounds, and surgical reconstruction	• Grafts can incorporate cells like keratinocytes and fibroblasts to enhance healing • Use of biocompatible scaffolds such as collagen and hyaluronic acid	• Engineered skin grafts: used for burn patients and chronic ulcers • Cosmetic surgery: tissue-engineered skin for reconstruction after injuries or surgeries	(Berthiaume *et al.*, 2011; Zhao *et al.*, 2013)
Bone and cartilage repair	Regenerates bone and cartilage tissues to treat fractures, joint injuries, and degenerative diseases like osteoarthritis	• Materials like hydroxyapatite and calcium phosphate are used in bone regeneration • Cartilage repair often involves chondrocytes and growth factors in hydrogels	• Bone grafts: repair bone defects from trauma, tumors, or congenital conditions • Cartilage scaffolds: used in joint repair for conditions like osteoarthritis	(Rijal & Li, 2018)
Cardiovascular tissue engineering	Develops tissue-engineered heart valves and vascular grafts to replace or repair damaged cardiovascular tissues	• Biodegradable polymers like polyglycolic acid (PGA) and polycaprolactone (PCL) are used in valve and vessel engineering • Endothelial cells are often seeded on grafts to promote integration	• Heart valve replacement: engineered valves that mimic natural heart valve function • Vascular grafts: artificial blood vessels to replace damaged arteries, particularly in cardiovascular disease patients	(Stoltz *et al.*, 2012)

(*Continued*)

Table 12.2 Applications in tissue engineering. (*Continued*)

Application area	Description	Uses	Examples	Reference
Neural tissue engineering	Aims to restore nervous system function by regenerating neural tissues after injuries such as spinal cord damage or peripheral nerve damage	• Use of neurotrophic factors like BDNF and NGF to stimulate neural regeneration • Hydrogels and conduits made of materials like collagen and polylactic acid	• Spinal cord repair: biomaterial scaffolds used to support the regrowth of neural tissues • Peripheral nerve regeneration: tubular scaffolds guide nerve growth	(Lu *et al.*, 2015)
Organ regeneration	Develops functional tissue constructs for organs like the liver, kidney, and pancreas, potentially for transplantation or supporting failing organs	• 3D bioprinting techniques used to create complex organ structures • Incorporation of hepatocytes for liver tissue and insulin-producing cells for pancreatic islets	• Liver tissue engineering: constructs used to support liver function in liver failure • Pancreatic islets: engineered islets as a therapy for diabetes	(Webber *et al.*, 2015)
Drug screening and disease modeling	Utilizes engineered tissues to create accurate *in vitro* models of human organs and diseases for drug testing and personalized medicine	• Organoids and microfluidic systems (organ-on-a-chip) are increasingly used to simulate organ function • Enables high-throughput screening of drug candidates and reduces reliance on animal models	• *In vitro* models: tissue constructs used to model diseases like cancer or fibrosis • Personalized drug testing: patient-specific tissue models to tailor treatments	(O'Donnell *et al.*, 2019)

12.6 Clinical Applications of Tissue Engineering

Table 12.3 shows the list of clinical applications of the tissue engineering.

Table 12.3 Clinical applications of tissue engineering.

Application area	Current status	Challenges	Successful applications	Reference
Skin substitutes	• Products like Integra and Apligraf are FDA-approved and widely used for burns and chronic wounds.	• Immune response to non-autologous cells • High production costs	• Integra: Extensively used in burn treatment, with significant reductions in scarring • Apligraf: Effective in healing diabetic foot ulcers and venous leg ulcers	(Boehler *et al.*, 2011; O'Donnell *et al.*, 2019)
Bone and cartilage repair	• INFUSE Bone Graft is clinically approved for bone defects. • ACI and MACI are used for cartilage repair.	• Difficulty replicating the complex structure and function of natural bone and cartilage • High cost and limited availability of cell-based therapies	• INFUSE Bone Graft: widely used in spinal fusion and bone repair surgeries, showing enhanced bone regeneration • ACI: Successfully used in thousands of patients for knee cartilage repair	(O'Donnell *et al.*, 2019)

(Continued)

Table 12.3 Clinical applications of tissue engineering. (*Continued*)

Application area	Current status	Challenges	Successful applications	Reference
Cardiovascular repair	• Tissue-engineered heart valves and vascular grafts are in advanced stages of clinical trials.	• Ensuring long-term durability and functionality of engineered tissues • Risk of thrombosis, calcification, and immune rejection	• TRIC Valve: emerging as a potential alternative to mechanical valves in clinical studies • Gore-Tex Vascular Graft: successfully used in bypass surgeries and as shunts	(Webber *et al.*, 2015)
Corneal reconstruction	• Tissue-engineered corneas and limbal stem cell therapies are being tested in clinical trials.	• Challenges in maintaining transparency and biomechanical properties • Risk of immune-mediated rejection	• Limbal stem cell therapy: demonstrated success in restoring corneal surface and vision in patients with severe ocular surface disorders	(Lu *et al.*, 2015)
Bladder augmentation	• Tissue-engineered bladders have been successfully implanted in human clinical trials.	• Ensuring proper vascularization and long-term functionality • Risk of infection, rejection, and bladder dysfunction	• Wake Forest Bladders: successfully implanted, showing promise in treating congenital bladder anomalies and neurogenic bladder conditions	(Stoltz *et al.*, 2012)

(*Continued*)

Table 12.3 Clinical applications of tissue engineering. (*Continued*)

Application area	Current status	Challenges	Successful applications	Reference
Tracheal replacement	• Bioengineered tracheas have been used in compassionate-use cases and early-stage clinical trials.	• Maintaining structural integrity and airway patency over time • Challenges with tissue integration and vascularization	• Bioengineered trachea: used in emergency surgeries, showing potential as a life-saving option for severe tracheal stenosis and trauma	(Rijal & Li, 2018)
Pancreatic Islet Transplantation	• Experimental stage with clinical trials underway for type 1 diabetes.	• Long-term viability and function of islet cells post-transplant • Overcoming immune rejection without long-term immunosuppression	• Islet transplants: early successes in restoring endogenous insulin production, reducing dependence on exogenous insulin in type 1 diabetes patients	(Zhao *et al.*, 2013)
Liver assist devices	• Bioengineered liver tissues used in extracorporeal devices are in the clinical testing phases.	• Replicating complex liver functions in engineered tissues • Preventing device-related complications and infections	• ELAD (Extracorporeal Liver Assist Device): Used in clinical trials to support liver function in patients with acute liver failure, showing potential to bridge patients to liver transplantation	(Berthiaume *et al.*, 2011)

(*Continued*)

Table 12.3 Clinical applications of tissue engineering. (*Continued*)

Application area	Current status	Challenges	Successful applications	Reference
Nerve regeneration	• Nerve conduits and scaffolds are in use and being further developed for peripheral nerve injuries.	• Achieving full functional recovery and proper alignment of regenerating nerves • Preventing scar tissue formation and neuroma development	• NeuraGen Nerve Guide: approved for clinical use in peripheral nerve repair, showing improved outcomes in nerve regeneration and functional recovery	(O'Donnell *et al.*, 2019)
Cartilage regeneration in joints	• Advanced techniques such as matrix-induced autologous chondrocyte implantation (MACI) are in clinical use.	• Replicating the biomechanical properties of natural cartilage, particularly in weight-bearing joints • Preventing degradation over time	• MACI: Successfully used for knee cartilage defects, improving joint function, and reducing pain with long-term positive outcomes	(Webber *et al.*, 2015)
Spinal cord repair	• Early-stage clinical trials are exploring the use of engineered scaffolds and stem cells for spinal cord injuries.	• Overcoming the inhibitory environment of the injured spinal cord • Ensuring proper integration and functional recovery of neural connections	• Neuro-spinal Scaffold: under investigation in clinical trials, showing promise in promoting regeneration and recovery in spinal cord injury patients	(Lu *et al.*, 2015)

12.7 Regulatory and Ethical Considerations

Table 12.4 represents the details about the regulatory and ethical considerations in tissue engineering.

Table 12.4 Regulatory and ethical considerations in tissue engineering.

Area	Description	Challenges	Current status	Examples	Reference
Regulatory approval	Ensuring that tissue-engineered products meet the safety, efficacy, and quality standards for clinical use	• Complex regulatory pathways due to product diversity (cells, biomaterials, bioactive molecules) • Regulatory variation across countries	• FDA (US): regulates tissue-engineered products under CBER • EMA (Europe): manages under ATMP framework	• Carticel: approved by the FDA as the first autologous cell therapy for cartilage repair • Holoclar: EMA-approved, the first stem cell-based product for corneal repair	(Baker *et al.*, 2016; O'Donnell *et al.*, 2019)

(*Continued*)

Table 12.4 Regulatory and ethical considerations in tissue engineering. (*Continued*)

Area	Description	Challenges	Current status	Examples	Reference
Quality control and manufacturing	Standardizing production to ensure the consistent quality, safety, and efficacy of products	• Reproducibility and scalability in manufacturing • Maintaining sterility and biological activity	• GMP standards: enforced globally to ensure product consistency • Continuous process verification: adopted in some regions to monitor production	• Novartis CAR-T Manufacturing: advanced process controls in place for consistent product quality • Lonza's Cell Therapy Manufacturing: leading example of large-scale GMP-compliant production	(Kashi & Saha, 2013)

(*Continued*)

Table 12.4 Regulatory and ethical considerations in tissue engineering. (*Continued*)

Area	Description	Challenges	Current status	Examples	Reference
Ethical sourcing of cells and tissues	Addressing ethical issues related to the sourcing of human cells and tissues, including consent	• Obtaining informed consent and ensuring donor rights • Ethical concerns over embryonic stem cells	• Informed consent protocols: strict requirements globally, especially for stem cell sourcing • Ethical oversight: institutional review boards (IRBs) monitor sourcing practices	• iPSCs: developed to bypass ethical concerns with embryonic stem cells, with widespread adoption in research • Umbilical Cord Blood Banks: provide ethically sourced stem cells for therapeutic use	(Boehler *et al.*, 2011)
Long-term monitoring and patient safety	Monitoring patients to assess long-term safety and effectiveness	• Tracking potential adverse effects (e.g., immune reactions, tumor formation) • Developing long-term follow-up protocols	• Post-market surveillance: required for approved products • Patient registries: used in various countries to track long-term outcomes	• Biovigilance programs: implemented by the FDA to monitor adverse events post-therapy • MACI (Vericel): long-term studies conducted to assess safety and efficacy in cartilage repair	(Zhang *et al.*, 2021)

(*Continued*)

Table 12.4 Regulatory and ethical considerations in tissue engineering. (*Continued*)

Area	Description	Challenges	Current status	Examples	Reference
Intellectual property (IP) and patents	Protecting innovations while ensuring accessibility and affordability	• Balancing IP protection with scientific collaboration • Ethical concerns over patenting living tissues	• Patent protections: strong in developed countries, providing incentives for innovation • Licensing agreements: commonly used to share technology while retaining IP rights	• Organogenesis: holds multiple patents on their Apligraf product for wound healing • Bluebird Bio: licensed technologies to other companies while retaining IP on gene therapies	(Gokcekuyu *et al.*, 2024)
Cost and accessibility	Addressing high costs to ensure broader patient access to therapies	• High R&D costs lead to expensive therapies • Inequitable access, especially in low-income regions	• Insurance coverage: varies significantly by country and product • Government subsidies/grants: offered in some regions to improve access	• Kymriah (Novartis): high cost but covered by insurance in some regions, including the US • NHS Funding (UK): certain cell therapies, like Strimvelis, are funded to ensure accessibility	(Bagherpour *et al.*, 2024)

(*Continued*)

Table 12.4 Regulatory and ethical considerations in tissue engineering. (*Continued*)

Area	Description	Challenges	Current status	Examples	Reference
Ethical use of emerging technologies	Addressing ethical issues related to CRISPR, gene editing, and advanced biotechnologies	• Risk of off-target effects and unintended consequences • Ethical debates over germline modification	• Strict oversight: regulations for gene editing are stringent in most countries • Ethical guidelines: issued by organizations like the ISSCR to guide research	• CRISPR Therapeutics: leading the way in developing CRISPR-based therapies with rigorous ethical oversight • Chinese Clinical Trials: early and controversial use of CRISPR, highlighting regulatory gaps	(Auger *et al.*, 2013)
Informed consent in emerging technologies	Ensuring patients and donors are fully aware of the risks and benefits of novel therapies	• Communicating complex scientific risks effectively • Ensuring voluntary and fully informed participation	• Enhanced consent protocols: required for high-risk therapies • Ethical committees: oversee consent processes in clinical trials	• CAR-T therapies: use detailed consent processes due to the novel and high-risk nature of treatments • Gene therapy trials: require thorough patient education and informed consent due to potential risks	(de Isla *et al.*, 2010)

(*Continued*)

Table 12.4 Regulatory and ethical considerations in tissue engineering. (*Continued*)

Area	Description	Challenges	Current status	Examples	Reference
Cultural and religious considerations	Addressing cultural and religious beliefs that may influence acceptance of tissue-engineered products	• Respecting cultural and religious values while advancing science • Ethical dilemmas in using animal-derived or human tissues	• Cultural Sensitivity Training: Increasingly required for researchers and clinicians • Alternative therapies: developed to accommodate cultural and religious preferences	• Halal-certified biomaterials: developed to address concerns in Muslim-majority countries • Kosher certification: obtained for certain products to align with Jewish dietary laws	(Zhang *et al.*, 2021)

(*Continued*)

Table 12.4 Regulatory and ethical considerations in tissue engineering. (*Continued*)

Area	Description	Challenges	Current status	Examples	Reference
Global regulatory harmonization	Ensuring consistent regulatory standards across countries to facilitate global access	• Harmonizing regulations in different jurisdictions • Addressing disparities in regulatory stringency	• International collaboration: organizations like ICH work towards global regulatory harmonization • WHO guidelines: provide a framework for consistent global standards	• ICH guidelines: widely adopted to standardize regulatory practices across countries • EMA–FDA collaborations: Joint initiatives to streamline approvals across the US and Europe	(Boehler *et al.*, 2011)

(*Continued*)

Table 12.4 Regulatory and ethical considerations in tissue engineering. (*Continued*)

Area	Description	Challenges	Current status	Examples	Reference
Transparency and public engagement	Engaging the public and maintaining transparency in the development and application of new therapies	• Building public trust in novel technologies • Addressing misinformation and ethical concerns proactively	• Public outreach programs: increasingly implemented to educate and involve the public • Transparency initiatives: required by many regulatory bodies for clinical trials	• NIH public engagement: extensive efforts to educate the public on stem cell research • EMA's PRIME scheme: enhances transparency and public involvement in accelerated approval pathways	(J. L. Guo *et al.*, 2023)
Environmental impact	Considering the environmental implications of tissue engineering, particularly in manufacturing and disposal	• Minimizing waste and managing disposal of biological materials • Reducing the environmental footprint of production facilities	• Green manufacturing practices: adopted by some companies to reduce environmental impact • Biodegradable materials: development of environmentally friendly scaffolds and biomaterials	• Organovo: uses sustainable practices in bio-printing processes • Regenovo Biotech: Pioneers in using biodegradable materials to reduce environmental impact	(Auger *et al.*, 2013)

12.8 Future Directions in Tissue Engineering

Table 12.5 shows the details about the future directions in tissue engineering.

Table 12.5 Future directions in tissue engineering.

Area	Description	Potential developments	Examples	Reference
Advanced biomaterials and scaffolds	Innovations in the design and functionality of biomaterials and scaffolds used in tissue engineering	• Smart biomaterials: responsive materials that release therapeutic agents • Biodegradable scaffolds: environmentally friendly materials	• Responsive hydrogels: materials that change properties based on stimuli • Natural polymer scaffolds: made from materials like collagen and alginate	(Bajpai & Andreadis, 2012)
Integration of stem cell technology	Advances in stem cell science to improve regenerative treatments	• Enhanced stem cell reprogramming: more efficient methods for creating iPSCs • Customized cell therapies: personalized treatments based on patient-specific cells	• iPSC-derived cardiomyocytes: used in cardiac disease models • Patient-specific iPSC therapies: personalized treatments for genetic disorders	(Bagherpour *et al.*, 2024)

(Continued)

Table 12.5 Future directions in tissue engineering. (*Continued*)

Area	Description	Potential developments	Examples	Reference
3D bioprinting and biofabrication	Use of advanced printing technologies to create complex tissue structures and organs	• Complex tissue printing: development of tissues with functional vasculature • Microfluidic integration: combining with 3D printing for advanced tissue models	• Organovo bioprinting: printing of liver and kidney tissues • Lab-on-a-chip devices: microfluidic systems for tissue engineering	(Higuchi *et al.*, 2013)
Regenerative medicine innovations	New approaches and technologies in regenerative medicine for treating complex diseases	• Gene editing: application of CRISPR for correcting genetic defects • Complex disease models: engineered tissues for diseases like cancer and neurodegenerative conditions	• CRISPR-Cas9 for genetic disorders: used to correct mutations in iPSCs • Cancer organoids: engineered tissues that model tumor growth for drug testing	(Zhang *et al.*, 2021)
Personalized medicine and tailored therapies	Customizing treatments based on individual patient data for improved outcomes	• Omics data integration: using genomics and proteomics to tailor therapies • Advanced diagnostics: real-time monitoring of engineered tissues	• Pharmacogenomics: tailoring drug treatments based on genetic information • Wearable sensors: monitoring physiological responses to therapies	(O'Donnell *et al.*, 2019)

(*Continued*)

Table 12.5 Future directions in tissue engineering. (*Continued*)

Area	Description	Potential developments	Examples	Reference
Ethical and regulatory evolution	Evolving regulations and ethical frameworks to address new challenges in tissue engineering	• Global regulatory harmonization: aligning standards across countries • Ethical frameworks: developing guidelines for new technologies	• ICH guidelines: international collaboration for regulatory standards • ISSCR ethical guidelines: frameworks for stem cell research and application	(Baker *et al.*, 2016)
Integration with digital health technologies	Combining tissue engineering with digital technologies for enhanced patient care	• Wearable biotechnologies: devices for monitoring health and therapy effectiveness • Digital twins: virtual models of patients for personalized treatment planning	• Wearable health monitors: tracking physiological parameters in real time • Digital twin simulations: used for predicting outcomes of therapeutic interventions	(Gokcekuyu *et al.*, 2024; J. L. Guo *et al.*, 2023)
Emerging technologies	New technologies advancing the field of tissue engineering	• Gene editing: precision editing of genetic material using tools like CRISPR • Smart biomaterials: materials with programmable responses to environmental changes	• CRISPR-Cas9: used for targeted gene modification in tissue engineering • Hydrogel-based smart materials: adaptable materials used for controlled drug release	(Zhang *et al.*, 2021)

(*Continued*)

Table 12.5 Future directions in tissue engineering. (*Continued*)

Area	Description	Potential developments	Examples	Reference
Vascularization: engineering blood supply	Strategies to develop functional blood vessels in engineered tissues	• Vascular networks: creating complex networks to support tissue survival and function • Bioprinted blood vessels: integrating vascular structures with engineered tissues	• *In vivo* vascularization models: using animal models to study vascular integration • 3D bioprinted vasculature: producing vascular structures to improve tissue integration	(Auger *et al.*, 2013)
Cell sourcing: autologous vs. allogeneic stem cells	Comparison of using patient-derived (autologous) versus donor-derived (allogeneic) stem cells	• Autologous cells: cells derived from the patient, reducing the risk of immune rejection • Allogeneic cells: donor cells that may offer broader applicability but require immunosuppression	• Autologous chondrocyte implantation: using patient's cells for cartilage repair • Allogeneic mesenchymal stem cells: used in clinical trials for various regenerative therapies	(Bajpai & Andreadis, 2012; J. Li *et al.*, 2021)

12.9 Conclusion and Future Perspectives

Tissue engineering offers a hopeful method for producing functioning tissue substitutes, which has the potential to transform the treatment of several illnesses. Important areas of concentration are enhancing the growth and specialization of stem cells, creating sophisticated materials for medical use, and effectively using tissue-engineered products in clinical settings. Stem cell treatment has already achieved substantial advancements in tissue engineering, facilitating the development of intricate and highly functioning tissues. With ongoing research advancements, the influence of stem cell therapy is anticipated to expand, providing novel therapeutic alternatives for individuals with previously incurable illnesses (X. Guo *et al.*, 2023; Johnson *et al.*, 2007; Xue *et al.*, 2022). The field of tissue engineering has great promise for the future, as a multitude of developing technologies are positioned to propel further progress. Gene editing, advanced biomaterials, and artificial intelligence are some of the advances that will likely have a significant impact on the future of the profession. These advancements will allow for the development of more complex and customized tissue-engineered products.

To fully harness the potential of tissue engineering, it is necessary to confront and overcome many problems that arise throughout its ongoing development. An important obstacle is the capacity of tissue-engineered products to be scaled up effectively. Still it is difficult to generate large and sophisticated tissues capable of replacing whole organs because blood vessels are not formed well or the cells deep in the tissue cannot access nutrients. Additionally, our understanding of how long-lasting these tissues are as well as their usefulness in a living organism is quite limited. The other hindrance here lies on an intricate and continually changing regulatory climate and a shifting economic landscape. Successful commercialization of engineered tissues involves navigating complex regulations and demonstrating cost effectiveness as well as safety and efficacy. The high cost of developing or producing tissue-engineered products may limit their use especially for low-resource settings. Ongoing moral concerns are associated with stem cell therapy, gene editing, and the potential for creating tissues that blur the line between human beings and non-human animals. These challenges also make the greatest need for continuous cross-disciplinary research and the proactive communication of research findings and information among scientists, physicians, regulators, and the public (X. Guo *et al.*, 2023). It will take a concerted effort to overcome these tissue engineering challenges and advance the field.

It is estimated that the use of artificial intelligence, machine learning, and high-throughput screening applications within biotechnology will play a major role in accelerating the discovery time and development of novel, advanced tissue engineering strategies. The development of gene-edited universal donor cells will thus resolve the issue of immune rejection and make tissue-engineered products more widely applicable. Public and commercial support for this tissue engineering research will prove vital for the breakthroughs but, most importantly, in its mainstream implementation. Clear regulatory channels and ethical principles would have to be developed that ensure the responsible advancement of new technologies and the benefits in general that technology brings to the world. In the end, the success of tissue engineering will hinge upon the development of safe, efficient, and commercially accessible therapies for addressing unmet medical needs that can truly change the paradigm of current medicine (X. Guo *et al.*, 2023; Johnson *et al.*, 2007).

References

Akter, F., *Tissue Eng. Made Easy*, 3–16, 2016. https://doi.org/10.1016/B978-0-12-805361-4.00002-3.AU: Please provide article title and volume number.

Auger, F.A., Gibot, L., Lacroix, D., The Pivotal Role of Vascularization in Tissue Engineering. *Annu. Rev. Biomed. Eng.*, 15, 1, 177–200, 2013. https://doi.org/10.1146/annurev-bioeng-071812-152428.

Bagherpour, R., Bagherpour, G., Mohammadi, P., Application of Artificial Intelligence in Tissue Engineering. *Tissue Eng. Part B Rev.*, 31, 1, 2024. https://doi.org/10.1089/ten.teb.2024.0022.

Bajpai, V.K. and Andreadis, S.T., Stem Cell Sources for Vascular Tissue Engineering and Regeneration. *Tissue Eng. Part B Rev.*, 18, 5, 405–425, 2012. https://doi.org/10.1089/ten.teb.2011.0264.

Baker, H.B., McQuilling, J.P., King, N.M.P., Ethical considerations in tissue engineering research: Case studies in translation. *Methods*, 99, 135–144, 2016. https://doi.org/10.1016/j.ymeth.2015.08.010.

Baptista, L.S., Chapter 13 - Adipose Tissue Resident Stem Cells. in: R. Coeli dos Santos Goldenberg and A. Campos de Carvalho (Eds.), *Resident Stem Cells and Regenerative Therapy Second Edition*, pp. 339–355, Academic Press, Rio de Janeiro, Brazil, 2024, https://doi.org/https://doi.org/10.1016/B978-0-443-15289-4.00008-1.

Berthiaume, F., Maguire, T.J., Yarmush, M.L., Tissue Engineering and Regenerative Medicine: History, Progress, and Challenges. *Annu. Rev. Chem. Biomol. Eng.*, 2, 1, 403–430, 2011. https://doi.org/10.1146/annurev-chembioeng-061010-114257.

Boehler, R.M., Graham, J.G., Shea, L.D., Tissue Engineering Tools For Modulation of the Immune Response. *BioTechniques*, 51, 4, 239–254, 2011. https://doi.org/10.2144/000113754.

Brown, K.S., Rao, M.S., Brown, H.L., The Future State of Newborn Stem Cell Banking. *J. Clin. Med.*, 8, 1, 117, 2019. https://doi.org/10.3390/jcm8010117.

Castells-Sala, C., Alemany-Ribes, M., Fernandez-Muiños, T., Recha-Sancho, L., Lopez-Chicon, P., *et al.*, Current Applications of Tissue Engineering in Biomedicine. *J. Biochip Tissue chip*, S2, 004, 2013, doi:10.4172/2153-0777.S2-004.

de Isla, N., Huseltein, C., Jessel, N., Pinzano, A., Decot, V., Magdalou, J., Bensoussan, D., Stoltz, J.-F., Introduction to tissue engineering and application for cartilage engineering. *Bio-Med. Mater. Eng.*, 20, 3–4, 127–133, 2010. https://doi.org/10.3233/BME-2010-0624.

Doblado, L.R., Martínez-Ramos, C., Pradas, M.M., Biomaterials for Neural Tissue Engineering. *Front. Nanotechnol.*, 3, 1–17, 2021. https://doi.org/10.3389/fnano.2021.643507.

Esmaeili, Y., Bidram, E., Bigham, A., Atari, M., Nasr Azadani, R., Tavakoli, M., Salehi, S., Mirhaj, M., Basiri, A., Mirzavandi, Z., Boshtam, M., Rafienia, M., Zargar Kharazi, A., Karbasi, S., Shariati, L., Zarrabi, A., Exploring the evolution of tissue engineering strategies over the past decade: From cell-based strategies to gene-activated matrix. *Alex. Eng. J.*, 81, 137–169, 2023. https://doi.org/10.1016/j.aej.2023.08.080.

Gheitasi, M., Safdel, S., Kumar Patra, S., Zandvakili, R., Nemati, M., Saha, B., Jafarzadeh, A., Generation of immune cells from induced pluripotent stem cells (iPSCs): Their potential for adoptive cell therapy. *Hum. Immunol.*, 85, 5, 110836, 2024. https://doi.org/10.1016/j.humimm.2024.110836.

Goetzl, E.J., Alpert, J.S., Chen, Q.M., Human Stem Cells in Regenerative Medicine. *Am. J. Med.*, 137, 9, 805–809, 2024 Sep., doi: 10.1016/j.amjmed.2024.05.024. Epub 2024 May 23. PMID: 38795938.

Gokcekuyu, Y., Ekinci, F., Guzel, M.S., Acici, K., Aydin, S., Asuroglu, T., Artificial Intelligence in Biomaterials: A Comprehensive Review. *Appl. Sci.*, 14, 15, 6590, 2024. https://doi.org/10.3390/app14156590.

Guo, J.L., Januszyk, M., Longaker, M.T., Machine Learning in Tissue Engineering. *Tissue Eng. Part A*, 29, 1–2, 2–19, 2023. https://doi.org/10.1089/ten.tea.2022.0128.

Guo, X., Ma, Y., Min, Y., Sun, J., Shi, X., Gao, G., Sun, L., Wang, J., Progress and prospect of technical and regulatory challenges on tissue-engineered cartilage as therapeutic combination product. *Bioact. Mater.*, 20, 501–518, 2023. https://doi.org/10.1016/j.bioactmat.2022.06.015.

Gustin, P. and Prasad, A., EnduroBone: A 3D printed bioreactor for extended bone tissue culture. *HardwareX*, 18, e00535, 2024. https://doi.org/10.1016/j.ohx.2024.e00535.

Higuchi, A., Ling, Q.-D., Chang, Y., Hsu, S.-T., Umezawa, A., Physical Cues of Biomaterials Guide Stem Cell Differentiation Fate. *Chem. Rev.*, 113, 5, 3297–3328, 2013. https://doi.org/10.1021/cr300426x.

Hovland, A.S., Bhattacharya, D., Azambuja, A.P., Pramio, D., Copeland, J., Rothstein, M., Simoes-Costa, M., Pluripotency factors are repurposed to shape the epigenomic landscape of neural crest cells. *Dev. Cell*, 57, 19, 2257–2272.e5, 2022. https://doi.org/10.1016/j.devcel.2022.09.006.

Johnson, P.C., Mikos, A.G., Fisher, J.P., Jansen, J.A., Strategic Directions in Tissue Engineering. *Tissue Eng.*, 13, 12, 2827–2837, 2007. https://doi.org/10.1089/ten.2007.0335.

Kharkar, P.S., Cancer Stem Cell (CSC) Inhibitors in Oncology—A Promise for a Better Therapeutic Outcome: State of the Art and Future Perspectives. *J. Med. Chem.*, 63, 24, 15279–15307, 2020. https://doi.org/10.1021/acs.jmedchem.0c01336.

Kwon, S.G., Kwon, Y.W., Lee, T.W., Park, G.T., Kim, J.H., Recent advances in stem cell therapeutics and tissue engineering strategies. *Biomater. Res.*, 22, 36, 1–8, 2018. https://doi.org/10.1186/s40824-018-0148-4.

Li, C., Zhou, Y., Wei, R., Napier, D.L., Sengoku, T., Alstott, M.C., Liu, J., Wang, C., Zaytseva, Y.Y., Weiss, H.L., Wang, Q., Evers, B.M., Glycolytic Regulation of Intestinal Stem Cell Self-Renewal and Differentiation. *Cell. Mol. Gastroenterol. Hepatol.*, 15, 4, 931–947, 2023. https://doi.org/10.1016/j.jcmgh.2022.12.012.

Li, J., Liu, Y., Zhang, Y., Yao, B., Li, Z., Song, W., Wang, Y., Duan, X., Yuan, X., Fu, X., Huang, S., Biophysical and Biochemical Cues of Biomaterials Guide Mesenchymal Stem Cell Behaviors. *Front. Cell Dev. Biol.*, 9, 1–13, 2021. https://doi.org/10.3389/fcell.2021.640388.

Li, Y.-R., Zhou, Y., Kramer, A., Yang, L., Engineering stem cells for cancer immunotherapy. *Trends Cancer*, 7, 12, 1059–1073, 2021. https://doi.org/10.1016/j.trecan.2021.08.004.

Li, Z., Liu, Q., Zhao, K., Cao, D., Cao, Z., Zhao, K., Ma, Q., Zhai, G., Hu, S., Li, Z., Wang, K., Gong, F., Ma, X., Zhang, X., Ren, R., Qiu, D., Zhao, Y., Yin, D., Dynamic DNA methylation modification in peanut seed development. *IScience*, 26, 7, 107062, 2023. https://doi.org/10.1016/j.isci.2023.107062.

Liu, W., Zheng, L., Wang, C., Yin, H., Raffaella, A., Apicella, A., Ji, P., Zhang, H., Fan, Y., Additively manufactured bioceramic scaffolds with 3D architecture for vertical bone augmentation: A proof-of-concept study. *Mater. Des.*, 239, 112749, 2024. https://doi.org/10.1016/j.matdes.2024.112749.

Lu, L., Arbit, H.M., Herrick, J.L., Segovis, S.G., Maran, A., Yaszemski, M.J., Tissue Engineered Constructs: Perspectives on Clinical Translation. *Ann. Biomed. Eng.*, 43, 3, 796–804, 2015. https://doi.org/10.1007/s10439-015-1280-0.

Muenst, S., Muraro, M.G., Spagnoli, G.C., Oertli, D., Weber, W.P., Soysal, S.D., P021 Primary tissue culture in a bioreactor system: a new tool for personalized medicine? *Breast*, 24, S33, 2015. https://doi.org/10.1016/S0960-9776(15)70071-8.

Niu, X., Xu, Z., Di, M., Huang, D., Li, X., Bioreactor strategies for tissue-engineered osteochondral constructs: Advantages, present situations and future trends. *Composites, Part B*, 259, 110736, 2023. https://doi.org/10.1016/j.compositesb.2023.110736.

O'Donnell, B.T., Ives, C.J., Mohiuddin, O.A., Bunnell, B.A., Beyond the Present Constraints That Prevent a Wide Spread of Tissue Engineering and Regenerative Medicine Approaches. *Front. Bioeng. Biotechnol.*, 7, 1–12, 2019. https://doi.org/10.3389/fbioe.2019.00095.

Pan, Y., Qin, H., Zheng, L., Guo, Y., Liu, W., Disturbance in transcriptomic profile, proliferation and multipotency in human mesenchymal stem cells caused by hexafluoropropylene oxides. *Environ. Pollut.*, 292, 118483, 2022. https://doi.org/10.1016/j.envpol.2021.118483.

Papantoniou, I., Nilsson Hall, G., Loverdou, N., Lesage, R., Herpelinck, T., Mendes, L., Geris, L., Turning Nature's own processes into design strategies for living bone implant biomanufacturing: a decade of Developmental Engineering. *Adv. Drug Deliv. Rev.*, 169, 22–39, 2021. https://doi.org/10.1016/j.addr.2020.11.012.

Pérez, L.M., de Lucas, B., Gálvez, B.G., Unhealthy Stem Cells: When Health Conditions Upset Stem Cell Properties. *Cell. Physiol. Biochem.*, 46, 5, 1999–2016, 2018. https://doi.org/10.1159/000489440.

Rijal, G. and Li, W., Native-mimicking *in vitro* microenvironment: an elusive and seductive future for tumor modeling and tissue engineering. *J. Biol. Eng.*, 12, 1, 20, 2018. https://doi.org/10.1186/s13036-018-0114-7.

Stoltz, J.-F., Huselstein, C., Bensoussan, D., Isla, N., Decot, V., Tissue Engineering: From Bench to Bed Side Exemples of Applications. *Engineering*, 04, 10, 43–45, 2012. https://doi.org/10.4236/eng.2012.410B011.

Tian, J., Zhu, J., Fan, Q., Luo, X., Nie, Q., Yu, J., Wu, X., Tang, Y., Liu, T., Yin, H., Interleukin-33 improves the neurogenesis of neural stem cells in perinatal brain after hypoxia–ischemia. *Int. Immunopharmacol.*, 123, 110778, 2023. https://doi.org/10.1016/j.intimp.2023.110778.

Tsiapalis, D. and O'Driscoll, L., Mesenchymal Stem Cell Derived Extracellular Vesicles for Tissue Engineering and Regenerative Medicine Applications. *Cells*, 9, 4, 991, 2020. https://doi.org/10.3390/cells9040991.

Ude, C.C., Miskon, A., Idrus, R.B.H., Abu Bakar, M.B., Application of stem cells in tissue engineering for defense medicine. *Mil. Med. Res.*, 5, 1, 7, 2018. https://doi.org/10.1186/s40779-018-0154-9.

Vanaei, S., Parizi, M.S., Vanaei, S., Salemizadehparizi, F., Vanaei, H.R., An Overview on Materials and Techniques in 3D Bioprinting Toward Biomedical Application. *Eng. Regener.*, 2, 1–18, 2021. https://doi.org/10.1016/j.engreg.2020.12.001.

Webber, M.J., Khan, O.F., Sydlik, S.A., Tang, B.C., Langer, R., A Perspective on the Clinical Translation of Scaffolds for Tissue Engineering. *Ann. Biomed. Eng.*, 43, 3, 641–656, 2015. https://doi.org/10.1007/s10439-014-1104-7.

Xu, Y., Chen, C., Hellwarth, P.B., Bao, X., Biomaterials for stem cell engineering and biomanufacturing. *Bioact. Mater.*, 4, 366–379, 2019. https://doi.org/10.1016/j.bioactmat.2019.11.002.

Xue, Y., Baig, R., Dong, Y., Recent advances of biomaterials in stem cell therapies. *Nanotechnology*, 33, 13, 132501, 2022. https://doi.org/10.1088/1361-6528/ac4520.

Yuan, W., Ferreira, L., de A.Q., Yu, B., Ansari, S., Moshaverinia, A., Dental-derived stem cells in tissue engineering: the role of biomaterials and host response. *Regen. Biomater.*, 11, 2024. https://doi.org/10.1093/rb/rbad100.

Zhang, H., Qin, C., An, C., Zheng, X., Wen, S., Chen, W., Liu, X., Lv, Z., Yang, P., Xu, W., Gao, W., Wu, Y., Application of the CRISPR/Cas9-based gene editing technique in basic research, diagnosis, and therapy of cancer. *Mol. Cancer*, 20, 1, 126, 2021. https://doi.org/10.1186/s12943-021-01431-6.

Zhao, C., Tan, A., Pastorin, G., Ho, H.K., Nanomaterial scaffolds for stem cell proliferation and differentiation in tissue engineering. *Biotechnol. Adv.*, 31, 5, 654–668, 2013. https://doi.org/10.1016/j.biotechadv.2012.08.001.

13

Personalized Medicine Frontiers: Interrelating Stem Cell Microenvironments with Stem Cell Therapies for Individual Patients

Sagar Mondal, Swati Priya, Jutishna Bora, Smita Lata, Ajay Kumar Mahalka* and Sumira Malik†

Amity Institute of Biotechnology, Amity University Jharkhand, Ranchi, India

Abstract

The intrinsic ability of stem cells to self-renew and differentiate into many cell types makes them a significant asset for tissue engineering and regenerative medicine applications. The interaction of physical, chemical, and biological signals within the stem cell niche, both temporally and spatially, profoundly influences their capacity for differentiation and self-renewal. Embryonic and adult stem cells possess potential for cell-based therapeutics; nevertheless, regulating stem cell behavior presents a substantial obstacle to their clinical application. Numerous current approaches for influencing stem cell fate insufficiently consider the complex signaling networks that dictate stem cell behaviors in their natural microenvironments. To overcome this limitation, researchers are developing a novel generation of biomaterials that function as three-dimensional synthetic microenvironments, mimicking the regulatory characteristics of natural extracellular matrix (ECM) proteins and ECM-associated growth factors. This chapter examines synthetic microenvironments as substrates for the surface immobilization of bioactive compounds via controlled release mechanisms to affect stem cell fates in the advancement of personalized medicine, which currently lacks variable treatments.

Keywords: Stem cells, extracellular matrix (ECM), personalized medicine disease microenvironment, regenerative medicine

**Corresponding author*: akmahalka@rnc.amity.edu

†*Corresponding author*: smalik@rnc.amity.edu

Ayush Madan, Sumel Ashique, Deepika Arora and Mantosh Kumar Satapathy (eds.) Stem Cell Therapeutics, (575–598) © 2025 Scrivener Publishing LLC

13.1 Introduction

Embryonic stem cells possess self-renewal as well as multipotential differentiation capabilities. Stem cells are characterized by their potential to generate different fully developed progeny while retaining the ability to regenerate themselves. These cells are essential for organ development during embryogenesis and their function offsets the regular cell loss in adulthood. SCs are essential or crucial for maintain the health of a long-lived organisms. While all types of cells can be generated in an organism with the help of embryonic stem cells (ES), adult SCs were once believed to possess a restricted ability for differentiation, bound to its specific cell types. New observations have shown that adult SCs are surprisingly prevalent and even suggest that they may possess a greater degree of flexibility in its capacity to develop into cell types from all the three germinal layers than formerly acknowledged [1]. Formerly apparent distinctions between embryonic and adult SCs have been significantly diminished to the extent that their existence is now rather disputed. The stem cells possess inherent abilities for self-renewal along with its multipotential differentiation. The existence of multipotent SCs in adulthood could potentially provide novel therapeutic possibilities for the replacement of tissues and organs. Hence, the accurate delineation of SCs and the capacity to establish their isolation are of utmost significance. Nevertheless, despite the diligent work of several researchers aiming to ascertain the characteristics of stem cells, a conclusive identification of their nature is still absent. In contrast to the anticipated selective expression of genes by stem cells, these cells have a remarkably promiscuous gene expression program.

The fields of regenerative medicine as well as tissue engineering endeavor to generate synthetic organ or tissues for therapeutic applications, pharmacological testing, disease modeling, and cell-based biochips and sensors. Regardless of the chosen technique for tissue engineering, such as scaffold which is made up of biomaterial, use of decellularized organic substances, or scaffold-free procedures, the components of the cell is indispensable [2]. Scientific progress in several biological disciplines has shown the remarkable responsiveness of cells to their surroundings, emphasizing the essential need of regulating the cell microenvironment in regenerative medicine along with tissue engineering. The microenvironment of SCs is of great significance in tissue engineering along with regenerative therapies; they are presently the most viable alternative for generating the requisite number of cells to produce clinically meaningful quantities of tissue. Vast reserves of SCs exist to regenerate tissues within the human body. Engineered tissues must adhere to the standard of preserving the microenvironmental regulation of these cells' plasticity [2], which pertains to their capacity to remain inactive and only utilize them when required in appropriate circumstances. Moreover, the improper regulation of the microenvironment of stem cells can lead to harmful consequences, including dedifferentiation and subsequent proliferation of tumors. Contemporary techniques in industrial cell culture frequently employ immortalized, tumor-inducing cell lines for the production of pure organic goods. In contrast, SC-based approaches essentially possess the capability to repeatedly generate a large quantity of functioning cells or tissues. The multitude of procedures involved will need the implementation of intricate and carefully regulated systems. In order to accomplish these objectives, it is therefore necessary to utilize the interactions between SCs and the surroundings. This section will examine the essential components of SCs and the processes related to their behavior in the disease microenvironment (Figure 13.1).

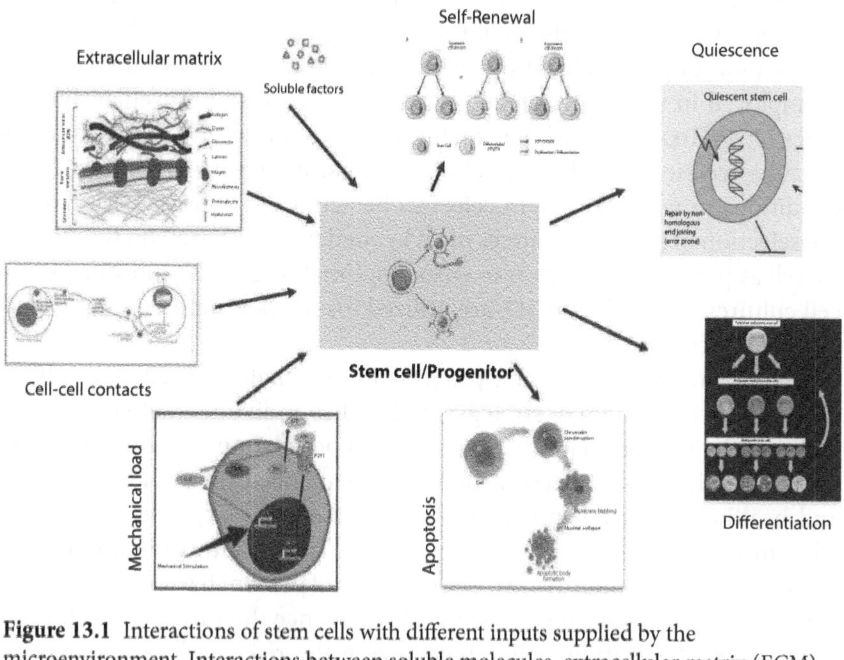

Figure 13.1 Interactions of stem cells with different inputs supplied by the microenvironment. Interactions between soluble molecules, extracellular matrix (ECM), intercellular connections, and biophysical forces collectively shape the destiny of stem cells. The inherent adaptability of progenitor cells allows them to undergo self-renewal, differentiation, quiescence, or death.

13.2 Stem Cells and their Microenvironments as Personalized Medicine

Personalized medicine, as an emerging discipline, pertains to the tailored prescription of treatment interventions for an individual. This technique is founded on pharmacogenetic along with pharmacogenomic data as well as information. The terms precision medicine and personalized medicine are occasionally used synonymously. There has been a transition from "personalized medicine" to "precision medicine." While personalized medicine originated from pharmacogenetics, it today encompasses other domains of healthcare. Regenerative medicine and cellular therapy, as emerging disciplines, utilize cell-based products to provide personalized treatments. Various sources of stem cells, particularly embryonic stem cells, mesenchymal stem cells, and especially induced pluripotent stem cells (iPSCs), have been evaluated in targeted therapies, offering numerous advantages. Induced pluripotent stem cells (iPSCs) have emerged as suitable candidates for personalized cell treatments. Cellular therapies offer a tailored methodology. Due to individual and demographic variations in the outcomes of stem cell therapy, personalized cellular therapy must be tailored to the exact profile of the patient to optimize therapeutic effects and outcomes. Application of personalized medicine and induced pluripotent stem cells is employed in the treatment of Parkinson's disease and muscular dystrophies [3, 4]. To accomplish personalized stem cell therapy, it is essential to consider recipient variables and donor factors, along with the entire bodily conditions in which the stem cells can be active and effective. Moreover, the selection of stem cell sources must be meticulously based on functional as well as physical parameters to ensure optimal results [5]. Intestinal stem cell cultures, both 2D and 3D, are utilized for the development of personalized medicine [6].

Notwithstanding the recent progress in personalized medicine, which relies on focused therapy strategies tailored to individual patient characteristics, the predominant therapies remain population-based techniques. Consequently, personalized medicine has enhanced the standards of life and healthcare while simultaneously decreasing the expenses associated with unsuccessful therapies, thereby assisting health authorities in managing healthcare expenditures [5]. The objective of personalized medicine is to develop optimal treatment strategies that enhance the therapeutic efficacy of medical interventions while minimizing the risks and adverse effects associated with standard drugs. Significant advancements in personalized medicine, as a novel medical strategy, have led to the deployment of robust techniques for the genetic analysis of various cancer types. Molecular diagnostics aid in predicting responses and identifying suitable treatments, thereby enabling the assessment of therapy efficacy across many diseases, seen as an essential element of personalized medicine. Conversely, cellular therapies require a tailored strategy since interactions between the donor and host can affect treatment outcomes [5]. Consequently, due to variations among population members regarding the outcomes of stem cell therapy, personalized cellular therapy needs to be aligned with a patient-specific profile to attain the optimal therapeutic strategy. Consequently, various elements must be evaluated in the domain of personalized stem cell therapies, including recipient specifications, donor attributes, and the overarching environment wherein the stem cells function. The origin of stem cells must be meticulously selected based on functional as well as physical characteristics to get the best outcomes. Consequently, personalized medicine is the most promising approach to tailor healthcare, enhancing the standard of medical services by addressing patients' specific differences in treatment [5], as mentioned in Figure 13.2.

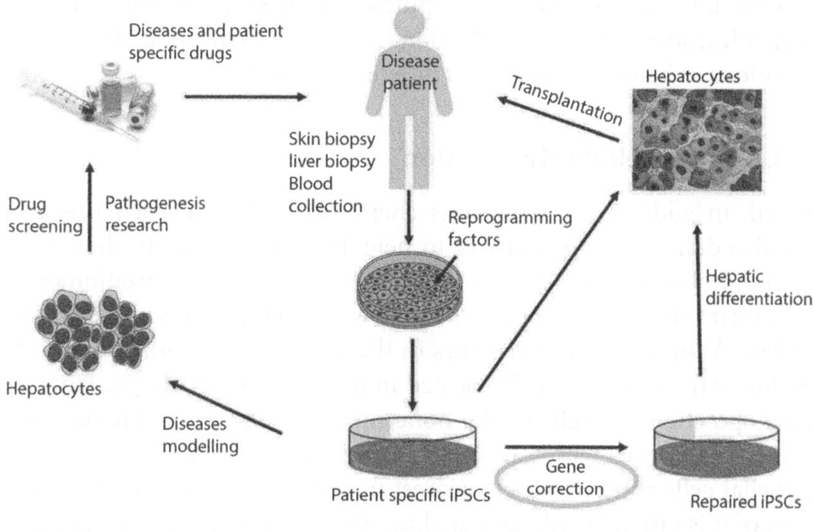

Figure 13.2 Personalized patient therapy using stem cells and gene editing.

13.3 The Disease Microenvironment: A Complex Ecosystem

The disease microenvironment, particularly in relation to stem cells, characterizes the intricate connection of stromal cells and cancer cells along with extracellular matrix (ECM) constituents within a tumor. The dynamic nature of this environment has a significant impact on the actions and development along with the function of SCs, namely, mesenchymal stem cells (MSCs), and plays a major part in the advancement, spread, as well as resistance to treatment for tumors.

The constituents of the disease include the microenvironment involving stromal cells—non-malignant tissue cells, including fibroblasts, endothelial cells, as well as antibody-producing cells—that engage through cancer cells and stem cells, therefore exerting an impact on their behavior and destiny. Extracellular matrix (ECM) proteins such as collagen, fibronectin, and laminin offer structural support and signaling signals to stem cells, ultimately influencing their differentiation, adhesion, and migratory processes. Also, the regulatory signaling factors—cytokines, growth factors (such as TGF-β, VEGF, and PDGF), and hormones (such as EGF, NGF) that control the behavior of stem cells, including their differentiation, proliferation, and survival. Cell–cell interactions refer to the direct contacts between cancer cells, stromal cells, and stem cells by which their activity is modulated and tumor development is influenced. Physical factors—mechanical stress, hypoxia, and inflammation—have the potential to cause alterations in behavior along with differentiation of SCs.

13.3.1 Intercellular Interaction

Indeed, in addition to exogenous chemicals in the microenvironment, SCs also demonstrate sensitivity to neighboring cells within their niche. Recent *in vivo* as well as *in vitro* studies on the spatial organization of SCs demonstrate the various mechanisms by which cell contacts influence stem cell fate. Mature neural precursors in the subventricular zone retain connection with ependymal cells located in the lateral ventricle [7]. However, hematopoietic stem cells in the bone marrow commence differentiation upon losing contact with osteoblasts. Adhesion junctions are formed when epithelial cells adhere to their nearest neighboring cells. Certain components of this intricate cellular signaling system regulate cell activity via the Lef1/β-catenin or else MAPK pathways [7]. Consequently, changes in the number of cells or abnormalities resulting in the dysfunction of adherens junction components (α-catenin and β-catenin, along with E-cadherin) may initiate cell division, promoting growth or potentially leading to tumorigenesis. Previous *in vitro* experiments have shown that embryonic stem cells (ESCs) can be effectively differentiated into hematopoietic progenitors when cocultured using an epithelial cell population. Furthermore, spatial investigation of mESC clusters has uncovered radial variations in the level of expression for SC markers. These findings have helped identify an autoregulatory mechanism that is mediated by an autocrine gp130 ligand. Moreover, the promotion of cell-to-cell contact through pellet culture has been demonstrated to enhance the specialization of MSCs into chondrocytes, indicating that the culture of a three-dimensional matrix and intercellular communication are important factors in controlling cell differentiation. Despite the characterization of several cell-mediated regulatory mechanisms, the challenge of managing these interactions hinders their application in stem cell cultivation technologies. Certain ESC development strategies have utilized coculture, while the incorporation of tumorigenic cell lines is considered undesirable. Recently, innovative techniques for creating spatially orientated cocultures have been devised by depositing bioactive molecules layer by layer. These systems have the potential to be highly beneficial for the targeted variation of SCs within non-therapeutic purposes. It is essential that any techniques involving several cell lineages facilitate following the separation of the target cells. Although the practical implementation of intercellular interactions in engineering solutions may be limited, it is crucial to recognize that any differentiation model involving cell colonies or aggregates, such as embryoid structures, inherently encompasses intercellular relationships as distinct cell lines develop.

13.3.2 Extracellular Matrix Interaction

ECM elements are essential components that significantly influence the instructional efficacy of the niche. Upon their formation as well as 3-D organization, these extracellular macromolecules create a milieu that effectively integrates signals from cell-ECM interactions, together with soluble and ECM-bound substances, to sustain stem cell homeostasis [8]. The diminished capacity of modified or aged stem cell niches to preserve the stem cell characteristics yield *in vivo* data that underscore the importance of the extracellular matrix, also known as the ECM, in SC behavior. A further and direct confirmation of the major function of the extracellular matrix (ECM) in regulating SC characteristics is provided by experiments conducted with decellularized tissue types, where the ECM is maintained.

Research have shown that natural extracellular matrix (ECM) scaffolds, obtained from decellularized tissues, direct the development of stem cells within the specific cellular lineages observed in the piece of tissue in which the extracellular matrix was derived. Decellularized tissues have been employed in the development of tissues and the enhancement of cell therapy techniques based on these properties [8].

The interactions between extracellular matrix (ECM) as well as SCs can be clearly facilitated by several cellular receptors, such as integrins along with carious other receptors. Although there are many potential receptors involved in the communication between extracellular matrix (ECM) and stem cells, a limited number of studies have particularly investigated the role for non-integrin receptors. CD44 is essential for directing hematopoietic stem cells (HSCs) after obtaining a transplant, whereas Robo4, an axon guidance receptor, plays a role in promoting HSC adherence to the niche in competitive repopulation experiments. Conversely, an expanding corpus of research has demonstrated that integrins serve a vital function as receptors in the relationships between the extracellular matrix (ECM) along with stem cells, as well as in the processes of adhesion, anchoring, and homing of stem cells. The family of heterodimeric transmembrane receptors known as integrins serves to link the external environment with the intracellular cytoskeleton, therefore facilitating cell motility, proliferation, and survival along with differentiation [8]. Diverse categories of integrins participate in the direct interaction with several extracellular matrix (ECM) components or with other molecules and receptors on the cell surface. Integrins can directly activate subsequent signaling pathways via focal adhesion kinase (FAK) as well as phosphoinositide 3-kinase (PI3K), hence regulating the self-renewal along with proliferation of several stem cell types. Integrin-mediated contact is essential for the anchorage of follicular stem cells in the *Drosophila* ovary within the niche, facilitating their proper self-renewal and uneven cell division. The α6β1 integrin, which binds to the extracellular matrix component laminin, is essential for the recruitment of spermatogonial SCs into the testicular niche alongside enabling the adherence of neural stem cells (NSCs) into the vascular niche [9]. The α9 integrin band linked to the extracellular matrix protein tenascin C plays a role in the growth of hematopoietic stem cells (HSC) along with the neural stem cell (NSC) niche. Furthermore, the α4, α6, α9, as well as β1 integrin chains serve as crucial for guiding hematopoietic stem cells (HSCs) to the BM niche in mice subjected to radiation exposure. Additionally, αvβ3 integrin regulates the homing and proliferation of hematopoietic stem cells (HSCs). The cytokine ligand thrombopoietin facilitates its expression. Bulge stem cells in the hair follicle synthesize the extracellular matrix

(ECM) protein nephronectin. This protein interacts via the α8β1 integrin receptor on the arrector pili muscle mass aiding in the maintenance of the correct positioning and functionality of hair follicle stem cells (HFSCs). In skeletal muscle, muscle SCs, known as cell satellites, interact with the basal lamina on one side of their niche through α7β1 integrin, in addition to the myofiber cellular membrane on the opposite side via M-cadherin. Furthermore, β1 integrins are essential for sustaining diverse SC populations by modulating the balance between symmetric as well as asymmetric divisions, as shown in skin and brain, in addition to influencing stem cell self-renewal as well as differentiation. Integrins can influence signaling pathways in response to growth factors as well as cytokines, including IL-3 as well as TGF-β, whereas signaling pathways can also regulate integrin expression. It is necessary to establish the distinct functions of these extracellular matrix (ECM) receptors in each specific niche for specific types

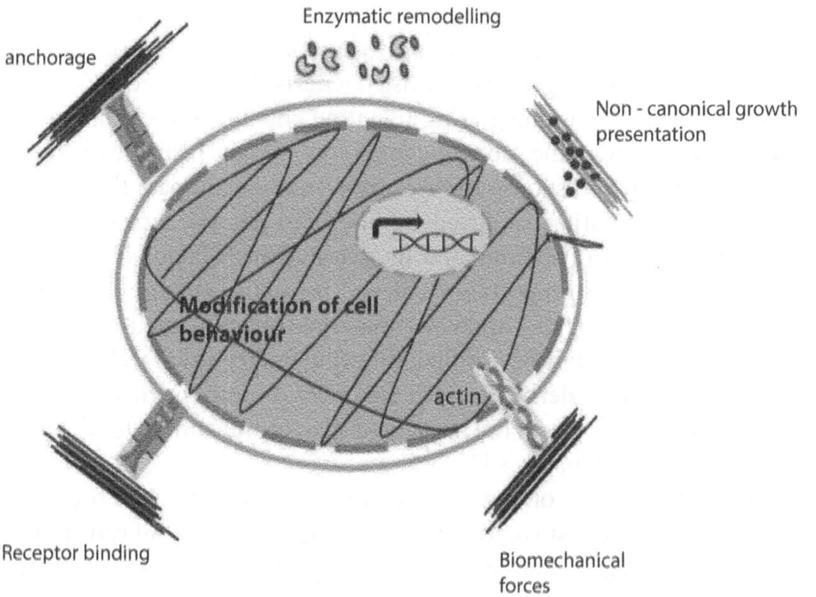

Figure 13.3 ECM-mediated control of cellular behavior. Regulation of cellular behavior by ECM. The effects exerted on cells by the extracellular matrix (ECM) may be facilitated through many mechanisms. The extracellular matrix (ECM) demonstrates direct binding affinity for several cell surface receptors as well as co-receptors (red, orange, black), hence promoting cell adhesion including regulating numerous pathways linked to intracellular signaling including mechanotransduction. Additionally, the extracellular matrix (ECM) can operate through the presentation of noncanonical growth factors (cyan) as well as undergo remodeling by enzymes (yellow pie), which can release functional pieces (green).

of stem cells as mentioned in Figure 13.3. Numerous occurrences of the relationship between integrins along with additional signaling molecules were identified in neural stem cells (NSCs) including mammary SCs. β1 integrins were demonstrated to regulate self-renewal as well as differentiation via influencing the functions of Notch with the EGF receptor. The essential function of β1 integrins in regulating the growth of intestinal stem cells (ISCs) through the mediation of Hedgehog signaling has been demonstrated.

13.3.3 Role of Soluble Bioactive Agents

Alternative methods for regulating the cellular microenvironment include the administration of bioactive substances, including drugs, hormones, or growth factors. Deviation from the normal signaling microenvironment can disrupt the signaling pathways, which are fundamental to several disorders such as cancer, diabetes, and several others [2]. The regulation of cell activation, growth, proliferation, migration, along with differentiation by growth factors, is of utmost importance for numerous biological processes, including angiogenesis and osteogenesis. Recent studies have predominantly focused on integrating signaling chemicals, including growth factors as well as cytokines, forming into biomaterials. Instances of altering cellular behavior in response to these molecules encompass the enhancement of vascularization (induced by fibroblast growth factor 2 FGF-2), neuronal regeneration (stimulated by nerve growth factor NGF), maintenance of stem cell properties (achieved through cytokine immobilization on maleic anhydride copolymer material thin-film coatings), and the establishment of a supportive milieu for cellular survival as well as growth. The integration of epidermal growth factor (EGF) into the matrix material enhances cell adherence to the implanted matrix and stimulates the proliferation of mesenchymal SCs [2].

Nevertheless, these compounds are prone to face enzymatic breakdown or deactivation in physiological fluids because of their significant instability and brief biological half-lives. Moreover, growth factors or medications require targeted delivery to have their desired effects. Consequently, systematic introduction is not a viable approach to regulate the amount they focus in designated target areas [2]. Consequently, it was essential to develop means of delivery that include growth hormones or pharmaceuticals into scaffolds for tissue engineering to overcome these limitations. Studies demonstrate that the sustained and targeted delivery of growth factors like VEGF via biodegradable hydrogels seems more effective than an intramuscular injection in stimulating angiogenesis. Upon administration with a bolus injection,

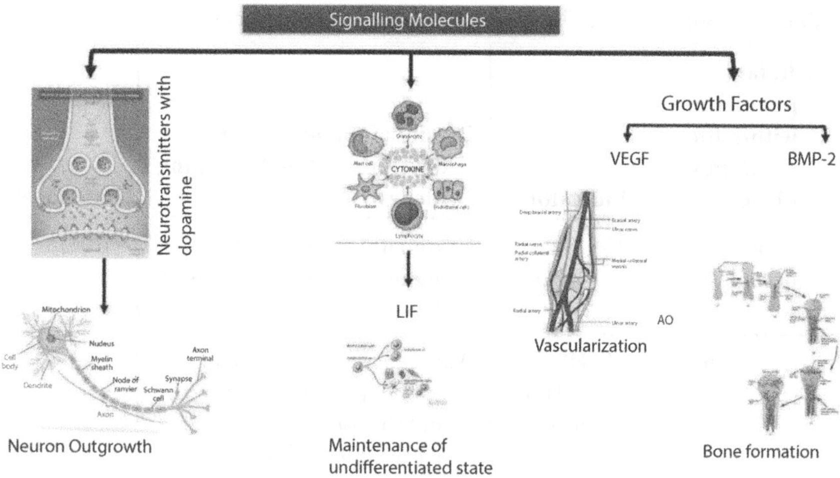

Figure 13.4 Different categories of soluble molecules exert unique impacts on cellular activity at both the individual cell and tissue levels. In tissue engineering research, the precise distribution of these components and their coordinated presence within the cellular niche is a crucial instrument.

VEGF failed to remain localized in the targeted region and persisted there for merely 72 h. Nonetheless, when administered via a renewable alginate hydrogel, 95% of the growth component remained at the ischemic site, leading to enhanced biodistribution. It persisted at that site for almost 15 days and had shown enhanced bioactivity, possibly due to protection against denaturation, as mentioned in Figure 13.4 and Table 13.1.

13.4 Stem Cell Plasticity and Adaptation

SC adaptability denotes the capacity of SCs to alter their lineage commitment, morphology, and function in response to various stimuli. Numerous studies have demonstrated that adult SCs derived from one organ can differentiate into cell types normally found in completely different organs following transplantation. Specifically, adoptive transfer has shown the involvement of bone marrow cells in the development of muscle, pulmonary, gastric, intestinal, and hepatic cells, together with neuronal stem cells, contribute to the development of hematopoietic, muscular, and neural tissues. Contemporary research indicates that stem cell plasticity is a highly uncommon occurrence. In the majority of the participants, the observed process of donor SC differentiation consisted essentially a fusion event

Table 13.1 Supporting cellular kinds in the developing niche.

Cellular populations within the developing niche	**Indicators**	**Possible functions in the development of developing niches**	**Reference**
Dermal fibroblasts	TWIST2, KRT19	Chromatin remodeling and maturation, including fascia development	[10]
Fetal myofibers	MYH8, ACTC1, spectrin	Adherence to fetal skeletal muscle progenitor cells and the formation of extracellular matrix necessary to maintain skeletal muscle progenitor cells	
Limb mesenchyme	DUSP6	Regulation of Wnt signaling as well as limb development	
Macrophages	SRGN, AIF1	Modulates muscle regeneration, stromal cell activation, including maturation	
Neural crest	SOX10, Wnt1	Prevention of early differentiation of fetal skeletal muscle progenitor cells	
Prechondrocyte	SOX5, SOX6, SOX9	Specification along with support of muscular development, BMP synthesis	
Preosteoblast	SHOX2	Specification as well as reinforcement of musculature	
Schwann cells	CDH19	Formation and maintenance of the neuromuscular junction	

(*Continued*)

Table 13.1 Supporting cellular kinds in the developing niche. (*Continued*)

Cellular populations within the developing niche	Indicators	Possible functions in the development of developing niches	Reference
Stromal cells	CD73, PDGFRα	Regulation of the immune system as well as SMPC activation via cytokine secretion and the differentiation of adipocytes vs. fibrosis	
Tendon precursors	SOX9, TNMD, SCX	Patterning of skeletal muscle as well as tensile forces promoting structural maturation	
Vascular and endothelial cells	ESAM, AP	Support of skeletal muscle through metabolites and growth factors	

between monocytes, macrophages, along with epithelial cells from the recipient tissues. The present optimism for medicinal multi-tissue regeneration in ill patients via the injection of a singular population of multipotent SCs encounters markedly waned.

13.4.1 Influence of Signaling Pathways

Niche cells secrete chemicals that influence stem cell fate through paracrine or neuroendocrine mechanisms. Molecular pathways, including angiopoietin-1 and growth factors, regulate stem cell maintenance and function in various ecological niches. Both invertebrates and mammals conserve BMP and Wnt signaling, regulating self-renewal and lineage commitment. The consequences of Wnt/beta-catenin signaling can differ according to the specific tissue involved. The Wnt/beta-catenin pathway is essential for promoting the self-renewal and multiplication of SCs in the hematopoietic system and gastrointestinal tract. In the epidermis, it stimulates the development of hair follicle precursors. The expansion of neural SC groups in the mammalian brain cells is stimulated by the upregulation of beta-catenin through Wnt signaling.

Bone morphogenic protein (BMP) signaling is crucial for regulating the proliferation of germinal centre stem (GCS) cells, the hematopoietic system, skin, and central nervous system. It suppresses bam, a cystoblast growth regulator, and promotes epidermal cell differentiation. Notch signaling pathways in these systems also facilitate cell proliferation. Stem cell behavior is influenced by reactive oxygen species (ROS), oxygen, and ions. The absence of calcium sensing receptor (CaR) in HSCs can lead to dysfunctional activity, compromising bone marrow engraftment. Additional investigation reveals that the preferential localization of HSCs occurs along a hypoxic gradient as well as that reactive oxygen species might cause the premature senescence of HSCs [11]. Table 13.2 provides a summary of these components.

13.5 Stem Cell Dysfunction in Disease

The essential importance of SCs in tissue homeostasis and repair is due to their unique ability for self-renewal as well as differentiation into various cellular types. Nevertheless, their participation extends beyond these regenerative processes, since some diseases have implicated them in their onset and advancement.

An exemplary instance is cancer. Cancer stem cells (CSCs) have a distinct subset of cancer cells that possess the characteristic traits of stem cells, including regenerative capacity, differentiation potential, and resistance to treatment. These cells are believed that they're responsible for the initiation, proliferation, and metastasis of tumors. Oncogenic stem cells (CSCs) can elude traditional cancer therapies, resulting in disease recurrence and unfavorable prognosis. Furthermore, scientists have linked stem cells to the pathogenesis of chronic illnesses, extending beyond cancer. Neuronal loss and disease development in neurodegenerative disorders such as Parkinson's and Alzheimer's disease may be recognized to the malfunctioning of neural stem cells. Abnormal cardiac stem cell activity in cardiovascular disorders can result in compromised heart regeneration and heightened susceptibility to heart failure [12].

Furthermore, the disease microenvironment, which consists of an intricate network of cells, extracellular matrix components, along with signaling chemicals, can impact stem cells. The microenvironment encompasses several stimuli that can either facilitate or inhibit the proliferation, development, and survival of stem cells. For example, in conditions characterized by inflammation, the inflammatory milieu might stimulate stem cells to follow pathogenic pathways.

Table 13.2 Overview of the cellular and molecular constituents identified in SC environments.

Tissue	Stem cell	Support cell	Signaling pathways	Adhesion	Reference
D. melanogaster ovary	GSC	Cap cells, ESCs	DPP-BMP	DE-cadherin, β-catenin	[11]
Mouse small intestine	CBC	Crypt fibroblasts, Paneth cells	Wnt, BMP	β-catenin	
Mouse skin	Follicular bulge stem cells	Dermal fibroblasts	Wnt, BMP	β-catenin, β-1 integrin	
C. elegans gonad	GSC	Distal tip cell	Notch	NI	
D. melanogaster testis	GSC	Hub cells	JAK-STAT	DE-cadherin, β-catenin	
D. melanogaster testis	CPC	Hub cells	JAK-STAT	DE-cadherin, β-catenin	
D. melanogaster ovary	ESC	NI	JAK-STAT	NI	
D. melanogaster ovary	FSC	NI	Hedgehog	DE-cadherin, β-catenin	
Mouse skeletal muscle	Satellite cells	NI	Notch	β-1 integrin	
Mouse skin	Interfollicular keratinocyte	NI	Wnt, Shh, Notch	E-cadherin, β-catenin, β-1 integrin	
Mouse bone marrow	HSC	Osteoblastic cells, endothelial cells	Wnt, Notch, ANG1, OPN	β-1 integrin	
Mouse brain (lateral ventricle)	SVZ stem cells	Vascular cells, astrocytes	Shh, BMP	N-cadherin, β-catenin	
Rat brain (hippocampus)	SVZ stem cells	Vascular cells, astrocytes	Shh, Wnt	N-cadherin, β-catenin	

13.5.1 Stem Cell-Derived Cancer Stem Cells

Some decades ago, researchers identified that barely a limited number of cancer cells could undergo significant multiplication, both in laboratory settings and in living organisms. Moreover, although malignant neoplasms are characterized by a single lineage, tumors can display a diverse range of distinct cell types, reflecting the hierarchical structure of normal regenerating tissues. Moreover, similar to healthy tissues, malignancies contain a population of cancer cells capable of unlimited division and proliferation, which is another distinguishing feature of stem cells [13]. These cells are designated as cancer stem cells (CSCs). The recently reintroduced idea of cancer stem cells (CSC), initially proposed over 40 years ago, is predicated on the CSC's ability to begin and sustain the formation of the neoplastic duplicate *in vivo*. Furthermore, despite the long-term self-renewal ability of the CSC, it produces a varied assortment of aberrantly differentiated tumor cells that comprise almost all of the tumor. Importantly, these cancerous cells are not essential for continued survival, expansion, and spread of the tumor and, as a result, typically do not have the ability to initiate cancer when transplanted into a capable host. Normal stem cells as well as cancer stem cells (CSCs) exhibit significant similarities, notably their ability to undergo lifelong self-renewal following division. Researchers have identified leukemic stem cells that exhibit a similar organizational structure to that of normal hematopoietic stem cells (HSCs) [13]. These cells have the ability to undergo periods of self-renewal, either of short duration, long duration, or during periods of inactivity. Conventional SCs are located within the SC "niche", a specialized microenvironment that consists of cell types that are different from the SC itself. The close proximity of indigenous SCs as well as niche cells enhances direct cell-to-cell and extracellular matrix communication, essential for maintaining SC identity. The balance between self-renewal as well as differentiation of SCs is governed by stimulation from certain habitats, together with internal and external signals. This equilibrium regulates the amount of SCs and the generation of more specialized progenitors during homeostasis or subsequent to tissue injury. Extensive research has demonstrated that the interactions among cancer cells, vascular components, mesenchymal cells, along with inflammatory cells substantially affect the progression as well as development of human malignancies. The interactions between stem cell niches and cancer stroma exhibit a high degree of overlap. This implies that the CSC should possess the capability to modify its own surroundings. For instance, it has the potential to transform a typical stem cell environment into a distinct tissue known as tumor stroma, therefore facilitating the growth and survival of the toxic clone. Moreover, the presence of somatic stem cells in seed colonies and the frequent migration of cancer cells spreading from the primary tumor suggest that normal niches may contribute to the metastatic process. Preferred sites for the selective introduction of metastatic cancer stem cells (CSCs) are SC niches located in the BM, liver, or even in brain. The hematopoietic cell niches within the BM release chemicals including SDF-1 (CXCL12), and that activate the CXCR4 receptor found on hematopoietic SCs. This signals hematopoietic stem cells (HSCs) present in the bloodstream to migrate to the bone marrow and initiate a fresh existence there. CSCs can develop sensitivity to certain secreted cues from the stem cell niche and produce metastases from advantageous microenvironments far from the primary cancer location. Although the comprehension of the signaling networks associated with cancer stem cells (CSCs) is limited, these atypical cells are anticipated to rely on pathways similar to those utilized by regular SCs to regulate cell proliferation, self-renewal, including cellular expansion. Numerous studies have shown that malignant cells frequently interfere with many signaling pathways critical for the proper functioning for SCs. The processes included in this list are Wnt-β-catenin signaling, hedgehog-patched-Gli signaling, c-Kit signaling, the PTEN/PI3-kinase connection, as well as Notch non-expression. Long-term self-renewal maintenance requires the polycomb protein Bmi-1, which inhibits the INK4a tumor suppressor gene. The absence of Bmi-1 inside the abdominal region leads to the gradual depletion of adult hematopoietic stem cells, neuronal stem cells, including leukemic stem cells, attributable to compromised self-renewal capabilities. Epigenetic regulation, particularly DNA methylation as well as histone acetylation, plays a crucial role in cancer formation [13]. Normal SCs along with their neoplastic equivalents are intricately connected, and an enhanced comprehension of stem cells may substantially aid in devising novel strategies to target the mysterious cancer SC.

13.5.2 Therapeutic Implications

Significant consequences emerge for the research, diagnosis, as well as therapy of cancer when identical communities of cancer cells demonstrate tumorigenicity in both individuals as well as the xenograft model. The primary goal of treatment should be to identify the small fraction of cancer stem cells responsible for tumor formation and to develop therapies that particularly target this group. The existing methodology for cancer treatment neglects to account for the possible variations in drug response as well as target expression among tumor-infiltrating cells (TICs) as well as non-tumorigenic cells. Development of medicines that reliably eliminate solid tumors, excluding testicular cancer, may provide challenges due to the differential treatment of tumorigenic as well as non-tumorigenic populations.

Conventionally, malignant cell therapies have depended on the capacity to indiscriminately eliminate rapidly multiplying cells. In both the AML as well as breast research, the tumorigenic cells represent a relatively minor fraction of the total tumor volume. Hence, screening techniques that depend on the fast reduction of tumor size are prone to disregarding drugs that specifically target and eliminate cancer stem cells. Recurrence is probable if an agent spares the tumor-infiltrating carcinomas (TICs), as the remaining TICs stimulate the growth of the tumor. Thus far, only a limited number of research have examined possible variations in drug sensitivity between the populations with tumors and those without tumors. In 2000, Costello and teammates showed that CD34+ CD38- leukemic cells exhibited much reduced sensitivity to daunorubicin compared to the more dedicated CD34- CD38+ cells [14]. Jordan and colleagues expanded upon these findings in a recent investigation, showing that lymphoid cancer cells (LICs) display a certain level of resistance to cytarabine, while the other leukemic blasts readily survived this drug [15]. The introduction of this unique cancer model is anticipated to profoundly impact our understanding of the processes behind treatment resistance. A variety of transporters, including those from the ABC transporter family, have been identified in normal stem cells. Numerous transporters have been definitively characterized in their roles in drug efflux. Prior studies have shown that leukemic progenitors can expel mitoxantrone along with daunorubicin, two frequently utilized drugs for therapeutic purposes of acute myeloid leukemia (AML). Consequently, the finding that a limited proportion of cancer stem cells contributes to carcinogenesis could significantly impact our understanding of drug sensitivity as well as the development of resistance to drugs in tumors [16].

The primary objective of cancer research is to uncover treatments that specifically target crucial genes or pathways that are responsible for the progression of tumor cells. The stem cell idea for cancer posits that therapies aimed against particular macromolecules present on or by tumor infiltrating cells (TICs) are actually more probable to provide efficacy. Gleevec selectively targets the ATP-binding domain that comprises Abl kinase in chronic myeloid leukemia (CML). Most patients who get treatment with Gleevec have full cytogenetic responses. Nevertheless, even when employing sensitive techniques such as PCR to monitor the condition, most of patients continue to have favorable outcomes for the combined transcript [16].

Due to the ongoing existence of these atypical documents, this treatment may prove ineffective in resolving the problem. The cancer stem cell model offers a plausible rationale regarding this discovery. Bhatia *et al.* established that the BCR/ABL combination was identified in 6.5% up to 13% of CD34+ precursor cells sourced from a group of patients in relapse [17]. These cells did not exhibit any translocation when subjected to FISH analysis on unfractionated bone marrow samples. Previous research indicates that the aberrant gene, although found in the hematopoietic stem cells (HSC), does not become evident until the initial attached progenitors. The inability to eradicate the disease may be ascribed to the lack of target expression within the atypical SC type. Conversely, intrinsic properties of the LIC in CML may confer immunity to Gleevec, whereas the observed responses result from the targeted targeting of distinct non tumorigenic population.

Furthermore, apart from influencing our comprehension of the effectiveness of existing treatments, SC illustration of cancer probably influences the identification of potential treatment targets. The pattern of expression of pluripotent SCs along with their more specialized derivatives can vary significantly. Tumor DNA along with tissue microarrays having thus far failed to account for cellular heterogeneity and changes in the regenerative capacity of different populations. Enhancing expression studies to focus on specific clusters of cancer cells with a greater likelihood of tumorigenesis might aid in the discovery of new indicators of cancer along with therapeutic targets. Furthermore, it is becoming increasingly clear that therapies aimed at the pathways involved in the regeneration of malignant SCs are more likely to provide favorable results [16].

Ultimately, this novel cancer model has substantial ramifications for the development of ongoing research endeavors focused on enhancing our capacity to detect cancer and pinpoint persons who are susceptible to metastasis. Prior studies have demonstrated the detection of dispersed cytokeratin-positive cancerous breast cells within the BM for non-recurrent patients. A probable explanation for this outcome is that the cancer tissues remain dormant until an undetermined event triggers them to resume proliferation. Alternate scenario is that the malignant cells that proliferated in this cohort of patients originated from non-tumorous cells. Metastatic cancers arise only from the dissemination and subsequent self-renewal of cancer stem cells. Hence, conducting more research on the tumorigenic population could provide significant insights into the metastasis process and enable the creation of diagnostic reagents that can accurately forecast which individuals will progress to metastatic disease [16]. This would allow clinicians to identify patients that would be probably to benefit from adjuvant therapy, so potentially exempting numerous patients from unnecessary medical procedures.

13.6 Stem Cell Therapy for Human Diseases

The regulatory consent for cancer treatment targets cancer immunotherapies and mesenchymal SC-based therapies in Europe as a prominent therapeutic approach [18]. Cellular therapy, specifically stem cells, offers renewed optimism for individuals afflicted with terminal illnesses, when therapeutic strategies prioritize disease control rather than definitive cure. SC based treatment is a crucial domain of regenerative healing that seeks to enhance the internal repair processes by activating, altering, along with regulating the current SC number and/or increasing the overall cell pool to sustain tissue equilibrium and foster regeneration. Since the inception of stem cells, their distinctive characteristics of differentiation as well as self-renewal have prompted extensive fundamental study and clinical investigations, solidifying their potential as therapeutic agents. Cellular replacement and tissue regeneration are the main objectives of regenerative medicine, which employs many types of SCs, particularly human pluripotent stem cells (hPSCs), multipurpose SCs, as well as progenitor cells [19]. Nonetheless, the emergence of unregulated along with unverified clinics claiming the effectiveness from the use of SC as "mystical cells" caused significant concerns about the safety of such treatments [20]. Therefore, in order to dispel the misconception of the "magic" cells and advance the field of regenerative medicine, this review offers a concise summary of SC-based treatment for the management of individual pathologies.

SC therapy uses stem cells' regenerative and differentiative properties to repair or replace damaged tissues in the body, similar to myocardial infarction, which destroys cardiomyocytes, leading to heart failure [21]. Stem cell sources for cell-based therapy include autologous and allogeneic cells, sourced from healthy donors. Ernst Haeckel coined the term in 1868 to describe the essential properties of a fertilized egg [22]. In 1888, German animal scientists Boveri and Haecker discovered embryonic cells capable of differentiation, marking the beginning of stem cell therapy. Neumann and Maximov's research on bone marrow revealed hematopoiesis and introduced the notion of polyblasts, which were subsequently designated as stem cells by Ernst Haeckel due to their processes of proliferation and differentiation [23–25]. From the 1960s to the 1970s, Friendenstein along with other researchers conducted systematic investigations of BM aspirates to determine the relationship between osteogenic differentiation as well as a minor subset of cells derived from BM [26]. Originally pioneered by BM transplantation six decades ago, SC therapy progressed into a groundbreaking regenerative medicine treatment for an inclusive variety of chronic illnesses which includes diseases related to the nervous system, respiratory dysfunctions, metabolic/endocrine-related diseases, reproductive disorders, and skin blisters along with cardiac-related illnesses [21]. Stem cells and their products are widely used in regenerative medicine, which investigates diseases in pharmacological contexts. They are also used in tissue engineering for tissue replacement. The growing array of stem cell products is driven by market demand for their therapeutic benefits [27].

13.7 Conclusion

While stem cells and the microenvironment possess separate attributes, they are inherently interconnected both physically and functionally. Stem cells exert several effects on their milieu, including immunomodulation, hematopoietic proliferation, and tissue regeneration. Additionally, the niche commands or utilizes differentiation and proliferation along with the function of SCs. Within the pathological microenvironment, the presence of SCs can either worsen or improve the medical condition. Ongoing debate persists regarding the use of SCs due to the possibility of conflicting effects on SCs resulting from the same microenvironment. Even though many clinical trials have been done, the use of SC treatment is still limited due to its unclear efficacy as well as security. Moreover, altering the microenvironment might either benefit or hinder the effectiveness of SC treatment and the advancement of the disease. A complicated entity, the microenvironment presents difficulties in its research and use. A more comprehensive investigation of the alterations in functions, composition, and location within the associated diseased process will elucidate the mechanism of illness onset and progression, thereby facilitating the development of more groundbreaking therapies.

References

1. Zipori, D., The nature of stem cells: state rather than entity. *Nat. Rev. Genet.*, 5, 11, 873–878, 2004.
2. Barthes, J., *et al.*, Cell microenvironment engineering and monitoring for tissue engineering and regenerative medicine: the recent advances. *Biomed. Res. Int.*, 2014, 1, 921905, 2014.

3. Stoddard-Bennett, T. and Reijo Pera, R., Treatment of Parkinson's disease through personalized medicine and induced pluripotent stem cells. *Cells*, 8, 1, 26, 2019.
4. Ortiz-Vitali, J.L. and Darabi, R., iPSCs as a platform for disease modeling, drug screening, and personalized therapy in muscular dystrophies. *Cells*, 8, 1, 20, 2019.
5. Arjmand, B., *et al.*, *Personalized regenerative medicine*, pp. 144–149, Acta Medica Iranica, 2017.
6. Liu, Y. and Chen, Y.-G., 2D-and 3D-based intestinal stem cell cultures for personalized medicine. *Cells*, 7, 12, 225, 2018.
7. Metallo, C.M., *et al.*, Engineering the stem cell microenvironment. *Biotechnol. Progr.*, 23, 1, 18–23, 2007.
8. Gattazzo, F., Urciuolo, A., Bonaldo, P., Extracellular matrix: a dynamic microenvironment for stem cell niche. *Biochim. Biophys. Acta (BBA)-General Subj.*, 1840, 8, 2506–2519, 2014.
9. Kanatsu-Shinohara, M., *et al.*, Homing of mouse spermatogonial stem cells to germline niche depends on β1-integrin. *Cell Stem Cell*, 3, 5, 533–542, 2008.
10. Hicks, M.R. and Pyle, A.D., The emergence of the stem cell niche. *Trends Cell Biol.*, 33, 2, 112–123, 2023.
11. Ferraro, F., Celso, C.L., Scadden, D., *Adult stem cells and their niches*, pp. 155–168, The Cell Biology of Stem Cells, 2010.
12. Farini, A., *et al.*, Clinical applications of mesenchymal stem cells in chronic diseases. *Stem Cells Int.*, 2014, 1, 306573, 2014.
13. Trumpp, A. and Wiestler, O.D., Mechanisms of disease: cancer stem cells—targeting the evil twin. *Nat. Clin. Pract. Oncol.*, 5, 6, 337–347, 2008.
14. Costello, R.T., *et al.*, Human acute myeloid leukemia CD34+/CD38– progenitor cells have decreased sensitivity to chemotherapy and Fas-induced apoptosis, reduced immunogenicity, and impaired dendritic cell transformation capacities. *Cancer Res.*, 60, 16, 4403–4411, 2000.
15. Guzman, M.L., *et al.*, Preferential induction of apoptosis for primary human leukemic stem cells. *Proc. Natl. Acad. Sci.*, 99, 25, 16220–16225, 2002.
16. Al-Hajj, M., *et al.*, Therapeutic implications of cancer stem cells. *Curr. Opin. Genet. Dev.*, 14, 1, 43–47, 2004.
17. Bhatia, R., *et al.*, Persistence of malignant hematopoietic progenitors in chronic myelogenous leukemia patients in complete cytogenetic remission following imatinib mesylate treatment. *Blood*, 101, 12, 4701–4707, 2003.
18. Hoang, D.M., *et al.*, Stem cell-based therapy for human diseases. *Signal Transduction Targeted Ther.*, 7, 1, 1–41, 2022.
19. Mousaei Ghasroldasht, M., *et al.*, Stem cell therapy: from idea to clinical practice. *Int. J. Mol. Sci.*, 23, 5, 2850, 2022.
20. Kuriyan, A.E., *et al.*, Vision loss after intravitreal injection of autologous "stem cells" for AMD. *N. Engl. J. Med.*, 376, 11, 1047–1053, 2017.

21. Malik, S. and Dhasmana, A., Differentiation of Stem Cells into Cardiomyocyte Lineage: *In Vitro* Cell Culture, *In Vivo* Transplantation in Animal Models, in: *Advances in Application of Stem Cells: From Bench to Clinics*, pp. 103–121, 2021.
22. Ramalho-Santos, M. and Willenbring, H., On the origin of the term "stem cell". *Cell Stem Cell*, 1, 1, 35–38, 2007.
23. Konstantinov, I.E., In search of Alexander A. Maximow: the man behind the unitarian theory of hematopoiesis. *Perspect. Biol. Med.*, 43, 2, 269–276, 2000.
24. Dröscher, A., Images of cell trees, cell lines, and cell fates: the legacy of Ernst Haeckel and August Weismann in stem cell research. *Hist. Philos. Life Sci.*, 36, 157–186, 2014.
25. Cheng, M., H artmann S tahelin (1925–2011) and the contested history of cyclosporin A. *Clin. Transplant.*, 27, 3, 326–329, 2013.
26. Friedenstein, A., Chailakhyan, R., Gerasimov, U., Bone marrow osteogenic stem cells: in vitro cultivation and transplantation in diffusion chambers. *Cell Proliferation*, 20, 3, 263–272, 1987.
27. Malik, S., Commercialization, IPR, and Market of Stem Cell Products, in: *Advances in Application of Stem Cells: From Bench to Clinics*, pp. 299–314, 2021.

14

Stem Cell Therapy for Cancer Treatment: Current Advances and Future Directions

Anand Kumar Shukla[1,2], Sindhu D. Bali[1,2], Varsha A. Mahadik[1,2], Rosaleen Sahoo[1,2] and Narendra Kadoo[1,2*]

[1]*Biochemical Sciences Division, CSIR-National Chemical Laboratory, Pune, India*
[2]*Academy of Scientific and Innovative Research (AcSIR), Ghaziabad, India*

Abstract
Stem cell therapy is emerging as a promising approach in the fight against cancer, offering a potential solution to the limitations of traditional cancer treatments. This chapter provides an in-depth review of the current state of stem cell therapy for cancer treatment, highlighting the latest advances and future directions in this field. The chapter begins by discussing the fundamental principles of stem cell therapy, including various types of stem cells and their unique properties. It then explores the current clinical applications of stem cell therapy for various types of cancer, including the use of mesenchymal stem cells, neural stem cells, and hematopoietic stem cells. The chapter also delves into the mechanisms by which stem cells modulate the immune response and promote tumor targeting, as well as the potential challenges and limitations of this therapeutic approach. Finally, the chapter concludes by discussing the future directions for stem cell therapy in cancer treatment, including integrating stem cell therapy with other therapeutic techniques and developing new stem cell sources.

***Keywords*:** Stem cell therapy, tumor targeting mechanisms, immune modulation in cancer therapy, stem cell immuno-oncology challenges, oncogenesis

14.1 Introduction

Cancer has become one of the leading causes of death globally. Moreover, demographic trends predict a rise in cancer incidence, with over 20 million

**Corresponding author*: ny.kadoo@ncl.res.in

Ayush Madan, Sumel Ashique, Deepika Arora and Mantosh Kumar Satapathy (eds.) Stem Cell Therapeutics, (599–628) © 2025 Scrivener Publishing LLC

new annual cases expected by 2025 [1, 2]. Cancer treatment has always been a complicated procedure. Although significant advancements have been made in treatment, such as targeted therapy, ablation therapy, radionics, chemodynamic therapy, sonodynamic therapy, etc., traditional approaches such as surgery, chemotherapy, and radiotherapy are still being routinely employed [3]. Conventional cancer treatment plans often involve surgically excising tumors, followed by chemotherapy or radiation therapy using X-rays [4]. While surgery is most effective when the disease is in its early stages, radiation therapy can damage organs, tissues, and healthy cells. Similarly, chemotherapy, although it has reduced morbidity and mortality, often harms healthy cells, particularly those that are fast-dividing and developing [5]. Moreover, drug resistance is one of the main challenges with chemotherapy, in which cancer cells that initially responded to an anti-cancer drug eventually become resistant to it [6]. This resistance is primarily due to increased drug efflux and decreased drug absorption. Limitations of traditional chemotherapy include challenging dose selection, poor selectivity, rapid drug metabolism, and predominantly adverse side effects [3]. Conventional therapies such as surgery, radiation, and chemotherapy often have limited effectiveness, especially against metastatic disease, and can result in severe side effects. Metastatic cancer cells usually resist traditional therapies, making recurrence highly probable. Consequently, developing new and effective treatments that are either nontoxic or have very little toxicity for normal cells is essential.

Cancer treatment encompasses a variety of strategies tailored to target specific characteristics of different cancer types. These treatments aim to eliminate or control the growth of cancer cells without damaging healthy tissues. Table 14.1 provides an overview of the primary cancer treatment modalities, including their descriptions, examples, and mechanisms of action. This highlights the diverse approaches employed in fighting cancer and underscores the importance of personalized treatment plans to improve patient outcomes.

Recently, stem cell treatment has emerged as a potential cancer treatment, with the ability to target tumor cells specifically and sparing healthy tissues [20, 21]. Stem cells have unique properties, such as self-renewal, migration to tumor sites, secretion of bioactive factors, and differentiation into various cell types, making them ideal for cancer therapy. Multiple stem cell types, such as neural, induced pluripotent, and mesenchymal stem cells, are being explored for their potential in cancer therapy. It is possible to modify these cells to express cytotoxic substances, which can serve as delivery vehicles for therapeutic compounds and may alter immune responses to target cancer cells [22]. Preclinical research has advanced

Table 14.1 Overview of cancer treatment strategies

Treatment strategy	Description	Examples	Mechanism of action	References
Surgery	Physical removal of the tumor from the body	Tumor resection, debulking surgery	Removes cancerous tissues to eliminate or reduce cancer burden	[7]
Radiation therapy	Uses high-energy radiation to inhibit, damage, or kill cancer cells	External beam radiation, brachytherapy	It damages the DNA of cancer cells, inhibiting their ability to replicate and grow	[8]
Chemotherapy	Using drugs to kill cancer cells	Cisplatin, doxorubicin, paclitaxel	Targets and kills rapidly dividing cells by interfering with cell division and DNA replication	[9]
Immunotherapy	Stimulates the body's immune system to recognize and attack cancer cells	Checkpoint inhibitors (nivolumab), CAR-T cell therapy	Enhances the immune response against cancer cells by targeting specific proteins or modifying immune cells	[10]
Targeted therapy	Uses drugs designed to target specific molecular targets associated with cancer	Imatinib (Gleevec), trastuzumab (Herceptin)	Targets specific proteins or pathways involved in the growth and survival of cancer cells	[11]

(Continued)

Table 14.1 Overview of cancer treatment strategies (*Continued*)

Treatment strategy	Description	Examples	Mechanism of action	References
Hormone therapy	Suppresses or eliminates hormones that promote certain cancers, such as breast and prostate cancers	Tamoxifen, aromatase inhibitors, androgen deprivation therapy	Interferes with hormone production or receptor activity to slow or stop cancer growth	[12]
Stem cell transplant	Replaces the diseased bone marrow cells with healthy stem cells	Autologous and allogeneic stem cell transplants	Healthy bone marrow is restored after chemotherapy or radiation therapy	[13]
Photodynamic therapy	Cancer cells are destroyed using light-sensitive drugs and a light source	Porfimer sodium (Photofrin)	Activates light-sensitive compounds that produce reactive oxygen species, killing cancer cells	[14]
Hyperthermia	It uses heat to damage or kill cancer cells	Local hyperthermia, whole-body hyperthermia	Increases the temperature of cancer cells to damage proteins and structures within the cells, causing cell death	[15]
Cryotherapy	Cancer cells are destroyed using extreme cold treatment	Cryoablation	Freezes cancer cells, leading to cell death and destruction of the tumor	[16]

(*Continued*)

Table 14.1 Overview of cancer treatment strategies (*Continued*)

Treatment strategy	Description	Examples	Mechanism of action	References
Gene therapy	Modifies genes to treat or prevent cancer	Viral vector delivery of therapeutic genes	Replaces, inactivates, or introduces genes to correct genetic defects or enhance cancer cell destruction	[17]
Nanotechnology-based therapy	Uses nanoparticles to deliver drugs or other therapeutic agents directly to cancer cells	Liposomal doxorubicin, gold nanoparticles	It enhances drug delivery, reduces side effects, and improves the targeting of cancer cells	[18]
Precision medicine	Customizes treatment based on the genetic profile of the patient's cancer	Personalized cancer vaccines, targeted gene sequencing	Tailored treatment according to individual genetic mutations and characteristics of the cancer	[19]

significantly, demonstrating the capability of stem cell-based therapies to reduce tumor burden and extend survival in animal models. However, several challenges remain, including the possibility of tumor formation, immune rejection, and disease relapse. Additionally, the complex biology of cancer stem cells (CSCs) driving tumor initiation, progression, and recurrence requires further examination [23].

Resistance and recurrence are frequently caused by standard cancer therapy's inadequate and non-specific targeting. Recent developments in stem cell biology have provided new therapeutic options for cancer patients. Unique biological properties of stem cells include self-renewal, migration in a specific direction, differentiation, and the ability to modulate other cells. These characteristics can be used for immunological cell production, medication targeting, therapeutic administration, and regenerative medicine [21].

Understanding cancer biology necessitates in-depth research on how microenvironments affect the maintenance of CSCs. Stem cells require specific microenvironments to sustain their self-renewal capabilities, and peritumoral tissues affect tumor maintenance. Two fundamental properties of stem cells that impact tumor progression are their high expression of multidrug resistance proteins and low proliferation rate. These characteristics suggest that conventional chemotherapy primarily targeting proliferating cells would not effectively eradicate these stem cells. Consequently, the discovery of CSCs with unique molecular pathways and an improved understanding of the tumor microenvironment could contribute to developing targeted treatments [24, 25]. These treatments are expected to be less invasive and more successful in eliminating the cells contributing to tumor growth.

14.1.1 Importance of Stem Cell Therapy in Cancer Treatment

Stem cell therapy is emerging as a pivotal approach in the fight against cancer, addressing many of the limitations associated with traditional cancer treatments. The unique properties of stem cells, such as their ability to self-renew, differentiate into various cell types, and modulate immune responses, make them invaluable in enhancing therapeutic efficacy and improving patient outcomes.

14.1.1.1 Enhanced Targeting of Tumors

One of the primary advantages of stem cell therapy is its potential for targeted delivery of therapeutic agents directly to tumor sites. Traditional cancer treatments often lack specificity, leading to significant off-target effects and damage to healthy tissues. In contrast, stem cells can be engineered to home in on tumors, thereby increasing the concentration of therapeutic agents at the site of cancer while minimizing systemic toxicity. This targeted approach can improve the efficacy of the treatment and also reduce adverse effects associated with conventional therapies [21, 26].

14.1.1.2 Overcoming Therapy Resistance

The CSCs are a subset of cells found within tumors. They can self-renew and promote tumor growth. These cells are often resistant to standard treatments like chemotherapy and radiation, which can lead to tumor recurrence and the spreading of cancer. Stem cell therapy presents a new approach to target and remove CSCs, potentially addressing the primary cause of treatment failure. By concentrating on these resilient cells, stem cell-based therapies can improve overall treatment responses and reduce the chances of disease relapse [27].

14.1.1.3 Regenerative Potential

Stem cells are crucial for tissue regeneration and repair. Aggressive treatments like chemotherapy and radiation often cause significant damage to patients' blood and immune systems. Stem cell transplants, especially hematopoietic stem cell transplants, can help the body produce healthy blood cells, improving recovery and quality of life. This regenerative capability is especially beneficial for patients with blood cancers like leukemia and lymphoma, where restoring blood cell production is critical [28].

14.1.1.4 Immune Modulation

Stem cells have immunomodulatory properties and can boost the body's immune response against cancer. They can generate immune effector cells or modify immune pathways, helping the body better recognize and attack cancer cells. This not only aids in direct anti-tumor effects but also promotes a more effective and long-lasting immune response, crucial for long-term cancer control [29]. To enhance the effectiveness of stem cell therapy, the stem cells can be genetically engineered to express a drug or a protein that binds to a specific target in a cancer cell. For instance, MSCs from bone marrow can be engineered to deliver cancer medications directly to cancer cells as they naturally home in on tumors. Additionally, stem cells can be used to create organoid models and miniature copies of organs to test potential cancer treatments and study the fundamental processes underlying cancer.

14.1.2 Fundamental Principles of Stem Cell Therapy

Stem cell therapy for cancer treatment utilizes the unique properties and potential of various stem cell types, such as mesenchymal stem cells (MSCs), neural stem cells (NSCs), and hematopoietic stem cells (HSCs). These stem cells can self-renew, differentiate into multiple cell types, and migrate toward tumor sites, making them promising candidates for targeted cancer therapy.

14.1.3 Types of Stem Cells

14.1.3.1 Mesenchymal Stem Cells (MSCs)

MSCs are also known as mesenchymal stromal cells. These multipotent stromal cells can differentiate into various cell types, such as osteoblasts (bone cells), chondrocytes (cartilage cells), myocytes (muscle cells), and adipocytes (Figure 14.1). They have been identified in nearly all tissues of the human body and can be isolated from bone marrow, adipose tissue, and umbilical cord blood [30].

Upon receiving appropriate signals from their microenvironment, MSCs can differentiate into multiple lineages. The differentiation process is highly regulated and can be influenced by various growth factors, cytokines, and extracellular matrix components. The following are the primary lineages that MSCs can differentiate into: myocytes (muscle cells), adipocytes (fat cells), osteoblasts (bone cells), chondrocytes (cartilage cells), and also neuron-like cells under specific conditions [32, 33].

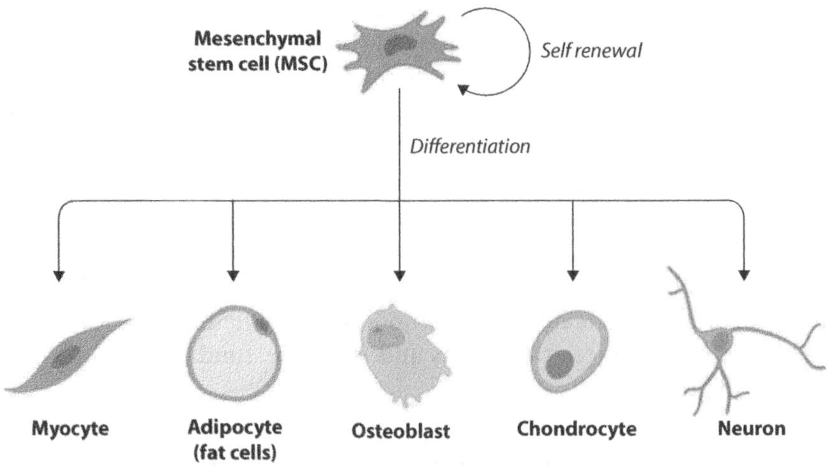

Figure 14.1 Differentiation process of mesenchymal stem cells (MSCs) from their self-renewing state into specialized cell types, such as myocytes, adipocytes, osteoblasts, chondrocytes, and neurons. This differentiation process is affected by both intrinsic and extrinsic factors, enabling MSCs to play crucial roles in tissue regeneration and repair [31]. The image is rendered using Biorender (https://www.biorender.com/).

14.1.3.2 *Neural Stem Cells (NSCs)*

Neural stem cells (NSCs) are the self-renewing and multipotent somatic stem cells that give rise to neurons, astrocytes, and oligodendrocytes (Figure 14.2). These stem cells are primarily located in the subventricular and subgranular zones of the hippocampus [34].

14.1.3.3 *Hematopoietic Stem Cells (HSCs)*

HSCs are the precursors to all blood cell types, including white blood cells, red blood cells, and platelets. These stem cells are primarily present in the bone marrow. However, they can also be isolated from peripheral and umbilical cord blood [36].

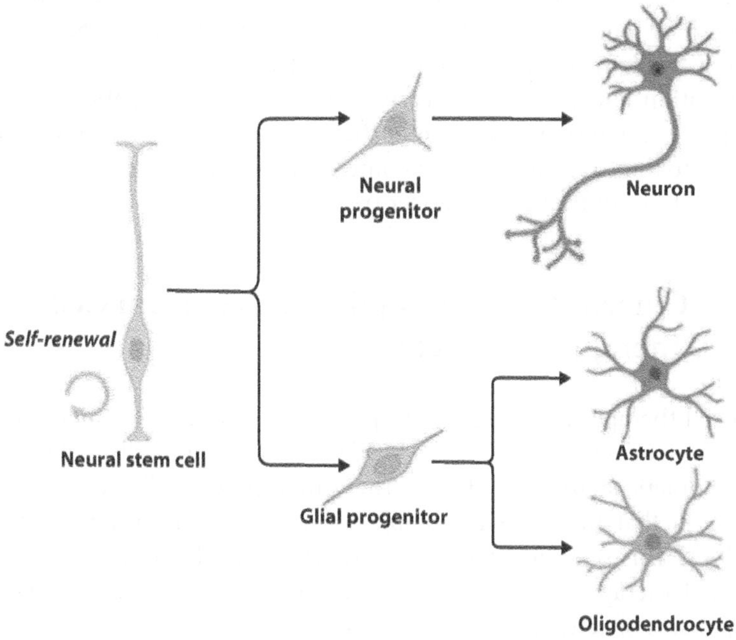

Figure 14.2 Differentiation process of neural stem cells (NSCs) from their self-renewing state into various neural cell types, such as neurons, astrocytes, and oligodendrocytes. The development and maintenance of the central nervous system (CNS) relies on this critical differentiation pathway [35]. The image is rendered using Biorender (https://www.biorender.com/).

14.1.4 Unique Properties of Stem Cells

Stem cells possess several unique properties that make them valuable tools for cancer therapy. One of their key attributes is self-renewal, which allows stem cells to divide and maintain their undifferentiated state, ensuring the continuous replenishment of the stem cell pool. This property is crucial for their long-term maintenance and regenerative potential. Additionally, stem cells can differentiate into specialized cell types under specific conditions, enabling them to replace damaged or diseased cells. The differentiation potential varies among different types of stem cells, with embryonic stem cells being the most versatile and adult stem cells being more lineage-restricted [37]. Another significant property of stem cells is their homing ability. They can migrate toward sites of injury or disease, including tumor sites, in response to various chemokines and growth factors. This tumor-tropic property allows stem cells to selectively target and deliver therapeutic agents to cancer cells while minimizing off-target effects [38, 39].

These fundamental properties, combined with the potential for genetic modification and targeted delivery, make stem cells valuable for cancer therapy. By exploiting their unique characteristics, stem cell-based therapies aim to improve treatment outcomes, reduce side effects, and overcome the limitations of traditional cancer treatments.

14.2 Current Clinical Applications of Stem Cell Therapy for Cancer

Stem cell therapy shows significant promise for cancer treatment, mainly using mesenchymal stem cells (MSCs), neural stem cells (NSCs), and hematopoietic stem cells (HSCs). These multipotent cells possess unique properties that make them suitable for various clinical applications in oncology. This section explores the current clinical applications of these stem cell therapies for cancer, emphasizing their therapeutic potential, mechanisms of action, and ongoing clinical trials.

14.2.1 Mesenchymal Stem Cells (MSCs)

MSCs exhibit immunomodulatory properties and can migrate to sites of inflammation and tumors, making them attractive candidates for cancer therapy. These cells have various therapeutic mechanisms in cancer treatment. MSCs are known for their tumor tropism, allowing them to home in on tumor sites and enable targeted delivery of therapeutic agents. This property is crucial to enhance the effectiveness of cancer treatments while minimizing damage to surrounding healthy tissues [40]. Additionally, MSCs can modulate the immune response by creating an immunosuppressive environment, which can help reduce tumor growth and metastasis. They achieve this by secreting various cytokines and growth factors that influence immune cell behavior, promoting tolerance and reducing inflammation [41]. The gene therapy potential of MSCs is another promising aspect. Genetic modification of MSCs to express therapeutic genes, such as those encoding pro-apoptotic factors or oncolytic viruses, has shown promise in preclinical studies. These engineered MSCs can directly target and kill cancer cells or enhance the immune response against tumors [42].

The application of MSCs in cancer therapy has been explored in numerous clinical trials, yielding varying degrees of success. MSCs have been utilized as therapeutic agents in direct cancer treatment. For example, in a phase I/II clinical trial, the metastatic and refractory solid tumors in pediatric and adult patients were treated using autologous MSCs infected with an oncolytic adenovirus (CELYVIR). The results demonstrated excellent safety and promising anti-tumor effects, including complete tumor remission in some cases [30]. Moreover, MSCs are commonly used to treat graft-versus-host disease (GVHD) succeeding hematopoietic stem cell transplantation. By mitigating the immune response that can lead to GVHD, MSCs help improve patient outcomes. MSCs have also been investigated with other cancer therapies, such as chemotherapy and immunotherapy, to enhance therapeutic efficacy. They can modulate the tumor microenvironment and support the immune system, a property that can synergize with conventional treatments, potentially leading to improved patient responses [42].

14.2.2 Neural Stem Cells (NSCs)

NSCs have emerged as a promising therapeutic approach to treat various cancers, particularly brain tumors like glioblastoma. NSCs have unique properties that make them attractive candidates for cancer therapy, including their ability to home to tumor sites and their potential for genetic modification to deliver therapeutic agents. NSCs have an inherent ability to migrate toward tumor sites, including distant metastases, which makes them suitable for cancer therapy. This tumor tropism is facilitated by cancer cells secreting chemokines and growth factors that attract NSCs to the tumor microenvironment. This capability enables NSCs to access both primary and invasive tumor sites, making them an ideal platform for delivering therapeutic agents. NSCs can also be genetically engineered to express cytotoxic agents, such as pro-apoptotic proteins, oncolytic viruses, and suicide genes. In preclinical studies, these modified NSCs have shown significant anti-tumor effects, reducing tumor volumes and prolonging survival in animal models [43]. A phase I clinical trial is currently assessing the safety and feasibility of human NSCs engineered to express a modified form of the cytokine IL-12 for treating recurrent high-grade gliomas [35]. As research progresses, several challenges must be addressed, including improving NSC targeting and retention within tumors, optimizing therapeutic payload delivery, and ensuring these treatments' safety and efficacy. Ongoing preclinical studies and clinical trials will provide valuable insights into the potential of NSC-based therapies for cancer management [44].

14.2.3 Hematopoietic Stem Cells (HSCs)

HSC transplantation is a pivotal treatment for hematologic malignancies such as leukemia, lymphoma, and multiple myeloma. This procedure involves infusing healthy HSCs into patients whose blood-forming cells have been depleted by intensive chemotherapy or radiation. HSCs regenerate the ability to produce new blood cells, aiding recovery from cancer treatments and enhancing survival rates [21, 28]. Allogeneic HSC transplantation offers a significant advantage through the graft-versus-leukemia (GVL) effect, wherein the donor immune cells attack the residual cancer cells, potentially leading to long-term remission and reducing the risk of relapse, particularly in leukemia cases. In contrast, in autologous HSC transplantation, the patient's own previously harvested stem cells are used after intensive chemotherapy, primarily benefiting conditions like multiple myeloma and certain lymphomas [45].

Recent advancements in gene therapy utilizing HSCs show promise for treating genetic blood disorders and enhancing anti-tumor responses in cancer. Correcting genetic defects in autologous HSCs aims to provide lasting cures for diseases such as sickle cell anemia and thalassemia. Improving HSC mobilization techniques, donor matching, and refining conditioning regimens are required to enhance transplant efficacy and safety. Additionally, exploring HSC-derived immune cells for adoptive cell therapy is a promising frontier in cancer treatment, underscoring the enduring role of stem cell transplantation in managing hematologic malignancies and fostering ongoing advancements in oncology [46].

14.2.4 Induced Pluripotent Stem Cells (iPSCs)

The iPSCs can differentiate into any cell type, providing a versatile platform for developing personalized treatments. They can be generated from a patient's somatic cells, thus offering a patient-specific approach that minimizes immune rejection. iPSCs are being explored in oncology due to their potential to generate immune cells, such as T cells and natural killer (NK) cells, which can be engineered to target and destroy cancer cells. iPSCs can also be differentiated into dendritic cells, which can be used in cancer vaccines to stimulate a robust anti-tumor immune response. The ability to create iPSCs from cancerous cells also allows for the development of disease models to study tumor biology and screen for novel therapeutics in a patient-specific context. However, some challenges, such as ensuring the genomic stability of iPSCs, preventing tumorigenicity, and scaling up production for clinical use, are significant hurdles that should be addressed [47, 48].

14.2.5 Embryonic Stem Cells (ESCs)

ESCs are the pluripotent stem cells derived from the inner cell mass of the blastocyst. They can differentiate into any cell type. This unique character makes ESCs a promising tool in cancer therapy, particularly for regenerative approaches and developing cancer models for drug screening. ESCs can be used to generate immune cells, such as cytotoxic T lymphocytes or natural killer cells, which can target and destroy cancer cells. Additionally, ESC-derived dendritic cells can be employed in cancer vaccines to enhance the body's immune response against tumors. Despite their potential, the use of ESCs in clinical applications faces significant ethical and technical challenges, including the risk of teratoma formation, immune rejection, and ethical concerns regarding the source of ESCs. Efforts are underway to overcome these obstacles by improving differentiation protocols, ensuring controlled cell proliferation, and developing immunocompatible ESC lines. Advances in gene editing technologies, such as CRISPR/Cas9, will be promising to enhance the safety and efficacy of ESC-based therapies [49, 50].

14.3 Molecular Mechanisms Underlying the Action of Stem Cells

The molecular mechanisms responsible for the action of stem cells in cancer therapy may involve direct and indirect pathways. For example, stem cells such as MSCs and NSCs exhibit inherent tumor tropism, enabling them to migrate toward tumor sites. Once at the tumor site, stem cells can exert anti-cancer effects through several mechanisms. Genetically engineered stem cells can deliver therapeutic agents, such as pro-apoptotic proteins, oncolytic viruses, or cytotoxic drugs, directly to the cancer cells, enhancing targeted therapy and minimizing systemic toxicity. Additionally, stem cells can modulate the tumor microenvironment by secreting cytokines and growth factors that inhibit angiogenesis, promote immune cell infiltration, or induce cancer cell apoptosis. Stem cells can also be used to generate immune effector cells, such as T cells or natural killer cells, which enhance the body's anti-tumor immune response. Furthermore, the immunomodulatory properties of stem cells can be harnessed to reduce adverse effects and improve the efficacy of conventional cancer therapies [51, 52]. Some of these molecular mechanisms are discussed briefly below.

1. Homing and migration: Stem cells, particularly MSCs, and NSCs, exhibit tumor-tropic behavior guided by chemokine receptors (e.g., CXCR4) and adhesion molecules (e.g., integrins) that respond to chemokines and signals released by the tumor microenvironment.
2. Secretion of cytokines and growth factors: Stem cells can secrete various cytokines and growth factors, such as interleukins (e.g., IL-10), tumor necrosis factor-alpha (TNF-α), and vascular endothelial growth factor (VEGF), which can modulate the tumor microenvironment by inhibiting angiogenesis, promoting immune cell infiltration, or inducing cancer cell apoptosis.
3. Paracrine signaling: Through paracrine signaling, stem cells release exosomes and microvesicles containing microRNAs, proteins, and lipids that can influence tumor cell behavior, suppress tumor growth, and modulate immune responses.
4. Immunomodulation: Stem cells, particularly MSCs, exhibit immunomodulatory properties by secreting immunosuppressive factors such as transforming growth factor-beta (TGF-β) and prostaglandin E2 (PGE2), which can reduce inflammation and modulate the activity of immune cells, such as T cells and natural killer cells.
5. Genetic engineering: Stem cells can be genetically engineered to express therapeutic genes, such as pro-apoptotic proteins (e.g., TRAIL), oncolytic viruses, or cytotoxic drugs, allowing targeted delivery of anti-cancer agents to the tumor site.

6. Modulation of tumor microenvironment: Stem cells can alter the tumor microenvironment by affecting the extracellular matrix composition, reducing fibrosis, and creating an anti-tumor milieu that inhibits tumor progression and metastasis.
7. Angiogenesis inhibition: Stem cells can secrete factors that inhibit angiogenesis, such as thrombospondin-1 and endostatin, depriving the tumor of the necessary blood supply and nutrients.
8. Stem cell differentiation: Directed differentiation of stem cells into immune effector cells, such as cytotoxic T lymphocytes or dendritic cells, can enhance anti-tumor immune responses and contribute to the elimination of cancer cells.

Stem cells have a complex role in cancer biology, affecting tumor progression, metastasis, and treatment response. Gaining insights into the interactions between stem cells and cancer cells is essential for creating effective therapeutic approaches. This section delves into four crucial mechanisms: bone marrow homing, tumor-tropic effects, paracrine factor secretion, and the differentiation capacity and signaling pathways in cancer stem cells (Table 14.2).

Table 14.2 Molecular mechanisms responsible for the action of stem cells in cancer

Mechanism	Description
Homing to bone marrow	Migration of stem cells to bone marrow to restore blood cell production post-therapy
Tumor-tropic effect	Attraction of stem cells to tumor sites *via* chemokines and cytokines, enhancing tumor growth
Paracrine factor secretion	Releasing cytokines and growth factors to modulate the tumor microenvironment and promote growth
Signaling in cancer stem cells	Activation of pathways (Wnt, Notch, Hedgehog) that maintain stemness and contribute to tumor progression

14.3.1 Homing to Bone Marrow

Homing describes the process of migration of stem cells to specific body sites, such as the bone marrow. This is vital for HSCs during transplantation, as they must reach the bone marrow to restore blood cell production. The homing process involves:

1. Initial tethering: Selectins bind stem cells to blood vessel walls.
2. Activation: Cytokines prepare the cells for migration.
3. Arrest: Integrins help stem cells adhere to the extracellular matrix.
4. Diapedesis: Stem cells pass through the endothelial barrier and move into the bone marrow following chemokine gradients [53, 54].

14.3.2 Tumor-Tropic Effect

Stem cells, especially MSCs, exhibit a tumor-tropic effect, meaning that they are drawn to tumor sites. This property is utilized in cancer therapies to deliver treatments directly to tumors. Key factors driving this effect include the following:

1. Chemokines and cytokines: Tumor cells release these substances to which the stem cells get attracted.
2. Hypoxic conditions: Low oxygen levels within tumors stimulate stem cell migration.
3. Reciprocal interactions: MSCs can enhance cancer cell proliferation, invasion, and angiogenesis, promoting tumor growth and metastasis [55].

14.3.3 Secretion of Paracrine Factors and Differentiation Capacity

Stem cells secrete various paracrine factors that affect the tumor microenvironment and support tumor growth. These factors include the following:

1. Cytokines, growth factors, and extracellular vesicles: These substances modulate immune responses, creating an immunosuppressive environment that shields tumor cells from immune detection.

2. Angiogenesis: In this process, new capillaries are generated from the existing blood vessels to supply nutrients to tumors.
3. Survival and proliferation of tumor cells: Growth factor stem cells' release enhances cancer cell growth and survival [56].

14.3.4 Signaling in Cancer Stem Cells

Cancer stem cells (CSCs) are the cancer cells found within tumors or hematological cancers that can self-renew and differentiate, driving tumor heterogeneity and resistance to treatment. Key signaling pathways regulating CSC properties include the following:

1. Wnt, Notch, and Hedgehog pathways: These pathways maintain stemness and promote self-renewal.
2. PI3K/AKT/mTOR and JAK/STAT pathways: These are involved in cell survival and proliferation.
3. TGF-β signaling: This pathway is responsible for the epithelial-to-mesenchymal transition (EMT), increasing invasiveness and metastasis.

These pathways often interact with and are disrupted by the tumor microenvironment, presenting potential targets for therapeutic intervention [57].

14.4 Challenges and Limitations of Stem Cell Therapy for Cancer

Off-target effects associated with stem cell therapy for treating cancer refer to unintended impacts where stem cells, designed to target and eliminate cancer cells, inadvertently affect non-target tissues or cells. These unintended consequences can manifest as differentiation into non-specific cell types, immune responses that may result in inflammation or rejection, and the risk of tumor formation if undifferentiated stem cells remain in the body. Furthermore, off-target effects can occur due to the migration of cells to unintended locations, interactions with the surrounding microenvironment, or genetic alterations. One example of an off-target effect in stem cell therapy for cancer is the potential for cardiotoxicity associated with certain cancer treatments, such as anthracyclines. These agents can induce arrhythmias through increased production of reactive oxygen species (ROS) and alterations in calcium handling within cardiac cells. When stem cells are employed with these therapies, the risk of such off-target effects may increase, leading to adverse cardiac events. Understanding these mechanisms is crucial to developing safer cancer therapies that minimize such risks while effectively targeting cancer cells [21]. To reduce these risks, it is essential to implement improved targeting methods, thorough characterization of stem cells, ongoing patient monitoring, preclinical assessments, and strict regulation of stem cell proliferation. Effectively managing these off-target effects is vital for ensuring the safety and efficacy of cancer treatments utilizing stem cells [58].

Immune responses present significant challenges in stem cell therapy for cancer, as introducing stem cells into a patient's body can lead to immune activation and potential rejection. This reaction may occur when the immune system recognizes the stem cells as foreign, resulting in inflammation, tissue damage, or elimination of the transplanted cells. The risk of immune rejection is heightened when stem cells are sourced from a donor, often requiring immunosuppressive therapies to prevent graft-versus-host disease. In contrast, using autologous stem cells derived from the patient's tissues necessitates careful management of autoimmunity risks. Due to their immune evasion properties, the immune response to CSCs can get frequently compromised. CSCs can impair antigen processing and presentation, downregulate tumor-associated antigens, and inhibit cytolytic granules released from cytotoxic T lymphocytes (CTLs) and natural killer (NK) cells. Additionally, they secrete immunosuppressive factors like TGF-β and IL-6 and upregulate immune checkpoint molecules. These crucial mechanisms provide critical elements for the development of novel tumor immunotherapies targeting CSCs [59]. Effective strategies to mitigate immune responses include optimizing tissue matching, utilizing advanced immunomodulation techniques, and establishing rigorous pre-treatment protocols to minimize the likelihood of adverse immune reactions. Addressing these immune-related challenges is crucial to enhance the safety and success of stem cell therapies in cancer treatment [60, 61].

Tumor heterogeneity poses a complex challenge in stem cell therapy for cancer, as the diverse characteristics of tumors can complicate treatment outcomes. Tumor heterogeneity encompasses variations in cell types, genetic mutations, and microenvironmental conditions within a single tumor or among different tumors in the same patient. This variability can lead to differential responses to stem cell therapy, with certain cancer cell subpopulations potentially exhibiting resistance to treatment or evading therapeutic effects. Additionally, heterogeneous CSCs within tumors can further complicate therapeutic effectiveness, as these cells may have unique properties and mechanisms of drug resistance. For instance, in breast cancer, distinct populations of CSCs identified by surface markers such as CD44 and CD24 demonstrate different tumorigenic potentials and responses to treatment. This highlights how intra-tumor heterogeneity can significantly affect therapeutic efficacy, as some CSC subpopulations may evade treatment and lead to tumor regrowth after therapy [59, 62, 63]. Addressing tumor heterogeneity necessitates personalized therapeutic strategies, such as targeting multiple tumor subtypes, employing combination therapies, and continuously adapting treatment plans based on real-time tumor profiling. Effectively managing tumor heterogeneity is essential for improving the efficacy and precision of stem cell therapies in the fight against cancer [57, 64].

Stem cell exhaustion refers to the reduction in the regenerative capacity and therapeutic potential of stem cells due to prolonged or repeated use in cancer therapy. This phenomenon can arise when stem cells undergo continuous or intensive therapeutic interventions, resulting in decreased functionality, reduced proliferation, and impaired differentiation capacity. In cancer treatments, stem cell exhaustion can undermine the effectiveness of therapies, as the therapeutic stem cells may fail to perform their intended roles, such as repairing damaged tissues or targeting cancer cells. Factors contributing to stem cell exhaustion include cellular stress, genetic mutations, and a hostile tumor microenvironment. This phenomenon is particularly evident in HSC transplantation for hematological cancers. For example, repeated cycles of chemotherapy can lead to HSC depletion and functional impairment, which compromises the reconstitution of the hematopoietic system and increases the risk of complications such as infections and graft failure. It has been shown that the hostile tumor microenvironment and cellular stress contribute significantly to this exhaustion, limiting the effectiveness of stem cell therapies [65]. To counteract this exhaustion, it is crucial to optimize stem cell culture conditions, minimize exposure to stressors, and develop strategies to rejuvenate or replace exhausted stem cells. Addressing stem cell exhaustion is vital for maintaining the efficacy and safety of stem cell-based cancer therapies [61, 66].

14.5 Future Directions for Stem Cell Therapy in Cancer Treatment

Stem cell therapy is being explored as a promising route in cancer treatment, unfolding the unique properties of stem cells, elevating the therapeutic efficiency, and addressing the limitations associated with conventional methods for cancer treatment. With progress in research, stem cell therapy is getting new directions in the field of oncology.

As mentioned, the stem cells showcase remarkable features like self-renewal, differentiation, and the ability to migrate to the tumor sites. These features make them function as carriers for the therapeutic agents, elevate the immune responses, and potentially differentiate into the cancer-fighting cells. For example, as discussed, the MSCs show tumor tropic properties, which allow them to deliver the anti-cancer agents directly to the tumor microenvironment, thereby increasing the concentration of the therapeutic agents at the site of action and reducing the system toxicity. Stem cell therapy may be combined with other existing treatment techniques, such as immunotherapies and targeted therapies, to enhance the overall treatment efficacy. For instance, stem cell therapy in integration with oncolytic viruses can help target and act on tumor cells, specifically avoiding off-target effects.

Thus, advancements in areas such as genomics and proteomics can lead to improvements in personalized stem cell therapies. By enhancing stem cell treatments based on the genetic profile of an individual's tumor, clinicians could optimize therapeutic outcomes and minimize the adverse effects [21].

14.5.1 Integration with Other Therapeutic Approaches

For better patient outcomes, stem cell therapy can be integrated with conventional treatments, such as chemotherapy, radiotherapy, and immunotherapy. Hematopoietic stem cell (HSC) transplantation is a well-known approach for treating hematological malignancies. After chemotherapy with high levels of doses, which results in severe myelosuppression, the restoration of hematopoiesis can be done using HSC transplantation. This technique is effective in conditions such as leukemia and lymphoma, where chemotherapy alone cannot meet the requirements [67]. Radiotherapy damages cancer cells and alters the tumor microenvironment, increasing the tumor antigens' visibility to the immune system. As it is well studied that the interaction between radiation and the immune system has unique features, the combination of radiotherapy and stem cell therapy is gaining attention now. Furthermore, this effect can be improved by using immune checkpoint inhibitors. The timing and sequencing of these therapies are critical; for instance, administering immunotherapy post-radiotherapy can capitalize on the immune activation induced by radiation, potentially leading to improved systemic anti-tumor responses [68, 69].

14.5.2 Development of New Stem Cell Sources

Developing new stem cell sources, mainly induced pluripotent stem cells (iPSCs) and embryonic stem cells (ESCs), shows significant advancement in regenerative medicine and therapeutic research. This evolution has expanded the possibilities for treating various diseases and understanding the fundamental biological processes. iPSCs can be created by reprogramming the somatic cells, like the skin or blood cells, to achieve the pluripotent state. iPSCs share similar properties with embryonic stem cells, including the capability to differentiate into any cell type and unlimited self-renewal. However, despite their advantages, iPSCs face challenges, including the risk of tumor formation and the need for efficient differentiation protocols. Ongoing research aims to address these issues to improve the safety and efficacy of iPSC applications in clinical settings [70].

In contrast, the ESCs are derived from the inner cell mass of blastocysts, which are early-stage embryos. These cells are naturally pluripotent and can differentiate into any cell type. ESCs have been extensively studied for their potential in regenerative medicine, particularly tissue repair and organ regeneration. However, the use of ESCs faces significant ethical issues, as they are derived by destroying embryos. Due to this, many countries have imposed strict regulations and limitations on embryonic stem cell research. The controversy surrounding ESCs has prompted scientists to seek alternative sources of pluripotent cells, such as iPSCs [71].

14.5.3 Gene Editing and Gene Therapy

Integrating gene editing and gene therapy with stem cell therapy is a transformative approach in regenerative medicine. It opens new avenues to treat genetic diseases and enhance therapeutic efficacy. This combination leverages the unique properties of stem cells alongside advanced gene editing technologies, such as CRISPR/Cas9, for addressing various medical challenges. Therapies for conditions like diabetes can be developed using gene-edited stem cells to regenerate damaged tissues or organs, where gene editing can enhance the functionality of transplanted cells and improve their survival against immune responses [72, 73]. Gene editing can be employed to modify immune cells derived from stem cells, which can enhance their ability to target and destroy cancer cells. Thus, gene editing is particularly relevant in the development of CAR T-cell therapies, where T cells are genetically engineered to express receptors that specifically recognize tumor antigens [74]. Models of various diseases can be created using iPSCs, allowing the study of disease mechanisms and testing new therapies. Gene editing can facilitate the introduction of specific mutations into these models, providing insights into the effects of genetic changes on cell behavior and disease progression.

14.5.4 Clinical Trials and Regulatory Mechanisms

These are crucial for advancing stem cell therapy for cancer treatment. These processes ensure that new therapies are safe and effective before they become widely available to patients. The safety and efficacy of stem cell therapies in cancer treatment can be evaluated through clinical trials. They typically progress through several phases, namely:

- Phase I trials: These focus on the safety and dosage of the treatments. Small groups of patients receive the treatment to assess treatment safety and identify any side effects.
- Phase II trials: These aim to evaluate the efficacy of the treatment, in which a larger group of patients receives the treatment to determine its effectiveness and further assess safety.
- Phase III trials: These are large-scale trials that compare the new treatment against standard therapies. These trials provide comprehensive data on the treatment's effectiveness, benefits, and potential risks.

Transplantation of hematopoietic stem cells is currently the most established stem cell therapy approved by the U.S. Food and Drug Administration (FDA). The treatment is primarily used to treat blood cancers and disorders affecting the blood and immune system. Other stem cell therapies for various cancers are still in experimental stages and undergoing clinical trials to evaluate their safety and efficacy [75]. The regulations for stem cell therapies are complex and vary by country. For example, in the United States, the FDA oversees the development and approval of stem cell therapies [76]. Key considerations include the following:

- Safety and efficacy: The FDA requires robust data from clinical trials demonstrating that a therapy is safe and effective for its intended use.
- Manufacturing standards: To ensure product quality and safety, stem cell therapies must be produced in compliance with Good Manufacturing Practices (GMP).

- Ethical guidelines: Regulatory bodies enforce ethical standards regarding the sourcing of stem cells, particularly concerning embryonic stem cells, which raises significant ethical considerations.
- Post-market surveillance: Once a therapy is approved, ongoing monitoring is essential to track long-term safety and efficacy in the general population [77].

14.6 Summary and Conclusions

Stem cell therapy for cancer treatment is advancing rapidly, with significant potential for improving patient outcomes. Continued research and clinical trials are essential for addressing existing challenges and optimizing these innovative therapeutic approaches. The chapter discussed the significance of stem cell therapy and its application in the field. Future therapies may target CSCs to prevent tumor recurrence and metastasis, utilizing anticancer vaccines derived from CSC antigens. Combining stem cell therapies with chemotherapy, radiotherapy, and immunotherapy can enhance treatment efficacy and reduce resistance. Furthermore, gene editing technologies (e.g., CRISPR) in conjunction with stem cell therapy could allow for precise corrections of genetic defects in patient-derived cells, leading to personalized treatment options. Ongoing research must navigate complex regulatory landscapes for ensuring the safety and efficacy of stem cell therapies, while addressing ethical concerns, especially regarding the use of embryonic stem cells. In summary, stem cell therapy for cancer treatment is advancing rapidly, with significant potential for improving patient outcomes. Continued research and clinical trials are essential to address existing challenges and optimize these innovative therapeutic approaches.

Acknowledgments

VAM acknowledges the University Grants Commission (UGC), New Delhi, India, for junior and senior research fellowships. AKS, SDB, VAM, and RS acknowledge the Academy of Scientific and Innovative Research (AcSIR), Ghaziabad, Uttar Pradesh, India, for the M.Tech. and Ph.D. programs. All authors thank the CSIR-National Chemical Laboratory, Pune, India, for providing the necessary facilities to undertake this study.

References

1. Siegel, R.L., Miller, K.D., Jemal, A., Cancer statistics, 2018. *CA. Cancer J. Clin.*, 68, 1, 7–30, 2018. doi: 10.3322/caac.21442.
2. Zugazagoitia, J., Guedes, C., Ponce, S., Ferrer, I., Molina-Pinelo, S., Paz-Ares, L., Current Challenges in Cancer Treatment. *Clin. Ther.*, 38, 7, 1551–1566, Jul. 2016. doi: 10.1016/j.clinthera.2016.03.026.
3. Debela, D.T., *et al.*, New approaches and procedures for cancer treatment: Current perspectives. *SAGE Open Med.*, 9, 20503121211034366, Jan. 2021. doi: 10.1177/20503121211034366.
4. Arruebo, M., *et al.*, Assessment of the Evolution of Cancer Treatment Therapies. *Cancers*, 3, 3, 3279–3330, Sep. 2011. doi: 10.3390/cancers3033279.
5. Moses, M.A., Brem, H., Langer, R., Advancing the field of drug delivery: Taking aim at cancer. *Cancer Cell*, 4, 5, 337–341, Nov. 2003. doi: 10.1016/S1535-6108(03)00276-9.
6. Shapira, A., Livney, Y.D., Broxterman, H.J., Assaraf, Y.G., Nanomedicine for targeted cancer therapy: Towards the overcoming of drug resistance. *Drug Resist. Updat.*, 14, 3, 150–163, Jun. 2011. doi: 10.1016/j.drup.2011.01.003.
7. Wyld, L., Audisio, R.A., Poston, G.J., The evolution of cancer surgery and future perspectives. *Nat. Rev. Clin. Oncol.*, 12, 2, 115–124, Feb. 2015. doi: 10.1038/nrclinonc.2014.191.
8. Kaliberov, S.A. and Buchsbaum, D.J., Cancer treatment with gene therapy and radiation therapy. *Adv. Cancer Res.*, 115, 221–263, 2012.
9. Behranvand, N., *et al.*, Chemotherapy: a double-edged sword in cancer treatment. *Cancer Immunol., Immunother.*, 71, 3, 507–526, Mar. 2022. doi: 10.1007/s00262-021-03013-3.
10. Martin, J.D., Cabral, H., Stylianopoulos, T., Jain, R.K., Improving cancer immunotherapy using nanomedicines: progress, opportunities and challenges. *Nat. Rev. Clin. Oncol.*, 17, 4, 251–266, 2020.
11. Tsimberidou, A.-M., Targeted therapy in cancer. *Cancer Chemother. Pharmacol.*, 76, 6, 1113–1132, Dec. 2015. doi: 10.1007/s00280-015-2861-1.
12. Mørch, L.S., Løkkegaard, E., Andreasen, A.H., Krüger-Kjær, S., Lidegaard, Ø., Hormone Therapy and Ovarian Cancer. *Jama*, 302, 3, 298–305, 2009, doi: 10.1001/jama.2009.1052.
13. Przepiorka, D., *et al.*, Allogeneic blood stem cell transplantation in advanced hematologic cancers. *Bone Marrow Transplant.*, 19, 5, 455–460, 1997.
14. dos Santos, A.F., de Almeida, D.Q., Terra, L.F., Baptista, M.S., Labriola, L., Photodynamic therapy in cancer treatment-an update review. *J. Cancer Metastasis Treat.*, 5, 25, 10–20517, 2019.
15. Field, S.B. and Bleehen, N.M., Hyperthermia in the treatment of cancer. *Cancer Treat. Rev.*, 6, 2, 63–94, 1979.
16. Theodorescu, D., Cancer cryotherapy: evolution and biology. *Rev. Urol.*, 6, Suppl 4, S9, 2004.

17. Cross, D. and Burmester, J.K., Gene therapy for cancer treatment: past, present and future. *Clin. Med. Res.*, 4, 3, 218–227, 2006.
18. Singh, R., Nanotechnology based therapeutic application in cancer diagnosis and therapy. *3 Biotech.*, 9, 11, 415, Nov. 2019. doi: 10.1007/s13205-019-1940-0.
19. Tsimberidou, A.M., Fountzilas, E., Nikanjam, M., Kurzrock, R., Review of precision cancer medicine: Evolution of the treatment paradigm. *Cancer Treat. Rev.*, 86, 102019, 2020.
20. Zhang, C.-L., Huang, T., Wu, B.-L., He, W.-X., Liu, D., Stem cells in cancer therapy: opportunities and challenges. *Oncotarget*, 8, 43, 75756–75766, Sep. 2017. doi: 10.18632/oncotarget.20798.
21. Chu, D.-T., *et al.*, Recent Progress of Stem Cell Therapy in Cancer Treatment: Molecular Mechanisms and Potential Applications. *Cells*, 9, 3, 563, Feb. 2020. doi: 10.3390/cells9030563.
22. Patel, M.S. and Harris, R.A., Metabolic Regulation, in: *Encyclopedia of Cell Biology*, Second, R.A. Bradshaw, G.W. Hart, P.D. Stahl (Eds.), pp. 353–365, Academic Press, Oxford, 2023, doi: 10.1016/B978-0-12-821618-7.00147-4.
23. Alnasser, S.M., Advances and Challenges in Cancer Stem Cells for Onco-Therapeutics. *Stem Cells Int.*, 2023, 8722803, Dec. 2023. doi: 10.1155/2023/8722803.
24. Collins, A.T., Berry, P.A., Hyde, C., Stower, M.J., Maitland, N.J., Prospective Identification of Tumorigenic Prostate Cancer Stem Cells. *Cancer Res.*, 65, 10946–10951, 23Dec. 2005. doi: 10.1158/0008-5472.CAN-05-2018.
25. Cox, C.V., Evely, R.S., Oakhill, A., Pamphilon, D.H., Goulden, N.J., Blair, A., Characterization of acute lymphoblastic leukemia progenitor cells. *Blood*, 104, 9, 2919–2925, Nov. 2004. doi: 10.1182/blood-2004-03-0901.
26. Sagar, J., Chaib, B., Sales, K., Winslet, M., Seifalian, A., Role of stem cells in cancer therapy and cancer stem cells: a review. *Cancer Cell Int.*, 7, 1, 9, Jun. 2007. doi: 10.1186/1475-2867-7-9.
27. Yin, W., Wang, J., Jiang, L., James Kang, Y., Cancer and stem cells. *Exp. Biol. Med.*, 246, 16, 1791–1801, 2021, doi: 10.1177/15353702211005390.
28. Li, H.C., Stoicov, C., Rogers, A.B., Houghton, J., Stem cells and cancer: evidence for bone marrow stem cells in epithelial cancers. *World J. Gastroenterol.*, 12, 3, 363–371, 2006, doi: 10.3748/wjg.v12.i3.363.
29. Huang, Y., Wu, Q., Tam, P.K.H., Immunomodulatory mechanisms of mesenchymal stem cells and their potential clinical applications. *Int. J. Mol. Sci.*, 23, 17, 10023, 2022, doi: 10.3390/ijms231710023.
30. Lan, T., Luo, M., Wei, X., Mesenchymal stem/stromal cells in cancer therapy. *J. Hematol. Oncol.*, 14, 1, 195, Nov. 2021. doi: 10.1186/s13045-021-01208-w.
31. Hernández, R., Jiménez-Luna, C., Perales-Adán, J., Perazzoli, G., Melguizo, C., Prados, J., Differentiation of Human Mesenchymal Stem Cells towards Neuronal Lineage: Clinical Trials in Nervous System Disorders. *Biomol. Ther.*, 28, 1, 34–44, Jan. 2020. doi: 10.4062/biomolther.2019.065.

32. Pittenger, M.F., Discher, D.E., Péault, B.M., Phinney, D.G., Hare, J.M., Caplan, A., II, Mesenchymal stem cell perspective: cell biology to clinical progress. *Npj Regen. Med.*, 4, 1, 22, Dec. 2019. doi: 10.1038/s41536-019-0083-6.
33. Almalki, S.G. and Agrawal, D.K., Key Transcription Factors in the Differentiation of Mesenchymal Stem Cells. *Differ. Res. Biol. Divers.*, 92, 1–2, 41–51, 2016. doi: 10.1016/j.diff.2016.02.005.
34. Chu, D.T., Nguyen, T.T., Tien, N.L.B., Tran, D.K., Jeong, J.H., Anh, P.G., Thanh, V.V., Truong, D.T., Dinh, T.C., Recent Progress of Stem Cell Therapy in Cancer Treatment: Molecular Mechanisms and Potential Applications. *Cells*, 9, 3, 563, 2020, doi: 10.3390/cells9030563.
35. Tang, Y., Yu, P., Cheng, L., Current progress in the derivation and therapeutic application of neural stem cells. *Cell Death Dis.*, 8, 10, e3108, 2017, doi: 10.1038/cddis.2017.504.
36. Eaves, C.J., Hematopoietic stem cells: concepts, definitions, and the new reality. *Blood J. Am. Soc. Hematol.*, 125, 17, 2605–2613, 2015.
37. Choi, Y., Lee, H.K., Choi, K.C., Engineered adult stem cells: a promising tool for anti-cancer therapy. *BMB Rep.*, 56, 2, 71–77, 2023, doi: 10.5483/BMBRep.2022-0091.
38. Lapidot, T., Dar, A., Kollet, O., How do stem cells find their way home? *Blood*, 106, 6, 1901–1910, 2005.
39. Poliwoda, S., Noor, N., Downs, E., Schaaf, A., Cantwell, A., Ganti, L., Kaye, A. D., Mosel, L.I., Carroll, C.B., Viswanath, O., Urits, I., Stem cells: a comprehensive review of origins and emerging clinical roles in medical practice. *Orthop. Rev.*, 14, 3, 37498, 2022, doi: 10.52965/001c.37498.
40. Kim, N. and Cho, S.-G., Clinical applications of mesenchymal stem cells. *Korean J. Intern. Med.*, 28, 4, 387–402, Jul. 2013. doi: 10.3904/kjim.2013.28.4.387.
41. Margiana, R., *et al.*, Clinical application of mesenchymal stem cell in regenerative medicine: a narrative review. *Stem Cell Res. Ther.*, 13, 1, 366, Dec. 2022. doi: 10.1186/s13287-022-03054-0.
42. TomyTomcy, A., Sindhu, E.R., Mesenchymal stem cells-an excellent therapeutic agent for cancer. *Asia Pac. J. Clin. Oncol.*, 20, 1, 7–15, 2024, doi: 10.1111/ajco.13969.
43. Bagó, J.R., Sheets, K.T., Hingtgen, S.D., Neural stem cell therapy for cancer. *Methods San Diego Calif.*, 99, 37, Apr. 2016. doi: 10.1016/j.ymeth.2015.08.013.
44. Benmelouka, A.Y., Munir, M., Sayed, A., Attia, M.S., Ali, M.M., Negida, A., Alghamdi, B.S., Kamal, M.A., Barreto, G.E., Ashraf, G.M., Meshref, M., Bahbah, E.I., Neural Stem Cell-Based Therapies and Glioblastoma Management: Current Evidence and Clinical Challenges. *Int. J. Mol. Sci.*, 22, 5, 2258, 2021, doi: 10.3390/ijms22052258.
45. Morgan, R.A., Gray, D., Lomova, A., Kohn, D.B., "Hematopoietic Stem Cell Gene Therapy–Progress and Lessons Learned,". *Cell Stem Cell*, 21, 5, 574–590, Nov. 2017. doi: 10.1016/j.stem.2017.10.010.

46. Hawsawi, Y.M., Al-Zahrani, F., (Harris) Mavromatis, C., Baghdadi, M.A., Saggu, S., Oyouni, A.A.A., Stem Cell Applications for Treatment of Cancer and Autoimmune Diseases: Its Promises, Obstacles, and Future Perspectives. *Technol. Cancer Res. Treat.*, 17, 1533033818806910, 2018. doi: 10.1177/1533033818806910.
47. Takahashi, K. and Yamanaka, S., A decade of transcription factor-mediated reprogramming to pluripotency. *Nat. Rev. Mol. Cell Biol.*, 17, 3, 183–193, Mar. 2016. doi: 10.1038/nrm.2016.8.
48. Stadtfeld, M. and Hochedlinger, K., Induced pluripotency: history, mechanisms, and applications. *Genes Dev.*, 24, 20, 2239–2263, Oct. 2010. doi: 10.1101/gad.1963910.
49. Villa-Diaz, L.G., Ross, A.M., Lahann, J., Krebsbach, P.H., The evolution of human pluripotent stem cell culture: from feeder cells to synthetic coatings. *Stem Cells Dayt. Ohio*, 31, 1, 1–7, Jan. 2013. doi: 10.1002/stem.1260.
50. Zakrzewski, W., Dobrzyński, M., Szymonowicz, M., Rybak, Z., Stem cells: past, present, and future. *Stem Cell Res. Ther.*, 10, 1, 68, Feb. 2019. doi: 10.1186/s13287-019-1165-5.
51. Sun, Y., Jiang, X., Gao, J., Stem cell-based ischemic stroke therapy: Novel modifications and clinical challenges. *Asian J. Pharm. Sci.*, 19, 1, 100867, Feb. 2024. doi: 10.1016/j.ajps.2023.100867.
52. Liu, J., *et al.*, Mesenchymal stem cells and their microenvironment. *Stem Cell Res. Ther.*, 13, 1, 429, Aug. 2022. doi: 10.1186/s13287-022-02985-y.
53. Ullah, M., Liu, D.D., Thakor, A.S., Mesenchymal Stromal Cell Homing: Mechanisms and Strategies for Improvement. *iScience*, 15, 421–438, May 2019. doi: 10.1016/j.isci.2019.05.004.
54. Liesveld, J.L., Sharma, N., Aljitawi, O.S., Stem cell homing: From physiology to therapeutics. *Stem Cells*, 38, 10, 1241–1253, 2020. doi: 10.1002/stem.3242.
55. Cuiffo, B.G. and Karnoub, A.E., Mesenchymal stem cells in tumor development. *Cell Adhes. Migr.*, 6, 3, 220–230, May 2012. doi: 10.4161/cam.20875.
56. Mirotsou, M., Jayawardena, T.M., Schmeckpeper, J., Gnecchi, M., Dzau, V.J., Paracrine mechanisms of stem cell reparative and regenerative actions in the heart. *J. Mol. Cell. Cardiol.*, 50, 2, 280–289, Feb. 2011. doi: 10.1016/j.yjmcc.2010.08.005.
57. Yang, L., *et al.*, Targeting cancer stem cell pathways for cancer therapy. *Signal Transduct. Target. Ther.*, 5, 1, 1–35, Feb. 2020. doi: 10.1038/s41392-020-0110-5.
58. Leiva, O., Bohart, I., Ahuja, T., Park, D., Off-target effects of cancer therapy on development of therapy-induced arrhythmia: A review. *Cardiology*, 148, 4, 324–334, 2023.
59. Tsuchiya, H. and Shiota, G., Immune evasion by cancer stem cells. *Regen. Ther.*, 17, 20–33, Mar. 2021. doi: 10.1016/j.reth.2021.02.006.
60. Phi, L.T.H., *et al.*, Cancer Stem Cells (CSCs) in Drug Resistance and their Therapeutic Implications in Cancer Treatment. *Stem Cells Int.*, 2018, 5416923, Feb. 2018. doi: 10.1155/2018/5416923.

61. Alnasser, S.M., Stem cell challenge in cancer progression, oncology and therapy. *Gene*, 840, 146748, Oct. 2022. doi: 10.1016/j.gene.2022.146748.
62. Singh, A.K., *et al.*, Tumor heterogeneity and cancer stem cell paradigm: Updates in concept, controversies and clinical relevance. *Int. J. Cancer*, 136, 9, 1991–2000, 2015. doi: 10.1002/ijc.28804.
63. Ruiu, R., *et al.*, Cancer stem cell immunology and immunotherapy: Harnessing the immune system against cancer's source, in: *Progress in Molecular Biology and Translational Science*, D.B. Teplow (Ed.), Cancer Immunotherapy, vol. 164, 119–188, Academic Press, 2019, doi: 10.1016/bs.pmbts.2019.03.008.
64. Veschi, V., Turdo, A., Stassi, G., Novel insights into cancer stem cells targeting: CAR-T therapy and epigenetic drugs as new pillars in cancer treatment. *Front. Mol. Med.*, 3, 1120090, May 2023. doi: 10.3389/fmmed.2023.1120090.
65. Porada, C.D., Atala, A.J., Almeida-Porada, G., The Hematopoietic System in the Context of Regenerative Medicine. *Methods San Diego Calif.*, 99, 44–61, Apr. 2016. doi: 10.1016/j.ymeth.2015.08.015.
66. Zhou, H.-M., Zhang, J.-G., Zhang, X., Li, Q., Targeting cancer stem cells for reversing therapy resistance: mechanism, signaling, and prospective agents. *Signal Transduct. Target. Ther.*, 6, 1, 1–17, Feb. 2021. doi: 10.1038/s41392-020-00430-1.
67. Donini, C., Rotolo, R., Proment, A., Aglietta, M., Sangiolo, D., Leuci, V., Cellular immunotherapy targeting cancer stem cells: preclinical evidence and clinical perspective. *Cells*, 10, 3, 543, 2021.
68. Liu, Y., Yang, M., Luo, J., Zhou, H., Radiotherapy targeting cancer stem cells 'awakens' them to induce tumour relapse and metastasis in oral cancer. *Int. J. Oral Sci.*, 12, 1, 19, 2020.
69. El-Derany, M.O., Said, R.S., El-Demerdash, E., Bone marrow-derived mesenchymal stem cells reverse radiotherapy-induced premature ovarian failure: emphasis on signal integration of TGF-β, Wnt/β-catenin and hippo pathways. *Stem Cell Rev. Rep.*, 17, 1429–1445, 2021.
70. Medvedev, S.P., Shevchenko, A., II, Zakian, S.M., Induced Pluripotent Stem Cells: Problems and Advantages when Applying them in Regenerative Medicine. *Acta Naturae*, 2, 2, 18–28, 2010.
71. Kim, J., Orkin, S.H., Embryonic stem cell-specific signatures in cancer: insights into genomic regulatory networks and implications for medicine. *Genome Med.*, 3, 11, 75, 2011, doi: 10.1186/gm291.
72. Hosseinkhani, H., Domb, A.J., Sharifzadeh, G., Nahum, V., Gene Therapy for Regenerative Medicine. *Pharmaceutics*, 15, 3, 856, Mar. 2023. doi: 10.3390/pharmaceutics15030856.
73. Sneddon, J.B., Tang, Q., Stock, P., Bluestone, J.A., Roy, S., Desai, T., Hebrok, M., Stem cell therapies for treating diabetes: progress and remaining challenges. *Cell Stem Cell*, 22, 6, 810–823, 2018, doi: 10.1016/j.stem.2018.05.016.
74. Jogalekar, M.P., Rajendran, R.L., Khan, F., Dmello, C., Gangadaran, P., Ahn, B.C., CAR T-Cell-Based gene therapy for cancers: new perspectives,

challenges, and clinical developments. *Front. Immunol.*, 13, 925985, 2022, doi: 10.3389/fimmu.2022.925985

75. Şanlı, N.M., Timuroğlu, A., Current Advances in Stem Cell-Based Therapies: Adult Stem Cell Applications. Stem Cell Transplantation. *Intech Open* [In Press], 2024, doi: 10.5772/intechopen.1007380.
76. Mousaei Ghasroldasht, M., Seok, J., Park, H.S., Liakath Ali, F.B., Al-Hendy, A., Stem cell therapy: from idea to clinical practice. *Int. J. Mol. Sci.*, 23, 5, 2850, 2022, doi: 10.3390/ijms23052850.
77. Elstner, A., *et al.*, The changing landscape of European and international regulation on embryonic stem cell research. *Stem Cell Res.*, 2, 2, 101–107, 2009.

15

Challenges and Breakthroughs: Lessons from Stem-Cell-Based Cancer Therapies

Priya Chaudhary[1], Ayush Madan[2,3], Divya Kumari[4], Apurva[1] and Pracheta Janmeda[4*]

[1]*Department of Biotechnology, School of Applied and Life Sciences, Uttaranchal University, Dehrdaun, Uttarakhand, India*
[2]*Department of Biotechnology, School of Research & Technology, People's University, Bhanpur, Bhopal, Madhya Pradesh, India*
[3]*Institute of Tropical Aquaculture and Fisheries (AKUATROP), Universiti Malaysia Terengganu, Kuala Nerus, Terengganu, Malaysia*
[4]*Department of Bioscience and Biotechnology, Banasthali Vidyapith, Banasthali, Rajasthan, India*

Abstract
Although medical research has made great strides, a permanent cure for cancer has not materialized. This has led to the exploration of a variety of therapeutic approaches, one of which is stem cell therapy. Furthermore, the use of cancer stem cells as a therapeutic option presents significant promise for the cure of incurable illnesses. Because stem cells can locate and target both metastatic and primary tumor foci, they can serve as innovative delivery systems. In experimental models of animals, stem cells modified to consistently express different cytotoxic drugs reduce tumor sizes and increase survival. They have also been used to reduce adverse reactions and improve primary medicinal effectiveness by acting as carriers of viruses and nanoparticles. Stem cells can also be utilized in cancer stem cell-directed therapy, regenerative medicine, and immunotherapy. Though it is technically feasible to treat human tumors with stem cells, issues like tumorigenesis and treatment durability demand for additional research to enhance therapeutic efficacy and applicability. The importance of cancer stem cells in oncotherapies and their capacity to support early cancer prognosis are highlighted in this study. An endless supply of drugs to fight cancer is emerging with the growth of tissue banks and their assortment of cancerous tissue types. Cancer stem cell-based

**Corresponding author*: pracheta@banasthali.in

Ayush Madan, Sumel Ashique, Deepika Arora and Mantosh Kumar Satapathy (eds.) Stem Cell Therapeutics, (629–672) © 2025 Scrivener Publishing LLC

therapy offers previously unheard-of possibilities to prolong human life, but it also poses complex and serious problems.

***Keywords*:** Cancer, stem cell, targeted therapy, regenerative medicine, onco-therapeutics

15.1 Introduction

As the world's population ages and grows, cancer is becoming a more common cause of mortality in both industrialized and developing nations. Chemotherapy, fractionated radiation, and removal by surgery are the primary techniques used for treating cancer. Nevertheless, many therapy choices' outcomes are limited by resistance to drugs, adverse reactions, and treatment-related complications. In addition, recurrence is extremely probable in cases when standard medicines are unable to eradicate metastatic cancer cells. Consequently, scientists are trying to create novel, potent treatments that are either completely safe or less hazardous to healthy cells [1].

Stem cells possess special abilities like motility to cancer cells, bioactive factor release, and immunosuppression that facilitate targeting tumors and get around present barriers to gene therapy approaches. Numerous factors including a mix of stochastic and hierarchical factors are considered essential in the "transformation" of stem cells into cancer stem cells (CSCs). Because stem cells may self-regenerate and specialize into different kinds of cells, they tend to collect epigenetic effects and mutations as they age. Natural selection has designed them to thrive in hostile situations, so they can proliferate and become more harmful no matter where they are [2].

Because malignant stem cells are resistant to traditional anticancer drugs, cancers derived from them are more dangerous because even one live CSC can cause a tumor to recur. Moreover, it has been suggested that, in contrast to the typical stem cell niche, these tumor cells might be shielded by modified microenvironmental ecosystems. By focusing on CSCs, cancer therapy could be revolutionized and patient outcomes could be much improved [3].

Eliminating the cells that cause metastasis and resistance to cancer therapy increases the likelihood of long-term remission and raises survival rates. The creation of treatments which target CSCs in particular has the power to revolutionize cancer care and give people battling this aggressive illness fresh hope. Application of personalized anti-cancer therapy plays a major function for the discovery of various approaches related to preclinical stem cell-based research. However, there is still requirement of more scientific validation to confirm the healing applicability of stem cells in the therapy of cancer [4]. Thus, this chapter aims to discuss the possible hurdles, applications, and associated side effects of latest research including anti-cancer stem cell therapies.

15.2 Current Hurdles in Stem-Cell-Based Treatment for Cancer

15.2.1 Ethical Considerations

The last 20 years are dedicated to the growth of stem cell research. However, questions of ethics and policy are raised due to these emerging stem cell study opportunities. Among various cell types, embryonic stem cells are mostly considered with the embryos moral issues, whereas adult and induced pluripotent stem cells (iPSCs) do not hold these moral. Awareness about ownership, how to obtain patients' authorization, the risk of stem cell therapies, and the usual harm that unproven stem cell therapies might trigger are the major ethical values that need to be considered for different types of stem cells [5].

For instance, medical investigators utilized an amalgam of autologous chondrons and allogeneic mesenchymal stem cells in an initial-in-animal experiment between 2013 and 2014 to direct individual cartilage transformation in the knee, with uplifting outcomes. There were problems with some individual's recuperation during the clinical trials. Their recovery procedure was unintentionally hampered by their failure to follow the researcher's advice to exercise caution when bearing too much weight on their knee [6]. It is possible that a few patients thought the stem cell treatment worked better than it actually did. The disadvantage of the treatment did not arise directly from the administration of stem cell [7]. Thus, this study shows how the limited perspective on ethical concern is unable to completely address the ramifications associated with the application and incorporation of stem cells into the therapeutic society.

According to [5], this point of view is mainly concerned with the issues of exotic stem cell therapies and their after effects. A more comprehensive investigation of ethical problems could aid in the development of practical measures that can maximize the advantages of stem cells application while minimizing their negative impacts [8].

15.2.2 Technical Constraints

The difficulty of transferring academic procedures into an industrial medicinal production facility should not be underestimated. This can involve difficulties with reagents, specialized tools, unclear user-dependent actions, and, lastly, scalability. There may also be technical challenges in characterizing the cells and connecting this to their therapeutic efficacy. This necessitates knowing their targeted MoA in detail. Lastly, according to Kolkundkar *et al.* [9], therapeutic cells must exhibit constant results one batch to another and upon distribution into the clinic.

Mass production techniques differ based on the particular item and its clinical administration method. While allogeneic goods can be manufactured at a central location, autologous products might need to have cellular modification done near the clinic. When trying to offer more dosages or across several locations, the number of manufacturing facilities also becomes a factor. To guarantee product safety, all actions deemed to be "substantial manipulation" must, nevertheless, take place in cleanliness and in a facility accredited by the GMP. The type of culture systems utilized will determine the precise grade of clean room required, such as open spaces or closed systems [10].

The variety of facilities that manufacture industrial cells has expanded along with the development of therapeutic possibilities. Quick links and advice for interacting with industrial production associates are offered by organizations like the UK's Cell and Gene Therapy Catapult. Comparable networks also operate on a European scale, such as the Restore program financed by the European Commission [11].

15.2.3 Regulatory Restrictions

The regulatory process has a significant role in dictating how quickly clinical trials advance. Since each biological product is different, figuring out which therapeutic category to use is a crucial initial phase. However, the degree of manipulation of cells and whether it qualifies as "substantial"—that is, a manufacturing step—determine whether transplant and cell therapy differ from one another. Gene therapies, sCTMPs, and TEPs are examples of advanced therapy medical products (ATMPs) as reported by the 2001/83/EC Directive in Europe.

As was mentioned, the product's margin of appreciation (MoA) determines how sCTMPs and TEPs differ from one another [12]. This definition has been updated to cover combination-ATMPs, where medical equipment (such as a biomaterial) is an essential part, as a result of the 1394/2007

ATMP regulation. As previously mentioned, based on the degree of alteration and purposeful application of the implanted stuff, the FDA in the United States of America controls cell treatments through their HCT/P rules [13].

There are regulatory pathways that can simplify novel product criteria and expedite the research study process in both nations. When it comes to diseases with tiny populations of patients, the classification of orphan can significantly lower the number of clinical trials needed to obtain market authorization [14].

Naturally, regulators have difficulties due to the wide range of ATMPs, and when a carrier substance is utilized, there are frequently regulatory overlaps with those pertaining to medical devices. When employing autologous cells, they could be totally uncontrolled in some countries (like Australia). Consequently, there are even less research studies for 3D printed ATMPs [15]. A case study of a 3D manufactured nasal implant that might also include biological elements was published by Hourd *et al.* [16]. The regulatory obstacles associated with bringing this kind of medicine into the clinic are examined in this investigation. The need to show GMP assurance of the implant which is inevitably going to be unique for every patient is fundamental to this.

15.2.4 Unanticipated Clinical Results

In vitro proof of cell function and phenotypic should be followed by *in vivo* models demonstrating the therapeutic's safety and efficacy. For information to be as relevant as feasible, models of animals should closely mimic the human state that is being studied. But the accessibility of these models, whether from academic institutions or through contract research organizations (CROs), frequently dictates one's capacity to do so [17].

Production variables are optimized for the particular therapeutic objective (e.g., culture circumstances, administering approach). Therefore, minimizing the requirement for *in vivo* additional research and the possibility of another non-success can be achieved by taking advantage of chemicals and techniques [18].

Furthermore, effectiveness levels that are shown *in vivo* ought to go beyond mere statistical relevance and exhibit therapeutic value. This is especially important when contrasting the novel treatment with controls that replicate the accepted standard practice at the moment. One important factor to take into account is the research's design. It is important to make sure that studies have enough power and adhere to procedures that maximize repeatability [19].

Many treatments move on to bigger models of animals if effectiveness is proven in small experimental animals such rats. When there is doubt about the therapy's capacity to be scaled up (as in the case of bigger tissue repairs), this is frequently the option used. Large animal research studies may only sometimes be required though, as it depends on the clinical studies, the ailment being managed, and the clinical application of related cell item in the past. The significance of this kind of research regarding human translation is still up for discussion [20].

In fact, it is questionable whether it is preferable to produce a cell line akin to the animal under test or to test a human cell line via immune modulation of an animal when using these models. It is doubtful if either situation is relevant to the end use by humans. Because of this, interest in humanized *ex vivo* or *in vitro* models is rising [21].

Any novel cell product must be shown to be harmless *in vivo* in addition to being effective before it can be used in clinical studies. While safety signs might be gathered in early investigations, critical testing for safety will be necessary to conduct a thorough safety evaluation of the final therapeutic composition. To show equivalency with the end result, this is typically done using GMP-manufactured or GMP-like cells [22].

The structure of critical safety investigations will be unique to the treatment, and an approach based on risk is recommended in the European Union guidelines. This attempts to pinpoint possible negative consequences, after which pivotal investigations are planned to assess how likely they are to occur [23]. There are several frequent concerns associated with MSCs that need to be taken into account, including as immunogenicity, instability of chromosomes, tumorgenicity, and hazardous biological distribution profiles. It will be necessary to defend the suitability of the model from the point of view of regulation, both in respect of the actual situation and length. Additionally, it is customary to conduct these investigations in accordance with good laboratory practice (GLP) guidelines, which include analyzing the data. The readily accessible nature of GLP models may impede advancement and necessitate the joint development of a fresh model in conjunction with a for-profit testing facility [24].

15.3 Stem Cell Classification and their Properties

The classification of different stem cells is discussed in Table 15.1 below.

Table 15.1 Classification of stem cells.

Stem cell type	Property	References
Embryonic stem cell (ESCs)	Capable of self-regeneration and differentiation into three primordial germlayer cells	[25]
Induced pluripotent stem cell (iPSCs)	Retroviral upregulation of the stem cell proteins and Klf4 reprograms somatic cells. According to their shape, expression of gene characteristics, and cell surface markers, these cells are identical to ES cells	
Neural stem cell (NSCs)	Oligodendrocytes, astrocytes, and neurons are the primary cell types that can be produced by neural stem cells	[26]
Mesenchymal stem cells (MSCs)	Multipotent cells called mesenchymal stem cells (MSCs) can be cultured and altered outside of the body. Three biological traits exhibited by these cells make them suitable for application in regenerative therapy: (1) differentiation capacity, (2) trophic factor release, and (3) immunoregulatory characteristics	[27]
Hematopoietic stem cells (HSCs)	Generate platelets, white blood cells, and red blood cells, at any point in their lifespan. Two unique biological traits are present in HSCs: distinct multilineage relationships and renewal themselves	[28]
Endothelial progenitor cells (EPCs)	Essential for angiogenesis, metastasis, and tumor formation. Considerable study has been conducted on potential EPCs produced from bone marrow at CFU-Hill	[29]
Cancer stem cells (CSCs)	Have traits similar to those of normal stem cells, majorly the capability to be transformed into every kind of cell that is available in a given cancer sample	[30]

15.4 Mechanism Underlying the Action of Stem Cells in Cancer

15.4.1 Homing to Bone Marrow

Chemotherapy at highly concentrated doses to eradicate all cancer cells either entirely or partially damages blood-forming cells and leukocytes in resistant tumors. People in those circumstances require an intravenous injection of autologous or allogeneic HSCs. It is expected that these HSCs would use a homing strategy to move quickly into specific bone marrow (BM) stem cell environment. The engrafted HSCs go through the engraftment procedure after coming into contact with BM before discovering into differentiated blood cells. The dynamic relationship among the gradient SDF-1 released by cells surrounding the BM stem cell habitats and the stem cell CXCR4 receptor is the primary molecular principle underlying the HSC migration process. Extracellular UTP or ATP and the ions H^+ and Ca^{2+} are involved in additional biochemical signalings. Moreover, HSCs were found to be associated with endothelial cells via CD44 and LFA-1, resulting in the release of matrix-degrading enzymes or the movement via blood artery [31].

15.4.2 Tumor-Tropic Effect

The development and invasion of tumors is characterized by the extracellular matrix (ECM) which results in the production of paracrine substances that encourage the movement of different kinds of cells, involving MSCs, and invading immune cells. Chronic wound tissue including tumors is characterized by persistent hypoxia, inflammation, and free radical generation. In addition, the migration of MSC into the tumor microenvironment was determined to be identical with the migration of MSC toward ischemia or damaged locations [32]. Cancer cells and its associated immune system cells can directly participate in the procedure of migration by the secretion of different chemoattractant. For instance, MSC movement toward the tumor microenvironment is achieved by CXCL16 released from osteosarcoma, breast carcinoma, multiple myeloma, and prostate cancer cells, respectively [33].

15.4.3 Paracrine Factor Secretion and Differentiation Capacity

Some research showcased that the stem cells release soluble compounds and extracellular vesicles (EVs) that affected the tumor spread, development, and survival. Nano-sized exosomes are one of the types of extracellular vesicles (EVs) that majorly regulate the cell–cell contact. They are formed by the intracellular multivesicular vesicles by the help of exocytosis and contain different types of biological substances including miRNAs, mRNAs, and proteins [34]. Exosomes can target cancer niches after being released from MSCs via integrating themself into cancer cells and then discharge their cargo into the recipient cells. MSCs are also able to release multiple other substances like immunosuppressive (TGF-β1), and anti-inflammatory (TSG-6) [35]. Transplanted HSCs, for instance, have the potential to generate all blood cell types. Consequently, clinical outcome and the blood system's capacity to recover would base on the caliber and quantity of transplanted HSCs. Similarly, in cases of brain cancer, NSCs can replace damaged neuronal and glial cells. iPSCs and ESCs can be used as cell types to induce effector immune system cells that seek out a tumor in a particular way [36].

15.4.4 Signaling in Cancer Stem Cells

The control of typical stem cell multiplication is widely known to be mediated by cellular signaling systems such as NF-κB, JAK/STAT, PI3K/PTEN, Hedgehog, and Notch. The formation of CSCs and subsequently cancer cells will result from the ongoing alteration in those pathways of signaling. Because of their great potential for renewal themselves and distinction, CSCs can promote the development, recurrence, and migration of tumors. These cells are also in charge of the malignancies' ability to resist standard treatment. Thus, it is crucial to do studies on CSCs in order to create a cancer therapy regimen that works [37].

Leukemia, brain, breast, lung, and gastrointestinal malignancies are among the tumor forms where CSCs have been shown to exist. These cells are usually separated and characterized utilizing a range of techniques, such as surface protein indicators and metabolic/functional characteristics. Originating from regular stem cells, CSCs' essential traits and biological mechanism are somewhat comparable to those of the latter. Although they can also be found in normal tissues, several surface markers like EpCAM, Lgr5, CD24, CD44, and CD133 are commonly utilized to distinguish CSCs from widely diverse kinds of cells. It was possible to effectively separate CD133+ cells of the brain, colon, and stomach; these cells showed a greater ability for regeneration and the *in vivo* formation of the initial tumor. Similarly, CD44 is frequently utilized to identify the CSC population in a variety of tumors when combined with CD133 or CD24. Although CD44+CD24−/low cells are rare in the SCID/NOD mice model of breast carcinoma, they can promote tumor development and occasionally form the entire primordial tumor tissues [38].

The assessment of CSC metabolic and functional activity yields additional characterization. Aldehyde dehydrogenase (ALDH) enzymes are usually evaluated in this regard since they catalyze the process of oxidation of aldehyde in a variety of biological poisons and substrates. Using a flow cytometry-based ALDEFLUOR assay, it is possible to detect high ALDH concentrations in cells that have elevated metabolic mechanism, such as CSCs and typical stem cells. Furthermore, CSCs have been shown to exhibit increased immunological response, improper division of cells, high metabolic rate, and improved resistance against treatment [38].

15.5 Potential Application of Stem Cell Therapy in Cancer Treatment

The potential applications of stem cell in cancer therapy are discussed as follows and also mentioned in Table 15.2 and Figure 15.1.

15.5.1 HSC Transplantation

Following cycles of greater-dose radiotherapy or treatment with chemotherapeutic agents, HSC implantation has mostly been utilized as a usual approach for the management of lymphomas, leukemia, and multiple myeloma. Furthermore, this process is currently being extensively studied in research studies to treat different cancer types, including breast cancer (NCT00003927), brain tumors (NCT00528437), and neuroblastoma, sarcomas (NCT01807468), alongside chemotherapy and immunotherapy. Nevertheless, graft-versus-host ailment (GVHD) is still a problem when employing allogeneic donor HSCs; immune-suppressive drugs are frequently used to treat GVHD, but they have serious negative effects and are less successful [39].

15.5.2 MSC Transplantation After Cancer Treatment

When aggressive malignancies are treated with intrusive tumor excision and high-dose chemotherapy, healthy tissues and the hematological system are frequently harmed. Empirical data has demonstrated that the introduction of MSCs contributes to the preservation of the undifferentiated

Table 15.2 Application of stem cell in cancer therapy.

Stem cell type	Stem cell application	Cancer type	References
Mesenchymal stem cell	Tumor reduction resulted in using a suicide gene that is dependent on proteasomes	Breast cancer	[61]
Mesenchymal stem cells and cancer-associated fibroblast	The discovery of mesenchymal stromal/stem cell-based anti-tumor medicines that inhibit primary cancer growth and metastasis is particularly relevant given the cells' inherent tumor tropism and immunomodulatory abilities	Breast cancer	[62]
Mesenchymal stem cells	MSC promoted apoptosis and reduced multiplication in colorectal cancer cells with elevated p53, caspase 3, and p21 expression	Colorectal cancer	[63]
Hematopoietic stem cell	Aid in the remission and control of certain tumors that are solid and hematological malignancies	Hematological malignancies	[64]
Induced pleuripotent stem cell	iPSC-derived NK cells are a possible candidate for ovarian cancer therapy since they caused antiovarian cancer death	Ovarian cancer	[65]
Neural stem cell	A successful therapy method for brain metastases of malignancies involves the systemic delivery of the prodrug 5-FC together with brain implantation of human NSCs expressing the suicide enzyme CD	Breast cancer brain metastasis	[66]
Endothelial progenitor stem cell	Greater following the initial chemotherapeutic cycle	Cervix, sarcoma, head and neck, prostrate, esophagus, ovarian, colorectal, and breast	[67]

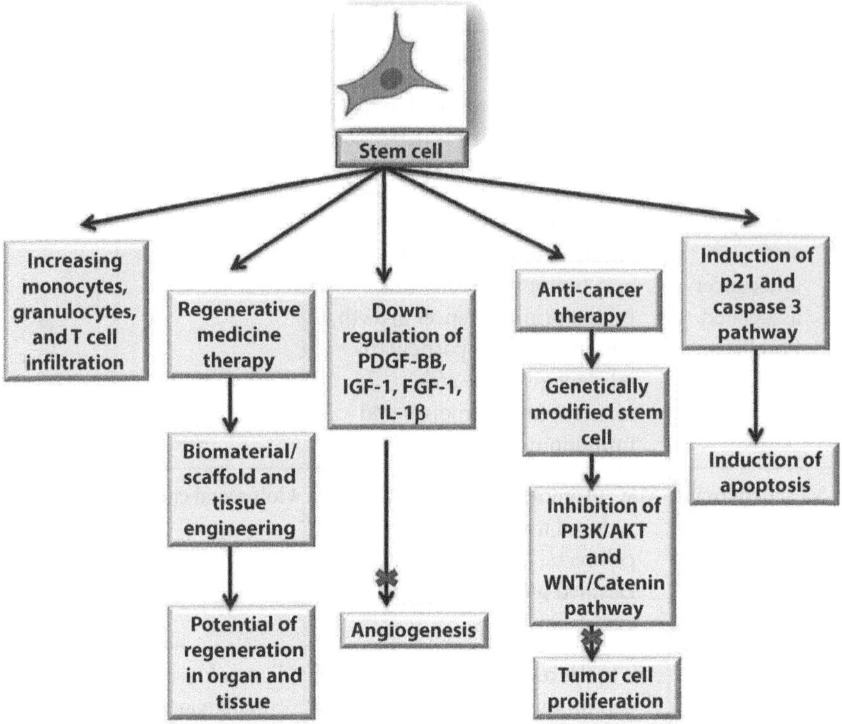

Figure 15.1 Application of stem cell in cancer therapy.

phase and the multiplication of HSCs, thereby improving the therapy's overall effect. Furthermore, MSCs with immunomodulatory activity may be able to successfully lessen severe immunological reactions in GVHD patients who are refractory. Current clinical experiments involving the co-engraftment of both MSCs and HSCs have demonstrated encouraging results with no associated side effect [40].

15.5.3 Enzyme Therapy

It is possible to modify NSCs and MSCs to produce the enzymes needed to change harmless prodrugs into toxic final products. Following transplantation into mice with tumors, the transformed stem cells relocate to the tumor cells, where an external enzyme alter the prodrug into a cytotoxic agent that eventually kills the tumor cells. Consequently, exact control over the dosage, timing, and site of drug release is possible. Suicide gene therapy, often known as enzyme/prodrug therapy, was the first designed NSC for medicinal purposes to go through clinical trials [40].

At present, one important enzyme utilized in enzyme/prodrug therapy is cytosine deaminase (CD). 5-Fluorocytosine (5-FC) is changed into 5-fluorouracil, a hazardous version, by CD. GBM cell development was suppressed by an amalgam of CD-bearing mouse 5-FC and NSCs. Using 5-FC to inject CD-expressing MSCs also inhibited the formation of tumors. A popular cytotoxic therapy involves the expression of CD (HB1.F3.CD) in human HB1.F3 cells. The therapy was extremely safe and effective. The findings of this finished research have not yet been made public. In case of suicide gene therapeutic treatment, herpes simplex virus-thymidine kinase (HSV-TK) has been utilized. During cellular division, GCV-TP incorporates into the DNA of surrounding cells, inhibiting DNA polymerase and causing death of cells. Intratumoral HSV-TK-transduced NSC (NSC-TK) injection and intraperitoneal GCV injection twice a day for 10 days (each containing 15 mg/kg) were shown to be a successful treatment for C6 gliomas in rats. Nine rats were injected, and six of them lived for 100 days with no tumor visible. According to Zhao *et al.* [41], there was evidence in another investigation that NSCs-TK injection moved into the brain, then co-localized with U87 cells, and gave GCV-treated mice a long lifespan.

15.5.4 Secreted Agents

Stem cells can behave as *in situ* agent manufacturing facilities, releasing anticancer drugs for prolonged periods of time and getting over several barriers to cancer treatment, like high systemic reactivity and brief drug half-lives. One of the most frequently employed curative drugs that is released is TNF-α-associated death-inducing ligand (TRAIL), which causes tumor cells to undergo apoptosis. Its brief half-life, however, lessens its *in vivo* therapeutic effectiveness. At the margins of resection, the encapsulated cells may constantly discharge therapeutic chemicals. This method improves mice's longevity by delaying the establishment of invasive and malignant brain tumors [42].

Additionally, growth-inhibiting proteins, such as IFN-β, can be specifically delivered by modified stem cells, making the surrounding tissue unsuitable for the formation of tumors. It was observed that tumor cell proliferation was inhibited and pulmonary and hepatic metastases were lessened when MSCs expressing IFN-β migrated and engrafted into primary breast tumor locations. Within the tumor microenvironment, MSCs produced large amounts of IFN-β, but not into the bloodstream [43].

15.5.5 Viral Therapy

In cancer cells, oncolytic viruses (OVs) show selective reproduction. It is hypothesized that OVs could cause tumor cell death and produce alarm signals to trigger the immune system to attack and eliminate tumor cells more efficiently. Nonetheless, immune cells can quickly identify and eliminate bare OVs from the body. One possible and promising vehicle for delivering and safeguarding OVs to tumor locations is stem cells [44]. Similarly, it was discovered that MSCs could efficiently load weakened oncolytic HSV and oncolytic OMV, which inhibited the development of GBM and hepatocellular carcinoma in mice, respectively. Remarkably, MSCs obtained from patients with ovarian cancer showed excellent promise for OMV carriers and similar rates of multiplication as those from donors who were healthy. After thawing the frozen transduced MSCs, this activity persisted, suggesting that they could be used for the treatment of cancer [45].

15.5.6 Nanoparticle Carrier

Nanoparticle carrier-based delivery systems (NPs) shield highly concentrated insoluble chemotherapy chemicals from deterioration in challenging biological environments. Using stem cells as NP carriers helps get around issues including ineffective dispersion in solid tumors, inability to strike micrometastatic lesions, and other restrictions. Additionally, stem cells can shield medications from host immunosurveillance and limit mononuclear cells unrrestricted reception of nanoparticles [46]. MSCs effectively ingested NPs, and Roger *et al.* [47] discovered that they might serve as NP transporters in brain malignancies.

15.5.7 Exosomes as Therapeutic Carrier

Exosomes enclosed a wide variety of proteins, tiny medicines, and mi-RNAs, among other anti-cancer medicinal substances. In comparison to the synthetic nanoparticles, these organic carriers have various advantages including safety, loading capacity, biological compatibility, and absorption ability into the tumor cells. Furthermore, they can be easily functionalized by the incorporation of particular ligands or proteins to their surface which can enhance their ability to penetrate in the tumor microenvironment. Genetic materials for instance anti-tumor mRNAs or siRNAs were beneficially loaded into the exosomes produced from stem cells by the help of conventional transfection technique. For instance, exosomes produced from bone marrow stromal cells carrying miR-146b were collected by

Katakowski *et al.* [48]. This exosome was directly infused into tumors and caused a significant decrease in the development of glioma xenograft in the animal model with initial brain tumor. Similarly, Greco *et al.* [49] reported that MSC-derived exosomes effectively transported siRNA to suppress the polo-like kinase 1 gene in case of bladder cancer cells.

15.6 Other Applications of Stem-Cell-Based Cancer Therapy

15.6.1 Inhibition of Angiogenesis

One important process that keeps tumor cells growing and spreading is angiogenesis. One possible method of delivering anti-angiogenetic drugs to restrict tumor proliferation and metastasis is through the use of MSC carriers. These altered MSCs can supply antiangiogenic substances without causing unintended adverse reactions and have a tropism to tumor tissue. Despite this, systemic and prolonged administration of anti-angiogenic medicines has been linked to toxicity and impaired blood flow, which lessens the ability of the medications used in chemotherapy to reach the cancer cells [50].

It has been determined that growth factors are generated in the micro-environment of tumor and a disparity in the control of pro- and antiangiogenic variables are the root causes of cancer-associated angiogenesis. Endostatin is a potent endogenous regulator of angiogenesis in tumors that has been widely used to block antiangiogenic agents in the treatment of various cancers. In an investigation conducted on animals, endostatin was delivered into nude mice by means of modulating human placenta-derived MSCs using adenoviral transmission [51].

These MSCs, which produced human endostatin, colonized the cancerous tissue and effectively reduced the size of the tumor without causing significant systemic harmful effects. Benefits of tumor treatment included increased cancer cell death and disrupted neovascularization, which reduced tumor cell development and slowed tumor growth. Additionally, administering the anti-angiogenic drugs in a phase II clinical trial led to the aberrant constructions normalizing and the blood vessels' poor performance decreasing, which in turn produced a notable reduction in tumor-related vasogenic brain edema [52]. Vascular normalization was mostly caused by decreased vascular permeability and diameter. Therefore, encouraged mural cell coating of the tiny vessels was an additional factor in achieving normalization of the vessel. Research indicates that MSCs may be able to localize to the tumor vasculature after intratumoral injection, which could be advantageous for the targeted administration of anti-angiogenic drugs, mainly in vascularized cancers [53].

15.6.2 Regenerative Medicine

Stem cells have the ability to regenerate themselves and divide, which makes them useful for repairing tissues that have been damaged by chemotherapy. After treating cancers with high-dose radiation or chemotherapy, transferring HSCs has been used extensively in clinical practice to promote a patient's hematopoietic restoration for the rest of their life. By delivering stem cells that specialize into the specialized kind of blood cell, this therapy works to address blood cell genetic disorders and rebuild bone marrow under circumstances of marrow failure. Recipients can regain hematopoiesis with transplantation and the successful engraftment of a single HSC [54].

It is potentially possible to use healthy iPSCs produced from patient cells to rebuild tissues harmed by tumors or treatments. iPSCs can be used in the field of regeneration to develop a variety of tissues. However, strong *in vivo* transplantation of iPSC-derived tissues is necessary for regeneration therapy achieved through human iPSCs. According to Cerneckis *et al.* [55], only a small number of human iPSC-derived types of cells have been effectively implanted in animal experiments to date.

15.6.3 Immunotherapy

Some blood-related cancers may be curable with allogeneic HSC transplantation alone due to an immune-related anticancer impact. HSCs become appealing for application in cancer immunotherapy when genes producing chimeric antigen receptors (CARs) or T-cell receptors (TCRs) targeted against antigens related with tumors are administered [56].

Immunotherapy techniques could profit from patient-specific iPSCs. Human iPSCs generated from T lymphocytes balance the pre-rearranged TCR gene, and they can be further stimulated to develop into T cells with functional capabilities. Through the process of reprogramming particular T cells into iPSCs, which subsequently develop back into T lymphocytes for patient injection, it is possible to manufacture functional, tumor antigen-specific T lymphocytes *in vitro*. But detailed investigation is required to assure the security of human iPSCs generated from T cells [57].

15.6.4 Anticancer Drug Screening

iPSCs can be utilized to screen novel anticancer medicines alongside to being utilized for treating cancer effectively. iPSCs produced from patient cancer tissue can be differentiated to produce various kinds of cells that may be extensively genetically similar to animal tumors than existing drug analysis techniques, which include mice xenograft models, conventional cancer cell lines, and mouse tumors. Furthermore, hepatocytes generated from human iPSCs with diverse genetic histories can be utilized to screen for hepatotoxicity, which hinders the practical application of several possible anticancer medicines [58]. Table 15.2 shows a list of the uses of stem cells in the treatment of cancer.

15.6.5 Targeting Cancer Stem Cells (CSCs)

Because of their multipotency, ability to regenerate themselves, and high regenerative potential, CSCs help trigger tumor growth and dissemination quickly. Thus, in order to guarantee high treatment effectiveness and avoid cancer recurrence, attacking CSCs is essential. Normal stem cells may be able to go after CSCs in cancer treatment since CSCs can draw normal stem cells. CSCs and normal stem cells combine to inhibit angiogenesis, metastasis, tumor growth, and inflammation as well as apoptosis [59]. As HSCs are less favored than NSCs to cause neoplastic changes in brain cancer, Bryukhovetskiy *et al.* [60] evaluated the possibility of HSCs and NSCs in anti-glioblastoma medication and came to the conclusion that HSCs might be suitable for discovering strategies meant to limit glioblastoma CSC activity. In a similar vein, modified HSCs might help create cell subsystems that can cause CSC apoptosis that is specifically targeted.

15.7 Factors Influencing Stem Cell Therapy

15.7.1 Stem Cell Type

Although stem cells have many features, their therapeutic actions might not be the same. Intracellular adenoviral replication was maintained by both types of stem cells; however, NSCs discharged a log greater amount of virus in comparison to MSCs. Moreover, in an animal experiment involving orthotopic glioma, mainly cerebral delivery of virus-enclosed NSCs increased existence, despite having similar migratory capacities, NSCs demonstrated greater therapeutic efficiency in intracranial tumors than

MSCs, indicating that the degree of association among malignant and carrier natural cells may be a significant factor in the effectiveness of carrier trafficking [68].

The selection of stem cell type in anti-cancer treatment is contingent upon the unique properties of the cell as well as its medicinal needs. After high-dose treatment with chemotherapy, autologous HSC engraftment is widely employed to maintain hematopoiesis in the cure of non-hematologic and hematologic cancers. This method is also frequently applied to consistently meet the needs for making mature blood cells in the condition of acquired marrow and congenital failures. Additionally, while evaluating the toxicities of potential anticancer drugs, iPSCs outperform other stem cells [69].

15.7.2 Transplantation Route

Anti-tumor treatment relies heavily on the mode of stem cell administration. Effective injection of NSCs in mouse models of GBM is accomplished by contralateral injection into the cancer location. Intracranial injections are not recommended for multiple injections due to their invasive nature. When administered intravenously, NSCs can still move to the tumor site with efficiency, enabling repeated delivery. Additionally, according to Balyasnikova *et al.* [70], this method can avoid problems associated with intravascular dispension.

Semisolid components, as opposed to suspended cells injections, may increase transplantation activity by reducing the stress of metabolism and offering mechanical assistance. Currently, stem cell transplant utilizing biocompatible technologies provides a provisional solution to the lower survival of NSC grafts. A 3-D extracellular matrix-associated substrate (3DECM), extracted from engineered skin cultures, was reported by Hansen *et al.* [71] as potentially offering a clinically effective method of administering cell grafts. The *in vitro* growth of implanted NSCs with uncontrolled differentiation was made possible by 3DCEM.

15.7.3 Cell Number and Transplanting Time

The quantity and timing of transplanted cells have an impact on the outcomes of treatment. Patients with oncohematological disorders who receive inadequate HSC transplants experience ineffective restoration of hematopoietic components and a high rate of disease recurrence.

Nevertheless, an excessive amount of transplanted cells can raise the possibility of ectopic engraftment or teratoma development. Accordingly, the ideal cell count for a specific therapy should be determined [72].

When stem cell treatment is administered is crucial to its effectiveness. According to Tobias *et al.* [44], rats with GBM43 xenografts that administered with loaded NSCs before XRT-TMZ treatment had a 9-day increase in median survival compared to those that administered with the opposite schedule ($p < 0.05$).

Prior to being used in oncolytic virotherapy, carrier cells need to proliferate in tumor beds. Subsequently, progeny of the virus becomes available to enable the virus's targeted distribution. Greater viral offspring were produced from NSCs 7 days after loading *in vitro*. Once implanted, the majority of NSCs moved to the locations of tumors in 24 to 48 h. According to Ahmed *et al.* [73], oncolytic viruses administered through NSCs should therefore have multiplication cycles suitable for NSC tumor-homing capacities.

15.8 Diapause and Hibernation Mechanism in Cancer Stem Cells

Chemotherapy causes cancer cells to go into hibernation or senescence. Administration of ATR protein inhibitors prior to chemotherapy prevents the cancer cells from going into hibernation in AML organoids and mice models. This strategy can be utilized to increase the chemotherapeutic efficacy for cancers of the breast, prostate, gastrointestinal tract, and other types. Moreover, studies have shown that the some of the cancer cell subtypes that are able to hibernate are also involved in the relapse of cancer [74]. This emphasizes how crucial is to study about these cancer cells type along with their hibernation in order to prevent the disease from relapsing and for the enhancement of treatment results. Various studies investigated this biological phenomenon to enhance the cancer therapeutics; however more detailed research is still required to fully understand the complicated processes of diapause and its relation in cancer studies [75]. In the battle of cancer, researchers are aiming to conquer treatment associated resistance, avoid the chances of tumor recurrence, and eventually enhance the lives of patients by understanding the cellular processes that control diapause and developing a methods to specifically target hibernating cancer cell type [76].

15.9 Cancer Stem Cell Models

15.9.1 Cell-Line-Derived Models

CSC models can be created by the application of known cell lines or cancerous cell lines that are obtained from tumor biopsies. In these models, stem-like cells that possess capabilities of differentiation and self-renewal are isolated and purified. Through the examination of these cell communities, scientists can learn more about the biology of CSCs, crucial markers, and find out how these cells may react to different therapies [77]. Many different types of research facilities can use them because of their affordability and ease of upkeep. Numerous cell lines grow quickly, which makes high-throughput research possible. Additionally, some cell lines can be genetically altered, which makes it possible to study the roles of particular genes. Furthermore, using cell lines eliminates the moral dilemmas related to studying humans or animals, which makes them a useful option for a variety of research contexts. Cell-line-derived models have drawbacks despite their benefits [78].

There may be differences between the physiological reality and the results of experiments because they fall short of capturing the intricate *in vivo* milieu and interactions that exist throughout the whole body. A restricted variety of tissue types constitutes the source of many cell lines, which reduces their representativeness.

Furthermore, cell lines may experience some genetic and phenotypic drift that cause divergence of these cells from acquiring the initial features of the tumor or tissue. These cells do not make interaction with the extracellular matrix and have no physiological significance as they are maintaining in an external environment. Lastly, their side reactions to therapeutic medications might not be precisely replicate *in vivo*, which could lead to false positive results during the testing and development of new therapies [79].

15.9.2 Patient-Derived Xenografts

Immunodeficient mice are implanted with tumor tissues or a CSC population which was obtained from the patients in PDX models. These models enable the investigation of tumor development, metastatic disease, and response to therapy *in vivo* and provide a most accurate representation of the tumor's microenvironment.

Customized treatment strategies and medication testing can greatly benefit from PDX models as they maintain the genetic and heterogeneous features of the actual tumor [80]. PDX models are quite advantageous in the stimulation of human cancer and then evaluation of medication in the same model. They do not have a human immune system, and are resource-intensive, and might not be useful for all types o0f tumors. Investigators should carefully reflect on the benefits and drawbacks related to the selection of PDX model for the particular research goal [81].

15.9.3 3D Culture Systems

Tumor spheroids and organoids which are examples of 3D culture systems are designed to replicate the three-dimensional structure and interactions between cells found in the tumor. Wilson has carried out a ground-breaking experiment in 1907 that revealed the extraordinary capacity of regeneration of sponge cells. Since then, stem cell studies have made a great stride in the creation of organoids from stem cells including pancreatic ductal adenocarcinoma, colon, lung, and breast [82].

To research stomach cancer, there are currently stem cell translation models and induced pluripotent stem cell organoids available. These models offer an atmosphere that is more physically realistic for researching the behavior of CSCs, such as self-regeneration, differentiation, thereby and therapy response [83]. Because of this characteristic, they are potentially useful instruments for translation studies and precise medicine, providing a strong foundation for investigating the genetics of cancer and individualized treatment approaches.

Existing cancer cell lines or samples from the patients can be utilized to create a 3D culture system. For example, whole-genome duplication, tandem-duplicator phenotype, and homologous recombination deficiency are some of the mutational aftereffects that have been successfully studied using patient-derived organoids in a high-grade severe ovarian cancer [58]. These organoids direct the creation of précised that can be utilized for the evaluation of chemical sensitivity. These organoids have the ability to foresee the treatment response and assist in the creation of specialized cancer therapy when combined with other methods like mass spectrometry and RNA-seq [84].

15.9.4 Genetically Engineered Mouse Models (GEMMs)

GEMMs direct genetic modification in the bio-composition of the mice in order to induce particular cancer subtypes or to target specific genes related to CSCs. These models enable scientists to investigate a particular genetic mutation that involves in the development of tumors, and their progression. The effectiveness of targeted treatments in a more intricate and challenging framework can also be assessed by the help of GEMMs. They are crucial for numerous scientific applications, but due to their resource-intensive property, they might not be appropriate for every illness, and caused moral dilemmas [85].

15.9.5 Spheroids

Spheroids are the 3D cell culture models that closely match the morphology and activity of real tumors. Usually, an extracellular matrix is used to enclose the cancer cell which is cultured in the suspended medium for the formation of spheres. CSCs or large numbers of different types of cancer cells are utilized to create spheroids. This large network is utilized to study drug reaction behavior, invasion, metastasis, self-renewal, differentiation, and interactions among CSCs and various other tumor cells in a three-dimensional (3D) milieu [86]. Different kinds of spheroids, including neurospheres, mammospheres, and colonospheres have been produced from distinct forms of cancer. The biology, gene-expression pattern, and genotype of all spheroids are highly similar with the parent malignancies. Mammospheres on the other side produced from ductal carcinoma and metastatic cells *in situ* have been applied to study signaling mechanism, and the impact of chemicals on CSCs. On the implantation of colonospheres into the mice, the histopathological characteristics of colonospheres resembles with the original tumor and have been widely used to discover single-cell aspects of CSC properties including chemoresistance, metastatic potential, and tumorigenicity [87].

15.10 Biomarkers Associated with Cancer

Recently, various complete genomes and tissue-specific methods for the molecular analysis of complex genes of tumors and cancer cell lines have been discovered. These methods have led to the discovery of particular gene sets related with certain types of cancer, which may anticipate the likelihood of metastatic activity at specific organ and tissues as well as the overall prognosis of cancer individuals [88]. Furthermore, their major consequences in disease relapse, treatment resistance, metastases, and tumor development have been showed by the *in vivo* and *in vitro* analysis of stem cell-like functional characteristics and markers of cancer stem/progenitor cells in tumor-related specimens from individuals suffering from cancer and cancer cell lines. Here we provide the stem cell-like biomarkers and gene signatures of cancer cells linked to specific invasive cancer variants, such as hormone-independent breast and prostate cancers [89] as shown in Table 15.3.

Table 15.3 Biomarkers of cancer stem cells in human cancer.

Cancer type	Biomarkers	Functions	References
Hematological cancers	CD44	Connected to increased metastasis, self-regeneration, and proliferation	[90]
Colorectal cancer	CD44/CD166	Propose anchorage-independent growth	[91]
	CD29	Regulates the adherence of external proteins; has a role in immune system processes, homeostasis, tissue repair, and embryogenesis	[92]
	Lgr5	Involved in the growth and upkeep of postnatal intestinal stem cells	
Breast cancer	CD49f	Crucial for signaling and cell surface attachment	[93]
Prostate cancer	CD133, CD44, integrin-α_2, nestin, ALDH1, and CD49f	Connected to both their resistance to therapy and their aggressive and invasive nature	[94]
Head and neck carcinomas	Sox2 and Oct 4	Its connection to the suppression of neuronal development. Additionally, Oct4 has been identified as a transcription factor that regulates ESC pluripotency and self-renewal in a crucial way. Acting in this capacity with Sox2 is known	[95]

(*Continued*)

Table 15.3 Biomarkers of cancer stem cells in human cancer. (*Continued*)

Cancer type	Biomarkers	Functions	References
Renal cell carcinoma	CD133/CD24	Chemoresistance and clonogenic	[96]
	CXCR4	Chemoresistance, tumorigenicity, and sphere formation	
	ABCB1$^+$/ ABCC1$^-$/ ABCG2$^-$	Resistance to radio and chemotherapy, tumorigenicity, and clonogenic	[97]
	CD73$^+$	Resistance of tumorigenicity to radiotherapy	
Hepatocellular carcinoma	CD133$^+$	Traditional chemotherapy cannot cure HCC due to the expression of CD133+ cells	[98]
	CD90$^+$/CD45$^-$ and CD44$^+$	CD45–, CD44+, and CD90+ cells are responsible of tumor metastasis	
	EpCAM	A cell surface protein called EpCAM (CD326) has been found to be produced by nearly all epithelial malignancies and may be a possible early biomarker for HCC	
Ovarian cancer	CA125, ALDH1, miRNA, Kallikrein, FOLR, and HE4	Greater sensitivity at early prognosis of ovarian carcinoma	[99]
Cervical cancer	GDA, AGR2, KRT7, CD63, MMP7	These cell populations express particular markers. There is evidence to suggest that these could be used as cancer targets	[100]

15.11 Side Effects Related with Stem Cell Therapy

Factors affecting the cancer therapy and its associated side effects are discussed as follows and are shown in Figure 15.2.

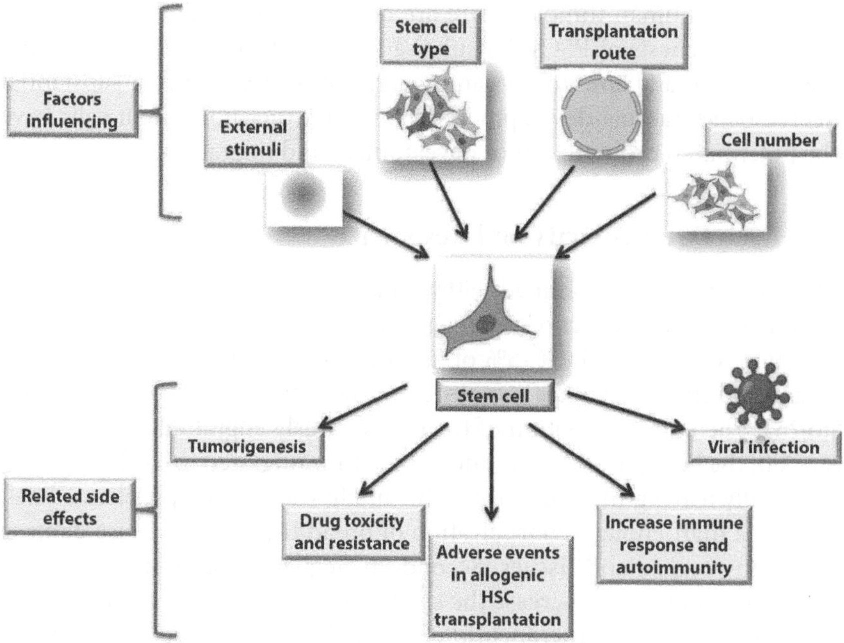

Figure 15.2 Factors influencing stem cell therapy and its related side effects.

15.11.1 Tumorigenesis

Normal stem cells and CSCs have a common mechanism for signaling. When the surrounding environment of stem cells changes negatively, normal stem cells can transition into CSCs, resulting in the creation of complete tumor tissue. The term endogenous stem cells in humans are closely tracked by other cells, so they continue to function normally. Yet, transplanted stem cells that are impacted by external circumstances during growth prior to administration may experience changes in expression of genome and consequent phenotype. The longer stem cells are cultured, the more likely they are to become cancerous. It was discovered that 45.8% of MSCs naturally changed into cancerous cells after a month in culture. Although this phenomena is still debated, it is vital to stress that the conditions of culture should be strictly managed in order to prevent detrimental changes in stem cells [101].

PSCs are reported to be more tumor-causing than ASCs. However, it should distinguish among "tetratocarcinoma" and "tetratoma" production. The first word frequently suggests that PSCs have innate traits that allow them to become a "normal cancer," whereas the second one denotes an "unusual human tumor". There are numerous methods for preventing the carcinogenesis of PSCs. Katsukawa *et al.* [102] observed that treating iPSCs with a fatal amount of gamma irradiation resulted in a substantial decrease in teratoma development in mice. Furthermore, stem cells have the potential to promote tumor development and migration.

15.11.2 Drug Toxicity and Resistance

The amount of cells confined within the tumor undoubtedly affects how well stem cells work as carriers of genes and drugs. Actually, upon systemic injection, only around 2%–5% of all of the stem cells entered the tumor tissue, and this percentage remained constant during the course of monitoring. The large population of the intravenously engrafted cells was first stuck in the lung parenchyma, after which they migrated to the liver before eventually leaving the body. Thus, this might cause some problems. First, there is a high-enough concentration of broad therapies to cause damage to healthy organs and tissues. Second, inadequate local medication concentrations in tumors increase the chances of drug resistance while simultaneously decreasing therapeutic efficacy. By figuring out how to more effectively target stem cell transplants to tumors, these hazards could be avoided and this therapeutic approach could become dependable [103].

15.11.3 Adverse Events in Allogenic HSC Transplantation

Hematologic and lymphoid malignancies can now be effectively treated with allogeneic HSC transplant. On the other hand, a significant proportion of individuals may experience permanent adverse outcomes from it, such as persistent GVHD, organ/tissue malfunction, infections linked to aberrant immune system responses, recurrence of cancer and secondary malignancies, and decreased individual quality of life. In actuality, the objective of treatment should now encompass both social ties and the full restoration of health rather than just extending survival. Clinical research on HSC sources for transplantation should be prioritized in order to enhance results. It is known that using homologous umbilical cord blood may lessen the severity and incidence of long-term GVHD [104].

15.11.4 Increased Immune Responses and Autoimmunity

Allogeneic stem cells from distant donors may cause strong host immunological reactions when applied. Indeed strong proof supporting the development of donor antigen-specific cellular immune responses as well as humoral antibodies produced by B cells has been presented [105]. Another concern is the possibility of autoimmune resulting from the administration of whole-cell vaccinations derived from autologous iPSCs. Both natural tissue-associated antigens and CSC-specific antigens are present in these vaccinations. Consequently, immunological reactions against normal tissues can be induced. Although autoimmunity was not observed by Kooreman *et al.* [106] following treatment with autologous iPSC vaccination, more research is necessary to verify the efficacy of the strategy.

15.11.5 Viral Infection

One popular and useful technique for transforming stem cells into gene delivery vehicles is viral transfection. On the other hand, it can increase the receivers' risk of contracting a viral infection. The primary issues that arise from this are the high immunogenicity of the virus, which may trigger unfavorable immune reactions that release toxins, eliminate transduced cells, reduce the extent of the transgenic capacity, or even result in mortality. Consequently, targeted motifs for an anti-cancer impact should be added to viral vectors, while certain sequences implicated in intrinsic toxicity among individuals should be removed [107].

15.12 Challenges to Stem Cell Therapy

15.12.1 Treatment Durability

Despite powerful early treatment results, tumors frequently return. A single drug used in stem cell therapy typically is not able to eradicate malignancies, just like most chemotherapies do. As a result, choosing the best medicine combo should be done with reason. Tests have been conducted on numerous combinations of treatments to increase therapy persistence [108]. Tumor cells exposed to radiation can produce substances that promote MSC penetration through intrinsic basement membranes, hence boosting the MSC population within the tumor. According to Kim *et al.* [109], TMZ raised the curative properties of MSC-TRAIL gene therapy by sensitizing glioma cells to TRAIL-directed apoptosis via changing the apoptotic mechanism. Treatment results were improved when TRAIL and immunoconjugates of EGFR-specific nanobodies supplied by stem cells were coupled.

15.12.2 Potential Tumorigenesis Concerns

Certain traits, such as self-regeneration, distinctions, and the ability to shift from epithelial to mesenchymal tissue, are shared by typical stem cells and CSCs. Therefore, more investigation is needed to determine how stem cell transplants can inhibit growth of tumor. It is unclear, nevertheless, if stem cells actually cause cancers or just encourage their proliferation. In subcutaneous transplanted cells, tumor growth was made possible by the addition of bone-marrow-derived MSCs to human breast carcinoma cells that were otherwise only marginally metastatic. This was shown by Karnoub *et al.* [110]. The chemokine CCL5, produced by MSCs in response to stimulation by the breast cancer cells, increased the motility of cancer cells through a paracrine effect.

15.12.3 Epithelial Mesenchymal Transition (EMT)

It has been shown that stem cell and epithelial mesenchymal transition (EMT) markers are mixed in tumor cells that circulate from individuals who have metastasized, and that cancer cells can acquire stem-like characteristics by EMT activation or through the expression of EMT transcription factors (TFs). Specifically, EMT-induced cells exhibit comparable markers to both healthy and neoplastic human breast stem-like cells. EMT also causes extremely differentiated neoplastic cells to produce a nearly limitless amount of cancer stem cells. In the meanwhile, resistances to medication and EMT stimulation are related. For example, stimulation of Notch-1 signaling results in the acquisition of EMT phenotype in gefitinib-induced resistant lung cancer cells [111]. In another direction to this, mesenchymal cell lines exhibit a greater number of genes associated with metastasis are considerably less vulnerable to EGFR restriction, such as cetuximab, gefitinib, and erlotinib; this is likely due in part to mesenchymal cell excessive expression of integrin-linked kinase (ILK) [112].

15.12.4 High Levels of Multidrug Resistance or Detoxification Proteins

A broad range of cancers that are solid, such as glioblastoma, gastrointestinal cancer, neuroblastoma, and retinoblastoma, have side population (SP) cells, which have an appearance similar to cancer stem cells. Their increased production of drug-transporter proteins, such as ABCB1, ABCG2, and MDR1, not only serves to exclude Hoechst dye but also eliminates cytotoxic medications, resulting in elevated resistance to chemotherapy medications along with better existence of cells and recurrence of the disease [113]. Furthermore, among individuals with resectable esophageal cancer, higher levels of ALDH1 overexpression indicate an inadequate response or rejection of preoperative chemoradiotherapy [114].

15.12.5 Tumor Microenvironment

It has been observed that the protection and control of typical stem cells require a specific microenvironment with a range of cellular makeup. Similar conditions were also discovered in the CSCs, wherein the cells were well supported in a histological niche known as the CSC microcosm, which included vascular and connective stromal components [115]. Because of the accelerated division processes occurring in the surroundings, CSCs are able to distinguish progenitor daughter cells, regenerate themselves, and remain in the basic developmental stage. According to Pires *et al.* [116], the cells in the CSC microenvironment have the potential to activate pathways of signaling which may help CSCs proliferate, avoid anoikis, and alter divisional kinetics in order to repopulate by symmetrical division [117]. Consequently, CSC ability to resist therapy may also be influenced by hypoxia and the HIF signaling system. Hypoxia and hypoxia-induced factor HIF1-α signaling, irrespective of BCR-ABL1 kinase function, boost CML cell survival in the hypoxic microenvironment primarily through the activation of hypoxia-inducible factor 1α (HIF1-α). In EGFR mutation-positive NSCLC, hypoxia likewise promotes gefitinib-resistant lung CSCs through stimulating IGF1 receptor (IGF1R) and raising insulin-like growth factor 1 (IGF1) production through HIF1α [118].

15.13 Strategies to Improve the Anticancer Efficacy of Stem-Cell-Based Therapy

There are several approaches that could be used to increase the medicinal capacity of SCs. The enhancement of immunotolerance and the tumor tropism/homing of stem cells, in addition to the best method of delivery and combinational methods, are all tactics meant to boost the cancer-fighting abilities of stem cells.

15.13.1 Improvement of Stem Cell Immunotolerance

Even in immunocompetent individuals, allogenic SC resistance remains a partially unresolved problem with SC-based treatment. HSC survival is increased when the human leukocyte antigen class I gene is knocked down because it stops HSCs from being rejected. Autologous NSCs are thought to be the finest SCs for treating gliomas despite being difficult to get as they can prevent immune system-based rejections [119]. According to the evaluation of these SCs in GB models of animals, they were able to prevent immunological rejection, which may maximize therapy duration in subsequent human studies [120].

15.13.2 Enhancement of the Stem Cell Tumor Tropism

The increase of SC tumor tropism/homing is made possible by the upregulation of the SC migratory components. For example, chemokine-directed movement into intracranial gliomas has been successfully increased by inducing receptors for chemokine in SCs. According to Pulukuri *et al.* [121], MSCs that are treated with histone deacetylase inhibitors exhibit an increased production of the urokinase plasminogen activator, which facilitates the cells' migration toward the cancer site.

Malignant tumors are characterized by hypoxia, which is associated with both inadequate therapy results and a rise in tumor recurrence. Surprisingly, these CXCR4-overexpressing hADSCs showed distant migration toward GB and selectively entered the hypoxic tumour core when implanted in the contralateral brain in a mouse intracranial GB xenograft [122].

15.13.3 Optimal Choice of the Stem Cell Administration Route

The most effective way to transport stem cells to the tumor location determines how effective stem cell-based treatment is in treating cancer. Garofalo *et al.* [123] assert that the intraventricular route of administration is preferable to the intravenous or intraperitoneal ones. Intranasal delivery is a unique approach that provides an outstanding way to overcome the present barriers to SC administration for the treating of intracerebral gliomas [70]. Another innovative tactic to increase the curative value of SCs is the transplantation of encapsulated SCs within the resection cavity. Hyaluronic acid, alginate, agarose, and other artificial substances, together with biodegradable hydrogels, are used to encapsulate stem cells (SCs) in biocompatible and semi-permeable scaffolds. The primary benefits of SC encapsulation are increased cell viability and a reduction in the significant and quick "washout" of cells by the brain fluid [42].

15.13.4 Combinational Approaches

The diverse groups of cells that make up glioblastoma (GB) are unpredictable both physically and genetically. Therefore, when GB is viewed as a heterogeneous structure, combinational techniques may be able to combat cancer more successfully. Research on stem cells (SCs) designed to release diverse medicinal products that target multiple signaling pathways in malignant cells are becoming more and more numerous. In animal studies with GB, combination of the immune-stimulatory cytokines IL-18 and IFN-β by medicinal SCs was found to be an effective tactic [124].

The immunological-conjugate Enb-TRAIL was created by fusing the pro-apoptotic TRAIL protein with a nanobody that targets EGFR. Enb-TRAIL inhibits the growth of tumors by activating death receptors and suppressing EGFR signaling, which leads to caspase-mediated cell death. According to Zhu *et al.* [125], in mice models of GB, the expression of ENb-TRAIL by NSCs results in significant tumor shrinkage and extension of survival.

Numerous research have demonstrated how various medications work in concert with stem cell therapy. In intracranial gliomas, valproic acid amplifies the antitumor impact of MSC-mediated HSV-TK gene therapy. In the same way, regressing GB was more successful when temozolomide was administered in conjunction with MSCs that expressed INF-β or TRAIL. It was discovered lately that the histone deacetylase inhibitor panobinostat enhances the therapeutic benefits of MSCs displaying TRAIL. Mice receiving xenografts generated from GB patients were shown to have a higher survival rate when treated with radiation, temozolomide, and NSCs expressing OV [44].

15.14 Conclusion

The utilization of stem cells has made a remarkable stride in the zone of malignant tissue studies, with constant breakthroughs in terms of testing. Despite various limitations, each new research advances the comprehension of stem cells in the fields of regeneration and transplantation. The opportunity for cancer stem cell-based therapy to be utilized for the cure of formerly incurable illnesses is quite interesting. The potential to use malignant tissues from individuals, thanks to the discovery of pluripotency, has resulted in the formation of tissue banks which used as a useful resource for the regenerative medicinal therapy in the battle against different ailments. Cancer stem cell therapy has had an unparalleled impact in prolonging the human life, and it still holds an exciting potential for the future. Thus, therapy that utilizes cancer stem cells is currently one of the most intriguing and exciting fields of cancer study.

References

1. Zhang, C.L., Huang, T., Wu, B.L., He, W.X., Liu, D., Stem cells in cancer therapy: opportunities and challenges. *Oncotarget*, 8, 43, 75756–75766, 2017. doi: 10.18632/oncotarget.20798.
2. Wan Kamarul Zaman, W.S., Nurul, A.A., Nordin, F., Stem cells and cancer stem cells: the Jekyll and Hyde scenario and their implications in stem cell therapy. *Biomedicines*, 9, 9, 1245, 2021. https://doi.org/10.3390/biomedicines9091245, 1245.
3. Zhou, H.M., Zhang, J.G., Zhang, X., Li, Q., Targeting cancer stem cells for reversing therapy resistance: mechanism, signaling, and prospective agents. *Signal Transduct. Target. Ther.*, 6, 1, 62, 2021. doi: 10.1038/s41392-020-00430-1.
4. Zheng, X., Yu, C., Xu, M., Linking tumor microenvironment to plasticity of cancer stem cells: mechanisms and application in cancer therapy. *Front. Oncol.*, 11, 678333, 2021. https://doi.org/10.3389/fonc.2021.678333, 678333.
5. MacPherson, A. and Kimmelman, J., Ethical development of stem-cell-based interventions. *Nat. Med.*, 25, 7, 1037–1044, 2019. doi: 10.1038/s41591-019-0511-6.
6. de Windt, T.S., Vonk, L.A., Slaper-Cortenbach, I.C., van den Broek, M.P., Nizak, R., van Rijen, M.H., de Weger, R.A., Dhert, W.J.A., Saris, D.B., Allogeneic mesenchymal stem cells stimulate cartilage regeneration and are safe for single-stage cartilage repair in humans upon mixture with recycled autologous chondrons. *Stem Cells*, 35, 1, 256–264, 2017. https://doi.org/10.1002/stem.2475.
7. Habets, M.G., van Delden, J.J., Bredenoord, A.L., Studying the lay of the land: views and experiences of professionals in the translational pluripotent stem cell field. *Regener. Med.*, 11, 1, 63–71, 2016. doi: 10.2217/rme.15.78.
8. Swierstra, T., Identifying the normative challenges posed by technology's 'soft' impacts. *Nord. J. Appl. Ethics*, 9, 1, 5–20, 2015. https://doi.org/10.5324/eip.v9i1.1838.
9. Kolkundkar, U., Gottipamula, S., Majumdar, A., Cell therapy manufacturing and quality control: current process and regulatory challenges. *J. Stem Cell Res. Ther.*, 4, 2, 2014. https://doi.org/10.4172/2157-7633.1000230.
10. Fekete, N., Béland, A.V., Campbell, K., Clark, S.L., Hoesli, C.A., Bags versus flasks: a comparison of cell culture systems for the production of dendritic cell-based immunotherapies. *Transfusion*, 58, 7, 1800–1813, 2018. https://doi.org/10.1111/trf.14621.

11. Sensebé, L., Gadelorge, M., Fleury-Cappellesso, S., Production of mesenchymal stromal/stem cells according to good manufacturing practices: a review. *Stem Cell Res. Ther.*, 4, 3, 66, 2013. https://doi.org/10.1186/scrt217.
12. Committee for Advanced Therapies, *EMA/CAT/600280/2010 - Reflection paper on classification of advanced therapy medicinal products,* European Medicines Agency, 2012, Available from: https://www.ema.europa.eu/en/documents/regulatory-procedural-guideline/reflection-paper-classification-advanced-therapy-medicinal-products_en.pdf.
13. US Food and Drug Administration, *FDA regulation of human cells, tissues, and cellular and tissue-based products (HCT/P's) product list,* Docket Number: FDA-2022-D-0563; Issued by: Center for Biologics Evaluation and Research, 2018. Available from: https://www.fda.gov/vaccines-blood-biologics/tissue-tissue-products/fda-regulation-human-cells-tissues-and-cellular-and-tissue-based-products-hctps-product-list.
14. Vaggelas, A. and Seimetz, D., Expediting drug development: FDA's New regenerative medicine advanced therapy designation. *Ther. Innov. Regul. Sci.*, 53, 3, 364–373, 2019. https://doi.org/10.1177/2168479018779373.
15. Gilbert, F., O'Connell, C.D., Mladenovska, T., Dodds, S., Print me an organ? Ethical and regulatory issues emerging from 3D bioprinting in medicine. *Sci. Eng. Ethics*, 24, 1, 73–91, 2018. https://doi.org/10.1007/s11948-017-9874-6.
16. Hourd, P., Medcalf, N., Segal, J., Williams, D.J., A 3D bioprinting exemplar of the consequences of the regulatory requirements on customized processes. *Regener. Med.*, 10, 863–883, 2015. https://doi.org/10.2217/rme.15.52.
17. Liebschner, M.A., Biomechanical considerations of animal models used in tissue engineering of bone. *Biomaterials*, 25, 9, 1697–1714, 2004. https://doi.org/10.1016/s0142-9612(03)00515-5.
18. Archambault, J., Moreira, A., McDaniel, D., Winter, L., Sun, L., Hornsby, P., Therapeutic potential of mesenchymal stromal cells for hypoxic ischemic encephalopathy: A systematic review and meta-analysis of preclinical studies. *PLoS One*, 12, 12, e0189895, 2017. https://doi.org/10.1371%2Fjournal.pone.0189895.
19. Ioannidis, J.P.A., Kim, B.Y.S., Trounson, A., Design preclinical studies for reproducibility. *Nat. Biomed. Eng.*, 2, 789–790, 2018. https://doi.org/10.1038/s41551-018-0322-y.
20. Epstein, S.E., Luger, D., Lipinski, M.J., Large animal model efficacy testing Is needed prior to launch of a stem cell clinical trial: An evidence-Lacking conclusion based on conjecture. *Circ. Res.*, 121, 5, 496–498, 2017. https://doi.org/10.1161/circresaha.117.311562.
21. Liguori, G.R., Jeronimus, B.F., de Aquinas Liguori, T.T., Moreira, L.F.P., Harmsen, M.C., Ethical issues in the Use of animal models for tissue engineering: Reflections on legal aspects, moral theory. three Rs strategies, and harm–Benefit analysis. *Tissue Eng. Part C Methods*, 23, 12, 850–862, 2017. https://doi.org/10.1089/ten.tec.2017.0189.

22. Baghbaderani, B.A., Tian, X., Neo, B.H., Burkall, A., Dimezzo, T., Sierra, G., Zeng, X., Warren, K., Kovarcik, D.P., Fellner, T., Rao, M.S., cGMP-Manufactured Human Induced Pluripotent Stem Cells Are Available for Pre-clinical and Clinical Applications. *Stem Cell Rep.*, 5, 4, 647–59, 2015. doi: 10.1016/j.stemcr.2015.08.015.
23. Wallenborn, M., Petters, O., Rudolf, D., Hantmann, H., Richter, M., Ahnert, P., Rohani, L., Smirk, J.J., Bulwin, G.C., Krupp, W., Schulz, R.M., Holland, H., , Comprehensive high-resolution genomic profiling and cytogenetics of human chondrocyte cultures by GTG-banding SKY and SNP array. *Eur. Cell Mater.*, 35, 225–241, 2018. https://doi.org/10.22203/ecm.v035a16.
24. Barkholt, L., Flory, E., Jekerle, V., Lucas-Samuel, S., Ahnert, P., Bisset, L.D., Fibbe, W., Foussat, A., Kwa, M., Lantz, O., Maciulaitis, R., Polamaki, T., Schneider, C.K., Sensebe, L., Tachdjian, G., Tarte, K., Tosca, L., Salmikangas, P., Risk of tumorigenicity in mesenchymal stromal cell–based therapies—bridging scientific observations and regulatory viewpoints. *Cytotherapy*, 15, 753–759, 2013. https://doi.org/10.1016/j.jcyt.2013.03.005.
25. Zakrzewski, W., Dobrzyński, M., Szymonowicz, M., Rybak, Z., Stem cells: past, present, and future. *Stem Cell Res. Ther.*, 10, 68, 2019. https://doi.org/10.1186/s13287-019-1165-5.
26. Zhao, X. and Moore, D.L., Neural stem cells: developmental mechanisms and disease modeling. *Cell Tissue Res.*, 371, 1, 1–6, 2018. doi: 10.1007/s00441-017-2738-1.
27. Vasanthan, J., Gurusamy, N., Rajasingh, S., Sigamani, V., Kirankumar, S., Thomas, E.L., Rajasingh, J., Role of Human Mesenchymal Stem Cells in Regenerative Therapy. *Cells*, 10, 1, 54, 2020. doi: 10.3390/cells10010054.
28. Pinho, S. and Frenette, P.S., Haematopoietic stem cell activity and interactions with the niche. *Nat. Rev. Mol. Cell Biol.*, 20, 303–320, 2019. https://doi.org/10.1038/s41580-019-0103-9.
29. Salybekov, A.A., Kobayashi, S., Asahara, T., Characterization of Endothelial Progenitor Cell: Past, Present, and Future. *Int. J. Mol. Sci.*, 23, 14, 7697, 2022. doi: 10.3390/ijms23147697.
30. Bisht, S., Nigam, M., Kunjwal, S.S., Sergey, P., Mishra, A.P., Sharifi-Rad, J., Cancer Stem Cells: From an Insight into the Basics to Recent Advances and Therapeutic Targeting. *Stem Cells Int.*, 2022, 9653244, 2022. doi: 10.1155/2022/9653244.
31. Liesveld, J.L., Sharma, N., Aljitawi, O.S., Stem cell homing: From physiology to therapeutics. *Stem Cells*, 38, 10, 1241–1253, 2020. https://doi.org/10.1002/stem.3242.
32. Pattabiraman, D. and Weinberg, R.A., Tackling the cancer stem cells—What challenges do they pose? *Nature Rev. Drug Discov.*, 13, 497–512, 2014. doi: 10.1038/nrd4253.

33. Uchibori, R., Tsukahara, T., Mizuguchi, H., Saga., Y., Urabe, M., Mizukami, H., Kume, A., Ozawa, K., NF-κB Activity Regulates Mesenchymal Stem Cell Accumulation at Tumor Sites. *Cancer Res.*, 73, 364–372, 2012. doi: 10.1158/0008-5472.CAN-12-0088.
34. Vakhshiteh, F., Atyabi, F., Ostad, S.N., Mesenchymal stem cell exosomes: A two-edged sword in cancer therapy. *Int. J. Nanomed.*, 14, 2847–2859, 2019. doi: 10.2147/IJN.S200036.
35. Lee, R.H., Oh, J.Y., Choi, H., Bazhanov, N., Therapeutic factors secreted by mesenchymal stromal cells and tissue repair. *J. Cell. Biochem.*, 112, 3073–3078, 2011. doi: 10.1002/jcb.23250.
36. Reddy, R.L., Mobilization and collection of peripheral blood progenitor cells for transplantation. *Transfus. Apheresis Sci.*, 32, 63–72, 2005. doi: 10.1016/j.transci.2004.10.007.
37. Batlle, E. and Clevers, H., Cancer stem cells revisited. *Nat. Med.*, 23, 1124–1134, 2017. doi: 10.1038/nm.4409.
38. Codd, A.S., Kanaseki., T., Torigo., T., Tabi, Z., Cancer stem cells as targets for immunotherapy. *Immunology*, 153, 304–314, 2017. doi: 10.1111/imm.12866.
39. Casper, J., Wolff, D., Knauf, W., Blau., I.W., Ruutu., T., Volin., L., Wandt., H., Schäfer-Eckart, K., Holowiecki., J., Giebel, S., *et al.*, Allogeneic Hematopoietic Stem-Cell Transplantation in Patients With Hematologic Malignancies After Dose-Escalated Treosulfan/Fludarabine Conditioning. *J. Clin. Oncol.*, 28, 3344–3351, 2010. doi: 10.1200/JCO.2009.23.3429.
40. Aboody, K.S., Najbauer, J., Metz, M.Z., D'Apuzzo, M., Gutova, M., Annala, A.J., Synold, T.W., Couture, L.A., Blanchard, S., Moats, R.A., Garcia, E., Aramburo, S., Valenzuela, V.V., Frank, R.T., Barish, M.E., Brown, C.E., Kim, S.U., Badie, B., Portnow, J., Neural stem cell-mediated enzyme/prodrug therapy for glioma: preclinical studies. *Sci. Transl. Med.*, 5, 184, 159r–184r, 2013. https://doi.org/10.1126%2Fscitranslmed.3005365.
41. Zhao, Y., Lam, D.H., Yang, J., Lin, J., Tham, C.K., Ng, W.H., Wang, S., Targeted suicide gene therapy for glioma using human embryonic stem cell-derived neural stem cells genetically modified by baculoviral vectors. *Gene Ther.*, 19, 189–200, 2012. https://doi.org/10.1038/gt.2011.82.
42. Kauer, T.M., Figueiredo, J.L., Hingtgen, S., Shah, K., Encapsulated therapeutic stem cells implanted in the tumor resection cavity induce cell death in gliomas. *Nat. Neurosci.*, 15, 197–204, 2011. https://doi.org/10.1038/nn.3019.
43. Ling, X., Marini, F., Konopleva, M., Schober, W., Shi, Y., Burks, J., Clise-Dwyer, K., Wang, R.Y., Zhang, W., Yuan, X., Lu, H., Caldwell, L., Andreeff, M., Mesenchymal Stem Cells Overexpressing IFN-beta Inhibit Breast Cancer Growth and Metastases through Stat3 Signaling in a Syngeneic Tumor Model. *Cancer Microenvironment*, 3, 1, 83–95, 2010. https://doi.org/10.1007/s12307-010-0041-8.

44. Tobias, A.L., Thaci, B., Auffinger, B., Rincón, E., Balyasnikova, I.V., Kim, C.K., Han, Y., Zhang, L., Aboody, K.S., Ahmed, A.U., Lesniak, M.S., The timing of neural stem cell-based virotherapy is critical for optimal therapeutic efficacy when applied with radiation and chemotherapy for the treatment of glioblastoma. *Stem Cells Transl. Med.*, 2, 2013, 655–666, 2013. 10.5966/sctm.2013-0039.
45. Mader, E.K., Butler, G.W., Dowdy, S.C., Mariani, A., Knutson, K.L., Federspiel, M.J., Russell, S.J., Galanis, E., Dietz, A.B., Peng, K.-W., Optimizing patient derived mesenchymal stem cells as virus carriers for a Phase I clinical trial in ovarian cancer. *J. Transl. Med.*, 11, 20, 2013. doi: 10.1186/1479-5876-11-20.
46. Auffinger, B., Morshed, R., Tobias, A., Cheng, Y., Ahmed, A.U., Lesniak, M.S., Drug-loaded nanoparticle systems and adult stem cells: a potential marriage for the treatment of malignant glioma? *Oncotarget*, 4, 378–396, 2013. https://doi.org/10.18632%2Foncotarget.937.
47. Roger, M., Clavreul, A., Venier-Julienne, M.C., Passirani, C., Sindji, L., Schiller, P., Montero-Menei, C., Menei, P., Mesenchymal stem cells as cellular vehicles for delivery of nanoparticles to brain tumors. *Biomaterials*, 31, 8393–8401, 2010. https://doi.org/10.1016/j.biomaterials.2010.07.048.
48. Katakowski, M., Buller, B., Zheng, X., Lu, Y., Rogers, T., Osobamiro, O., Shu, W., Jiang, F., Chopp, M., Exosomes from marrow stromal cells expressing miR-146b inhibit glioma growth. *Cancer Lett.*, 335, 201–204, 2013. doi: 10.1016/j.canlet.2013.02.019.
49. Greco., K.A., Franzen, C., Foreman., K.E., Flanigan, R.C., Kuo, P.C., Gupta, G., PLK-1 Silencing in Bladder Cancer by siRNA Delivered With Exosomes. *Urology*, 91, 241.e1–241.e7, 2016. doi: 10.1016/j.urology.2016.01.028.
50. Zheng, L., Zhang, D., Chen, X., Yang, L., Wei, Y., Zhao, X., Antitumor activities of human placenta-derived mesenchymal stem cells expressing endostatin on ovarian cancer. *PLoS One*, 7, 7, e39, 2012. https://doi.org/10.1371/journal.pone.0039119.
51. Saman, H., Raza, S.S., Uddin, S., Rasul, K., Inducing Angiogenesis, a Key Step in Cancer Vascularization, and Treatment Approaches. *Cancers (Basel)*, 12, 5, 1172, 2020. doi: 10.3390/cancers12051172.
52. Aravindhan, S., Ejam, S.S., Lafta, M.H., Markov, A., Yumashev, A.V., Ahmadi, M., Mesenchymal stem cells and cancer therapy: insights into targeting the tumour vasculature. *Cancer Cell Int.*, 21, 1, 158, 2021. doi: 10.1186/s12935-021-01836-9.
53. Martin, J.D., Seano, G., Jain, R.K., Normalizing Function of Tumor Vessels: Progress, Opportunities, and Challenges. *Annu. Rev. Physiol.*, 81, 505–534, 2019. doi: 10.1146/annurev-physiol-020518-114700.
54. Mansouri, V., Beheshtizadeh, N., Gharibshahian, M., Sabouri, L., Varzandeh, M., Rezaei, N., Recent advances in regenerative medicine strategies for cancer treatment. *Biomed. Pharmacother.*, 141, 111875, 2021. https://doi.org/10.1016/j.biopha.2021.111875.

55. Cerneckis, J., Cai, H., Shi, Y., Induced pluripotent stem cells (iPSCs): molecular mechanisms of induction and applications. *Signal Transduction Targeted Ther.*, 9, 112, 2024. https://doi.org/10.1038/s41392-024-01809-0.
56. Gschweng, E., De Oliveira, S., Kohn, D.B., Hematopoietic stem cells for cancer immunotherapy. *Immunol. Rev.*, 257, 1, 237–249, 2014. doi: 10.1111/imr.12128.
57. Netsrithong, R. and Wattanapanitch, M., Advances in Adoptive Cell Therapy Using Induced Pluripotent Stem Cell-Derived T Cells. *Front. Immunol.*, 12, 759558, 2021. doi: 10.3389/fimmu.2021.759558.
58. Chehelgerdi, M., Behdarvand Dehkordi, F., Chehelgerdi, M., Kabiri, H., Salehian-Dehkordi, H., Abdolvand, M., Salmanizadeh, S., Rashidi, M., Niazmand, A., Ahmadi, S., Feizbakhshan, S., Kabiri, S., Vatandoost, N., Ranjbarnejad, T., Exploring the promising potential of induced pluripotent stem cells in cancer research and therapy. *Mol. Cancer*, 22, 1, 189, 2023. doi: 10.1186/s12943-023-01873-0.
59. Sun, H.R., Wang, S., Yan, S.C., Zhang, Y., Nelson, P.J., Jia, H.L., Qin, L.X., Dong, Q.Z., Therapeutic Strategies Targeting Cancer Stem Cells and Their Microenvironment. *Front. Oncol.*, 9, 1104, 2019. doi: 10.3389/fonc.2019.01104.
60. Bryukhovetskiy, I., Mischenko, P.V., Tolok, E.V., Khotimchenko, Y., Xaitsev, S., Bryukhovetskiy, A.S., Hematopoietic stem cells with induced apoptosis effectively inhibit glioma cell growth *in vitro*, but started new mechanism of tumor stem cells. *Genes Cells*, 9, 4, 90–95, 2016.
61. Madjd, Z., Gheytanchi, E., Erfani, E., M. Asadi-Lari, M., Application of Stem Cells in Targeted Therapy of Breast Cancer: A Systematic Review. *Asian Pac. J. Cancer Prev.*, 14, 5, 2789–2800, 2013. DOI:http://dx.doi.org/10.7314/APJCP.2013.14.5.2789.
62. Borzone, F.R., Giorello, M.B., Martinez, L.M., Sanmartin, M.C., Feldman, L., Dimase, F., Batagelj, E., Yannarelli, G., Chasseing, N.A., Senescent mesenchymal stem/stromal cells in pre-metastatic bone marrow of untreated advanced breast cancer patients. *Oncol. Res.*, 31, 3, 361–374, 2023. doi: 10.32604/or.2023.028104.
63. Aldoghachi, A., Chong, Z.X., Yeap, S.K., Cheong, S.K., Ho, W.Y., Ong, A.H.K., Stem Cells for Cancer Therapy: Translating the Uncertainties and Possibilities of Stem Cell Properties into Opportunities for Effective Cancer Therapy. *Int. J. Mol. Sci.*, 24, 2, 1012, 2023. https://doi.org/10.3390/ijms24021012.
64. Hoang, D.M., Pham, P.T., Bach, T.Q., Ngo, A.T.L., Nguyen, Q.T., Phan, T.T.K., Nguyen, G.H., Le, P.T.T., Hoang, V.T., Forsyth, N.R., Heke, M., Nguyen, L.T., Stem cell-based therapy for human diseases. *Signal Transduction Targeted Ther.*, 7, 272, 2022. https://doi.org/10.1038/s41392-022-01134-4.
65. Hermanson, D.L., Bendzick, L., Pribyl, L., McCullar, V., Vogel, R.I., Miller, J.S., Geller, M.A., Kaufman, D.S., Induced Pluripotent Stem Cell-Derived Natural Killer Cells for Treatment of Ovarian Cancer. *Stem Cells*, 34, 1, 93–101, 2016. https://doi.org/10.1002/stem.2230.

66. Joo, K.M., Park, I.H., Shin, J.Y., Jin, J., Kang, B.G., Kim, M.H., Lee, S.J., Jo, M.-Y., Kim, S.U., Nam, D.-H., Human Neural Stem Cells Can Target and Deliver Therapeutic Genes to Breast Cancer Brain Metastases. *Mol. Ther.*, 17, 3, 570–575, 2009. https://doi.org/10.1038/mt.2008.290.
67. Flamini, V., Jiang, W.G., Lane, J., Cui, Y.-X., Significance and therapeutic implications of endothelial progenitor cells in angiogenic-mediated tumour metastasis. *Crit. Rev. Incology/Hematology*, 100, 177–189, 2016. https://doi.org/10.1016/j.critrevonc.2016.02.010.
68. Miska, J. and Lesniak, M.S., Neural Stem Cell Carriers for the Treatment of Glioblastoma Multiforme. *EBioMedicine*, 2, 774–775, 2015. https://doi.org/10.1016%2Fj.ebiom.2015.08.022.
69. Sharkis, S.J., Jones, R.J., Civin, C., Jang, Y.Y., Pluripotent stem cell-based cancer therapy: promise and challenges. *Sci. Transl. Med.*, 4, 127–129, 2012. https://doi.org/10.1126/scitranslmed.3003920.
70. Balyasnikova, I.V., Prasol, M.S., Ferguson, S.D., Han, Y., Ahmed, A.U., Gutova, M., Tobias, A.L., Mustafi, D., Rincon, E., Zhang, L., Aboody, K.S., Lesniak, M.S., Intranasal delivery of mesenchymal stem cells significantly extends survival of irradiated mice with experimental brain tumors. *Mol. Ther.*, 22, 140–148, 2014. https://doi.org/10.1038/mt.2013.199.
71. Hansen, K., Muller, F.J., Messing, M., Zeigler, F., Loring, J.F., Lamszus, K., Westphal, M., Schmidt, N.O., A 3-dimensional extracellular matrix as a delivery system for the transplantation of glioma-targeting neural stem/progenitor cells. *Neuro Oncol.*, 12, 645–654, 2010. https://doi.org/10.1093/neuonc/noq002.
72. Golubeva, V., Mikhalevich, J., Novikova, J., Tupizina, O., Trofimova, S., Zueva, Y., Novel cell population data from a haematology analyzer can predict timing and efficiency of stem cell transplantation. *Transfus. Apheresis Sci.*, 50, 39–45, 2014. https://doi.org/10.1016/j.transci.2013.12.004.
73. Ahmed, A.U., Thaci, B., Alexiades, N.G., Han, Y., Qian, S., Liu, F., Balyasnikova, I.V., Ulasov, I.Y., Aboody, K.S., Lesniak, M.S., Neural stem cell-based cell carriers enhance therapeutic efficacy of an oncolytic adenovirus in an orthotopic mouse model of human glioblastoma. *Mol. Ther.*, 19, 1714–1726, 2011. https://doi.org/10.1038/mt.2011.100.
74. Duy, C., Li., M., Teater., M., Meydan., C., Garrett-Bakelman., F.E., Lee, T.C., Chin, C.R., Durmaz, C., Kawabata, K.C., Dhimolea, E., Chemotherapy induces senescence-like resilient cells capable of initiating AML recurrence senescence-like cells contribute to relapse of AML. *Cancer Discovery*, 11, 6, 1542–1561, 2021. https://doi.org/10.1158/2159-8290.CD-20-1375.
75. Park, S.-Y. and Nam, J.-S., The force awakens: metastatic dormant cancer cells. *Exp. Mol. Med.*, 52, 4, 569–581, 2020. https://doi.org/10.1038/s12276-020-0423-z.

76. Sabisz, M. and Skladanowski, A., Cancer stem cells and escape from drug-induced premature senescence in human lung tumor cells: implications for drug resistance and *in vitro* drug screening models. *Cell Cycle*, 8, 19, 3208–3217, 2009. https://doi.org/10.4161/cc.8.19.9758, 2-s2.0-70349912085.
77. Yeung, T.M., Gandhi, S.C., Bodmer, W.F., Hypoxia and lineage specification of cell line-derived colorectal cancer stem cells. *Proc. Natl. Acad. Sci.*, 108, 11, 4382–4387, 2011. https://doi.org/10.1073/pnas.1014519107, 2-s2.0-79952717492.
78. Ding, K., Banerjee, A., Tan, S., Zhao, J.S., Zhuang, Q., Li, R., Qian, P., Liu, S., Wu, Z.-S., Lobie, P.E., Zhu, T., Artemin, a member of the glial cell line-derived neurotrophic factor family of ligands, is HER2-regulated and mediates acquired trastuzumab resistance by promoting cancer stem cell-like behavior in mammary carcinoma cells. *J. Biol. Chem.*, 289, 23, 16057–16071, 2014. https://doi.org/10.1074/jbc.M113.529552, 2-s2.0-84902182083.
79. Alnasser, S.M., Advances and Challenges in Cancer Stem Cells for Onco-Therapeutics. *Stem Cells Int.*, 2023, 8722803, 2023. doi: 10.1155/2023/8722803.
80. Yoshida, G.J., Applications of patient-derived tumor xenograft models and tumor organoids. *J. Hematol. Oncol.*, 13, 1, 4, 2020. doi: 10.1186/s13045-019-0829-z.
81. Liu, Y., Wu., W., Cai, C., Zhang, H., Shen, H., Han, Y., Patient-derived xenograft models in cancer therapy: technologies and applications. *Signal Transduction Targeted Ther.*, 8, 160, 2023. https://doi.org/10.1038/s41392-023-01419-2.
82. Cacciamali, A., Villa, R., Dotti, S., 3D Cell Cultures: Evolution of an Ancient Tool for New Applications. *Front. Physiol.*, 13, 836480, 2022. doi: 10.3389/fphys.2022.836480. PMID: 35936888; PMCID: PMC9353320.
83. Wuputra, K., Ku, C.C., Kato, K., Wu, D.C., Saito, S., Yokoyama, K.K., Translational models of 3-D organoids and cancer stem cells in gastric cancer research. *Stem Cell Res. Ther.*, 12, 1, 492, 2021. doi: 10.1186/s13287-021-02521-4.
84. Papaccio, F., García-Mico, B., Gimeno-Valiente, F., Cabeza-Segura, M., Gambardella, V., Gutiérrez-Bravo, M.F., Alfaro-Cervelló, C., Martinez-Ciarpaglini, C., Rentero-Garrido, P., Zúñiga-Trejos, S., Carbonell-Asins, J.A., Fleitas, T., Roselló, S., Huerta, M., Sánchez Del Pino, M.M., Sabater, L., Roda, D., Tarazona, N., Cervantes, A., Castillo, J., Proteotranscriptomic analysis of advanced colorectal cancer patient derived organoids for drug sensitivity prediction. *J. Exp. Clin. Cancer Res.*, 42, 1, 8, 2023. doi: 10.1186/s13046-022-02591-z. PMID: 36604765; PMCID: PMC9817273.
85. Menezes, M.R., Das, S.K., Emdad, L., Windle, J.J., Wang, X.-Y., Sarkar, D., Fisher, P.B., Chapter Eight - Genetically Engineered Mice as Experimental Tools to Dissect the Critical Events in Breast Cancer. *Adv. Cancer Res.*, 121, 331–382, 2014. https://doi.org/10.1016/B978-0-12-800249-0.00008-1.

86. Pinto, B., Henriques, A.C., Silva, P.M.A., Bousbaa, H., Three-Dimensional Spheroids as *In Vitro* Preclinical Models for Cancer Research. *Pharmaceutics*, 2, 12, 1186, 2020. doi: 10.3390/pharmaceutics12121186.
87. Ishiguro, T., Ohata, H., Sato, A., Yamawaki, K., Enomoto, T., Okamoto, K., Tumor-derived spheroids: Relevance to cancer stem cells and clinical applications. *Cancer Sci.*, 108, 3, 283–289, 2017. doi: 10.1111/cas.13155.
88. Goodspeed, A., Heiser, L.M., Gray, J.W., Costello, J.C., Tumor-Derived Cell Lines as Molecular Models of Cancer Pharmacogenomics. *Mol. Cancer Res.*, 14, 1, 3–13, 2016. doi: 10.1158/1541-7786.MCR-15-0189.
89. Qin, S., Jiang, J., Lu, Y., Nice, E.C., Huang, C., Zhang, J., He, W., Emerging role of tumor cell plasticity in modifying therapeutic response. *Signal Transduction Targeted Ther.*, 5, 1, 228, 2020. doi: 10.1038/s41392-020-00313-5.
90. Walcher, L., Kistenmacher, A.-K., Suo, H., Kitte, R., Dluczek, S., Straub, A., Blaudszun, A.-R., Yevsa, T., Fricke, S., Kossatz-Boehlert, U., Cancer Stem Cells—Origins and Biomarkers: Perspectives for Targeted Personalized Therapies. *Front. Immunol.*, 11, 1280, 2020. https://doi.org/10.3389/fimmu.2020.01280.
91. Manhas, J., Bhattacharya, A., Agrawal, S.K., Gupta, B., Das, P., Deo, S.V., Pal, S., Sen, S., Characterization of cancer stem cells from different grades of human colorectal cancer. *Tumour Biol.*, 37, 10, 14069–14081, 2016. doi: 10.1007/s13277-016-5232-6.
92. Abetov, D., Mustapova, Z., Saliev, T., Bulian, D., Biomarkers and signaling pathways of colorectal cancer stem cells. *Tumor Biol.*, 36, 1339–1353, 2015. https://doi.org/10.1007/s13277-015-3198-4.
93. Vassilopoulos, A., Chisholm, C., Lahusen, T., Zheng, H., Deng, C.X., A critical role of CD29 and CD49f in mediating metastasis for cancer-initiating cells isolated from a Brca1-associated mouse model of breast cancer. *Oncogene*, 33, 5477–5482, 2014. https://doi.org/10.1038/onc.2013.516.
94. Mimeault, M. and Batra, S.K., Molecular Biomarkers of Cancer Stem/Progenitor Cells Associated with Progression, Metastases, and Treatment Resistance of Aggressive Cancers. *Cancer Epidemiology Biomarkers Prev.*, 23, 2, 234–254, 2014. https://doi.org/10.1158/1055-9965.EPI-13-0785.
95. Curtarelli, R.B., Gonçalves, J.M., dos Santos, L.G.P., Savi, M.G., Nör, J.E., Mezzomo, L.A.M., Cordeiro, M.M.R., Expression of Cancer Stem Cell Biomarkers in Human Head and Neck Carcinomas: a Systematic Review. *Stem Cell Rev. Rep.*, 14, 769–784, 2018. https://doi.org/10.1007/s12015-018-9839-4.
96. Lasorsa, F., Rutigliano, M., Milella, M., Ferro, M., Pandolfo, S.D., Crocetto, F., Autorino, R., Battaglia, M., Ditonno, P., Lucarelli, G., Cancer Stem Cells in Renal Cell Carcinoma: Origins and Biomarkers. *Int. J. Mol. Sci.*, 249, 17, 13179, 2023. https://doi.org/10.3390/ijms241713179.
97. Corró, C. and Moch, H., Biomarker discovery for renal cancer stem cells. *J. Pathology Clin. Res.*, 4, 1, 3–18, 2017. https://doi.org/10.1002/cjp2.91.

98. Romano, M., De Francesco, F., Pirozzi, G., Gringeri, E., Boetto, R., Di Domenico, M., Zavan, B., Ferraro, G.A., Cillo, U., Expression of cancer stem cell biomarkers as a tool for a correct therapeutic approach to hepatocellular carcinoma. *Oncoscience*, 2, 5, 443–456, 2015. doi: 10.18632/oncoscience.163.
99. Muinao, T., Boruah, H.P.D., Pal, M., Diagnostic and Prognostic Biomarkers in ovarian cancer and the potential roles of cancer stem cells – An updated review. *Exp. Cell. Res.*, 362, 1, 1–10, 2018. https://doi.org/10.1016/j.yexcr.2017.10.018.
100. Sudhalkar, N., Rathod, N.P., Mathews, A., Chopra, S., Sriram, H., Shrivastava, S.K., Goda, J.S., Potential role of cancer stem cells as biomarkers and therapeutic targets in cervical cancer. *Cancer Rep.*, 2, c1144, 2018. Doi: 10.1002/cnr2.1144.
101. Chu, D.T., Nguyen, T.T., Tien, N.L.B., Tran, D.K., Jeong, J.H., Anh, P.G., Thanh, V.V., Truong, D.T., Dinh, T.C., Recent Progress of Stem Cell Therapy in Cancer Treatment: Molecular Mechanisms and Potential Applications. *Cells*, 9, 3, 563, 2020. doi: 10.3390/cells9030563.
102. Katsukawa, M., Nakajima, Y., Fukumoto, A., Doi, D., Takahashi, J., Katsukawa, M.M., Fail-Safe Therapy by Gamma-Ray Irradiation Against Tumor Formation by Human-Induced Pluripotent Stem Cell-Derived Neural Progenitors. *Stem Cells Dev.*, 25, 815–825, 2016. doi: 10.1089/scd.2015.0394.
103. Albarenque, S.M., Zwacka, R., Mohr, A., Both human and mouse mesenchymal stem cells promote breast cancer metastasis. *Stem Cell Res.*, 7, 163–171, 2011. doi: 10.1016/j.scr.2011.05.002.
104. Mohty, B. and Mohty, M., Long-term complications and side effects after allogeneic hematopoietic stem cell transplantation: An update. *Blood Cancer J.*, 1, e16, 2011. doi: 10.1038/bcj.2011.14.
105. Lohan, P., Treacy, O., Griffin, M.D., Ritter, T., Ryan, A., Anti-Donor Immune Responses Elicited by Allogeneic Mesenchymal Stem Cells and Their Extracellular Vesicles: Are We Still Learning? *Front. Immunol.*, 8, 1626, 2017. doi: 10.3389/fimmu.2017.01626.
106. Kooreman, N.G., Kim, Y., De Almeida, P., Termglinchan, V., Diecke, S., Shao, N., Wei, T., Yi, H., Dey, D., Nelakanti, R., Brouwer, T.P., Paik, D.T., Sagiv-Barfi, I., Han, A., Quax, P.H.A., Hamming, J.F., Levy, R., Davis, M.M., Wu, J.C., Autologous iPSC-Based Vaccines Elicit Anti-tumor Responses *In Vivo. Cell Stem Cell*, 22, 501–513.e7, 2018. doi: 10.1016/j.stem.2018.01.016.
107. Nayerossadat, N., Maedeh, T., Ali, P.A., Viral and nonviral delivery systems for gene delivery. *Adv. Biomed. Res.*, 1, 27, 2012. doi: 10.4103/2277-9175.98152.
108. Yi, B.R., Park, M.A., Lee, H.R., Kang, N.H., Choi, K.J., Kim, S.U., Choi, K.C., Suppression of the growth of human colorectal cancer cells by therapeutic stem cells expressing cytosine deaminase and interferon-beta via their tumor-tropic effect in cellular and xenograft mouse models. *Mol. Oncol.*, 7, 543–554, 2013. https://doi.org/10.1016%2Fj.molonc.2013.01.004.

109. Kim, S.M., Woo, J.S., Jeong, C.H., Ryu, C.H., Jang, J.D., Jeun, S.S., Potential application of temozolomide in mesenchymal stem cell-based TRAIL gene therapy against malignant glioma. *Stem Cells Transl. Med.*, 3, 2, 172–182, 2014. https://doi.org/10.5966/sctm.2013-0132.
110. Karnoub, A.E., Dash, A.B., Vo, A.P., Sullivan, A., Brooks, M.W., Bell, G.W., Richardson, A.L., Polyak, K., Tubo, R., Weinberg, R.A., Mesenchymal stem cells within tumour stroma promote breast cancer metastasis. *Nature*, 449, 557–563, 2007. https://doi.org/10.1038/nature06188.
111. Xie, M., Zhang, L., He., C.S., Xu, F., Liu, J.L., Hu, Z.H., Zhao, L.P., Tian, Y., Activation of Notch-1 enhances epithelial–mesenchymal transition in gefitinib-acquired resistant lung cancer cells. *J. Cell. Biochem.*, 113, 5, 1501–1513, 2012. https://doi.org/10.1002/jcb.24019.
112. Fucha, B.C., Fujii, T., Dorfman, J.D., Goodwin, J.M., Zhu, A.X., Lanuti, M., Tanabe, K.K., Epithelial-to-mesenchymal transition and integrin-linked kinase mediate sensitivity to epidermal growth factor receptor inhibition in human hepatoma cells. *Cancer Res.*, 68, 7, 2391–2399, 2008. https://doi.org/10.1158/0008-5472.CAN-07-2460, 2-s2.0-42049116522, 18381447.
113. Hirschmann-Jax, C., Foster, A.E., Wulf, G.G., Nuchtern, J.G., Jax, T.W., Gobel, U., Goodell, M.A., Brenner, M.K., A distinct "side population" of cells with high drug efflux capacity in human tumor cells. *PNAS*, 101, 39, 14228–14233, 2004. https://doi.org/10.1073/pnas.0400067101, 2-s2.0-4644306516, 15381773.
114. Ajani, J.A., Wang, X., Song, S., Suzuki, A., Taketa, T., Sudo, K., Wadhwa, R., Hofstetter, W.L., Komaki, R., Maru, D.M., Lee, J.H., Bhutani, M.S., Weston, B., Baladandayuthapani, V., Yao, Y., Honjo, S., Scott, A.W., Skinner, H.D., Johnson, R.L., Berry, D., ALDH-1 expression levels predict response or resistance to preoperative chemoradiation in resectable esophageal cancer patients. *Mol. Oncol.*, 8, 1, 142–149, 2014. https://doi.org/10.1016/j.molonc.2013.10.007, 2-s2.0-84893006227, 24210755.
115. Krishnamurthy, S., Warner, K.A., Dong, Z., Imai, A., Nör, C., Ward, B.B., Helman, J.I., Taichman, R.S., Bellile, E.L., McCauley, L.K., Polverini, P.J., Prince, M.E., Wicha, M.S., Nör, J.E., Endothelial interleukin-6 defines the tumorigenic potential of primary human cancer stem cells. *Stem Cells*, 32, 11, 2845–2857, 2014. https://doi.org/10.1002/stem.1793, 2-s2.0-84907887777, 25078284.
116. Pires, B.R.B., De Amorim, ISS., Souza., L.D.E., Rodrigues, J.A., Mencalha, A.L., Targeting cellular signaling pathways in breast cancer stem cells and its implication for cancer treatment. *Anticancer Res.*, 36, 11, 5681–5692, 2016. https://doi.org/10.21873/anticanres.11151, 2-s2.0-84993965457, 27793889.
117. Bao, B., Azmi, A.S., Ali, S., Ahmad, A., Li, Y., Banerjee, S., Kong, D., Sarkar, F.H., The biological kinship of hypoxia with CSC and EMT and their relationship with deregulated expression of miRNAs and tumor aggressiveness. *Biochim. Biophys. Acta (BBA) - Rev. Cancer*, 1826, 2, 272–296, 2012. https://doi.org/10.1016/j.bbcan.2012.04.008, 2-s2.0-84861828232, 22579961.

118. Phi, L.T.H., Sari, I.N., Yang, Y.-G., Lee, S.-H., Jun, N., Kim, K.S., Lee, Y.K., Kwon, H.Y., Cancer Stem Cells (CSCs) in Drug Resistance and their Therapeutic Implications in Cancer Treatment. *Stem Cells Int.*, 2018, 1, 5416923, 2018. https://doi.org/10.1155/2018/5416923.
119. Aboody, K., Capela, A., Niazi, N., Stern, J.H., Temple, S., Translating stem cell studies to the clinic for CNS repair: current state of the art and the need for a Rosetta stone. *Neuron*, 70, 2011, 597–613, 2011. 10.1016/j.neuron.2011.05.007.
120. Rong, Z., Wang, M., Hu, Z., Stradner, M., Zhu, S., Kong, H., Yi, H., Goldrath, A., yang, Y.-G., Xu, Y., Fu, X., An effective approach to prevent immune rejection of human ESC-derived allografts. *Cell Stem Cell*, 14, 2014, 121–130, 2014. 10.1016/j.stem.2013.11.014.
121. Pulukuri, S.M.K., Gorantla, B., Dasari, V.R., Gondi, C.S., Rao, J.S., Epigenetic upregulation of urokinase plasminogen activator promotes the tropism of mesenchymal stem cells for tumor cells. *Mol. Cancer Res.*, 8, 2010, 1074–1083, 2010. 10.1158/1541-7786.MCR-09-0495.
122. Jiang, X., Wang, C., Fitch, S., Yang, F., Targeting tumor hypoxia using nanoparticle-engineered CXCR4-overexpressing adipose-derived stem cells. *Theranostics*, 8, 2018, 1350–1360, 2018. 10.7150/thno.22736.
123. Garofalo, M., Villa, A., Rizzi, N., Kuryk, L., Mazzaferro, V., Ciana, P., Systemic administration and targeted delivery of immunogenic oncolytic adenovirus encapsulated in extracellular vesicles for cancer therapies. *Viruses*, 10, 2018, 558, 2018. 10.3390/v10100558.
124. Choi, S.H., Tamura, K., Khajuria, R.K., Bhere, D., Nesterenko, I., Lawler, J., Shah, K., Antiangiogenic variant of TSP-1 targets tumor cells in glioblastomas. *Mol. Ther.*, 23, 2015, 235–243, 2015. 10.1038/mt.2014.214.
125. Zhu, Y., Bassoff, N., Reinshagen, C., Bhere, D., Nowicki, M.O., Lawler, S.E., Roux, J., Shah, K., Bi-specific molecule against EGFR and death receptors simultaneously targets proliferation and death pathways in tumors. *Sci. Rep.*, 7, 2017, 2602, 2017. 10.1038/s41598-017-02483-9.

16

Clinical Trials and Triumphs: Navigating the Path from Bench to Bedside

Abdelmonem Siddiq[1], Al-Hassan Soliman Wadan[2], Abdalla Ali[1], Shubham Shrestha[3] and Ejaz Ahmad Khan[4*]

[1]*Faculty of Pharmacy, Mansoura University, Mansoura, Egypt*
[2]*Faculty of Dentistry, Sinai University, Arish, North Sinai, Egypt*
[3]*School of Medicine, Patan Academy of Health Sciences, Lagankhel, Lalitpur, Nepal*
[4]*Department of Public Health, Health Services Academy, Islamabad, Pakistan*

Abstract

Stem cells have emerged as an option for regenerative therapies for multiple diseases. Their mechanism of action involves the regeneration of human parts such as bone, fat, cartilage, tendon, and others. Moreover, studies have shown that stem cells can induce the secretion of multiple pro-regenerative factors that help in the regeneration of injured tissues. Multiple clinical and pre-clinical studies have shown their efficacy in the management of human disorders. We aimed here to summarize the available evidence regarding the efficacy of different types of stem cells in the management of different disorders by elaborating on different conducted clinical trials of all phases. We summarized approved stem cells such as Queencell, Cellgram-AMI, Cartistem, Cupistem, and others used in disease management. Outlining these FDA- and EMA-approved stem cells is essential in identifying the current landscape of stem cell usage with a focus on different challenges facing the research, industrial, and regulatory companies while testing and approving them.

**Corresponding author*: ejaz@hsa.edu.pk; ORCID: https://orcid.org/0000-0002-7072-8035
Abdelmonem Siddiq: ORCID: https://orcid.org/0000-0003-3826-9702
Al-Hassan Soliman Wadan: ORCID: https://orcid.org/0000-0003-2282-4046
Abdalla Ali: ORCID: https://orcid.org/0000-0003-4237-9541
Shubham Shrestha: ORCID: https://orcid.org/0009-0006-3599-1264

Ayush Madan, Sumel Ashique, Deepika Arora and Mantosh Kumar Satapathy (eds.) Stem Cell Therapeutics, (673–704) © 2025 Scrivener Publishing LLC

***Keywords*:** Hematopoietic stem cells, mesenchymal stem cells, induced pluripotent stem cells, Neuronata-R, prochymal, stempeucel

16.1 Introduction

A. Overview of stem cell therapeutics in disease management
Stem cells are at the forefront of regenerative medicine due to their ability to divide and differentiate into various cell types. They have their roles in repairing tissues and organs affected by congenital abnormalities, diseases, and age related changes [1]. Stem cells are building blocks for all tissue and organ systems in our body. Also, they are fundamental in how diseases develop and how tissue heals. Depending on how they can change, stem cells fall into four categories: unipotent, multipotent, pluripotent, and totipotent [2].

When we talk about stem cell therapies, we mean using living human stem cells like embryonic stem cells (ESCs), induced pluripotent stem cells (iPSCs), and adult stem cells for patient themselves (autologous) and donor-derived (allogeneic) treatments [3]. These therapies are especially helpful in transplants since stem cells can change into whatever cell type is needed to fix damaged tissues.

Cell-based therapies that use stem cells open new doors for patients with diseases that cannot be cured otherwise. Mostly, past treatments have just been about managing symptoms instead of curing the problem. This field of regenerative medicine aims to boost the body's own healing abilities by stimulating or replenishing its own stem cell pool. The goal is to get tissue homeostasis again and promote healing [4]. This groundbreaking technology offers substantial potential for creating transformative and potentially curative treatments for some of the most challenging diseases affecting humanity.

B. Importance of clinical trials in translating research to treatments
Studies showing success with stem cell therapies give hope that these methods might help tackle many challenging diseases [5]. In fact, over the past few years, clinical trials for stem cell therapies have hit significant milestones with promising effects on different illnesses [6].

As stem cells can become specific types of cells needed for repair, there is a shiny glimmer of hope for transplantable tissues and organs since the demand far outweighs what we have. Diseases like macular degeneration, strokes, osteoarthritis, neurodegenerative issues, and diabetes have seen benefits from these therapies [7–9]. For example, scientists are getting closer to making healthy heart muscle cells for people with heart disease. One standout case involved treating a person with epidermolysis bullosa where they saw their skin started to heal after getting cultured epidermal stem cells [10].

With private clinics offering unproven stem cell treatments popping up everywhere, a serious concern is raised about how safe these therapies really are. A scary incident happened when fat-derived cells were injected into three patients' eyes for macular degeneration which ended up with them losing their vision [11]. So, as regenerative medicine keeps moving forward, it is super important to do lots of scientific trials to prove these therapies actually work and are safe. We need to bust the myth that some "magic" cell can solve every problem.

Since working with stem cell therapies is complex, researchers have to find safe and stable sources of these cells that can turn into different cell types. So selecting the right type of stem cells for clinical uses is not just important; it is crucial [12].

C. Key milestones and successes in stem cell clinical trials

The journey of stem cell therapy got its foundation after German biologists who noticed that certain special cells could turn into specialized type way back in the late 19th and early 20th centuries [13]. Some first victories came from discovering hematopoietic stem cells and performing successful bone marrow transplants which were game-changers against conditions like aplastic anemia and leukemia. The field really took off after identifying mesenchymal stem cells which these can change into various types like osteoblasts, adipocytes, and chondrocytes [14]. In more recent times, breakthroughs like finding human embryonic stem cells opened new doors for therapy applications. Nowadays, we are seeing this therapy shine as a hopeful option for numerous tough-to-treat diseases like neurological disorders, lung problems, metabolic issues (like diabetes), reproductive disorders, skin burns, and heart issues. These milestones have firmly pushed stem cell therapy as an essential part of regenerative medicine, providing fresh hope for innovative treatments.

D. Role of laboratory studies and animal models in the initiation of clinical trials

Laboratory studies and animal models are absolutely key when launching clinical trials for stem cell treatments. Using animals has been crucial ever since around the 6th century BCE; the practice has helped push forward medical knowledge which includes vaccines and antibiotics and improved our grasp on human illness [15]. These studies usually check if new treatments are safe before testing human patients.

In the realm of stem cell research, tons of studies investigated into animal models, especially those involving companion animals, to explore how well these treatments might work out. They mostly look at whether doing these treatments is feasible or not while checking things like safety levels and ways to administer them properly and biological responses. Those initial animal model tests provide important data guiding how human trials get designed and initiated.

16.2 Current and Ongoing Clinical Trials

Stem cells promise to address unmet medical needs in conditions ranging from neurodegenerative and cardiovascular diseases to cancer, autoimmune disorders, and diabetes. Stem cell research has progressed from the lab to clinical use through important achievements, prompting multiple trials to confirm the safety and effectiveness of stem cell therapies. This section outlines present clinical trials, emphasizing their potential in treating human ailments. Current studies are focusing on neurodegenerative conditions like Parkinson's disease (PD) and amyotrophic lateral sclerosis (ALS), utilizing embryonic and mesenchymal stem cells to ease symptoms and delay disease advancement [16]. Stem cells have been used in conducting robust clinical trials addressing management of post cardiovascular events across many countries. These trails use mesenchymal and induced pluripotent stem cells in repairing the damaged cardiac tissues, mostly occurring in post myocardial infarction cases, and consequently failing hearts. Stem cells also see a promising future in the field of oncology where hematopoietic stem cells transplants and iPSC-derived natural killer (NK) cells are being investigated and potentially tested for improved clinical outcomes of the treatments [17]. This chapter also looks into stem cells' therapeutic application and use in certain resistant antibody mediated disabling diseases such as multiple sclerosis and, very commonly, rheumatoid arthritis (RA). The mesenchymal stem cells, therefore, may play an important role in the management of such diseases through their immunomodulatory properties [18]. This chapter will also discuss potential of stem cell therapy in one of the most prevalent chronic diseases worldwide, i.e., diabetes mellitus (DM), which could be a game changer having clinical trials designed to test alternates to beta cells of pancreas using IPSCs to restoring optimum insulin production [17]. Overall, this chapter will try to give an insight into the development and potentiality of the clinical trials changing the landscape of medical treatment in the near future [19].

Neurodegenerative Diseases

Progressive degeneration of the nerve cells leads to disability among the populations. The neurodegenerative diseases are resistant to complete cure due to lack of medical treatment, which could result in the regeneration of the nerve cells. However, a few recent and the latest clinical trials have been promising in finding a role of stem cell therapies in possible definite treatment of the neurodegenerative diseases [20, 21].

Parkinson's Disease (PD)

One of the most prevalent neurological disorders is Parkinson's disease (PD). Its etiology lies in gradual decline in dopamine production by the dopamine-producing neurons of brain cells. During 2020, a phase I/II trial used human embryonic stem cell (hESC)-derived dopamine neurons in patients with Parkinson's disease. The trial evaluated the safety, tolerability, and probable efficacy of the treatment. The results were encouraging,

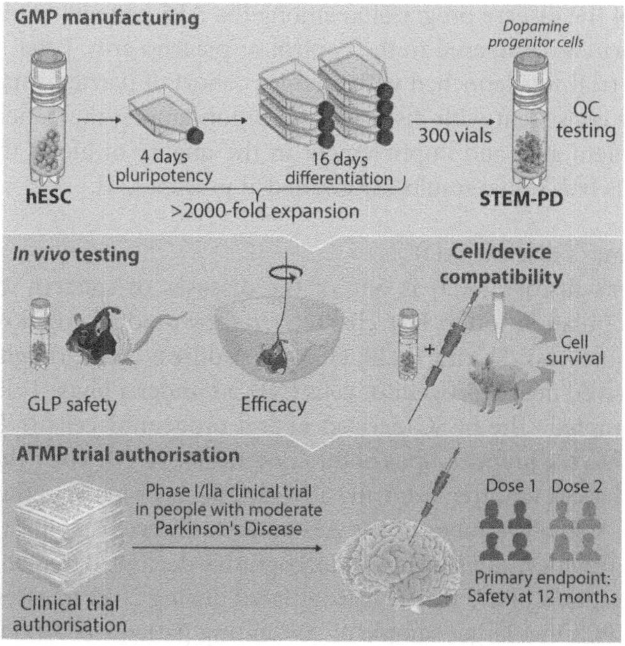

Figure 16.1 Process of stem cell technology used in the treatment of PD (adapted from Kirkeby *et al.* 2023).

showing the survival of the transplanted cells which were later engrafted and produced optimum levels of dopamine without any observable adverse effects in a considerable number of trial participants. Similarly, another clinical trial commenced in 2022, involving induced pluripotent cells (iPSCs), which were derived from the patients (trials treatment arm participants) and later differentiated into dopamine neurons involved using induced pluripotent stem cells (iPSCs) derived from patients' cells, which are then differentiated into dopamine neurons and transplanted into the patients' brains. The very initial results have been affirmative, indicating a good tolerance toward the treatment, and a promise to minimize the PD symptoms [22, 23]. Figure 16.1 gives a graphical summary of such processes.

1. Amyotrophic lateral sclerosis
Recent advances in stem cell therapy's research have shown the use of mesenchymal stem cells (MSCs), derived from bone marrow, in improving the quality of life among patients with Amyotrophic Lateral Sclerosis (ALS), which is a debilitating motor neuron disease. A phase II trial conducted in 2021 aimed to evaluate the safety and effectiveness of intrathecal MSC transplantation. The results of the trial showed immune modulation and slowing of the disease progression among the ALS patients or of the intervention arm as compared to the control or placebo arm. Later, in 2023, a phase III trial was launched with a larger cohort of participants. The trial data were consistent with disease stabilization among the patients within the treatment arm and improvement in the quality of life of the participants. This trial's final results are awaited at present [24].

2. Alzheimer's disease (AD)
Alzheimer's disease (AD) is one of the diseases of concern among the elderly population today. It is characterized by gradual cognitive decline and memory loss. During 2020, efforts to address AD through injection hESC-derived neural progenitor cells (NPCs) under a phase I clinical trial were attempted. The hESC-derived neural progenitor cells (NPCs) were injected into the hippocampus of the consenting AD patients. The trial was found to be safe and free from any adverse effects or harm to the trial participants. A small number of the AD patients showed stabilization of the cognitive function and slowing of the cognitive decline. Based on the phase I trial's results, a phase II trial was initiated during 2022 to assess the efficacy of NPCs in a larger sample of consenting patients of AD. The results showed stability of NPCs, their successful host brain cell integration, and

Table 16.1 Recent clinical trials in neurodegenerative diseases of concern.

Disease	Trial description	Year	Phase	Key results	Reference
PD	hESC-derived dopamine neuron transplantation	2020	I/II	Improved motor function, and cell survival	[25, 26]
	iPSC-derived dopamine neurons transplantation	2022	I	Well-tolerated, symptom alleviation	[27]
ALS	Bone marrow-derived MSC transplantation	2021	II	Slower motor function decline, immune modulation	[28]
	Bone marrow-derived MSC transplantation	2023	III	Disease stabilization, improved quality of life	[29]
AD	hESC-derived neural progenitor cells (NPCs) injection into the hippocampus	2020	I	Safety confirmed, stabilized cognitive function	[30]
	NPC transplantation	2022	II	Cell survival, neuroprotective effects	[31]

neuroprotection along with improvement in cognitive function of the recipients. Table 16.1 shows the details of some of the important clinical trials since 2020 until today.

Diabetes Mellitus

Diabetes mellitus (DM) happens when the human body fails to utilize its consumed sugars appropriately. This leads to a progressive decline in the function of vital organs. There are many types of diabetes. However, the most prevalent among those are type 1 and type 2 diabetes. Type 1 is due to the destruction of the insulin-producing beta cells of the pancreas, while type 2 is mainly due to insulin resistance at the cellular level. Stem cell research has focused on developing therapies to counter beta cells' destruction by the immune response. Clinical trials conducted between

Table 16.2 Selected clinical trials investigating the use of stem cells in type 1 and type 2 diabetes mellitus.

Disease	Trial description	Year	Phase	Key results	References
T1D	Encapsulated hESC-derived pancreatic progenitor cell transplantation	2020	I/II	Insulin production and reduced insulin requirements	[32]
	iPSC-derived beta cells	2022	I	Engraftment, improved blood glucose regulation	[33]
T2D	MSC treatment to improve insulin sensitivity and reduce inflammation	2021	I	Improved insulin sensitivity, glycemic control	[34]

2020 and 2024 focused on beta cells' replacement and, through immunomodulation, attempted to restore the beta cell function by production of insulin by them. This may lead to better blood glucose level control and an improved quality of life of the type 1 diabetes patients. This chapter reviews these trials and their potential impacts on diabetes treatment, as outlined in Table 16.2.

1. Type 1 diabetes (T1D)

Type 1 diabetes requires lifelong insulin administration to manage blood glucose levels, often making it difficult to achieve optimal control and avoid complications. Stem cell therapies aim to overcome these challenges by replacing lost beta cells or modulating the immune system to preserve the remaining cells. During 2020, an important phase I/II clinical trial was initiated. The trial examined the safety and effectiveness of encapsulated hESC-derived pancreatic progenitor cells. Once matured, those cells could develop into insulin-producing beta cells after being transplanted into the type 1 diabetes patients. The cells were found to be responding to the blood glucose level by producing insulin, leading to less need of exogenous insulin with better blood glucose level control without any major adverse effects observed. Later, in 2022, a second trial but with iPSCs derived from the patients' cells was commenced. Those iPSCs were differentiated into beta cells and thereafter transplanted into the recipients' pancreas.

The early data from the ongoing trial suggest that iPSC-derived beta cells can engraft, produce insulin, and enhance blood glucose regulation, with an added advantage of reducing the risk of immune rejection due to their autologous origin.

2. Type 2 diabetes (T2D)
Despite the new knowledge of managing type 2 DM with latest medications, healthy eating and lifestyle changes, stem cell therapies have been continuously explored as a better and definite treatment option. Type 2 diabetes (T2D) is typically managed with lifestyle changes and medication; stem cell therapies are being explored as additional treatment options. A few years ago, in 2021, a phase I clinical trial investigated the use of MSC to enhance insulin sensitivity and reduce inflammation among type 2 diabetes patients. The trial concluded that treatment with MSC was not only safe but also improved insulin sensitivity and hence glycemic control. This underscores its potential benefit in better management of type 2 diabetes mellitus.

Blood Disorders
Recent stem cell research brought profound advancements in the treatment of blood disorders, offering promising therapeutic options for conditions

Table 16.3 Clinical trials investigating the use of stem cells in blood disorders.

Disease	Trial description	Year	Phase	Key results	References
Leukemia	Umbilical cord blood-derived stem cell transplantation	2020	II	Similar survival rates, reduced GVHD	[34]
Lymphoma	Haploidentical HSCT	2022	III	Comparable outcomes to fully matched transplants	[35]
Sickle cell disease	Gene-edited hematopoietic stem cells using CRISPR/Cas9	2021	I/II	Reduced disease complications, increased healthy hemoglobin	[36]
	Autologous stem cells modified with lentiviral vectors	2023	I	Safe, increased functional hemoglobin production	[37]

such as leukemia, lymphoma, and sickle cell disease. Certain clinical trials, conducted during 2020 to 2024, focused on hematopoietic stem cell transplantation (HSCT) and the use of gene editing techniques to improve patient outcomes. Table 16.3 of this part of the chapter reviews such trials, highlighting their potential impacts and summarizing critical findings as discussed.

1. Leukemias and lymphomas
There are two most common types of blood cancers in children and adults: leukemias and lymphomas, which have traditionally been treated with HSCT. The recent clinical trials have been focusing on optimization of established treatments along with exploring new stem cell resources. During 2020, a phase II trial evaluated the use of umbilical cord blood-derived stem cells in patients with acute myeloid leukemia (AML). The comparison trial was designed to compare the safety and efficacy of the cord blood transplants with traditional bone marrow transplants. The inferences drawn were supportive of similar survival rates with comparatively reduced graft-versus-host disease (GVHD) among the cord blood transplant recipients, suggesting their potential as an alternative stem cell source. A phase III trial conducted 2 years after this trial (2022) used haploidentical (half-matched) donors in lymphoma patients receiving HSCT to measure the patients' outcome. The approach expanded the donor pool and increased the likelihood of finding a suitable match. This approach expands the donor pool and increases the likelihood of finding a suitable match. Preliminary data indicated that haploidentical HSCT provided comparable outcomes to fully matched donor transplants, with manageable levels of GVHD, and promising survival rates.

2. Sickle cell disease (SCD)
Sickle cell disease is more primarily common in Africa. This disease is a genetic blood disorder characterized by abnormal hemoglobin that leads to disfigured and distorted red blood cells (RBCs). These RBCs cause pain and damage while passing through the vital organs. Over the years, stem cell transplantation and gene editing have been in practice as tested treatment approaches. During 2021, a novel phase I/II trial used gene-edited hematopoietic stem cells to treat SCD. The CRISPR/Cas9 technology was employed to correct the genetic mutation responsible for SCD. Early results showed that treated patients experienced significant reductions in disease-related complications and increased healthy hemoglobin production, demonstrating the potential of gene editing for curative therapy. Another notable trial started in 2023 and investigated the safety and efficacy of

using autologous (patient's own) stem cells modified with lentiviral vectors to express a functional hemoglobin gene. Preliminary findings from this ongoing study suggest that the treatment is safe and can effectively increase the production of functional hemoglobin, reducing the frequency of sickle cell crises.

Cardiovascular Diseases

Cardiovascular diseases (CVD) continue to be a leading cause of morbidity and mortality globally, driving extensive research into novel treatments. Stem cell therapy has shown promise in repairing and regenerating damaged heart tissue. Between 2020 and 2024, clinical trials focused on using MSCs, iPSCs, and cardiac progenitor cells (CPCs) to treat conditions like myocardial infarction (MI) and heart failure (HF). Table 16.4 summarizes the outcomes and potential impacts of these trials conducted for cardiovascular disease.

1. Myocardial infarction

Heart attack, as commonly known, is technically myocardial infarction (MI). Mostly, it results in the substantial loss of heart muscle tissue. Stem cell therapies play a pivotal role in regenerating damaged tissues and improving the muscle's function. Stem cell therapies aim to regenerate this damaged tissue and improve heart function. In 2020, allogenic MSC was used by a phase II clinical trial among those who suffered a recent MI. The

Table 16.4 Clinical trials using stem cell therapies in cardiovascular diseases.

Disease	Trial description	Year	Phase	Key results	References
MI	Allogeneic MSC transplantation *via* intracoronary infusion	2020	II	Improved LVEF, reduced infarct size	[38]
	iPSC-derived cardiomyocytes injection	2022	I/II	Engraftment, improved cardiac function	[39]
HF	Autologous MSC injection *via* catheter-based delivery	2021	II	Improved LVEF, exercise capacity, quality of life	[40]
	CPC administration *via* transendocardial injection	2023	III	Improved cardiac function, reduced hospitalizations	[41]

trial evaluated the safety and efficacy of MSC transplantation delivered *via* intracoronary infusion. The results indicated that MSC treatment was safe and showed association with an improved left ventricular ejection fraction along with reduced infarct size as compared to the control group. Besides the use of MSC, a 2022 phase I/II trial explored using iPSC-derived cardiomyocytes in MI patients. The iPSCs were differentiated into cardiac cells and injected directly into the damaged heart tissue. Preliminary findings promised that the iPSC-derived cardiomyocytes could engraft, integrate with the host tissue, and improve the heart function, depicting its potential as a regenerative therapy for MI.

2. Heart failure
Cardiac failure or heart failure (HF) is characterized by the heart's inability to pump blood effectively. Considering the failing heart's gravity and impact on the quality of life of a patient, stem cell research has been having a focus on it since long. A number of clinical trials have examined the potential role of stem cells therapy in enhancing the heart function focusing enhancing the heart's capacity and capability to pump blood effectively. During 2021, a phase II trial was conducted to evaluate the effects of autologous MSC injected into the myocardium or the heart muscle *via* catheter in patients with chronic heart failure (CHF). The results were promising, showing their safety and significant improvement in left ventricular function (LVF), exercise capacity, and ultimately the quality of life. Later, during 2023, a phase III trial was initiated to assess the efficacy of cardiac progenitor cells (CPCs) in patients with advanced heart failure. The CPCs were delivered through a transendocardial injection. The early data suggested that CPC therapy improved cardiac function and reduced hospitalization for heart failure exacerbations.

Oncology

Cancer remains one of the most challenging and devastating diseases globally, affecting millions of lives each year. Traditional cancer treatments such as chemotherapy, radiation, and surgery, while often effective, can lead to significant side effects and are not always curative. Stem cell therapy and regenerative medicine have emerged as promising fields offering new avenues for cancer treatment. This part of the chapter explores the various clinical trials conducted from 2020 to 2024 that focus on stem cell-based therapies in oncology, highlighting their potential impacts, mechanisms of action, and challenges. Table 16.5 shows a summary of the stem cell trials referred in this section.

Table 16.5 Clinical trials investigating the use of stem cells in oncology.

Disease	Trail description	Year	Phase	Key results	References
Acute lymphoblastic leukemia	Haploidentical HSCT	2020	II	Comparable survival rates, manageable GVHD	[42]
Multiple myeloma	UCB HSCT	2021	II	Lower relapse rates, improved overall survival	[43]
Metastatic breast cancer	MSCs loaded with oncolytic viruses	2021	I/II	Well-tolerated, anti-tumor activity	[39]
Glioblastoma	MSCs delivering therapeutic genes	2022	I	Selective targeting, tumor shrinkage	[44]
Refractory solid tumors	iPSC-derived NK cells	2023	I/II	Safe, significant anti-tumor responses	[45]
HER2-positive breast cancer	iPSC-derived CAR-T cells targeting HER2	2024	I/II	Safe, anti-tumor activity, partial/ complete responses	[16]

1. Hematopoietic stem cell transplantation (HSCT)

Hematopoietic stem cell transplantation (HSCT) has been a cornerstone in the treatment of hematologic malignancies such as leukemia and lymphoma. Recent clinical trials have sought to optimize this procedure and expand its applicability.

1.1. Leukemia and lymphoma

In 2020, a Phase II trial investigated using haploidentical (half-matched) donors for HSCT in acute lymphoblastic leukemia (ALL) patients. This approach aimed to increase the donor pool and improve outcomes. The trial results demonstrated that haploidentical HSCT provided comparable survival rates to fully matched donor transplants with manageable levels of GVHD. A Phase II trial in 2022 focused on improving HSCT outcomes in multiple myeloma (MM) patients using umbilical cord blood (UCB) stem cells. Preliminary data indicated that UCB transplants were associated with improved overall survival.

2. MSCs and solid tumors

MSCs have shown potential in treating solid tumors due to their ability to home to tumor sites and their immunomodulatory properties. Clinical trials have explored their use in various cancers, including breast cancer, glioblastoma, and lung cancer.

2.1. Breast cancer

A phase I/II trial in 2021 investigated the safety and efficacy of MSCs loaded with oncolytic viruses in metastatic breast cancer patients. The MSCs were engineered to deliver the viruses directly to the tumor site, enhancing their anti-tumor effects. The trial results indicated that the treatment was well tolerated and showed promising anti-tumor activity, with some patients experiencing partial tumor regression and prolonged stable disease.

2.2. Glioblastoma

Glioblastoma, an aggressive brain tumor with limited treatment options and a poor prognosis, has been the focus of innovative research. In 2022, a phase I trial investigated the use of MSCs to deliver therapeutic genes to glioblastoma cells. The genetically modified MSCs were engineered to produce a prodrug-activating enzyme, which converts a non-toxic prodrug into a cytotoxic agent directly within the tumor. Early findings suggested that this approach could target glioblastoma cells selectively, leading to tumor shrinkage and extended survival in some patients.

3. IPSC and immunotherapy

IPSCs have revolutionized cancer treatment by providing a source for generating various immune cells, including NK cells and T cells, which can be used in immunotherapy.

3.1. iPSC-derived NK cells

A groundbreaking phase I/II trial launched in 2023 investigated the use of iPSC-derived NK cells in patients with refractory solid tumors, including ovarian and pancreatic cancers. These NK cells were engineered to enhance their anti-tumor activity and persistence. Early findings showed that the treatment was safe and led to significant anti-tumor responses, including tumor shrinkage and extended progression-free survival in several patients.

3.2. CAR-T cell therapy

Chimeric antigen receptor (CAR)-T cell therapy has remarkably succeeded in treating hematologic malignancies. Recent trials have aimed to extend its efficacy to solid tumors. A notable phase I/II trial in 2024 investigated using iPSC-derived CAR-T cells targeting HER2-positive breast cancer. The iPSC-derived CAR-T cells were engineered to express a receptor specific for the HER2 antigen, allowing them to target and destroy HER2-expressing tumor cells. Preliminary data suggested that the treatment was safe and demonstrated anti-tumor activity, with some patients achieving partial or complete responses.

16.3 Approved Stem Cell Therapies

Stem cell therapeutics are a rapidly growing treatment trend which exploits the differentiation potency and paracrine activity of the stem cells to make many major illnesses curable. They include the processes of collection, isolation, infusion, and finally migration of stem cells to the desired organ or tissue. Conventional therapies, either pharmacologic or surgical, have some drawbacks. For instance, a considerable percentage of diseases eventually develop resistance and tolerance after long-term drug treatment. Moreover, some conditions such as severe osteoarthritis do not have a definite pharmaceutical cure, and their management rely only on delaying the progression and disability [46]. Accordingly, stem cell-based therapies represent an alternative strategy to overcome the multi-factorial ineffectiveness of a drug or surgical therapy, so they are now incorporated in a variety of clinical guidelines.

Hematopoietic Stem Cell Transplantation

Hematopoietic stem cell transplantation (HSCT) is considered a standard of care for many cancerous and non-cancerous conditions [47]. It is a multi-step clinical procedure where sufficient healthy hematopoietic stem cells are harvested from peripheral blood (other sources include bone marrow and umbilical cord) with the aid of colony-stimulating factors like filgrastim (mobilization phase). Then, the diseased hematopoietic tissue is destroyed using a suitable conditioning regimen involving high-dose chemotherapy and/or radiation with selective hematological toxicity (conditioning phase). Finally, the collected hematopoietic stem cells are infused into the patient (expressed as day 0), and subsequent cellular migration to the bone marrow paves the way to the restoration of hematopoietic function (engraftment and post-engraftment phases) [47]. The source of the harvested cells determines the type of transplantation as either autologous or allogeneic HSCT. Table 16.6 shows a detailed comparison between the two sources regarding definition, therapeutic goals, advantages, and complications.

Human leukocyte antigens (HLAs) are a group of cell surface markers through which the immune system can distinguish between foreign and

Table 16.6 Comparison between autologous and allogeneic HSCT.

Items	Autologous HSCT	Allogenic HSCT
Definition	The patient's stem cells are collected and infused after a proper myeloablative regimen.	Hematopoietic stem cells are collected from HLA-matched donors either related or unrelated and infused after a myeloablative, reduced-intensity, or non-myeloablative regimen.
Therapeutic goals	Aims to completely eradicate malignancy or hematopoietic and immunologic dysfunction through depending solely on high-dose chemotherapy, total-body radiation, or both. Therefore, the primary role of the transplanted stem cells is to reconstitute the aplastic bone marrow.	Conditioning therapy is utilized to achieve total or partial disease elimination along with suppression of the recipient's immunity to prevent potential host-versus-graft reactions. Since the transplanted stem cells have immunologic activity, they contribute to the elimination of the residual cancer cells besides hematopoiesis recovery.
Advantages	Transplanted stem cells are syngeneic, hence the absence of graft rejection or graft-versus-host disease (GvHD) and higher survival rates.	Transplanted stem cells can produce an immune response known as graft-versus-tumor (GvT) effect which aids in disease eradication and allows lower-intensity conditioning therapy [64].
Complications and limitations	Higher risk of chemotherapeutic toxicity especially for elderly patients and in the presence of comorbidities due to intense conditioning therapy. Others: infection and pancytopenia.	Acute and chronic GvHD and graft failure necessitate long-term immunosuppressive therapy and regular monitoring up to 100 days post-engraftment [65]. Others: same as autologous.

non-foreign antigens [48]. In allogeneic HSCT, HLA-matching is the standard tool that governs the selection of a suitable donor that may be related or unrelated. Generally, matched-related donors are the best choice [47].

HSCT Guidelines and Indications

The extent of how HSCT will be effective for certain conditions is assessed according to the European Society for Blood and Marrow Transplantation (EBMT) guidelines [49] into four grades:

1- Standard of care (S): HSCT has equal or superior positive outcomes over other therapeutic approaches.
2- Clinical option (CO): Only small studies show efficacy.
3- Developmental (D): Limited research is available.
4- Generally, not recommended (GNR): HSCT is irrational.

a) Leukemias

Allogenic HSCT represents a standard of care for all leukemia subtypes, especially for refractory or drug-resistant cases [50].

b) Hodgkin and non-Hodgkin lymphoma

Relapsed or poor prognosis patients are the best candidates for autologous HSCT [51, 52].

c) Multiple myeloma

Autologous HSCT combined with chemotherapy can improve survival without a cure [53].

d) Neuroblastoma

Autologous HSCT is indicated for high-risk patients in their first complete remission [54].

e) Hemoglobinopathies

Allogenic transplantation is curative and increases the survival rates for sickle cell anemia and thalassemia major [55, 56]. However, the rarity of matched sibling donors is a limitation in sickle cell anemia [55].

f) Autoimmune diseases

Multiple sclerosis patients with relapsing re-emitting disease who underwent autologous HSCT experienced slower disability progression and longer disease-free state [57]. In Crohn's disease, several studies reported clinical remission after autologous HSCT [58, 59].

g) Acquired-immunodeficiency syndrome (AIDS)
The rationale behind the use of allogeneic HSCT in AIDS is established on finding donors with a homozygous CCR5 receptor gene mutation which prevents its expression and further HIV attachment and cellular entry [60]. Although cure or sustained remission was achieved in some HIV patients with blood malignancies who received mutated CCR5 donor stem cells [61, 62], applying this procedure for non-malignant patients carries few constraints that outweigh its benefits—for example, multi-organ toxicity due to unprovoked high-dose chemotherapy exposure as a myeloablative regimen. Furthermore, an HIV patient may be co-infected with HIV variants that are not CCR5-tropic and depend on other chemokine receptors where the infection may persist after the procedure [63]. These challenges highlighted the necessity of developing novel drug molecules targeting the CCR5 receptors as a safer and more reasonable treatment option for the largest portion of HIV cases.

Mesenchymal Stem Cell Transplantation
Mesenchymal stem cells (MSCs) are multipotent, non-hematopoietic, plastic-adherent stem cells present in the stroma of a wide range of body and extra-embryonic tissues [66]. Their importance arises from their superior proliferative ability and differentiation potency to numerous types of somatic cells including chondrocytes, adipocytes, myocytes, osteoblasts, and other stromal cells [66]. Despite bone marrow MSCs being the most clinically used type, umbilical cord MSCs have the highest yield and multiplication tendency [67]. Currently, MSCs are constituting a promising research area in regenerative medicine, and several approved MSCs products are now used in various clinical applications. Figure 16.2 shows details on the possible use of MSCs in clinical trials.

Worldwide-Approved MSCs Products
1. Products approved by the Korean Ministry of Food and Drug Safety (MFDS)
In 2010, Queencell was the first approved product globally for the treatment of subcutaneous tissue disorders [68]. A year later, Cellgram-AMI was marketed for improving the ejection fraction in acute myocardial infarction after intracoronary infusion [69]. Plausible mechanisms are myocardium regeneration and angiogenesis [70]. In 2012, Cartistem and Cupistem were approved for use in osteoarthritis-induced knee cartilage degeneration and

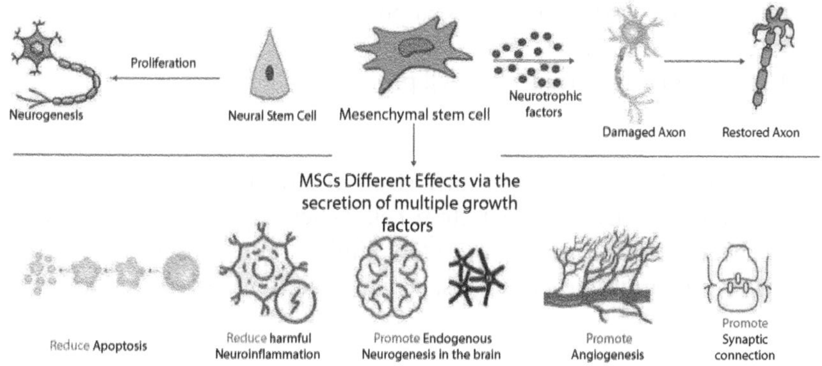

Figure 16.2 Process of neurorestoration by MSCs (adapted from Najafi S. *et al.* 2023).

fistulizing Crohn's disease, respectively [68, 71]. Cartistem alleviates osteoarthritic degenerative changes through the inhibition of cartilage-degrading enzymes, induction of chondrocyte differentiation, and interruption of the inflammatory processes. Cupistem efficacy is attributed to the promotion of local healing and reversion of Crohn's disease pathogenesis through regulatory T cell upregulation [72]. In addition, fibrin glue, an adhesive material commonly mixed with MSC products, exhibited a synergistic action in fistula healing [73]. Ultimately, Neuronata-R was approved in 2014 as the first stem cell therapy for amyotrophic lateral sclerosis (ALS) where it led to slower disease progression due to motor neuron protection, immunomodulatory, and anti-inflammatory effects [74]. All of the mentioned products are autologous bone marrow-derived MSCs except for Cartistem (allogenic umbilical cord-derived MSCs), Queencell, and Cupistem (autologous adipose tissue-derived MSCs).

2. Products approved by the Japanese Pharmaceuticals and Medical Devices Agency (PMDA)

Temcell HS was revealed in 2015 as an allogenic bone marrow-derived MSCs for treatment of acute GvHD. It activates regulatory T cells, resulting in tissue repair and immunomodulation [75]. Another product, Stemirac (autologous bone marrow-derived MSCs), was approved in 2018 as a promising treatment for traumatic spinal cord injury where it acts by boosting neurogenesis, secretion of neurotrophic factors, and immunomodulatory properties [76].

3. Products approved by other global pharmaceutical authorities are summarized in Table 16.7.

Miscellaneous approved stem cell products:

1. Holoclar

It is a stem-cell-based therapy authorized in 2015 by the EMA for use in patients suffering from limbal stem cell deficiency (LSCD) [81]. LSCD occurs because of damaged unipotent limbal stem cells in the corneal limbus due to multiple etiologies, which results in severe reduction in

Table 16.7 Other approved mesenchymal stem cell products.

	Approving authority/ country/year	**Origin of the MSCs**	**Indications/mechanism**
Prochymal [77]	Health Canada/2012	Allogenic bone marrow	Steroid-refractory acute GvHD/diminishes pro-inflammatory cytokines and promotes tissue repair through secretion of growth factors
Stempeucel [78, 79]	Central Drugs Standard Control Organization (CDSCO)/ India/2016	Allogenic bone marrow	Critical limb ischemia associated with Buerger's and peripheral arterial diseases/knee osteoarthritis/anti-inflammatory and pro-angiogenic properties
Alofisel [80]	European Medicine Agency (EMA)/ Belgium/2018	Allogenic adipose tissue	Complex perianal fistulas in adults with Crohn's disease/ reduction of pro-inflammatory cytokines release allows fistula healing

visual acuity and corneal inflammation [82]. Holoclar involves amplified autologous limbal stem cells taken from an intact portion of the corneal limbus and cultured over a fibrin layer to be implanted surgically [81]. Consequently, the corneal epithelium begins to regenerate after differentiation and multiplication of the implanted limbal cells.

2. Strimvelis
It is a genetically modified hematopoietic stem cell product accepted for marketing in 2016 by the EMA for the treatment of adenosine deaminase (ADA)-deficient severe combined immunodeficiency in the absence of a compatible donor [83]. Normally, ADA exists with high concentrations in B and T lymphocytes and is responsible for preventing the accumulation of the lymphotoxic phosphorylated deoxyadenosine metabolites by converting deoxyadenosine into deoxyinosine, which explains the severe lymphocytopenia in those patients [84]. The hematopoietic stem cells are harvested and undergo viral transduction using ADA-encoding retrovirus [39]. After transplantation, the ADA expression is restored with cellular count improvement.

16.4 Challenges in Clinical Trials

A. Scientific and technical challenges

1. Complex mechanisms of action: For cellular deficiency-based diseases, stem cell therapy seems more logical. Stem cell-based therapies can involve direct cellular replacement or paracrine signaling. For effective therapies, these mechanisms of action are needed to be understood along with integration of cells or signaling within tissues [85]. The biggest challenge is the fear of the unknown mechanism which needs to be overcome for global acceptance of the procedure [86].
2. Preclinical evidence requirements: Proper safety and efficacy preclinical data must demonstrate that it is safe and effective in the species so that human trials do not become open experiments [87]. It is a potential risk to skip this step, but many of the early-stage trials lack this foundational evidence.
3. Quality control and manufacturing: Stem cell products need to be made in compliance with Good Manufacturing Practice, such that all reagents

are of high quality and all products are tested. Here we need to overcome issues of standardizing cell products and delivery methods [85, 86].

4. Tumorigenicity and genetic stability: Tumorigenicity and genetic instability could pose challenges for clinical applications of iPSCs. There are significant shortcomings and challenges associated with the reprogramming technique, which could result in the presence of unwanted mutations and increase the risk of cancer [86].

5. Immunological tolerance and rejection: The stem cells must gain immunological tolerance with the recipient body by overcoming the barriers of immune rejection and being able to function within the patient [86].

B. Ethical and regulatory hurdles

1. Ethical problems with ESCs: When embryonic stem cells were isolated in the lab, it led to the destruction of embryos. These embryos had the potential to develop into a full-grown human being. Using embryonic stem cells can lead to ethical problems due to the embryos being destroyed. This is one of the main ethical problems with the use of ESCs. In recent years, progress has been made toward developing alternatives that do not destroy potential human life.

2. Unproven therapies and direct-to-consumer pitfalls: Stem cell "therapies" developed without adequately supported basic or pre-clinical data, when sometimes sold directly to consumers, can raise serious ethical concerns [85].

3. Risk evaluation and informed consent: The stem cell trials and experiments are new. The patients need better risk perception of harm and benefits of such therapies. Much awareness of the researchers and their target groups is required for the trials' safety and better compliance and consenting to the trials, which could prove more benefit to the society.

4. Trials' oversighting committees and boards: More technical as well as diverse trial oversight is required. Strict compliance to the healthcare research and medical ethics is required. Stem cell research may need more brain storming among the experts as it is going to be the future of therapeutic research.

C. Logistic and financial constraints

1. Funding and financial barriers: The trials of stem cells often have financial constraints, with certain trials relying on the contributions of patients rather than the conventional funding ways. This can lead to a reduction in the scope and quality of the research.

2. Trial design and execution: The planning and implementation of stem cell trials depend on the utilization of the most modern techniques in the field of logistics, such as patient monitoring, cell delivery, and sequentially designed trials. These trials are expensive from the material point of view, and they need careful planning [85].
3. Infrastructure and production costs: The setting up of facilities for GMP-compliant cell production and testing is a huge investment. This involves solving of the problems of mass production while maintaining a high standard of outcomes.
4. Scale and complexity of procedures: The intricate formation of the entire laboratory-grown organs by stem cell therapy is still challenging. The large-scale production of such biologically accurate cells requires interdisciplinary collaboration.

D. Case examples of overcoming these challenges

1. Structured review frameworks: The ISSCR has published a framework for the evaluation of stem cell trials that is helpful in addressing key scientific, ethical and regulatory issues. This helps to see to it that trials are evidence-based and ethically conducted [88].
2. Guidelines and practice tools: Developing standards as well as relevant checklists for ethics committees or institutional review boards is a prerequisite for more informed decisions on the specific use of stem cell therapies [85].
3. New and improved reprogramming methods: The efforts to ameliorate the reprogramming technique of induced pluripotent stem cells (iPSCs), with emphasis on reducing tumorigenicity while improving efficiency, pose important scientific advances for overcoming scientific hurdles [89].
4. Patient-specific uses: The use of patient-specific cells to bypass the possibility of allorejection and reduce ethical problems is an enormous advancement in improving the safety and effectiveness of stem cell treatments.

16.5 Conclusion

Stem cell therapy is a revolutionary approach in medicine, having a noteworthy impact on regenerative medicine and transplantology as it has potential for the treatment of currently untreatable neurodegenerative diseases. It has the ability to address present and future diseases with advancement of induced pluripotent stem cells from a patient's own cells and concept of tissue banks. Despite its lots of successes, it faces many ethical concerns, immunological compatibility, risk of tumor formation, and instability when re-programmed. Clinical trials and research which are carefully designed and ethically sound are important for advancing this field of regenerative medicine. Development of new therapeutics technology and its widespread clinical application are possible with collaboration among clinicians, academics, industry, and regulatory authorities for overcoming existing barriers.

References

1. Mason, C. and Dunnill, P., A brief definition of regenerative medicine. *Regen. Med.* [Internet], 3, 1–5, 2008, Available from: http://www.ncbi.nlm.nih.gov/pubmed/18154457.
2. Fortier, L.A., Stem cells: classifications, controversies, and clinical applications. *Vet. Surg.* [Internet], 34, 415–23, 2005, Available from: http://www.ncbi.nlm.nih.gov/pubmed/16266332.
3. Zhang, F.-Q., Jiang, J.-L., Zhang, J.-T., Niu, H., Fu, X.-Q., Zeng, L.-L., Current status and future prospects of stem cell therapy in Alzheimer's disease. *Neural Regen. Res.* [Internet], 15, 242–50, 2020, Available from: http://www.ncbi.nlm.nih.gov/pubmed/31552889.
4. O'Brien, T. and Barry, F.P., Stem cell therapy and regenerative medicine. *Mayo Clin. Proc.* [Internet], 84, 859–61, 2009, Available from: http://www.ncbi.nlm.nih.gov/pubmed/19797773.
5. Madl, C.M., Heilshorn, S.C., Blau, H.M., Bioengineering strategies to accelerate stem cell therapeutics. *Nature* [Internet], 557, 335–42, 2018, Available from: http://www.ncbi.nlm.nih.gov/pubmed/29769665.
6. Pérez López, S. and Otero Hernández, J., Advances in stem cell therapy. *Adv. Exp. Med. Biol.* [Internet], 741, 290–313, 2012, Available from: http://www.ncbi.nlm.nih.gov/pubmed/22457117.
7. Sun, S., Li, Z., Glencer, P., Cai, B., Zhang, X., Yang, J., *et al.*, Bringing the age-related macular degeneration high-risk allele age-related maculopathy susceptibility 2 into focus with stem cell technology. *Stem Cell Res. Ther.* [Internet], 8, 135, 2017, Available from: http://www.ncbi.nlm.nih.gov/pubmed/28583181.
8. Liu, J., Induced pluripotent stem cell-derived neural stem cells: new hope for stroke? *Stem Cell Res. Ther.* [Internet], 4, 115, 2013, Available from: http://www.ncbi.nlm.nih.gov/pubmed/24067059.
9. Shahjalal, H.M., Abdal Dayem, A., Lim, K.M., Jeon, T.-I., Cho, S.-G., Generation of pancreatic β cells for treatment of diabetes: advances and challenges. *Stem Cell Res. Ther.* [Internet], 9, 355, 2018, Available from: http://www.ncbi.nlm.nih.gov/pubmed/30594258.

10. Hirsch, T., Rothoeft, T., Teig, N., Bauer, J.W., Pellegrini, G., De Rosa, L., *et al.*, Regeneration of the entire human epidermis using transgenic stem cells. *Nature* [Internet], 551, 327–32, 2017, Available from: http://www.ncbi.nlm.nih.gov/pubmed/29144448.
11. Kuriyan, A.E., Albini, T.A., Townsend, J.H., Rodriguez, M., Pandya, H.K., Leonard, II R.E., *et al.*, Vision Loss after Intravitreal Injection of Autologous "Stem Cells" for AMD. *N. Engl. J. Med.* [Internet], 376, 1047–53, 2024, Available from: https://www.nejm.org/doi/full/10.1056/NEJMoa1609583.
12. Bobba, S., Di Girolamo, N., Munsie, M., Chen, F., Pébay, A., Harkin, D., *et al.*, The current state of stem cell therapy for ocular disease. *Exp. Eye Res.* [Internet], 177, 65–75, 2018, Available from: http://www.ncbi.nlm.nih.gov/pubmed/30029023.
13. Ramalho-Santos, M. and Willenbring, H., On the origin of the term "stem cell". *Cell Stem Cell* [Internet], 1, 35–8, 2007, Available from: http://www.ncbi.nlm.nih.gov/pubmed/18371332.
14. Caplan, Mesenchymal stem cells. *J. Orthop. Res.* -Wiley Online Lib. [Internet], 9, 5, 641–50, 1991, Available from: https://onlinelibrary.wiley.com/doi/abs/10.1002/jor.1100090504.
15. Ericsson, A.C., Crim, M.J., Franklin, C.L., A brief history of animal modeling. *Mo. Med.* [Internet], 110, 201, 2013, Available from: https://www.ncbi.nlm.nih.gov/pmc/articles/PMC3979591/.
16. Swain, S.M., Shastry, M., Hamilton, E., Targeting HER2-positive breast cancer: advances and future directions. *Nat. Rev. Drug Discov.*, 22, 101–26, 2023.
17. Teale, M.A., Schneider, S., Eibl, D., van den Bos, C., Neubauer, P., Eibl, R., Mesenchymal and induced pluripotent stem cell–based therapeutics: a comparison. *Appl. Microbiol. Biotechnol.*, 107, 4429–45, 2023.
18. Sarsenova, M., Issabekova, A., Abisheva, S., Rutskaya-Moroshan, K., Ogay, V., Saparov, A., Mesenchymal stem cell-based therapy for rheumatoid arthritis. *Int. J. Mol. Sci.*, 22, 11592, 2021.
19. Karamini, A., Bakopoulou, A., Andreadis, D., Gkiouras, K., Kritis, A., Therapeutic potential of mesenchymal stromal stem cells in rheumatoid arthritis: a systematic review of *in vivo* studies. *Stem Cell Rev. Rep.*, 16, 276–87, 2020.
20. Tolba, M.M., Jabbar, A., Afzal, S., Mahmoud, M., Zulfiqar, F., El-Soudany, I., *et al.*, A promising RNA nanotechnology in clinical therapeutics: a future perspective narrative review. *Future Sci. OA*, 9, FSO883, 2023.
21. Ellakwa, D.E.-S., Mushtaq, N., Khan, S., Jabbar, A., Abdelmalek, M.A., Wadan, A.-H.S., *et al.*, Molecular functions of microRNAs in colorectal cancer: recent roles in proliferation, angiogenesis, apoptosis, and chemoresistance. *Naunyn Schmiedebergs Arch. Pharmacol.*, 397, 5617–30, 2024.
22. Mohamed, W., Kumar, J., Alghamdi, B.S., Soliman, A.-H., Toshihide, Y., Neurodegeneration and inflammation crosstalk: Therapeutic targets and perspectives. *IBRO Neurosci. Rep.*, 14, 95–110, 2023.

23. Alshial, E.E., Abdulghaney, M.I., Wadan, A.-H.S., Abdellatif, M.A., Ramadan, N.E., Suleiman, A.M., *et al.*, Mitochondrial dysfunction and neurological disorders: a narrative review and treatment overview. *Life Sci.*, 334, 122257, 2023.
24. Wadan, A.S., Raza, M.L., Moradikor, N., Synaptic modulation by coffee compounds: Insights into neural plasticity. *Prog. Brain Res.*, 289, 181–91, 2024.
25. Kirkeby, A., Nelander, J., Hoban, D.B., Rogelius, N., Bjartmarz, H., Storm, P., *et al.*, Preclinical quality, safety, and efficacy of a human embryonic stem cell-derived product for the treatment of Parkinson's disease, STEM-PD. *Cell Stem Cell*, 30, 1299–1314.e9, 2023.
26. Cha, Y., Park, T.-Y., Leblanc, P., Kim, K.-S., Current status and future perspectives on stem cell-based therapies for Parkinson's disease. *J. Mov. Disord.*, 16, 22, 2023.
27. Park, S., Park, C.W., Eom, J.H., Jo, M.Y., Hur, H.J., Choi, S.K., *et al.*, Preclinical and dose-ranging assessment of hESC-derived dopaminergic progenitors for a clinical trial on Parkinson's disease. *Cell Stem Cell*, 31, 25–38.e8, 2024.
28. Najafi, S., Najafi, P., Farkhad, N.K., Torshizi, G.H., Darban, R.A., Boroumand, A.R., *et al.*, Mesenchymal stem cell therapy in amyotrophic lateral sclerosis (ALS) patients: A comprehensive review of disease information and future perspectives. *Iran. J. Basic Med. Sci.*, 26, 872, 2023.
29. Nabavi, S.M., Karimi, S.K., Arab, L., Sanjari, L., Mardpour, S., Azimian, V., *et al.*, Safety and efficacy of allogeneic adipose tissue mesenchymal stromal cells in amyotrophic lateral sclerosis patients, Single-center, Prospective, Open-label, Single-arm clinical trial, long-term follow-up. *Cell J.*, 23, 772, 2021.
30. Chen, X., Jiang, S., Wang, R., Bao, X., Li, Y., Neural stem cells in the treatment of Alzheimer's disease: Current status, challenges, and future prospects. *J. Alzheimers Dis.*, 94, S173–86, 2023.
31. Jiang, S., Wang, H., Yang, C., Feng, F., Xu, D., Zhang, M., *et al.*, Phase 1 study of safety and preliminary efficacy of intranasal transplantation of human neural stem cells (ANGE-S003) in Parkinson's disease. *J. Neurol. Neurosurg. Psychiatry*, 95, 12, 1102–11, 2024.
32. Krentz, N.A.J., Shea, L.D., Huising, M.O., Shaw, J.A.M., Restoring normal islet mass and function in type 1 diabetes through regenerative medicine and tissue engineering. *Lancet Diabetes Endocrinol.*, 9, 708–24, 2021.
33. Gerace, D., Zhou, Q., Kenty, J.H.-R., Veres, A., Sintov, E., Wang, X., *et al.*, Engineering human stem cell-derived islets to evade immune rejection and promote localized immune tolerance. *Cell Rep. Med.*, 4, 1, 2023.
34. Gao, S., Zhang, Y., Liang, K., Bi, R., Du, Y., Mesenchymal stem cells (MSCs): a novel therapy for type 2 diabetes. *Stem Cells Int.*, 2022, 8637493, 2022.
35. Thakar, M.S., Browning, M., Hari, P., Charlson, J.A., Margolis, D.A., Logan, B., *et al.*, Phase II trial using haploidentical hematopoietic cell transplantation (HCT) followed by donor natural killer (NK) cell infusion and sirolimus

maintenance for patients with high-risk solid tumors. *J. Clin. Oncol.*, 38, e23551–e23551, 2020.
36. Sharma, A., Boelens, J.-J., Cancio, M., Hankins, J.S., Bhad, P., Azizy, M., *et al.*, CRISPR-Cas9 Editing of the HBG1 and HBG2 Promoters to Treat Sickle Cell Disease. *N. Engl. J. Med.*, 389, 820–32, 2023.
37. Spencer Chapman, M., Cull, A.H., Ciuculescu, M.F., Esrick, E.B., Mitchell, E., Jung, H., *et al.*, Clonal selection of hematopoietic stem cells after gene therapy for sickle cell disease. *Nat. Med.*, 29, 3175–83, 2023.
38. Attar, A., Farjoud Kouhanjani, M., Hessami, K., Vosough, M., Kojuri, J., Ramzi, M., *et al.*, Effect of once versus twice intracoronary injection of allogeneic-derived mesenchymal stromal cells after acute myocardial infarction: BOOSTER-TAHA7 randomized clinical trial. *Stem Cell Res. Ther.*, 14, 264, 2023.
39. Cheng, Y.C., Hsieh, M.L., Lin, C.J., Chang, C.M.C., Huang, C.Y., Puntney, R., *et al.*, Combined Treatment of Human Induced Pluripotent Stem Cell-Derived Cardiomyocytes and Endothelial Cells Regenerate the Infarcted Heart in Mice and Non-Human Primates. *Circulation*, 148, 1395–409, 2023.
40. Yan, W., Xia, Y., Zhao, H., Xu, X., Ma, X., Tao, L., Stem cell-based therapy in cardiac repair after myocardial infarction: Promise, challenges, and future directions. *J. Mol. Cell. Cardiol.*, 188, 1–14, 2024.
41. Carvalho, A.B., Kasai-Brunswick, T.H., de Carvalho, A.C.C., Advanced cell and gene therapies in cardiology. *EBioMedicine*, 103, 105125, 2024.
42. Fierro-Pineda, J.C., Tsai, H.L., Blackford, A., Cluster, A., Caywood, E., Dalal, J., *et al.*, Prospective PTCTC trial of myeloablative haplo-BMT with post-transplant cyclophosphamide for pediatric acute leukemias. *Blood Adv.*, 7, 5639–48, 2023.
43. Srour, S.A., Mehta, R.S., Shah, N., Qazilbash, M.H., Im, J., Bashir, Q., *et al.*, Phase II study of umbilical cord blood–derived natural killer (CB-NK) cells with elotuzumab, lenalidomide, and high-dose melphalan followed by autologous stem cell transplantation (ASCT) for patients with high-risk multiple myeloma (HRMM). *J. Clin. Oncol.*, 40, 16_suppl, 8009, 2022, doi:10.1200/jco.2022.40.16_suppl.8009.
44. Varela, M.L., Comba, A., Faisal, S.M., Argento, A., Franson, A., Barissi, M.N., *et al.*, Gene therapy for high grade glioma: the clinical experience. *Expert Opin. Biol. Ther.*, 23, 145–61, 2023.
45. Zhang, H., Yang, L., Wang, T., Li, Z., NK cell-based tumor immunotherapy. *Bioact. Mater.*, 31, 63–86, 2024.
46. Block, J.A. and Cherny, D., Management of Knee Osteoarthritis: What Internists Need to Know. *Med. Clin. North Am.* [Internet], 105, 367–85, 2021, Available from: https://pubmed.ncbi.nlm.nih.gov/33589109/.
47. Balassa, K., Danby, R., Rocha, V., Haematopoietic stem cell transplants: principles and indications. *Br. J. Hosp. Med. (Lond.)* [Internet], 80, 33–9, 2019, Available from: https://pubmed.ncbi.nlm.nih.gov/30592675/.

48. Hunt, J.S. and Orr, H.T., HLA and maternal-fetal recognition. *FASEB J.* [Internet], 6, 2344–8, 1992, Available from: https://pubmed.ncbi.nlm.nih.gov/1544544/.
49. Snowden, J.A., Sánchez-Ortega, I., Corbacioglu, S., Basak, G.W., Chabannon, C., de la Camara, R., *et al.*, Indications for haematopoietic cell transplantation for haematological diseases, solid tumours and immune disorders: current practice in Europe, 2022. *Bone Marrow Transplant.*, 57, 8, 1217–39, 2022, [Internet]. Available from: https://www.nature.com/articles/s41409-022-01691-w.
50. Board PDQATE, Adult Acute Myeloid Leukemia Treatment (PDQ®): Health Professional Version, in: *PDQ Cancer Inf Summ* [Internet], 2002, Available from: https://www.cancer.gov/types/leukemia/hp/adult-aml-treatment-pdq.
51. Van Den Neste, E., Casasnovas, O., André, M., Touati, M., Senecal, D., Edeline, V., *et al.*, Classical Hodgkin's lymphoma: the Lymphoma Study Association guidelines for relapsed and refractory adult patients eligible for transplant. *Haematologica* [Internet], 98, 1185, 2013, Available from: file:///pmc/articles/PMC3729898/.
52. Ayers, E.C., Li, S., Medeiros, L.J., Bond, D.A., Maddocks, K.J., Torka, P., *et al.*, Outcomes in patients with aggressive B-cell non-Hodgkin lymphoma after intensive frontline treatment failure. *Cancer*, 126, 293–303, 2020.
53. Shah, N., Callander, N., Ganguly, S., Gul, Z., Hamadani, M., Costa, L., *et al.*, Hematopoietic Stem Cell Transplantation for Multiple Myeloma: Guidelines from the American Society for Blood and Marrow Transplantation. *Biol. Blood Marrow Transplant.* [Internet], 21, 1155–66, 2015, Available from: http://www.astctjournal.org/article/S1083879115001743/fulltext.
54. Fish, J.D. and Grupp, S.A., Stem cell transplantation for neuroblastoma. *Bone Marrow Transplant.* [Internet], 41, 159, 2008, Available from: file:///pmc/articles/PMC2892221/.
55. Khemani, K., Katoch, D., Krishnamurti, L., Curative Therapies for Sickle Cell Disease. *Ochsner J.* [Internet], 19, 131–7, 2019, Available from: https://pubmed.ncbi.nlm.nih.gov/31258425/.
56. Algeri, M., Lodi, M., Locatelli, F., Hematopoietic Stem Cell Transplantation in Thalassemia. *Hematol. Oncol. Clin. North Am.* [Internet], 37, 413–32, 2023, Available from: https://pubmed.ncbi.nlm.nih.gov/36907612/.
57. Msheik, A., Assi, F., Hamed, F., Jibbawi, A., Nakhl, A.-M., Khoury, A., *et al.*, Stem Cell Transplantation for Multiple Sclerosis: A 2023 Review of Published Studies. *Cureus* [Internet], 15, 10, 2023, Available from: https://pubmed.ncbi.nlm.nih.gov/38034162/.
58. Craig, R.M., Traynor, A., Oyama, Y., Burt, R.K., Hematopoietic stem cell transplantation for severe Crohn's disease. *Bone Marrow Transplant.* [Internet], 32, Suppl 1, S57–9, 2003, Available from: https://pubmed.ncbi.nlm.nih.gov/12931244/.
59. Scimè, R., Cavallaro, A.M., Tringali, S., Santoro, A., Rizzo, A., Montalbano, L., *et al.*, Complete clinical remission after high-dose immune suppression

and autologous hematopoietic stem cell transplantation in severe Crohn's disease refractory to immunosuppressive and immunomodulator therapy. *Inflamm. Bowel Dis.* [Internet], 10, 892–4, 2004, Available from: https://pubmed.ncbi.nlm.nih.gov/15626911/.

60. Faivre, N., Verollet, C., Dumas, F., The chemokine receptor CCR5: multifaceted hook for HIV-1. *Retrovirology* [Internet], 21, 2, 2024, Available from: https://pubmed.ncbi.nlm.nih.gov/38263120/.
61. Hütter, G., Nowak, D., Mossner, M., Ganepola, S., Müßig, A., Allers, K., *et al.*, Long-term control of HIV by CCR5 Delta32/Delta32 stem-cell transplantation. *N. Engl. J. Med.* [Internet], 360, 692–8, 2009, Available from: https://pubmed.ncbi.nlm.nih.gov/19213682/.
62. Henrich, T.J., Hu, Z., Li, J.Z., Sciaranghella, G., Busch, M.P., Keating, S.M., *et al.*, Long-term reduction in peripheral blood HIV type 1 reservoirs following reduced-intensity conditioning allogeneic stem cell transplantation. *J. Infect. Dis.* [Internet], 207, 1694–702, 2013, Available from: https://pubmed.ncbi.nlm.nih.gov/23460751/.
63. Verheyen, J., Thielen, A., Lübke, N., Dirks, M., Widera, M., Dittmer, U., *et al.*, Rapid Rebound of a Preexisting CXCR4-tropic Human Immunodeficiency Virus Variant After Allogeneic Transplantation With CCR5 Δ32 Homozygous Stem Cells. *Clin. Infect. Dis.* [Internet], 68, 684–7, 2019, Available from: https://pubmed.ncbi.nlm.nih.gov/30020413/.
64. Barisic, S. and Childs, R.W., Graft-Versus-Solid-Tumor Effect: From Hematopoietic Stem Cell Transplantation to Adoptive Cell Therapies. *Stem Cells* [Internet], 40, 556–63, 2022, Available from: https://pubmed.ncbi.nlm.nih.gov/35325242/.
65. Paczesny, S., Raiker, N., Brooks, S., Mumaw, C., Graft-versus-host disease biomarkers: omics and personalized medicine. *Int. J. Hematol.* [Internet], 98, 275–92, 2013, Available from: https://pubmed.ncbi.nlm.nih.gov/23959582/.
66. Ding, D.C., Shyu, W.C., Lin, S.Z., Mesenchymal stem cells. *Cell Transplant.* [Internet], 20, 5–14, 2011, Available from: https://pubmed.ncbi.nlm.nih.gov/21396235/.
67. Reyhani, S., Abbaspanah, B., Mousavi, S.H., Umbilical cord-derived mesenchymal stem cells in neurodegenerative disorders: from literature to clinical practice. *Regen. Med.* [Internet], 15, 1561–78, 2020, Available from: https://pubmed.ncbi.nlm.nih.gov/32479211/.
68. (Cell therapy products) ANTEROGEN [Internet]. [cited 2024 Aug 4]. Available from: https://www.mfds.go.kr/eng/brd/m_30/view.do?seq=71337&srchFr=&srchTo=&srchWord=&srchTp=&itm_seq_1=0&itm_seq_2=0&multi_itm_seq=0&company_cd=&company_nm=&page=1
69. Stem Cell Drug Hearticellgramâ“‡-AMI -PHARMICELL [Internet]. Available from: http://www.pharmicell.com/eng/biz/medicine_cellgram.html.
70. Kim, S.H., Cho, J.H., Lee, Y.H., Lee, J.H., Kim, S.S., Kim, M.Y., *et al.*, Improvement in Left Ventricular Function with Intracoronary Mesenchymal

Stem Cell Therapy in a Patient with Anterior Wall ST-Segment Elevation Myocardial Infarction. *Cardiovasc. Drugs Ther.* [Internet], 32, 329, 2018, Available from: file:///pmc/articles/PMC6133167/.
71. 식품의약품안전처 의약품통합정보시스템 의약품제품정보 상세보기-카티스템(동종제대혈유래중간엽줄기세포) [Internet]. Available from: https://nedrug.mfds.go.kr/pbp/CCBBB01/getItemDetailCache?cacheSeq=202106062aupdateTs2024-07-15 14:17:06.09794b.
72. Salem, H.K. and Thiemermann, C., Mesenchymal Stromal Cells: Current Understanding and Clinical Status. *Stem Cells* [Internet], 28, 585, 2010, Available from: file:///pmc/articles/PMC2962904/.
73. Cheng, F., Huang, Z., Li, Z., Efficacy and Safety of Mesenchymal Stem Cells in Treatment of Complex Perianal Fistulas: A Meta-Analysis. *Stem Cells Int.* [Internet], 2020, 8816737, 2020, Available from: https://pubmed.ncbi.nlm.nih.gov/33299423/.
74. First licensed stem cell therapy for ALS [Internet]. Available from: https://www.nature.com/articles/d43747-020-00727-8.
75. PMDA, Temcell-Report on the Deliberation Results [Internet], 2015, Available from: https://www.pmda.go.jp/files/000215658.pdf.
76. PMDA, Stemirac for Injection-Report on the Deliberation Results [Internet], 2018, Available from: https://www.pmda.go.jp/files/000231946.pdf.
77. Register D and HP, Prochymal-Produuct Monograph [Internet], 2012, Available from: https://pdf.hres.ca/dpd_pm/00024994.PDF.
78. Gupta, P.K., Krishna, M., Chullikana, A., Desai, S., Murugesan, R., Dutta, S., *et al.*, Administration of Adult Human Bone Marrow-Derived, Cultured, Pooled, Allogeneic Mesenchymal Stromal Cells in Critical Limb Ischemia Due to Buerger's Disease: Phase II Study Report Suggests Clinical Efficacy. *Stem Cells Transl. Med.* [Internet], 6, 689, 2017, Available from: file:///pmc/articles/PMC5442769/.
79. Gupta, A., StemOneTM/Stempeucel®: CDSCO Approved, Adult Human Bone Marrow-Derived, Cultured, Pooled, Allogenic Mesenchymal Stem Cells for Knee Osteoarthritis. *Biomedicines* [Internet], 11, 2894, 2023, Available from: file:///pmc/articles/PMC10669254/.
80. Alofisel | European Medicines Agency (EMA) [Internet]. Available from: https://www.ema.europa.eu/en/medicines/human/EPAR/alofisel.
81. Holoclar | European Medicines Agency (EMA) [Internet]. Available from: https://www.ema.europa.eu/en/medicines/human/EPAR/holoclar.
82. Kate, A. and Basu, S., A Review of the Diagnosis and Treatment of Limbal Stem Cell Deficiency. *Front. Med.* [Internet], 9, 836009, 2022, Available from: file:///pmc/articles/PMC9175008/.
83. Strimvelis | European Medicines Agency (EMA) [Internet]. Available from: https://www.ema.europa.eu/en/medicines/human/EPAR/strimvelis.
84. Bradford, K.L., Moretti, F.A., Carbonaro-Sarracino, D.A., Gaspar, H.B., Kohn, D.B., Adenosine Deaminase (ADA)-Deficient Severe Combined Immune Deficiency (SCID): Molecular Pathogenesis and Clinical Manifestations.

J. Clin. Immunol. [Internet], 37, 626–37, 2017, Available from: https://pubmed.ncbi.nlm.nih.gov/28842866/.

85. Barker, R.A., Carpenter, M.K., Forbes, S., Goldman, S.A., Jamieson, C., Murry, C.E., *et al.*, The Challenges of First-in-Human Stem Cell Clinical Trials: What Does This Mean for Ethics and Institutional Review Boards? *Stem Cell Rep.* [Internet], 10, 1429–31, 2018, Available from: https://www.ncbi.nlm.nih.gov/pmc/articles/PMC5995446/.
86. Zakrzewski, W., Dobrzyński, M., Szymonowicz, M., Rybak, Z., Stem cells: past, present, and future. *Stem Cell Res. Ther.* [Internet], 10, 68, 2019, Available from: https://www.ncbi.nlm.nih.gov/pmc/articles/PMC6390367/.
87. Kimmelman, J. and Federico, C., Consider drug efficacy before first-in-human trials. *Nature* [Internet], 542, 25–7, 2017, Available from: http://www.ncbi.nlm.nih.gov/pubmed/28150789.
88. Clinical Resources [Internet]. *Int. Soc. Stem Cell Res.*, 2024, Available from: https://www.isscr.org/clinicians.
89. Quinlan, A.R., Boland, M.J., Leibowitz, M.L., Shumilina, S., Pehrson, S.M., Baldwin, K.K., *et al.*, Genome sequencing of mouse induced pluripotent stem cells reveals retroelement stability and infrequent DNA rearrangement during reprogramming. *Cell Stem Cell* [Internet], 9, 366–73, 2011, Available from: http://www.ncbi.nlm.nih.gov/pubmed/21982236.

17

Beyond Borders: Global Perspectives on Stem Cell Therapeutics

N.L. Swathi[1*], Praveen Mallari[2], Mihir Dabhi[3] and Shweta Shrivastava[4†]

[1]*Pharm D, JNTUA University, Department of Pharmacy Practice, Andhra Pradesh, India*
[2]*Department of Zoology, Indira Gandhi National Tribal University, Amarkantak, Madhya Pradesh, India*
[3]*MBBS, Civil Hospital, Ahmedabad, Gujarat, India*
[4]*School of Pharmacy, School of Health and Allied Sciences, ARKA JAIN University, Jharkhand, India*

Abstract

Treatments for numerous diseases, including cancer, neurological disorders, and cardiovascular ailments, are being made possible by stem cell therapies, which are evolving medicine. The forefront research and clinical trials from North America, Europe, Asia-Pacific, Latin America, and Africa are the main focus of this study, which examines the global developments, difficulties, and prospects for stem cell applications and research. The research and application of stem cell therapies are shaped by the legal and ethical contexts that it analyzes, taking into account public opinion and acceptance in various nations. By presenting case studies of fruitful international collaborations that get over financial, logistical, and scientific obstacles, the chapter highlights the significance of international collaborations and partnerships in advancing stem cell research. Emphasizing the value of global perspectives in promoting innovation and improving patient outcomes, the study also examines cutting-edge technology and emerging trends that have the potential to change stem cell research. The present analysis highlights the significance of a cooperative worldwide strategy in the progress of stem cell therapies, opening up new avenues for medical discovery and better healthcare options for various demographics.

**Corresponding author*: nlswathi1999@gmail.com
†*Corresponding author*: shweta.shrivastwa@gmail.com

Ayush Madan, Sumel Ashique, Deepika Arora and Mantosh Kumar Satapathy (eds.) Stem Cell Therapeutics, (705–738) © 2025 Scrivener Publishing LLC

Keywords: Stem cell therapies, global health, medical innovation, clinical trials, ethical issues, regulatory frameworks, international collaboration, public perception, biotechnology, regenerative medicine

17.1 Introduction

17.1.1 Overview of Stem Cell Therapies

Globally, stem cell therapies are developing quickly and are showing promise in a number of medical specialties, such as neurology, cancer, and orthopedics. Even though the use of mesenchymal stem cells (MSCs) to treat ailments including spinal cord injury and osteoarthritis has advanced significantly, issues with long-term safety and standards still exist. The field of stem cell therapy is broad, including applications in orthopedics, neurology, cardiology, and dermatology, among other specialties. Tang *et al.* (2024) conducted a systematic review and meta-analysis examining the effectiveness of mesenchymal stem cells (MSCs) obtained from different sources in the treatment of knee osteoarthritis (KOA) [1]. Visual Analogue Scale [VAS] SMD 0.97; 95% CI 0.37–1.57 showed that adipose-derived MSCs (AD-MSCs) were significantly more effective than placebo in decreasing pain and functional ratings. This network meta-analysis included 16 randomized controlled trials (RCTs) with 1,005 participants. The results indicate a strong degree of evidence in favor of AD-MSCs as a potential KOA therapeutic option despite ongoing worries about MSC therapy's side effects [1].

Alessandrini *et al.* (2019) provide a summary of stem cell uses in the treatment of neurological illnesses, including multiple sclerosis and spinal cord injuries. They noted that there are now over 200 registered clinical studies, but only a small number of them have advanced to pivotal trials, which are required for regulatory approval [2]. This emphasizes how many treatments must first undergo extensive clinical research to confirm their safety and effectiveness before being widely used.

Lotfy *et al.* (2023) examined clinical trials employing MSC-derived exosomes (MSC-Exos), which are showing promise as a substitute for direct stem cell therapies because of their decreased risk of side effects. This further supports the potential of MSCs [3]. The analysis highlighted MSC-Exos' promise as a safer treatment option by pointing out that they have demonstrated success in immunomodulation and regenerative medicine across a range of disorders.

Rikhtegar *et al.* (2019) addressed the use of various stem cell types, including induced pluripotent stem cells (iPSCs), embryonic stem cells (ESCs), cardiac stem cells (CSCs), and skeletal myoblasts, in the therapy of heart disease in cardiology [4]. They observed that despite the fact that a number of trials have shown promise in terms of restoring cardiac function following a MI, there is still a great deal of variation in the treatment approaches used in different research projects.

Furthermore, a review of stem cell therapy for spinal cord injuries was carried out by Yamazaki *et al.* (2020). While pointing out a number of clinical trials, they also emphasized the lack of agreement regarding the best transplantation techniques, such as the best cell type and delivery method, highlighting a crucial subject for further study [5]. It is more difficult to reach firm findings regarding the efficacy of stem cell treatments for SCI because of this contradiction.

Wang *et al.* (2023) investigated the application of human umbilical cord MSC-derived exosomes for the treatment of melasma, a difficult-to-treat skin disorder, highlighting their promise in regenerative medicine. According to their research, using these exosomes in conjunction with cutting-edge delivery techniques may improve treatment results [6].

In spite of these encouraging advancements, stem cell therapy still faces a number of obstacles. Mazzini *et al.* (2016) explored the challenges associated with using stem cells to treat ALS. They noted that although mesenchymal and neural fetal stem cells exhibit potential, a thorough understanding through carefully planned clinical trials is necessary to advance these therapies into routine care procedures [7].

A historical perspective on stem cell research was also provided by Zakrzewski *et al.* (2019), who also addressed modern concerns like public perception and regulatory constraints. In order to integrate stem cell treatments into traditional medicine, they stressed that overcoming these challenges is crucial [8].

17.2 Global Advances in Stem Cell Research

17.2.1 Breakthroughs in North America

North America achieved greater research in stem cell research in the past decades as there was ample funding from the government and other private sectors. It was at the beginning of the 1980s when Evans and Kaufman (1981) isolated mouse embryonic stem cells and supported their culture, which indicated the possibility of the given approach [9]. Subsequently,

Table 17.1 Global advances in stem cell research.

Country	Type of stem cell	Outcome	Methodology used	Challenges faced	References
North America	Mouse embryonic stem cells	Derived and maintained embryonic stem cells in culture	Mouse embryonic stem cells were derived and cultured	N/A	Evans and Kaufman, 1981 [9]
	Human embryonic stem cells	Isolated the first human embryonic stem cell line	Derived from blastocysts to study human development and diseases	Ethical issues surrounding the use of embryos	Thomson *et al.*, 1998 [10]
	iPSCs	Converted adult cells into iPSCs using transcription factors	Yamanaka's method uses four transcription factors	Technical challenges in reprogramming	Takahashi and Yamanaka, 2006 [11]
	Embryonic stem cells	Initial clinical trial using embryonic stem cells to treat spinal cord injury	FDA-endorsed trial	The trial was cut short	Lebkowski, 2011 [12]
		Treated patients with age-related macular degeneration	Used embryonic stem cell-derived retinal pigment epithelium	Technical and ethical challenges	Liu *et al.*, 2021 [13]

(*Continued*)

Table 17.1 Global advances in stem cell research. (*Continued*)

Country	Type of stem cell	Outcome	Methodology used	Challenges faced	References
	Various stem cells	Developed methods for reprogramming cells for clinical use	Cost-effective reprogramming and disease modeling	Financial and technical challenges	Fatehullah *et al.*, 2016 [14]
Europe	Human embryonic stem cells	Derivation of hESC lines	Derivation from human blastocysts	Ethical issues	Vazin and Freed, 2010 [15]
	MSCs	Isolated MSCs from human liposuction procedures	Isolation of MSCs from liposuction	Ethical and sourcing issues	Zuk *et al.*, 2001 [16]
	iPSCs	Generated human iPSC lines	Reprogramming skin cells into iPSCs	Technical and ethical challenges	Park *et al.*, 2008 [17]
	Various stem cells	Over 400 active clinical trials by 2010	Various methodologies for clinical trials	Regulatory and ethical challenges	Trounson *et al.*, 2011 [18]
	Human embryonic stem cells	First trial using hESCs in the treatment of macular degeneration	Transplantation of new eye cells	Ethical issues and technical challenges	da Cruz *et al.*, 2018 [19]

(*Continued*)

Table 17.1 Global advances in stem cell research. (*Continued*)

Country	Type of stem cell	Outcome	Methodology used	Challenges faced	References
Asia-Pacific	Various stem cells	Significant growth in publications, patents, and trials in China	Investment in research infrastructure and trials	Ethical concerns (China)	Chen *et al.*, 2021 [20]
	Adult stem cells	Focused on adult stem cell research in India	Research and trials focusing on adult stem cells	Funding and infrastructural support	Bagcchi, 2014 [21]
	iPSCs	Advanced iPSC research and applications in Japan	Center for iPS Cell Research and Application	Ethical and technical challenges	Enosawa, 2022 [22]
	Various stem cells	Collaboration through International Consortium of Stem Cell Networks	Regional collaboration and annual conferences	Ethical issues, funding, and infrastructure	Chen *et al.*, 2022 [23]

(*Continued*)

Table 17.1 Global advances in stem cell research. (*Continued*)

Country	Type of stem cell	Outcome	Methodology used	Challenges faced	References
Latin America	Embryonic and adult stem cells	Development of hESC lines and clinical trials in Brazil	Networked research institutions and partnerships	Ethical issues and regulatory challenges	Martins de Oliveira *et al.*, 2023 [24]
	Mesenchymal stem cells	Phase III trials using MSCs for diabetes patients in Mexico	Use of MSCs for diabetes, spinal cord injury, multiple sclerosis, and Parkinson's disease		Anand, 2024 [25]
	Various stem cells	Developed public stem cell systems and clinical applications in Cuba	Use of hematopoietic stem cells and mesenchymal stem cells for various treatments		Fernández-Montequín *et al.*, 2009 [26]
Africa and Middle East	Human embryonic stem cells	Creation of hESC lines and iPSCs in South Africa	Establishment of SAn Stem Cell Institute	Ethical and regulatory challenges	Azeez *et al.*, 2022 [27].
	Various stem cells	Coordination of research activities and training in 11 African countries	Development of stem cell banks, institutes, and regulatory authorities	Funding, ethical, and infrastructural challenges	Aboalola *et al.*, 2024 [28]

Thomson *et al.* (1998) derived the first embryonic stem cell line from human blastocysts for understanding human development and diseases [10]. The use of iPSCs came to a significant milestone in 2006 when Takahashi and Yamanaka (2006) changed ordinary somatic cells into iPSCs using only four proteins called transcription factors, a technology that earned the 2012 Nobel Prize in Physiology/Medicine [11]. Refer Table 17.1 as it provides an overview of global advances in stem cell research, summarizing key developments, methodologies, and challenges in various regions.

Stem cell treatments have also progressed; they have improved in a way that they are now able to be used to heal wounds and other injuries. Stem cells have the potential to transform into other differentiated cells at some point in early life or as the organs and tissues grow. In 2010, the first FDA-approved clinical trial involving embryonic stem cells began with the firm Geron Corporation conducting a trial to treat spinal cord injuries [12]. Liu *et al.* (2021), at the University of California, San Francisco used embryonic stem cell-derived RPE to restore vision in patients with AMD and improve their visual acuity by providing them with a new RPE layer [13].

Another milestone in this field was achieved by Chad Cowan of Harvard University who has come up with a cheaper technique of reprogramming cells for use in clinical applications and to generate cell types that capture diseases and potentially cure them. The mouse and human long-term organoids of Fatehullah *et al.* (2016) offered the best model to investigate disease and development [14]. Refer Figure 17.1 as it illustrates the major breakthroughs in stem cell research in North America, including early stem cell derivations, clinical applications, and recent advancements in regenerative medicine.

17.2.2 Innovations in Europe

For the last two decades, Europe has been at the forefront of several key stem cell technologies such as embryonic stem cells (hESCs), adult stem cells, induced pluripotent stem cells (iPSCs), and stem cell regenerative medicine. The discovery of the first hESCs was carried out at the University of Wisconsin in 1998 from human blastocysts. Next, Sweden was the first country to generate European hESC lines in the year 2000, while the UK generated in the year 2001 [15]. Refer Figure 17.2 as it presents the advancements in European stem cell research, highlighting key milestones such as the derivation of hESC lines and the rapid adoption of iPSC technology.

European researchers have advancements in adult stem cells such as the mesenchymal stem cells (MSCs) found in bone marrow and other organs. For instance, Italian researchers are credited for the isolation of MSCs from

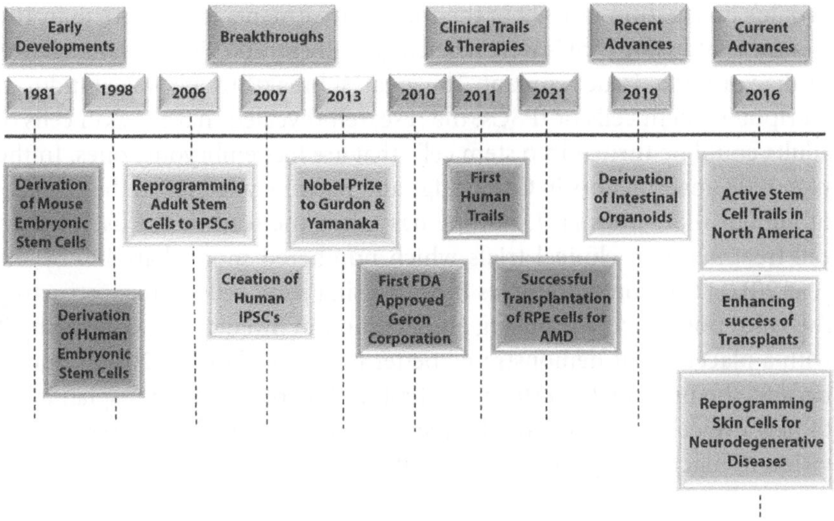

Figure 17.1 Global advances in stem cell research in North America. Stem cell research timeline highlights key milestones from 1981 to 2021 in five categories. Some topics include early stem cell discovery, mouse/human embryonic stem cells, human iPSCs, Nobel Prizes, stem cell clinical trials/therapies, the most recent in stem cell therapy.

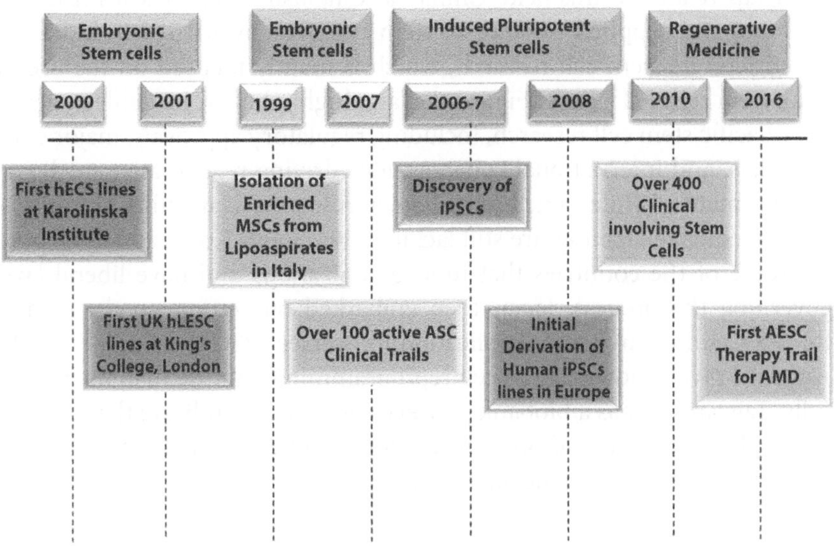

Figure 17.2 Innovations of stem cell research in Europe. Stem cell research has undergone immense progress throughout the years 1999–2016: human embryonic stem cells in 2000, discovery of induced pluripotent stem cells in 2006, over 400 stem cell trials by 2010, more embryonic stem cell milestones in Europe, and advancements in stem-cell-derived regenerative medicine.

people's liposuction in 1999 in their use in treatments [16]. More than 100 clinical trials using MSCs were labeled globally by 2007 in domains such as cardiac or bone tissue remnant. The greatest breakthrough was of induced pluripotent stem cells or iPSCs; the research revealed that the skin cells of adults could be turned into stem cells that are like embryonic ones. In the year 2008, investigators in Cambridge and Barcelona launched the creation of human iPSC lines [17]. Europe has taken the lead in the use of stem cell treatments in clinical trials which numbered more than 400 by the year 2010 [18]. Some examples include using stem cells in reconstructing the nasal bone/cartilage during an operation in Sweden and comparing it with a placebo that demonstrated better outcomes [29]. In 2016, London became the first city to perform the first trial involving the transplantation of new eye cells derived from human embryonic stem cells to restore lost vision in macular degeneration patients [19].

17.2.3 Progress in Asia-Pacific

In recent years, nations of the Asia-Pacific region have gradually emerged as leaders in stem cell research and regenerative medicine. This country alone has one-third of the global total production of publications, patents, and trials related to this field. China has established well-known research institutes and supplied 1,000 or so stem cell trials by 2015 [20]. India has also driven its stem cell research, which is mainly focused on the use of adult stem cells [21]. Refer Figure 17.3 as it highlights the key milestones in Asia-Pacific stem cell research, including regulatory approvals, major clinical trials, and international collaborations driving regenerative medicine advancements. While India has laid down rules, the availability of funds as well as requisite facilities are still lacking [30]. South Korea and Singapore are some of the countries that fund organizations and have liberal laws governing the process. Japan then embarked on setting up the Center for iPS Cell Research and Application to oversee iPSC development [22]. Australia gives knowledge in cell reprogramming and the manufacture of cells. Taiwan still has a biobank. Observable patterns indicate that elites in China, Japan, India, and Singapore integrate internationally primarily to learn from one another and share assets [23].

17.2.4 Developments in Latin America

In 2008, a group of researchers created the Brazilian National Laboratory of Embryonic Stem Cells which was devoted to the isolation, archiving,

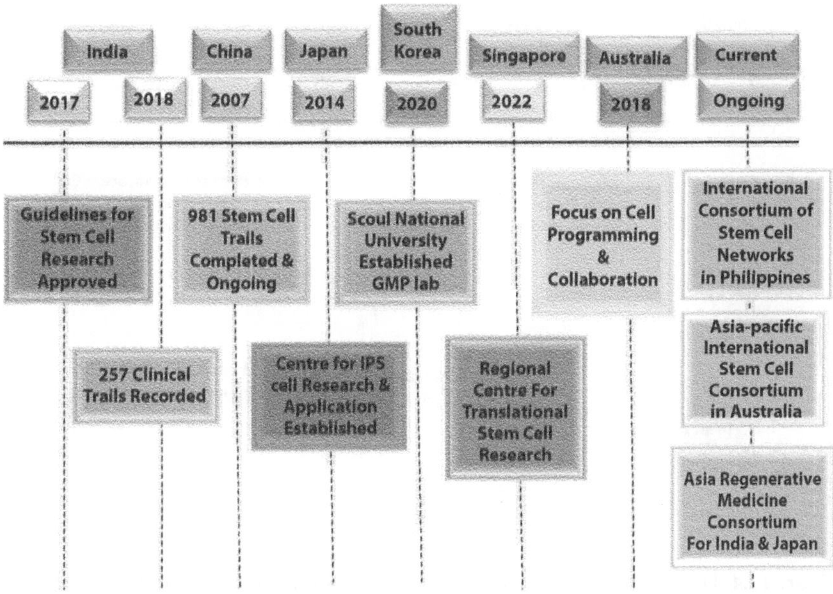

Figure 17.3 Progress in Asia-Pacific regions on stem cell research. Stem cell research timeline across the Asia Pacific from 2007 to 2022: India's guidelines 2017, China's clinical trials, Japan and South Korea's developments, Singapore and Australia's consortium, and international collaborative efforts in the Asia-Pacific region.

and description of embryonic stem cells for other scientists to use. By 2018, this lab has successfully produced 17 human embryonic stem cell lines, all by the standards of the global community [24]. Refer Figure 17.4 as it highlights the development in stem cell research. The University of São Paulo's Regenerative Medicine Center is now among the leaders in stem cell therapy clinical trials. In the past 5 years, the center's investigators have devised numerous phase II trials utilizing stem cells to regenerate heart function in heart disease patients [25]. In 2007, Mexico looked for further stem cell research by establishing the Mexican Stem Cell Network offered funds for stem cell research and partnerships and technology exchange [31].

Cuba has established its official public stem cell system focusing on hematopoietic stem cells derived from umbilical cord blood to address blood-related diseases and mesenchymal stem cells for regeneration. Other Cuban clinics have also pioneered the application of patients' bone marrow-derived mesenchymal stem cells for orthopedic diseases and wounds and neurological disorders [32].

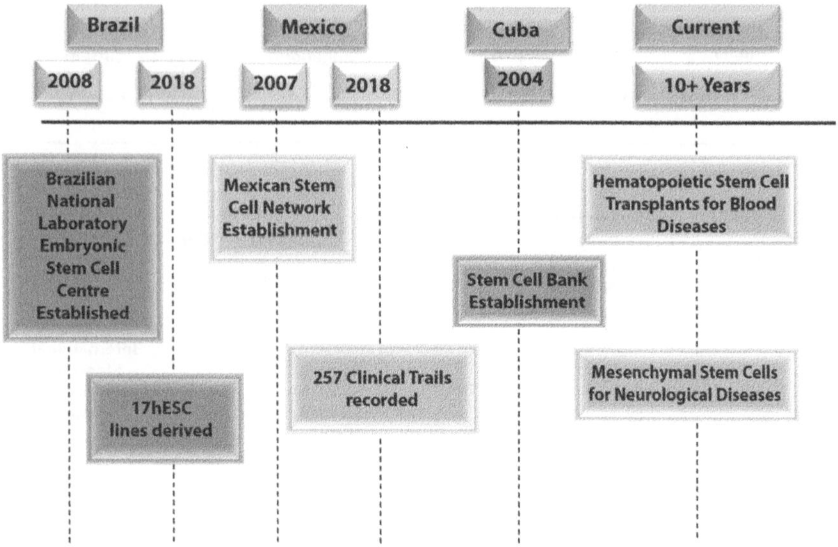

Figure 17.4 Developments in stem cell research in Latin America. Major milestones in stem cell research in three Latin American countries. It also emphasizes the establishment of research institutes and networks, as well as the principal clinical uses, thereby providing an understanding of the development from the initial science to today's complex cell therapies that are specific to each country.

17.2.5 Research in Africa and the Middle East

The research on stem cells has risen significantly within the African and Middle Eastern regions over the last 20 years. South Africa was the first country in Africa to generate human embryonic stem cell lines in 2004 and the first to make induced pluripotent stem cells in 2010 as Jackson and Pepper noted. While South Africa started the SAn Stem Cell Institute in 2005 to foster research and provide a platform for research collaboration with other countries in the African continent, Egypt, Nigeria, Ghana, and Tanzania have only recently initiated stem cell research at the basic level as was stated by Azeez *et al.* (2022) [27]. Refer Figure 17.5 as it outlines the achievements, ongoing research, and future trends in stem cell research across Africa and the Middle East, highlighting ethical policies, funding challenges, and key research hubs.

The following new institutions have emerged over the past 5 years to coordinate stem cell banking, institutes, and regulatory authorities in Africa. The Pan-African Cell Therapy Immunotherapy Alliance has

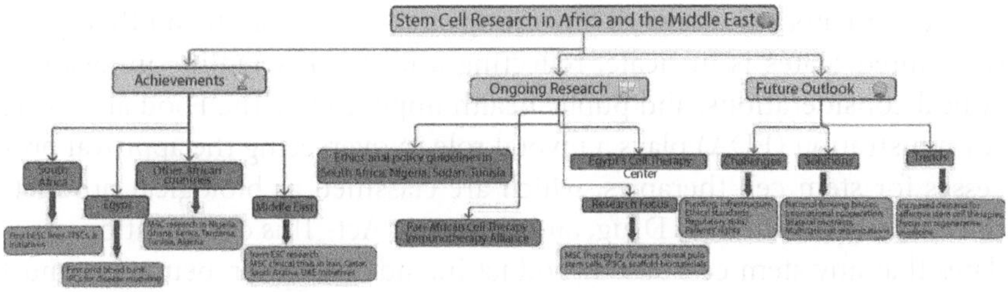

Figure 17.5 Flowchart diagram for stem cell research in Africa and the Middle East. Africa and the Middle East are actively progressing in the field of stem cell research through partnerships, codes of ethical conduct, and addressing challenges in funding and base infrastructure. Some progress has been made but the issue is still far from being solved completely. Emphasizing collaboration and agreement to enhance the capability of the field to its full capacity.

expanded and is offering training, policy, and networking services in 11 African countries according to Aboalola *et al.* (2024) [28]. Main research areas of interest include the use of mesenchymal stem cells to address tissue repair for diseases like diabetes, cardiovascular diseases, and autoimmune disorders [33].

In this case, scientists have developed some practical ways of sustainable development, but there are constraints on the practice of the concept. These are the factors such as inadequate funding and infrastructural development attributed to low national research expenditure as well as limited industry players. Some of the areas that require progress include ethical standards for deriving embryos and cell lines, supervision of gene editing processes, and ownership or patents for embryonic stem cells [28].

17.3 Regional Regulatory Frameworks

17.3.1 Regulatory Environment in the USA

The regulatory environment for stem cell research and therapy in the USA is characterized by a complex interplay of federal and state regulations, ethical considerations, and evolving scientific standards. Key agencies such as the FDA oversee the approval processes for stem cell therapies, emphasizing safety and efficacy while navigating public health needs and ethical concerns. The landscape is marked by significant variability across states, leading to disparities in access and application of stem cell treatments.

The regulatory framework governing stem cell research and therapy in the United States is intricate, reflecting a blend of scientific innovation, ethical considerations, and public health imperatives. The Food and Drug Administration (FDA) plays a pivotal role in overseeing the approval processes for stem cell therapies, which are classified as biological products under the Federal Food, Drug, and Cosmetic Act. This classification mandates that any stem cell-based product intended for therapeutic use must undergo rigorous testing to demonstrate its safety and efficacy before it can be marketed [34].

One critical aspect of the FDA's regulatory approach is its emphasis on distinguishing between minimally manipulated cells and those that are more significantly altered or combined with other substances. The former may qualify for more lenient regulation under Section 361 of the Public Health Service Act, while the latter falls under stricter scrutiny as a drug or biologic product [35]. This distinction is crucial as it influences the pathway through which stem cell therapies can be developed and made available to patients.

Moreover, state-level regulations add another layer of complexity to the regulatory landscape. States like California have enacted more permissive laws that encourage stem cell research through funding initiatives and supportive policies. In contrast, other states may impose stricter limitations or outright bans on certain types of stem cell research or therapies [36]. This patchwork regulatory environment creates significant disparities in access to innovative treatments across different regions of the country.

Ethical considerations also play a fundamental role in shaping the regulatory framework. The National Institutes of Health (NIH) has established guidelines for stem cell research that emphasize the importance of informed consent when using human embryonic stem cells. These guidelines aim to address ethical concerns surrounding the sourcing of embryonic cells while promoting transparency in research practices [37]. Furthermore, ongoing debates about the moral implications of using fetal tissue for research continue to influence public perception and policy decisions regarding stem cell therapies.

Recent advancements in regenerative medicine have prompted discussions about potential updates to existing regulations. For instance, there is growing interest in gene editing technologies like CRISPR-Cas9, which raise questions about their integration into stem cell therapies. As these technologies evolve, regulatory bodies will need to adapt their frameworks to ensure patient safety while fostering innovation [5].

Overall, the regulatory environment for stem cell research and therapy in the USA is dynamic and multifaceted. It reflects ongoing efforts to balance scientific progress with ethical considerations and public health needs. As new technologies emerge and our understanding of stem cells deepens, continuous evaluation and adaptation of regulatory frameworks will be essential to ensure patient safety while promoting innovation.

The regional regulatory frameworks governing stem cell research and therapy in the USA are shaped by a combination of federal oversight, state-specific regulations, ethical considerations, and evolving scientific knowledge. The FDA's stringent approval processes are critical for ensuring that only safe and effective therapies reach patients; however, disparities across states create significant variability in access to these innovative treatments. As advancements continue within regenerative medicine, ongoing dialogue among stakeholders—including researchers, regulators, ethicists, and patients—will be vital for navigating this complex landscape effectively.

17.4 Ethical Considerations and Public Perception

17.4.1 Ethical Dilemmas in Stem Cell Research

Apart from the challenges and barriers that halt the progress of stem cell therapy, there are various ethical dilemmas that also need to be dealt with concurrently. Cultural factors such as religion and society also have a huge influence on the public image and acceptability of this therapy.

17.4.1.1 Embryonic Stem Cells (ESCs)

ESCs are pluripotent and have the potential to virtually differentiate into any type of adult cell, which makes them highly valuable in various applications of stem cell therapy. However, their use is riddled with certain ethical issues.

ESCs are generally derived from embryos, which are around 5 to 6 days old. Their extraction involves inevitable destruction of the embryo and this is considered morally unacceptable since embryos have the potential to give rise to human life. The union of male and female gametes, to form an embryo, is considered to be the moment of creation of new life according to many religious and philosophical beliefs.

The embryos are usually obtained from the surplus embryos that are formed through *in vitro* fertilization techniques. People usually assign special emotions and moral value to their gametes in contrast to other tissues/products of their body. This raises an ethical issue of taking consent from the donor parents, who should be thoroughly and carefully explained about all the potential use of their gametes. This is particularly important because some donors might find it as a violation of their privacy in case their genetic sequence is decoded by scientists through research on their gametes/embryos because there is a possibility of re-identification of donor through their genomic sequence, and future genetic susceptibility of donors to various diseases can be done [38, 39]. Some donors might have an objection with their cells injected into animals or with mixing their cells with those of animals to give rise to chimeras [40].

Confidentiality is another concern regarding donors, which must be safeguarded at any costs, because a breach in confidentiality can lead to the donors facing unwanted publicity and opposition from those against stem cell therapy or from opponent companies. There have been cases of break in confidentiality through break-ins by staff, hacking of computers containing such information, or loss/theft of such computers. In order to ensure that such incidents do not happen in the future, security must be increased and improved so that such volatile information does not get jeopardized [41].

17.4.1.2 Safety and Efficacy of Therapy

Stem cell therapy, though promising, is not very effective as of yet. Combined with the issues of its safety in terms of tumorigenicity, there are ethical issues regarding its efficacy and hence, implementation on a larger scale. Even with various safeguards in place, due to the relatively newer advent of the therapy and less human knowledge and familiarity with this therapy, unforeseen serious events might happen. In an instance, some trials involving the transfer of fetal dopaminergic neurons in patients of Parkinson's disease resulted in failure to improve clinical outcomes. Moreover, in 15% of patients, late disabling dyskinesias developed as a side effect, with some patients requiring surgery to get relief from such side effects [42, 43].

To be implemented on a large scale for the welfare of the society, stem cell therapy has a lot of ground to cover in safety and efficacy. The commercialization of stem cell therapy also raises issues such as exploitation of patients by giving them false hopes and marketing unproven treatments to charge exorbitant fees from patients. Often, such therapies lack any scientific backing, which leads to loss of money and mistrust in the minds of people who avail such therapies. The International Society of Stem Cell Research supports medical innovations although in limited circumstances and opposes the use of unproven stem cell transplantation which charge patients exorbitant fees for therapies without any scientific backing, patient protections, or transparency [44]. Healthcare providers and regulatory bodies should therefore ensure that guidelines are strictly enforced and patient welfare is kept as the prime priority.

Stem cell therapy is potentially life-saving, but due to the high cost and uncertainty in the efficiency of therapy, few people can actually avail the benefits of this therapy. This can lead to disparity in healthcare accessibility particularly to financially lower segments of the society, by creating inequality in access to stem cell therapy.

17.4.1.3 Animal Models and Chimeras

Stem cell research inevitably includes creation of animal models that correspond to different human diseases and even chimeras (organisms that contain cells from two or more different species) to study the pathophysiology of diseases and test the safety and efficacy of various therapies. This use of animals for research is a concern from the view of animal welfare and with newer stricter regulations in place regarding animal use and to minimize suffering of animals; it has become difficult for researchers to adhere to regulatory guidelines. Also, using chimeras involving human and animal cells raises a completely new set of ethical and moral dilemmas regarding the human-ness of such chimeras and their possibility of developing human like consciousness or characteristics.

17.4.1.4 Responsible and Ethical Research

Studies have been found in which regulations are disregarded and unethical research has been conducted. For instance, the Hwang Scandal (in South Korea) featured claims of scientific fraud, inadequate compensation for cell donors, irregularities in obtaining fully informed consent, undue pressure on researchers and staff to donate blood, and a high rate of adverse medical effects from the oocyte donation procedure [45–47]. It has been suggested that such adverse effects should be brought under the cover of a health insurance but it could not be realized due to difficulties in estimating the risk associated with this therapy over a long term and also in determining all the intervening factors that can directly or indirectly contribute to the adverse events.

17.4.2 Cultural Influences on Public Perception

The public's perception of in order to advance and gain acceptance of stem cell therapy, it is imperative that more and more people learn about it through education and communication. Any misunderstandings should be cleared up in order to promote mutual trust. They should be educated on the possible advantages, dangers, and ethical ramifications of thc treatment as part of this. Public perception is influenced by a range of religious, cultural, and societal issues. These factors subsequently impact laws, regulations, financing for research, and the acceptance of stem cell therapy. Stem cell therapy advancement and progress are therefore influenced both directly and indirectly by public perception.

17.4.2.1 Religious Viewpoints

Various religions have differing viewpoints regarding research in stem cell therapy and its use. There are various viewpoints within Christianity which are contradictory. The Catholic Church strongly opposes the destruction of embryos, considering it as morally unacceptable, while Protestant followers have a more open perspective and have an optimistic belief about the medical advancement that can be made through stem cell therapy. Jehovah's witnesses, on the other hand, might not accept stem cell therapy since they are strictly against the introduction of animal or human products in their body, even for life-saving procedures.

The acceptability of stem cell research in Islam is debated among Islamic scholars, with some authorities considering it acceptable within scientific limits, whereas their conservative counterparts are being more restrictive about stem cell therapy.

The principle of "pikuach nefesh" (which prioritizes saving a life) guides the religious stance of Judaism, with Jewish ethicists supporting the use of embryos in research and its subsequent use if it turns out to be life-saving [48].

17.4.3 Case Studies of Public Engagement

17.4.3.1 UK Stem Cell Network (UKSCN)

Established in 2003, the UKSCN aims to coordinate and promote stem cell research across the UK. It organized various public dialogues and consultations to know public opinion about stem cell therapy and address its ethical concerns. These dialogues involved many stakeholders such as scientists, ethicists, patients, religious groups, and people from the general public, and the insights that were gathered from them helped to shape policies regarding research and development of stem cell therapy. UKSCN has also mentioned the importance of increasing public understanding of stem cell science and its implications through educational outreaching. They collaborated with schools and universities to involve this topic in their curriculum and developed educational resources and lectures on the topic.

The Nuffield Council on Bioethics and Human Fertilisation and Embryology Authority are some examples of the robust ethical framework that the UK has established to tackle the ethical concerns. These institutions constantly work to guide researchers and organizations on ethical issues, ensure that regulations are being complied, and increase awareness among the general public [49].

17.4.3.2 Stem Cell Network (SCN) Canada

SCN Canada has played a significant role in furthering awareness and public engagement in Canada since its establishment in 2001 by organizing various public lectures, workshops and interactive exhibitions at science museums. SCN has also fostered collaboration among scientists, clinicians, and biotech organizations to accelerate research in the field and its application in medicine. This has led to a more supportive environment for innovation and research [50].

17.4.3.3 California Institute for Regenerative Medicine (CIRM)

With a staggering allocation of $3 billion for stem cell research in California, CIRM was established in 2004. It is an example of state-level initiatives in increasing public engagement and stem cell research. It was established after the approval for Proposition 71, which was an initiative that gathered significant support from the public through a state wide voting. This proposition was designed to address federal restrictions in the field of SCT, with an aim to make the state of California a leader in the field. This was made possible through extensive public engagement campaigns, involving educational programmes, forums for public discussion and other workshops with endorsements from leading scientists and research groups in the field.

CIRM has kept transparency and accountability as the core values in its workings and regularly publishes its progress reports, research outcomes, and updates as well as funding decisions. It also holds public meetings and stakeholder meetings from time to time to know public opinion and inputs and maintain public trust. CIRM has always kept patients and general public actively involved in its decision making and governance. Review committees and advisory panels always involve public and patients advisors to ensure that the viewpoints of patients who are affected by the disease and are likely to be affected by therapy are always heard [51].

17.5 Challenges and Barriers

The field of stem cell therapy (SCT) is a relatively new area of medicine with great promise to manage and treat a wide range of diseases that were previously thought to be incurable. Nevertheless, due to a number of obstacles and challenges, the field of SCT is not yet as widely applicable as predicted.

17.5.1 Scientific and Technical Challenges

17.5.1.1 Cell Source and Differentiation

Broadly, stem cells are of three types: embryonic (ESC), induced pluripotent (iPSCs), and adult stem cells. Each of them presents unique challenges, and the selection of a desirable and ethically acceptable cell type among them remains one of the initial dilemmas.

ESCs are stem cells that originate from embryos. They can differentiate into any cell type, making them highly valuable for therapeutic purposes. However, if not fully differentiated before transplantation, they can give rise to teratoma (which is a tumor composed of different types of tissues such as skin, bone, hairs, and even teeth). iPSCs, on the other hand, are not derived from embryos but instead they are differentiated cells that have been re-programmed to the embryo-like state. They do bypass the ethical issues associated with ESCs but their reprogramming has the possibility of inducing genetic mutations which is not desirable. Also, their efficiency in differentiating into a desired cell type is not satisfactory and this remains a challenge in their use. Adult stem cells, by definition, are multipotent (instead of pluri-potent) and have the lowest potential to differentiate into different types of cells. They are harvested from a human donor, which makes the process invasive and at times painful for the donor. In certain cases, acquisition of gametes from donors can cause medical complications for the donor. For example, oocyte retrieval poses the risk of ovarian hyperstimulation syndrome, excessive blood loss, infections, and complications due to the delivery of anesthesia. Even though that can be reduced through processes such as exclusion of donors that have an increased susceptibility to develop these reactions and adjusting the dose of ovulation-inducing drugs, the threat still looms. Hence, the use of adult stem cells is not free from challenges as well [52].

17.5.1.2 Cell Homing, Survival, and Integration

Homing can be termed as the process where stem cells migrate from exogenous or endogenous sources, through blood stream or connective tissues, to a place where they differentiate and replace the injured tissue by taking up its function or repairing it. Inappropriate homing of stem cells to their destined place can lead to failure of therapy. Another significant challenge is to ensure that transplanted stem cells survive in the recipient, their integration with the host cells, and their proper functioning in the host tissue. After integration, there are various issues such as transplant rejection, inflammation, and lack of vascularization. All of these can lead to the destruction of the transplanted stem cells and subsequent failure of therapy. Overcoming these setbacks would require an in-depth understanding of the host environment and pathways involving stem cell differentiation, integration, and survival [53].

17.5.1.3 Immune Tolerance and Tumorigenicity

One of the major hurdles in achieving successful stem cell integration and function is immunological tolerance of stem cells. There is a possibility that the immune system of the recipient can recognize the stem cells as "foreign" and initiate an immune attack against them, causing their death. This phenomenon, known as "immune rejection" is more common with ESCs and iPSCs (adult stem cells have a relatively low immune rejection rate since they are derived from the host cells, but their use is not always feasible or effective). Another big concern in using stem cells, particularly pluripotent types, is the inability to control the degree of their proliferation. Though infinite proliferation is considered an advantage, this can be a double-edged sword because there is a possibility that these cells might multiply uncontrollably even after transplantation, leading to the formation of tumors. Retention of immature/undifferentiated cells in final cell product, activation of reprogramming factors in certain stem cells that promote tumorigenesis, and defects in the multiplication of stem cells (*in vitro* or post-transplantation) due to genetic instability, in the form of mutations or improper chromosomal distribution/disjunction, can all lead to the formation of tumors [54].

17.5.2 Logistical and Financial Barriers

17.5.2.1 Cost of Therapy

A huge non-scientific barrier hindering the progress of stem cell therapy is the substantially high amount of financial investment that is needed to develop it. From research, trials, and approvals to the personnel, facilities, and equipment that are a pre-requisite, expenses can be enormous, which limit its accessibility to patients and, hence, its adoption on a large scale. With the lack of a standardized cost of treatment, a massive increase in the number of clinics set up solely for the purpose of profit which often offer unproven therapies adds to the financial uncertainties and contributes to disparities in access to safe and ethical stem cell therapy [55].

Rules and regulations surrounding new therapies regarding their safety and efficacy are usually very stringent, and obtaining regulatory approvals from agencies such as US FDA and European Medicines Agency usually involves extensive research in the form of preclinical trials and clinical studies, which can be very time consuming and expensive. Adding to this issue is the fact that regulations regarding stem cell therapy are constantly evolving, which can further add to the uncertainties and delays in getting approved. Furthermore, due to differences in regulations across different countries and regions across the world, it becomes harder to work together in harmony to develop and commercialize stem cell therapy to a global level. For example, in an effort to decrease the number of oocytes, embryos, and stem cells utilized for research, they might be shared across different institutions. However, ethical issues can arise if the researchers at an institution work on cells or tissues that are obtained in other legal jurisdictions through processes and techniques that would not have been allowed in the jurisdiction of their home institution. Conversely, such loopholes can be used by some institutions to incentivize the breaking of such codes of ethics by people in a jurisdiction where it is permitted [56].

17.5.2.2 Copyright and Competition

Copyright of knowledge, methods, and technology related to stem cell therapy can create issues to the development of this therapy by restricting its access and distribution. This leads to a slower rate of progress of stem cell therapy by halting innovation and collaboration. Navigating through these issues further increases the time and cost that it takes to create a breakthrough in this field. Competition among bio-tech companies also prevents a smoother exchange of information between one another, which can be one reason for less collaboration and subsequent progress in the field.

17.5.3 Overcoming Obstacles: Best Practices

Although riddled with various challenges, SCT is constantly advancing to overcome barriers. New sources of stem cells like MSCs (mesenchymal stem cells) and disease/organ-specific cells (such as cardiac stem cells to treat MI and other cardiac disorders) are being discovered to potentially overcome the limitations of currently available conventional sources of stem cells. Furthermore, the genotypic and phenotypic characteristics of cells can be maintained and identified by characterizing them with various cell-type-specific markers. For example, cardiac stem cells have cardiomyocyte-specific markers such as troponin T2 (TNNT2), troponin I3 (TNNI3), etc., and their functionality can be ensured by checking their conductance, ion current, cell structure, and morphology, including cell membrane proteins [57].

Homing of stem cells can be improved by preconditioning the stem cells. In an instance, a group of researchers from Egypt have reported that certain stem cells that are treated with sodium hydrosulfide *in vitro* before transplantation showed improvement in homing compared to non-preconditioned cells [57]. Problems with rejection of stem cells can be dealt with by processing the stem cells before transplantation. By preconditioning the stem cells, their survival chances within a hostile recipient can be enhanced. Novel techniques related to cell delivery, timing, dosage, and co-administration of stem-cell-supportive agents can maximize cell survival, integration, differentiation, and immune response modulation. Nanotechnology also has a role in improving these techniques by allowing the targeted delivery of growth factors. Such advancements promise a highly personalized, targeted, and effective stem cell therapy in the future [58].

The chances of immune rejection can also be reduced by using immunosuppressive drugs both pre- and post-transplantation procedure. This, however, can predispose the host to various opportunistic infections (including fungal and viral infections) and flaring up of latent infections such as TB. Hence, the use of immunosuppressive drugs can be similar to walking on a tightrope, with an abyss on either side in the form of immune rejection and serious infections.

Methods to ensure the genetic stability of stem cells and their constant monitoring post-transplantation is vital to decrease the chances of these cells giving rise to cancer. As an example, stem cells were GFP (green fluorescent protein)-labeled and injected into a lab animal and then continuously monitored by qPCR, NMR (nuclear magnetic resonance), and MRI scans [59]. Such innovative monitoring methods are important to understand how stem cells behave and develop into either ectopic tissue or tumor after transplantation.

A novel approach is the use of engineered supporting scaffold which bio-mimics the host tissue in both structure and function. They are artificial support systems which have the capacity to maintain the proliferating and differentiating abilities of stem cells. It has been observed that conventional cell cultures which consist of "static tissue culture dish" support cell growth in two dimensions (2D). Even though the initial cell growth rate is higher in such mediums, once the cells reach overgrow and fill up the surface, the proliferation of the implanted cells stops. On the other hand, bio-engineered porous materials are 3D structures which provide a comparatively bigger surface area for cellular attachment and subsequent proliferation compared to the 2D counterparts used earlier. Vascularization and supply of nutrients, which are crucial for the survival of stem cells, are also facilitated with the use of such models where the distribution of nutrients, oxygen, and other vital cellular factors is enough to sustain a large number of cells over an extended period of time [60, 61].

Collaboration between individuals, governing bodies, and countries is vital for advancing stem cell therapy. Integrating research from various fields such as medicine and engineering can foster innovative solutions to problems. This can be helpful in tackling complex challenges, be they scientific or logistical. Harmony among different countries and regions on working toward this therapy can help in making getting approvals easier and, in turn, foster collaboration. International organizations can play a role in making this happen on a global level. Ethical issues can be dealt with by establishing clear and consistent policies and guidelines. Involvement of various stakeholders such as scientists, patients, and policymakers can help to better shape the policies and regulations regarding the therapy.

Innovations in technology can drastically change the manufacture and transportation of stem cells, overcoming current issues. Developments in bio-reactors and automated cell cultures, which can produce stem cells on a large scale while also maintaining their quality, seem promising in this direction. Techniques such as cryopreservation increase the chances of stem cells remaining viable and potent during their distribution and storage. Newer designs in manufacturing, which have quality maintenance as the key principle, can ensure the reproducibility of stem cells on a large scale. Such improvements in manufacturing can also cut down costs and increase accessibility to stem cell therapy.

Combined with constantly improving computational approaches and modeling, accurate predictions can be made pertaining to the vital conditions that enable efficient cell engineering that is free from any errors and can be easily reproduced [62].

Government and private funding in the form of grants, tax credits, and subsidies can encourage research and interest in the field of SCT. Philanthropic support and public–private partnership can also help ensure that there is maximum availability of funds for research and development of SCT. Also, the accessibility of SCT can be increased by providing reimbursement and health insurance coverage of this therapy so that more and more people can avail it.

In conclusion, stem cell therapy has the potential to become a game changer in medicine, with its capabilities growing with each experiment and innovation. However, there are many obstacles to overcome still.

17.6 Future Directions

17.6.1 Emerging Technologies in Stem Cell Research

Three areas of technologies that add value to stem cells and their applications include gene editing, 3D bioprinting, and organoids. Examples of methodologies for targeted genetic sequence modification are CRISPR and TALENs [63]. This can be utilized to create disease models using the patients' stem cells for drug testing or for correction of the mutation for the cell therapies [64]. Tissue and organ generation involve stem cells, biomaterials, and signaling factors to print tissues and organs using 3D printing. The benefit of this is that it can print cells in a pattern that replicates that of the tissues in the body regarding structure and function. 3D bioprinting involves the use of bioinks to develop tissues for uses such as developing tissues, testing drugs, and creating organs on demand [65].

17.6.2 Predictions for the Next Decade

Organoids develop branched cell cultures that assemble organ structure and functionality through stem-cell-based self-organization. Refer Figure 17.6 as it discusses about emerging technologies in stem cell research. Organoids are more effective in modeling human development

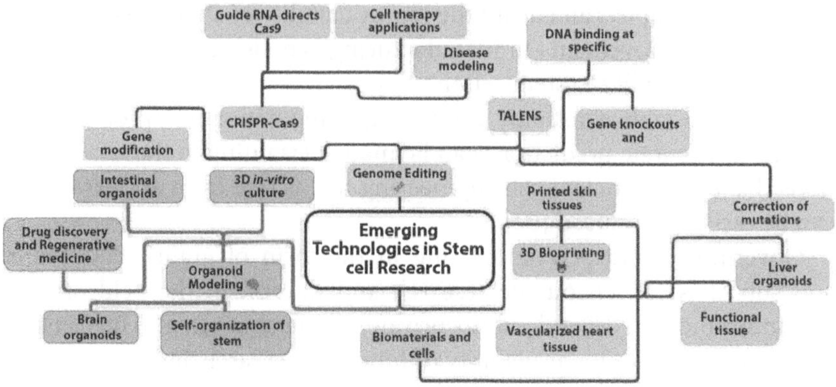

Figure 17.6 Emerging technologies in stem cell research. Interlinked circles of technologies which are genome editing, stem cells, organoids, and printing technologies, all of which are both enablers and drivers. It is important to note that editing techniques occupy a significant position in stem cell progression. Organoids and bioprinting build complex human tissue for disease modeling, drug testing, and maybe even medicine. 3D organoids are self-organized, branch-like cell cultures that mimic the architecture and physiology of organs. Through stem cell self-organization, organoids can mimic complex organs such as the central nervous system, liver, and gut [66].

and diseases than cultures in two dimensions and animal models [67]. Organoids reproduce the function of organ systems like the central nervous system, the liver, and the gastrointestinal tract.

In addition to the discussed applications of CRISPR technologies, it is also important to note that scientists can advance stem cell technologies and systems utilizing these novel tools to differentiate patient-specific induced pluripotent stem cells (iPSCs) and reintroduce them to patients for autologous transplantation or treat congenital diseases by using early gene targeting [68]. Two fields that are particularly closely related to the present study and are expected to witness a quantum leap in the next decade include stem cell biology and regenerative medicine due to enhanced knowledge of cell reprogramming, 3D culture, and gene editing concepts. Refer Figure 17.7 as it discusses about advancements of stem cell research for next decade. However, although models of research are evolving and there is increased creativity regarding clinical practice, restrictions still exist [69].

17.6.3 Global Impact and Potential Developments

Recent developments in stem cell studies have been made especially after the onset of the new century and the techniques of differentiation between

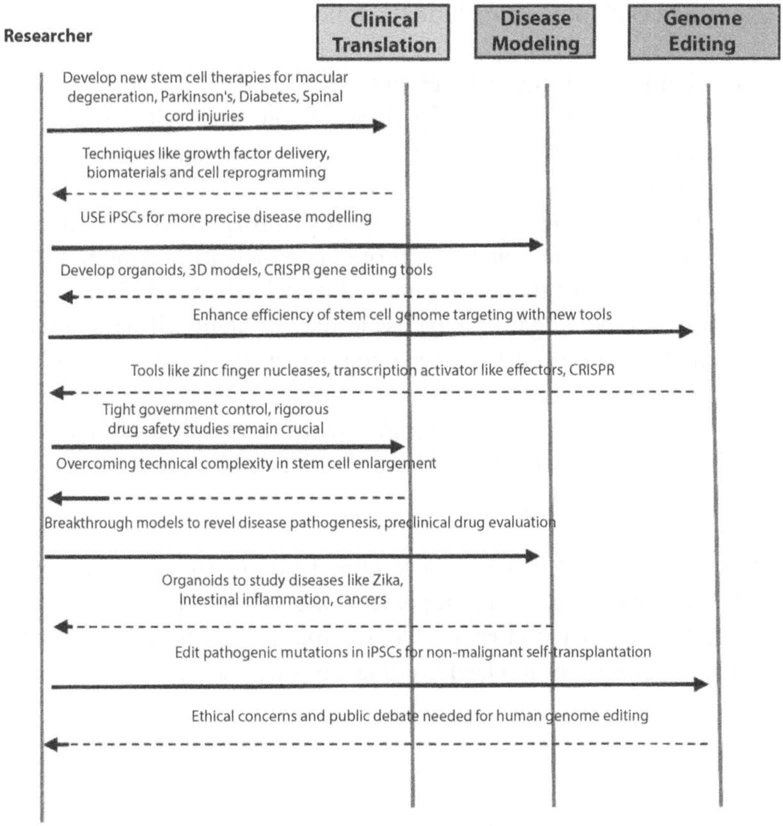

Figure 17.7 Fortune advancements in stem cell research for the next decade. Through an intricate link between the three areas of focus which include clinical application in the development of new therapies, disease modeling using organoids and genome editing using CRISPR, and dealing with ethical considerations.

embryonic and adult stem cells and the technique of propagation of these stem cells have also been developed. Some potential applications of the therapy involve its application on diabetes, heart diseases, neurological disorders, spinal cord injuries, and visual impairments, among others. Another important application is precision medicine where patients are treated according to their specific cell lines [70]. Some emerging economies have adopted the development and expansion of stem cell research to address the demand of patients locally and prevent the outsourcing of stem cell treatments to developed economies [71]. In the future, the results of the international team will be revealed, and they will present their discoveries and the possible stem cell therapy uses to the public as available worldwide [72].

Expected developments include the following:

- Genome editing: Applications such as CRISPR enable the specific manipulation of patient-derived stem cells for modeling diseases or treating conditions with autologous cell therapies [73].
- 3D bioprinting: This allows for the printing of cell constructs along with the matrix components which will facilitate the fabrication of tissues/organs and vessel formation [74].
- Organoids: Organ-like or tissue-like 3D stem cell structures that can be used to foster drug screening and tissue engineering [75].
- AI and machine learning: Prediction methods and heuristics for mechanistically driven understanding and improving production [76].

Opportunities for further work to effectively realize stem cell potential include the following: regulating safety and efficacy using rigorous but protracted and costly controlled trials, explaining stem cell self-renewal and tissue engraftment to clarify biological processes, creating large-scale

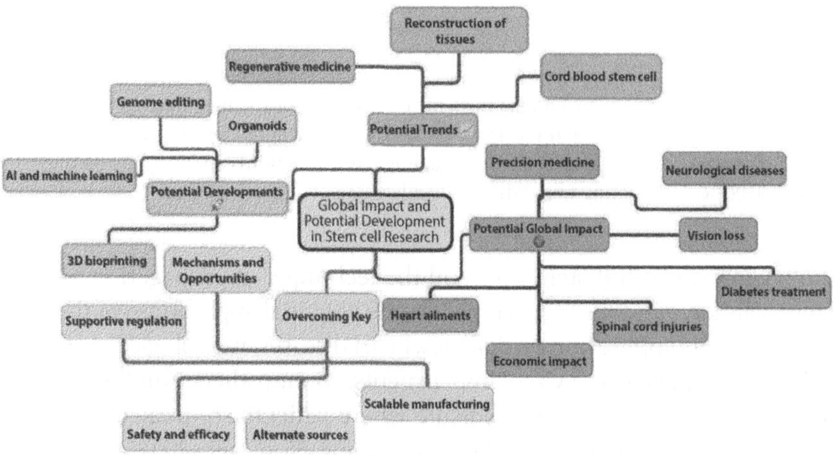

Figure 17.8 Global impact and potential development in stem cell research. With advancements in genome editing, organoids, and AI in stem cell research, there is support for mechanisms and opportunities for regulation and safety, such as new trends that are on the horizon like regenerative medicine stem cells and cord blood stem cells, as well as the impact of stem cell research on precision medicine, economic benefits, and diseases including neurological illnesses and diabetes.

manufacture of potent yet safe stem-cell-based products, and exploring other stem cell concepts like adult stem cells to avoid ethical concerns associated with using embryonic stem cells and dealing with regulations [77]. Refer Figure 17.8 as it discusses about potential impact of stem cell research.

17.7 Conclusion

Global medical reform could be greatly enhanced by stem cell therapy. This chapter has covered the fascinating developments that are occurring in North America, Europe, Asia-Pacific, Latin America, and Africa, among other regions. A vast array of diseases have been the focus of innovative research and clinical trials that we have witnessed.

The chapter did, however, also address the difficulties preventing stem cell treatments from being used widely. There are many obstacles due to legal and ethical issues, which necessitate constant debate and global consensus building. Diverse societies also have different public perceptions, which makes ongoing public engagement and education initiatives necessary.

The chapter stressed the importance of international cooperation in removing financial, logistical, and scientific obstacles. Case studies have demonstrated how well international alliances work to advance development. The field of stem cell research has exciting prospects for the future thanks to new trends and innovative technologies.

To sum up, achieving the maximum benefit from stem cell therapies requires a cooperative, international approach. We can bring in a new era of medical advancements and better healthcare solutions for a variety of populations across the world by establishing international partnerships, addressing ethical issues, and advancing public understanding. The secret to achieving the transformative potential of stem cell therapy on a global scale is this spirit of cooperation combined with never-ending innovation.

References

1. Tang, X., Huang, H., Hao, L., Decadal analysis of efficacy and safety profiles of mesenchymal stem cells from varied sources in knee osteoarthritis patients: A systematic review and network meta-analysis. *Exp. Gerontol.*, 192, 112460, 2024 Jul.

2. Alessandrini, M., Preynat-Seauve, O., De Bruin, K., Pepper, M.S., Stem cell therapy for neurological disorders. *S. Afr. Med. J.*, 109, 8b, 70–7, 2019 Sep 10.
3. Lotfy, A., AboQuella, N.M., Wang, H., Mesenchymal stromal/stem cell (MSC)-derived exosomes in clinical trials. *Stem Cell Res. Ther.*, 14, 1, 66, 2023 Apr 7.
4. Rikhtegar, R., Pezeshkian, M., Dolati, S., Safaie, N., Afrasiabi Rad, A., Mahdipour, M., *et al.*, Stem cells as therapy for heart disease: iPSCs, ESCs, CSCs, and skeletal myoblasts. *Biomed. Pharmacother.*, 109, 304–13, 2019 Jan.
5. Yamazaki, K., Kawabori, M., Seki, T., Houkin, K., Clinical Trials of Stem Cell Treatment for Spinal Cord Injury. *Int. J. Mol. Sci.*, 21, 11, 3994, 2020 Jun 2.
6. Wang, T., Gao, H., Wang, D., Zhang, C., Hu, K., Zhang, H., *et al.*, Stem cell-derived exosomes in the treatment of melasma and its percutaneous penetration. *Lasers Surg. Med.*, 55, 2, 178–89, 2023 Feb.
7. Mazzini, L., Vescovi, A., Cantello, R., Gelati, M., Vercelli, A., Stem cells therapy for ALS. *Expert Opin. Biol. Ther.*, 16, 2, 187–99, 2016.
8. Zakrzewski, W., Dobrzyński, M., Szymonowicz, M., Rybak, Z., Stem cells: past, present, and future. *Stem Cell Res. Ther.*, 10, 1, 68, 2019 Feb 26.
9. Evans, M.J. and Kaufman, M.H., Establishment in culture of pluripotential cells from mouse embryos. *Nature*, 292, 5819, 154–6, 1981 Jul.
10. Thomson, J.A., Itskovitz-Eldor, J., Shapiro, S.S., Waknitz, M.A., Swiergiel, J.J., Marshall, V.S., *et al.*, Embryonic Stem Cell Lines Derived from Human Blastocysts. *Science*, 282, 5391, 1145–7, 1998 Nov 6.
11. Takahashi, K. and Yamanaka, S., Induction of Pluripotent Stem Cells from Mouse Embryonic and Adult Fibroblast Cultures by Defined Factors. *Cell*, 126, 4, 663–76, 2006 Aug.
12. Lebkowski, J., GRNOPC1: The World's First Embryonic Stem Cell-Derived Therapy. *Regen. Med.*, 6, sup6, 11–3, 2011 Nov.
13. Liu, Z., Parikh, B.H., Tan, Q.S.W., Wong, D.S.L., Ong, K.H., Yu, W., *et al.*, Surgical Transplantation of Human RPE Stem Cell-Derived RPE Monolayers into Non-Human Primates with Immunosuppression. *Stem Cell Rep.*, 16, 2, 237–51, 2021 Feb.
14. Fatehullah, A., Tan, S.H., Barker, N., Organoids as an *in vitro* model of human development and disease. *Nat. Cell Biol.*, 18, 3, 246–54, 2016 Mar.
15. Vazin, T. and Freed, W.J., Human embryonic stem cells: Derivation, culture, and differentiation: A review. *Restor. Neurol. Neurosci.*, 28, 4, 589–603, 2010.
16. Zuk, P.A., Zhu, M., Mizuno, H., Huang, J., Futrell, J.W., Katz, A.J., *et al.*, Multilineage Cells from Human Adipose Tissue: Implications for Cell-Based Therapies. *Tissue Eng.*, 7, 2, 211–28, 2001 Apr.
17. Park, I.H., Zhao, R., West, J.A., Yabuuchi, A., Huo, H., Ince, T.A., *et al.*, Reprogramming of human somatic cells to pluripotency with defined factors. *Nature*, 451, 7175, 141–6, 2008 Jan.
18. Trounson, A., Thakar, R.G., Lomax, G., Gibbons, D., Clinical trials for stem cell therapies. *BMC Med.*, 9, 1, 52, 2011 Dec.

19. Da Cruz, L., Fynes, K., Georgiadis, O., Kerby, J., Luo, Y.H., Ahmado, A., *et al.*, Phase 1 clinical study of an embryonic stem cell–derived retinal pigment epithelium patch in age-related macular degeneration. *Nat. Biotechnol.*, 36, 4, 328–37, 2018 Apr.
20. Chen, J., Luo, L., Tian, R., Yu, C., A review and update for registered clinical studies of stem cells for non-tumorous and non-hematological diseases. *Regen. Ther.*, 18, 355–62, 2021 Dec.
21. Bagcchi, S., Authorities in India ban use of stem cells in patients except in clinical trials. *BMJ*, 348, mar04 14, g1901–g1901, 2014 Mar 4.
22. Enosawa, S., Clinical Trials of Stem Cell Therapy in Japan: The Decade of Progress under the National Program. *JCM*, 11, 23, 7030, 2022 Nov 28.
23. Chen, G., Yin, S., Zeng, H., Li, H., Wan, X., Regulation of Embryonic Stem Cell Self-Renewal. *Life*, 12, 8, 1151, 2022 Jul 29.
24. Martins De Oliveira, M.L., Tura, B.R., Meira Leite, M., Melo Dos Santos, E.J., Pôrto, L.C., Pereira, L.V., *et al.*, Creating an HLA-homozygous iPS cell bank for the Brazilian population: Challenges and opportunities. *Stem Cell Rep.*, 18, 10, 1905–12, 2023 Oct.
25. Affordable Stem Cell Therapy in Mexico [Internet], [cited 2024 Aug 5]. Available from: https://www.medicaltourismco.com/stem-cell-therapy-mexico-3/.
26. Fernández-Metzler, C., Ackermann, B., Garofolo, F., Arnold, M.E., DeSilva, B., Gu, H., *et al.*, Biomarker Assay Validation by Mass Spectrometry. *AAPS J.*, 24, 3, 66, 2022 May.
27. Azeez, I.A., Awogbindin, I.O., Olayinka, J.N., Folarin, R.O., Adamu, A.S., Ior, L.D., *et al.*, Neural stem cell research in Africa: current realities and future prospects. *Biol. Open*, 11, 11, bio059574, 2022 Nov 1.
28. Aboalola, D., Ramadan, M., Baadhaim, M., Alsiary, R., Badraiq, H., Alghamdi, T., *et al.*, Public awareness and understanding of stem cell treatments available in Saudi Arabia and their trust in hospitals and research centers involved in stem cell research—a cross sectional study. *Front. Public Health*, 12, 1364809, 2024 Apr 2.
29. Rosenthal, J.C., Wisotzky, E.L., Matuschek, C., Hobl, M., Hilsmann, A., Eisert, P., *et al.*, Endoscopic measurement of nasal septum perforations. *HNO*, 70, S1, 1–7, 2022 Feb.
30. Mittal, S., Stem cell research: The India perspective. *Perspect. Clin. Res.*, 4, 1, 105, 2013.
31. Mamaeva, A., Krasnova, O., Khvorova, I., Kozlov, K., Gursky, V., Samsonova, M., *et al.*, Quality Control of Human Pluripotent Stem Cell Colonies by Computational Image Analysis Using Convolutional Neural Networks. *IJMS*, 24, 1, 140, 2022 Dec 21.
32. Fernández-Montequín, J.I., Betancourt, B.Y., Leyva-Gonzalez, G., Mola, E.L., Galán-Naranjo, K., Ramírez-Navas, M., *et al.*, Intralesional administration of epidermal growth factor-based formulation (Heberprot-P) in chronic diabetic foot ulcer: treatment up to complete wound closure. *Int. Wound J.*, 6, 1, 67–72, 2009 Feb.

33. De Klerk, E. and Hebrok, M., Stem Cell-Based Clinical Trials for Diabetes Mellitus. *Front. Endocrinol.*, 12, 631463, 2021 Feb 26.
34. Hu, L., Zhao, B., Wang, S., Stem-Cell Therapy Advances in China. *Hum. Gene Ther.*, 29, 2, 188–96, 2018 Feb.
35. Mousaei Ghasroldasht, M., Seok, J., Park, H.S., Liakath Ali, F.B., Al-Hendy, A., Stem Cell Therapy: From Idea to Clinical Practice. *Int. J. Mol. Sci.*, 23, 5, 2850, 2022 Mar 5.
36. Andrade, R., Vasta, S., Pereira, R., Pereira, H., Papalia, R., Karahan, M., *et al.*, Knee donor-site morbidity after mosaicplasty - a systematic review. *J. Exp. Orthop.*, 3, 1, 31, 2016 Dec.
37. Robijns, J., Nair, R.G., Lodewijckx, J., Arany, P., Barasch, A., Bjordal, J.M., *et al.*, Photobiomodulation therapy in management of cancer therapy-induced side effects: WALT position paper 2022. *Front. Oncol.*, 12, 927685, 2022.
38. Lowrance, W.W. and Collins, F.S., Ethics. Identifiability in genomic research. *Science*, 317, 5838, 600–2, 2007 Aug 3.
39. McGuire, A.L. and Gibbs, R.A., Genetics. No longer de-identified. *Science*, 312, 5772, 370–1, 2006 Apr 21.
40. Greely, H.T., Cho, M.K., Hogle, L.F., Satz, D.M., Thinking about the human neuron mouse. *Am. J. Bioethics*, 7, 5, 27–40, 2007 May.
41. Lo, B., Zettler, P., Cedars, M.I., Gates, E., Kriegstein, A.R., Oberman, M., *et al.*, A new era in the ethics of human embryonic stem cell research. *Stem Cells*, 23, 10, 1454–9, 2005.
42. Olanow, C.W., Goetz, C.G., Kordower, J.H., Stoessl, A.J., Sossi, V., Brin, M.F., *et al.*, A double-blind controlled trial of bilateral fetal nigral transplantation in Parkinson's disease. *Ann. Neurol.*, 54, 3, 403–14, 2003 Sep.
43. Freed, C.R., Greene, P.E., Breeze, R.E., Tsai, W.Y., DuMouchel, W., Kao, R., *et al.*, Transplantation of embryonic dopamine neurons for severe Parkinson's disease. *N. Engl. J. Med.*, 344, 10, 710–9, 2001 Mar 8.
44. Hyun, I., Allowing innovative stem cell-based therapies outside of clinical trials: ethical and policy challenges. *J. Law Med. Ethics*, 38, 2, 277–85, 2010.
45. Holden, C., Korean stem cell scandal. Schatten: Pitt panel finds "misbehavior" but not misconduct. *Science*, 311, 5763, 928, 2006 Feb 17.
46. Chong, S., Scientific misconduct. Investigations document still more problems for stem cell researchers. *Science*, 311, 5762, 754–5, 2006 Feb 10.
47. Chong, S. and Normile, D., Stem cells. How young Korean researchers helped unearth a scandal. *Science*, 311, 5757, 22–5, 2006 Jan 6.
48. Fadel, H.E., Developments in stem cell research and therapeutic cloning: Islamic ethical positions, a review. *Bioethics*, 26, 3, 128–35, 2012 Mar.
49. Lewis, A.J., Working for a cure for Type 1 diabetes. An interview. *Regen. Med.*, 4, 2, 177–80, 2009 Mar.
50. Lyall, D., Stem Cell Network: world's first stem cell network remains true to mission. *Stem Cells Transl. Med.*, 1, 8, 579–80, 2012 Aug.
51. McCormack, K., Regen Med California Institute for Regenerative Medicine Profile CIRM 2.0. *Regen. Med.*, 11, 8, 759–62, 2016 Dec.

52. Read, Assessing the Medical Risks of Human Oocyte Donation for Stem Cell Research: Workshop Report, NAP.edu [Internet]. [cited 2024 Aug 5]. Available from: https://nap.nationalacademies.org/read/11832/chapter/3.
53. Boushell, M.K., Mosher, C.Z., Suri, G.K., Doty, S.B., Strauss, E.J., Hunziker, E.B., *et al.*, Polymeric mesh and insulin-like growth factor 1 delivery enhance cell homing and graft-cartilage integration. *Ann. N. Y. Acad. Sci.*, 1442, 1, 138–52, 2019 Apr.
54. Talib, S., Millan, M.T., Jorgenson, R.L., Shepard, K.A., Proceedings: Immune Tolerance and Stem Cell Transplantation: A CIRM Mini-Symposium and Workshop Report. *Stem Cells Transl. Med.*, 4, 1, 4–9, 2015 Jan.
55. Bahari, M., Mokhtari, H., Yeganeh, F., Stem Cell Therapy, the Market, the Opportunities and the Threat. *Int. J. Mol. Cell. Med.*, 12, 3, 310–9, 2023.
56. Lo, B. and Parham, L., Ethical issues in stem cell research. *Endocr. Rev.*, 30, 3, 204–13, 2009 May.
57. Sarathkumar, E., Victor, M., Menon, J.A., Jibin, K., Padmini, S., Jayasree, R.S., Nanotechnology in cardiac stem cell therapy: cell modulation, imaging and gene delivery. *RSC Adv.*, 11, 55, 34572–88, 2021 Oct 25.
58. Abubakar, M., Masood, M.F., Javed, I., Adil, H., Faraz, M.A., Bhat, R.R., *et al.*, Unlocking the Mysteries, Bridging the Gap, and Unveiling the Multifaceted Potential of Stem Cell Therapy for Cardiac Tissue Regeneration: A Narrative Review of Current Literature, Ethical Challenges, and Future Perspectives. *Cureus*, 15, 7, e41533, 2023 Jul.
59. Noaksson, K., Zoric, N., Zeng, X., Rao, M.S., Hyllner, J., Semb, H., *et al.*, Monitoring differentiation of human embryonic stem cells using real-time PCR. *Stem Cells*, 23, 10, 1460–7, 2005.
60. Polverini, P.J., The pathophysiology of angiogenesis. *Crit. Rev. Oral Biol. Med.*, 6, 3, 230–47, 1995.
61. Kim, J.A., Choi, J.H., Kim, M., Rhee, W.J., Son, B., Jung, H.K., *et al.*, High-throughput generation of spheroids using magnetic nanoparticles for three-dimensional cell culture. *Biomaterials*, 34, 34, 8555–63, 2013 Nov.
62. Apostolou, E., Blau, H., Chien, K., Lancaster, M.A., Tata, P.R., Trompouki, E., *et al.*, Progress and challenges in stem cell biology. *Nat. Cell Biol.*, 25, 2, 203–6, 2023 Feb.
63. Cong, L., Ran, F.A., Cox, D., Lin, S., Barretto, R., Habib, N., *et al.*, Multiplex Genome Engineering Using CRISPR/Cas Systems. *Science*, 339, 6121, 819–23, 2013 Feb 15.
64. Li, H.L., Fujimoto, N., Sasakawa, N., Shirai, S., Ohkame, T., Sakuma, T., *et al.*, Precise Correction of the Dystrophin Gene in Duchenne Muscular Dystrophy Patient Induced Pluripotent Stem Cells by TALEN and CRISPR-Cas9. *Stem Cell Rep.*, 4, 1, 143–54, 2015 Jan.
65. Zhang, L., Stokes, N., Polak, L., Fuchs, E., Specific MicroRNAs Are Preferentially Expressed by Skin Stem Cells To Balance Self-Renewal and Early Lineage Commitment. *Cell Stem Cell*, 8, 3, 294–308, 2011 Mar 4.

66. Camp, J.G., Sekine, K., Gerber, T., Loeffler-Wirth, H., Binder, H., Gac, M., *et al.*, Multilineage communication regulates human liver bud development from pluripotency. *Nature*, 546, 7659, 533–8, 2017 Jun.
67. Sloan, S.A., Darmanis, S., Huber, N., Khan, T.A., Birey, F., Caneda, C., *et al.*, Human Astrocyte Maturation Captured in 3D Cerebral Cortical Spheroids Derived from Pluripotent Stem Cells. *Neuron*, 95, 4, 779–790.e6, 2017 Aug.
68. Mali, P., Yang, L., Esvelt, K.M., Aach, J., Guell, M., DiCarlo, J.E., *et al.*, RNA-Guided Human Genome Engineering via Cas9. *Science*, 339, 6121, 823–6, 2013 Feb 15.
69. Musunuru, K., Genome editing of human pluripotent stem cells to generate human cellular disease models. *Dis. Models Mech.*, 6, 896–904, 2013 Jan 1, dmm.012054.
70. El-Kadiry, A.E.H., Rafei, M., Shammaa, R., Cell Therapy: Types, Regulation, and Clinical Benefits. *Front. Med.*, 8, 756029, 2021 Nov 22.
71. Gottweis, H., Salter, B., Waldby, C., *The Global Politics of Human Embryonic Stem Cell Science* [Internet], Palgrave Macmillan UK, London, 2009, [cited 2024 Aug 1]. Available from: http://link.springer.com/10.1057/9780230594364.
72. Kimmelman, J., Hyun, I., Benvenisty, N., Caulfield, T., Heslop, H.E., Murry, C.E., *et al.*, Policy: Global standards for stem-cell research. *Nature*, 533, 7603, 311–3, 2016 May.
73. Drost, J., Van Boxtel, R., Blokzijl, F., Mizutani, T., Sasaki, N., Sasselli, V., *et al.*, Use of CRISPR-modified human stem cell organoids to study the origin of mutational signatures in cancer. *Science*, 358, 6360, 234–8, 2017 Oct 13.
74. Gao, G., Yonezawa, T., Hubbell, K., Dai, G., Cui, X., Inkjet-bioprinted acrylated peptides and PEG hydrogel with human mesenchymal stem cells promote robust bone and cartilage formation with minimal printhead clogging. *Biotechnol. J.*, 10, 10, 1568–77, 2015 Oct.
75. Hofer, M. and Lutolf, M.P., Engineering organoids. *Nat. Rev. Mater.*, 6, 5, 402–20, 2021 Feb 19.
76. Ching, T., Himmelstein, D.S., Beaulieu-Jones, B.K., Kalinin, A.A., Do, B.T., Way, G.P., *et al.*, Opportunities and obstacles for deep learning in biology and medicine. *J. R. Soc. Interface*, 15, 141, 20170387, 2018 Apr.
77. Aguirre, M., Escobar, M., Forero Amézquita, S., Cubillos, D., Rincón, C., Vanegas, P., *et al.*, Application of the Yamanaka Transcription Factors Oct4, Sox2, Klf4, and c-Myc from the Laboratory to the Clinic. *Genes*, 14, 9, 1697, 2023 Aug 26.

18

Ethical Considerations in Stem Cell Therapeutics: Striking a Balance

Rahul Kumar[1*], Ayush Madan[2,3], Debasis Mitra[1†] and Jyotsana Singh Chandravanshi[2]

[1]*Department of Microbiology, Graphic Era (Deemed to be University), Clement Town, Dehradun, Uttarakhand, India*
[2]*Department of Biotechnology, School of Research & Technology, People's University, Bhanpur, Bhopal, Madhya Pradesh, India*
[3]*Institute of Tropical Aquaculture and Fisheries (AKUATROP), Universiti Malaysia Terengganu, Kuala Nerus, Terengganu, Malaysia*

Abstract
This chapter explores the ethical dilemmas surrounding stem cell therapeutics, particularly the use of embryonic stem cells. It discusses autonomy, beneficence, non-maleficence, and justice issues and provides a global perspective on the regulatory and policy environments in different countries. The chapter highlights the cultural, religious, and societal values that shape ethical expectations in stem cell research and application, offering hope for the potential of these therapies despite their complex ethical challenges. The chapter explores ethical issues in clinical translation, emphasizing the need for balance between innovation and ethical oversight. It highlights the responsibilities of researchers and clinicians in patient safety and informed consent. The chapter also examines the ethical considerations in different countries and the complex challenges faced by stakeholders in stem cell therapeutics. It aims to guide researchers, policymakers, and practitioners in making informed decisions.

***Keywords*:** Stem cell therapy, bioethics, medical ethics, biotechnology ethics

**Corresponding author*: rkpma.dpmi@gmail.com
†*Corresponding author*: debasismitra3@gmail.com

Ayush Madan, Sumel Ashique, Deepika Arora and Mantosh Kumar Satapathy (eds.) Stem Cell Therapeutics, (739–756) © 2025 Scrivener Publishing LLC

18.1 Introduction

While in the past most medical procedures could only treat the symptoms of an illness, today's sophisticated diagnosis and treatment techniques allow efforts to eliminate the root causes of illnesses, and precisely treating them has gained prominence. Research on stem cells is therefore a "revolutionary" approach to medical advancement (Wade, 2023). Stem cell therapies based on the idea of swapping out unhealthy cells in the body with functioning cells provide enormous promises for the treatment of chronic illnesses or long-term impairments. Regenerative medicine and stem cell research have raised several ethical concerns, many of which have been centered on science itself (Reed, 2020). Despite enduring differences of view, rigorous discussions have produced valuable public discourse and established a regulatory framework within which significant science has been undertaken (Arda, and Güvercin, 2013).

The practical ethical issues surrounding stem cell research are far more widespread and socially significant now that we are in a different era. Certain arguments are acknowledged as heralding a new era in medicine; these include the use of stem cell technologies to create new organs in place of non-functioning ones, such as in liver failure, and to treat conditions like blindness, deafness, cancer, infertility, muscular disease, rheumatoid arthritis, osteoarthritis, spinal cord injury, Alzheimer's disease, Parkinson's disease, myocardial infarction, and diabetes. In the near future, providing customized cellular therapies is anticipated (Arda, and Güvercin, 2013) (Sandel, 2004). Even while knowledge in the field of stem cells is rapidly expanding, research has not kept up with the high expectations that have been placed on the field, particularly with regard to therapeutic outcomes. However, with regard to practice guidelines and standards for the therapeutic use of stem cells, bone marrow transplantation is now the most dependable therapy available.

Significant advancements in skin and corneal healing have also been made recently. Although promising, experimental research on alternative therapies is still in the early stages of development (Funderud, 2008). The topic of ongoing ethical discussions over stem cell research has shifted as chances to benefit from research become available for treating specific ailments. The need to develop non-embryonic stem cells, the philosophical significance of the embryo, the issue of embryo rights, spare embryos for therapy and research, embryo experimentation, artificial tissue, and organ production have been highlighted. A novel therapeutic strategy in medicine has emerged as a result of stem cells' ability to differentiate into the necessary tissue (Sapna *et al.*, 2023). Regenerative medicine, which focuses on the regeneration of tissues and organs in the treatment of illnesses originating from cellular degeneration or tissue injury, is a wide field of debate and practice within the mainstream medical paradigm of today.

For such disorders, for which there is typically no cure and which necessitate a lifetime of therapy, regenerative medicine has proven to be a promising treatment option. A significant new chapter in biomedical research began in 1998 when James Alexander Thomson and his colleagues at Wisconsin University successfully isolated human stem cells from human embryos and maintained them alive in a lab setting (JA, T. 1998). A lot of common knowledge about cell growth, maturation, and roles has altered as a result of the investigations that have been done. For example, the idea that cell differentiation is an unchangeable, one-way process vanished; it was demonstrated instead that differentiation may be reversed, returning adult stem cells to their undifferentiated state (Gorelik, and Flavell, 2000).

18.2 Properties of Stem Cells

Parent cells called stem cells give rise to all bodily tissues and organs. Undifferentiated cells have the ability to undergo infinite divisibility, regenerate, and differentiate into organs and tissues through the production of differentiated cells (Arda, and Güvercin, 2013). While certain organ cells, including heart and nerve cells, have specific tasks, stem cells are not differentiated and stay that way until they are stimulated to become a particular kind of cell. Undifferentiated cells are special because of their capacity to proliferate and change into other types of cells with specific tasks, such as muscle or nerve cells (Sandel, 2004).

The process of a single-celled zygote developing into a multicellular creature with the ability to regenerate, differentiate, and develop is known as human biological development. Fertilized ovum makes up the first ring in this chain. During this procedure, the cells' developmental capacity varies from one another but typically decreases. When a zygote is able to produce a gestation product (totipotent), it changes into over 200 cells that become a blastocyst fetus (pluripotent). When these cells are inserted into the wall of the uterus, they can change into fewer cells (multipotent) (Rousseaux, and Bolon, 2013) (Figure 18.1).

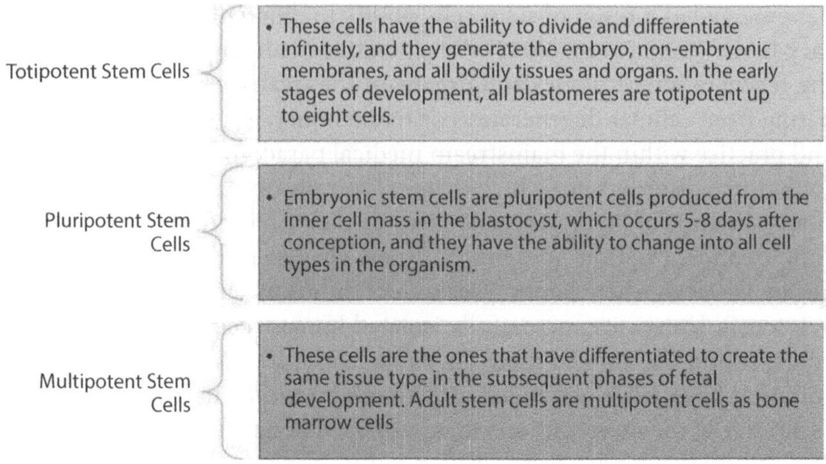

Figure 18.1 Three kinds of stem cells are identified by this developmental process: totipotent, pluripotent, and multipotent (Mitalipov, and Wolf, 2009).

18.3 Sources of Stem Cell

The use of embryos with abnormal chromosomal configurations or dead organisms for embryonic stem cell generation is a topic of discussion, using single blastomere biopsy for preimplantation genetic diagnosis. One very intriguing technique that attempts to reverse the cell's developmental process is called somatic cell dedifferentiation. Research on adult somatic cells that have finished differentiating and are returning to their pluripotent phase presents a vast field of possibilities for stem cell studies aimed at therapeutic applications. Technically, the rejection of tissue response will be overcome, and the ethical debate over embryos will conclude (Arda, and Güvercin, 2013) (Figure 18.2). Research on human stem cells is crucial for the clinical and fundamental medical sciences. The anticipated benefits of these studies include elucidating the processes governing cell differentiation and converting stem cells into unique types of cells that can be employed to treat certain illnesses for which there is now no effective treatment (Siegel, 2008) (Figure 18.2).

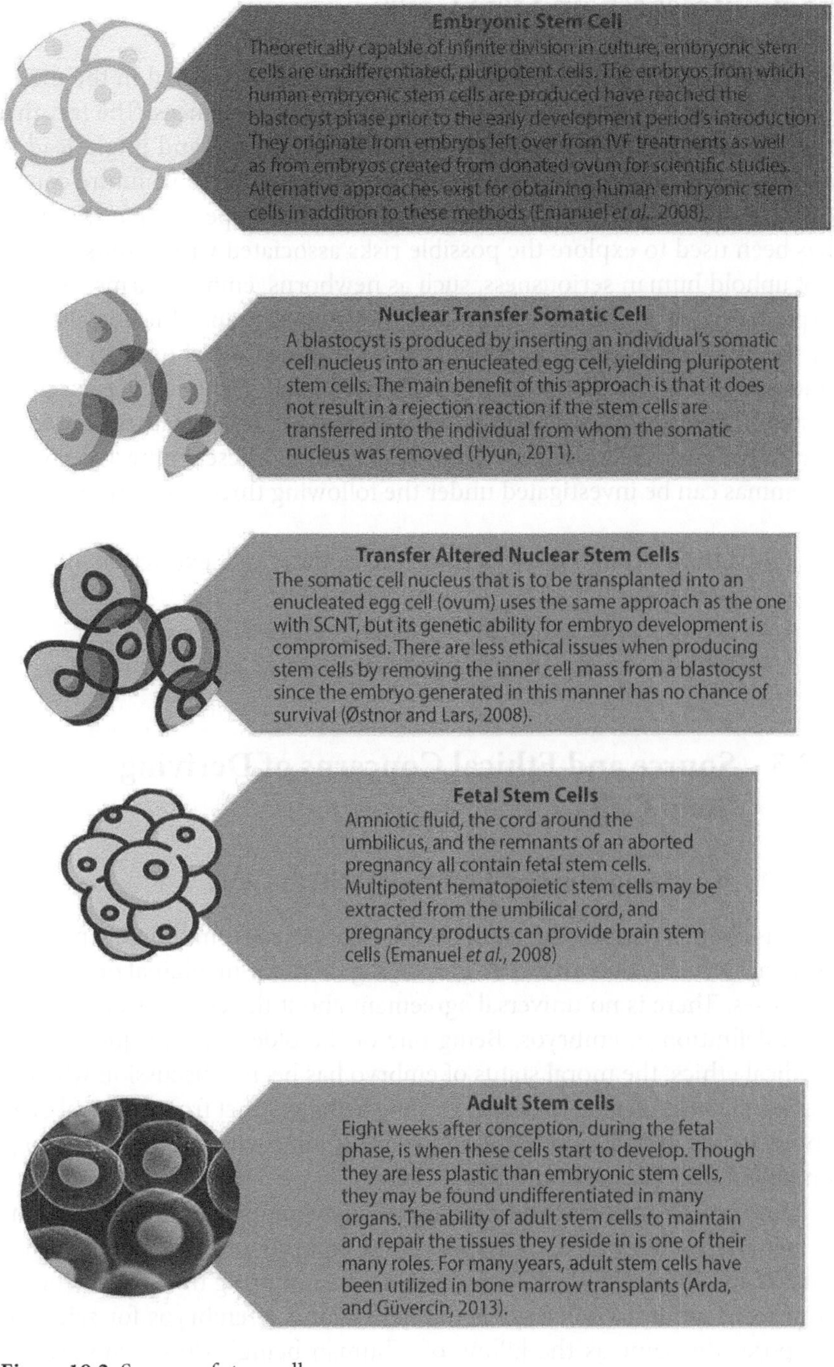

Figure 18.2 Sources of stem cells.

18.4 Research on Stem Cells

The fundamental difficulties in stem cell research are rooted in the methodologies used in the study and the potential consequences. The fact that these studies entail the destruction of human embryos and might lead to actions that could possibly be inhumane with unpredictable future consequences are the most hotly contested topics. The slippery slope argument has been used to explore the possible risks associated with results that do not uphold human seriousness, such as newborns, embryo farms, and the exploitation of fetuses as spare parts. Additionally, an ethical evaluation of the study procedures is conducted. Is it ethical to choose to employ undesirable means like violating someone's basic rights to get excellent results? There are several issues in human stem cell research that need to be carefully considered ethically (Reya *et al.*, 2001). These connected ethical dilemmas can be investigated under the following three categories:

- Ethical concerns regarding the source and extraction of stem cells
- Ethical concerns about the research process
- Ethical concerns regarding the clinic phase and beyond

18.5 Source and Ethical Concerns of Deriving Stem Cells

18.5.1 Utilizing Human Embryonic Stem Cells

Because human embryonic stem cells are generated from human embryos, there are ethical concerns when employing them for medicinal or research purposes. There is no universal agreement about the ethical standing and legal definition of embryos. Being one of the oldest sets of questions in medical ethics, the moral status of embryo has been a discussion which is essentially held on abortus issue. Despite being distinct from the goal of an abortion, the destruction of the embryo is at issue when obtaining embryonic stem cells (Bongso, and Fong, 2009).

Diverse sociocultural and theological viewpoints give rise to differing values attached to human embryos. Is the embryo a unique being? When can we speak of an embryo or fetus as a person during the whole process, from conception to delivery? Is the destruction of embryos for scientific purposes the same as the killing of a human being? Those who ground these debates in religious doctrine contend that all human beings begin as embryos and that God endows embryos with souls through fertilization. As a result, everyone, regardless of developmental stage, deserves respect for their lives and privileges, just as adults do. Those who hold this perspective consider stem cell research on embryos to be unethical. Opponents counter that the fact that each person begins life as an embryo does not demonstrate their uniqueness. They argue that although the embryo has potential, human life develops gradually. Furthermore, it is stressed that environmental factors play a significant role in this process since the embryo does not have a chance to survive outside of the uterus (Liew *et al.*, 2005).

It is suggested that the moral standing of an embryo in a petri dish would differ from that of an individual. Given that stem cell research and embryo morality are incompatible, evaluating stem cell research in relation to embryo morality would inevitably bring ethical discussions to a standstill. Therefore, it is thought to be beneficial to investigate other choices that are currently on the table for the source of stem cells.

18.5.2 Utilizing Extra/Spare Embryos Created During IVF Applications

The need for extra protection for these embryos that are cell masses which are not going to be able to reach human status will vanish when the realities of IVF treatments, that is, the process of creating an embryo that takes place *in vitro*, are considered. Additionally, excess embryos that are not placed in the uterus are used during research. Since embryos extracted from the uterus cannot develop into adult individuals, using these embryos appears to be less morally dubious. It may turn out that using embryos that would otherwise be destroyed for the benefit of suffering patients and mankind is beneficial and appropriate in the sense of preventing waste. To ensure the ethical use of embryos for stem cell research, it is crucial to obtain informed consent from the women or parents (Franklin, 2006).

18.5.3 Utilization of Fetal Tissue Received From Termination of Pregnancy (Intentionally Induced or Miscarriages)

Abortion is the international or spontaneous removal of the gestational product (embryo or fetus) from the mother before viability (Neal *et al.*, 2016). This makes the ethical debates around embryos pointless because, at the end of the process, the result of gestation is dead. It is legitimate to utilize fetal tissue which would be destroyed anyhow for research purposes after obtaining the agreement of the woman and following the guidelines of waste elimination and benefit-sharing. However, it is unethical to create a pregnancy and then use abortion to end it for medical study or therapeutic objectives. The process may go in very dangerous directions, as in the cases of forcing women to have abortions in order to make money, selling embryos obtained in some countries, removing babies from their mothers shortly after delivery and forcing them to become pregnant, and removing newborns from these women and destroying them in order to trade stem cells (Lee *et al.*, 2020). Organ and tissue trade is prohibited by the Oviedo Convention (a. 22) (Ismayilova, 2023).

18.5.4 Producing Embryos for Use in Stem Cell Studies

In light of the embryo's and humankind's basic instrumentalization, the production of embryos for scientific purposes is an undesirable practice. It is unacceptable to use an embryo as a tool or instrument because it violates human dignity (Hansen, 2002). The Oviedo Convention (a.18) permits research on embryos; nevertheless, the production of embryos for scientific purposes is expressly forbidden (Petersen, 2019).

18.5.5 Somatic Cell Uses

The difficulties associated with employing embryonic or fetal tissues and the morality of embryos appear to be resolved using somatic stem cells. It is advised to focus the study on this axis. It is possible to create embryonic stem cells that are compatible with the person who is the nucleus donor by using the nucleus transfer procedure. In light of this circumstance, patients in the future requiring tissue and organ transplantation may find hope in the use of customized tissues and organs, which reduce the immune system's rejection problem. If this is realized, living donors will no longer be required for organ donation, which will eliminate any associated ethical concerns (Arda, and Güvercin, 2013). Dedifferentiation or the process of turning a differentiated somatic cell back into a pluripotent cell is also the final step that has to be completed. This will address ethical issues with embryonic stem cells, rejection, histocompatibility, and the requirement for a living donor. These kinds of studies are seen to be the most appropriate ones going forward (Arda, and Güvercin, 2013).

18.6 Research Ethics Concerns and Process

18.6.1 Concerns of Ethics in Relation to Women's Reproductive Rights

The woman's status becomes important since the ovum is employed in studies using embryonic stem cells. Owing to the crucial responsibilities that women play as a vulnerable group throughout research procedures, studies may raise concerns about women's health risks, oppression, and classification as "meta" because they donate or give ovum (Beall and DeCherney, 2012). Women's sensitivity should be taken into account when conducting these studies and allowing them to make their own decisions about abortion, IVF applications, embryo donation, and ovum donation without coercion, forcing, financial incentive, or deceit (Beall and DeCherney, 2012). Women must receive hormone therapy for overstimulation prior to ovum removal. Following ovum removal surgery, there are concerns about infertility and even deadly consequences (Beall, and DeCherney, 2012).

"Every woman has the right to be included in the decision-making mechanism (diagnosis and treatment) which will affect herself and the fetus," states the Declaration on Rights of Mother and Newborn. Expectant mothers should be informed about the matter and each decision should be taken on its own (Guimarães *et al.*, 2012). Women's sociocultural and economic circumstances, her educational background, her social standing and societal gender disparity all have an impact on their ability to access healthcare services and exercise control over their bodies (Şimşek, 2011).

Women are at danger from the potential side effects of stem cell research, both for therapeutic purposes and for research purposes, as long as these parameters continue to shift negatively. Due to financial concerns, a woman may engage in stem cell research projects that have ethical committee permission in addition to participating in illicit stem cell applications. It would appear that doctors' and researchers' primary ethical responsibility is to defend the rights of all participants (Arda, and Güvercin, 2013). The United Nations Declaration on Human Cloning emphasizes the prevention of women's abuse, urging member nations to take measures to achieve this goal.

18.6.2 Informed Consent During the Research Process to Volunteers

The use and preservation of every component of the human body requires the patient's informed permission, which is a requirement for all medical procedures (Phelan, 2012). The Amsterdam Declaration's universal precepts apply to stem cell research as well. Even though the research is extremely intricate and sophisticated, volunteers should be fully informed about it before giving their agreement. The freedom to make decisions about oneself at the cellular and tissue levels belongs to the person. Stem cells can occasionally be exposed to undesirable genetic alterations during *in vitro* cultivation, which could be detrimental to the organism (Arda, and Güvercin, 2013). The possible health risks associated with cells that are replicated in the presence of serum, chemicals, and medium, all of which are utilized in research, are not well understood (Arda, and Güvercin, 2013). All of these hazards, along with any potential benefits, should be openly disclosed throughout the informed consent period, and voluntariness should be guaranteed. The participants should give their informed permission once more if a novel circumstance arises throughout stem cell therapies (Arda, and Güvercin, 2013).

18.6.3 Privacy Confidentiality and Information Masking During the Research Process

Volunteers' private life should be respected and properly protected during the study process, and information belonging to them should be kept secret from outside parties. These computers should not be connected to the Internet, and authorized users should only be able to access them with a unique password in order to safeguard the information sent to an electronic environment. The Oviedo Convention's Article 23 emphasizes the need to protect personal data and ensure its security (Porxas, 2017).

18.7 Ethical Concerns Throughout the Clinical Phase and After

18.7.1 Problems Caused by Business Interests Taking the Initiative and Persistently Pursuing Research

The predominance of commercial expectancies in stem cell research ought to be acknowledged as an ethical issue. Research directions are determined by economic interests rather than scientific truths, which causes outcomes to be assessed using profit and loss criteria. However, the outcomes of these studies, which the whole human race eagerly awaits, should not be sacrificed to corporate gain.

18.7.2 Promotion, Deception of the Public, Amazing Therapies, and Sharing of Research Findings with Unproven Theories

It is widely acknowledged that only information that is protective and promotes health may be advertised or promoted; neither unfair competition nor demand-generating campaigns for healthcare services are permitted (Ersoy, and Güler, 2008). Exaggerated and inaccurate media presentations of stem cell research raise expectations in the field and facilitate the issue's commercialization. As stem cell therapies proliferate, this topic—which needs to be tackled in terms of media ethics as well—will become increasingly significant. Public dissemination of study findings without predicted outcomes is just as important as highlighting accomplishments on this topic. By honestly and clearly presenting associations' and scientists' perspectives—who are authorized in the field of stem cells—we can bring society's unrealistic expectations down to earth.

18.7.3 During the Research Process, Several Treatment Methods That Have Not Been Proven to be Effective Were Reflected as Miraculous Treatments

This strategy could result in iatrogenic consequences as well as the moral and financial abuse of people. With little evidence of efficacy, patients, particularly those nearing the end of their lives, might readily consent to receiving stem cell therapy. Patients may suffer needless discomfort and inconvenience as well as potentially fatal hazards from these non-standardized therapeutic studies. These occurrences might generate a hostile atmosphere that would bolster the arguments of those who are against stem cell research and make it more difficult to carry out actual research (Jiwa, 2012).

18.7.4 Copyright and Patent Issues

When asked if there was a patent for the polio vaccine he developed, Jonas Edward Salk replied, "There isn't." Is it possible for you to patent the sun? Could the patent cover a beneficial "thing," like humanity's shared heritage? From this perspective, the stem cell problem is a little more nuanced and does not fit the conventional patent idea. The Biomedicine Convention (a. 21) forbids using human bodies and bodily parts for financial or comparable benefit (Ersoy, and Güler, 2008). It is unclear what the European Patent Office (EPO) believes about human embryonic stem cells not being patentable (de Miguel Beriain, 2017). Patent issues in stem cell research are still relevant today.

18.7.5 Issues with Treatment Accessibility and Costs to be Developed as a Result of Stem Cell Research

It is said that stem cell research primarily aims to address illnesses associated with aging, but it also holds potential for expensive therapies that a select few wealthy people may benefit from (Føllesdal, 2008). While there are limitations on the usage of potential remedies to be created in terms of patients, the whole society participates to the research process through volunteers, medical professionals, and public monies. The sense of fairness principle may be used to discuss this circumstance. At the societal level, there must be an equitable and fair distribution of the costs and benefits. Since healthcare is seen as a birthright, it should be one of the services that society requires to be supplied to all members equitably and in accordance with social fairness (Rothstein, 2002). All patients who require stem cell therapies ought to be considered from this angle when they are brought up in the future. Geographical, cultural, and socioeconomic variables should not limit who may use the services.

18.7.6 Disregarding Other Regional and Worldwide Health Issues Unrelated to Stem Cell Research

The fields that require assistance should receive the resources designated for healthcare difficulties, and they should be dispersed equitably. However, fundamental healthcare services may be neglected as a result of the high expense of stem cell research and the focus on this field of medicine. The United Nations Declaration on Human Cloning calls on member states to consider the major problems that global emergencies like HIV/AIDS, TB, and malaria pose to their research budgets, especially for poor nations (Arda, and Güvercin, 2013).

18.8 Fundamental Methods of Stem Cell Research in Different Nations

Regarding the topic of stem cell research, many nations throughout the world have different perspectives. Still these may be broadly divided into three categories: moderate, permissive, and restricted (Figure 18.3).

Restrictive option

Forbids research on human embryos; does not specifically allow study on hESC lines that are already in existence. Ireland, Poland, Austria, Slovakia, Latvia, Germany, and Italy are the nations included in this choice.

Permissive option

Acknowledges the use of nuclear transfer or *in vitro* fertilization to create human embryos for scientific purposes (cloning). Spain, the UK, Northern Ireland, Belgium, Sweden, Finland, South Africa, Israel, Japan, China. India, South Korea, and Australia are the nations included in this choice.

Moderate option

Allows for the creation of new hESC lines, but only when leftover embryos from infertility clinics are used. France, Portugal, the Netherlands, Denmark, Iceland, Norway, Russia, Czech Republic, Hungary, Slovenia, Estonia, Lithuania, Greece, Turkey, Iran, New Zealand, Brazil, and Canada are the nations included in this other choice (Walters, 2008).

Figure 18.3 Stem cell research method.

18.8.1 International Law and National Strategies for Research on Stem Cells

Regarding stem cell research, there is currently no clear consensus on a global scale. The United Nations Declaration on Human Cloning appears to be the product of lengthy discussions over the stem cell issue held under the auspices of the UN (Arda, and Güvercin, 2013). In the use of life sciences, member states are urged to take all essential precautions to preserve human life. It is requested of member states to take action to stop the exploitation of women in the life sciences. Insofar as human dignity and the preservation of human life are concerned, member states are urged to outlaw all kinds of human cloning.

Compared to the 84 nations that supported the declaration, the total number of countries casting counter and abstention votes (37 and 34) is almost equal to the number of countries listed above. This finding highlights the stark disparities in viewpoints around the globe in this field. The proclamation forbids human cloning in any form and includes a crucial allusion to the preservation of human life. During this time, it places particular focus on preventing the mistreatment of women and allocating financial resources fairly.

Member states are further called upon to consider the pressing global issues like HIV/AIDS, tuberculosis, and malaria, which affect the developing countries, in particular, in their financing of medical research, including life sciences research. One may argue that the Council of Europe's Oviedo Convention, another international organization similar to the United Nations, likewise allots a significant amount of space for this subject. It is necessary to mention Article 18 of the convention in this context since it has also been incorporated into the laws of every member state of the council.

18.8.2 Oviedo Convention on Human Rights and Biomedicine, Article 18 (Oviedo, 1997)

In cases when *in vitro* embryo research is permitted by law, the embryo must be adequately protected. It is forbidden to create human embryos for scientific reasons. The only thing that is forbidden is the production of embryos for scientific study. Research on *in vitro* embryos is permitted as long as sufficient protection for the embryo is offered. The convention authorized research on the topic and led to the creation of national rules.

The ISSCR Guideline (Daley, *et al.*, 2007) is an additional international intervention. Scientists, ethicists, and legal specialists from 14 different nations contributed to its preparation. The purpose of this guideline is to promote worldwide collaboration in the field of human embryonic stem cell research. It was believed that creating a collection of common applications would inspire institutions and researchers.

The recommendation states that while doing research using human embryonic stem cells, three primary aspects should be considered: operating in a transparent manner, taking care of social affairs, and preserving society's feeling of trust. The following are the fundamental ideas that the guidelines specify: demand supervision, acceptable and unacceptable research, need for expressed permission, monetary factors, promoting adherence, and research materials that are easily accessible.

Within this framework, the guidelines (Guidelines for the Conduct of Human Embryonic Stem Cell Research, Version 1; Dec. 21, 2006) comprise 14 subsections addressing justification, task force mission, commentary on scientific terminology, guidelines' scope, conduct responsibility, statement on reproductive cloning, issues relevant to international collaborations, supervision recommendations, enforcement mechanisms, research categories, material procurement, derivation principles, banking, distribution of human stem cell lines, dispute resolution, and continuous guidelines review.

The part of the guideline's annexes contains five examples of informed consent forms:

1. Sample consent form for study; purchase of eggs for SCR (p. 22)
2. Sample consent form for study; donation of eggs for SCR (p. 36)
3. Sample consent form for research; donation of embryos for SCR (p. 38)
4. Sample consent form for study; somatic cell donation for stem cell research (p. 45)
5. Consent form for sample study; sperm donation for SCR (p. 52)

Moreover, there is also an agreement form on the transmission of materials relevant to human embryonic stem cells (pp. 59–64). As a result, the guideline appears to be designed to ensure that research on human embryonic stem cells, a novel area in the fields of biological sciences and medicine, is carried out within the bounds of ethical acceptability while also taking social responsibility into consideration.

References

Arda, B. and Güvercin, C.H., Stem cell research: Ethical considerations, in: *Stem cells: Current challenges and new directions*, pp. 341–358, 2013.

Beall, S.A. and DeCherney, A., History and challenges surrounding ovarian stimulation in the treatment of infertility. *Fertil. Steril.*, *97*, 4, 795–801, 2012.

Bongso, A. and Fong, C.Y., Human embryonic stem cells: their nature, properties, and uses, in: *Trends in stem cell biology and technology*, pp. 1–17, 2009.

Daley, G.Q., Richter, L.A., Auerbach, J.M., Benvenisty, N., Charo, R.A., Chen, G., Zoloth, L., The ISSCR guidelines for human embryonic stem cell research. *Science*, *315*, 5812, 603–604, 2007.

de Miguel Beriain, I., Patenting human embryonic stem cells in the European Union context: an updated analysis of a complex issue, in: *Safety, Ethics and Regulations*, pp. 263–273, 2017.

Emanuel, E.J., Grady, C., Crouch, R.A., Lie, R.K., Miller, F.G., Wendler, D. (Eds.), *The Oxford Textbook of Clinical Research Ethics*, pp. 488–499, Oxford University Press, Oxford, New York, 2008.

Ersoy, V. and Güler, M., Advertisement and promotion in offering healthcare service, Turkish Medical Association Publications, Ankara, 2008.

Føllesdal, D., The potentiality argument and stem cell research, in: *Stem Cells, Human Embryos and Ethics: Interdisciplinary Perspectives*, pp. 137–147, 2008.

Franklin, S., Embryonic economies: the double reproductive value of stem cells. *BioSocieties*, *1*, 1, 71–90, 2006.

Funderud, S., Stem Cells: Sources and Clinical Applications, in: *Stem Cells, Human Embryos and Ethics: Interdisciplinary Perspectives*, pp. 21–30, 2008.

Gorelik, L. and Flavell, R.A., Abrogation of TGFβ signaling in T cells leads to spontaneous T cell differentiation and autoimmune disease. *Immunity*, *12*, 2, 171–181, 2000.

Guimarães, H., Rocha, G., Bellieni, C., Buonocore, G., Rights of the newborn and end-of-life decisions. *J. Matern.-Fetal Neonatal Med.*, *25*, sup1, 76–78, 2012.

Hansen, B., Opinion on the Ethical Aspects of Patenting Inventions Involving Human Stem Cells: Verslag van de ronde-tafelconferentie van de European Group on Ethics in Science and New Technologies to the European Commission (EGE). *Ethische Perspectieven*, *12*, 1, 43–48, 2002.

Hyun, I., Moving human SCNT research forward ethically. *Cell Stem Cell*, *9*, 4, 295–297, 2011.

Ismayilova, P., International standards and domestic legislation on human organ donation and transplantation. *Amazon. Investig.*, *12*, 71, 175–187, 2023.

Ja, T., Embryonic stem cell lines, derived from human blastocysts. *Science*, *282*, 1145–1147, 1998.

Jiwa, M., Doctors and the media. *Australas. Med. J.*, *5*, 11, 603, 2012.

Lee, T.S., Feeney, M.B., Schmainda, K.M., Sherley, J.L., Prentice, D.A., Human fetal tissue from elective abortions in research and medicine: Science, ethics, and the law. *Issues L. Med.*, *35*, 3, 2020.

Liew, C.G., Moore, H., Ruban, L., Shah, N., Cosgrove, K., Dunne, M., Andrews, P., Human embryonic stem cells: possibilities for human cell transplantation. *Ann. Med.*, *37*, 7, 521–532, 2005.

Mitalipov, S. and Wolf, D., Totipotency, pluripotency and nuclear reprogramming, in: *Engineering of stem cells*, pp. 185–199, 2009.

Neal, S., Mahendra, S., Bose, K., Camacho, A.V., Mathai, M., Nove, A., Matthews, Z., The causes of maternal mortality in adolescents in low and middle income countries: a systematic review of the literature. *BMC Pregnancy Childbirth*, *16*, 1–18, 2016.

Østnor, L. (Ed.), *Stem Cells, Human Embryos and Ethics: Interdisciplinary Perspectives*, pp. 205–220, Springer, New York, 2008.

Petersen, T., Legislation on Biotechnology in the Nordic Countries:-an overview 2019, 2019.

Phelan, M., Medical interpreting and the law in the European Union. *Eur. J. Health Law*, *19*, 4, 333–353, 2012.

Porxas Roig, M.A., A Bioethics Approach to the Convention against Trafficking in Human Organs. *Rev. Bioet. Derecho*, *40*, 141, 2017.

Reed, D.C., *Revolutionary Therapies: How the California Stem Cell Program Saved Lives, Eased Suffering-and Changed the Face of Medicine Forever*, World Scientific Publishing Co. Pte. Ltd., Singapore, 2020.

Reya, T., Morrison, S.J., Clarke, M.F., Weissman, I.L., Stem cells, cancer, and cancer stem cells. *Nature*, *414*, 6859, 105–111, 2001.

Rothstein, M.A., Rethinking the meaning of public health. *J. Law Med. Ethics*, *30*, 2, 144–149, 2002.

Rousseaux, C.G. and Bolon, B., Embryo and fetus, in: *Haschek and Rousseaux's Handbook of Toxicologic Pathology*, pp. 2695–2759, Academic Press, Inc., San Diego, California, USA, 2013. https://doi.org/10.1016/C2010-1-67850-9.

Sandel, M.J., Embryo ethics–the moral logic of stem-cell research. *N. Engl. J. Med.*, *351*, 3, 207–209, 2004.

Sapna, F.N.U., Raveena, F.N.U., Chandio, M., Bai, K., Sayyar, M., Varrassi, G., Mohamad, T., Advancements in heart failure management: a comprehensive narrative review of emerging therapies. *Cureus*, 15, 10, e46486, 2023.

Siegel, A., Ethics of stem cell research, 2008.

Şimşek, H., Effects of gender inequalities on women's reproductive health: The case of Turkey. *Dokuz Eylül Üniv. Tıp Fak. Derg.*, *25*, 2, 119–26, 2011.

Wade, J., A Historical Review of Hematopoietic Stem Cell Transplantation: Developmental Advancements Then and Now. *Senior Honors Theses*, 2023.

Walters, L., An intercultural perspective on human embryonic stem cell research, in: *Stem cells, human embryos and ethics: Interdisciplinary perspectives*, pp. 91–109, 2008.

Index

β-Globin gene, 65
β-Thalassemia, 52, 55

2D cultures, 18

3D bioprinting, 732
3D culture, 730
3D culture system, 649
3D human brain models, 370

Abortion, 745, 746, 747
Actions of stem cells
homing, 614
Acute and chronic coronary syndromes, 317
Acute myeloid leukemia (AML), 429, 592
Acute myeloid leukemia; CD_{34}^{+} / CD_{38}^{+}, 413
Adenine base editors, 80
Adipocytes, 329
Adipose-derived SCs (ADSCs), 417
Adipose-derived stem and progenitor cells, 314
Adrenoleukodystrophy, 429
Adult (somatic) stem cells, 42, 53
Adult stem cells (ASCs), 111, 317, 355
Advanced therapy medicinal products, 368
Allogeneic HSCT, 59
Alzheimer's disease, 59, 69, 362, 475, 740
Amniotic fluid SCs, 417
Amyotrophic lateral sclerosis (ALS), 59, 68, 364, 432, 678
Angiogenesis, 254–260, 269–272, 643
Animal models, 272–277
Anti-apoptotic, 328
Antigen presentation, 425
Antigenic stimulation, 425
Anti-inflammatory, 330
Antitumor, 420
Aplastic anemia, 432
APOE4, 69
Apoptosis, 420, 424
Asthma, 470
Atherosclerosis (ASVD), 229, 237, 238
Atrial fibrillation, 317
Autoimmune diseases, 411
Autoimmune disorders, 410, 417
Autologous HSCT, 57
Autologous SCs, 429
Autologous stem cells modified with lentiviral vectors, 682
Autologous tissue transplantation, 507
Autologous transplantation, 730

B cell receptor editing, 425
Bacteria, 469
Base editing, 80
Base editors, 80
Basic fibroblast growth factor (bFGF), 144
BCL11A, 68
Bioartificial pancreas (BAP), 148

Biocompatibility, 164, 165, 167, 168, 206, 207
Biocompatible, 545
Bioengineering strategies, 336
Biomarkers, 160, 181, 209, 213, 237–241, 650
Biomaterials, 541
Biomaterials and scaffolding, 360
Biomimetric strategies, 338
Bioreactor system, 547
Biotransformation, 485
Blindness, 740
BM-derived endothelial progenitor cells, 312
BM-derived mesenchymal stem cells, 313
Bone marrow hematopoiesis, 424
Bone marrow mononuclear cells, 418
Bone marrow stem cells (BMSCs), 137
Bone tissue, 508, 509, 514
Bone-marrow-derived mononuclear cells, 312
Brain tumor; CD_{133} $^{+}$, 413
Breast cancer; CD_{44} $^{+}$ / CD^{24} $^{+}$, 413

Calcium sensing receptor (CaR), 588
Cancer, 740
Cancer (immunotherapy), 67
Cancer and its treatment, 599–605
Cancer immunotherapy, 209
Cancer stem cells (CSCs), 14, 19–21, 23–26, 591, 615, 629
Cardiac failure, 310
Cardiac progenitor cells (CPCs), 684
Cardiac rehabilitation (CR), 146
Cardiac SCs, 418
Cardiac stem and progenitor cells, 315
Cardiac stem cells, 727
Cardiomyocytes, 175, 310
Cardiotoxins, 315
Cardiovascular diseases (CVDs), 228, 229, 237–241, 269–272, 277, 278
Cardiovascular illness, 309
Cardiovascular regeneration, 227–309
CAR-T cell therapy, 159, 161, 209, 211, 212, 214
Cartistem and Cupistem, 690
CCR5, 68
CD105, 434
CD14, 434
CD19, 434
CD34, 434
CD45, 434
CD73, 434
CD79α, 434
CD90, 434
Cell fusion, 327
Cell migration, 506
Cell retension, 337
Cell survival and proliferation, 261–263
Cell transplants, 366
Cell-based therapy, 659
Cellgram-AMI, 690
Cell-line derived models, 648
Cellular plasticity, 507
Cellular signaling, 637
CEP290, 69
Cerebral delivery, 646
CFTR gene, 65, 66
Challenges and future directions, 277, 278
Challenges and limitations of stem cell therapy
  immune responses, 397
  off-target effects, 396
  stem cell exhaustion, 398
  tissue heterogeneity, 397
Chemokine gradients, 425
Chemokines, 416, 509
Chemotherapy, 630
Chimeric antigen receptor (CAR)-T cell therapy, 686
Chondrocytes, 329
Chondrocytes (cartilage cells), adipocytes (fat cells), and osteoblasts (bone cells), 117
Chronic illnesses, 508, 516, 740

Circadian rhythms, 425
Clinical trials, 277, 278, 365, 477, 620, 621, 705
Clonal expansion, 424
Clustered Regularly Interspaced Short Plaindromic Repeats associated protein 9 (CRISPR-Cas9) 40, 82
Collagen, 165, 175
Colloid osmotic pressure (COP), 148
Colon cancer; CD_{133} $^{+}$, 413
Combination therapy, 400
Combinational approach, 659
Communities, 481
Complement system activation, 425
Confocal microscopy, 16, 17
Continuous renal replacement treatment (CRRT), 147
Contract research organizations, 633
Contralateral injection, 646
Controlled release systems, 166, 169, 213
Converting stem cells, 742
Corneal healing, 740
Corneal limbal SCs, 419
Coronory blockage, 324
COVID-19, 478
COVID-19 pneumonia, 56
Cranioplasty, 509
CRISPR and gene editing, 359
CRISPR/Cas9, 23, 29
CRISPR-Cas12, 79
CRISPR-Cas13, 79
CRISPR-Cas9, 167, 169, 177, 718
CRISPR-Cas9 gene editing, 361
CRISPR-Cas9 technology, 432
Crohn's disease, 428
Cryopreservation, 728
Crypts, 118
Cyclin-dependent kinase, 506
Cystic fibrosis (CF), 51, 65
Cytokine base editors, 80
Cytokine-mediated differentiation, 424
Cytokines, 522, 524
Cytotoxic drugs, 629
Deafness, 740
Dedifferentiated cells, 519
Defensive system, 410
Dendritic cell maturation, 426
Dendritic cells, 414
Dental pulp SCs (DPSCs), 417
Diabetes, 717, 740
Diapause, 647
Dielectrophoresis (DEP), 128
Differentiation, 168, 169, 172–175, 211, 213
Differentiation and Integration, 357
Disability-adjusted life years, 352
Dishevelled protein, 508, 529
Donation of eggs, 753
Donation of embryos, 753
Dopaminergic neurons, 440
Double-strand break (DSB), 42
Drug delivery, 164–166, 168, 183, 206, 208, 209, 211, 213
Drug toxicity, 654
Duchenne muscular dystrophy (DMD), 56, 65, 66
Dystrophin protein, 65, 66

Electroporation, 73
Electrospinning, 169
Embronic stem cells, 635
Embryonic and mesenchymal stem cells, 676
Embryonic stem (ES) cell, 3, 5
Embryonic stem cells (ESCs), 42, 53, 119, 316, 354, 417, 509, 524, 525, 576, 707
Encapsulated hESCderived pancreatic progenitor cell transplantation, 680
Endothelial progenitor cells, 418, 635
Engraftment, 336
Environment, 471
Enzyme therapy, 640
Epidermal SCs, 418
Epigenetic modifications, 425
Epithelial cells, 465

Epithelial-mesenchymal transition (EMT), 20, 21, 656
eSpCas9, 49
Essential, 472
Ethical advantage, 543
Ethical considerations, 367, 631
European Medicines Agency (EMA), 62, 88, 89
European Union (EU), 88
Ex vivo expansion, 335
Exosomes, 159, 179–185, 211, 227–309, 506, 527, 637, 707
Extracellular complex (ECM), 4, 8–10, 14–16, 19, 20, 22, 26, 29
Extracellular matrix (ECM), 142, 511, 515, 521, 541, 575
Extracellular vesicle-derived stem cells, 327
Extracellular vesicles (EVs), 123, 228, 237, 238, 242, 243
Extramedullary hematopoiesis, 426

F8, 70
Feedback inhibition, 425
Fermentation, 467
Fertilization, 745
Fertilized ovum, 741
Fibrosis reduction, 263–268, 270, 271
Field flow fractionation (FFF), 126
Flow cytometry, 335
Flow cytometry (FACS), 128
Flow cytometry based assay, 638
FokI nuclease, 45, 81, 82
Food and Drug Administration (FDA), 62, 88, 89

Gastric cancer; $CD_{44}{}^{+}$, 413
Gastrointestinal tract, 474
Gene editing, 718
Gene modified stem cells, 328
Gene therapies, 632
Gene therapy, 169, 182, 214, 619
Genetically engineered mouse model, 649
Germline, 85
Gingival mesenchymal SCs, 419
Global health, 706
Gold nanoparticles, 73
Good laboratory practices, 634
Good manufacturing practice, 693
Graft-versus-host disease (GVHD), 59
Graft-versus-leukemia, 432
Granulopoiesis, 426
Growth factors, 524, 525
Guide RNA (gRNA), 42
Gut health, 464
Gut microbiome interactions, 425

Hair follicle SCs, 419
Harmful pathogens, 410
HBB, 66, 67, 68
Healing process, 508
Health Canada, 90
Heart and nerve cells, 741
Heart disease, 411
Heart disorder, 310
Heart failure (HF), 228, 238, 239, 271
Heart-on-a-chip, 325
Hedgehog signaling, 4, 13, 14, 21
Hematology, 428
Hematopoietic stem cell transplantation (HSCT), 57, 682
Hematopoietic stem cells (HSCs), 3, 5, 7, 8, 13, 53, 109, 417, 430, 509, 578, 635
Hemophilia A, 70
Hemophilia B, 68
Hepatocyte growth factor, 420
Hepatocytes, 175, 176
hESC-derived neural progenitor cells (NPCs) , 679
Heterogeneity, 331, 486
HEXA, 70
Histocompatibility, 746

HIV infection, 56
HIV/AIDS, 68, 750, 750
Hodgkin, 429
Holoclar, 692
Homeostasis, 424
Homing of stem cell *in vivo*, 332
Homing to bone marrow, 636
Homology-directed repair (HDR), 42
Hormonal influences, 425
Hormone therapy, 747
HSC transplantation, 638
HTT, 67
Human cloning, 751
Human embryonic stem cells (ESCs), 110, 752, 753
Human embryos, 741, 744
Human endothelial precursor cells (EPCs), 132
Human Fertilisation and Embryology Authority (HFEA), 90
Human induced pluripotent stem cells (hiPSCs), 64
Human leucocyte antigens (HLA), 138
Human papillomavirus, 483
Human pluripotent stem cells (PSCs), 51
Human rights, 752
Human umbilical cord (HUC), 108
Huntington's disease, 51, 67
Hurler syndrome, 429
Hydrodynamic injection, 75
Hydrogels, 164, 169, 178, 250, 272
Hypoxia, 11, 20, 21, 23, 25, 26
Hypoxia preconditioning, 237, 262

Immune checkpoint inhibition, 426
Immune modulation, 169, 170, 181, 183
Immune tolerance, 725
Immunogenicity, 313, 444, 470
Immunomodulation, 235, 236, 251–254, 358, 410, 411, 414, 416
Immunosuppresive substances, 333
Immunosuppressive, 420
Immunotherapy, 629, 644, 716
Immunotolerance, 657
In vitro fertilization, 720
In vitro models, 18
Indian Council of Medical Research (ICMR), 91
Induced pleuripotent stem cell, 316, 631, 635
Induced pluripotent stem cells (iPSCs), 2, 5, 23, 24, 27, 29, 42, 53, 119, 355, 356, 417, 578, 713
Infections, 479
Infectious pathogens, 410
Infertility, 740
Inflammation, 3, 9, 10, 22–24, 26, 30, 465, 470
Inflammation modulation, 251–254
Inflammatory response, 508
Insulin sensitivity, 681
Interactions, 474
Interferon signaling, 426
Interleukin-10, 420, 451
Intestinal SCs, 418
Intestinal stem cells (ISCs), 466, 583
Intracranial hemorrhage (ICH), 108
Intravenous injection, 335
Ischemia-reperfusion injury, 322
Ischemic cardiomyopathy, 323
Ischemic heart failure, 324

Keratinocytes, 118

Leber congenital amaurosis, 56, 69
Left ventricular assist device (LVAD), 146
Lipid nanoparticles, 72
Liposuction, 714
Liver cancer; $CD_{133}{}^{+}$, 413
Liver failure, 740
Liver SCs, 418

Lung cancer; $CD_{133}{}^{+}$, 413
Lung SCs, 419
Lymph node trafficking, 426
Lymphoblastic leukemia, 429
Lymphocytes, 482
Lymphoid cancer cells (LICs), 592

Macrophage polarization, 426
Macrophages, 414
Magnetic activated cell sorting (MACS), 128
Main histocompatibility complex, 313
Malaria, 750
Mammary SCs, 419
Melanoma; $CD_4{}^{+}$ / $CD_{25}{}^{+}$ / Fox $P_3{}^{+}$, 413
Memory cell formation, 426
Menstrual blood-derived SCs, 418
Mesenchymal, 116, 415
Mesenchymal stem cell (MSCs), 3, 5, 8, 15, 16, 24, 27, 28, 53, 108, 172–175, 181, 208, 230–236, 244, 245, 251–253, 269–271, 273–277, 417, 635, 678, 712
Mesenchymal stem cell therapy, 325
Mesoderm, 116
Metabolic disorder, 382–388
Metabolic disorder signalling pathway
  AMP-activated protein kinase (AMPK), 395
  c-Jun N-terminal kinase (JNK), 395
  mechanistic target of rapamycin (mTOR), 394, 398
Metabolic rates, 506
Metabolic regulation, 425
Metalloproteinases, 414
Metformin, 438
Microbiome, 464, 476, 487
Microenvironment, 507, 526, 531
Microfluidics based on negative DEP (nDEP), 132
Microinjection, 74
Microvascular damage, 316
Microvesicles, 417
Migration, 329
Ministry of Food and Drug Safety (MFDS), 91
Ministry of Health, Labour and Welfare (MHLW), 90
miRNAs, 238–241, 252–254, 258–260, 264–268
Miscarriages, 745
Mobilized stem and progenitor cells, 313
Monoclonal antibodies, 446
MSC transplantation, 638
Multigene-modified MSC exploration, 330
Multilineage property, 329
Multiple myeloma, 55
Multiple myeloma; $CD_{138}{}^{+}$, 413
Multiple sclerosis, 363, 439
Muscle satellite cells, 418
Muscle-derived progenitors, 311
Muscular disease, 740
Muscular dystrophy, 52
MYBPC3, 69
Myeloid-derived suppressor cells, 424
Myocardial infarction (MI), 228, 237, 238, 269û271, 273–277, 322, 740
Myocardial restoration, 317
Myocardium, 175
Myocardium carriers, 339

Nanobiomaterials, 159, 163–167, 206–208, 213–215
Nanocarriers, 162, 166, 169, 183, 209, 211
Nanoparticle carrier, 642
Nanoparticles, 162–164, 166, 167, 169, 177, 183, 207, 209
National Health and Medical Research Council (NHMRC), 90
National Health Commission (NHC), 89
National Health Surveillance Agency (ANVISA), 91
Natural killer cell education, 426

Neural fetal stem cells, 707
Neural stem cells (NSCs), 8, 22, 23, 109, 175, 356, 417, 583, 635
Neurological deterioration, 429
Neurological disorders, 351
Neurology, 372
Neuroprotection, 357
Neuroprotective effects, 452
Neurotransmission, 438
Neurotransmitter, 475
Neurotropic factors, 549
Neutrophil extracellular traps (NETs), 425
Non-homologous end joining (NHEJ), 42
Notch signaling, 4, 5, 13–25
N-succinimidyl ester, 521

Olfactory ensheathing cells, 418
Oncology, 428
Orchestrating effective, 422
Organ function, 507, 508
Organism, 487
Organoids, 60, 712
Organoids and 3D bioprinting, 359
Osteoarthritis, 706, 740
Osteoblast differentiation, 334
Osteoblasts, adipocytes, and chondrocytes, 675
Osteocytes, 329
Ovary cancer; $CD_{44}{}^{+}$ / $CD_{117}{}^{+}$, 413
Ovum removal surgery, 747

Pancreatic cancer; $CD_{44}{}^{+}$ / $CD_{24}{}^{+}$, 413
Pancreatic progenitor cells (PPCs), 138
Pancreatic SCs, 419
Paracrine effect, 326, 358
Paracrine factor secretion, 636
Paracrine factors, 507
Paracrine signaling, 228, 237, 269, 434
Parkinson's disease, 59, 70, 362, 677, 740
Pathological conditions, 506
Patient consent, 369
Patient-derived iPSCs, 52, 54
PCSK9, 69
Penicillin-binding proteins, 438
Pentylenetetrazol (PTZ), 148
Peptide-based delivery, 74
Percutaneous coronary intervention, 313
Periodontal ligament SCs, 418
Peripheral artery disease (PAD), 271, 272
Permeability, 480
Personalized medicine, 159–161, 208–210, 212, 214, 400, 401
Phosphoinositide 3-kinase (PI3K), 582
Placenta derived SCs, 418
Plasma cell differentiation, 426
Pluripotent, 543
Polio vaccine, 749
Polymeric nanoparticles, 75
Post-infarction wounds, 311
Precision therapeutics, 159, 213, 214
Preclinical studies, 272–277
Preventive therapies, 371
Prime editing guide RNA (pegRNA), 81
Probiotics, 468, 476, 477
Progenitor stem cells specific to tissues (TSPSCs), 137
Prostate cancer; $CD_{133}{}^{+}$ / $CD_{44}{}^{+}$, 413
Protospacer adjacent motif (PAM), 43
Purchase of eggs, 753

Queencell, 690

Regenerative medicine, 137, 159, 162, 171, 172, 174, 211, 213, 229, 230, 236, 249, 250, 644, 674, 706
Regulatory pathways, 633
Regulatory restrictions, 632
Regulatory T cell suppression, 424
Retinal diseases, 67
Rheumatoid arthritis, 740

Scaffold based method, 546
Scaffold design, 169, 170, 178, 211
Scaffold-based implantation, 335
SCs, 576
SCs of umbilical cord, 417
Secreted agents, 641
Self regeneration, 542
Self-renewal, 4
Sequencing, 484
Sickle cell anemia, 51, 55, 66
Side effects, 652
Signaling molecules, 472
Single-cell RNA sequencing, 17
Single-cell sequencing, 360
Skeletal myoblast, 311
SNCA, 70
SOD1, 68
Somatic cell donation, 753
Somatic stem cells, 114
SpCas9-HF1, 49
Sperm donation, 753
Spermatogonial SCs, 419
Spheroids, 650
Spinal cord injuries, 364, 411
Spinal cord injury, 740, 708
Stability, 473
Static tissue culture dish, 728
Stem cell action
  homing, 388, 392
  immune response modulation, 393
  modulation of immune response, 393
  tissue repair and regeneration, 392
Stem cell classification, 634
Stem cell properties, 608–611
Stem cell therapies, 309
Stem cell therapy, 159, 162, 168, 171, 172, 208, 211, 512, 526, 599, 706
  embryonic stem cell, 387, 391
  hematopoietic stem cell, 387, 388, 390
  mesenchymal stem cell, 386, 389, 393, 399
  neural stem cell, 387, 391
  puripotent stem cell, 386, 387, 389, 396, 399
Stem cell therapy: challenges, 615–619
Stem cell transplantation, 507, 508, 511
Stem cell tropism, 658
Stem cell types, 606–608
Stem cells (SCs), 227–309, 353, 471
Stemirac, 691
Strimvelis, 693
Stroke, 363
Surface modification, 165, 168, 170, 177, 183, 207
Symptoms, 479
Synovial mesenchymal SCs, 419
Systemic lupus erythematosus, 427
Systolic pump, 311

TALENs, 81
Targeted delivery, 159, 162, 166, 182, 183, 211, 213
Tay–Sachs disease, 70
T-cell acute lymphoblastic leukemia (T-ALL), 56
Technical constraints, 632
Temcell HS, 691
Testicular SCs, 419
Therapeutic applications, 269–272
Therapeutic efficacy, 507
Thymic involution, 426
Thymic selection, 424
Tissue differentiation, 509
Tissue engineering, 141,162, 169, 170, 213, 542, 732
Tissue injection, 335
Tissue regeneration, 505, 506, 508, 509
Tissue repair, 230–235, 249, 250
TNF-alpha-associated death-inducing ligand, 641
Transcription activator-like effector nucleases (TALENs), 40, 81, 82
Transcriptional programs, 506

Transforming growth factor beta (TGF-β), 6, 8, 15, 20
Transfusion-dependent β-thalassemia, 56
Translational application, 339
Translocation, 480
Transplantation, 412
Trauma, 508
Truncated gRNAs (tru-gRNAs), 49
Tumor, 482
Tumor microenvironment (TME), 19–21, 26, 657
Tumorigenesis, 444, 653
Tumor-infiltrating cells (TICs), 591
Tumor-tropic effect, 636
Two-photon microscopy, 16, 17
Tyrosine kinase inhibitors, 316

Urine-derived SCs, 418

Vaccinations, 410
Vascular endothelial growth factor (VEGF), 20, 24–26
VEGF signaling, 332
Viral infection, 655
Viral therapy, 642
Viral vectors (lentivirus, AAV), 72

Wnt signaling, 4, 12, 13, 15
Wnt, notch, and hedgehog, 3, 4, 12
World Health Organization (WHO), 51, 92
Wound healing, 508, 531

Xenograft, 648

Yamanaka factors, 439

Zinc finger nucleases (ZFN's), 40, 81, 82
Zygote developing, 741

Also of Interest

Check out these other related titles from Scrivener Publishing

Design and Forecasting Models for Disease Management, Edited by Pijush Dutta, Sudip Mandal, Korhan Cengiz, Arindam Sadhu, and Gour Gopal Jana, ISBN: 9781394234042. The book provides an essential overview of AI techniques in disease management and how these computational methods can lead to further innovations in healthcare.

Smart Hospitals: 5G, 6G, and Moving Beyond Connectivity, Edited by Arun Kumar, Manoj Gupta, Sanjeev Sharma, Er. Himanshu Sharma, and Khursheed Aurangzeb, ISBN 9781394275441. Discover the potential of 5G, 6G, and smart hospitals beyond connectivity in Smart Hospitals: 5G, 6G, and Moving Beyond Connectivity and learn how these advancements are revolutionizing healthcare and the digital world.

Sustainable Nanomaterials for Treatment and Diagnosis of Infectious Diseases, Edited by Rajasekhar Chokkareddy, Suvardhan Kanchi and Gan G. Redhi, ISBN: 9781394200016. The book focuses on the design and novel synthetic routes of sustainable nanomaterials in diagnosing and treating infectious diseases offering potential benefits in terms of efficiency, biocompatibility, and environmental impact.

Federated Intelligent System for Healthcare: A Practical Guide, Edited by S. Rakesh Kumar, N. Gayathri and Seifedine Kadry, ISBN: 9781394271351. This practical guide gives valuable insights for integrating advanced technologies in healthcare, empowering researchers to effectively navigate and implement federated systems to enhance patient care.

Internet of Medicine for Smart Healthcare, Edited by Abhishek Kumar, Narayan Vyas, Pramod Singh Rathore, Abhineet Anand, and Pooja Dixit, ISBN: 9781394272235. This book provides in-depth explanations and discussions of the latest applications of AI, machine learning, and the Internet of Medicine, offering readers the cutting edge on this rapidly growing technology that has the potential to transform healthcare and improve patient outcomes.

Viral Vectors for Vaccine Delivery, Edited by Vivek Chavda and Vasso Apostolopoulos, ISBN: 9781394271535. The book is essential for anyone interested in vaccine development, as it highlights the unique advantages of viral vector vaccines in triggering robust, long-lasting immunity and provides an in-depth exploration of the technology and advancements shaping the future of healthcare.

Artificial Intelligence and Cybersecurity in Healthcare, Edited by Rashmi Agrawal, Pramod Singh Rathore, Ganesh Gopal Devarajan, and Rajiva Ranjan Divivedi, ISBN: 9781394229796. *Artificial Intelligence and Cybersecurity in Healthcare* offers a vital exploration of the intersection of artificial intelligence and cybersecurity within healthcare Cyber Physical Systems, equipping readers with knowledge on how to navigate the transformative yet complex technological landscape shaping modern patient care and data protection.

Explainable Artificial Intelligence in the Healthcare Industry, Edited by Abhishek Kumar, T. Ananth Kumar, Prasenjit Das, Chetan Sharma, and Ashutosh Kumar Dubey, ISBN: 9781394249268. Discover the essential insights and practical applications of explainable AI in healthcare that will empower professionals and enhance patient trust with Explainable AI in Healthcare, a must-have resource for anyone looking to navigate the future of medical practices with confidence.

Role of Flavonoids in Chronic Metabolic Diseases: From Bench to Clinic, Edited by Neeraj Mishra, Sumel Ashique, B.H. Jaswanth Gowda, Arshad Farid, Ashish Garg, ISBN: 9781394238040. This comprehensive volume covers the entire field of flavonoids by explaining their complex functions in reducing chronic metabolic illnesses, from the early stages of laboratory research to the development of therapeutic uses.

Printed and bound by CPI Group (UK) Ltd, Croydon, CR0 4YY

17/09/2026

14964264-0008